INTERMEDIATE ALGEBRA
EIGHTH EDITION

STUDENT'S
SOLUTIONS MANUAL

JEFFERY A. COLE
Anoka-Ramsey Community College

Margaret L. Lial
American River College

John Hornsby
University of New Orleans

 ADDISON-WESLEY

An imprint of Addison Wesley Longman, Inc.

Reading, Massachusetts • Menlo Park, California • New York • Harlow, England
Don Mills, Ontario • Sydney • Mexico City • Madrid • Amsterdam

Reproduced by Addison Wesley Longman from camera-ready copy supplied by the author.

Copyright © 2000 Addison Wesley Longman, Inc.

ISBN: 0-321-06200-0

2 3 4 5 6 7 8 9 10 PHC 02

Preface

This *Student's Solutions Manual* contains solutions to selected exercises in the text *Intermediate Algebra, Eighth Edition* by Margaret L. Lial and E. John Hornsby, Jr. It contains solutions to the odd-numbered exercises in each section, as well as solutions to all the exercises in the review sections, the chapter tests, and the cumulative review sections.

This manual is a text supplement and should be read along *with* the text. You should read all exercise solutions in this manual because many concept explanations are given and then used in subsequent solutions. All concepts necessary to solve a particular problem are not reviewed for every exercise. If you are having difficulty with a previously covered concept, refer back to the section where it was covered for more complete help.

A significant number of today's students are involved in various outside activities, and find it difficult, if not impossible, to attend all class sessions; this manual should help meet the needs of these students. In addition, it is my hope that this manual's solutions will enhance the understanding of all readers of the material and provide insights to solving other exercises.

I appreciate feedback concerning errors, solution correctness or style, and manual style. Any comments may be sent directly to me at the address below or in care of the publisher: Addison Wesley Longman, One Jacob Way, Reading, MA 01867.

I would like to thank Ken Grace, of Anoka–Ramsey Community College, for typesetting the manuscript and providing invaluable help with many features of the manual; Marv Riedesel and Mary Johnson, of Inver Hills Community College, for their careful accuracy checking; my wife, Joan, for proofreading the manuscript; Ruth Berry, of Addison Wesley Longman, for facilitating the production process; and the authors and Jennifer Crum, of Addison Wesley Longman, for entrusting me with this project.

Jeffery A. Cole
Anoka-Ramsey Community College
11200 Mississippi Blvd. NW
Coon Rapids, MN 55433

Table of Contents

CHAPTER 1   REVIEW OF THE REAL NUMBER SYSTEM

Section 1.1

1. *True*; division of a nonzero number by zero is undefined.

3. The statement "Every number has a positive additive inverse" is *false*. The additive inverse of a positive number is negative, and the additive inverse of 0 is 0.

5. *True*; the absolute value of a negative number is its additive inverse. For example, the absolute value of -5 is 5, which is also its additive inverse.

7. $\{y \mid y$ is a natural number greater than 5$\}$
The set of natural numbers is $\{1, 2, 3, \dots\}$, so the set of natural numbers greater than 5 is $\{6, 7, 8, \dots\}$.

9. $\{z \mid z$ is an integer less than or equal to 4$\}$
The set of integers is $\{\dots, -3, -2, -1, 0, 1, 2, 3, \dots\}$, so the set of integers less than or equal to 4 is $\{\dots, -1, 0, 1, 2, 3, 4\}$.

11. $\{a \mid a$ is an even integer greater than 8$\}$
The set of even integers is $\{\dots, -2, 0, 2, 4, \dots\}$, so the set of even integers greater than 8 is $\{10, 12, 14, 16, \dots\}$.

13. $\{x \mid x$ is an irrational number that is also rational$\}$
Irrational numbers cannot also be rational numbers. The set of irrational numbers that are also rational is $\emptyset$.

15. $\{p \mid p$ is a number whose absolute value is 4$\}$
This is the set of numbers that lie a distance of 4 units from 0 on the number line. Thus, the set of numbers whose absolute value is 4 is $\{-4, 4\}$.

17. Yes, the choice of variables is not important. The sets have the same description so they represent the same set.

19. $\{4, 8, 12, 16, \dots\}$
One way to describe this set is $\{x \mid x$ is a multiple of 4 greater than 0$\}$.

21. $\{2, 4, 6, 8\}$
One way to describe this set is $\{x \mid x$ is an even natural number less than or equal to 8$\}$.

23.
$$\left\{-8, -\sqrt{5}, -.6, 0, \frac{1}{0}, \frac{3}{4}, \sqrt{3}, 4, 5, \frac{13}{2}, 17, \frac{40}{2}\right\}$$

(a) The elements $4, 5, 17,$ and $\frac{40}{2}$ (or 20) are natural numbers.

(b) The elements $0, 4, 5, 17,$ and $\frac{40}{2}$ are whole numbers.

(c) The elements $-8, 0, 4, 5, 17,$ and $\frac{40}{2}$ are integers.

(d) The elements $-8, -.6, 0, \frac{3}{4}, 4, 5, \frac{13}{2}, 17,$ and $\frac{40}{2}$ are rational numbers.

(e) The elements $-\sqrt{5}$ and $\sqrt{3}$ are irrational numbers.

(f) All the elements are real numbers except $\frac{1}{0}$.

(g) The only element that is undefined is $\frac{1}{0}$.

25. Graph $\{-3, -1, 0, 4, 6\}$.
Place dots for $-3, -1, 0, 4,$ and 6 on a number line.

27. Graph $\left\{-\frac{2}{3}, 0, \frac{4}{5}, \frac{12}{5}, \frac{9}{2}, 4.8\right\}$.
Place dots to indicate the points
$-\frac{2}{3} = -.\overline{6}, 0, \frac{4}{5} = .8, \frac{12}{5} = 2.4, \frac{9}{2} = 4.5,$ and 4.8 on a number line.

29. The graph of a number is a point on the number line. The coordinate of a point on the number line is the number that corresponds to the point.

31. $|-8|$
Use the definition of absolute value.
$$|-8| = -(-8) = 8$$

33. $-|5| = -(5) = -5$

35. $-|-2| = -[-(-2)] = -(2) = -2$

37. $-|4.5| = -(4.5) = -4.5$

39. $|-2| + |3| = 2 + 3 = 5$

41. $|-9| - |-3| = 9 - 3 = 6$

43. $|-9| + |-13| = 9 + 13 = 22$

45. $|-1| + |-2| - |-3| = 1 + 2 - 3$
$$= 3 - 3$$
$$= 0$$

47. We want the greatest change, without regard to whether the change is an increase or a decrease. Look for the number in the list with the largest absolute value. That number, 10.2, is associated with the year 1988.

49. True; since -6 is to the left of -2 on a number line, -6 is less than -2.

51. False; since $-.32$ is to the right of $-\frac{4}{3}$ on a number line, $-.32$ is *greater* than $-\frac{4}{3}$, not less.

53. Since 3 is to the right of -2 on a number line, the statement $3 > -2$ is true and the calculator would return a 1.

55. Since -3 is equal to -3, the statement $-3 \geq -3$ is true and the calculator would return a 1.

57. The inequality $-3 < 2$ can also be written $2 > -3$. In each case, the inequality symbol points toward the smaller number so both inequalities are true.

59. "6 is less than 11" can be written as $6 < 11$.

61. "4 is greater than x" can be written as $4 > x$.

63. "$3t - 4$ is less than or equal to 10" can be written as $3t - 4 \leq 10$.

65. "5 is greater than or equal to 5" can be written as $5 \geq 5$.

67. "t is between -3 and 5" can be written as $-3 < t < 5$.

69. "$3x$ is between -3 and 4, including -3 and excluding 4" can be written as $-3 \leq 3x < 4$.

71. "$5x + 3$ is not equal to 0" can be written as $5x + 3 \neq 0$.

73. $3 \not< 2$ is read "3 is not less than 2," is equivalent to "3 is greater than or equal to 2," written $3 \geq 2$.

75. $-3 \not> -3$ is read "-3 is not greater than -3," is equivalent to "-3 is less than or equal to -3," written $-3 \leq -3$.

77. $5 \geq 3$ is read "5 is greater than or equal to 3," is equivalent to "5 is not less than 3," written $5 \not< 3$.

79. Compare the depths of the bodies of water. The deepest, that is, the body of water whose depth has the greatest absolute value, is the Pacific Ocean ($|-12,925| = 12,925$) followed by the Indian Ocean, the Caribbean Sea, the South China Sea, and the Gulf of California.

81. True; the absolute value of the depth of the Pacific Ocean is

$$|-12,925| = 12,925$$

which is greater than the absolute value of the depth of the Indian Ocean,

$$|-12,598| = 12,598.$$

83. $\{x \mid x > -2\}$ includes all numbers greater than -2, written $(-2, \infty)$. Place a parenthesis at -2 since -2 is not an element of the set. The graph extends from -2 to the right.

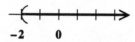

85. $\{x \mid x \leq 6\}$ includes all numbers less than or equal to 6, written $(-\infty, 6]$. Place a bracket at 6 since 6 is an element of the set. The graph extends from 6 to the left.

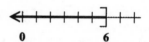

87. $\{x \mid 0 < x < 3.5\}$ includes all numbers between 0 and 3.5, written $(0, 3.5)$. Place parentheses at 0 and 3.5 since these numbers are not elements of the set. The graph goes from 0 to 3.5.

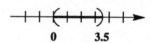

89. $\{x \mid 2 \leq x \leq 7\}$ includes all numbers from 2 to 7, written $[2, 7]$. Place brackets at 2 and 7 since these numbers are elements of the set. The graph goes from 2 to 7.

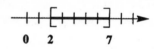

91. $\{x \mid -4 < x \leq 3\}$ includes all numbers between -4 and 3, excluding -4, but including 3, written $(-4, 3]$. Place a parenthesis at -4 and a bracket at 3 to show that -4 is not included, but 3 is. The graph goes from -4 to 3.

93. $\{x \mid 0 < x \leq 3\}$ includes all numbers between 0 and 3, excluding 0, but including 3, written $(0, 3]$. Place a parenthesis at 0 and a bracket at 3 to show that 0 is not included, but 3 is. The graph goes from 0 to 3.

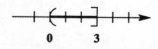

95. Draw a horizontal line through $18{,}000$. List all years whose graphs go above this line: 1992, 1993, and 1994.

97. The number of charges in 1991 is greater than $16{,}000$ and the number of charges in 1997 is less than $16{,}000$, so $x > y$.

99. The *natural numbers* are the numbers with which we count. Some examples are 1, 2, and 3. The *whole numbers* are formed by including 0 with the natural numbers. Three examples are 0, 10, and 100. The *integers* are formed by including the negatives of the natural numbers with the whole numbers, such as -1, -2, and -3. The *rational numbers*, such as $\frac{1}{2}$, $.75$, and $-\frac{3}{4}$, are formed by quotients of integers (nonzero denominator). The *irrational numbers* are positive or negative numbers that are not rational, such as π, $\sqrt{2}$, and $-\sqrt{5}$. The real numbers include all positive numbers, negative numbers, and 0, such as $-\pi$, 0, and $\sqrt{2}$.

Section 1.2

1. The sum of a positive number and a negative number is 0 if _the numbers are additive inverses_. For example, $4 + (-4) = 0$.

3. The sum of two negative numbers is a _negative_ number. For example, $-7 + (-21) = -28$.

5. The sum of a negative number and a positive number is positive if _the positive number has a larger absolute value_. For example, $15 + (-2) = 13$.

7. The difference between two negative numbers is negative if _the one with the smaller absolute value is subtracted from the one with the larger absolute value_. For example, $-15 - (-3) = -12$.

9. The product of two numbers with unlike signs is _negative_. For example, $(-5)(15) = -75$.

11. $13 + (-4) = 9$

13. $-6 + (-13) = -(6 + 13) = -19$

15. $-\dfrac{7}{3} + \dfrac{3}{4} = -\dfrac{28}{12} + \dfrac{9}{12} = -\dfrac{19}{12}$

17. $-.125 + .312 = .187$

19. $\begin{aligned} -8 - (-12) - (2 - 6) &= -8 + 12 - (-4) \\ &= -8 + 12 + 4 \\ &= 4 + 4 \\ &= 8 \end{aligned}$

21. $\begin{aligned} \left(-\dfrac{5}{4} - \dfrac{2}{3}\right) + \dfrac{1}{6} &= \left(-\dfrac{15}{12} - \dfrac{8}{12}\right) + \left(\dfrac{2}{12}\right) \\ &= -\dfrac{23}{12} + \dfrac{2}{12} \\ &= -\dfrac{21}{12} \text{ or } -\dfrac{7}{4} \end{aligned}$

23. $\begin{aligned} (-.382) + (4 - .6) &= (-.382) + (3.4) \\ &= 3.018 \end{aligned}$

25. $A = -4$ and $B = 2$. The distance between A and B is the absolute value of their difference.
$$|-4 - 2| = |-6| = 6$$

27. $D = -6$ and $F = \dfrac{1}{2}$. The distance between D and F is the absolute value of their difference.
$$\begin{aligned} \left|-6 - \dfrac{1}{2}\right| &= \left|-\dfrac{12}{2} - \dfrac{1}{2}\right| \\ &= \left|-\dfrac{13}{2}\right| \\ &= \dfrac{13}{2} \text{ or } 6\dfrac{1}{2} \end{aligned}$$

29. Answers will vary. One example is
$$-4 - (-9) = -4 + 9 = 5.$$
The sign is determined by choosing the sign of the number with the larger absolute value after subtraction has been changed to addition. In this example, 9 has a larger absolute value than -4, so the sign of the answer is the same as the sign of 9, which is positive.

31. It is true for multiplication (and division). It is false for addition and for subtraction when the number to be subtracted has the smaller absolute value. A more precise statement is, "The product or quotient of two negative numbers is positive."

33. The product of two numbers with the same sign is positive, so
$$(-15)(-3) = 45.$$

35. $\dfrac{3}{4}(-20)(-12) = (-15)(-12) = 180$

37. $-3.45(-2.14) = 7.383$

39. $\dfrac{-100}{-25} = -100\left(-\dfrac{1}{25}\right) = 4$

41.
$$\cfrac{\cfrac{12}{13}}{-\cfrac{4}{3}} = \frac{12}{13} \div \left(-\frac{4}{3}\right) = \frac{12}{13}\left(-\frac{3}{4}\right)$$

$$= -\frac{9}{13}$$

43. Division by 0 is undefined, so $\dfrac{5}{0}$ is undefined.

45. The statement "$(-2)^7$ is a negative number" is *true*. $(-2)^7$ gives an odd number of negative factors, so the expression is negative.

47. The statement "The product of 8 positive factors and 8 negative factors is positive" is *true*. The product of an even number of negative factors is positive.

49. $-4^6 = (-4)^6$ is *false*.
$$-4^6 = -(4\cdot4\cdot4\cdot4\cdot4\cdot4) = -4096$$
whereas
$$(-4)^6 = (-4)(-4)(-4)(-4)(-4)(-4)$$
$$= 4096.$$

51. The statement "$\sqrt{16}$ is a positive number" is *true*. The symbol $\sqrt{}$ always gives a positive square root.

53. The statement "In the exponential -3^5, -3 is the base" is *false*. The base is 3, not -3. If the problem were written $(-3)^5$, then -3 would be the base.

55. $\sqrt{121} = 11$ since 11 is positive and $11^2 = 121$.

57. $.28^3 = (.28)(.28)(.28) = .021952$

59. $\left(-\dfrac{7}{10}\right)^2 = \left(-\dfrac{7}{10}\right)\left(-\dfrac{7}{10}\right) = \dfrac{49}{100}$

61. $-\sqrt{900} = -\left(\sqrt{900}\right) = -(30) = -30$

63. $\sqrt{16}$ is the positive (or principal) square root of 16, namely 4; -4 is the negative square root of 16, written $-\sqrt{16}$.

65. $\sqrt{18,499} \approx 136.011029$

67. $\sqrt{93.26} \approx 9.657121724$

69. (a) If a is a positive number, then $-a$ is a negative number. Therefore, $\sqrt{-a}$ is not a real number, so $-\sqrt{-a}$ *is not a real number*.

(b) If a is a positive number, then $\sqrt{a}$ is also a positive number, and $-\sqrt{a}$ is a *negative* number.

71. $-7(-3) - (-2)^3$

$= -7(-3) - (-8)$	*Simplify power.*
$= 21 - (-8)$	*Multiply.*
$= 21 + 8$	*Use definition of subtraction.*
$= 29$	*Add.*

73. $|-6 - 5|(-8) + 3^2$

$= |-11|(-8) + 3^2$	*Simplify within absolute value bars.*
$= 11(-8) + 3^2$	*Take absolute value.*
$= 11(-8) + 9$	*Simplify power.*
$= -88 + 9$	*Multiply.*
$= -79$	*Add.*

75. $(-8 - 5)(-2 - 1)$

$= (-13)(-3)$	*Simplify within each parenthesis.*
$= 39$	*Multiply.*

77. $\dfrac{(-6 + 3)\cdot(-4)}{-5 - 1}$

$= \dfrac{(-3)\cdot(-4)}{-5 - 1}$	*Work inside parentheses.*
$= \dfrac{12}{-6}$	*Simplify numerator and denominator separately.*
$= -2$	*Divide.*

79. $\dfrac{2(-5) + (-3)(-2^2)}{-6 + 5 + 1}$

$= \dfrac{2(-5) + (-3)(-4)}{-6 + 5 + 1}$	*Take powers and roots.*
$= \dfrac{-10 + 12}{0}$	*Work in numerator and denominator separately.*
$= \dfrac{2}{0}$	

Since division by 0 is not possible, the given expression is undefined.

81. $-\dfrac{1}{4}[3(-5) + 7(-5) + 1(-2)]$

$= -\dfrac{1}{4}[-15 - 35 - 2]$	*Multiply inside brackets.*
$= -\dfrac{1}{4}[-52]$	*Add inside brackets.*
$= 13$	*Multiply.*

83.
$$\dfrac{-4\left(\dfrac{12-(-8)}{3\cdot2+4}\right)-5(-1-7)}{-9-(-7)-[-5-(-8)]}$$

$$=\dfrac{-4\left(\dfrac{20}{10}\right)-5(-8)}{-9+7-[3]}$$

*Simplify numerator and denominator
separately using order of operations.*

$$=\dfrac{-4(2)-5(-8)}{-2-3}$$

$$=\dfrac{-8+40}{-5}$$

$$=\dfrac{32}{-5}=-\dfrac{32}{5}$$

In Exercises 85–91, $a = -3$, $b = 64$, and $c = 6$.

85. $\quad 3a + \sqrt{b} = 3(-3) + \sqrt{64}$
$$= 3(-3) + 8$$
$$= -9 + 8$$
$$= -1$$

87. $\quad \sqrt{b} + c - a = \sqrt{64} + 6 - (-3)$
$$= 8 + 6 + 3 = 17$$

89. $\quad \dfrac{2c + a^3}{4b + 6a} = \dfrac{2(6) + (-3)^3}{4(64) + 6(-3)}$
$$= \dfrac{12 + (-27)}{256 + (-18)}$$
$$= \dfrac{-15}{238} = -\dfrac{15}{238}$$

91. $\quad 4A^3 + 2C = 4(-3)^3 + 2(6)$
$$= 4(-27) + 12$$
$$= -108 + 12$$
$$= -96$$

93. To find the difference between these two
temperatures, subtract the lowest temperature
from the highest temperature.
$$90° - (-22°) = 90° + 22°$$
$$= 112°$$

The difference is $112°$ F.

95. 282 feet below sea level is denoted -282. Thus,
$$11{,}049 - (-282) = 11{,}049 + 282$$
$$= 11{,}331.$$

The difference is $11{,}331$ ft.

97. The difference between the January and February
changes is

$$-6{,}439 - 5{,}039$$
$$= -(6{,}439 + 5{,}039) = -11{,}478.$$

99. Choose values for a and b such that $a < b$, for
example, $a = 1$ and $b = 2$. Then the difference
$a - b$ is
$$1 - 2 = 1 + (-2) = -1.$$

100. The answer in Exercise 99, -1, is less than 0.

101. This time let $a = -2$ and $b = -1$. Then the
difference $a - b$ is
$$-2 - (-1) = -2 + 1 = -1.$$

102. Again, the answer, -1, is less than 0. Based on
these exercises, if $a < b$, then $a - b \underline{\ <\ } 0$.

Section 1.3

1. The identity element for addition is 0 since, for
any real number a, $a + 0 = 0 + a = a$.
Choice (b)

3. The coefficient in the term $-8yz^2$ is -8 since the
number that is multiplied times the variables in
the term is -8. Choice (a)

5. The distinction between the commutative and
associative properties is that <u>the commutative
property says that the *order* in which terms are
added or multiplied does not matter, while the
associative property says that the *grouping* of the
terms to be added or multiplied does not matter.</u>

7. Using the second form of the distributive property,
$$5k + 3k = (5 + 3)k$$
$$= 8k.$$

9. $\quad -9r + 7r = (-9 + 7)r$
$$= -2r$$

11. $\quad -8z + 4w$
Since there is no common variable factor here, we
cannot use the distributive property to simplify the
expression.

13. Using the identity property, then the distributive
property,
$$-a + 7a = -1a + 7a$$
$$= (-1 + 7)a$$
$$= 6a.$$

15. Using the distributive property,
$$2(m + p) = 2m + 2p.$$

17. Using the distributive property,
$$-12(x - y) = -12[x + (-y)]$$
$$= -12(x) + (-12)(-y)$$
$$= -12x + 12y.$$

19. Using the distributive property,

$$-5(2d + f) = -5(2d) + (-5)(f)$$
$$= (-5 \cdot 2)d + (-5)f$$
$$= -10d - 5f.$$

21. $4x + 3x + 7 + 19$
$$= (4 + 3)x + (7 + 19)$$
$$= 7x + 26$$

23. $-12y + 4y + 3 + 2y$
$$= -12y + 4y + 2y + 3$$
$$= (-12 + 4 + 2)y + 3$$
$$= -6y + 3$$

25. $3(k + 2) - 5k + 6 + 3$
$$= 3k + 6 - 5k + 6 + 3$$
$$= 3k - 5k + 6 + 6 + 3$$
$$= (3 - 5)k + 6 + 6 + 3$$
$$= -2k + 15$$

27. $.25(8 + 4p) - .5(6 + 2p)$
$$= 2 + p - 3 - p$$
$$= p - p + 2 - 3$$
$$= (1 - 1)p + 2 - 3$$
$$= 0p - 1$$
$$= -1$$

29. $-(2p + 5) + 3(2p + 4) - 2p$
$$= -2p - 5 + 6p + 12 - 2p$$
$$= (-2 + 6 - 2)p + (-5 + 12)$$
$$= 2p + 7$$

31. $2 + 3(2z - 5) - 3(4z + 6) - 8$
$$= 2 + 6z - 15 - 12z - 18 - 8$$
$$= 6z - 12z + 2 - 15 - 26$$
$$= (6 - 12)z - 13 - 26$$
$$= -6z - 39$$

33. $-6P + 11P - 4P + 6 + 5$
$$= (-6 + 11 - 4)P + 6 + 5$$
$$= P + 11$$
For $P = 2$, $P + 11 = 13$.

35. $-2(M + 1) + 3(M - 4)$
$$= -2M - 2 + 3M - 12$$
$$= (-2 + 3)M - 2 - 12$$
$$= M - 14$$
For $M = -2$, $M - 14 = -16$.

37. $5x + 8x = (5 + 8)x = 13x$ *Distributive property*

39. $5(9r) = (5 \cdot 9)r = 45r$ *Associative property*

41. $5x + 9y = 9y + 5x$ *Commutative property*

43. $1 \cdot 7 = 7$ *Identity property*

45. $-\dfrac{1}{4}ty + \dfrac{1}{4}ty = 0$ *Inverse property*
A number plus its opposite equals 0.

47. $8(-4 + x) = 8(-4) + 8x$ *Distributive property*
$$= -32 + 8x$$

49. $0(.875x + 9y - 88z) = 0$ *Multiplication property of 0*
Zero times any quantity equals 0.

51. Answers will vary. Commutative: one example is washing your face and brushing your teeth. The activities can be carried out in either order. Not commutative: one example is putting on your socks and putting on your shoes.

53. $2 + 6x = 2 + 6 \cdot 5$ *Replace x with 5.*
$$= 2 + 30$$
$$= 32$$
$$8x = 8 \cdot 5 \qquad \text{\textit{Replace x with 5.}}$$
$$= 40$$
Since $32 \neq 40$, $2 + 6x \neq 8x$.

55. $96 \cdot 19 + 4 \cdot 19$
$$= (96 + 4)19$$
$$= (100)19$$
$$= 1900$$

57. $58 \cdot \dfrac{3}{2} - 8 \cdot \dfrac{3}{2}$
$$= (58 - 8)\dfrac{3}{2}$$
$$= (50)\dfrac{3}{2} = \dfrac{50}{1} \cdot \dfrac{3}{2}$$
$$= \dfrac{150}{2} = 75$$

59. $4.31(69) + 4.31(31)$
$$= (69 + 31)4.31$$
$$= (100)4.31$$
$$= 431$$

61. The terms have been grouped using the associative property of addition.

62. The terms have been regrouped using the associative property of addition.

63. The order of the terms inside the parentheses has been changed using the commutative property of addition.

64. The terms have been regrouped using the associative property of addition.

65. The common factor, x, has been distributed over the expression in parentheses using the distributive property.

66. The numbers in parentheses have been added to simplify the expression.

67. Answers will vary but should include the following properties: distributive, inverse (2), identity (2), associative (2), commutative (2), and multiplying by 0. See the summary in the Quick Review for Section 1.3 on page 39 of the text.

69. Is $a + (b \cdot c) = (a + b)(a + c)$?
No. One example is

$$7 + (5 \cdot 3) = (7 + 5)(7 + 3),$$

which is false since

$$7 + (5 \cdot 3) = 7 + 15 = 22$$

and

$$(7 + 5)(7 + 3) = 12(10) = 120.$$

Chapter 1 Review Exercises

1. $\left\{ -4, -1, 2, \dfrac{9}{4}, 4 \right\}$

Place dots for $-4, -1, 2, \dfrac{9}{4} = 2.25$, and 4 on a number line.

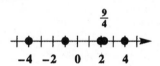

2. $\left\{ -5, -\dfrac{11}{4}, -.5, 0, 3, \dfrac{13}{3} \right\}$

Place dots for $-5, -\dfrac{11}{4} = -2.75, -5, 0, 3$, and $\dfrac{13}{3} = 4.\overline{3}$ on a number line.

$$\begin{array}{c} -\frac{11}{4} \quad\; \frac{13}{3} \\[-2pt] -.5 \end{array}$$

$$-5 \;\; -3 \;\; -1\;0\;1 \quad 3 \quad 5$$

3. $|-16| = -(-16) = 16$

4. $|23| = 23$

5. $-|-4| = -[-(-4)] = -(4) = -4$

In Exercises 6–9,

$$S = \left\{ -9, -\dfrac{4}{3}, -\sqrt{10}, 0, \dfrac{5}{3}, \sqrt{7}, \dfrac{12}{3} \right\}.$$

Simplified, $\dfrac{12}{3} = 4$.

6. 0 and 4 are whole numbers.

7. $-9, 0$, and 4 are integers.

8. $-9, -\dfrac{4}{3}, 0, \dfrac{5}{3}$, and 4 are rational numbers.

9. All the elements in the set are real numbers.

10. $\{ x \mid x$ is a natural number between 3 and 9 $\}$
The natural numbers between 3 and 9 are $4, 5, 6, 7$, and 8. Therefore, the set is $\{ 4, 5, 6, 7, 8 \}$.

11. $\{ y \mid y$ is a whole number less than 4 $\}$
The whole numbers less than 4 are $0, 1, 2$, and 3. Therefore, the set is $\{ 0, 1, 2, 3 \}$.

12. $4 \cdot 2 \le |12 - 4|$
The statement is true.
Since $4 \cdot 2 = 8$, and $|12 - 4| = 8$, the statement becomes $8 \le 8$, which is true.

13. $2 + |-2| > 4$
The statement is false.
Since $2 + |-2| = 2 + 2 = 4$, the statement becomes $4 > 4$, which is false.

14. **(a)** Look for the entry with the largest absolute value. This number is $-33,865$, which is associated with Japan.

(b) Look for the entry with the smallest absolute value. This number is 72, which is associated with South Korea.

15. $\{ x \mid x < -5 \}$
In interval notation, $x < -5$ is written as $(-\infty, -5)$. The parenthesis at -5 indicates that -5 is not included. The graph extends from -5 to the left.

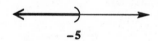

-5

16. $\{ x \mid -2 < x \le 3 \}$
In interval notation, $-2 < x \le 3$ is written as $(-2, 3]$. The parenthesis indicates that -2 is not included, while the bracket indicates that 3 is included. The graph goes from -2 to 3.

$-2 \qquad\qquad 3$

17. $-\dfrac{5}{8} - \left(-\dfrac{7}{3} \right) = -\dfrac{5}{8} + \dfrac{7}{3}$

$$= -\dfrac{15}{24} + \dfrac{56}{24} = \dfrac{41}{24}$$

18. $-\dfrac{4}{5} - \left(-\dfrac{3}{10} \right) = -\dfrac{4}{5} + \dfrac{3}{10}$

$$= -\dfrac{8}{10} + \dfrac{3}{10}$$

$$= -\dfrac{5}{10} = -\dfrac{1}{2}$$

19. $-5 + (-11) + 20 - 7$
$= -16 + 20 - 7$
$= 4 - 7$
$= -3$

20. $-9.42 + 1.83 - 7.6 - 1.9$
$= -7.59 - 7.6 - 1.9$
$= -15.19 - 1.9$
$= -17.09$

21. $-15 + (-13) + (-11)$
$= -28 + (-11)$
$= -39$

22. $-1 - 3 - (-10) + (-7)$
$= -4 + 10 + (-7)$
$= 6 + (-7)$
$= -1$

23. $\dfrac{3}{4} - \left(\dfrac{1}{2} - \dfrac{9}{10}\right) = \dfrac{3}{4} - \left(\dfrac{5}{10} - \dfrac{9}{10}\right)$
$= \dfrac{3}{4} - \left(-\dfrac{4}{10}\right)$
$= \dfrac{3}{4} + \dfrac{4}{10}$
$= \dfrac{15}{20} + \dfrac{8}{20} = \dfrac{23}{20}$

24. If the signs of the numbers being added are the same, the sign of the sum matches them. If the signs are different, the sign of the sum matches the one with the larger absolute value.

25. To subtract $a - b$, write it as the addition problem $a + (-b)$, and add.

26. $2(-5)(-3)(-3) = (-10)(-3)(-3)$
$= (30)(-3)$
$= -90$

27. $-\dfrac{3}{7}\left(-\dfrac{14}{9}\right) = \dfrac{3}{7} \cdot \dfrac{2 \cdot 7}{3 \cdot 3} = \dfrac{2}{3}$

28. $\dfrac{-38}{-19} = -38\left(\dfrac{1}{-19}\right) = 2$

29. $\dfrac{-2.3754}{-.74} = 3.21$

30. False; the percent change for Latinos (-6.9%) was less than that for African Americans (-17.6%).

31. True; $-5.4\% + .04\% = -5.36\%$, which is negative.

32. False; the largest percent change shown is $+191.0\%$.

33. $\dfrac{5}{7 - 7} = \dfrac{5}{0}$, which is undefined.
$\dfrac{7 - 7}{5} = \dfrac{0}{5} = 0$

34. $\left(\dfrac{3}{7}\right)^3 = \dfrac{3}{7} \cdot \dfrac{3}{7} \cdot \dfrac{3}{7} = \dfrac{27}{343}$

35. $(1.7)^2 = (1.7)(1.7) = 2.89$

36. $\sqrt{400} = 20$, because 20 is positive and $20^2 = 400$.

37. $\sqrt{-64}$ is not a real number.

38. $-14\left(\dfrac{3}{7}\right) + 6 \div 3 = -2(3) + (6 \div 3)$
$= -6 + 2 = -4$

39. $\dfrac{-5(3^2) + 9\left(\sqrt{4}\right) - 5}{6 - 5\left(-\sqrt{4}\right)}$
$= \dfrac{-5(9) + 9(2) - 5}{6 - 5(-2)}$
$= \dfrac{-45 + 18 - 5}{6 + 10}$
$= \dfrac{-32}{16} = -2$

In Exercises 40–41, let $k = -4$, $m = 2$, and $n = 16$.

40. $4k - 7m = 4(-4) - 7(2)$
$= -16 - 14$
$= -30$

41. $-3\sqrt{n} + m + 5k$
$= -3\left(\sqrt{16}\right) + 2 + 5(-4)$
$= -3(4) + 2 - 20$
$= -12 - 18$
$= -30$

42. In order to evaluate $(3 + 2)^2$, you should work within the parentheses first.

43. Let $a = 4$ and $b = 6$.
$(a + b)^2 = (4 + 6)^2 = 10^2 = 100$
$a^2 + b^2 = 4^2 + 6^2 = 16 + 36 = 52$
Since $100 \neq 52$, this shows that $(a + b)^2 \neq a^2 + b^2$.

44. $2q + 19q$
$= (2 + 19)q$ *Distributive property*
$= 21q$

45. $13z - 17z$

$= (13 - 17)z$ *Distributive property*

$= -4z$

46. $-m + 6m$

$= -1m + 6m$ *Identity property*

$= (-1 + 6)m$ *Distributive property*

$= 5m$

47. $5p - p$

$= 5p + (-1)p$ *Identity property*

$= [5 + (-1)]p$ *Distributive property*

$= 4p$

48. $-2(k + 3)$

$= -2(k) + (-2)(3)$ *Distributive property*

$= -2k - 6$

49. $6(r + 3)$

$= 6(r) + 6(3)$ *Distributive property*

$= 6r + 18$

50. $-3y + 6 - 5 + 4y$

$= -3y + 4y + 6 - 5$

$= y + 1$

51. $2a + 3 - a - 1 - a - 2$

$= 2a - a - a + 3 - 1 - 2$

$= 0$

52. $-2(k - 1) + 3k - k$

$= -2k + 2 + 3k - k$

$= -2k + 3k - k + 2$

$= 2$

53. $-3(4m - 2) + 2(3m - 1) - 4(3m + 1)$

$= -12m + 6 + 6m - 2 - 12m - 4$

$= -12m + 6m - 12m + 6 - 2 - 4$

$= -18m$

54. $2x + 3x = (2 + 3)x = 5x$ *Distributive property*

55. $-4 \cdot 1 = -4$ *Identity property*

56. $2(4x) = (2 \cdot 4)x = 8x$ *Associative property*

57. $-3 + 13 = 13 + (-3) = 10$ *Commutative property*

58. $-3 + 3 = 0$ *Inverse property*

59. $5(x + z) = 5x + 5z$ *Distributive property*

60. $0 + 7 = 7$ *Identity property*

61. $8 \cdot \dfrac{1}{8} = 1$ *Inverse property*

62. $\dfrac{9}{28} \cdot 0 = 0$ *Multiplication property of 0*

63. $\left(-\dfrac{4}{5}\right)^4 = \left(-\dfrac{4}{5}\right)\left(-\dfrac{4}{5}\right)\left(-\dfrac{4}{5}\right)\left(-\dfrac{4}{5}\right) = \dfrac{256}{625}$

64. $-\dfrac{5}{8}(-40) = -\dfrac{5}{8} \cdot \dfrac{-40}{1}$

$= 5 \cdot 5 = 25$

65. $\dfrac{75}{-5} = 75\left(\dfrac{1}{-5}\right) = -15$

66. $9(2m + 3n)$

$= 9(2m) + 9(3n)$ *Distributive property*

$= 18m + 27n$

67. $-25\left(-\dfrac{4}{5}\right) + 3^3 - 32 \div \sqrt{4}$

$= -25\left(-\dfrac{4}{5}\right) + 27 - 32 \div 2$

$= 20 + 27 - 16$

$= 31$

68. $(-5)^3 = (-5)(-5)(-5) = -125$

69. $-(3k - 4h)$

$= -1(3k - 4h)$ *Identity property*

$= -1(3k) + (-1)(-4h)$

$= -3k + 4h$

70. $-8 + |-14| + |-3| = -8 + 14 + 3$

$= 9$

71. $-\sqrt{25} = -(5) = -5$

72. $\dfrac{6 \cdot \sqrt{4} - 3 \cdot \sqrt{16}}{-2 \cdot 5 + 7(-3) - 10}$

$= \dfrac{6 \cdot 2 - 3 \cdot 4}{-2 \cdot 5 + 7(-3) - 10}$

$= \dfrac{12 - 12}{-10 - 21 - 10}$

$= \dfrac{0}{-41} = 0$

73. $-4.6(2.48) = -11.408$

74. $-\dfrac{10}{21} \div -\dfrac{5}{14} = -\dfrac{10}{21} \cdot -\dfrac{14}{5}$

$$= \dfrac{2 \cdot 5}{3 \cdot 7} \cdot \dfrac{2 \cdot 7}{5}$$

$$= \dfrac{2 \cdot 2}{3} = \dfrac{4}{3}$$

75. $-\dfrac{2}{3}[5(-2) + 8 - 4^3]$

$$= -\dfrac{2}{3}[5(-2) + 8 - 64]$$

$$= -\dfrac{2}{3}[-10 + 8 - 64]$$

$$= -\dfrac{2}{3}[-66] = 44$$

76.

$-(-p + 6q) - (2p - 3q)$
$= -1(-p + 6q) + (-1)(2p - 3q)$
$= -1(-p) + (-1)(6q) + (-1)(2p) + (-1)(-3q)$
$= p - 6q - 2p + 3q$
$= p - 2p - 6q + 3q$
$= -p - 3q$

77. $-2(3k^2 + 5m) = -2[3(-4)^2 + 5(2)]$

Let k= -4, m=2

$$= -2[3(16) + 5(2)]$$

$$= -2[48 + 10]$$

$$= -2[58]$$

$$= -116$$

78. $.8 - 4.9 - 3.2 + 1.14$

$$= -4.1 - 3.2 + 1.14$$

$$= -7.3 + 1.14$$

$$= -6.16$$

79. $-|-8| + |-3| = -8 + 3 = -5$

80. $-3^2 = -(3 \cdot 3) = -9$

81. $-\dfrac{4.64}{.16} = -29$

82. $-\dfrac{2}{3} - \left(\dfrac{1}{6} - \dfrac{5}{9}\right) = -\dfrac{2}{3} - \left(\dfrac{3}{18} - \dfrac{10}{18}\right)$

$$= -\dfrac{2}{3} - \left(-\dfrac{7}{18}\right)$$

$$= -\dfrac{2}{3} + \dfrac{7}{18}$$

$$= -\dfrac{12}{18} + \dfrac{7}{18} = -\dfrac{5}{18}$$

83. $-2x + 5 - 4x + 1$

$$= -2x - 4x + 5 + 1$$

$$= -6x + 6$$

84. $\dfrac{4m^3 - 3n}{7k^2 - 10} = \dfrac{4(2)^3 - 3(16)}{7(-4)^2 - 10}$

Let m=2, n=16, k= -4

$$= \dfrac{4(8) - 3(16)}{7(16) - 10}$$

$$= \dfrac{32 - 48}{112 - 10}$$

$$= \dfrac{-16}{102} = -\dfrac{8}{51}$$

For Exercises 85–94, evaluate the following expression. Let $x = 5$, $y = -4$, and $z = 1$.

$$\dfrac{2}{3}x - y^2 - 3z = \dfrac{2}{3}(5) - (-4)^2 - 3(1)$$

$$= \dfrac{10}{3} - 16 - 3$$

$$= \dfrac{10}{3} - \dfrac{48}{3} - \dfrac{9}{3} = -\dfrac{47}{3}$$

85. The value of the expression for $x = 5$, $y = -4$, and $z = 1$ is $-\dfrac{47}{3}$.

86. Since $-16 = -\dfrac{48}{3}$ and $-\dfrac{48}{3} < -\dfrac{47}{3}$, the value of the expression is greater than -16.

87. The set of integers is $\{\ldots, -2, -1, 0, 1, 2, \ldots\}$ so the value of the expression, $-\dfrac{47}{3}$, is not an integer. The value of the expression is a rational number since it is in the form $\dfrac{p}{q}$ with p and q integers and $q \neq 0$. The answers are (a) no and (b) yes.

88. $\left|-\dfrac{47}{3}\right| = -\left(-\dfrac{47}{3}\right) = \dfrac{47}{3}$

89. $\sqrt{-\dfrac{47}{3}}$ is *not a real number* since the square root of a negative number is not real.

90. The square of the expression is

$$\left(-\dfrac{47}{3}\right)^2 = \left(-\dfrac{47}{3}\right)\left(-\dfrac{47}{3}\right) = \dfrac{2209}{9}$$

91. The additive inverse of a number is its opposite.

$$-\left(-\dfrac{47}{3}\right) = \dfrac{47}{3}$$

The additive inverse is $\dfrac{47}{3}$.

92. A number times its multiplicative inverse equals one.

$$-\frac{47}{3}\left(-\frac{3}{47}\right) = 1$$

The multiplicative inverse is $-\frac{3}{47}$.

93. Placing parentheses around the first two terms of the expression should not change the answer since the original expression was added and subtracted from left to right.

$$\left(\frac{2}{3}x - y^2\right) - 3z = \left[\frac{2}{3}(5) - (-4)^2\right] - 3(1)$$

$$= \left(\frac{10}{3} - 16\right) - 3$$

$$= \left(\frac{10}{3} - \frac{48}{3}\right) - \frac{9}{3}$$

$$= -\frac{38}{3} - \frac{9}{3} = -\frac{47}{3}$$

The answer is the same.

94. Placing parentheses around the last two terms of the expression will change the answer since the work inside the parentheses with the two terms must be done before the final subtraction occurs.

$$\frac{2}{3}x - \left(y^2 - 3z\right) = \frac{2}{3}(5) - \left[(-4)^2 - 3(1)\right]$$

$$= \frac{10}{3} - (16 - 3)$$

$$= \frac{10}{3} - 13$$

$$= \frac{10}{3} - \frac{39}{3} = -\frac{29}{3}$$

The new answer is $-\frac{29}{3}$.

Chapter 1 Test

1. $\left\{-3, .75, \frac{5}{3}, 5, 6.3\right\}$

Place dots at $-3, .75, \frac{5}{3} = 1.\overline{6}, 5,$ and 6.3.

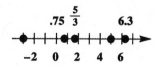

In Exercises 2–5,

$$A = \left\{-\sqrt{6}, -1, -.5, 0, 3, 7.5, \frac{24}{2}\right\}.$$

Simplified, $\frac{24}{2} = 12$.

2. $0, 3,$ and 12 are whole numbers.

3. $-1, 0, 3,$ and 12 are integers.

4. $-1, -.5, 0, 3, 7.5,$ and 12 are rational numbers.

5. All the elements in the set are real numbers.

6. $\{x \mid x < -3\}$
In interval notation, $x < -3$ is written as $(-\infty, -3)$. The parenthesis at -3 indicates that -3 is not included.. The graph extends from -3 to the left.

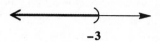

7. $\{y \mid -4 < y \leq 2\}$
In interval notation, $-4 < y \leq 2$ is written as $(-4, 2]$. The parenthesis indicates that -4 is not included, while the bracket indicates that 2 is included. The graph goes from -4 to 2.

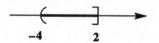

8. $-6 + 14 + (-11) - (-3)$
$= 8 + (-11) + 3$
$= -3 + 3 = 0$

9. $10 - 4 \cdot 3 + 6(-4)$
$= 10 - 12 + (-24)$
$= -2 + (-24) = -26$

10. $7 - 4^2 + 2(6) + (-4)^2$
$= 7 - 16 + 12 + 16$
$= 19$

11. $\dfrac{10 - 24 + (-6)}{\sqrt{16}(-5)}$
$= \dfrac{-14 + (-6)}{4(-5)}$
$= \dfrac{-20}{-20} = 1$

12. $\dfrac{-2[3 - (-1 - 2) + 2]}{\sqrt{9}(-3) - (-2)}$
$= \dfrac{-2[3 - (-3) + 2]}{3(-3) - (-2)}$
$= \dfrac{-2[8]}{-9 - (-2)}$
$= \dfrac{-16}{-7} = \dfrac{16}{7}$

13. $\dfrac{8\cdot 4 - 3^2\cdot 5 - 2(-1)}{-3\cdot 2^3 + 1}$

$= \dfrac{8\cdot 4 - 9\cdot 5 - 2(-1)}{-3\cdot 8 + 1}$

$= \dfrac{32 - 45 + 2}{-24 + 1}$

$= \dfrac{-11}{-23} = \dfrac{11}{23}$

14. The largest change (rightmost on a number line) is 8.0. The smallest change (leftmost on a number line) is -11.4.

15. The change with the largest absolute value is -11.4. The change with the smallest absolute value is 1.3.

16. The difference between the change for the Postal Service and the change for the Internal Revenue Service is

$$-6.8\% - 8.0\% = -14.8\%,$$

which is *negative*.

17. $\sqrt{196} = 14$, because 14 is positive and $14^2 = 196$.

18. $-\sqrt{225} = -\left(\sqrt{225}\right) = -(15) = -15$

19. $\sqrt{-16}$ is not a real number.

20. **(a)** If a is positive, then $\sqrt{a}$ will represent a positive number.

(b) If a is negative, then $\sqrt{a}$ will not represent a real number.

(c) If a is 0, then $\sqrt{a}$ will be 0.

21. $\dfrac{8k + 2m^2}{r - 2} = \dfrac{8(-3) + 2(-3)^2}{25 - 2}$

$\qquad$ *Let k= –3, m= –3, r=25*

$= \dfrac{8(-3) + 2(9)}{23}$

$= \dfrac{-24 + 18}{23}$

$= \dfrac{-6}{23} \text{ or } -\dfrac{6}{23}$

22. $-3(2k - 4) + 4(3k - 5) - 2 + 4k$

$= -3(2k) + (-3)(-4) + 4(3k)$
$\quad + 4(-5) - 2 + 4k$

$= -6k + 12 + 12k - 20 - 2 + 4k$

$= -6k + 12k + 4k + 12 - 20 - 2$

$= 10k - 10$

23. When simplifying

$$(3r + 8) - (-4r + 6),$$

the subtraction sign in front of $(-4r + 6)$ changes the sign of the terms $-4r$ and 6.

$$(3r + 8) - (-4r + 6)$$
$$= 3r + 8 - (-4r) - 6$$
$$= 3r + 8 + 4r - 6$$
$$= 3r + 4r + 8 - 6$$
$$= 7r + 2$$

24. $6 + (-6) = 0$

The answer is *B. Inverse Property*.
The sum of 6 and its inverse, -6, equals zero.

25. $-2 + (3 + 6) = (-2 + 3) + 6$

The answer is *D. Associative Property*.
The order of the terms is the same, but the grouping has changed.

26. $5x + 15x = (5 + 15)x$

The answer is *A. Distributive Property*.
This is the second form of the distributive property.

27. $13\cdot 0 = 0$

The answer is *F. Multiplication Property of* 0.
Multiplication by 0 always equal 0.

28. $-9 + 0 = -9$

The answer is *C. Identity Property*.
The addition of 0 to any number does not change the number.

29. $4\cdot 1 = 4$

The answer is *C. Identity Property*.
Multiplication of any number by 1 does not change the number.

30. $(a + b) + c = (b + a) + c$

The answer is *E. Commutative Property*.
The order of the terms a and b is reversed.

CHAPTER 2 LINEAR EQUATIONS AND INEQUALITIES

Section 2.1

1. **(a)** $3x + x - 1 = 0$ can be written as $4x = 1$, so it is linear.

(c) $6x + 2 = 9$ can be written as $6x = 7$, so it is linear.

3.
$$3(x + 4) = 5x \qquad \textit{Original equation}$$
$$3(6 + 4) = 5 \cdot 6 \text{ ? } \quad \textit{Let x=6.}$$
$$3(10) = 30 \quad \text{ ? } \quad \textit{Add.}$$
$$30 = 30 \qquad \textit{True}$$

Since a true statement is obtained, 6 is a solution.

5. If two equations are equivalent, they have the same solution set .

7. $.06(10 - x)(100)$
$$= .06(100)(10 - x)$$
$$= 6(10 - x)$$
$$= 60 - 6x \quad \text{Choice (b)}$$

9.
$$7k + 8 = 1$$
$$7k + 8 - 8 = 1 - 8 \qquad \textit{Subtract 8.}$$
$$7k = -7$$
$$\frac{7k}{7} = \frac{-7}{7} \qquad \textit{Divide by 7.}$$
$$k = -1$$

Solution set: $\{-1\}$

11.
$$8 - 8x = -16$$
$$8 - 8x - 8 = -16 - 8 \qquad \textit{Subtract 8.}$$
$$-8x = -24$$
$$\frac{-8x}{-8} = \frac{-24}{-8} \qquad \textit{Divide by -8.}$$
$$x = 3$$

Solution set: $\{3\}$

13.
$$7y - 5y + 15 = y + 8$$
$$2y + 15 = y + 8 \qquad \textit{Combine terms.}$$
$$2y = y - 7 \qquad \textit{Subtract 15.}$$
$$y = -7 \qquad \textit{Subtract y.}$$

Solution set: $\{-7\}$

15.
$$12w + 15w - 9 + 5 = -3w + 5 - 9$$
$$27w - 4 = -3w - 4 \qquad \textit{Combine terms.}$$
$$30w - 4 = -4 \qquad \textit{Add 3w.}$$
$$30w = 0 \qquad \textit{Add 4.}$$
$$w = 0 \qquad \textit{Divide by 30.}$$

Solution set: $\{0\}$

17.
$$2(x + 3) = -4(x + 1)$$
$$2x + 6 = -4x - 4 \qquad \textit{Remove parentheses.}$$
$$6x + 6 = -4 \qquad \textit{Add 4x.}$$
$$6x = -10 \qquad \textit{Subtract 6.}$$
$$x = \frac{-10}{6} = -\frac{5}{3} \qquad \textit{Divide by 6.}$$

Solution set: $\left\{-\dfrac{5}{3}\right\}$

19.
$$3(2w + 1) - 2(w - 2) = 5$$
$$6w + 3 - 2w + 4 = 5 \qquad \textit{Remove parentheses.}$$
$$4w + 7 = 5 \qquad \textit{Combine terms.}$$
$$4w = -2 \qquad \textit{Subtract 7.}$$
$$w = \frac{-2}{4} \qquad \textit{Divide by 4.}$$
$$w = -\frac{1}{2}$$

Solution set: $\left\{-\dfrac{1}{2}\right\}$

21.
$$2x + 3(x - 4) = 2(x - 3)$$
$$2x + 3x - 12 = 2x - 6$$
$$5x - 12 = 2x - 6$$
$$3x = 6$$
$$x = \frac{6}{3} = 2$$

Solution set: $\{2\}$

23.
$$6p - 4(3 - 2p) = 5(p - 4) - 10$$
$$6p - 12 + 8p = 5p - 20 - 10$$
$$14p - 12 = 5p - 30$$
$$9p = -18$$
$$p = -2$$

Solution set: $\{-2\}$

25. $-[2z - (5z + 2)] = 2 + (2z + 7)$
   $-[2z - 5z - 2] = 2 + 2z + 7$
   $-2z + 5z + 2 = 2 + 2z + 7$
   $3z + 2 = 2z + 9$
   $z = 7$

Solution set: $\{7\}$

27. $-3m + 6 - 5(m - 1)$
   $= 4m - (2m - 4) - 9m + 5$
   $-3m + 6 - 5m + 5$
   $= 4m - 2m + 4 - 9m + 5$
   $-8m + 11 = -7m + 9$
   $-m = -2$
   $m = 2$

Solution set: $\{2\}$

29. $-[3y - (2y + 5)] = -4 - [3(2y - 4) - 3y]$
   $-[3y - 2y - 5] = -4 - [6y - 12 - 3y]$
   $-[y - 5] = -4 - [3y - 12]$
   $-y + 5 = -4 - 3y + 12$
   $-y + 5 = -3y + 8$
   $2y = 3$
   $y = \dfrac{3}{2}$

Solution set: $\left\{\dfrac{3}{2}\right\}$

31.

$-(9 - 3a) - (4 + 2a) - 3 = -(2 - 5a) + (-a) + 1$
   $-9 + 3a - 4 - 2a - 3 = -2 + 5a - a + 1$
   $a - 16 = 4a - 1$
   $-15 = 3a$
   $-5 = a$

Solution set: $\{-5\}$

33. $2(-3 + m) - (3m - 4)$
   $= -(-4 + m) - 4m + 6$
   $-6 + 2m - 3m + 4 = 4 - m - 4m + 6$
   $-m - 2 = -5m + 10$
   $4m = 12$
   $m = 3$

Solution set: $\{3\}$

35. Yes, the coefficients will be larger, but you will get the correct solution. As long as you multiply both sides of the equation by the *same* nonzero number, the resulting equation is equivalent and the solution does not change.

37. $\dfrac{8y}{3} - \dfrac{2y}{4} = -13$

Multiply both sides by the least common denominator, 12.

$12\left(\dfrac{8y}{3} - \dfrac{2y}{4}\right) = 12(-13)$
   $32y - 6y = -156$
   $26y = -156$
   $y = \dfrac{-156}{26} = -6$

Solution set: $\{-6\}$

39. $\dfrac{2r - 3}{7} + \dfrac{3}{7} = -\dfrac{r}{3}$

Multiply both sides by the least common denominator, 21.

$21\left(\dfrac{2r - 3}{7} + \dfrac{3}{7}\right) = 21\left(-\dfrac{r}{3}\right)$
   $3(2r - 3) + 3(3) = 7(-r)$
   $6r - 9 + 9 = -7r$
   $6r = -7r$
   $13r = 0$
   $r = \dfrac{0}{13} = 0$

Solution set: $\{0\}$

41. $\dfrac{2x + 5}{5} = \dfrac{3x + 1}{2} + \dfrac{-x + 7}{2}$

Multiply both sides by the least common denominator, 10.

$10\left(\dfrac{2x + 5}{5}\right) = 10\left(\dfrac{3x + 1}{2} + \dfrac{-x + 7}{2}\right)$
   $2(2x + 5) = 5(3x + 1) + 5(-x + 7)$
   $4x + 10 = 15x + 5 - 5x + 35$
   $4x + 10 = 10x + 40$
   $-6x = 30$
   $x = \dfrac{30}{-6} = -5$

Solution set: $\{-5\}$

43. $.09k + .13(k + 300) = 61$

Multiply both sides by 100.

$100[.09k + .13(k + 300)] = 100(61)$
$100(.09k) + 100(.13)(k + 300) = 6100$
   $9k + 13(k + 300) = 6100$
   $9k + 13k + 3900 = 6100$
   $22k = 2200$
   $k = \dfrac{2200}{22} = 100$

Solution set: $\{100\}$

45. $.20(14,000) + .14t = .18(14,000 + t)$

Multiply both sides by 100.

$100[.20(14,000) + .14t] = 100[.18(14,000 + t)]$

$20(14,000) + 14t = 18(14,000 + t)$

$280,000 + 14t = 252,000 + 18t$

$28,000 = 4t$

$t = 7000$

Solution set: $\{7000\}$

47. $.08x + .12(260 - x) = .48x$

Multiply both sides by 100.

$8x + 12(260 - x) = 48x$

$8x + 3120 - 12x = 48x$

$-4x + 3120 = 48x$

$3120 = 52x$

$x = \dfrac{3120}{52} = 60$

Solution set: $\{60\}$

49. Solve each equation.

(a) $2x + 1 = 3$

$2x = 2$

$x = 1$

The solution set, $\{1\}$, has only one element, so $2x + 1 = 3$ is a conditional equation.

(b) $x = 3x - 2x$

$x = x$

The equation is an identity, not a conditional equation.

(c) $2(x + 2) = 2x + 2$

$2x + 4 = 2x + 2$

$4 = 2$ *False*

The equation is a contradiction, not a conditional equation.

(d) $5x - 3 = 4x + x - 5 + 2$

$5x - 3 = 5x - 3$

The equation is an identity, not a conditional equation.

Therefore, the correct answer is (a).

51. $-2p + 5p - 9 = 3(p - 4) - 5$

$3p - 9 = 3p - 12 - 5$

$3p - 9 = 3p - 17$

$-9 = -17$ *False*

The equation is a contradiction.

Solution set: $\emptyset$

53. $6x + 2(x - 2) = 9x + 4$

$6x + 2x - 4 = 9x + 4$

$8x - 4 = 9x + 4$

$-8 = x$

This is a conditional equation.

Solution set: $\{-8\}$

55. $-11m + 4(m - 3) + 6m = 4m - 12$

$-11m + 4m - 12 + 6m = 4m - 12$

$-m - 12 = 4m - 12$

$0 = 5m$

$0 = m$

This is a conditional equation.

Solution set: $\{0\}$

57. $7[2 - (3 + 4r)] - 2r = -9 + 2(1 - 15r)$

$7[2 - 3 - 4r] - 2r = -9 + 2 - 30r$

$7[-1 - 4r] - 2r = -7 - 30r$

$-7 - 28r - 2r = -7 - 30r$

$-7 - 30r = -7 - 30r$

The equation is an identity.

Solution set: $\{\text{all real numbers}\}$

59. $5x = 10$ and $\dfrac{5x}{x + 2} = \dfrac{10}{x + 2}$

Solve the first equation.

$$5x = 10$$
$$x = 2$$

Solution set: $\{2\}$

Solve the second equation.

$$\dfrac{5x}{x + 2} = \dfrac{10}{x + 2}$$

Since the denominators are the same, the numerators must be equal.

$$5x = 10$$
$$x = 2$$

Solution set: $\{2\}$

Since these two equations have the same solution set, they are equivalent equations.

61. $y = -3$ and $\dfrac{y}{y + 3} = \dfrac{-3}{y + 3}$

For $y = -3$, the solution set is $\{-3\}$.

Since -3 would cause the denominators in the second equation to become zero, it cannot be used as a solution. Therefore, the equations are not equivalent.

63. $k = 4$ and $k^2 = 16$

The solution set of the first equation is $\{4\}$. Since $(-4)^2 = 16$ and $4^2 = 16$, the solution set of the second equation is $\{-4, 4\}$.

Since these two equations do not have the same solution set, they are not equivalent equations.

65. **(a)** $y = .55x - 42.5$

$\qquad y = .55(95) - 42.5 \qquad$ *Let x=95.*

$\qquad\quad = 52.25 - 42.5$

$\qquad\quad = 9.75$ million tickets

(b) $\qquad y = .55x - 42.5$

$\qquad 7.9 = .55x - 42.5 \qquad$ *Let y=7.9.*

$\qquad 50.4 = .55x$

$\qquad\quad x = \dfrac{50.4}{.55} \approx 91.6$

The answer represents the 1991–92 season.

67. $y = .1x - 8.5$

$\qquad y = .1(93) - 8.5 \qquad$ *Let x=93.*

$\qquad\quad = 9.3 - 8.5 = .8$ million tickets

$\qquad y = .1x - 8.5$

$\qquad .75 = .1x - 8.5 \qquad$ *Let y=.75.*

$\qquad 9.25 = .1x$

$\qquad\quad x = \dfrac{9.25}{.1} = 92.5$

The answer represents the 1992–93 season.

Section 2.2

1. **(a)** $x = \dfrac{5x + 8}{3}$

$\qquad 3x = 3\left(\dfrac{5x + 8}{3}\right)$

$\qquad 3x = 5x + 8$

(b) $t = \dfrac{bt + k}{c} \ (c \neq 0)$

$\qquad ct = c\left(\dfrac{bt + k}{c}\right)$

$\qquad ct = bt + k$

2. **(a)** $3x - 5x = 5x + 8 - 5x$

$\qquad\quad 3x - 5x = 8$

(b) $ct - bt = bt + k - bt$

$\qquad\quad ct - bt = k$

3. Use the distributive property in each case.

(a) $-2x = 8 \qquad$ **(b)** $(c - b)t = k$

4. **(a)** $\dfrac{-2x}{-2} = \dfrac{8}{-2} \qquad$ **(b)** $\dfrac{(c - b)t}{c - b} = \dfrac{k}{c - b}$

$\qquad\quad x = -4 \qquad\qquad\qquad\quad t = \dfrac{k}{c - b}$

5. The restriction $b \neq c$ must be applied. If $b = c$, the denominator becomes 0 and division by 0 is undefined.

6. The first equation doesn't have any variables in the denominator.

7. $\qquad A = \dfrac{1}{2}bh$

$\qquad 2A = 2\left(\dfrac{1}{2}bh\right) \qquad$ *Multiply by 2.*

$\qquad 2A = bh$

$\qquad \dfrac{2A}{b} = \dfrac{bh}{b} \qquad$ *Divide by b.*

$\qquad \dfrac{2A}{b} = h$

$\qquad \dfrac{2A}{b} = \dfrac{2}{1} \cdot \dfrac{A}{b}$

$\qquad\quad = 2\left(\dfrac{A}{b}\right) \qquad$ *This choice is (a).*

$\qquad\quad = 2A\left(\dfrac{1}{b}\right) \qquad$ *This is choice (b).*

To get choice (c), multiply $\dfrac{2A}{b}$ by $1 = \dfrac{\frac{1}{2}}{\frac{1}{2}}$.

$$\dfrac{\frac{1}{2}}{\frac{1}{2}} \cdot \dfrac{2A}{b} = \dfrac{A}{\frac{1}{2}b}$$

Choice (d), $h = \dfrac{\frac{1}{2}A}{b}$, can be multiplied by $\dfrac{2}{2}$ on the right side to get $h = \dfrac{A}{2b}$, so it is *not* equivalent to $h = \dfrac{2A}{b}$. Therefore, the correct answer is (d).

9. Solve $d = rt$ for t.

$\qquad \dfrac{d}{r} = \dfrac{rt}{r} \qquad$ *Divide by r.*

$\qquad \dfrac{d}{r} = t$ or $t = \dfrac{d}{r}$

11. Solve $A = bh$ for b.

$\qquad \dfrac{A}{h} = b \qquad$ *Divide by h.*

13. Solve $P = a + b + c$ for a.

$$P - (b + c) = a + b + c - (b + c)$$

  Subtract $(b + c)$.

$$P - b - c = a$$

15. Solve $A = \frac{1}{2}bh$ for h.

$$\frac{2A}{b} = \frac{2}{b}\left(\frac{1}{2}bh\right) \quad \text{Multiply by 2;}$$
  divide by b.

$$\frac{2A}{b} = h$$

17. Solve $S = 2\pi rh + 2\pi r^2$ for h.

$$S - 2\pi r^2 = 2\pi rh \quad \text{Subtract } 2\pi r^2.$$

$$\frac{S - 2\pi r^2}{2\pi r} = \frac{2\pi rh}{2\pi r} \quad \text{Divide by } 2\pi r.$$

$$\frac{S - 2\pi r^2}{2\pi r} = h$$

or $\quad \dfrac{S}{2\pi r} - r = h$

19. Solve $C = \frac{5}{9}(F - 32)$ for F.

$$\frac{9}{5}C = \frac{9}{5} \cdot \frac{5}{9}(F - 32) \quad \text{Multiply by } \frac{9}{5}.$$

$$\frac{9}{5}C = F - 32$$

$$\frac{9}{5}C + 32 = F \qquad \text{Add 32.}$$

21. Solve $A = 2HW + 2LW + 2LH$ for H.

$$A - 2LW = 2HW + 2LH \quad \text{Subtract } 2LW.$$

$$A - 2LW = (2W + 2L)H \quad \text{Distributive property}$$

$$\frac{A - 2LW}{2W + 2L} = H \quad \text{Divide by } 2W + 2L.$$

23. Solve $2k + ar = r - 3y$ for r.
Get the "r-terms" on one side and the other terms on the other side.

$$ar - r = -2k - 3y$$

$$(a - 1)r = -2k - 3y \quad \text{Distributive property}$$

$$r = \frac{-2k - 3y}{a - 1} \quad \text{Divide by } a-1.$$

The answer can also be written as

$$r = \frac{2k + 3y}{1 - a},$$

which would occur if you took the r-terms to the right side in your first step.

25. Solve $w = \dfrac{3y - x}{y}$ for y.

$$wy = 3y - x \quad \text{Multiply by } y.$$

$$x = 3y - wy \quad \text{Get } y\text{-terms on one side.}$$

$$x = (3 - w)y \quad \text{Distributive property}$$

$$\frac{x}{3 - w} = y \quad \text{Divide by } 3-w.$$

Equivalently, we have

$$y = \frac{-x}{w - 3}.$$

27. $2x - mx = z - m$ First, get all terms with m on one side of the equation and all terms without m on the other side. The result should be $m - mx = z - 2x$ (or an equivalent expression). Then use the distributive property on the left, so m is a factor. This step gives $m(1 - x) = z - 2x$. Finally, divide both sides by the coefficient of m, $1 - x$, to get $m = \dfrac{z - 2x}{1 - x}$.

29. Solve $d = rt$ for t.

$$t = \frac{d}{r}$$

To find t, substitute $d = 500$ and $r = 134.5$.

$$t = \frac{500}{134.5} \approx 3.7$$

His time was about 3.7 hr.

31. Use the formula $F = \dfrac{9}{5}C + 32$, and substitute -40 for C.

$$F = \frac{9}{5}(-40) + 32$$
$$= -72 + 32$$
$$= -40$$

The temperature was $-40°$ F.

33. Solve $P = 4s$ for s.

$$s = \frac{P}{4}$$

To find s, substitute 920 for P.

$$s = \frac{920}{4} = 230$$

The length of each side is 230 m.

35. Solve $d = rt$ for r.

$$r = \frac{d}{t}$$

$$r = \frac{520}{10} = 52 \quad \textit{Let d=520, t=10.}$$

Her rate was 52 mph.

37. Solve $S = 2\pi rh + 2\pi r^2$ for h.

$$S - 2\pi r^2 = 2\pi rh$$

$$\frac{S - 2\pi r^2}{2\pi r} = h$$

$$h = \frac{86.125\pi - 2\pi(3.25)^2}{2\pi(3.25)} \quad \textit{Let S=86.125}\pi,$$
$$\textit{r=3.25.}$$

$$= \frac{86.125\pi - 21.125\pi}{6.5\pi}$$

$$= \frac{65\pi}{6.5\pi} = 10 \text{ cm}$$

39. Use $V = \pi r^2 h$ and replace r with $2r$ to see the effect of doubling the radius.

$$V_1 = \pi(2r)^2 h$$
$$= \pi(4r^2)h$$
$$= 4(\pi r^2 h)$$

We see that V_1 is 4 times V, so the volume is 4 times as large.

41. Solve $V = LWH$ for L.

$$L = \frac{V}{WH}$$

To find L, substitute $V = 128$, $W = 4$, and $H = 4$.

$$L = \frac{128}{(4)(4)} = 8$$

The stack is 8 ft long.

43. The distance around the yard is the perimeter so you would need to use the perimeter to decide how much fencing to buy.

45. The mixture is 36 oz and that part which is alcohol is 9 oz. Thus, the percent of alcohol is

$$\frac{9}{36} = \frac{1}{4} = \frac{25}{100} = 25\%.$$

The percent of water is

$$100\% - 25\% = 75\%.$$

47. Find what percent $6300 is of $210,000, that is,

$$\frac{6300}{210,000} = .03 = 3\%$$

The agent received a 3% rate of commission.

49. In 1982, 3% of the television audience watched cable.

$$3\% \text{ of } 50,000$$
$$= .03(50,000)$$
$$= 1,500$$

51. In 1997, 35% of the television audience watched cable.

$$35\% \text{ of } 35,000$$
$$= .35(35,000)$$
$$= 12,250$$

In Exercises 53–56, use the rule of 78.

$$u = f \cdot \frac{k(k+1)}{n(n+1)}$$

53. Substitute 700 for f, 4 for k, and 36 for n.

$$u = 700 \cdot \frac{4(4+1)}{36(36+1)}$$

$$= 700 \cdot \frac{4(5)}{36(37)} \approx 10.51$$

The unearned interest is $10.51.

55. Substitute 380.50 for f, 8 for k, and 24 for n.

$$u = (380.50) \cdot \frac{8(8+1)}{24(24+1)}$$

$$= (380.50) \cdot \frac{8(9)}{24(25)} \approx 45.66$$

The unearned interest is $45.66.

57. Use $A = \dfrac{24f}{b(p+1)}$.

Substitute 200 for f, 1920 for b, and 24 for p.

$$A = \frac{24(200)}{1920(24+1)} = .1 = 10\%$$

The approximate annual interest is 10%.

59.
$$2L + 2W = 28$$
$$2L + 2(4) = 28 \quad \textit{Let W=4.}$$
$$2L + 8 = 28$$
$$2L = 20$$
$$L = 10$$

Section 2.3

1. The phrase "the sum of a number and 6" is an *expression* and can be translated as $x + 6$.

3. The sentence "$\dfrac{2}{3}$ of a number is 12" is an *equation* and can be translated as $\dfrac{2}{3}x = 12$.

5. The phrase "the ratio of a number and 5" is an *expression* and can be translated as $\dfrac{x}{5}$.

7. If x is the amount invested at 6%, then $40,000 - x$ is the amount invested at 4% and the equation becomes

$$.04(40,000 - x) + .06x = 2040.$$

Solve the equation.

$$1600 - .04x + .06x = 2040$$
$$.02x = 440$$
$$x = \frac{440}{.02} = 22,000$$
$$40,000 - x = 18,000$$

These are the same answers as given in Example 5.

9. "Decreased by" indicates subtraction. "A number decreased by 18" translates as

$$x - 18.$$

11. The phrase "the product of 9 less than a number and 6 more than the number" is translated $(x - 9)(x + 6)$.

13. The phrase "the ratio of 12 and a nonzero number" is translated

$$\frac{12}{x},\ (x \neq 0).$$

15. Decide what you are asked to find, choose a variable and write down what it represents. If there are other unknown quantities, express them using the same variable. Write an equation from the wording in the problem and solve it. Answer the question(s) of the problem. Check the answers in the wording of the problem.

17. The sentence "if the quotient of a number and 6 is added to twice the number, the result is 8 less than the number" can be translated as

$$2x + \frac{x}{6} = x - 8.$$

19. The sentence "when $\dfrac{2}{3}$ of a number is subtracted from 12, the result is 10" can be translated as

$$12 - \frac{2}{3}x = 10.$$

21. *Step 1*
Let $x =$ the number of shoppers at large chain bookstores.

Step 2
$x - 70 =$ the number of shoppers at small chain/independent bookstores.

Step 3
A total of 442 book buyers shopped at these two types of stores, so

$$x + (x - 70) = 442.$$

Step 4

$$2x - 70 = 442$$
$$2x = 512$$
$$x = 256$$

Step 5
There were 256 large chain bookstore shoppers and $256 - 70 = 186$ small chain/independent bookstore shoppers.

Step 6
The number of large chain shoppers was 70 more than the number of small chain/independent shoppers, and the total number of these two bookstore types was $256 + 186 = 442$.

23. *Step 1*
Let $x =$ sales for TLC Beatrice International.

Step 2
$x + 9.3 =$ sales for Ingram Industries (in billions).

Step 3
Together their sales totaled 13.7 billion, so

$$x + (x + 9.3) = 13.7.$$

Step 4

$$2x + 9.3 = 13.7$$
$$2x = 4.4$$
$$x = 2.2$$

Step 5
The sales for TLC Beatrice International were $2.2 billion. The sales for Ingram Industries were $2.2 + \$9.3 = \$11.5$ billion.

Step 6
$11.5 billion is $9.3 billion more than $2.2 billion and the total is $2.2 + \$11.5 = \$13.7$ billion.

25. *Step 1*
Let $x =$ the number of hits Ruth got.

Step 2
$x + 57 =$ the number of hits Hornsby got.

Step 3
Their base hits totaled 5803, so

$$x + (x + 57) = 5803.$$

Step 4

$$2x + 57 = 5803$$
$$2x = 5746$$
$$x = 2873$$

Step 5
Ruth got 2873 base hits, and Hornsby got
$2873 + 57 = 2930$ base hits.

Step 6
2930 is 57 more than 2873 and the total is
$2873 + 2930 = 5803$.

27. *Step 1*
Let $x =$ video sales revenue in 1990.

Step 2
$2x + .27 =$ video rental revenue (in billions).

Step 3
Together these sales were $9.81 billion, so

$$x + (2x + .27) = 9.81.$$

Step 4

$$3x + .27 = 9.81$$
$$3x = 9.54$$
$$x = 3.18$$
$$2x + .27 = 6.63$$

Step 5
In 1990, video sales revenue was $3.18 billion
and video rental revenue was $6.63 billion.

Step 6
The total sales were $3.18 + $6.63 = $9.81
billion and $6.63 is $.27 more than twice $3.18.

29. The values in Exercise 27, $3.18 billion and
$6.63 billion, and the values in Exercise 28, $7.34
billion and $7.49 billion, correspond well with the
information depicted in the graph.

31. Let $P =$ the percent increase.
The amount of increase was

$$20.9 - 20.6 = .3.$$

The base score was 20.6.

$$P = \frac{.3}{20.6} \approx .015 = 1.5\%$$

The increase was about 1.5%.

33. 227% of $1386
   $= 2.27(\$1386)$
   $= \$3146.22$
So tuition increased $3146.22 from 1985 to 1996.
Adding this amount to the base of $1386 gives us
about $4532 for the 1996 cost.

35. Let $x =$ the motel receipts before
         tax is added.
   $.08x =$ amount of sales tax
The total amount of receipts is $1650.78, so

$$x + .08x = 1650.78.$$
$$1x + .08x = 1650.78$$
$$1.08x = 1650.78$$
$$x = \frac{1650.78}{1.08} = 1528.50$$

This gives the motel receipts before the taxes. The
sales tax is 8% of 1528.50 or
$.08(1528.50) = \$122.28$.

37. Let $x =$ the number of liters of
         the 20% alcohol solution.
Complete the table.

Strength	Liters of solution	Liters of pure alcohol
12%	12	$.12(12) = 1.44$
20%	x	$.20x$
14%	$x + 12$	$.14(x + 12)$

From the last column, we can formulate an
equation that compares the number of liters of
pure alcohol.

$$\text{Alcohol in 12\%} + \text{alcohol in 20\%} = \text{alcohol in 14\%}.$$

$$1.44 + .20x = .14(x + 12)$$
$$1.44 + .20x = .14x + 1.68$$
$$.06x = .24$$
$$x = 4$$

4 L of the 20% alcohol solution are needed.

39. Let $x =$ the number of liters of
         pure alcohol.

Make a table.

Strength	Liters of solution	Liters of pure alcohol
70%	50	$.70(50) = 35$
100%	x	$1.00x = x$
78%	$50 + x$	$.78(50 + x)$

Note that pure alcohol has a strength of 100%.
The total number of liters of pure alcohol in the
two original containers must equal the number of
liters of pure alcohol in the final container. The
last column of the table gives the equation.

$$35 + x = .78(50 + x)$$
$$3500 + 100x = 78(50 + x) \qquad \textit{Multiply by 100.}$$
$$3500 + 100x = 3900 + 78x$$
$$22x = 400$$
$$x = \frac{400}{22} \text{ or } 18\frac{2}{11}$$

$18\frac{2}{11}$ L of pure alcohol should be added.

41. Let $x =$ the number of liters of the 20%
solution that must be drained.
The number of liters drained is equal to the
number of liters replaced. Therefore, x is also the
number of liters of 100% solution. The amount of
20% solution left will be $20 - x$. Make a table.

Strength	Liters of solution	Liters of pure antifreeze
20%	$20 - x$	$.20(20 - x)$
100%	x	$1.00x$
40%	20	$.40(20)$

The last column gives the equation.

$$.20(20 - x) + 1.00x = .40(20)$$
$$4 - .2x + x = 8$$
$$.8x = 4$$
$$x = \frac{4}{.8} = 5$$

5 L of the 20% antifreeze solution must be
drained and replaced.

43. Let $x =$ the amount invested at 3%.
$12,000 - x =$ the amount invested at 4%.
Complete the table.

Percent as a Decimal	Amount Invested	Interest in One Year
.03	x	$.03x$
.04	$12,000 - x$	$.04(12,000 - x)$
Total	$12,000$	440

The last column gives the equation.

$$\begin{array}{c} \text{Interest} \\ \text{at 3\%} \end{array} + \begin{array}{c} \text{interest} \\ \text{at 4\%} \end{array} = \begin{array}{c} \text{total} \\ \text{interest.} \end{array}$$

$$.03x + .04(12,000 - x) = 440$$
$$3x + 4(12,000 - x) = 44,000 \qquad \textit{Multiply by 100.}$$
$$3x + 48,000 - 4x = 44,000$$
$$-x = -4000$$
$$x = 4000$$

Jason should invest $4000 at 3% and
$12,000 - 4000 = \$8000$ at 4%.

45. Let $x =$ the amount invested at 4.5%.
$2x - 1000 =$ the amount invested at 3%.
The formula for interest is $I = prt$ with $t = 1$ yr.
The information is organized in the following
table.

Percent as a Decimal	Amount Invested	Interest in One Year
.045	x	$.045x$
.03	$2x - 1000$	$.03(2x - 1000)$
Total		1020

The last column gives the equation.

$$\begin{array}{c} \text{Interest} \\ \text{at 4.5\%} \end{array} + \begin{array}{c} \text{interest} \\ \text{at 3\%} \end{array} = \begin{array}{c} \text{total} \\ \text{interest.} \end{array}$$

$$.045x + .03(2x - 1000) = 1020$$
$$45x + 30(2x - 1000) = 1,020,000$$
$$\qquad\qquad \textit{Multiply by 1000.}$$
$$45x + 60x - 30,000 = 1,020,000$$
$$105x = 1,050,000$$
$$x = \frac{1,050,000}{105} = 10,000$$
$$2x - 1000 = 2(10,000) - 1000 = 19,000$$

He invested $10,000 at 4.5% and $19,000 at 3%.

47. Let $x =$ the amount invested at 2%.
$x + 29,000 =$ the amount invested at the
average rate of 3%.
Make a table to organize the information.

Percent as a Decimal	Amount Invested	Interest in One Year
.05	$29,000$	$.05(29,000)$
.02	x	$.02x$
.03	$x + 29,000$	$.03(x + 29,000)$

The last column gives the equation.

$$.05(29,000) + .02x = .03(x + 29,000)$$
$$5(29,000) + 2x = 3(x + 29,000)$$
$$\qquad\qquad \textit{Multiply by 100.}$$
$$145,000 + 2x = 3x + 87,000$$
$$58,000 = x$$

Ed should invest $58,000 at 2%.

49. The sum of the measures of the angles of any
triangle is 180 degrees.

51. If two angles are supplementary, the sum of their
measures is 180 degrees.

53. The sum of the measures of the three angles of a triangle is 180°.

$$(x+15)+(10x-20)+(x+5)=180$$
$$12x=180$$
$$x=15$$

With $x=15$, the three angles measures become

$$(15+15)°=30°,$$
$$(10\cdot15-20)°=130°,$$
and $\quad(15+5)°=20°.$

55. Supplementary angles have an angle measure sum of 180°.

$$(3x+5)+(5x+15)=180$$
$$8x+20=180$$
$$8x=160$$
$$x=20$$

With $x=20$, the two angle measures become

$$(3\cdot20+5)°=65°$$
and $\quad(5\cdot20+15)°=115°.$

57. **(a)** Let $\quad x=$ the amount invested at 5%.
$800-x=$ the amount invested at 10%.

(b) Let $\quad y=$ the amount of 5% acid used.
$800-y=$ the amount of 10% acid used.

58. Organize the information in a table.

(a)

Percent as a Decimal	Amount Invested	Interest in One Year
.05	x	$.05x$
.10	$800-x$	$.10(800-x)$
.0875	800	$.0875(800)$

The amount of interest earned at 5% and 10% is found in the last column of the table, $.05x$ and $.10(800-x)$.

(b)

Strength as a Decimal	Liters of solution	Liters of pure acid
.05	y	$.05y$
.10	$800-y$	$.10(800-y)$
.0875	800	$.0875(800)$

The amount of pure acid in the 5% and 10% mixtures is found in the last column of the table, $.05y$ and $.10(800-y)$.

59. Refer to the tables for Exercise 58. In each case, the last column gives the equation.

(a) $.05x+.10(800-x)=.0875(800)$

(b) $.05y+.10(800-y)=.0875(800)$

60. In both cases, multiply by $10,000$ to clear the decimals.

(a)
$$.05x+.10(800-x)=.0875(800)$$
$$500x+1000(800-x)=875(800)$$
$$500x+800,000-1000x=700,000$$
$$-500x=-100,000$$
$$x=200$$
$$800-x=800-200=600$$

Jack invested $200 at 5% and $600 at 10%.

(b)
$$.05y+.10(800-y)=.0875(800)$$
$$500y+1000(800-y)=875(800)$$
$$500y+800,000-1000y=700,000$$
$$-500y=-100,000$$
$$y=200$$
$$800-y=800-200=600$$

Jill used 200 L of 5% acid solution and 600 L of 10% acid solution.

61. The processes used to solve Problems A and B were virtually the same. Aside from the variables chosen, the problem information was organized in similar tables and the equations solved were the same.

Section 2.4

1. No, the answers must be whole numbers because they represent the number of area codes in 1947.

3. Most people will choose time, because that is the unknown.

5. The total amount is

$$21(.05)+14(.10)=1.05+1.40$$
$$=\$2.45.$$

7. Use $d=rt$, or $t=\dfrac{d}{r}$. Substitute 520 for d and 10 for t.

$$r=\frac{520}{10}=52$$

His rate was 52 mph.

9. The dot over 1975 is just under 10 million, so choice (c), 9.7 million is the best estimate.

11. Let $\quad x=$ the number of pennies.
$x=$ the number of dimes.
$44-2x=$ the number of quarters.

Denomination	Number of coins	Value
.01	x	$.01x$
.10	x	$.10x$
.25	$44-2x$	$.25(44-2x)$
Total	44	$\$4.37$

Write the equation from the last column of the table.

$$.01x + .10x + .25(44 - 2x) = 4.37$$
$$x + 10x + 25(44 - 2x) = 437$$
Multiply by 100.
$$x + 10x + 1100 - 50x = 437$$
$$-39x + 1100 = 437$$
$$-39x = -663$$
$$x = 17$$

There are 17 pennies, 17 dimes, and $44 - 2(17) = 10$ quarters.

13. Let $x =$ the number of student tickets sold.

$410 - x =$ the number of nonstudent tickets sold.

Make a table.

Cost of Ticket	Number Sold	Amount Collected
$3	x	$3x$
$7	$410 - x$	$7(410 - x)$
Total	410	$1650

Write the equation from the last column of the table.

$$3x + 7(410 - x) = 1650$$
$$3x + 2870 - 7x = 1650$$
$$-4x = -1220$$
$$x = 305$$

305 student tickets were sold; $410 - 305 = 105$ nonstudent tickets were sold.

15. Let $x =$ the number of Row 1 seats sold.

$105 - x =$ the number of Row 2 seats sold.

Cost of Ticket	Number Sold	Value
$35	x	$35x$
$30	$105 - x$	$30(105 - x)$
Total	105	$3420

Write the equation from the last column of the table.

$$35x + 30(105 - x) = 3420$$
$$35x + 3150 - 30x = 3420$$
$$5x = 270$$
$$x = 54$$

54 Row 1 seats and $105 - 54 = 51$ Row 2 seats were sold.

17. Let $t =$ the time it takes for the trains to be 315 km apart.

Use the formula $d = rt$. Complete the table.

	Rate	Time	Distance
First train	85	t	$85t$
Second train	95	t	$95t$
			315

The total distance traveled is the sum of the distances traveled by each train, since they are traveling in opposite directions. This total is 315 km. Therefore,

$$85t + 95t = 315$$
$$180t = 315$$
$$t = \frac{315}{180} = \frac{7}{4} = 1\frac{3}{4}.$$

It will take the trains $1\frac{3}{4}$ hr before they are 315 km apart.

19. Let $x =$ time for Nancy to commute to work. Since Mark leaves 15 minutes after Nancy, $\frac{15}{60} = \frac{1}{4}$ hr, and $x - \frac{1}{4} =$ time for Mark.

Make a table using the formula $rt = d$.

	r	t	d
Nancy	35	x	$35x$
Mark	40	$x - \frac{1}{4}$	$40\left(x - \frac{1}{4}\right)$

Since Nancy and Mark are going in opposite directions, we add their distances to get 140 mi.

$$35x + 40\left(x - \frac{1}{4}\right) = 140$$
$$35x + 40x - 10 = 140$$
$$75x = 150$$
$$x = 2$$

They will be 140 mi apart at 8 A.M. $+ 2$ hr $= 10$ A.M.

21. Let $x =$ Tri's speed when he drives his car.

Use the formula $d = rt$ with 30 min $= \dfrac{1}{2}$ hr and

45 min $= \dfrac{3}{4}$ hr.

	r	t	d
Travel by car	x	$\dfrac{1}{2}$	$\dfrac{1}{2}x$
Travel by bus	$x-12$	$\dfrac{3}{4}$	$\dfrac{3}{4}(x-12)$

The distance by car and the distance by bus are the same. Therefore,

$\dfrac{1}{2}x = \dfrac{3}{4}(x - 12).$

$2x = 3(x - 12)$ *Multiply by 4.*

$2x = 3x - 36$

$36 = x$

Tri's speed by car is 36 mph. The distance he travels to work is

$$\dfrac{1}{2}(36) = 18 \text{ mi.}$$

23. Let $x =$ the number of hours that each part of the trip took.

The distance traveled in the first part of the trip was $10x$ $(d = rt)$; the distance traveled in the second part of the trip was $15x$. The entire trip covered 100 miles, so

$$10x + 15x = 100$$
$$25x = 100$$
$$x = 4.$$

Each part of the trip took 4 hr, so the entire trip took 8 hr.

25. Let $x =$ the time for Paula to catch Janet.

$x + \dfrac{1}{6} =$ the time Janet runs

$\left(10 \text{ min} = \dfrac{10}{60} = \dfrac{1}{6} \text{ hr}\right).$

	r	t	d
Janet	5	$x + \dfrac{1}{6}$	$5\left(x + \dfrac{1}{6}\right)$
Paula	6	x	$6x$

The distances they run are the same.

$$6x = 5\left(x + \dfrac{1}{6}\right)$$

$$6x = 5x + \dfrac{5}{6}$$

$$x = \dfrac{5}{6}$$

It will take $\dfrac{5}{6}$ hr (or 50 min) for Paula to catch up with Janet.

27. Let $x =$ the width of the base.

$2x - 65 =$ the length of the base.

The perimeter is 860 ft. Use the formula $P = 2L + 2W$ and substitute 860 for P, $2x - 65$ for L and x for W.

$$P = 2L + 2W$$
$$860 = 2(2x - 65) + 2(x)$$
$$860 = 4x - 130 + 2x$$
$$990 = 6x$$
$$165 = x$$

The width is 165 ft, and the length is

$$2x - 65 = 2(165) - 65 = 265 \text{ ft.}$$

29. Let $x =$ the length of the middle side.

$x - 75 =$ the length of the shortest side.

$x + 375 =$ the length of the longest side.

The perimeter equals the sum of the lengths of the three sides. Since the perimeter is 3075 mi, an equation is

$$x + (x - 75) + (x + 375) = 3075.$$
$$3x + 300 = 3075$$
$$3x = 2775$$
$$x = 925$$

Then

$$x - 75 = 925 - 75 = 850$$

and

$$x + 375 = 925 + 375 = 1300.$$

The shortest side measures 850 mi, the middle side measures 925 mi, and the longest side measures 1300 mi.

31. Let $x =$ the width.

$2x =$ the length (twice the width).

Use $P = 2L + 2W$ and add one more width to cut the area into two parts. Thus, an equation is

$$2L + 2W + W = 210.$$
$$2(2x) + 2x + x = 210$$
$$7x = 210$$
$$x = 30$$

The width is 30 m, and the length is $2(30) = 60$ m.

33. Let x = the page number on one page.
$x + 1$ = the page number on the next page.
Since the sum of the page numbers is 153, an equation is

$$x + (x + 1) = 153.$$
$$2x + 1 = 153$$
$$2x = 152$$
$$x = 76$$
Then $\quad x + 1 = 77.$

The page numbers are 76 and 77.

35. Let x = the first consecutive integer.
$x + 1$ = the second consecutive integer.
$x + 2$ = the third consecutive integer.
The sum of the first and twice the second is 17 more than twice the third, so an equation is

$$x + 2(x + 1) = 2(x + 2) + 17.$$
$$x + 2x + 2 = 2x + 4 + 17$$
$$3x + 2 = 2x + 21$$
$$x = 19$$
Then $\quad x + 1 = 20,$
and $\quad x + 2 = 21.$

The three consecutive integers are 19, 20, and 21.

37. Let x = the regular price of the disc player.
The sale price after a discount of 40% (or .40) was $255, so an equation is

$$x - .40x = 255.$$
$$.60x = 255$$
$$x = 425$$

The regular price of the disc player was $425.

39. Let x = the width of the rectangle.
$x + 3$ = the length of the rectangle.
If the length was decreased by 2 in and the width was increased by 1 in, the perimeter would be 24 in. Use the formula $P = 2L + 2W$, and substitute 24 for P, $(x + 3) - 2$ or $x + 1$ for L, and $x + 1$ for W.

$$P = 2L + 2W$$
$$24 = 2(x + 1) + 2(x + 1)$$
$$24 = 2x + 2 + 2x + 2$$
$$24 = 4x + 4$$
$$20 = 4x$$
$$5 = x$$

The width of the rectangle is 5 in, and the length is $5 + 3 = 8$ in.

41. Let x = the number of heads he buys.
Of the lettuce he buys, 10%, or .10x, cannot be sold. That leaves 90%, or .90x, that is sold. If the grocer charges 40¢ or $.40 for each head he sells,

he gets $.40(.90x)$. This must equal what he pays for the lettuce, $5.20, plus a 10¢ or $.10 profit on each head he buys. Thus, an equation is

$$.40(.90x) = 5.20 + .10x.$$
$$.36x = 5.20 + .10x$$
$$.26x = 5.20$$
$$x = 20$$

There are 20 heads of lettuce in the crate.

Section 2.5

1. $x \le 3$ can be written in interval notation as $(-\infty, 3]$. Choice D

3. $x < 3$ represents all numbers less than 3, which is graphed in choice B.

5. $-3 \le x \le 3$ can be written in interval notation as $[-3, 3]$. Choice F

7. If an endpoint is an element of the solution set, then this is shown on the graph using a bracket. If an endpoint is *not* an element of the solution set, then this is shown on the graph using a parenthesis.

9.
$$4x + 1 \ge 21$$
Subtract 1 from both sides.
$$4x + 1 - 1 \ge 21 - 1$$
$$4x \ge 20$$
Divide both sides by 4.
$$\frac{4x}{4} \ge \frac{20}{4}$$
$$x \ge 5$$
Solution set: $[5, \infty)$

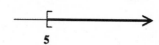

11.
$$\frac{3k - 1}{4} > 5$$
$$4\left(\frac{3k - 1}{4}\right) > 4(5) \qquad \textit{Multiply by 4.}$$
$$3k - 1 > 20$$
$$3k - 1 + 1 > 20 + 1 \quad \textit{Add 1.}$$
$$3k > 21$$
$$\frac{3k}{3} > \frac{21}{3} \qquad \textit{Divide by 3.}$$
$$k > 7$$
Solution set: $(7, \infty)$

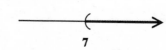

13. $-4x < 16$

Divide both sides by -4, and reverse the inequality sign.

$$\frac{-4x}{-4} > \frac{16}{-4}$$
$$x > -4$$

Solution set: $(-4, \infty)$

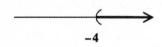

15. $-\frac{3}{4}r \geq 30$

Multiply both sides by $-\frac{4}{3}$, and reverse the inequality sign.

$$-\frac{4}{3}\left(-\frac{3}{4}r\right) \leq -\frac{4}{3}(30)$$
$$r \leq -40$$

Solution set: $(-\infty, -40]$

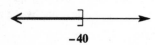

17. $-1.3m \geq -5.2$

Divide both sides by -1.3, and reverse the inequality sign.

$$\frac{-1.3m}{-1.3} \leq \frac{-5.2}{-1.3}$$
$$m \leq 4$$

Solution set: $(-\infty, 4]$

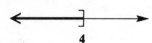

19. $\frac{2k-5}{-4} > 5$

Multiply both sides by -4, and reverse the inequality sign.

$$-4\left(\frac{2k-5}{-4}\right) < -4(5)$$
$$2k - 5 < -20$$
$$2k < -15 \qquad \text{Add 5.}$$
$$k < -\frac{15}{2} \qquad \text{Divide by 2.}$$

Solution set: $\left(-\infty, -\frac{15}{2}\right)$

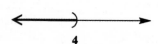

21. $y + 4(2y - 1) \geq y$

$$y + 8y - 4 \geq y \qquad \text{\textit{Clear parentheses.}}$$
$$9y - 4 \geq y \qquad \text{\textit{Combine terms.}}$$
$$8y \geq 4 \qquad \text{\textit{Add 4; Subtract y.}}$$
$$y \geq \frac{4}{8} = \frac{1}{2} \qquad \text{\textit{Divide by 8.}}$$

Solution set: $\left[\frac{1}{2}, \infty\right)$

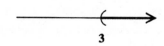

23. $-(4 + r) + 2 - 3r < -14$

$$-4 - r + 2 - 3r < -14 \qquad \text{\textit{Clear parentheses.}}$$
$$-4r - 2 < -14 \qquad \text{\textit{Combine terms.}}$$
$$-4r < -12 \qquad \text{\textit{Add 2.}}$$

Divide by -4, and reverse the inequality sign.

$$r > 3$$

Solution set: $(3, \infty)$

25. $-3(z - 6) > 2z - 2$

$$-3z + 18 > 2z - 2 \qquad \text{\textit{Clear parentheses.}}$$
$$-5z > -20 \qquad \text{\textit{Subtract 2z; subtract 18.}}$$

Divide by -5, and reverse the inequality sign.

$$z < 4$$

Solution set: $(-\infty, 4)$

27. $\frac{2}{3}(3k - 1) \geq \frac{3}{2}(2k - 3)$

Multiply both sides by 6 to clear the fractions.

$$6 \cdot \frac{2}{3}(3k - 1) \geq 6 \cdot \frac{3}{2}(2k - 3)$$
$$4(3k - 1) \geq 9(2k - 3)$$
$$12k - 4 \geq 18k - 27 \qquad \text{\textit{Clear parentheses.}}$$
$$-6k \geq -23 \qquad \text{\textit{Subtract 18k; add 4.}}$$

Divide by -6, and reverse the inequality sign.

$$k \leq \frac{23}{6}$$

Solution set: $\left(-\infty, \dfrac{23}{6}\right]$

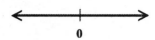

$\dfrac{23}{6}$

29. $-\dfrac{1}{4}(p+6) + \dfrac{3}{2}(2p-5) < 10$

Multiply each term by 4 to clear the fractions.

$$-1(p+6) + 6(2p-5) < 40$$
$$-p - 6 + 12p - 30 < 40$$
$$11p - 36 < 40$$
$$11p < 76$$
$$p < \dfrac{76}{11}$$

Solution set: $\left(-\infty, \dfrac{76}{11}\right)$

$\dfrac{76}{11}$

31. $3(2x - 4) - 4x < 2x + 3$
$$6x - 12 - 4x < 2x + 3$$
$$2x - 12 < 2x + 3$$
$$-12 < 3 \quad \textit{True}$$

The statement is true for all values of x.
Therefore, the original inequality is true for any real number.

Solution set: $(-\infty, \infty)$

0

33. $8\left(\dfrac{1}{2}x + 3\right) < 8\left(\dfrac{1}{2}x - 1\right)$
$$4x + 24 < 4x - 8$$
$$24 < -8 \quad \textit{False}$$

This is a false statement, so the inequality is a contradiction.

Solution set: $\emptyset$

35. It is incorrect. The inequality symbol should be reversed only when multiplying or dividing by a negative number. Since 5 is positive, the inequality symbol should not be reversed.

37. $-4 < x - 5 < 6$

Add 5 to each part of the inequality
to isolate the variable x.

$$-4 + 5 < x - 5 + 5 < 6 + 5$$
$$1 < x < 11$$

Solution set: $(1, 11)$

$1 \qquad 11$

39. $-9 \le k + 5 \le 15$

Subtract 5 from each part.

$$-9 - 5 \le k + 5 - 5 \le 15 - 5$$
$$-14 \le k \le 10$$

Solution set: $[-14, 10]$

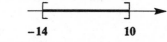

$-14 \qquad\qquad 10$

41. $-6 \le 2z + 4 \le 16$
$$-10 \le 2z \le 12 \qquad \textit{Subtract 4.}$$
$$-5 \le z \le 6 \qquad \textit{Divide by 2.}$$

Solution set: $[-5, 6]$

$-5 \qquad 6$

43. $-19 \le 3x - 5 \le 1$
$$-14 \le 3x \le 6 \qquad \textit{Add 5.}$$
$$-\dfrac{14}{3} \le x \le 2 \qquad \textit{Divide by 3.}$$

Solution set: $\left[-\dfrac{14}{3}, 2\right]$

$-\dfrac{14}{3} \qquad 2$

45. $-1 \le \dfrac{2x - 5}{6} \le 5$
$$-6 \le 2x - 5 \le 30 \quad \textit{Multiply by 6.}$$
$$-1 \le 2x \le 35 \qquad \textit{Add 5.}$$
$$-\dfrac{1}{2} \le x \le \dfrac{35}{2} \qquad \textit{Divide by 2.}$$

Solution set: $\left[-\dfrac{1}{2}, \dfrac{35}{2}\right]$

$-\dfrac{1}{2} \qquad \dfrac{35}{2}$

47. $4 \le 5 - 9x < 8$
$$-1 \le -9x < 3 \qquad\qquad \textit{Subtract 5.}$$

Divide each part by -9; reverse the inequalities.

$$\dfrac{1}{9} \ge x > -\dfrac{1}{3}$$

This inequality may be written

$$-\dfrac{1}{3} < x \le \dfrac{1}{9}.$$

Solution set: $\left(-\dfrac{1}{3}, \dfrac{1}{9}\right]$

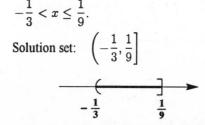

$-\dfrac{1}{3} \qquad\qquad \dfrac{1}{9}$

49. "Exceed" means greater than. From the graph, the percent of tornadoes was greater than 7.7% in April, May, June, and July.

51. A total of 17,252 tornadoes were reported. To find the months in which fewer than (or less than) 1500 were reported, find what percent 1500 is of 17,252.

$$\frac{1500}{17,252} \approx .087 \approx 8.7\%$$

The months where less then 8.7% of tornadoes were reported were January, February, March, August, September, October, November, and December.

53. Notice from the problem that the taxicab rates are assessed per $\frac{1}{5}$ mi. Therefore, let

$x =$ the number of $\frac{1}{5}$-mi distances

Dantrell can travel.
He must pay \$1.50 plus $.25x$, and this amount must be no more than \$3.75.

$$1.50 + .25x \leq 3.75$$
$$.25x \leq 2.25$$
$$x \leq 9$$

Thus, Dantrell can travel the first $\frac{1}{5}$ mi plus $\frac{9}{5}$ additional miles or

$$\frac{1}{5} + \frac{9}{5} = \frac{10}{5} = 2 \text{ mi.}$$

55. Let $x =$ her score on the third test.
Her average must be at least 84 (≥ 84). To find the average of three numbers, add them and divide by 3.

$$\frac{90 + 82 + x}{3} \geq 84$$
$$\frac{172 + x}{3} \geq 84 \qquad Add.$$
$$172 + x \geq 252 \qquad Multiply\ by\ 3.$$
$$x \geq 80 \qquad Subtract\ 172.$$

She must score at least 80 on her third test.

57. Let $x =$ the number of miles driven.
The cost of renting from Ford is \$35 plus the mileage cost of $.14x$, while the cost of renting from Chevrolet is \$34 plus the mileage cost of $.16x$.

Cost from Chevrolet > Cost from Ford
$$34 + .16x > 35 + .14x$$
$$3400 + 16x > 3500 + 14x$$
$$\qquad\qquad Multiply\ by\ 100.$$
$$2x > 100$$
$$x > 50$$

After 50 mi, the price to rent the Chevrolet exceeds the price to rent the Ford.

59. Cost $C = 20x + 100$; Revenue $R = 24x$
The business will show a profit only when $R > C$. Substitute the given expressions for R and C.

$$R > C$$
$$24x > 20x + 100$$
$$4x > 100$$
$$x > 25$$

The company will show a profit upon selling 26 tapes.

61. $5(x + 3) - 2(x - 4) = 2(x + 7)$
$$5x + 15 - 2x + 8 = 2x + 14$$
$$3x + 23 = 2x + 14$$
$$x = -9$$

Solution set: $\{-9\}$

The graph is the point -9 on a number line.

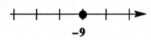
-9

62. $5(x + 3) - 2(x - 4) > 2(x + 7)$
$$5x + 15 - 2x + 8 > 2x + 14$$
$$3x + 23 > 2x + 14$$
$$x > -9$$

Solution set: $(-9, \infty)$

The graph extends from -9 to the right on a number line; -9 is not included in the graph.

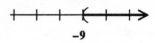
-9

63. $5(x + 3) - 2(x - 4) < 2(x + 7)$
$$5x + 15 - 2x + 8 < 2x + 14$$
$$3x + 23 < 2x + 14$$
$$x < -9$$

Solution set: $(-\infty, -9)$

The graph extends from -9 to the left on a number line; -9 is not included in the graph.

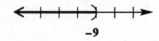
-9

64. If we graph all the solution sets from Exercises 61–63; that is, $\{-9\}$, $(-9, \infty)$, and $(-\infty, -9)$, on the same number line, we will have graphed the set of all real numbers.

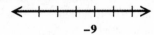

65. The solution set of the given equation is the point -3 on a number line. The solution set of the first inequality extends from -3 to the right (toward ∞) on the same number line. Based on Exercises 61–63, the solution set of the second inequality should then extend from -3 to the left (toward $-\infty$) on the number line. Complete the statement with $(-\infty, -3)$.

67. $4 < y < 1$ is an improper statement since $4 \not< 1$. There is no such number y.

69. From the graph, the solution is either $(.5, \infty)$ or $[.5, \infty)$. Since the symbol is $\geq$, we include .5 in the solution set. The solution set is $[.5, \infty)$ or, equivalently, $\left[\frac{1}{2}, \infty\right)$.

71. The symbol is $>$, so we don't include the endpoints. The solution set is $(-6, -4)$.

Section 2.6

1. This statement is true. The solution set of $x + 1 = 5$ is $\{4\}$. The solution set of $x + 1 > 5$ is $(4, \infty)$. The solution set of $x + 1 < 5$ is $(-\infty, 4)$. Taken together we have the set of real numbers. (See Section 2.5, Exercises 61–65, for a discussion of this concept.)

3. This statement is false. The union is $(-\infty, 8) \cup (8, \infty)$. The only real number that is *not* in the union is 8.

5. This statement is false since 0 is a rational number but not an irrational number. The sets of rational numbers and irrational numbers have no common elements so their intersection is $\emptyset$.

In Exercises 7–20, let

$A = \{1, 2, 3, 4, 5, 6\}$, $B = \{1, 3, 5\}$, $C = \{1, 6\}$, and $D = \{4\}$.

7. The intersection of sets B and A contains only those elements in both sets B and A.

$$B \cap A = \{1, 3, 5\} \text{ or set } B$$

9. The intersection of sets A and D is the set of all elements in both set A and D. Therefore,

$$A \cap D = \{4\} \text{ or set } D.$$

11. The intersection of set B and the set of no elements (empty set), $B \cap \emptyset$, is the set of no elements or $\emptyset$.

13. The union of sets A and B is the set of all elements that are in either set A or set B or both sets A and B. Since all numbers in set B are also in set A, the set $A \cup B$ will be the same as set A.

$$A \cup B = \{1, 2, 3, 4, 5, 6\} \text{ or set } A$$

15. The union of sets B and C contains all elements in either set B or set C or both sets B and C.

$$B \cup C = \{1, 3, 5, 6\}$$

17. The union of sets C and D is the set of all elements that are in either set C or set D or both sets C and D.

$$C \cup D = \{1, 4, 6\}$$

19. $B \cap C =$ the set of elements common to both B and $C = \{1\}$.

$$A \cap (B \cap C) = A \cap \{1\} = \{1\}.$$
$$A \cap B = \{1, 3, 5\}.$$
$$(A \cap B) \cap C = \{1, 3, 5\} \cap C = \{1\}.$$

Therefore,

$$A \cap (B \cap C) = (A \cap B) \cap C.$$

This illustrates the associative property of set intersection.

21. One example of how the concept of intersection can be applied to a real-life situation is in the grocery store. The set of red apples is the intersection of the set of red fruit and the set of apples.

23. The first graph represents the set $(-\infty, 2)$. The second graph represents the set $(-3, \infty)$. The intersection includes the elements common to both sets, that is, $(-3, 2)$.

25. The first graph represents the set $(-\infty, 5]$. The second graph represents the set $(-\infty, 2]$. The intersection includes the elements common to both sets, that is, $(-\infty, 2]$.

27. $x - 3 \leq 6$ and $x + 2 \geq 7$
$x \leq 9$ and $x \geq 5$

The graph of the solution set is all numbers that are both less than or equal to 9 and greater than or equal to 5. This is the intersection. The elements common to both sets are the numbers between 5 and 9, including the endpoints. The solution set is $[5, 9]$.

29. $-3x > 3$ and $x + 3 > 0$
$x < -1$ and $x > -3$

The graph of the solution set is all numbers that are both less than -1 and greater than -3. This is the intersection. The elements common to both sets are the numbers between -3 and -1, not including the endpoints. The solution set is $(-3, -1)$.

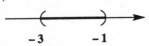

31. $3x - 4 \leq 8$ and $-4x + 1 \geq -15$
$3x \leq 12$ and $-4x \geq -16$
$x \leq 4$ and $x \leq 4$

Since both inequalities are identical, the graph of the solution set is the same as the graph of one of the inequalities. The solution set is $(-\infty, 4]$.

33. The first graph represents the set $(-\infty, 2]$. The second graph represents the set $[4, \infty)$. The union includes all elements in either set, or in both, that is, $(-\infty, 2] \cup [4, \infty)$.

35. The first graph represents the set $(-\infty, 1]$. The second graph represents the set $(-\infty, 8]$. The union includes all elements in either set, or in both, that is, $(-\infty, 8]$.

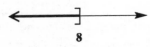

37. $x + 2 > 7$ or $1 - x > 6$
$-x > 5$
$x > 5$ or $x < -5$

The graph of the solution set is all numbers either greater than 5 or less than -5. This is the union. The solution set is

$$(-\infty, -5) \cup (5, \infty).$$

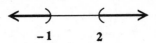

39. $x + 1 > 3$ or $-4x + 1 > 5$
$-4x > 4$
$x > 2$ or $x < -1$

The graph of the solution set is all numbers either less than -1 or greater than 2. This is the union. The solution set is $(-\infty, -1) \cup (2, \infty)$.

41. $(-\infty, -1] \cap [-4, \infty)$
The intersection is the set of numbers less than or equal to -1 and greater than or equal to -4. The numbers common to both original sets are between, and including, -4 and -1. The solution set is $[-4, -1]$.

43. $(-\infty, -6] \cap [-9, \infty)$
The intersection is the set of numbers less than or equal to -6 and greater than or equal to -9. The numbers common to both original sets are between, and including, -9 and -6. The solution set is $[-9, -6]$.

45. $(-\infty, 3) \cup (-\infty, -2)$
The union is the set of numbers that are either less than 3 or less than -2, or both. This is all numbers less than 3. The solution set is $(-\infty, 3)$.

47. $[3, 6] \cup (4, 9)$
The union is the set of numbers between, and including, 3 and 6, or between, but not including, 4 and 9. This is the set of numbers greater than or equal to 3 and less than 9. The solution set is $[3, 9)$.

49. $x < -1$ and $x > -5$
The word "and" means to take the intersection of both sets. $x < -1$ and $x > -5$ is true only when

$$-5 < x < -1.$$

The graph of the solution set is all numbers greater than -5 *and* less than -1. This is all numbers between -5 and -1, not including -5 or -1. The solution set is $(-5, -1)$.

51. $x < 4$ or $x < -2$

The word "or" means to take the union of both sets. The graph of the solution set is all numbers that are either less than 4 *or* less than -2, or both. This is all numbers less than 4. The solution set is $(-\infty, 4)$.

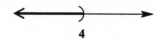

53. $-3x \le -6$ or $-3x \ge 0$

     $x \ge 2$ or $x \le 0$

The word "or" means to take the union of both sets. The graph of the solution set is all numbers that are either greater than or equal to 2 *or* less than or equal to 0. The solution set is $(-\infty, 0] \cup [2, \infty)$.

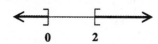

55. $x + 1 \ge 5$ and $x - 2 \le 10$

     $x \ge 4$ and $x \le 12$

The word "and" means to take the intersection of both sets. The graph of the solution set is all numbers that are both greater than or equal to 4 *and* less than or equal to 12. This is all numbers between, and including, 4 and 12. The solution set is $[4, 12]$.

57. From the graph, the number of stations sold exceeded 1200 in the years 1993, 1994, 1995, and 1996.

The value of transactions exceeded 3000 (in millions of dollars) in 1995 and 1996. Because of the word "and," we want the intersection of these two sets of years, which is 1995 and 1996.

For Exercises 59–64, find the area and perimeter of each of the given yards.

For Luigi's, Maria's, and Than's yards, use the formulas $A = LW$ and $P = 2L + 2W$.

Luigi's yard
$A = 50(30) = 1500$ ft^2
$P = 2(50) + 2(30) = 160$ ft

Maria's yard
$A = 40(35) = 1400$ ft^2
$P = 2(40) + 2(35) = 150$ ft

Than's yard
$A = 60(50) = 3000$ ft^2
$P = 2(60) + 2(50) = 220$ ft

continued

For Joe's yard, use the formulas $A = \frac{1}{2}bh$ and $P = a + b + c$.

Joe's yard
$A = \frac{1}{2}(40)(30) = 600$ ft^2
$P = 30 + 40 + 50 = 120$ ft

To be fenced, a yard must have a perimeter $P \le 150$ ft. To be sodded, a yard must have an area $A \le 1400$ ft^2.

59. Find "the yard can be fenced *and* the yard can be sodded."

A yard that can be fenced has $P \le 150$. Maria and Joe qualify.

A yard that can be sodded has $A \le 1400$. Again, Maria and Joe qualify.

Find the intersection. Maria's and Joe's yards are common to both sets, so Maria and Joe can have their yards both fenced and sodded.

60. Find "the yard can be fenced *and* the yard cannot be sodded."

A yard that can be fenced has $P \le 150$. Maria and Joe qualify.

A yard that cannot be sodded has $A > 1400$. Luigi and Than qualify.

Find the intersection. There are no yards common to both sets, so none of them qualify.

61. Find "the yard cannot be fenced *and* the yard can be sodded."

A yard that cannot be fenced has $P > 150$. Luigi and Than qualify.

A yard that can be sodded has $A \le 1400$. Maria and Joe qualify.

Find the intersection. There are no yards common to both sets, so none of the qualify.

62. Find "the yard cannot be fenced *and* the yard cannot be sodded."

A yard that cannot be fenced has $P > 150$. Luigi and Than qualify.

A yard that cannot be sodded has $A > 1400$. Again, Luigi and Than qualify.

Find the intersection. Luigi's and Than's yards are common to both sets, so Luigi and Than qualify.

63. Find "the yard can be fenced *or* the yard can be sodded." From Exercise 59, Maria's and Joe's yards qualify for both conditions, so the union is Maria and Joe.

64. Find "the yard cannot be fenced *or* the yard can be sodded." From Exercise 61, Luigi's and Than's yards cannot be fenced, and Maria's and Joe's yards can be sodded. The union includes all of them.

Section 2.7

1. (a) $|x| = 5$ has two solutions, $x = 5$ or $x = -5$. The graph is Choice E.

(b) $|x| < 5$ is written $-5 < x < 5$. Notice that -5 and 5 are not included. The graph is Choice C, which uses parentheses.

(c) $|x| > 5$ is written $x < -5$ or $x > 5$. The graph is Choice D, which uses parentheses.

(d) $|x| \le 5$ is written $-5 \le x \le 5$. This time -5 and 5 are included. The graph is Choice B, which uses brackets.

(e) $|x| \ge 5$ is written $x \le -5$ or $x \ge 5$. The graph is Choice A, which uses brackets.

3. (a) $|ax + b| = k,\ k = 0$
This means the distance from $ax + b$ to 0 is 0, so $ax + b = 0$, which has one solution.

(b) $|ax + b| = k,\ k > 0$
This means the distance from $ax + b$ to 0 is a positive number, so $ax + b = k$ or $ax + b = -k$. There are two solutions.

(c) $|ax + b| = k,\ k < 0$
This means the distance from $ax + b$ to 0 is a negative number, which is impossible because distance is always positive. There are no solutions.

5. $|x| = 12$
$x = 12$ or $x = -12$
Solution set: $\{-12, 12\}$

7. $|4x| = 20$
$4x = 20$ or $4x = -20$
$x = 5$ or $x = -5$
Solution set: $\{-5, 5\}$

9. $|y - 3| = 9$
$y - 3 = 9$ or $y - 3 = -9$
$y = 12$ or $y = -6$
Solution set: $\{-6, 12\}$

11. $|2x + 1| = 7$
$2x + 1 = 7$ or $2x + 1 = -7$
$2x = 6$ $\qquad$ $2x = -8$
$x = 3$ or $x = -4$
Solution set: $\{-4, 3\}$

13. $|4r - 5| = 17$
$4r - 5 = 17$ $\qquad$ or $\qquad$ $4r - 5 = -17$
$4r = 22$ $\qquad\qquad\qquad\qquad$ $4r = -12$
$r = \dfrac{22}{4} = \dfrac{11}{2}$ or $\qquad$ $r = -3$
Solution set: $\left\{-3, \dfrac{11}{2}\right\}$

15. $|2y + 5| = 14$
$2y + 5 = 14$ or $2y + 5 = -14$
$2y = 9$ $\qquad\qquad$ $2y = -19$
$y = \dfrac{9}{2}$ or $\qquad$ $y = -\dfrac{19}{2}$
Solution set: $\left\{-\dfrac{19}{2}, \dfrac{9}{2}\right\}$

17. $\left|\dfrac{1}{2}x + 3\right| = 2$

$\dfrac{1}{2}x + 3 = 2$ $\qquad$ or $\qquad$ $\dfrac{1}{2}x + 3 = -2$
$\dfrac{1}{2}x = -1$ $\qquad\qquad\qquad$ $\dfrac{1}{2}x = -5$
$x = -2$ or $\qquad\qquad$ $x = -10$
Solution set: $\{-10, -2\}$

19. $\left|1 - \dfrac{3}{4}k\right| = 7$

$1 - \dfrac{3}{4}k = 7$ $\qquad$ or $\qquad$ $1 - \dfrac{3}{4}k = -7$
Multiply all sides by 4.
$4 - 3k = 28$ $\qquad$ or $\qquad$ $4 - 3k = -28$
$-3k = 24$ $\qquad\qquad\qquad$ $-3k = -32$
$k = -8$ $\qquad$ or $\qquad\qquad$ $k = \dfrac{32}{3}$
Solution set: $\left\{-8, \dfrac{32}{3}\right\}$

21. When solving an absolute value equation or inequality of the form

(a) $|ax + b| = k$
(b) $|ax + b| < k$, or
(c) $|ax + b| > k$,

where k is a positive number, use

(a) *or* for the $=$ case,
(b) *and* for the $<$ case, and
(c) *or* for the $>$ case.

23. $|x| > 3$
  $x > 3$ or $x < -3$
  Solution set: $(-\infty, -3) \cup (3, \infty)$

-3   3

25. $|k| \geq 4$
  $k \geq 4$ or $k \leq -4$
  Solution set: $(-\infty, -4] \cup [4, \infty)$

-4   4

27. $|t + 2| > 10$
  $t + 2 > 10$ or $t + 2 < -10$
  $t > 8$ or $t < -12$
  Solution set: $(-\infty, -12) \cup (8, \infty)$

-12   8

29. $|3 - x| > 5$
  $3 - x > 5$ or $3 - x < -5$
  $-x > 2$ or $-x < -8$
  Multiply by -1,
  and reverse the inequality signs.
  $x < -2$ or $x > 8$
  Solution set: $(-\infty, -2) \cup (8, \infty)$

-2   8

31. $|x| \leq 3$
  $-3 \leq x \leq 3$
  Solution set: $[-3, 3]$

-3   3

33. $|k| < 4$
  $-4 < k < 4$
  Solution set: $(-4, 4)$

-4   4

35. $|t + 2| \leq 10$
  $-10 \leq t + 2 \leq 10$
  $-12 \leq t \leq 8$
  Solution set: $[-12, 8]$

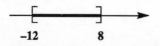

-12   8

37. $|3 - x| \leq 5$
  $-5 \leq 3 - x \leq 5$
  $-8 \leq -x \leq 2$
  Multiply by -1, and reverse the inequality signs.
  $8 \geq x \geq -2$ or $-2 \leq x \leq 8$
  Solution set: $[-2, 8]$

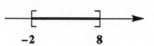

-2   8

39. $|-4 + k| > 9$
  $-4 + k > 9$ or $-4 + k < -9$
  $k > 13$ or $k < -5$
  Solution set: $(-\infty, -5) \cup (13, \infty)$

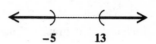

-5   13

41. $|r + 5| > 20$
  $r + 5 > 20$ or $r + 5 < -20$
  $r > 15$ or $r < -25$
  Solution set: $(-\infty, -25) \cup (15, \infty)$

-25   15

43. $|7 + 2z| = 5$
  $7 + 2z = 5$ or $7 + 2z = -5$
  $2z = -2$ $2z = -12$
  $z = -1$ or $z = -6$
  Solution set: $\{-6, -1\}$

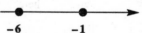

-6   -1

45. $|3r - 1| \leq 11$
  $-11 \leq 3r - 1 \leq 11$
  $-10 \leq 3r \leq 12$
  $-\dfrac{10}{3} \leq r \leq 4$

  Solution set: $\left[-\dfrac{10}{3}, 4\right]$

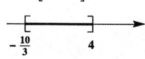

$-\dfrac{10}{3}$ 4

47. $|-6x - 6| \leq 1$

$-1 \leq -6x - 6 \leq 1$

$5 \leq -6x \leq 7$

Divide by -6. and reverse
the inequality signs.

$-\dfrac{5}{6} \geq x \geq -\dfrac{7}{6}$ or $-\dfrac{7}{6} \leq x \leq -\dfrac{5}{6}$

Solution set: $\left[-\dfrac{7}{6}, -\dfrac{5}{6}\right]$

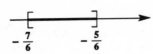

49. $|3x - 1| \geq 8$

$3x - 1 \geq 8$ or $3x - 1 \leq -8$

$3x \geq 9$ $3x \leq -7$

$x \geq 3$ or $x \leq -\dfrac{7}{3}$

Solution set: $\left(-\infty, -\dfrac{7}{3}\right] \cup [3, \infty)$

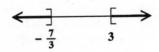

51. The distance between x and 4 equals 9 can be
written

$$|x - 4| = 9 \quad \text{or} \quad |4 - x| = 9.$$

53. $|x + 4| + 1 = 2$

$|x + 4| = 1$

$x + 4 = 1$ or $x + 4 = -1$

$x = -3$ or $x = -5$

Solution set: $\{-5, -3\}$

55. $|2x + 1| + 3 > 8$

$|2x + 1| > 5$

$2x + 1 > 5$ or $2x + 1 < -5$

$2x > 4$ $2x < -6$

$x > 2$ or $x < -3$

Solution set: $(-\infty, -3) \cup (2, \infty)$

57. $|x + 5| - 6 \leq -1$

$|x + 5| \leq 5$

$-5 \leq x + 5 \leq 5$

$-10 \leq x \leq 0$

Solution set: $[-10, 0]$

59. $|3x + 1| = |2x + 4|$

$3x + 1 = 2x + 4$ or $3x + 1 = -(2x + 4)$

$3x + 1 = -2x - 4$

$5x = -5$

$x = 3$ or $x = -1$

Solution set: $\{-1, 3\}$

61. $\left|m - \dfrac{1}{2}\right| = \left|\dfrac{1}{2}m - 2\right|$

$m - \dfrac{1}{2} = \dfrac{1}{2}m - 2$ or $m - \dfrac{1}{2} = -\left(\dfrac{1}{2}m - 2\right)$

Multiply by 2. $m - \dfrac{1}{2} = -\dfrac{1}{2}m + 2$

$2m - 1 = m - 4$ $2m - 1 = -m + 4$

$3m = 5$

$m = -3$ or $m = \dfrac{5}{3}$

Solution set: $\left\{-3, \dfrac{5}{3}\right\}$

63. $|6x| = |9x + 1|$

$6x = 9x + 1$ or $6x = -(9x + 1)$

$-3x = 1$ $6x = -9x - 1$

$15x = -1$

$x = -\dfrac{1}{3}$ or $x = -\dfrac{1}{15}$

Solution set: $\left\{-\dfrac{1}{3}, -\dfrac{1}{15}\right\}$

65. $|2p - 6| = |2p + 11|$

$2p - 6 = 2p + 11$ or $2p - 6 = -(2p + 11)$

$-6 = 11$ *False* $2p - 6 = -2p - 11$

$4p = -5$

No solution   or $p = -\dfrac{5}{4}$

Solution set: $\left\{-\dfrac{5}{4}\right\}$

67. $|12t - 3| = -8$

Since the absolute value of an expression can
never be negative, there are no solutions for this
equation.

Solution set: $\emptyset$

69. $|4x + 1| = 0$

The expression $4x + 1$ will equal 0 *only* for the
solution of the equation

$$4x + 1 = 0.$$

$4x = -1$

$x = \dfrac{-1}{4}$ or $-\dfrac{1}{4}$

Solution set: $\left\{-\dfrac{1}{4}\right\}$

71. $|2q - 1| < -6$

There are no numbers whose absolute value is
negative, so this inequality has no solution.

Solution set: $\emptyset$

73. $|x + 5| > -9$

Since the absolute value of an expression is always nonnegative (positive or zero), the inequality is true for any real number x.

Solution set: $(-\infty, \infty)$

75. $|7x + 3| \leq 0$

The absolute value of an expression is always nonnegative (positive or zero), so this inequality is true only when

$$7x + 3 = 0$$
$$7x = -3$$
$$x = -\frac{3}{7}.$$

Solution set: $\left\{-\dfrac{3}{7}\right\}$

77. $|5x - 2| \geq 0$

The absolute value of an expression is always nonnegative, so the inequality is true for any real number x.

Solution set: $(-\infty, \infty)$

79. $|10z + 7| > 0$

Since an absolute value expression is always nonnegative and $|10z + 7| \neq 0$, there is only one possible value of z that makes this statement false. The equation $10z + 7 = 0$ will give that value.

$$10z + 7 = 0$$
$$10z = -7$$
$$z = -\frac{7}{10}$$

Solution set: $\left(-\infty, -\dfrac{7}{10}\right) \cup \left(-\dfrac{7}{10}, \infty\right)$

81. Add the given heights with a calculator to get 4602. There are 10 numbers, so divide the sum by 10.

$$\frac{4602}{10} = 460.2$$

The average height is 460.2 ft.

82. $|x - k| < 50$

Substitute 460.2 for k and solve the inequality.
$$|x - 460.2| < 50$$
$$-50 < x - 460.2 < 50$$
$$410.2 < x < 510.2$$
The buildings with heights between 410.2 ft and 510.2 ft are the Federal Office Building, City Hall, Kansas City Power and Light, and the Hyatt Regency.

83. $|x - k| < 75$

Substitute 460.2 for k and solve the inequality.
$$|x - 460.2| < 75$$
$$-75 < x - 460.2 < 75$$
$$385.2 < x < 535.2$$
The buildings with heights between 385.2 ft and 535.2 ft are Southwest Bell Telephone, City Center Square, Commerce Tower, the Federal Office Building, City Hall, Kansas City Power and Light, and the Hyatt Regency.

84. **(a)** This would be the opposite of the inequality in Exercise 83, that is,

$$|x - 460.2| \geq 75.$$

(b) $|x - 460.2| \geq 75$

$\quad x - 460.2 \geq 75 \qquad$ or $\qquad x - 460.2 \leq -75$
$\qquad\quad x \geq 535.2 \quad$ or $\qquad\qquad x \leq 385.2$

(c) The buildings that are not within 75 ft of the average have height less than or equal to 385.2 or greater than or equal to 535.2. This would include Pershing Road Associates, AT&T Town Pavillion, and One Kansas City Place.

(d) The answer makes sense because it includes all the buildings *not* listed earlier which had heights within 75 ft of the average.

85. **(a)** For the weight x of the box to be within .5 ounce of 16 ounces, we must have

$$|x - 16| \leq .5.$$

(b) $-.5 \leq x - 16 \leq .5$
$\quad 15.5 \leq x \leq 16.5$

(c) The process is out of control if $|x - 16| \geq .5$, which is the same as $x \leq 15.5$ or $x \geq 16.5$.

Summary: Exercises on Solving Linear and Absolute Value Equations and Inequalities

1. $4z + 1 = 49$
$\quad\; 4z = 48$
$\qquad z = 12$
Solution set: $\{12\}$

3. $6q - 9 = 12 + 3q$
$\quad\; 3q = 21$
$\qquad q = 7$
Solution set: $\{7\}$

5. $|a + 3| = -4$

Since the absolute value of an expression is always nonnegative, there is no number that makes this statement true. Therefore, the solution set is $\emptyset$.

7. $8r + 2 \geq 5r$

  $3r \geq -2$

  $r \geq -\dfrac{2}{3}$

Solution set: $\left[-\dfrac{2}{3}, \infty\right)$

9. $2q - 1 = -7$

  $2q = -6$

  $q = -3$

Solution set: $\{-3\}$

11. $6z - 5 \leq 3z + 10$

  $3z \leq 15$

  $z \leq 5$

Solution set: $(-\infty, 5]$

13. $9y - 3(y + 1) = 8y - 7$

  $9y - 3y - 3 = 8y - 7$

  $6y - 3 = 8y - 7$

  $4 = 2y$

  $2 = y$

Solution set: $\{2\}$

15. $9y - 5 \geq 9y + 3$

  $-5 \geq 3$ *False*

This is a false statement, so the inequality is a contradiction.

Solution set: $\emptyset$

17. $|q| < 5.5$

  $-5.5 < q < 5.5$

Solution set: $(-5.5, 5.5)$

19. $\dfrac{2}{3}y + 8 = \dfrac{1}{4}y$

  $8y + 96 = 3y$ *Multiply by 12.*

  $5y = -96$

  $y = -\dfrac{96}{5}$

Solution set: $\left\{-\dfrac{96}{5}\right\}$

21. $\dfrac{1}{4}p < -6$

  $4\left(\dfrac{1}{4}p\right) < 4(-6)$

  $p < -24$

Solution set: $(-\infty, -24)$

23. $\dfrac{3}{5}q - \dfrac{1}{10} = 2$

  $6q - 1 = 20$ *Multiply by 10.*

  $6q = 21$

  $q = \dfrac{21}{6}$ or $\dfrac{7}{2}$

Solution set: $\left\{\dfrac{7}{2}\right\}$

25. $r + 9 + 7r = 4(3 + 2r) - 3$

  $8r + 9 = 12 + 8r - 3$

  $8r + 9 = 8r + 9$

  $0 = 0$ *True*

The last statement is true for any real number r.

Solution set: $(-\infty, \infty)$

27. $|2p - 3| > 11$

  $2p - 3 > 11$ or $2p - 3 < -11$

  $2p > 14$ $2p < -8$

  $p > 7$ or $p < -4$

Solution set: $(-\infty, -4) \cup (7, \infty)$

29. $|5a + 1| \leq 0$

The expression $|5a + 1|$ is never less than 0 since an absolute value expression must be nonnegative. However, $|5a + 1| = 0$ if $5a + 1 = 0$.

  $5a = -1$

  $a = \dfrac{-1}{5} = -\dfrac{1}{5}$

Solution set: $\left\{-\dfrac{1}{5}\right\}$

31. $-2 \leq 3x - 1 \leq 8$

  $-1 \leq 3x \leq 9$

  $-\dfrac{1}{3} \leq x \leq 3$

Solution set: $\left[-\dfrac{1}{3}, 3\right]$

33. $|7z - 1| = |5z + 3|$

  $7z - 1 = 5z + 3$ or $7z - 1 = -(5z + 3)$

  $2z = 4$ $7z - 1 = -5z - 3$

                $12z = -2$

  $z = 2$ or $z = \dfrac{-2}{12} = -\dfrac{1}{6}$

Solution set: $\left\{-\dfrac{1}{6}, 2\right\}$

35. $|1 - 3x| \geq 4$

$1 - 3x \geq 4 \qquad \text{or} \qquad 1 - 3x \leq -4$

$\qquad -3x \geq 3 \qquad\qquad\qquad -3x \leq -5$

$\qquad x \leq -1 \quad \text{or} \qquad\qquad x \geq \dfrac{5}{3}$

Solution set: $\left(-\infty, -1\right] \cup \left[\dfrac{5}{3}, \infty\right)$

37. $-(m + 4) + 2 = 3m + 8$

$\qquad -m - 4 + 2 = 3m + 8$

$\qquad\qquad -m - 2 = 3m + 8$

$\qquad\qquad\qquad -10 = 4m$

$\qquad\qquad\qquad m = \dfrac{-10}{4} = -\dfrac{5}{2}$

Solution set: $\left\{-\dfrac{5}{2}\right\}$

39. $-6 \leq \dfrac{3}{2} - x \leq 6$

$-12 \leq 3 - 2x \leq 12$

$-15 \leq -2x \leq 9$

$\dfrac{15}{2} \geq x \geq -\dfrac{9}{2} \text{ or } -\dfrac{9}{2} \leq x \leq \dfrac{15}{2}$

Solution set: $\left[-\dfrac{9}{2}, \dfrac{15}{2}\right]$

41. $|y - 1| \geq -6$

The absolute value of an expression is always nonnegative, so the inequality is true for any real number x.

Solution set: $(-\infty, \infty)$

43. $8q - (1 - q) = 3(1 + 3q) - 4$

$\qquad 8q - 1 + q = 3 + 9q - 4$

$\qquad\qquad 9q - 1 = 9q - 1 \quad \textit{True}$

This is an identity.

Solution set: $(-\infty, \infty)$

45. $|r - 5| = |r + 9|$

$r - 5 = r + 9 \qquad \text{or} \qquad r - 5 = -(r + 9)$

$\quad -5 = 9 \quad \textit{False} \qquad\qquad r - 5 = -r - 9$

$\qquad\qquad\qquad\qquad\qquad\qquad\qquad 2r = -4$

$\qquad \textit{No solution} \quad \text{or} \qquad\qquad r = -2$

Solution set: $\{-2\}$

47. $2x + 1 > 5 \quad \text{or} \quad 3x + 4 < 1$

$\qquad 2x > 4 \qquad\qquad 3x < -3$

$\qquad x > 2 \quad \text{or} \qquad x < -1$

Solution set: $(-\infty, -1) \cup (2, \infty)$

Chapter 2 Review Exercises

1. $-(8 + 3y) + 5 = 2y + 6$

$\qquad -8 - 3y + 5 = 2y + 6$

$\qquad\qquad -3y - 3 = 2y + 6$

$\qquad\qquad\qquad -5y = 9$

$\qquad\qquad\qquad y = -\dfrac{9}{5}$

Solution set: $\left\{-\dfrac{9}{5}\right\}$

2. $-\dfrac{3}{4}x = -12$

$\quad -3x = -48 \quad \textit{Multiply by 4.}$

$\quad x = 16$

Solution set: $\{16\}$

3. $\dfrac{2q + 1}{3} - \dfrac{q - 1}{4} = 0$

$4(2q + 1) - 3(q - 1) = 0 \qquad \textit{Multiply by 12.}$

$\qquad 8q + 4 - 3q + 3 = 0$

$\qquad\qquad 5q + 7 = 0$

$\qquad\qquad 5q = -7$

$\qquad\qquad q = -\dfrac{7}{5}$

Solution set: $\left\{-\dfrac{7}{5}\right\}$

4. $5(2x - 3) = 6(x - 1) + 4x$

$\quad 10x - 15 = 6x - 6 + 4x$

$\quad 10x - 15 = 10x - 6$

$\qquad -15 = -6 \quad \textit{False}$

This is a false statement, so the equation is a contradiction.

Solution set: $\emptyset$

5. $7r - 3(2r - 5) + 5 + 3r = 4r + 20$

$\quad 7r - 6r + 15 + 5 + 3r = 4r + 20$

$\qquad\qquad 4r + 20 = 4r + 20$

$\qquad\qquad\qquad 20 = 20 \quad \textit{True}$

This equation is an identity.

Solution set: $(-\infty, \infty)$

6. $8p - 4p - (p - 7) + 9p + 6 = 12p - 7$

$\quad 8p - 4p - p + 7 + 9p + 6 = 12p - 7$

$\qquad\qquad 12p + 13 = 12p - 7$

$\qquad\qquad\qquad 13 = -7 \quad \textit{False}$

This equation is a contradiction.

Solution set: $\emptyset$

7. $-2r + 6(r - 1) + 3r - (4 - r) = -(r + 5) - 5$

$\qquad -2r + 6r - 6 + 3r - 4 + r = -r - 5 - 5$

$\qquad\qquad\qquad 8r - 10 = -r - 10$

$\qquad\qquad\qquad\qquad 9r = 0$

$\qquad\qquad\qquad\qquad r = 0$

This equation is a conditional equation.

Solution set: $\{0\}$

8.   Solve $V = LWH$ for H.

$$\frac{V}{LW} = \frac{LWH}{LW}$$

$$\frac{V}{LW} = H$$

9.   Solve $A = \dfrac{1}{2}(B + b)h$ for b.

$\qquad 2A = (B + b)h$ *Multiply by 2.*

$\qquad \dfrac{2A}{h} = B + b$ *Divide by h.*

$\qquad \dfrac{2A}{h} - B = b$ or $b = \dfrac{2A - Bh}{h}$

10.   Solve $M = -\dfrac{1}{4}(x + 3y)$ for x.

$\qquad 4M = -1(x + 3y)$ *Multiply by 4.*

$\qquad 4M = -x - 3y$

$\qquad 4M + 3y = -x$

$\qquad x = -4M - 3y$

11.   Solve $P = \dfrac{3}{4}x - 12$ for x.

$\qquad 4P = 3x - 48$ *Multiply by 4.*

$\qquad 4P + 48 = 3x$ *Add 48.*

$\qquad \dfrac{4P + 48}{3} = x$ *Divide by 3.*

$\qquad \dfrac{4}{3}(P + 12) = x$ or $\dfrac{4}{3}P + 16 = x$

12.   Use the formula $A = LW$, and solve for W.

$$W = \frac{A}{L}$$

Substitute 132 for A and 16.5 for L.

$$W = \frac{132}{16.5} = 8$$

The width of the mural is 8 ft.

13.   Use the formula $I = prt$, and solve for t.

$$\frac{I}{pr} = \frac{prt}{pr}$$

$$\frac{I}{pr} = t \text{ or } t = \frac{I}{pr}$$

Substitute 1200 for p, .03 for r, and 126 for I.

$$t = \frac{126}{1200(.03)} = \frac{126}{36} = 3.5$$

It will take 3.5 yr for the deposit to earn \$126.

14.   Use the formula $I = prt$, and solve for r.

$$\frac{I}{pt} = \frac{prt}{pt}$$

$$\frac{I}{pt} = r$$

Substitute 30,000 for p, 7800 for I, and 4 for t.

$$r = \frac{7800}{30,000(4)} = \frac{7800}{120,000} = .065$$

The rate is 6.5%.

15.   Use the formula $F = \dfrac{9}{5}C + 32$.

Substitute 40 for C.

$$F = \frac{9}{5}(40) + 32 = 72 + 32 = 104$$

The child's temperature would be 104° F.

16.   To find the area of this trapezoid, I would first add the values of B and b. Then I would multiply the sum by the value of h, and finally, I would divide that answer by 2.

17.   "The difference between 9 and twice a number" is written

$$9 - 2x.$$

18.   "The product of 4 and a number, subtracted from 8" is written

$$8 - 4x.$$

19.   Let $x =$ the number.

$$x - .35x = 260$$

$$1x - .35x = 260$$

$$.65x = 260$$

$$x = \frac{260}{.65} = 400$$

The number is 400.

20. Let $x =$ the number of pairs in Washington.
$x + 289 =$ the number of pairs in California.
The total number of pairs is 1631, so an equation is

$$x + (x + 289) = 1631.$$
$$2x + 289 = 1631$$
$$2x = 1342$$
$$x = 671$$

There were 671 pairs in Washington, and
$671 + 289 = 960$ pairs in California.

21. The change (in billions) is
$\$17.76 - \$16.18 = \$1.58$.
The base is $\$16.18$, so the percent increase is

$$\frac{\$1.58}{\$16.18} \approx .0977 \approx 9.8\%.$$

22. Let $x =$ the amount invested at 6%.
$x - 4000 =$ the amount invested at 4%.
Make a table.

Percent as a Decimal	Amount Invested	Interest in One Year
.06	x	$.06x$
.04	$x - 4000$	$.04(x - 4000)$
Total		$\$840$

The last column gives the equation.

$.06x + .04(x - 4000) = 840$
$6x + 4(x - 4000) = 84{,}000$ *Multiply by 100.*
$6x + 4x - 16{,}000 = 84{,}000$
$10x = 100{,}000$
$x = 10{,}000$

Kevin should invest $\$10{,}000$ at 6% and
$\$10{,}000 - \$4000 = \$6000$ at 4%.

23. Let $x =$ the number of liters of the 20% solution.
Make a table.

Strength	Liters of Solution	Liters of Pure Chemical
20%	x	$.20x$
50%	15	$.50(15) = 7.5$
30%	$x + 15$	$.30(x + 15)$

The last column gives the equation.

$.20x + 7.5 = .30(x + 15)$
$.20x + 7.5 = .30x + 4.5$
$3 = .10x$
$30 = x$

30 L of the 20% solution should be used.

24. **(a)** The sum of the four systems is 5.4025 million. The percent using Primestar is

$$\frac{1.8995}{5.4025} \approx .3516 \approx 35.2\%.$$

(b) The system with the smallest market share at .133 million is USSB.

25. Let $x =$ the angle.
$90 - x =$ its complement.
$180 - x =$ its supplement.
Solve this equation.

$$90 - x = \frac{1}{5}(180 - x) - 10$$
$$450 - 5x = 180 - x - 50 \qquad \textit{Multiply by 5.}$$
$$450 - 5x = 130 - x$$
$$320 = 4x$$
$$80 = x$$

The measure of the angle is 80°.

26. The sum of the angles in a triangle is 180°. Therefore,

$$(3x + 7) + (4x + 1) + (9x - 4) = 180$$
$$16x + 4 = 180$$
$$16x = 176$$
$$x = 11.$$

The first angle is $3(11) + 7 = 40°$.
The second angle is $4(11) + 1 = 45°$.
The third angle is $9(11) - 4 = 95°$.

27. The marked angles are supplements which have a sum of 180°. Therefore,

$$(15x + 15) + (3x + 3) = 180$$
$$18x + 18 = 180$$
$$18x = 162$$
$$x = 9.$$

The angle measures are
$15(9) + 15 = 150°$ and $3(9) + 3 = 30°$.

28.   Let $x = $ the number of tickets sold in London.

$74,000,000 - x = $ the number of tickets sold in Frankfurt.

The total value of the tickets sold is \$686 million.

$$10.59(x) + 8.42(74,000,000 - x) = 686,000,000$$
$$1059(x) + 842(74,000,000 - x) = 68,600,000,000$$

Multiply by 100.

$$1059x + 62,308,000,000 - 842x = 68,600,000,000$$
$$217x = 6,292,000,000$$
$$x \approx 28,995,392$$
$$74,000,000 - x \approx 45,004,608$$

Rounded to the nearest million, there were $29,000,000$ tickets sold in London and $45,000,000$ tickets sold in Frankfurt.

29.   Let $x = $ the number of pounds of the \$3 candy.

$x + 30 = $ the number of pounds of the \$5 candy that will result from mixing.

Make a table.

	Amount of Candy	Cost per Pound	Value
First Candy	x	3	$3x$
Second Candy	30	6	$6(30)$
Mixture	$x + 30$	5	$5(x + 30)$

From the last column:

$$3x + 6(30) = 5(x + 30)$$
$$3x + 180 = 5x + 150$$
$$30 = 2x$$
$$15 = x$$

She should use 15 lb of the \$3 candy.

30.   Solve $d = rt$ for r and substitute.

$$r = \frac{d}{t} = \frac{405}{8.2} \approx 49.4$$

The best estimate is choice (a), 50 miles per hour.

31.   Let $x = $ average speed for the first hour.

$x - 7 = $ average speed for the second hour.

Using $d = rt$. The distance traveled for the first hour is $x(1)$, for the second hour is $(x - 7)(1)$, and for the whole trip, 85.

$$x + (x - 7) = 85$$
$$2x - 7 = 85$$
$$2x = 92$$
$$x = 46$$

The average speed for the first hour was 46 mph.

32.   Let $x = $ the speed of the slower car.

$x + 15 = $ the speed of the faster car.

Make a table.

	r	t	d
Slower car	x	2	$2x$
Faster car	$x + 15$	2	$2(x + 15)$

The total distance traveled, 230 km, is the sum of the distances traveled by each car.

$$2x + 2(x + 15) = 230$$
$$2x + 2x + 30 = 230$$
$$4x = 200$$
$$x = 50$$

The slower car travels at $50\dfrac{\text{km}}{\text{hr}}$, while the faster car travels at $50 + 15 = 65\dfrac{\text{km}}{\text{hr}}$.

33.   Let $x = $ the length of the shortest side.

$2x = $ the length of the middle side.

$3x - 2 = $ the length of the longest side.

The perimeter is 34 in. Add the lengths of the three sides.

$$x + 2x + (3x - 2) = 34$$
$$6x - 2 = 34$$
$$6x = 36$$
$$x = 6$$

The shortest side is 6 inches, the middle side is $2(6) = 12$ inches, and the longest side is $3(6) - 2 = 16$ inches long.

34.   Let $x = $ one integer.

$x + 1 = $ the other integer.

The sum of the integers is 105, so an equation is

$$x + (x + 1) = 105.$$
$$2x + 1 = 105$$
$$2x = 104$$
$$x = 52$$

Then $x + 1 = 53$.

The integers are 52 and 53.

35. $-\dfrac{2}{3}k < 6$

$-2k < 18$ *Multiply by 3.*

Divide by -2; reverse the inequality sign.

$k > -9$

Solution set: $(-9, \infty)$

36. $-5x - 4 \geq 11$

$\qquad -5x \geq 15$

Divide by -5; reverse the inequality sign.

$\qquad x \leq -3$

Solution set: $(-\infty, -3]$

37. $\dfrac{6a + 3}{-4} < -3$

Multiply by -4; reverse the inequality sign.

$\qquad 6a + 3 > 12$

$\qquad 6a > 9$

$\qquad a > \dfrac{9}{6} \text{ or } \dfrac{3}{2}$

Solution set: $\left(\dfrac{3}{2}, \infty\right)$

38. $5 - (6 - 4k) \geq 2k - 7$

$\qquad 5 - 6 + 4k \geq 2k - 7$

$\qquad 4k - 1 \geq 2k - 7$

$\qquad 2k \geq -6$

$\qquad k \geq -3$

Solution set: $[-3, \infty)$

39. $8 \leq 3y - 1 < 14$

$\qquad 9 \leq 3y < 15$

$\qquad 3 \leq y < 5$

Solution set: $[3, 5)$

40. $\dfrac{5}{3}(m - 2) + \dfrac{2}{5}(m + 1) > 1$

$\qquad 25(m - 2) + 6(m + 1) > 15$

$\qquad\qquad\qquad\quad$ *Multiply by 15.*

$\qquad 25m - 50 + 6m + 6 > 15$

$\qquad\qquad 31m - 44 > 15$

$\qquad\qquad\qquad 31m > 59$

$\qquad\qquad\qquad\quad m > \dfrac{59}{31}$

Solution set: $\left(\dfrac{59}{31}, \infty\right)$

41.   Let $x =$ the student's grade on the fifth test. The average of the five test grades must be at least 70. The inequality is

$$\frac{75 + 79 + 64 + 71 + x}{5} \geq 70.$$

$\qquad 75 + 79 + 64 + 71 + x \geq 350$

$\qquad\qquad\qquad\quad 289 + x \geq 350$

$\qquad\qquad\qquad\qquad\qquad x \geq 61$

The student will pass algebra if at least 61% on the fifth test is achieved.

42.   The result, $-8 < -13$, is a false statement. There are no real numbers that make this inequality true. The solution set is $\emptyset$.

For Exercises 43–46, let $A = \{a, b, c, d\}$, $B = \{a, c, e, f\}$, and $C = \{a, e, f, g\}$.

43. $A \cap B = \{a, b, c, d\} \cap \{a, c, e, f\}$

$\qquad\qquad = \{a, c\}$

44. $A \cap C = \{a, b, c, d\} \cap \{a, e, f, g\}$

$\qquad\qquad = \{a\}$

45. $B \cup C = \{a, c, e, f\} \cup \{a, e, f, g\}$

$\qquad\qquad = \{a, c, e, f, g\}$

46. $A \cup C = \{a, b, c, d\} \cup \{a, e, f, g\}$

$\qquad\qquad = \{a, b, c, d, e, f, g\}$

47. $x \leq 4$ and $x < 3$

The graph of the solution set is all numbers both less than or equal to 4 and less than 3. This is the intersection. The elements common to both sets are the numbers less than 3. The solution set is $(-\infty, 3)$.

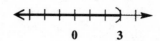

48. $x + 4 > 12$ and $x - 2 < 1$

$\qquad x > 8$ and $x < 3$

The graph of the solution set is all numbers both greater than 8 and less than 3. This is the intersection. Since there are no numbers that are both greater than 8 and less than 3, the solution set is $\emptyset$.

49. $x > 5$ or $x \leq -1$

The graph of the solution set is all numbers either greater than 5 or less than or equal to -1. This is the union. The solution set is $(-\infty, -1] \cup (5, \infty)$.

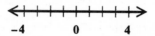

50. $x - 4 > 6$ or $x + 3 \leq 18$

$\qquad x > 10$ or $x \leq 15$

The graph of the solution set is all numbers either greater than 10 or less than or equal to 15, or both. This is the union. The solution set is the set of all real numbers, or $(-\infty, \infty)$.

51. $(-3, \infty) \cap (-\infty, 4)$

$(-3, \infty)$ includes all real numbers greater than -3.

$(-\infty, 4)$ includes all real numbers less than 4. Find the intersection. The numbers common to both sets are greater than -3 and less than 4.

$$-3 < x < 4$$

Solution set: $(-3, 4)$

52. $(-\infty, 6) \cap (-\infty, 2)$

$(-\infty, 6)$ includes all real numbers less than 6.

$(-\infty, 2)$ includes all real numbers less than 2.

Find the intersection. The numbers common to both sets are less than 2.

Solution set: $(-\infty, 2)$

53. $(4, \infty) \cup (9, \infty)$

$(4, \infty)$ includes all real numbers greater than 4.

$(9, \infty)$ includes all real numbers greater than 9.

Find the union. The numbers in the first set, the second set, or in both sets are all the real numbers that are greater than 4.

Solution set: $(4, \infty)$

54. $(1, 2) \cup (1, \infty)$

$(1, 2)$ includes the real numbers between 1 and 2, not including 1 and 2.

$(1, \infty)$ includes all real numbers greater than 1. Find the union. The numbers in the first set, the second set, or in both sets are all real numbers greater than 1.

Solution set: $(1, \infty)$

55. **(a)** The median earnings for men are less than $900 includes managerial and professional specialty, waiters, and bus drivers.

The median earnings for women are greater than $500 includes managerial and professional specialty and mathematical and computer scientists.

Find the intersection. The only occupation in both groups is managerial and professional specialty.

(b) The median earnings for men are greater than $900 includes mathematical and computer sciences.

The median earnings for women are greater than $600 includes mathematical and computer sciences and managerial and professional specialty, which is also the union.

56. $|x| = 7$

$x = 7$ or $x = -7$

Solution set: $\{-7, 7\}$

57. $|3k - 7| = 8$

$3k - 7 = 8$ or $3k - 7 = -8$

$\qquad 3k = 15$ $\qquad\qquad 3k = -1$

$\qquad\quad k = 5$ or $\qquad\quad k = -\dfrac{1}{3}$

Solution set: $\left\{-\dfrac{1}{3}, 5\right\}$

58. $|z - 4| = -12$

Since the absolute value of an expression can never be negative, there are no solutions for this equation.

Solution set: $\emptyset$

59. $|4a + 2| - 7 = -3$

$\quad |4a + 2| = 4$

$4a + 2 = 4$ or $4a + 2 = -4$

$\quad 4a = 2$ $\qquad\qquad 4a = -6$

$\qquad a = \dfrac{2}{4}$ $\qquad\qquad a = -\dfrac{6}{4}$

$\qquad a = \dfrac{1}{2}$ or $\qquad a = -\dfrac{3}{2}$

Solution set: $\left\{-\dfrac{3}{2}, \dfrac{1}{2}\right\}$

60. $|3p + 1| = |p + 2|$

$3p + 1 = p + 2$ or $3p + 1 = -(p + 2)$

$\quad 2p = 1$ $\qquad\qquad 3p + 1 = -p - 2$

$\qquad\qquad\qquad\qquad\qquad 4p = -3$

$\quad p = \dfrac{1}{2}$ or $\qquad p = -\dfrac{3}{4}$

Solution set: $\left\{-\dfrac{3}{4}, \dfrac{1}{2}\right\}$

61. $|2m - 1| = |2m + 3|$

$2m - 1 = 2m + 3$ or $2m - 1 = -(2m + 3)$

$\quad 0 = 4$ *False* $\qquad 2m - 1 = -2m - 3$

$\qquad\qquad\qquad\qquad\qquad 4m = -2$

No solution   or $\qquad m = -\dfrac{2}{4} = -\dfrac{1}{2}$

Solution set: $\left\{-\dfrac{1}{2}\right\}$

62. $|-y + 6| \le 7$

$-7 \le -y + 6 \le 7$

$-13 \le -y \le 1$

Multiply by -1; reverse the inequality signs.

$13 \ge y \ge -1$ or $-1 \le y \le 13$

Solution set: $[-1, 13]$

63. $|2p + 5| \le 1$

$-1 \le 2p + 5 \le 1$

$-6 \le 2p \le -4$

$-3 \le p \le -2$

Solution set: $[-3, -2]$

64. $|x + 1| \geq -3$

Since the absolute value of an expression is always nonnegative (positive or zero), the inequality is true for any real number x.

Solution set: $(-\infty, \infty)$

65. $|5r - 1| > 9$

$$5r - 1 > 9 \quad \text{or} \quad 5r - 1 < -9$$
$$5r > 10 \qquad\qquad 5r < -8$$
$$r > 2 \quad \text{or} \quad r < -\frac{8}{5}$$

Solution set: $\left(-\infty, -\frac{8}{5}\right) \cup (2, \infty)$

66. $|11x - 3| \leq -2$

There are no numbers whose absolute value is negative, so this inequality has no solution.

Solution set: $\emptyset$

67. $|11x - 3| \leq 0$

The absolute value of an expression is always nonnegative (positive or zero), so this inequality is true only when

$$11x - 3 = 0$$
$$11x = 3$$
$$x = \frac{3}{11}.$$

Solution set: $\left\{\frac{3}{11}\right\}$

68. The distance between x and 14 is greater than 12 can be written

$$|x - 14| > 12 \text{ or } |14 - x| > 12.$$

69. $5 - (6 - 4k) > 2k - 5$
$$5 - 6 + 4k > 2k - 5$$
$$-1 + 4k > 2k - 5$$
$$2k > -4$$
$$k > -2$$

Solution set: $(-2, \infty)$

70. Solve $S = 2HW + 2LW + 2LH$ for L.

Get the "L-terms" on one side.
$$S - 2HW = 2LW + 2LH$$
$$S - 2HW = L(2W + 2H)$$
$$\frac{S - 2HW}{2W + 2H} = L$$

71. $x < 3$ and $x \geq -2$

The real numbers that are common to both sets are the numbers greater than or equal to -2 and less than 3.

$$-2 \leq x < 3$$

Solution set: $[-2, 3)$

72. $-4(3 + 2m) - m = -3m$
$$-12 - 8m - m = -3m$$
$$-12 - 9m = -3m$$
$$-6m = 12$$
$$m = -2$$

Solution set: $\{-2\}$

73. Use the formula $V = LWH$, and solve for H.

$$\frac{V}{LW} = \frac{LWH}{LW}$$
$$\frac{V}{LW} = H \text{ or } H = \frac{V}{LW}$$

Substitute 1.5 for W, 5 for L, and 75 for V.

$$H = \frac{75}{5(1.5)} = \frac{75}{7.5} = 10$$

The height of the box is 10 ft.

74. $|3k + 6| \geq 0$

The absolute value of an expression is always nonnegative, so the inequality is true for any real number k.

Solution set: $(-\infty, \infty)$

75. Let $x =$ the smallest consecutive integer.
$x + 1 =$ the middle consecutive integer.
$x + 2 =$ the largest consecutive integer.
The sum of the smallest and largest integers is 47 more than the middle integer, so an equation is

$$x + (x + 2) = 47 + (x + 1).$$
$$2x + 2 = 48 + x$$
$$x = 46$$

Then $x + 1 = 47$,
and $x + 2 = 48$.

The integers are 46, 47, and 48.

76. $|2k - 7| + 4 = 11$
$$|2k - 7| = 7$$
$$2k - 7 = 7 \quad \text{or} \quad 2k - 7 = -7$$
$$2k = 14 \qquad\qquad 2k = 0$$
$$k = 7 \quad \text{or} \qquad k = 0$$

Solution set: $\{0, 7\}$

77. $.05x + .03(1200 - x) = 42$
$$5x + 3(1200 - x) = 4200$$
$$\textit{Multiply by 100.}$$
$$5x + 3600 - 3x = 4200$$
$$2x = 600$$
$$x = 300$$

Solution set: $\{300\}$

78. $|p| < 14$
$$-14 < p < 14$$

Solution set: $(-14, 14)$

79. $\dfrac{3}{4}(a - 2) - \dfrac{1}{3}(5 - 2a) < -2$
$$9(a - 2) - 4(5 - 2a) < -24$$
$$\textit{Multiply by 12.}$$
$$9a - 18 - 20 + 8a < -24$$
$$17a - 38 < -24$$
$$17a < 14$$
$$a < \dfrac{14}{17}$$

Solution set: $\left(-\infty, \dfrac{14}{17}\right)$

80. $-4 < 3 - 2k < 9$
$$-7 < -2k < 6$$

Divide by -2; reverse the inequality signs.
$$\dfrac{7}{2} > k > -3 \text{ or } -3 < k < \dfrac{7}{2}$$

Solution set: $\left(-3, \dfrac{7}{2}\right)$

81. $-.3x + 2.1(x - 4) \le -6.6$
$$-3x + 21(x - 4) \le -66$$
$$\textit{Multiply by 10.}$$
$$-3x + 21x - 84 \le -66$$
$$18x - 84 \le -66$$
$$18x \le 18$$
$$x \le 1$$

Solution set: $(-\infty, 1]$

82. Let $x =$ the degree measure of the angle.
$180 - x =$ the degree measure of its supplement.
$90 - x =$ the degree measure of its complement.
The supplement is 25 more than twice its complement, so an equation is
$$180 - x = 25 + 2(90 - x).$$
$$180 - x = 25 + 180 - 2x$$
$$x = 25 \qquad \textit{Add } 2x.$$

The measure of the angle is $25°$.

83. Use the rule of 78.
$$u = f \cdot \dfrac{k(k + 1)}{n(n + 1)}$$

Let $f = 450$, $k = 5$, and $n = 24$.
$$u = 450 \cdot \dfrac{5(6)}{24(25)} = 22.50$$

The unearned interest is \$22.50.

84. Let $x =$ amount of time spent averaging 45 miles per hour.

$4 - x =$ amount of time at 50 mph.

	Rate	Time	Distance
First Part	45	x	$45x$
Second Part	50	$4 - x$	$50(4 - x)$
Total			195

From the last column:
$$45x + 50(4 - x) = 195$$
$$45x + 200 - 50x = 195$$
$$-5x = -5$$
$$x = 1$$

The automobile averaged 45 mph for 1 hour.

85. $|5r - 1| > 14$
$$5r - 1 > 14 \quad \text{or} \quad 5r - 1 < -14$$
$$5r > 15 \qquad\qquad 5r < -13$$
$$r > 3 \quad \text{or} \qquad r < -\dfrac{13}{5}$$

Solution set: $\left(-\infty, -\dfrac{13}{5}\right) \cup (3, \infty)$

86. $x \ge -2 \text{ or } x < 4$

The solution set includes all numbers either greater than or equal to -2 or all numbers less than 4. This is the union and is the set of all real numbers. The solution set is $(-\infty, \infty)$.

87. Let $x =$ the number of liters of the 20% solution.

$x + 10 =$ the number of liters of the resulting 40% solution.

Make a table.

Strength	Liters of Solution	Liters of Mixture
20%	x	$.20x$
50%	10	$.50(10)$
40%	$x + 10$	$.40(x + 10)$

From the last column:

$$.20x + .50(10) = .40(x + 10)$$
$$.20x + 5 = .40x + 4$$
$$1 = .20x$$
$$5 = x$$

5 L of the 20% solution should be used.

88. $|m - 1| = |2m + 3|$

$m - 1 = 2m + 3$ or $m - 1 = -(2m + 3)$
$m - 1 = -2m - 3$
$3m = -2$
$-4 = m$ or $m = -\dfrac{2}{3}$

Solution set: $\left\{-4, -\dfrac{2}{3}\right\}$

89. $\dfrac{3y}{5} - \dfrac{y}{2} = 3$

$6y - 5y = 30$ *Multiply by 10.*
$y = 30$

Solution set: $\{30\}$

90. $|m + 3| \leq 1$

$-1 \leq m + 3 \leq 1$
$-4 \leq m \leq -2$

Solution set: $[-4, -2]$

91. $|3k - 7| = 4$

$3k - 7 = 4$ or $3k - 7 = -4$
$3k = 11$ $3k = 3$
$k = \dfrac{11}{3}$ or $k = 1$

Solution set: $\left\{1, \dfrac{11}{3}\right\}$

92. Let $x =$ the number of votes for Willkie.
$x + 367 =$ the number of votes for Roosevelt.
The total is 531, so

$$x + (x + 367) = 531$$
$$2x + 367 = 531$$
$$2x = 164$$
$$x = 82.$$

Willkie received 82 electoral votes, and Roosevelt received $82 + 367 = 449$ electoral votes.

93. $5(2x - 7) = 2(5x + 3)$
$10x - 35 = 10x + 6$
$-35 = 6$ *False*

This equation is a contradiction.

Solution set: $\emptyset$

94. $x > 6$ and $x < 8$
The graph of the solution set is all numbers both greater than 6 *and* less than 8. This is the

intersection. The elements common to both sets are the numbers between 6 and 8, not including the endpoints. The solution set is $(6, 8)$.

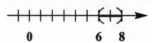

95. $-5x + 1 \geq 11$ or $3x + 5 \geq 26$
$-5x \geq 10$ $3x \geq 21$
$x \leq -2$ or $x \geq 7$

The graph of the solution set is all numbers either less than or equal to -2 *or* greater than or equal to 7. This is the union. The solution set is $(-\infty, -2] \cup [7, \infty)$.

96. (a) The endpoints on the number line are $-\dfrac{11}{3}$ and 1. The solution set of $|3x + 4| \geq 7$ would be

$$\left(-\infty, -\dfrac{11}{3}\right] \cup [1, \infty).$$

(b) The solution set of $|3x + 4| \leq 7$ would be

$$\left[-\dfrac{11}{3}, 1\right].$$

Chapter 2 Test

1. $3(2y - 2) - 4(y + 6) = 3y + 8 + y$
$6y - 6 - 4y - 24 = 4y + 8$
$2y - 30 = 4y + 8$
$-2y = 38$
$y = -19$

Solution set: $\{-19\}$

2. $.08x + .06(x + 9) = 1.24$
$8x + 6(x + 9) = 124$
Multiply by 100.
$8x + 6x + 54 = 124$
$14x + 54 = 124$
$14x = 70$
$x = 5$

Solution set: $\{5\}$

3. $\dfrac{x + 6}{10} + \dfrac{x - 4}{15} = \dfrac{x + 2}{6}$
$3(x + 6) + 2(x - 4) = 5(x + 2)$
Multiply by 30.
$3x + 18 + 2x - 8 = 5x + 10$
$5x + 10 = 5x + 10$ *True*

This is an identity.

Solution set: $(-\infty, \infty)$

4. $3x - (2 - x) + 4x + 2 = 8x + 3$
     $3x - 2 + x + 4x + 2 = 8x + 3$
               $8x = 8x + 3$
               $0 = 3$ *False*

The false statement indicates that the equation is a contradiction.

Solution set: $\emptyset$

5.   Solve $-16t^2 + vt - S = 0$ for v.

$$vt = S + 16t^2$$
$$v = \frac{S + 16t^2}{t}$$

6.   Solve $d = rt$ for t and substitute 450 for d and 140.9 for r.

$$t = \frac{d}{r} = \frac{450}{140.9} \approx 3.2$$

Petty's time was about 3.2 hr.

7.   Let $x =$ the population of the county.
Then, 63.1% of x or $.631x$ is the number of residents living in poverty.

$$.631x = 6118$$
$$x = \frac{6118}{.631} \approx 9695.7$$

There are about 9696 residents in the county.

8.   Let $x =$ Tanui's time.
$x + .3 =$ Pippig's time.
The distance they ran is the same. Use $d = rt$.

$$12.19x = 10.69(x + .3)$$
$$1219x = 1069(x + .3)$$
$$1219x = 1069x + 320.7$$
$$150x = 320.7$$
$$x = 2.138$$

So Tanui's time was about 2.14 hours and Pippig's time was about $2.14 + .3 = 2.44$ hours.

9.   Let $x =$ the amount invested at 3%.
$28,000 - x =$ the amount invested at 5%.

Percent as a Decimal	Amount Invested	Interest in One Year
.03	x	$.03x$
.05	$28,000 - x$	$.05(28,000 - x)$
Total	$28,000	$1240

From the last column:

$.03x + .05(28,000 - x) = 1240$
$3x + 5(28,000 - x) = 124,000$
                   Multiply by 100.
$3x + 140,000 - 5x = 124,000$
          $-2x = -16,000$
             $x = 8000$

Craig should invest $8000 at 3% and $28,000 - \$8000 = \$20,000$ at 5%.

10.   The sum of the three angle measures is 180°, so

$$(2x + 20) + x + x = 180$$
$$4x + 20 = 180$$
$$4x = 160$$
$$x = 40.$$

The three angle measures are 40°, 40°, and $(2 \cdot 40 + 20)° = 100°$.

11.   **(a)** $23.4 - .597 = 22.803$
The 1994 sales were 22.803 million more than the 1991 sales.

(b) Using 16.5 million as an estimate for the 1993 sales, we have

$$.43(16.5) = 7.095$$

million as an approximation for the number of reference books sold in CD-ROM format in 1993.

12.   When multiplying or dividing both sides of an inequality by a negative number, remember to reverse the direction of the inequality symbol.

13. $4 - 6(x + 3) \leq -2 - 3(x + 6) + 3x$
   $4 - 6x - 18 \leq -2 - 3x - 18 + 3x$
      $-6x - 14 \leq -20$
         $-6x \leq -6$
Divide by -6, and reverse the inequality sign.
           $x \geq 1$
Solution set: $[1, \infty)$

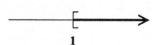

14. $-\frac{4}{7}x > -16$
   $-4x > -112$ *Multiply by 7.*
Divide by -4, and reverse the inequality sign.
      $x < 28$
Solution set: $(-\infty, 28)$

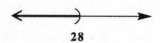

15. $-6 \leq \frac{4}{3}x - 2 \leq 2$

     $-18 \leq 4x - 6 \leq 6$ *Multiply by 3.*

     $-12 \leq 4x \leq 12$ *Add 6.*

     $-3 \leq x \leq 3$ *Divide by 4.*

Solution set: $[-3, 3]$

-3 3

16.   To find the inequalities equivalent to $x < -3$, solve each inequality for x.

  (a) $-3x < 9$

      $x > -3$ *Not equivalent*

  (b) $-3x > -9$

      $x < 3$ *Not equivalent*

  (c) $-3x > 9$

      $x < -3$ *Equivalent*

  (d) $-3x < -9$

      $x > 3$ *Not equivalent*

Of these four choices, only (c) is equivalent to $x < -3$.

17.   From the graph, the number of departures to Europe increased the most between 1992 and 1993 (about 1 million). The graph is flat between 1991 and 1992, so that is when the departures stayed the same.

18. **(a)** $A \cap B = \{1, 2, 5, 7\} \cap \{1, 5, 9, 12\}$
            $= \{1, 5\}$

  (b) $A \cup B = \{1, 2, 5, 7\} \cup \{1, 5, 9, 12\}$
            $= \{1, 2, 5, 7, 9, 12\}$

19. **(a)** $3k \geq 6$ and $k - 4 < 5$
      $k \geq 2$ and $k < 9$

The solution set is all numbers both greater than or equal to 2 *and* less than 9. This is the intersection. The numbers common to both sets are between 2 and 9, including 2 but not 9. The solution set is $[2, 9)$.

  (b) $-4x \leq -24$ or $4x - 2 < 10$
                   $4x < 12$
      $x \geq 6$ or $x < 3$

The solution set is all numbers less than 3 or greater than or equal to 6. This is the union. The solution set is $(-\infty, 3) \cup [6, \infty)$.

20. $|4x - 3| = 7$

     $4x - 3 = 7$ or $4x - 3 = -7$

         $4x = 10$ $4x = -4$

         $x = \frac{10}{4} = \frac{5}{2}$ or $x = -1$

Solution set: $\left\{ -1, \frac{5}{2} \right\}$

21. $|4x - 3| > 7$

     $4x - 3 > 7$ or $4x - 3 < -7$

         $4x > 10$ $4x < -4$

         $x > \frac{10}{4}$

         $x > \frac{5}{2}$ or $x < -1$

Solution set: $(-\infty, -1) \cup \left(\frac{5}{2}, \infty \right)$

Note that the solution sets for Exercises 21–22 could be obtained from the solution set in Exercise 20.

22. $|4x - 3| < 7$

     $-7 < 4x - 3 < 7$

     $-4 < 4x < 10$

     $-1 < x < \frac{10}{4}$

Solution set: $\left(-1, \frac{5}{2} \right)$

23. $|3 - 5x| = |2x + 8|$

   $3 - 5x = 2x + 8$ or $3 - 5x = -(2x + 8)$

     $-7x = 5$ $3 - 5x = -2x - 8$

                          $-3x = -11$

     $x = -\frac{5}{7}$ or $x = \frac{11}{3}$

Solution set: $\left\{ -\frac{5}{7}, \frac{11}{3} \right\}$

24. $|-3x + 4| - 4 < -1$

       $|-3x + 4| < 3$

     $-3 < -3x + 4 < 3$

     $-7 < -3x < -1$

     $\frac{7}{3} > x > \frac{1}{3}$ or $\frac{1}{3} < x < \frac{7}{3}$

Solution set: $\left(\frac{1}{3}, \frac{7}{3} \right)$

25. **(a)** $|5x + 3| < k$

If $k < 0$, then $|5x + 3|$ would be less than a negative number. Since the absolute value of an expression is always nonnegative (positive or zero), the solution set is $\emptyset$.

(b) $|5x + 3| > k$

If $k < 0$, then $|5x + 3|$ would be greater than a negative number. Since the absolute value of an expression is always nonnegative (positive or zero), the solution set is the set of all real numbers, $(-\infty, \infty)$.

(c) $|5x + 3| = k$

If $k < 0$, then $|5x + 3|$ would be equal to a negative number. Since the absolute value of an expression is always nonnegative (positive or zero), the solution set is $\emptyset$.

Cumulative Review Exercises Chapters 1–2

Exercises 1–6 refer to set A.

Let $A = \left\{ -8, -\dfrac{2}{3}, -\sqrt{6}, 0, \dfrac{4}{5}, 9, \sqrt{36} \right\}$.

Simplify $\sqrt{36} = 6$.

1. The elements 9 and 6 are natural numbers.

2. The elements 0, 9, and 6 are whole numbers.

3. The elements -8, 0, 9, and 6 are integers.

4. The elements -8, $-\dfrac{2}{3}$, 0, $\dfrac{4}{5}$, 9, and 6 are rational numbers.

5. The element $-\sqrt{6}$ is an irrational number.

6. All the elements in set A are real numbers.

7. $-\dfrac{4}{3} - \left(-\dfrac{2}{7} \right) = -\dfrac{4}{3} + \dfrac{2}{7}$

$= -\dfrac{28}{21} + \dfrac{6}{21}$

$= -\dfrac{22}{21}$

8. $|-4| - |2| + |-6| = 4 - 2 + 6$

$= 2 + 6$

$= 8$

9. $(-2)^4 + (-2)^3 = 16 + (-8) = 8$

10. $\sqrt{25} - 5(-1)^0 = 5 - 5(1) = 5 - 5 = 0$

11. $(-3)^5 = (-3)(-3)(-3)(-3)(-3) = -243$

12. $\left(\dfrac{6}{7} \right)^3 = \dfrac{6}{7} \cdot \dfrac{6}{7} \cdot \dfrac{6}{7} = \dfrac{216}{343}$

13. $\left(-\dfrac{2}{3} \right)^3 = \left(-\dfrac{2}{3} \right)\left(-\dfrac{2}{3} \right)\left(-\dfrac{2}{3} \right) = -\dfrac{8}{27}$

14. $-4^6 = -(4 \cdot 4 \cdot 4 \cdot 4 \cdot 4 \cdot 4) = -4096$

15. $-\sqrt{36} = -(6) = -6$

$\sqrt{-36}$ is not a real number.

16. $\dfrac{4-4}{4+4} = \dfrac{0}{8} = 0$

$\dfrac{4+4}{4-4} = \dfrac{8}{0}$, which is undefined.

For Exercises 17–20, let $a = 2$, $b = -3$, and $c = 4$.

17. $-3a + 2b - c = -3(2) + 2(-3) - 4$

$= -6 - 6 - 4$

$= -16$

18. $-2b^2 - 4c = -2(-3)^2 - 4(4)$

$= -2(9) - 4(4)$

$= -18 - 16$

$= -34$

19. $-8(a^2 + b^3) = -8[2^2 + (-3)^3]$

$= -8[4 + (-27)]$

$= -8(-23)$

$= 184$

20. $\dfrac{3a^3 - b}{4 + 3c} = \dfrac{3(2)^3 - (-3)}{4 + 3(4)}$

$= \dfrac{3(8) - (-3)}{4 + 3(4)}$

$= \dfrac{24 + 3}{4 + 12}$

$= \dfrac{27}{16}$

21. $-7r + 5 - 13r + 12$

$= -7r - 13r + 5 + 12$

$= (-7 - 13)r + (5 + 12)$

$= -20r + 17$

22. $-(3k + 8) - 2(4k - 7) + 3(8k + 12)$

$= -3k - 8 - 8k + 14 + 24k + 36$

$= -3k - 8k + 24k - 8 + 14 + 36$

$= 13k + 42$

23. $(a + b) + 4 = 4 + (a + b)$

The order of the terms $(a + b)$ and 4 have been reversed. This is an illustration of the commutative property.

24. $4x + 12x = (4 + 12)x$

The common variable, x, has been removed from each term. This is an illustration of the distributive property.

25. $-9 + 9 = 0$

The sum of a number and its opposite is equal to 0. This is an illustration of the inverse property.

26. The product of a number and its reciprocal must equal 1. Given the number $-\dfrac{2}{3}$, its reciprocal is $-\dfrac{3}{2}$, since

$$-\frac{2}{3}\left(-\frac{3}{2}\right) = 1.$$

27.
$$-4x + 7(2x + 3) = 7x + 36$$
$$-4x + 14x + 21 = 7x + 36$$
$$10x + 21 = 7x + 36$$
$$3x = 15$$
$$x = 5$$
Solution set: $\{5\}$

28.
$$-\frac{3}{5}x + \frac{2}{3}x = 2$$
$$3(-3x) + 5(2x) = 15(2) \quad \textit{Multiply by 15.}$$
$$-9x + 10x = 30$$
$$x = 30$$
Solution set: $\{30\}$

29.
$$.06x + .03(100 + x) = 4.35$$
$$6x + 3(100 + x) = 435 \quad \textit{Multiply by 100.}$$
$$6x + 300 + 3x = 435$$
$$9x = 135$$
$$x = 15$$
Solution set: $\{15\}$

30. Solve $P = a + b + c$ for b.
$$P - a - c = b \text{ or}$$
$$b = P - a - c$$

31.
$$3 - 2(x + 7) \leq -x + 3$$
$$3 - 2x - 14 \leq -x + 3$$
$$-2x - 11 \leq -x + 3$$
$$-x \leq 14$$
Multiply by -1, and reverse the inequality sign.
$$x \geq -14$$
Solution set: $[-14, \infty)$

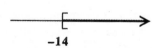
-14

32.
$$-4 < 5 - 3x \leq 0$$
$$-9 < -3x \leq -5$$
Divide by -3, and reverse the inequality sign.
$$3 > x \geq \frac{5}{3} \text{ or } \frac{5}{3} \leq x < 3$$
Solution set: $\left[\dfrac{5}{3}, 3\right)$

$\dfrac{5}{3}$ 3

33.
$$2x + 1 > 5 \quad \text{or} \quad 2 - x > 2$$
$$2x > 4 \qquad\qquad -x > 0$$
$$x > 2 \quad \text{or} \qquad x < 0$$
Solution set: $(-\infty, 0) \cup (2, \infty)$

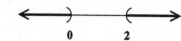
$0 \qquad 2$

34. $|-7k + 3| \geq 4$
$$-7k + 3 \geq 4 \quad \text{or} \quad -7k + 3 \leq -4$$
$$-7k \geq 1 \qquad\qquad -7k \leq -7$$
$$k \leq -\frac{1}{7} \quad \text{or} \qquad k \geq 1$$
Solution set: $\left(-\infty, -\dfrac{1}{7}\right] \cup [1, \infty)$

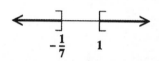
$-\dfrac{1}{7} \qquad 1$

35. Let x = the amount invested at 7% and at 10%. The total amount invested is $2x$, and the total interest is

$$.1(2x) - 150 = .2x - 150.$$

Interest at 7%		interest at 10%		total interest.
$.07x$	$+$	$.10x$	$=$	$.2x - 150$
		$.17x$	$=$	$.20x - 150$
		150	$=$	$.03x$
		5000	$=$	x

She invested \$5000 at each rate.

36. Let x = the amount of food C.
$$2x = \text{the amount of food A.}$$
$$5 = \text{the amount of food B.}$$
The total is at most 24 grams.

$$x + 2x + 5 \leq 24$$
$$3x \leq 19$$
$$x \leq \frac{19}{3} \text{ or } 6\frac{1}{3}$$

He may use at most $6\dfrac{1}{3}$ g of food C.

37. Let x = the grade the student must make on the third test.
To find the average of the three tests, add them and divide by 3. This average must be at least 80.

$$\frac{88 + 78 + x}{3} \geq 80$$
$$\frac{166 + x}{3} \geq 80$$
$$166 + x \geq 240$$
$$x \geq 74$$

She must earn at least 74 points on her third test.

38. Let $x =$ the time it takes for Jack to be $\frac{1}{4}$ mi ahead of Jill.

Use the formula $d = rt$. Make a table.

	r	t	d
Jack	7	x	$7x$
Jill	5	x	$5x$

Jack's distance must be $\frac{1}{4}$ mi more than Jill's distance.

$$7x = 5x + \frac{1}{4}$$
$$28x = 20x + 1 \quad \text{\textit{Multiply by 4.}}$$
$$8x = 1$$
$$x = \frac{1}{8}$$

It will take Jack $\frac{1}{8}$ hr.

39. Let $x =$ the amount of pure alcohol that should be added.

Strength	Liters of Solution	Liters of Pure Alcohol
100%	x	$1.00x$
10%	7	$.10(7)$
30%	$x + 7$	$.30(x + 7)$

From the last column:

$$1.00x + .10(7) = .30(x + 7)$$
$$10x + 1(7) = 3(x + 7) \quad \text{\textit{Multiply by 10.}}$$
$$10x + 7 = 3x + 21$$
$$7x = 14$$
$$x = 2$$

2 L of pure alcohol should be added to the solution.

40. Let $x =$ the number of nickels.

$x - 4 =$ the number of quarters.

The collection contains 29 coins, so the number of pennies is

$$29 - x - (x - 4) = 33 - 2x.$$

	Number of coins	Denomination	Value
Pennies	$33 - 2x$	.01	$.01(33 - 2x)$
Nickels	x	.05	$.05x$
Quarters	$x - 4$	.25	$.25(x - 4)$
Total	29		$2.69

From the last column:

$$.01(33 - 2x) + .05x + .25(x - 4) = 2.69$$
$$1(33 - 2x) + 5x + 25(x - 4) = 269$$
$$\text{\textit{Multiply by 100.}}$$
$$33 - 2x + 5x + 25x - 100 = 269$$
$$28x - 67 = 269$$
$$28x = 336$$
$$x = 12$$

There are $33 - 2(12) = 9$ pennies, 12 nickels, and $12 - 4 = 8$ quarters.

In Exercises 41 and 42, use Clark's rule.

$$\frac{\text{Weight of child in pounds}}{150} \times \frac{\text{adult}}{\text{dose}} = \frac{\text{child's}}{\text{dose}}$$

41. If the child weighs 55 lb and the adult dosage is 120 mg, then

$$\frac{55}{150} \times 120 = 44.$$

The child's dosage is 44 mg.

42. If the child weighs 75 lb and the adult dosage is 40 drops, then

$$\frac{75}{150} \times 40 = 20.$$

The child's dosage is 20 drops.

43. **(a)** From 1990 to 1995, the number of daily newspapers decreased by $1611 - 1532 = 79$.

(b) $\dfrac{79}{1611} \approx .049 = 4.9\%$

44. Use the BMI formula.

$$\text{BMI} = \frac{704 \, (\text{weight in pounds})}{(\text{height in inches})^2}$$

Ken Griffey, Jr., weighs 205 lb and is 6 ft, 3 in or 75 in (1 ft = 12 in) tall, so his

$$\text{BMI} = \frac{704(205)}{75^2} = \frac{144,320}{5625} \approx 25.7$$

CHAPTER 3   GRAPHS, LINEAR
EQUATIONS,
AND FUNCTIONS

Section 3.1

1. **(a)** The graph rises (so the numbers increase) in the consecutive years 1991–1992 and 1993–1994.

   (b) The biggest drop in the graph was between 1992 and 1993.

   (c) There were about 6000 solar collectors used for heating pools in 1993.

3. The data could be represented well with a bar graph.

5. The point with coordinates $(0, 0)$ is called the <u>origin</u> of a rectangular coordinate system.

7. The x-intercept is the point where a line crosses the x-axis. To find the x-intercept of a line, we let <u>y</u> equal 0 and solve for <u>x</u>.

9. To graph a straight line we must find a minimum of <u>two</u> points. A third point is sometimes found to check the accuracy of the first two points.

11. **(a)** The point $(1, 6)$ is located in quadrant I, since the x- and y-coordinates are both positive.

    (b) The point $(-4, -2)$ is located in quadrant III, since the x- and y-coordinates are both negative.

    (c) The point $(-3, 6)$ is located in quadrant II, since the x-coordinate is negative and the y-coordinate is positive.

    (d) The point $(7, -5)$ is located in quadrant IV, since the x-coordinate is positive and the y-coordinate is negative.

    (e) The point $(-3, 0)$ is located on the x-axis, so it does not belong to any quadrant.

13. **(a)** If $xy > 0$, then both x and y have the same sign.
    (x, y) is in quadrant I if x and y are positive.
    (x, y) is in quadrant III if x and y are negative.

    (b) If $xy < 0$, then x and y have different signs.
    (x, y) is in quadrant II if $x < 0$ and $y > 0$.
    (x, y) is in quadrant IV if $x > 0$ and $y < 0$.

    (c) If $\dfrac{x}{y} < 0$, then x and y have different signs.

    (x, y) is in either quadrant II or IV. (See part (b).)

    (d) If $\dfrac{x}{y} > 0$, then x and y have the same sign.

    (x, y) is in either quadrant I or III. (See part (a).)

For Exercises 15–24, see the rectangular coordinate system after Exercise 24.

15. To plot $(2, 3)$, go two units from zero to the right along the x-axis, and then go three units up parallel to the y-axis.

17. To plot $(-3, -2)$, go three units from zero to the left along the x-axis, and then go two units down parallel to the y-axis.

19. To plot $(0, 5)$, do not move along the x-axis at all since the x-coordinate is 0. Move five units up along the y-axis.

21. To plot $(-2, 4)$, go two units from zero to the left along the x-axis, and then go four units up parallel to the y-axis.

23. To plot $(-2, 0)$, go two units to the left along the x-axis. Do not move up or down since the y-coordinate is 0.

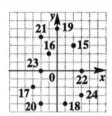

25. $x - y = 3$
    To complete the table, substitute the given values for x and y in the equation.
    For $x = 0$: $x - y = 3$
    $\qquad\qquad\quad 0 - y = 3$
    $\qquad\qquad\qquad\quad y = -3 \quad (0, -3)$

    For $y = 0$: $x - y = 3$
    $\qquad\qquad\quad x - 0 = 3$
    $\qquad\qquad\qquad\quad x = 3 \quad (3, 0)$

    For $x = 5$: $x - y = 3$
    $\qquad\qquad\quad 5 - y = 3$
    $\qquad\qquad\qquad -y = -2$
    $\qquad\qquad\qquad\quad y = 2 \qquad (5, 2)$

    For $x = 2$: $x - y = 3$
    $\qquad\qquad\quad 2 - y = 3$
    $\qquad\qquad\qquad -y = 1$
    $\qquad\qquad\qquad\quad y = -1 \quad (2, -1)$

Plot the ordered pairs and draw the line through them.

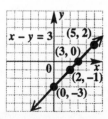

27. $x + 2y = 5$

To complete the table, substitute the given values for x or y in the equation.

For $x = 0$: $x + 2y = 5$
$$0 + 2y = 5$$
$$2y = 5$$
$$y = \frac{5}{2} \quad \left(0, \frac{5}{2}\right)$$

For $y = 0$: $x + 2y = 5$
$$x + 2(0) = 5$$
$$x + 0 = 5$$
$$x = 5 \quad (5, 0)$$

For $x = 2$: $x + 2y = 5$
$$2 + 2y = 5$$
$$2y = 3$$
$$y = \frac{3}{2} \quad \left(2, \frac{3}{2}\right)$$

For $y = 2$: $x + 2y = 5$
$$x + 2(2) = 5$$
$$x + 4 = 5$$
$$x = 1 \quad (1, 2)$$

Plot the ordered pairs and draw the line through them.

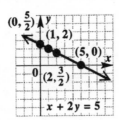

29. $4x - 5y = 20$

For $x = 0$: $4x - 5y = 20$
$$4(0) - 5y = 20$$
$$-5y = 20$$
$$y = -4 \quad (0, -4)$$

For $y = 0$: $4x - 5y = 20$
$$4x - 5(0) = 20$$
$$4x = 20$$
$$x = 5 \quad (5, 0)$$

For $x = 2$: $4x - 5y = 20$
$$4(2) - 5y = 20$$
$$8 - 5y = 20$$
$$-5y = 12$$
$$y = -\frac{12}{5} \quad \left(2, -\frac{12}{5}\right)$$

For $y = -3$: $4x - 5y = 20$
$$4x - 5(-3) = 20$$
$$4x + 15 = 20$$
$$4x = 5$$
$$x = \frac{5}{4} \quad \left(\frac{5}{4}, -3\right)$$

Plot the ordered pairs and draw the line through them.

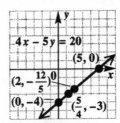

31. In quadrant III, both coordinates of the ordered pairs are negative. If $x + y = k$ and k is positive, then either x or y must be positive, because the sum of two negative numbers is negative.

33. $2x + 3y = 12$

To find the x-intercept, let $y = 0$.
$$2x + 3y = 12$$
$$2x + 3(0) = 12$$
$$2x = 12$$
$$x = 6$$

The x-intercept is $(6, 0)$.
To find the y-intercept, let $x = 0$.
$$2x + 3y = 12$$
$$2(0) + 3y = 12$$
$$3y = 12$$
$$y = 4$$

The y-intercept is $(0, 4)$.
Plot the intercepts and draw the line through them.

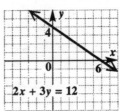

35. $x - 3y = 6$

To find the x-intercept, let $y = 0$.
$$x - 3y = 6$$
$$x - 3(0) = 6$$
$$x - 0 = 6$$
$$x = 6$$

The x-intercept is $(6, 0)$.
To find the y-intercept, let $x = 0$.

$$x - 3y = 6$$
$$0 - 3y = 6$$
$$-3y = 6$$
$$y = -2$$

The y-intercept is $(0, -2)$.
Plot the intercepts and draw the line through them.

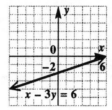

37. $\frac{2}{3}x - 3y = 7$

To find the x-intercept, let $y = 0$.

$$\frac{2}{3}x - 3(0) = 7$$
$$\frac{2}{3}x = 7$$
$$x = \frac{3}{2} \cdot 7 = \frac{21}{2}$$

The x-intercept is $\left(\frac{21}{2}, 0\right)$.

To find the y-intercept, let $x = 0$.

$$\frac{2}{3}(0) - 3y = 7$$
$$-3y = 7$$
$$y = -\frac{7}{3}$$

The y-intercept is $\left(0, -\frac{7}{3}\right)$.

Plot the intercepts and draw the line through them.

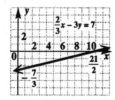

39. $y = 5$

This is a horizontal line. Every point has y-coordinate 5, so no point has y-coordinate 0. There is no x-intercept.
Since every point of the line has y-coordinate 5, the y-intercept is $(0, 5)$. Draw the horizontal line through $(0, 5)$.

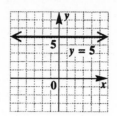

41. $x = 2$

This is a vertical line. Every point has x-coordinate 2, so the x-intercept is $(2, 0)$. Since every point of the line has x-coordinate 2, no point has x-coordinate 0. There is no y-intercept. Draw the vertical line through $(2, 0)$.

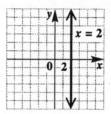

43. $x + 4 = 0 \quad (x = -4)$

This is a vertical line. Every point has x-coordinate -4, so the x-intercept is $(-4, 0)$. Since every point of the line has x-coordinate -4, no point has x-coordinate 0. There is no y-intercept. Draw the vertical line through $(-4, 0)$.

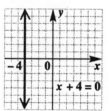

45. $x + 5y = 0$

To find the x-intercept, let $y = 0$.

$$x + 5y = 0$$
$$x + 5(0) = 0$$
$$x = 0$$

The x-intercept is $(0, 0)$, and since $x = 0$, this is also the y-intercept. Since the intercepts are the same, another point is needed to graph the line. Choose any number for y, say $y = -1$, and solve the equation for x.

$$x + 5y = 0$$
$$x + 5(-1) = 0$$
$$x = 5$$

This gives the ordered pair $(5, -1)$. Plot $(5, -1)$ and $(0, 0)$, and draw the line through them.

continued

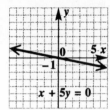

$x + 5y = 0$

47. $2x = 3y$

If $x = 0$, then $y = 0$, so the x- and y-intercepts are $(0, 0)$. To get another point, let $x = 3$.

$$2(3) = 3y$$
$$2 = y$$

Plot $(3, 2)$ and $(0, 0)$, and draw the line through them.

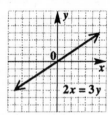

$2x = 3y$

49. $3y = -\dfrac{4}{3}x$

If $x = 0$, then $y = 0$, so the x- and y-intercepts are $(0, 0)$. To get another point, let $x = 9$.

$$3y = -\frac{4}{3}(9)$$
$$3y = -12$$
$$y = -4$$

Plot $(9, -4)$ and $(0, 0)$, and draw the line through them.

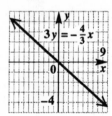

$3y = -\dfrac{4}{3}x$

51. $y = 1.22x + 118$

$y = 1.22(30) + 118$ *x=30 corresponds to 1995*
$= 36.6 + 118$
$= 154.6$

From the model, the approximate speed was 154.6 mph.

53. The graph goes through the point $(2, 6)$ which satisfies only equation (c). The correct equation is (c).

55. (a) From the table, the x-intercept is $(1.5, 0)$.

(b) The y-intercept is $(0, 3)$.

(c) The point $(0, 3)$ satisfies equations C and D. The point $(.5, 2)$ satisfies only equation D. Therefore, the correct equation is D.

57. The screen on the right is more useful because it shows the intercepts.

59. R has x-coordinate 6 and y-coordinate -2, so the ordered pair is $(6, -2)$.

60. The midpoint of $\overline{PR}$ has x-coordinate halfway from 4 to 6, which is 5, and y-coordinate -2, so it's the ordered pair $S = (5, -2)$.

61. The midpoint of $\overline{QR}$ has x-coordinate 6 and y-coordinate halfway from -2 to 2, so it's the ordered pair $T = (6, 0)$.

62. $M = (x\text{-coordinate of } S, y\text{-coordinate of } T)$
$= (5, 0)$

63. $P = (4, -2)$ and $Q = (6, 2)$. The average of the x-coordinates of P and Q is $\dfrac{4 + 6}{2} = \dfrac{10}{2} = 5$. The average of the y-coordinates of P and Q is $\dfrac{-2 + 2}{2} = \dfrac{0}{2} = 0$.

64. The x-coordinate of M is the average of the x-coordinates of P and Q. The y-coordinate of M is the average of the y-coordinates of P and Q.

Section 3.2

1. slope $= \dfrac{\text{change in vertical position}}{\text{change in horizontal position}}$

$= \dfrac{-30 \text{ feet}}{100 \text{ feet}}$

Choices (a) $-.3$, (b) $-\dfrac{3}{10}$, and (d) $-\dfrac{30}{100}$ are all correct.

3. To get to B from A, we must go up 2 units and move right 1 unit. Thus,

slope of $AB = \dfrac{\text{rise}}{\text{run}} = \dfrac{2}{1} = 2.$

5. slope of $CD = \dfrac{\text{rise}}{\text{run}} = \dfrac{-7}{0}$, which is undefined.

7. $m = \dfrac{6 - 2}{5 - 3} = \dfrac{4}{2} = 2$

9. $m = \dfrac{4 - (-1)}{-3 - (-5)} = \dfrac{4 + 1}{-3 + 5} = \dfrac{5}{2}$

11. $m = \dfrac{-5 - (-5)}{3 - 2} = \dfrac{-5 + 5}{1} = \dfrac{0}{1} = 0$

13. (b) and (d) are correct. Choice (a) is wrong because the order of subtraction must be the same in the numerator and denominator. Choice (c) is wrong because slope is defined as the change in y divided by the change in x.

15. Let $(x_1, y_1) = (-4, 3)$ and $(x_2, y_2) = (-3, 4)$. Then

$$m = \frac{y_2 - y_1}{x_2 - x_1} = \frac{4 - 3}{-3 - (-4)} = \frac{1}{1} = 1.$$

The slope is 1.

17. Let $(x_1, y_1) = (-3, -3)$ and $(x_2, y_2) = (5, 6)$.   Then

$$m = \frac{y_2 - y_1}{x_2 - x_1} = \frac{6 - (-3)}{5 - (-3)} = \frac{9}{8}.$$

The slope is $\frac{9}{8}$.

19. Let $(x_1, y_1) = (-6, 3)$ and $(x_2, y_2) = (2, 3)$.   Then

$$m = \frac{y_2 - y_1}{x_2 - x_1} = \frac{3 - 3}{2 - (-6)} = \frac{0}{8} = 0.$$

The slope is 0.

21. The points shown on the line are $(1, -1)$ and $(4, 5)$. The slope is

$$m = \frac{5 - (-1)}{4 - 1} = \frac{6}{3} = 2.$$

23. "The line has positive slope" means that the line goes up from left to right. This is line B.

25. "The line has slope 0" means that there is no vertical change; that is, the line is horizontal. This is line A.

27. To find the slope of

$$x + 2y = 4,$$

first find the intercepts. Replace y with 0 to find that the x-intercept is $(4, 0)$; replace x with 0 to find that the y-intercept is $(0, 2)$. The slope is then

$$m = \frac{2 - 0}{0 - 4} = -\frac{2}{4} = -\frac{1}{2}.$$

To sketch the graph, plot the intercepts and draw the line through them.

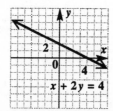

29. To find the slope of

$$-x + y = 4,$$

first find the intercepts. Replace y with 0 to find that the x-intercept is $(-4, 0)$; replace x with 0 to find that the y-intercept is $(0, 4)$. The slope is then

$$m = \frac{4 - 0}{0 - (-4)} = \frac{4}{4} = 1.$$

To sketch the graph, plot the intercepts and draw the line through them.

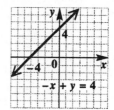

31. To find the slope of

$$6x + 5y = 30,$$

first find the intercepts. Replace y with 0 to find that the x-intercept is $(5, 0)$; replace x with 0 to find that the y-intercept is $(0, 6)$. The slope is then

$$m = \frac{6 - 0}{0 - 5} = \frac{6}{-5} = -\frac{6}{5}.$$

To sketch the graph, plot the intercepts and draw the line through them.

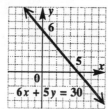

33. To find the slope of

$$5x - 2y = 10,$$

first find the intercepts. Replace y with 0 to find that the x-intercept is $(2, 0)$; replace x with 0 to find that the y-intercept is $(0, -5)$. The slope is then

$$m = \frac{-5 - 0}{0 - 2} = \frac{-5}{-2} = \frac{5}{2}.$$

To sketch the graph, plot the intercepts and draw the line through them.

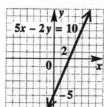

35. In the equation

$$y = 4x,$$

replace x with 0 and then x with 1 to get the ordered pairs $(0,0)$ and $(1,4)$, respectively. (There are other possibilities for ordered pairs.) The slope is then

$$m = \frac{4-0}{1-0} = \frac{4}{1} = 4.$$

To sketch the graph, plot the two points and draw the line through them.

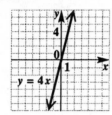

37. $x - 3 = 0$ $(x = 3)$

The graph of $x = 3$ is the vertical line with x-intercept $(3,0)$. The slope of a vertical line is undefined.

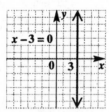

39. A vertical line has equation $\underline{x} = c$ for some constant c; a horizontal line has equation $\underline{y} = d$ for some constant d.

41. Locate the point on coordinate system. Write the slope as a fraction, if necessary. Move up or down the number of units in the numerator and left or right the number of units in the denominator to determine a second point. Draw the line through these two points.

43. To graph the line through $(-2,-3)$ with slope $m = \dfrac{5}{4}$, locate $(-2,-3)$ on the graph. To find a second point, use the definition of slope.

$$m = \frac{\text{change in } y}{\text{change in } x} = \frac{5}{4}$$

From $(-2,-3)$, go up 5 units. Then go 4 units to the right to get to $(2,2)$. Draw the line through $(-2,-3)$ and $(2,2)$.

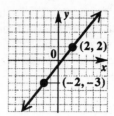

45. To graph the line through $(0,-4)$ with slope $m = -\dfrac{3}{2}$, locate the point $(0,-4)$ on the graph. To find a second point on the line, use the definition of slope, writing $-\dfrac{3}{2}$ as $\dfrac{-3}{2}$.

$$m = \frac{\text{change in } y}{\text{change in } x} = \frac{-3}{2}$$

From $(0,-4)$, move 3 units down and then 2 units to the right. Draw a line through this second point and $(0,-4)$. (Note that the slope could also be written as $\dfrac{3}{-2}$. In this case, move 3 units up and 2 units to the left to get another point on the same line.)

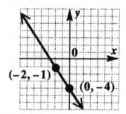

47. Locate $(-2,-4)$. Then use $m = 4 = \dfrac{4}{1}$ to go 4 units up and 1 unit right to $(-1,0)$.

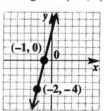

49. Locate $(-3,1)$. Since the slope is undefined, the line is vertical. The x-value of every point is -3.

51. Locate $(5,3)$. A slope of 0 means that the line is horizontal, so $y = 3$ at every point. Draw the horizontal line through $(5,3)$.

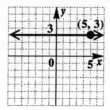

53. $3x = y$ and $2y - 6x = 5$

The slope of the first line is the coefficient of x, namely 3. Solve the second equation for y.

$$2y = 6x + 5$$
$$y = 3x + \frac{5}{2}$$

So the slope of the second line is also 3, and the lines are *parallel*.

55. $4x + y = 0$ and $5x - 8 = 2y$

Solve the equations for y.

$$y = -4x \qquad \frac{5}{2}x - 4 = y$$

The slopes are -4 and $\frac{5}{2}$. The lines are *neither parallel nor perpendicular*.

57. $4x - 3y = 8$ and $4y + 3x = 12$

Solve the equations for y.

$$-3y = -4x + 8 \qquad 4y = -3x + 12$$
$$y = \frac{4}{3}x - \frac{8}{3} \qquad y = -\frac{3}{4}x + 3$$

The slopes, $\frac{4}{3}$ and $-\frac{3}{4}$, are negative reciprocals of one another, so the lines are *perpendicular*.

59. $4x - 3y = 5$ and $3x - 4y = 2$

Solve the equations for y.

$$-3y = -4x + 5 \qquad -4y = -3x + 2$$
$$y = \frac{4}{3}x - \frac{5}{3} \qquad y = \frac{3}{4}x - \frac{1}{2}$$

The slopes are $\frac{4}{3}$ and $\frac{3}{4}$. The lines are *neither parallel nor perpendicular*.

61. The slope of the line through $(4, 6)$ and $(-8, 7)$ is

$$m = \frac{7 - 6}{-8 - 4} = \frac{1}{-12} = -\frac{1}{12}.$$

The slope of the line through $(7, 4)$ and $(-5, 5)$ is

$$m = \frac{5 - 4}{-5 - 7} = \frac{1}{-12} = -\frac{1}{12}.$$

Since the slopes are equal, the two lines are *parallel*.

63. The vertical change is 63 ft, and the horizontal change is $250 - 160 = 90$ ft.

The slope is $\dfrac{63}{90} = \dfrac{7}{10}$.

65. The traffic volume went from 18 billion min to 60 billion min during the 10-yr period from 1986 to 1995. Then

$$m = \frac{60 - 18}{10} = \frac{42}{10} = 4.2.$$

The average rate of change is about 4.2 billion $\dfrac{\text{min}}{\text{yr}}$.

67. For 1991 to 1992:

$$m = \frac{15.15 - 15.64}{1992 - 1991} = \frac{-.49}{1} = -.49$$

For 1992 to 1994:

$$m = \frac{14.17 - 15.15}{1994 - 1992} = \frac{-.98}{2} = -.49$$

For 1993 to 1995:

$$m = \frac{13.68 - 14.66}{1995 - 1993} = \frac{-.98}{2} = -.49$$

These answers suggest that the average rate of change has remained constant at $-.49$ quadrillion $\dfrac{\text{BTUs}}{\text{yr}}$.

69. **(a)** For 1991 through 1996:

$$m = \frac{22.9 - 26.2}{1996 - 1991} = \frac{-3.3}{5} = -.66$$

The average rate of change is $-.66$ million $\dfrac{\text{kilowatts}}{\text{year}}$.

(b) For 1993 through 1996:

$$m = \frac{22.9 - 26.2}{1996 - 1993} = \frac{-3.3}{3} = -1.1$$

The average rate of change is -1.1 million $\dfrac{\text{kilowatts}}{\text{year}}$.

(c) The graph is not a straight line, so the average rate of change varies for different pairs of years.

71. Let $(x_1, y_1) = (0, -4)$ and $(x_2, y_2) = (1, -1)$. Then

$$m = \frac{-1 - (-4)}{1 - 0} = \frac{-1 + 4}{1} = \frac{3}{1} = 3.$$

The slope is 3.

73. Line A has negative slope and line B has positive slope, so line A must be $Y_1 = -2x + 3$ and line B must be $Y_2 = 3x - 4$.

75. Label the points as shown in the figure.

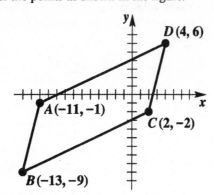

In order to determine whether $ABCD$ is a parallelogram, we need to show that the slope of $\overline{AB}$ equals the slope of $\overline{CD}$ and that the slope of $\overline{AD}$ equals the slope of $\overline{BC}$.

Slope of $\overline{AB} = \dfrac{-9 - (-1)}{-13 - (-11)} = \dfrac{-8}{-2} = 4$

Slope of $\overline{CD} = \dfrac{6 - (-2)}{4 - 2} = \dfrac{8}{2} = 4$

Slope of $\overline{AD} = \dfrac{6 - (-1)}{4 - (-11)} = \dfrac{7}{15}$

Slope of $\overline{BC} = \dfrac{-2 - (-9)}{2 - (-13)} = \dfrac{7}{15}$

Thus, the figure is a parallelogram.

77. For $A(3, 1)$ and $B(6, 2)$, the slope of $\overline{AB}$ is

$$m = \frac{1 - 2}{3 - 6} = \frac{-1}{-3} = \frac{1}{3}.$$

78. For $B(6, 2)$ and $C(9, 3)$, the slope of $\overline{BC}$ is

$$m = \frac{2 - 3}{6 - 9} = \frac{-1}{-3} = \frac{1}{3}.$$

79. For $A(3, 1)$ and $C(9, 3)$, the slope of $\overline{AC}$ is

$$m = \frac{1 - 3}{3 - 9} = \frac{-2}{-6} = \frac{1}{3}.$$

80. The slope of $\overline{AB} =$ slope of $\overline{BC}$
$$= \text{slope of } \overline{AC}$$
$$= \frac{1}{3}.$$

81. For $A(1, -2)$ and $B(3, -1)$, the slope of $\overline{AB}$ is

$$m = \frac{-1 - (-2)}{3 - 1} = \frac{1}{2}.$$

For $B(3, -1)$ and $C(5, 0)$, the slope of $\overline{BC}$ is

$$m = \frac{0 - (-1)}{5 - 3} = \frac{1}{2}.$$

For $A(1, -2)$ and $C(5, 0)$, the slope of $\overline{AC}$ is

$$m = \frac{0 - (-2)}{5 - 1} = \frac{2}{4} = \frac{1}{2}.$$

Since the three slopes are the same, the three points are collinear.

82. For $A(0, 6)$ and $B(4, -5)$, the slope of $\overline{AB}$ is

$$m = \frac{-5 - 6}{4 - 0} = \frac{-11}{4} = -\frac{11}{4}.$$

For $B(4, -5)$ and $C(-2, 12)$, the slope of $\overline{BC}$ is

$$m = \frac{12 - (-5)}{-2 - 4} = \frac{17}{-6} = -\frac{17}{6}.$$

Since these two slopes are not the same, the three points are not collinear.

Section 3.3

1. Choice (a) $3x - 2y = 5$ is in the form $Ax + By = C$ with $A \geq 0$ and integers A, B, and C having no common factor (except 1).

3. Choice (a) $y = 6x + 2$ is in the form $y = mx + b$.

5.
$$y = -3x + 10$$
$$3x + y = 10 \qquad \textit{Standard form}$$

7. $y = 2x + 3$

This line is in slope-intercept form with slope $m = 2$ and y-intercept $(0, b) = (0, 3)$. The only graph with positive slope and with a positive y-coordinate of its y-intercept is A.

9. $y = -2x - 3$

This line is in slope-intercept form with slope $m = -2$ and y-intercept $(0, b) = (0, -3)$. The only graph with negative slope and with a negative y-coordinate of its y-intercept is C.

11. $y = 2x$

This line has slope $m = 2$ and y-intercept $(0, b) = (0, 0)$. The only graph with positive slope and with y-intercept $(0, 0)$ is H.

13. $y = 3$

This line is a horizontal line with y-intercept $(0, 3)$. Its y-coordinate is positive. The only graph that has these characteristics is B.

15. Through $(-2, 4)$; slope $= -\dfrac{3}{4}$

Use the point-slope form with $(x_1, y_1) = (-2, 4)$ and $m = -\dfrac{3}{4}$.

$$y - y_1 = m(x - x_1)$$
$$y - 4 = -\frac{3}{4}[x - (-2)]$$
$$y - 4 = -\frac{3}{4}(x + 2)$$
$$y - 4 = -\frac{3}{4}x - \frac{3}{2}$$
$$y = -\frac{3}{4}x - \frac{3}{2} + \frac{8}{2}$$
$$y = -\frac{3}{4}x + \frac{5}{2}$$

This slope-intercept form gives the y-intercept $\left(0, \frac{5}{2}\right)$.

17. Through $(5, 8)$; slope $= -2$
Use the point-slope form with $(x_1, y_1) = (5, 8)$ and $m = -2$.

$$y - y_1 = m(x - x_1)$$
$$y - 8 = -2(x - 5)$$
$$y - 8 = -2x + 10$$
$$y = -2x + 18$$

19. Through $(-5, 4)$; slope $= \frac{1}{2}$
Use the point-slope form with $(x_1, y_1) = (-5, 4)$ and $m = \frac{1}{2}$.

$$y - y_1 = m(x - x_1)$$
$$y - 4 = \frac{1}{2}[x - (-5)]$$
$$y - 4 = \frac{1}{2}(x + 5)$$
$$y - 4 = \frac{1}{2}x + \frac{5}{2}$$
$$y = \frac{1}{2}x + \frac{5}{2} + \frac{8}{2}$$
$$y = \frac{1}{2}x + \frac{13}{2}$$

21. Through $(3, 0)$; slope $= 4$
Use the point-slope form with $(x_1, y_1) = (3, 0)$ and $m = 4$.

$$y - y_1 = m(x - x_1)$$
$$y - 0 = 4(x - 3)$$
$$y = 4x - 12$$

23. The point-slope form, $y - y_1 = m(x - x_1)$, is used when the slope and one point on a line or two points on a line are known. The slope-intercept form, $y = mx + b$, is used when the slope and y-intercept are known. The standard form, $Ax + By = C$, is not useful for writing the equation, but can be found from other forms of the equation. The form $y = d$ is used for a horizontal line through the point (c, d). The form $x = c$ is used for a vertical line through the point (c, d).

25. Through $(9, 5)$; slope 0
A line with slope 0 is a horizontal line. A horizontal line through the point (x, k) has equation $y = k$. Here $k = 5$, so an equation is $y = 5$.

27. Through $(9, 10)$; undefined slope
A vertical line has undefined slope and equation $x = c$. Since the x-value in $(9, 10)$ is 9, the equation is $x = 9$.

29. Through $(.5, .2)$; vertical
A vertical line through the point (k, y) has equation $x = k$. Here $k = .5$, so the equation is $x = .5$.

31. Through $(-7, 8)$; horizontal
A horizontal line through the point (x, k) has equation $y = k$, so the equation is $y = 8$.

33. $(3, 4)$ and $(5, 8)$
Find the slope.

$$m = \frac{8 - 4}{5 - 3} = \frac{4}{2} = 2$$

Use the point-slope form with $(x_1, y_1) = (3, 4)$ and $m = 2$.

$$y - y_1 = m(x - x_1)$$
$$y - 4 = 2(x - 3)$$
$$y - 4 = 2x - 6$$
$$y = 2x - 2$$

35. $(6, 1)$ and $(-2, 5)$
Find the slope.

$$m = \frac{5 - 1}{-2 - 6} = \frac{4}{-8} = -\frac{1}{2}$$

Use the point-slope form with $(x_1, y_1) = (6, 1)$ and $m = -\frac{1}{2}$.

$$y - y_1 = m(x - x_1)$$
$$y - 1 = -\frac{1}{2}(x - 6)$$
$$y - 1 = -\frac{1}{2}x + 3$$
$$y = -\frac{1}{2}x + 4$$

37. $\left(-\dfrac{2}{5}, \dfrac{2}{5}\right)$ and $\left(\dfrac{4}{3}, \dfrac{2}{3}\right)$

Find the slope.

$$m = \dfrac{\dfrac{2}{3} - \dfrac{2}{5}}{\dfrac{4}{3} - \left(-\dfrac{2}{5}\right)} = \dfrac{\dfrac{10-6}{15}}{\dfrac{20+6}{15}}$$

$$= \dfrac{\dfrac{4}{15}}{\dfrac{26}{15}} = \dfrac{4}{26} = \dfrac{2}{13}$$

Use the point-slope form with

$(x_1, y_1) = \left(-\dfrac{2}{5}, \dfrac{2}{5}\right)$ and $m = \dfrac{2}{13}$.

$$y - \dfrac{2}{5} = \dfrac{2}{13}\left[x - \left(-\dfrac{2}{5}\right)\right]$$

$$y - \dfrac{2}{5} = \dfrac{2}{13}\left(x + \dfrac{2}{5}\right)$$

$$y - \dfrac{2}{5} = \dfrac{2}{13}x + \dfrac{4}{65}$$

$$y = \dfrac{2}{13}x + \dfrac{4}{65} + \dfrac{26}{65}$$

$$y = \dfrac{2}{13}x + \dfrac{6}{13}\left(\dfrac{30}{65} = \dfrac{6}{13}\right)$$

39. $(2, 5)$ and $(1, 5)$

Find the slope.

$$m = \dfrac{5 - 5}{1 - 2} = \dfrac{0}{-1} = 0$$

A line with slope 0 is horizontal. A horizontal line through the point (x, k) has equation $y = k$, so the equation is $y = 5$.

41. $(7, 6)$ and $(7, -8)$

Find the slope.

$$m = \dfrac{-8 - 6}{7 - 7} = \dfrac{-14}{0} \quad \textit{Undefined}$$

A line with undefined slope is a vertical line. The equation of a vertical line is $x = k$, where k is the common x-value. So the equation is $x = 7$.

43. $(1, -3)$ and $(-1, -3)$

Find the slope.

$$m = \dfrac{-3 - (-3)}{-1 - 1} = \dfrac{0}{-2} = 0$$

A line with slope 0 is horizontal. A horizontal line through the point (x, k) has equation $y = k$, so the equation is $y = -3$.

45. $m = 5;\ b = 15$

Substitute these values in the slope-intercept form.

$$y = mx + b$$
$$y = 5x + 15$$

47. $m = -\dfrac{2}{3};\ b = \dfrac{4}{5}$

Substitute these values in the slope-intercept form.

$$y = mx + b$$
$$y = -\dfrac{2}{3}x + \dfrac{4}{5}$$

49. Slope $\dfrac{2}{5}$; y-intercept $(0, 5)$

Here, $m = \dfrac{2}{5}$ and $b = 5$. Substitute these values in the slope-intercept form.

$$y = mx + b$$
$$y = \dfrac{2}{5}x + 5$$

51. To get to the point $(3, 3)$ from the y-intercept $(0, 1)$, we must go up 2 units and to the right 3 units, so the slope is $\dfrac{2}{3}$. The slope-intercept form is

$$y = \dfrac{2}{3}x + 1.$$

53. $x + y = 12$

(a) Solve for y to get the equation in slope-intercept form.

$$x + y = 12$$
$$y = -x + 12$$

(b) The slope is the coefficient of x, -1.

(c) The y-intercept is the point $(0, b)$ or $(0, 12)$.

55. $5x + 2y = 20$

(a) Solve for y.

$$2y = -5x + 20$$
$$y = -\dfrac{5}{2}x + 10$$

(b) The slope is $-\dfrac{5}{2}$.

(c) The y-intercept is $(0, 10)$.

57. $2x - 3y = 10$

(a) Solve for y.

$$-3y = -2x + 10$$
$$y = \dfrac{2}{3}x - \dfrac{10}{3}$$

(b) The slope is $\dfrac{2}{3}$.

(c) The y-intercept is $\left(0, -\dfrac{10}{3}\right)$.

59. Through $(7, 2)$; parallel to $3x - y = 8$

Find the slope of $3x - y = 8$.

$$-y = -3x + 8$$
$$y = 3x - 8$$

The slope is 3, so a line parallel to it also has slope 3. Use $m = 3$ and $(x_1, y_1) = (7, 2)$ in the point-slope form.

$$y - y_1 = m(x - x_1)$$
$$y - 2 = 3(x - 7)$$
$$y - 2 = 3x - 21$$
$$y = 3x - 19$$

61. Through $(-2, -2)$; parallel to $-x + 2y = 10$

Find the slope of $-x + 2y = 10$.

$$2y = x + 10$$
$$y = \frac{1}{2}x + 5$$

The slope is $\dfrac{1}{2}$, so a line parallel to it also has slope $\dfrac{1}{2}$. Use $m = \dfrac{1}{2}$ and $(x_1, y_1) = (-2, -2)$ in the point-slope form.

$$y - y_1 = m(x - x_1)$$
$$y - (-2) = \frac{1}{2}[x - (-2)]$$
$$y + 2 = \frac{1}{2}(x + 2)$$
$$y + 2 = \frac{1}{2}x + 1$$
$$y = \frac{1}{2}x - 1$$

63. Through $(8, 5)$; perpendicular to $2x - y = 7$

Find the slope of $2x - y = 7$.

$$-y = -2x + 7$$
$$y = 2x - 7$$

The slope of the line is 2. Therefore, the slope of the line perpendicular to it is $-\dfrac{1}{2}$ since

$2\left(-\dfrac{1}{2}\right) = -1$. Use $m = -\dfrac{1}{2}$ and $(x_1, y_1) = (8, 5)$ in the point-slope form.

$$y - y_1 = m(x - x_1)$$
$$y - 5 = -\frac{1}{2}(x - 8)$$
$$y - 5 = -\frac{1}{2}x + 4$$
$$y = -\frac{1}{2}x + 9$$

65. Through $(-2, 7)$; perpendicular to $x = 9$

$x = 9$ is a vertical line so a line perpendicular to it will be a horizontal line. It goes through $(-2, 7)$ so its equation is

$$y = 7.$$

67. Since it costs \$15 plus \$3 per day to rent a chain saw, it costs $3x + 15$ dollars for x days. Thus, if y represents the charge to the user (in dollars), the equation is $y = 3x + 15$.

When $x = 1$, $y = 3(1) + 15 = 3 + 15 = 18$.

Ordered pair: $(1, 18)$

When $x = 5$, $y = 3(5) + 15 = 15 + 15 = 30$.

Ordered pair: $(5, 30)$

When $x = 10$, $y = 3(10) + 15 = 30 + 15 = 45$.

Ordered pair: $(10, 45)$

69. $(1, 80.1)$ and $(6, 89.4)$

$$m = \frac{89.4 - 80.1}{6 - 1} = \frac{9.3}{5} = 1.86$$

Use the point-slope form.

$$y - 80.1 = 1.86(x - 1)$$
$$y - 80.1 = 1.86x - 1.86$$
$$y = 1.86x + 78.24$$

71. (a) $(3, 21, 696)$ and $(7, 25, 050)$

$$m = \frac{25,050 - 21,696}{7 - 3} = \frac{3354}{4} = 838.5$$

Use the point-slope form.

$$y - 21,696 = 838.5(x - 3)$$
$$y - 21,696 = 838.5x - 2515.5$$
$$y = 838.5x + 19,180.5$$

(b) $x = 5$ represents 1995

$$y = 838.5(5) + 19,180.5$$
$$= 4192.5 + 19,180.5$$
$$= 23,373$$

This is close to the value $23,583$ that is given.

73. (a) $2x + 7 - x = 4x - 2$

$$x + 7 = 4x - 2$$
$$-3x + 9 = 0$$
$$-3x + 9 = y$$

(b) From the screen, we see that $x = 3$ is the solution.

(c) $2x + 7 - x = 4x - 2$

$$x + 7 = 4x - 2$$
$$9 = 3x$$
$$3 = x$$

Solution set: $\{3\}$

75. (a) $3(2x + 1) - 2(x - 2) = 5$

$$6x + 3 - 2x + 4 - 5 = 0$$
$$4x + 2 = 0$$
$$4x + 2 = y$$

(b) From the screen, we see that $x = -.5$ is the solution.

(c) $3(2x + 1) - 2(x - 2) = 5$

$$6x + 3 - 2x + 4 = 5$$
$$4x + 7 = 5$$
$$4x = -2$$
$$x = -\frac{1}{2} \text{ or } -.5$$

Solution set: $\{-.5\}$

77. The solution to the equation $Y_1 = 0$ is the x-coordinate of the x-intercept. In this case, the x-intercept is greater than 10, so the correct choice must be (d).

79. When $C = 0°$, $F = \underline{32°}$, and when $C = 100°$, $F = \underline{212°}$. These are the freezing and boiling temperatures for water.

80. The two points of the form (C, F) would be $(0, 32)$ and $(100, 212)$.

81. $m = \dfrac{212 - 32}{100 - 0} = \dfrac{180}{100} = \dfrac{9}{5}$

82. Let $m = \dfrac{9}{5}$ and $(x_1, y_1) = (0, 32)$.

$$y - y_1 = m(x - x_1)$$
$$F - 32 = \frac{9}{5}(C - 0)$$
$$F - 32 = \frac{9}{5}C$$
$$F = \frac{9}{5}C + 32$$

83. $F = \dfrac{9}{5}C + 32$

$$F - 32 = \frac{9}{5}C$$
$$\frac{5}{9}(F - 32) = C$$

84. A temperature of $50°$ C corresponds to a temperature of $122°$ F.

Section 3.4

1. The boundary of the graph of $y \leq -x + 2$ will be a _solid_ line (since the inequality involves $\leq$), and the shading will be _below_ the line (since the inequality sign is $\leq$ or $<$).

3. The boundary of the graph of $y > -x + 2$ will be a _dashed_ line (since the inequality involves $>$), and the shading will be _above_ the line (since the inequality sign is $\geq$ or $>$).

5. Change the inequality to an equation and graph it. For example, to graph $3x + 2y > 5$, begin by graphing the line with equation $3x + 2y = 5$. The line should be dashed if the inequality involves $>$ or $<$. (This is the case for our example.) It should be solid if the inequality involves $\geq$ or $\leq$. Decide which side of the line to shade by solving for y. We get $y > -\dfrac{3}{2}x + \dfrac{5}{2}$. Because of the $>$ symbol, shade above the line. For the $<$ symbol, we would shade below the line. The shaded region shows the solutions.

7. $x + y \leq 2$

Graph the line $x + y = 2$ by drawing a solid line (since the inequality involves $\leq$) through the intercepts $(2, 0)$ and $(0, 2)$.

Test a point not on this line, such as $(0, 0)$.

$$x + y \leq 2$$
$$0 + 0 \leq 2$$
$$0 \leq 2 \quad True$$

Shade that side of the line containing the test point $(0, 0)$.

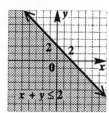

9. $4x - y < 4$

Graph the line $4x - y = 4$ by drawing a dashed line (since the inequality involves $<$) through the intercepts $(1, 0)$ and $(0, -4)$. Instead of using a test point, we will solve the inequality for y.

$$-y < -4x + 4$$
$$y > 4x - 4$$

Since we have "$y >$ " in the last inequality, shade the region *above* the boundary line.

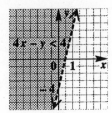

11. $x + 3y \geq -2$

Graph the solid line $x + 3y = -2$ (since the inequality involves $\geq$) through the intercepts $(-2, 0)$ and $\left(0, -\dfrac{2}{3}\right)$.

Test a point not on this line such as $(0, 0)$.

$$0 + 3(0) \geq -2$$
$$0 \geq -2 \quad \textit{True}$$

Shade that side of the line containing the test point $(0, 0)$.

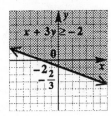

13. $x + y > 0$

Graph the line $x + y = 0$, which includes the points $(0, 0)$ and $(2, -2)$, as a dashed line (since the inequality involves $>$). Solving the inequality for y gives us

$$y > -x,$$

So shade the region above the boundary line.

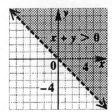

15. $x - 3y \leq 0$

Graph the solid line $x - 3y = 0$ through the points $(0, 0)$ and $(3, 1)$.

Solve the inequality for y.

$$-3y \leq -x$$
$$y \geq \frac{1}{3}x$$

Shade the region above the boundary line.

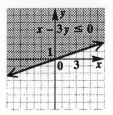

17. $y < x$

Graph the dashed line $y = x$ through $(0, 0)$ and $(2, 2)$. Since we have "$y <$ " in the inequality, shade the region *below* the boundary line.

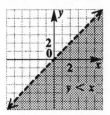

19. $x + y \leq 1$ and $x \geq 1$

Graph the solid line $x + y = 1$ through $(0, 1)$ and $(1, 0)$. The inequality $x + y \leq 1$ can be written as $y \leq -x + 1$, so shade the region below the boundary line.

Graph the solid vertical line $x = 1$ through $(1, 0)$ and shade the region to the right. The required graph is the common shaded area as well as the portions of the lines that bound it.

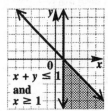

21. $2x - y \geq 2$ and $y < 4$

Graph the solid line $2x - y = 2$ through the intercepts $(1, 0)$ and $(0, -2)$. Test $(0, 0)$ to get $0 \geq 2$, a false statement. Shade that side of the graph not containing $(0, 0)$. To graph $y < 4$ on the same axes, graph the dashed horizontal line through $(0, 4)$. Test $(0, 0)$ to get $0 < 4$, a true statement. Shade that side of the dashed line containing $(0, 0)$.

The word "and" indicates the intersection of the two graphs. The final solution set consists of the region where the two shaded regions overlap.

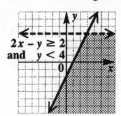

23. $x + y > -5$ and $y < -2$

Graph $x + y = -5$, which has intercepts $(-5, 0)$ and $(0, -5)$, as a dashed line. Test $(0, 0)$, which yields $0 > -5$, a true statement. Shade the region that includes $(0, 0)$.

Graph $y = -2$ as a dashed horizontal line. Shade the region below $y = -2$. The required graph of the intersection is the region common to both graphs.

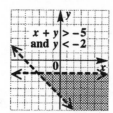

25. $|x| < 3$ can be rewritten as $-3 < x < 3$. The boundaries are the dashed vertical lines $x = -3$ and $x = 3$. Since x is between -3 and 3, the graph includes all points between the lines.

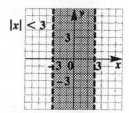

27. $|x + 1| < 2$ can be rewritten as

$$-2 < x + 1 < 2$$
$$-3 < x < 1.$$

The boundaries are the dashed vertical lines $x = -3$ and $x = 1$. Since x is between -3 and 1, the graph includes all points between the lines.

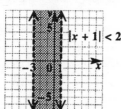

29. $x - y \geq 1$ or $y \geq 2$

Graph the solid line $x - y = 1$, which crosses the y-axis at -1 and the x-axis at 1. Use $(0, 0)$ as a test point, which yields $0 \geq 1$, a false statement. Shade the region that does not include $(0, 0)$. Now graph the solid line $y = 2$. Since the inequality is $y \geq 2$, shade above this line. The required graph of the union includes all the shaded regions, that is, all the points that satisfy either inequality.

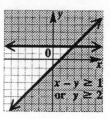

31. $x - 2 > y$ or $x < 1$

Graph $x - 2 = y$, which has intercepts $(2, 0)$ and $(0, -2)$, as a dashed line. Test $(0, 0)$, which yields $-2 > 0$, a false statement. Shade the region that does not include $(0, 0)$.

Graph $x = 1$ as a dashed vertical line. Shade the region to the left of $x = 1$.

The required graph of the union includes all the shaded regions, that is, all the points that satisfy either inequality.

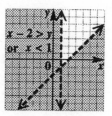

33. $3x + 2y < 6$ or $x - 2y > 2$

Graph $3x + 2y = 6$, which has intercepts $(2, 0)$ and $(0, 3)$, as a dashed line. Test $(0, 0)$, which yields $0 < 6$, a true statement. Shade the region that includes $(0, 0)$.

Graph $x - 2y = 2$, which has intercepts $(2, 0)$ and $(0, -1)$, as a dashed line. Test $(0, 0)$, which yields $0 > 2$, a false statement. Shade the region that does not include $(0, 0)$.

The required graph of the union includes all the shaded regions, that is, all the points that satisfy either inequality.

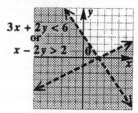

35. **(a)** The x-intercept is $(-4, 0)$, so the solution to $y = 0$ is $\{-4\}$.

(b) The solution to $y < 0$ is $(-\infty, -4)$, since the graph is below the x-axis for these values of x.

(c) The solution to $y > 0$ is $(-4, \infty)$, since the graph is above the x-axis for these values of x.

37. **(a)** The x-intercept is $(3.5, 0)$, so the solution to $y = 0$ is $\{3.5\}$.

(b) The solution to $y < 0$ is $(3.5, \infty)$, since the graph is below the x-axis for these values of x.

(c) The solution to $y > 0$ is $(-\infty, 3.5)$, since the graph is above the x-axis for these values of x.

39. $y \le 3x - 6$

The boundary line, $y = 3x - 6$, has slope 3 and y-intercept -6. This would be graph B or graph C. Since we want the region less than or equal to $3x - 6$, we want the region on or below the boundary line. The answer is graph C.

41. $y \le -3x - 6$

The slope of the boundary line $y = -3x - 6$ is -3, and the y-intercept is -6. This would be graph A or graph D. The inequality sign is $\le$, so we want the region on or below the boundary line. The answer is graph A.

43. **(a)** $5x + 3 = 0$
$$5x = -3$$
$$x = -\frac{3}{5} = -.6$$
Solution set: $\{-.6\}$

(b) $5x + 3 > 0$
$$5x > -3$$
$$x > -\frac{3}{5} \text{ or } -.6$$
Solution set: $(-.6, \infty)$

(c) $5x + 3 < 0$
$$5x < -3$$
$$x < -\frac{3}{5} \text{ or } -.6$$
Solution set: $(-\infty, -.6)$

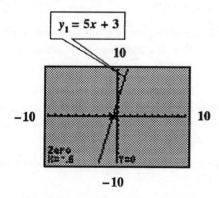

The x-intercept is $(-.6, 0)$, as in part (a). The graph is above the x-axis for $x > -.6$, as in part (b), and below the x-axis for $x < -.6$, as in part (c).

45. **(a)** $-8x - (2x + 12) = 0$
$$-8x - 2x - 12 = 0$$
$$-10x - 12 = 0$$
$$-10x = 12$$
$$x = -1.2$$
Solution set: $\{-1.2\}$

(b) $-8x - (2x + 12) \ge 0$
$$-8x - 2x - 12 \ge 0$$
$$-10x - 12 \ge 0$$
$$-10x \ge 12$$
$$x \le -1.2$$
Solution set: $(-\infty, -1.2]$

(c) $-8x - (2x + 12) \le 0$
$$-8x - 2x - 12 \le 0$$
$$-10x - 12 \le 0$$
$$-10x \le 12$$
$$x \ge -1.2$$
Solution set: $[-1.2, \infty)$

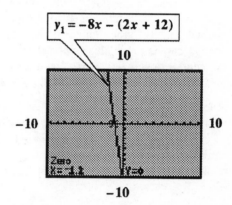

The x-intercept is $(-1.2, 0)$, as in part (a). The graph is on or above the x-axis for $(-\infty, -1.2]$, as in part (b), and on or below the x-axis for $[-1.2, \infty)$, as in part (c).

47. The inequality we want to solve is
$$y \ge 85.$$
$$1.86x + 78.24 \ge 85$$
$$1.86x \ge 6.76$$
$$x \ge \frac{6.76}{1.86} \approx 3.6$$

Rounding 3.6 to 4, this inequality represents the years 1994–1996.

Section 3.5

1. We give one of many possible answers here. A function is a set of ordered pairs in which each first element determines exactly one second element. For example, $\{(0, 1), (1, 2), (2, 3), (3, 4)\dots\}$ is a function.

3. In an ordered pair of a relation, the first element is the independent variable.

5. $\{(2, 5), (3, 7), (4, 9), (5, 11)\}$

The relation is a function since for each x-value, there is only one y-value. The domain is the set of x-values: $\{2, 3, 4, 5\}$. The range is the set of y-values: $\{5, 7, 9, 11\}$.

7. This relation is not a function since each input value corresponds to two output values. The domain is the set of positive real numbers; that is, $(0, \infty)$. The range is the set of positive real numbers along with their negatives; that is, $(-\infty, 0) \cup (0, \infty)$.

9. The relation is a function since each input value corresponds to exactly one output value. The domain is the set of inputs, {unleaded regular, unleaded premium, crude oil}. The range is the set of outputs, $\{1.22, 1.44, .21\}$.

11. Using the vertical line test, we find any vertical line will intersect the graph at most once. This indicates that the graph represents a function. This graph extends indefinitely to the left $(-\infty)$ and indefinitely to the right (∞). Therefore, the domain is $(-\infty, \infty)$. This graph extends indefinitely downward $(-\infty)$, and reaches a high point at $y = 4$. Therefore, the range is $(-\infty, 4]$.

13. Since a vertical line can intersect the graph of the relation in more than one point, the relation is not a function. The domain, the x-values of the points on the graph, is $[-4, 4]$. The range, the y-values of the points on the graph, is $[-3, 3]$.

15. $y = x^2$
Each value of x corresponds to one y-value. For example, if $x = 3$, then $y = 3^2 = 9$. Therefore, $y = x^2$ defines y as a function of x.
Since any x-value, positive, negative, or zero, can be squared, the domain is $(-\infty, \infty)$.

17. $x = y^2$
The ordered pairs $(4, 2)$ and $(4, -2)$ both satisfy the equation. Since one value of x, 4, corresponds to two values of y, 2 and -2, the relation does not define a function. Because x is equal to the square of y, the values of x must always be nonnegative. The domain is $[0, \infty)$.

19. $x + y < 4$
For a particular x-value, more than one y-value can be selected to satisfy $x + y < 4$. For example, if $x = 2$ and $y = 0$, then

$$2 + 0 < 4. \quad \textit{True}$$

Now, if $x = 2$ and $y = 1$, then

$$2 + 1 < 4. \quad \textit{Also true}$$

Therefore, $x + y < 4$ does not define y as a function of x.
The graph of $x + y < 4$ consists of the shaded region below the dashed line $x + y = 4$, which extends indefinitely from left to right. Therefore, the domain is $(-\infty, \infty)$.

21. $y = \sqrt{x}$
For any value of x, there is exactly one corresponding value for y, so this relation defines a function. Since the radicand must be a nonnegative number, x must always be nonnegative. The domain is $[0, \infty)$.

23. $xy = 1$
Rewrite $xy = 1$ as $y = \dfrac{1}{x}$. Note that x can never equal 0, otherwise the denominator would equal 0. The domain is $(-\infty, 0) \cup (0, \infty)$.
Each nonzero x-value gives exactly one y-value. Therefore, $xy = 1$ defines y as a function of x.

25. $y = \sqrt{4x + 2}$
To determine the domain of $y = \sqrt{4x + 2}$, recall that the radicand must be nonnegative. Solve the inequality $4x + 2 \geq 0$, which gives us $x \geq -\dfrac{1}{2}$.
Therefore, the domain is $\left[-\dfrac{1}{2}, \infty\right)$.
Each x-value from the domain produces exactly one y-value. Therefore, $y = \sqrt{4x + 2}$ defines a function.

27. $y = \dfrac{2}{x - 9}$
Given any value of x, y is found by subtracting 9, then dividing the result into 2. This process produces exactly one value of y for each x-value, so the relation represents a function. The domain includes all real numbers except those that make the denominator 0, namely 9. The domain is $(-\infty, 9) \cup (9, \infty)$.

29. (a) The number of gallons of water varies from 0 to 3000, so the possible values are in the set $[0, 3000]$.

(b) The graph rises for the first 25 hours, so the water level increases for 25 hours. The graph falls for $t = 50$ to $t = 75$, so the water level decreases for 25 hours.

(c) There are 2000 gallons in the pool when $t = 90$.

(d) $f(0)$ is the number of gallons in the pool at time $t = 0$. Here, $f(0) = 0$.

31. The amount of income tax you pay depends on your taxable income, so income tax is a function of taxable income.

33. $f(x) = -3x + 4$
$f(0) = -3(0) + 4$
$ = 0 + 4$
$ = 4$

35. $f(x) = -3x + 4$

$f(-x) = -3(-x) + 4$

$\qquad = 3x + 4$

37. $g(x) = -x^2 + 4x + 1$

$g(10) = -(10)^2 + 4(10) + 1$

$\qquad = -100 + 40 + 1$

$\qquad = -59$

39. $g(x) = -x^2 + 4x + 1$

$g\left(\dfrac{1}{2}\right) = -\left(\dfrac{1}{2}\right)^2 + 4\left(\dfrac{1}{2}\right) + 1$

$\qquad = -\dfrac{1}{4} + 2 + 1$

$\qquad = \dfrac{11}{4}$

41. $g(x) = -x^2 + 4x + 1$

$g(2p) = -(2p)^2 + 4(2p) + 1$

$\qquad = -4p^2 + 8p + 1$

43. $g[f(1)]$

First find $f(1)$.

$$f(x) = -3x + 4$$
$$f(1) = -3(1) + 4 = 1$$

Since $f(1) = 1$, $g[f(1)] = g(1)$.

$$g(x) = -x^2 + 4x + 1$$
$$g(1) = -1^2 + 4(1) + 1 = 4$$

Thus, $g[f(1)] = 4$.

45. From Exercise 42, $f[g(1)] = -8$.

From Exercise 43, $g[f(1)] = 4$.

In general, $f[g(x)] \neq g[f(x)]$.

47. (a) Solve the equation for y.

$$x + 3y = 12$$
$$3y = 12 - x$$
$$y = \frac{12 - x}{3}$$

Since $y = f(x)$,

$$f(x) = \frac{12 - x}{3}.$$

(b) $f(3) = \dfrac{12 - 3}{3} = \dfrac{9}{3} = 3$

49. (a) Solve the equation for y.

$$y + 2x^2 = 3$$
$$y = 3 - 2x^2$$

Since $y = f(x)$,

$$f(x) = 3 - 2x^2.$$

(b) $f(3) = 3 - 2(3)^2$

$\qquad = 3 - 2(9)$

$\qquad = -15$

51. (a) Solve the equation for y.

$$4x - 3y = 8$$
$$-3y = 8 - 4x$$
$$y = \frac{8 - 4x}{-3}$$

Since $y = f(x)$,

$$f(x) = \frac{8 - 4x}{-3}.$$

(b) $f(3) = \dfrac{8 - 4(3)}{-3} = \dfrac{8 - 12}{-3}$

$\qquad = \dfrac{-4}{-3} = \dfrac{4}{3}$

53. The equation $2x + y = 4$ has a straight line as its graph. To find y in $(3, y)$, let $x = 3$ in the equation.

$$2x + y = 4$$
$$2(3) + y = 4$$
$$6 + y = 4$$
$$y = -2$$

To use functional notation for $2x + y = 4$, solve for y to get

$$y = -2x + 4.$$

Replace y with $f(x)$ to get

$$f(x) = -2x + 4.$$
$$f(3) = -2(3) + 4 = -2$$

Because $y = -2$ when $x = 3$, the point $(3, -2)$ lies on the graph of the function.

55. $f(x) = -2x + 5$

The graph will be a line. The intercepts are $(0, 5)$ and $\left(\dfrac{5}{2}, 0\right)$.

The domain is $(-\infty, \infty)$.

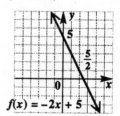

57. $h(x) = \frac{1}{2}x + 2$

The graph will be a line. The intercepts are $(0, 2)$ and $(-4, 0)$.

The domain is $(-\infty, \infty)$.

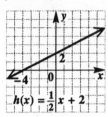

59. $G(x) = 2x$

This line includes the points $(0, 0)$, $(1, 2)$, and $(2, 4)$. The domain is $(-\infty, \infty)$.

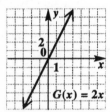

61. $g(x) = -4$

Using a y-intercept of $(0, -4)$ and a slope of $m = 0$, we graph the horizontal line. From the graph we see that the domain is $(-\infty, \infty)$.

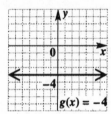

63. (a)

x	$f(x)$
0	0
1	\$1.50
2	\$3.00
3	\$4.50

(b) Since the charge equals the cost per mile, \$1.50, times the number of miles, x, $f(x) = \$1.50x$.

(c) To graph $f(x)$ for $x \in \{0, 1, 2, 3\}$, plot the points $(0, 0)$, $(1, 1.50)$, $(2, 3.00)$, and $(3, 4.50)$ from the chart.

65. Since the length of a man's femur is given, use the formula
$$h(r) = 69.09 + 2.24r.$$
Let $r = 56$.
$$h(56) = 69.09 + 2.24(56)$$
$$= 194.53$$
The man is 194.53 cm tall.

67. Since the length of a woman's femur is given, use the formula
$$h(r) = 61.41 + 2.32r.$$
Let $r = 50$.
$$h(50) = 61.41 + 2.32(50)$$
$$= 177.41$$
The woman is 177.41 cm tall.

69. $f(x) = (.91)(3.14)x^2$
$$f(.8) = (.91)(3.14)(.8)^2$$
$$= 1.828736$$
To the nearest hundredth, the volume of the pool is 1.83 m³.

71. $f(x) = (.91)(3.14)x^2$
$$f(1.2) = (.91)(3.14)(1.2)^2$$
$$= 4.114656$$
To the nearest hundredth, the volume of the pool is 4.11 m³.

73. $f(x) = -183x + 40,034$

(a) $f(1) = -183(1) + 40,034 = 39,851$

(b) $f(3) = -183(3) + 40,034 = 39,485$

(c) $f(5) = -183(5) + 40,034 = 39,119$

(d) $x = 2$ corresponds to 1992. In 1992, there were 39,668 post offices in the U.S.

75. The graph shows $x = 3$ and $y = 7$. In functional notation, this is
$$f(3) = 7.$$

77. Let $(x_1, y_1) = (-1, 8)$ and $(x_2, y_2) = (4, -7)$. Then
$$m = \frac{8 - (-7)}{-1 - 4} = \frac{15}{-5} = -3.$$

Use the point $(-1, 8)$ and $m = -3$ in the point-slope form.
$$y - y_1 = m(x - x_1)$$
$$y - 8 = -3[x - (-1)]$$
$$y - 8 = -3x - 3$$
$$y = -3x + 5$$
$$f(x) = -3x + 5$$

Section 3.6

1. The equation $y = \dfrac{3}{x}$ represents inverse variation. y varies inversely as x because x is in the denominator.

3. The equation $y = 10x^2$ represents direct variation. The number 10 is the constant of variation, and y varies directly as the square of x.

5. The equation $y = 3xz^4$ represents joint variation. y varies directly as x and z^4.

7. The equation $y = \dfrac{4x}{wz}$ represents combined variation. In the numerator, 4 is the constant of variation, and y varies directly as x. In the denominator, y varies inversely as w and z.

9. "x varies directly as y" means
$$x = ky$$
for some constant k.
Substitute $x = 9$ and $y = 3$ in the equation and solve for k.
$$x = ky$$
$$9 = k(3)$$
$$3 = k$$
So, $x = 3y$.
To find x when $y = 12$, substitute 12 for y in the equation.
$$x = 3y$$
$$x = 3(12)$$
$$x = 36$$

11. "z varies inversely as w" means
$$z = \dfrac{k}{w}$$
for some constant k. Since $z = 10$ when $w = .5$, substitute these values in the equation and solve for k.
$$z = \dfrac{k}{w}$$
$$10 = \dfrac{k}{.5}$$
$$5 = k$$
So, $z = \dfrac{5}{w}$.
To find z when $w = 8$, substitute 8 for w in the equation.
$$z = \dfrac{5}{w}$$
$$z = \dfrac{5}{8} \text{ or } .625$$

13. "p varies jointly as q and r^2" means
$$p = kqr^2$$
for some constant k. Given that $p = 200$ when $q = 2$ and $r = 3$, solve for k.
$$p = kqr^2$$
$$200 = k(2)(3)^2$$
$$200 = 18k$$
$$k = \dfrac{200}{18} = \dfrac{100}{9}$$
So, $p = \dfrac{100}{9}qr^2$.
Using $k = \dfrac{100}{9}$, $q = 5$, and $r = 2$, find p.
$$p = \dfrac{100}{9}qr^2$$
$$p = \dfrac{100}{9}(5)(2)^2$$
$$= \dfrac{100}{9}(20)$$
$$= \dfrac{2000}{9} \text{ or } 222\dfrac{2}{9}$$

15. For $k > 0$, if y varies directly as x (then $y = kx$), when x increases, y _increases_, and when x decreases, y _decreases_.

17. In a direct variation, both variables either increase or decrease.
In an inverse variation, one variable increases while the other decreases.

19. Let $x =$ the number of gallons he bought and $C =$ the cost.
C varies directly as x, so
$$C = kx.$$
Since $C = 8.79$ when $x = 8$,
$$8.79 = k(8)$$
$$k = \dfrac{8.79}{8} = 1.09875 \approx 1.099.$$
The price is $\dfrac{\$1.09\frac{9}{10}}{\text{gal}}$.

21. Let $y =$ the weight of an object on earth
and $x =$ the weight of the object on the moon.
y varies directly as x, so

$$y = kx$$

for some constant k. Since $y = 200$ when $x = 32$, substitute these values in the equation and solve for k.

$$y = kx$$
$$200 = k(32)$$
$$k = \frac{200}{32} = 6.25$$

So, $y = 6.25x$.
To find x when $y = 50$, substitute 50 for y in the equation

$$y = 6.25x.$$
$$50 = 6.25x$$
$$x = \frac{50}{6.25} = 8$$

The dog would weigh 8 lb on the moon.

23. Let $A =$ the amount of water emptied
by a pipe in one hour
and $d =$ the diameter of the pipe.
A varies directly as d^2, so

$$A = kd^2$$

for some constant k. Since $A = 200$ when $d = 6$, substitute these values in the equation and solve for k.

$$A = kd^2$$
$$200 = k(6)^2$$
$$200 = 36k$$
$$k = \frac{200}{36} = \frac{50}{9}$$

So, $A = \frac{50}{9}d^2$.
When $d = 12$,

$$A = \frac{50}{9}d^2$$
$$A = \frac{50}{9}(12)^2$$
$$= \frac{50}{9}(144) = 800.$$

A 12-inch pipe would empty 800 gal of water in one hour.

25. Let $d =$ the distance and
$t =$ the time.
d varies directly as the square of t, so $d = kt^2$.
Let $d = -576$ and $t = 6$. (You could also use $d = 576$, but the negative sign indicates the direction of the body.)

$$-576 = k(6)^2$$
$$-576 = 36k$$
$$-16 = k$$

So, $d = -16t^2$.
Let $t = 4$.

$$d = -16(4)^2 = -256$$

The object fell 256 ft in the first 4 seconds.

27. Let $I =$ the illumination produced
by a light source
and $d =$ the distance from the source.
I varies inversely as d^2, so

$$I = \frac{k}{d^2}$$

for some constant k. Since $I = 768$ when $d = 1$, substitute these values in the equation and solve for k.

$$I = \frac{k}{d^2}$$
$$768 = \frac{k}{1^2}$$
$$768 = k$$

So $I = \frac{768}{d^2}$.
When $d = 6$,

$$I = \frac{768}{d^2}$$
$$I = \frac{768}{6^2} = \frac{768}{36} = \frac{64}{3} \text{ or } 21\frac{1}{3}$$

The illumination produced by the light source is $21\frac{1}{3}$ footcandles.

29. Let $f =$ the frequency of a string
in cycles per second
and $s =$ the length in feet.
f varies inversely as s, so

$$f = \frac{k}{s}$$

for some constant k. Since $f = 250$ when $k = 2$, substitute these values in the equation and solve for k.

$$f = \frac{k}{s}$$
$$250 = \frac{k}{2}$$
$$500 = k$$

So, $f = \frac{500}{s}$.
When $s = 5$,

$$f = \frac{500}{5} = 100.$$

The string would have a frequency of 100 $\frac{\text{cycles}}{\text{sec}}$.

31. Let $d =$ the distance and
   $h =$ the height.
   d varies directly as the square root of h, so

$$d = k\sqrt{h}$$

for some constant k. Since $d = 18$ when $h = 144$, substitute these values in the equation and solve for k.

$$d = k\sqrt{h}$$
$$18 = k\sqrt{144}$$
$$k = \frac{18}{12} = \frac{3}{2}$$

So, $d = \frac{3}{2}\sqrt{h}$.
When $h = 1600$,

$$d = \frac{3}{2}\sqrt{1600} = \frac{3}{2}(40) = 60.$$

The distance is 60 km.

33. Let $F =$ the force of the wind,
   $A =$ the area,
   and $v =$ the velocity.
   Then $F = kAv^2$.

Let $F = 50$, $A = \frac{1}{2}$, and $v = 40$.

$$50 = k\left(\frac{1}{2}\right)(40)^2$$
$$50 = k\left(\frac{1}{2}\right)(1600)$$
$$50 = 800k$$
$$k = \frac{50}{800} = \frac{1}{16}$$

So, $F = \frac{1}{16}Av^2$.
Let $A = 2$ and $v = 80$.

$$F = \frac{1}{16}(2)(80)^2$$
$$= \frac{1}{16}(2)(6400) = 800$$

The force is 800 lb.

35. Let $P =$ the period of the pendulum,
   $L =$ the length, and
   $g =$ the acceleration due to gravity.

Then $P = \frac{k\sqrt{L}}{\sqrt{g}}$.

Let $P = 1.06\pi$, $L = 9$, and $g = 32$.

$$1.06\pi = \frac{k\sqrt{9}}{\sqrt{32}}$$
$$k = \frac{(1.06\pi)\sqrt{32}}{3}$$

So, $P = \frac{(1.06\pi)\sqrt{32}\sqrt{L}}{3\sqrt{g}}$.

Let $L = 4$ and $g = 32$.

$$P = \frac{(1.06\pi)\sqrt{32}\sqrt{4}}{3\sqrt{32}}$$
$$= \frac{2.12\pi}{3} \approx .71\pi$$

The period is about $.71\pi$ sec.

37. Let $L =$ the load,
   $w =$ the width,
   $h =$ the height, and
   $l =$ the length.

Then $L = \frac{kwh^2}{l}$.

Let $L = 360$, $l = 6$, $w = .1$, and $h = .06$.

$$360 = \frac{k(.1)(.06)^2}{6}$$
$$360 = .00006k$$
$$6,000,000 = k$$

So,

$$L = \frac{6,000,000wh^2}{l}.$$

Let $l = 16$, $w = .2$, and $h = .08$.

$$L = \frac{6,000,000(.2)(.08)^2}{16}$$
$$= \frac{7680}{16} = 480$$

The maximum load is 480 kg.

39. According to Example 6,

$$B = \frac{694w}{h^2}.$$

Let $w = 260$ and $h = 82$ (6 ft, 10 in = 82 in).

$$B = \frac{694(260)}{(82)^2} \approx 26.8 \approx 27$$

Chris Webber's BMI is about 27.

41. $f(x) = kx$
From the screen, let $x = 8$ and $f(x) = 2$.
Then
$$2 = k(8)$$
$$\frac{1}{4} = k.$$
So, $f(x) = \frac{1}{4}x$.
Therefore,
$$f(36) = \frac{1}{4}(36) = 9.$$

43. The ordered pairs in the form of (gallons, price) are $(0, 0)$ and $(1, 1.25)$.

44. Let $(x_1, y_1) = (0, 0)$ and $(x_2, y_2) = (1, 1.25)$.
Then
$$m = \frac{1.25 - 0}{1 - 0} = 1.25.$$
The slope is 1.25.

45. Since $m = 1.25$ and $b = 0$, the equation is
$$y = 1.25x + 0$$
$$\text{or} \quad y = 1.25x.$$

46. If $f(x) = ax + b$, then $a = 1.25$ and $b = 0$.

47. The value of a, 1.25, is the price in dollars per gallon of gasoline. It is also the slope of the line.

48. Since $f(x) = 1.25x$, it fits the form for a direct variation, that is, $y = kx$. The value of a, 1.25, is the constant of variation $(k = a)$.

49. The point $(4.6, 5.75)$ means that 4.6 gal of gasoline cost \$5.75.

50. $x = 12$ and $Y_1 = 15$ means that 12 gal of gas cost \$15.00.

Chapter 3 Review Exercises

1. $3x + 2y = 10$
For $x = 0$:
$$3(0) + 2y = 10$$
$$2y = 10$$
$$y = 5 \quad (0, 5)$$
For $y = 0$:
$$3x + 2(0) = 10$$
$$3x = 10$$
$$x = \frac{10}{3} \quad \left(\frac{10}{3}, 0\right)$$
For $x = 2$:
$$3(2) + 2y = 10$$
$$6 + 2y = 10$$
$$2y = 4$$
$$y = 2 \quad (2, 2)$$

For $y = -2$:
$$3x + 2(-2) = 10$$
$$3x - 4 = 10$$
$$3x = 14$$
$$x = \frac{14}{3} \quad \left(\frac{14}{3}, -2\right)$$

Plot the ordered pairs, and draw the line through them.

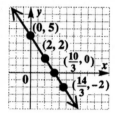

2. $x - y = 8$
For $x = 2$:
$$2 - y = 8$$
$$-y = 6$$
$$y = -6 \quad (2, -6)$$
For $y = -3$:
$$x - (-3) = 8$$
$$x + 3 = 8$$
$$x = 5 \quad (5, -3)$$
For $x = 3$:
$$3 - y = 8$$
$$-y = 5$$
$$y = -5 \quad (3, -5)$$
For $y = -2$:
$$x - (-2) = 8$$
$$x + 2 = 8$$
$$x = 6 \quad (6, -2)$$

Plot the ordered pairs, and draw the line through them.

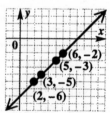

3. $4x - 3y = 12$
To find the x-intercept, let $y = 0$.
$$4x - 3y = 12$$
$$4x - 3(0) = 12$$
$$4x = 12$$
$$x = 3$$

The x-intercept is $(3, 0)$.
To find the y-intercept, let $x = 0$.

$$4x - 3y = 12$$
$$4(0) - 3y = 12$$
$$-3y = 12$$
$$y = -4$$

The y-intercept is $(0, -4)$.
Plot the intercepts and draw the line through them.

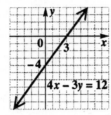

4. $5x + 7y = 28$
To find the x-intercept, let $y = 0$.

$$5x + 7y = 28$$
$$5x + 7(0) = 28$$
$$5x = 28$$
$$x = \frac{28}{5}$$

The x-intercept is $\left(\frac{28}{5}, 0\right)$.
To find the y-intercept, let $x = 0$.

$$5x + 7y = 28$$
$$5(0) + 7y = 28$$
$$7y = 28$$
$$y = 4$$

The y-intercept is $(0, 4)$.
Plot the intercepts and draw the line through them.

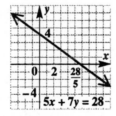

5. $2x + 5y = 20$
To find the x-intercept, let $y = 0$.

$$2x + 5y = 20$$
$$2x + 5(0) = 20$$
$$2x = 20$$
$$x = 10$$

The x-intercept is $(10, 0)$.
To find the y-intercept, let $x = 0$.

$$2x + 5y = 20$$
$$2(0) + 5y = 20$$
$$5y = 20$$
$$y = 4$$

The y-intercept is $(0, 4)$.
Plot the intercepts and draw the line through them.

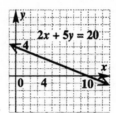

6. $x - 4y = 8$
To find the x-intercept, let $y = 0$.

$$x - 4y = 8$$
$$x - 4(0) = 8$$
$$x = 8$$

The x-intercept is $(8, 0)$.
To find the y-intercept, let $x = 0$.

$$0 - 4y = 8$$
$$-4y = 8$$
$$y = -2$$

The y-intercept is $(0, -2)$.
Plot the intercepts and draw the line through them.

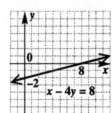

7. If both coordinates are positive, the point lies in quadrant I. If the first coordinate is negative and the second is positive, the point lies in quadrant II. To lie in quadrant III, the point must have both coordinates negative. To lie in quadrant IV, the first coordinate must be positive and the second must be negative.

8. Through $(-1, 2)$ and $(4, -5)$

$$m = \frac{\text{change in } y}{\text{change in } x} = \frac{2 - (-5)}{-1 - 4} = \frac{7}{-5} = -\frac{7}{5}$$

9. Through $(0, 3)$ and $(-2, 4)$
Let $(x_1, y_1) = (0, 3)$ and $(x_2, y_2) = (-2, 4)$.

$$m = \frac{y_2 - y_1}{x_2 - x_1} = \frac{4 - 3}{-2 - 0} = \frac{1}{-2} = -\frac{1}{2}$$

10. The slope of $y = 2x + 3$ is 2, the coefficient of x.

11. $3x - 4y = 5$
Write the equation in slope-intercept form.
$$-4y = -3x + 5$$
$$y = \frac{3}{4}x - \frac{5}{4}$$
The slope is $\frac{3}{4}$.

12. $x = 5$ is a vertical line and has undefined slope.

13. Parallel to $3y = 2x + 5$
Write the equation in slope-intercept form.
$$3y = 2x + 5$$
$$y = \frac{2}{3}x + \frac{5}{3}$$
The slope of $3y = 2x + 5$ is $\frac{2}{3}$; all lines parallel to it will also have a slope of $\frac{2}{3}$.

14. Perpendicular to $3x - y = 4$
Solve for y.
$$y = 3x - 4$$
The slope is 3; the slope of a line perpendicular to it is $-\frac{1}{3}$ since
$$3\left(-\frac{1}{3}\right) = -1.$$

15. Through $(-1, 5)$ and $(-1, -4)$
$$m = \frac{5 - (-4)}{-1 - (-1)} = \frac{9}{0} \quad Undefined$$
This is a vertical line; it has undefined slope.

16. Find the slope using two points from the table, say $(0, 2)$ and $(6, -6)$.
$$m = \frac{2 - (-6)}{0 - 6} = \frac{2 + 6}{-6} = \frac{8}{-6} = -\frac{4}{3}$$

17. Find the slope using the two points from the screens; that is, $(-2, -6)$ and $(4, 9)$.
$$m = \frac{-6 - 9}{-2 - 4} = \frac{-15}{-6} = \frac{5}{2}$$

18. The line goes up from left to right, so it has positive slope.

19. The line goes down from left to right, so it has negative slope.

20. The line is horizontal, so it has zero slope.

21. The line is vertical, so it has undefined slope.

22. The slope is $\frac{2}{10}$ which can be written as $.2$, 20%, $\frac{20}{100}$, or $\frac{1}{5}$.
The correct responses are (a), (b), (c), (d), and (f).

23. To rise 1 foot, we must move 4 feet in the horizontal direction. To rise 3 feet, we must move $3(4) = 12$ feet in the horizontal direction.

24. Let $(x_1, y_1) = (1970, 10,000)$ and $(x_2, y_2) = (1995, 41,000)$. Then
$$m = \frac{41,000 - 10,000}{1995 - 1970}$$
$$= \frac{31,000}{25} = 1240.$$
The average rate of change is $\frac{\$1240}{\text{yr}}$.

25. Slope $-\frac{1}{3}$, y-intercept $(0, -1)$
Use the slope-intercept form with $m = -\frac{1}{3}$ and $b = -1$.
$$y = mx + b$$
$$y = -\frac{1}{3}x - 1$$

26. Slope 0, y-intercept $(0, -2)$
Use the slope-intercept form with $m = 0$ and $b = -2$.
$$y = mx + b$$
$$y = (0)x - 2$$
$$y = -2$$

27. Slope $-\frac{4}{3}$, through $(2, 7)$
Use the point-slope form with $m = -\frac{4}{3}$ and $(x_1, y_1) = (2, 7)$.
$$y - y_1 = m(x - x_1)$$
$$y - 7 = -\frac{4}{3}(x - 2)$$
$$y - 7 = -\frac{4}{3}x + \frac{8}{3}$$
$$y = -\frac{4}{3}x + \frac{29}{3}$$

28. Slope 3, through $(-1, 4)$
Use the point-slope form with $m = 3$ and $(x_1, y_1) = (-1, 4)$.

$$y - y_1 = m(x - x_1)$$
$$y - 4 = 3[x - (-1)]$$
$$y - 4 = 3(x + 1)$$
$$y - 4 = 3x + 3$$
$$y = 3x + 7$$

29. Vertical, through $(2, 5)$

The equation of any vertical line is in the form $x = k$. Since the line goes through $(2, 5)$, the equation is $x = 2$. (Slope-intercept form is not possible.)

30. Through $(2, -5)$ and $(1, 4)$

Find the slope.

$$m = \frac{4 - (-5)}{1 - 2} = \frac{9}{-1} = -9$$

Use the point-slope form with $m = -9$ and $(x_1, y_1) = (2, -5)$.

$$y - y_1 = m(x - x_1)$$
$$y - (-5) = -9(x - 2)$$
$$y + 5 = -9x + 18$$
$$y = -9x + 13$$

31. Through $(-3, -1)$ and $(2, 6)$

Find the slope.

$$m = \frac{6 - (-1)}{2 - (-3)} = \frac{7}{5}$$

Use the point-slope form with $m = \frac{7}{5}$ and $(x_1, y_1) = (2, 6)$.

$$y - y_1 = m(x - x_1)$$
$$y - 6 = \frac{7}{5}(x - 2)$$
$$y - 6 = \frac{7}{5}x - \frac{14}{5}$$
$$y = \frac{7}{5}x + \frac{16}{5}$$

32. Parallel to $4x - y = 3$ and through $(7, -1)$

$y = 4x - 3$ has slope 4. Lines parallel to it will also have slope 4. The line with slope 4 through $(7, -1)$ is :

$$y - y_1 = m(x - x_1)$$
$$y - (-1) = 4(x - 7)$$
$$y + 1 = 4x - 28$$
$$y = 4x - 29$$

33. Perpendicular to $2x - 5y = 7$ and through $(4, 3)$

Write the equation in slope-intercept form.

$$2x - 5y = 7$$
$$-5y = -2x + 7$$
$$y = \frac{2}{5}x - \frac{7}{5}$$

$y = \frac{2}{5}x - \frac{7}{5}$ has slope $\frac{2}{5}$ and is perpendicular to lines with slope $-\frac{5}{2}$.

The line with slope $-\frac{5}{2}$ through $(4, 3)$ is

$$y - y_1 = m(x - x_1)$$
$$y - 3 = -\frac{5}{2}(x - 4)$$
$$y - 3 = -\frac{5}{2}x + 10$$
$$y = -\frac{5}{2}x + 13$$

34. From Exercise 16, $m = -\frac{4}{3}$.

From the table, the graph goes through $(0, 2)$, so $b = 2$. Therefore, the equation is

$$y = -\frac{4}{3}x + 2.$$

35. From Exercise 17, $m = \frac{5}{2}$.

Use $(x_1, y_1) = (4, 9)$ and $m = \frac{5}{2}$ in the point-slope form.

$$y - y_1 = m(x - x_1)$$
$$y - 9 = \frac{5}{2}(x - 4)$$
$$y - 9 = \frac{5}{2}x - 10$$
$$y = \frac{5}{2}x - 1$$

36. If $x = 0$ corresponds to 1980, then $x = 16$ corresponds to 1996. Substitute 16 for x in the equation and solve for y.

$$y = 2.1x + 230$$
$$y = 2.1(16) + 230$$
$$= 33.6 + 230 = 263.6$$

The population in 1996 was about 264 million.

37. Substitute 247 for y in the equation and solve for x.

$$y = 2.1x + 230$$
$$247 = 2.1x + 230$$
$$17 = 2.1x$$
$$x = \frac{17}{2.1} \approx 8$$

Since $x = 0$ corresponds to 1980, $x = 8$ corresponds to 1988. In 1988, the population reached 247 million.

38. $3x - 2y \leq 12$

Graph $3x - 2y = 12$ as a solid line through $(0, -6)$ and $(4, 0)$. Use $(0, 0)$ as a test point. Since $(0, 0)$ satisfies the inequality, shade the region on the side of the line containing $(0, 0)$.

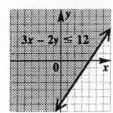

39. $5x - y > 6$

Graph $5x - y = 6$ as a dashed line through $(0, -6)$ and $\left(\frac{6}{5}, 0\right)$. Use $(0, 0)$ as a test point.

Since $(0, 0)$ does not satisfy the inequality, shade the region on the side of the line that does not contain $(0, 0)$.

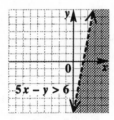

40. $x \geq 2$

Graph $x = 2$ as a solid vertical line, crossing the x-axis at 2. Since $(0, 0)$ does not satisfy the inequality, shade the region to the right of the line.

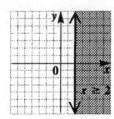

41. $2x + y \leq 1$ and $x \geq 2y$

Graph $2x + y = 1$ as a solid line through $\left(\frac{1}{2}, 0\right)$ and $(0, 1)$, and shade the region on the side containing $(0, 0)$ since it satisfies the inequality. Next, graph $x = 2y$ as a solid line through $(0, 0)$ and $(2, 1)$, and shade the region on the side containing $(2, 0)$ since $2 > 2(0)$ or $2 > 0$ is true. The intersection is the region where the graphs overlap.

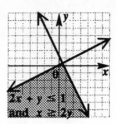

42. $\{(-4, 2), (-4, -2), (1, 5), (1, -5)\}$

The domain, the set of x-values, is $\{-4, 1\}$. The range, the set of y-values, is $\{2, -2, 5, -5\}$. Since each x-value has more than one y-value, the relation is not a function.

43. The domain, the x-values of the points on the graph, is $[-4, 4]$. The range, the y-values of the points on the graph, is $[0, 2]$. Since a vertical line intersects the graph of the relation in at most one point, the relation is a function.

44. The domain, the set of first components, is $\{$California, New York, Texas, Pennsylvania, Washington$\}$.
The range, the set of second components, is $\{71, 266, 50, 101, 48, 010, 42, 142, 38, 240\}$. Since each state corresponds to one number of small offices/home offices, the relation is a function.

45. **(a)** The independent variable must be the country, because, for each country, there is exactly one amount of power. On the other hand, for a specific amount of power, there would be more than one country generating that amount.

(b) The domain, the set of countries, is $\{$United States, France, Japan, Germany, Canada, Russia$\}$. The range, the amounts of nuclear power generated (in billion kilowatt-hours), is $\{101, 154, 286, 377, 706\}$.

46. The line graph passes the vertical line test, so it is the graph of a function; in fact, it is a linear function.
The end points give us two ordered pairs on the line: $(1991, 80.1)$ and $(1996, 89.4)$.
The horizontal axis values vary from 1991 to 1996, so the domain is $[1991, 1996]$. The vertical axis values vary from 80.1 to 89.4, so the range is $[80.1, 89.4]$.

47. $y = 3x - 3$

For any value of x, there is exactly one value of y, so the equation defines a function, actually a linear function. The domain is the set of all real numbers $(-\infty, \infty)$.

48. $y < x + 2$

For any value of x, there are many values of y. For example, $(1, 0)$ and $(1, 1)$ are both solutions of the inequality that have the same x-value but different y-values. The inequality does not define a function. The domain is the set of all real numbers $(-\infty, \infty)$.

49. $y = |x - 4|$

For any value of x, there is exactly one value of y, so the equation defines a function. The domain is the set of all real numbers $(-\infty, \infty)$.

50. $y = \sqrt{4x + 7}$

Given any value of x, y is found by multiplying x by 4, adding 7, and taking the square root of the result. This process produces exactly one value of y for each x-value, so the equation defines a function. Since the radicand must be nonnegative,

$$4x + 7 \geq 0$$
$$4x \geq -7$$
$$x \geq -\frac{7}{4}.$$

The domain is $\left[-\frac{7}{4}, \infty\right)$.

51. $x = y^2$

The ordered pairs $(4, 2)$ and $(4, -2)$ both satisfy the equation. Since one value of x, 4, corresponds to two values of y, 2 and -2, the equation does not define a function. Because x is equal to the square of y, the values of x must always be nonnegative. The domain is $[0, \infty)$.

52. $y = \dfrac{7}{x - 6}$

Given any value of x, y is found by subtracting 6, then dividing the result into 7. This process produces exactly one value of y for each x-value, so the equation defines a function. The domain includes all real numbers except those that make the denominator 0, namely 6. The domain is $(-\infty, 6) \cup (6, \infty)$.

53. If no vertical line intersects the graph in more than one point, then it is the graph of a function.

In Exercises 54–59, use

$$f(x) = -2x^2 + 3x - 6.$$

54. $f(0) = -2(0)^2 + 3(0) - 6 = -6$

55. $f(2.1) = -2(2.1)^2 + 3(2.1) - 6$
$$= -8.82 + 6.3 - 6 = -8.52$$

56. $f\left(-\dfrac{1}{2}\right) = -2\left(-\dfrac{1}{2}\right)^2 + 3\left(-\dfrac{1}{2}\right) - 6$
$$= -\frac{1}{2} - \frac{3}{2} - 6 = -8$$

57. $f(k) = -2k^2 + 3k - 6$

58. $f[f(0)]$

First find $f(0)$.

$$f(0) = -2(0)^2 + 3(0) - 6 = -6$$

Since $f(0) = -6$, $f[f(0)] = f(-6)$. Find $f(-6)$.

$$f(-6) = -2(-6)^2 + 3(-6) - 6$$
$$= -72 - 18 - 6 = -96$$

So, $f[f(0)] = -96$.

59. $f(2p) = -2(2p)^2 + 3(2p) - 6$
$$= -2\left(4p^2\right) + 6p - 6$$
$$= -8p^2 + 6p - 6$$

60. $\quad 2x^2 - y = 0$
$$-y = -2x^2$$
$$y = 2x^2$$

Since $y = f(x)$,
$$f(x) = 2x^2,$$
and $f(3) = 2(3)^2 = 2(9) = 18$.

61. Solve for y in terms of x.

$$2x - 5y = 7$$
$$2x - 7 = 5y$$
$$\frac{2x - 7}{5} = y$$

This is the same as choice (c),

$$f(x) = \frac{-7 + 2x}{5}.$$

62. No, because the equation of a line with undefined slope is $x = c$, so the ordered pairs have the form (c, y), where c is a constant and y is a variable. Thus, the number c corresponds to an infinite number of values of y.

63. The slope is negative since the line falls from left to right.

64. Use the points $(-1, 5)$ and $(3, -1)$.

$$m = \frac{5 - (-1)}{-1 - 3} = \frac{5 + 1}{-1 - 3} = \frac{6}{-4} = -\frac{3}{2}$$

65. $2y = -3x + 7$

To find the x-intercept, let $y = 0$.

$$2(0) = -3x + 7$$
$$3x = 7$$
$$x = \frac{7}{3}$$

The x-intercept is $\left(\frac{7}{3}, 0\right)$.

66. $2y = -3x + 7$

To find the y-intercept, let $x = 0$.

$$2y = -3(0) + 7$$
$$2y = 7$$
$$y = \frac{7}{2}$$

The y-intercept is $\left(0, \frac{7}{2}\right)$.

67. Solve $2y = -3x + 7$ for y.

$$y = -\frac{3}{2}x + \frac{7}{2}$$

Since $y = f(x)$,

$$f(x) = -\frac{3}{2}x + \frac{7}{2}.$$

68. $f(x) = -\frac{3}{2}x + \frac{7}{2}$

$$f(8) = -\frac{3}{2}(8) + \frac{7}{2}$$
$$= -\frac{24}{2} + \frac{7}{2} = -\frac{17}{2}$$

69. $f(x) = -\frac{3}{2}x + \frac{7}{2}$

Let $f(x) = -8$.

$$-8 = -\frac{3}{2}x + \frac{7}{2}$$
$$-16 = -3x + 7 \quad \textit{Multiply by 2.}$$
$$-23 = -3x$$
$$x = \frac{23}{3}$$

70.
$$f(x) \geq 0$$
$$-\frac{3}{2}x + \frac{7}{2} \geq 0$$
$$-\frac{3}{2}x \geq -\frac{7}{2}$$
$$x \leq \left(-\frac{7}{2}\right)\left(-\frac{2}{3}\right)$$
$$x \leq \frac{7}{3}$$

71. $f(x) = 0$ is equivalent to $y = 0$, which is the equation we solved in Exercise 65 to find the x-intercept.

Solution set: $\left\{\frac{7}{3}\right\}$

72. The graph is below the x-axis for $x > \frac{7}{3}$, so the solution set of $f(x) < 0$ is $\left(\frac{7}{3}, \infty\right)$.

73. The graph is above the x-axis for $x < \frac{7}{3}$, so the solution set of $f(x) > 0$ is $\left(-\infty, \frac{7}{3}\right)$.

74. Since $m = -\frac{3}{2}$, the slope of any line perpendicular to this line is $\frac{2}{3}$ since $\frac{2}{3}$ is the negative reciprocal of $-\frac{3}{2}$.

75. If y varies inversely as x, then $y = \frac{k}{x}$, for some constant k. This form fits choice (c).

76. Let $R =$ the resistance in ohms and $K =$ the temperature in degrees Kelvin. R varies directly as K, so

$$R = kK$$

for some constant k. Since $R = .646$ when $K = 190$, substitute these values in the equation and solve for k.

$$R = kK$$
$$.646 = k(190)$$
$$k = \frac{.646}{190} = .0034$$

So, $R = .0034K$.

When $K = 250$,

$$R = .0034(250) = .850.$$

The resistance is .850 ohm.

77. Let $v =$ the viewing distance and $e =$ the amount of enlargement. v varies directly as e, so

$$v = ke$$

for some constant k. Since $v = 250$ when $e = 5$, substitute these values in the equation and solve for k.

$$v = ke$$
$$250 = k(5)$$
$$50 = k$$

So, $v = 50e$.
When $e = 8.6$,

$$v = 50(8.6) = 430.$$

It should be viewed from 430 mm.

78. f varies directly as $\sqrt{t}$ and inversely as L, so

$$f = \frac{k\sqrt{t}}{L}$$

for some constant k. Since $f = 20$ when $t = 9$ and $L = 30$, substitute these values in the equation and solve for k.

$$f = \frac{k\sqrt{t}}{L}$$
$$20 = \frac{k\sqrt{9}}{30}$$
$$k = \frac{20(30)}{3} = 200$$

So, $f = \dfrac{200\sqrt{t}}{L}$.

When the tension is doubled and the length remains the same, $t = 18$ and $L = 30$, and

$$f = \frac{200\sqrt{18}}{30}$$
$$= \frac{20 \cdot 3\sqrt{2}}{3} = 20\sqrt{2}$$
$$\approx 28 \text{ (to the nearest unit).}$$

The frequency of vibration is about 28.

Chapter 3 Test

1. $2x - 3y = 12$
For $x = 1$:

$$2(1) - 3y = 12$$
$$2 - 3y = 12$$
$$-3y = 10$$
$$y = -\frac{10}{3} \quad \left(1, -\frac{10}{3}\right)$$

For $x = 3$:

$$2(3) - 3y = 12$$
$$6 - 3y = 12$$
$$-3y = 6$$
$$y = -2 \quad (3, -2)$$

For $y = -4$:

$$2x - 3(-4) = 12$$
$$2x + 12 = 12$$
$$2x = 0$$
$$x = 0 \quad (0, -4)$$

2. Through $(6, 4)$ and $(-4, -1)$

$$m = \frac{4 - (-1)}{6 - (-4)} = \frac{4+1}{6+4} = \frac{5}{10} = \frac{1}{2}$$

The slope of the line is $\dfrac{1}{2}$.

3. $3x - 2y = 20$
To find the x-intercept, let $y = 0$.

$$3x - 2(0) = 20$$
$$3x = 20$$
$$x = \frac{20}{3}$$

The x-intercept is $\left(\dfrac{20}{3}, 0\right)$.
To find the y-intercept, let $x = 0$.

$$3(0) - 2y = 20$$
$$-2y = 20$$
$$y = -10$$

The y-intercept is $(0, -10)$.
Draw the line through these two points.

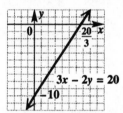

4. The graph of $y = 5$ is the horizontal line with slope 0 and y-intercept $(0, 5)$. There is no x-intercept.

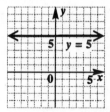

5. The graph of $x = 2$ is the vertical line with x-intercept at $(2, 0)$. There is no y-intercept.

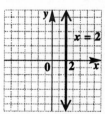

6. The graph of a line with undefined slope is the graph of a vertical line.

7. Find the slope of each line.

$$5x - y = 8$$
$$-y = -5x + 8$$
$$y = 5x - 8$$

The slope is 5.

$$5y = -x + 3$$
$$y = -\frac{1}{5}x + \frac{3}{5}$$

The slope is $-\frac{1}{5}$.

Since $5\left(-\frac{1}{5}\right) = -1$, the two slopes are negative reciprocals and the lines are perpendicular.

8. Find the slope of each line.

$$2y = 3x + 12$$
$$y = \frac{3}{2}x + 6$$

The slope is $\frac{3}{2}$.

$$3y = 2x - 5$$
$$y = \frac{2}{3}x - \frac{5}{3}$$

The slope is $\frac{2}{3}$.

The lines are neither parallel nor perpendicular.

9. Through $(4, -1)$; $m = -5$
Let $m = -5$ and $(x_1, y_1) = (4, -1)$ in the point-slope form.

$$y - y_1 = m(x - x_1)$$
$$y - (-1) = -5(x - 4)$$
$$y + 1 = -5x + 20$$
$$y = -5x + 19$$

10. Through $(-3, 14)$; horizontal
A horizontal line has equation $y = k$. Here $k = 14$, so the line has equation $y = 14$.

11. Through $(-7, 2)$ and parallel to $3x + 5y = 6$
To find the slope of $3x + 5y = 6$, write the equation in slope-intercept form by solving for y.

$$3x + 5y = 6$$
$$5y = -3x + 6$$
$$y = -\frac{3}{5}x + \frac{6}{5}$$

The slope is $-\frac{3}{5}$, so a line parallel to it also has

slope $-\frac{3}{5}$. Let $m = -\frac{3}{5}$ and $(x_1, y_1) = (-7, 2)$ in the point-slope form.

$$y - y_1 = m(x - x_1)$$
$$y - 2 = -\frac{3}{5}[x - (-7)]$$
$$y - 2 = -\frac{3}{5}(x + 7)$$
$$y - 2 = -\frac{3}{5}x - \frac{21}{5}$$
$$y = -\frac{3}{5}x - \frac{11}{5}$$

12. Through $(-7, 2)$ and perpendicular to $y = 2x$
Since $y = 2x$ is in slope-intercept form $(b = 0)$, the slope, m, of $y = 2x$ is 2. A line perpendicular to it has a slope that is the negative reciprocal of 2, that is, $-\frac{1}{2}$. Let $m = -\frac{1}{2}$ and $(x_1, y_1) = (-7, 2)$ in the point-slope form.

$$y - y_1 = m(x - x_1)$$
$$y - 2 = -\frac{1}{2}(x + 7)$$
$$y - 2 = -\frac{1}{2}x - \frac{7}{2}$$
$$y = -\frac{1}{2}x - \frac{3}{2}$$

13. From the graphs, find the equation of the line through $(-2, 3)$ and $(6, -1)$ First find the slope.

$$m = \frac{3 - (-1)}{-2 - 6} = \frac{3 + 1}{-8} = \frac{4}{-8} = -\frac{1}{2}$$

Use $m = -\frac{1}{2}$ and $(x_1, y_1) = (-2, 3)$ in the point-slope form.

$$y - y_1 = m(x - x_1)$$
$$y - 3 = -\frac{1}{2}[x - (-2)]$$
$$y - 3 = -\frac{1}{2}(x + 2)$$
$$y - 3 = -\frac{1}{2}x - 1$$
$$y = -\frac{1}{2}x + 2$$

14. Positive slope means that the line goes up from left to right. The only line that has positive slope and a negative y-coordinate for its y-intercept is choice (b).

15. For 1994, $x = 4$ (1994 $-$ 1990 $= 4$).

$$y = 1410x + 12,520$$
$$y = 1410(4) + 12,520 = 18,160$$

There were 18,160 cases.

16. (a) The number 1410 is the slope of the line.

(b) It is the annual rate of change in the number of cases served.

17. $3x - 2y > 6$

Graph the line $3x - 2y = 6$, which has intercepts $(2, 0)$ and $(0, -3)$, as a dashed line since the inequality involves $>$. Test $(0, 0)$, which yields $0 > 6$, a false statement. Shade the region that does not include $(0, 0)$.

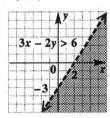

18. $y < 2x - 1$ and $x - y < 3$

First graph $y = 2x - 1$ as a dashed line through $(2, 3)$ and $(0, -1)$. Test $(0, 0)$, which yields $0 < -1$, a false statement. Shade the side of the line not containing $(0, 0)$.

Next, graph $x - y = 3$ as a dashed line through $(3, 0)$ and $(0, -3)$. Test $(0, 0)$, which yields $0 < 3$, a true statement.

Shade the side of the line containing $(0, 0)$. The intersection is the region where the graphs overlap.

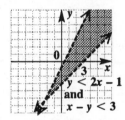

19. Choice (d) is the only graph that passes the vertical line test.

20. Choice (d) does not define a function, since its domain (input) element A is paired with two different range (output) elements, 1 and 2.

21. $f(x) = -x^2 + 2x - 1$

$f(1) = -(1)^2 + 2(1) - 1$

$\quad = -1 + 2 - 1$

$\quad = 0$

22. $f(x) = \dfrac{2}{3}x - 1$

This function represents a line with y-intercept $(0, -1)$ and x-intercept $\left(\dfrac{3}{2}, 0\right)$.

Draw the line through these two points. The domain is $(-\infty, \infty)$, and the range is $(-\infty, \infty)$.

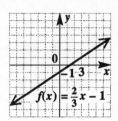

23. Choice (b) {(Year, Death rate)} defines a function since each year corresponds to one death rate. Choice (a) {(Death rate, year)} doesn't define a function since the values 8.6 and 8.8 would correspond to more than one year.

24. Let $I =$ the current and
$\quad R =$ the resistance.
The current is inversely proportional to the resistance, so

$$I = \dfrac{k}{R}.$$

Let $I = 80$ and $R = 30$. Find k.

$$80 = \dfrac{k}{30}$$

$$k = 80(30) = 2400$$

So, $I = \dfrac{2400}{R}$.

Let $R = 12$.

$$I = \dfrac{2400}{12} = 200$$

The current is 200 amps.

25. Let $I =$ the collision impact,
$\quad m =$ the mass, and
$\quad s =$ the speed.
The impact varies jointly as its mass and the square of its speed, so

$$I = kms^2.$$

Let $I = 6.1$, $m = 2000$, and $s = 55$. Find k.

$$6.1 = k(2000)(55)^2$$

$$6.1 = 6,050,000k$$

$$k \approx .000001008$$

So, $I = .000001008ms^2$.
Let $m = 2000$ and $s = 65$.

$$I = .000001008(2000)(65)^2$$

$$\approx 8.5$$

Cumulative Review Exercises Chapters 1–3

1. The absolute value of a negative number is a positive number and the additive inverse of the same negative number is the same positive number. For example, suppose the negative number is -5:

$$|-5| = -(-5) = 5$$
$$\text{and} \quad -(-5) = 5$$

The statement is *always true*.

2. The statement is *always true*; in fact, it is the definition of a rational number.

3. The sum of two negative numbers is another negative number, so the statement is *never true*.

4. The statement is *sometimes true*. For example,

$$3 + (-3) = 0,$$
$$\text{but} \quad 3 + (-1) = 2 \neq 0.$$

5. $-|-4| = -[-(-4)] = -[4] = -4$

6. $|-12| - |-3| = [-(-12)] - [-(-3)]$
$$= [12] - [3] = 9$$

7. $\dfrac{3\left(\sqrt{16}\right) - (-1)(7)}{4 + (-6)} = \dfrac{3(4) - (-7)}{-2}$
$$= \dfrac{12 + 7}{-2} = -\dfrac{19}{2}$$

8. $2p - (4 + 3p) - 1 - p = 2p - 4 - 3p - 1 - p$
$$= -2p - 5$$

9. $3x + 2(x - 4) = -(2x + 5)$
$$3x + 2x - 8 = -2x - 5$$
$$5x - 8 = -2x - 5$$
$$7x = 3$$
$$x = \dfrac{3}{7}$$
Solution set: $\left\{\dfrac{3}{7}\right\}$

10. $\dfrac{x - 1}{6} - \dfrac{x}{4} = \dfrac{2x + 4}{12}$
$$2(x - 1) - 3x = 2x + 4 \quad \textit{Multiply by 12.}$$
$$2x - 2 - 3x = 2x + 4$$
$$-x - 2 = 2x + 4$$
$$-3x = 6$$
$$x = -2$$
Solution set: $\{-2\}$

11. Solve $V = \dfrac{1}{3}\pi r^2 h$ for h.
$$3V = \pi r^2 h$$
$$\dfrac{3V}{\pi r^2} = h$$

12. $-4 < 3 - 2k < 9$
$$-7 < -2k < 6$$
Divide by -2; reverse the inequality signs.
$$\dfrac{7}{2} > k > -3 \text{ or } -3 < k < \dfrac{7}{2}$$
Solution set: $\left(-3, \dfrac{7}{2}\right)$

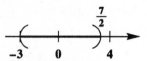

13. $-.3x + 2.1(x - 4) \leq -6.6$
$$-3x + 21(x - 4) \leq -66$$
$$\textit{Multiply by 10.}$$
$$-3x + 21x - 84 \leq -66$$
$$18x - 84 \leq -66$$
$$18x \leq 18$$
$$x \leq 1$$
Solution set: $(-\infty, 1]$

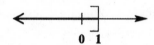

14. $x > 6 \quad \text{and} \quad x < 8$
The graph of the solution set is all numbers both greater than 6 *and* less than 8. This is the intersection. The elements common to both sets are the numbers between 6 and 8, not including the endpoints. The solution set is $(6, 8)$.

15. $-5x + 1 \geq 11 \quad \text{or} \quad 3x + 5 > 26$
$$-5x \geq 10 \qquad\qquad 3x > 21$$
$$x \leq -2 \quad \text{or} \qquad x > 7$$
The graph of the solution set is all numbers either less than or equal to -2 *or* greater than 7. This is the union. The solution set is $(-\infty, -2] \cup (7, \infty)$.

16. $|2k - 7| + 4 = 11$
$$|2k - 7| = 7$$
$$2k - 7 = 7 \quad \text{or} \quad 2k - 7 = -7$$
$$2k = 14 \qquad\qquad 2k = 0$$
$$k = 7 \quad \text{or} \qquad k = 0$$
Solution set: $\{0, 7\}$

17. $|3m + 6| \geq 0$

The absolute value of an expression is always nonnegative, so the inequality is true for any real number m.

Solution set: $(-\infty, \infty)$

18. The union of the three solution sets is $(-\infty, \infty)$; that is, the set of all real numbers.

19. $F = \dfrac{9}{5}C + 32$

$F = \dfrac{9}{5}(-55) + 32$ Let $C = -55$

$= -99 + 32 = -67$

$-55°$ C is equivalent to $-67°$ F.

20. The change is $11.7 - 10.9 = .8$.
The increase is

$$\dfrac{.8}{10.9} \approx .073 = 7.3\%.$$

21. Let s denote the side of the original square and $4s$ the perimeter. Now $s + 4$ is the side of the new square and $4(s + 4)$ is its perimeter.
"The perimeter would be 8 inches less than twice the perimeter of the original square" translates as

$$4(s + 4) = 2(4s) - 8.$$
$$4s + 16 = 8s - 8$$
$$24 = 4s$$
$$6 = s$$

The length of a side of the original square is 6 inches.

22. $3x + 5y = 12$
To find the x-intercept, let $y = 0$.

$$3x + 5(0) = 12$$
$$3x = 12$$
$$x = 4$$

The x-intercept is $(4, 0)$.
To find the y-intercept, let $x = 0$.

$$3(0) + 5y = 12$$
$$5y = 12$$
$$y = \dfrac{12}{5}$$

The y-intercept is $\left(0, \dfrac{12}{5}\right)$.

Plot the intercepts and draw the line through them.

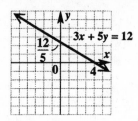

23. $A(-2, 1)$ and $B(3, -5)$
(a) The slope of line AB is

$$m = \dfrac{1 - (-5)}{-2 - 3} = \dfrac{6}{-5} = -\dfrac{6}{5}.$$

(b) The slope of a line perpendicular to line AB is the negative reciprocal of $-\dfrac{6}{5}$, which is $\dfrac{5}{6}$.

24. $-2x + y < -6$
Graph the line $-2x + y = -6$, which has intercepts $(3, 0)$ and $(0, -6)$, as a dashed line since the inequality involves $<$. Test $(0, 0)$, which yields $0 < -6$, a false statement. Shade the region that does not include $(0, 0)$.

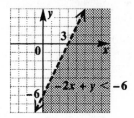

25. $f(x) = -3x + 6$
(a) The domain includes the set of all real numbers $(-\infty, \infty)$.

(b) $f(-6) = -3(-6) + 6 = 18 + 6 = 24$

26. The set $\{(\text{City, Percent})\}$ is a function. (The set $\{(\text{Percent, City})\}$ is not a function since 78, 79, and 85 would all correspond to more than one city.) The elements of the domain of the function are the names of the ten cities.

27. $2x - 7y = 14$
For any value of x, there is exactly one value of y, so the equation defines a function. Solve for y:

$$-7y = -2x + 14$$
$$y = \dfrac{2}{7}x - 2$$

Replace y with $f(x)$.

$$f(x) = \dfrac{2}{7}x - 2$$

28. $f(x) = 2x + \sqrt{x}$

$\quad f(4) = 2(4) + \sqrt{4}$

$\quad\quad = 8 + 2 = 10$

29. Choice (a), $2x + 3 = -4y$, is the only equation that can be written in the form $f(x) = mx + b$. Thus, (a) is the only equation that defines a linear function.

30. Let $C =$ the cost of a pizza

and $r =$ the radius of the pizza.

C varies directly as r^2, so

$$C = kr^2$$

for some constant k. Since $C = 6$ when $r = 7$, substitute these values in the equation and solve for k.

$$C = kr^2$$
$$6 = k(7)^2$$
$$6 = 49k$$
$$\frac{6}{49} = k$$

So, $C = \frac{6}{49}r^2$.

When $r = 9$,

$$C = \frac{6}{49}(9)^2$$
$$= \frac{6}{49}(81) = \frac{486}{49} \approx 9.92.$$

A pizza with a 9-inch radius should cost $9.92.

CHAPTER 4 SYSTEMS OF LINEAR EQUATIONS

Section 4.1

1. **(a)** In general, the graph falls from left to right, so the slope of an approximating line would be negative.

(b) The Giants' performance declined during those months, as indicated by the negative slope. As time passed, their winning percentage decreased.

(c) In general, the graph rises from left to right, so the slope of an approximating line would be positive.

(d) The Athletics' performance improved during those months, as indicated by the positive slope. As time passed, their winning percentage increased.

3. To determine whether $(5, 1)$ is a solution of the system

$$x + y = 6$$
$$x - y = 4,$$

replace x with 5 and y with 1 in each equation.

$$x + y = 6$$
$$5 + 1 = 6$$
$$6 = 6 \quad \text{True}$$

$$x - y = 4$$
$$5 - 1 = 4$$
$$4 = 4 \quad \text{True}$$

Since $(5, 1)$ makes both equations true, $(5, 1)$ is a solution of the system.

5. To determine whether $(5, 2)$ is a solution of the system

$$2x - y = 8$$
$$3x + 2y = 20,$$

replace x with 5 and y with 2 in each equation.

$$2x - y = 8$$
$$2(5) - 2 = 8$$
$$10 - 2 = 8 \quad \text{True}$$

$$3x + 2y = 20$$
$$3(5) + 2(2) = 20$$
$$15 + 4 = 20 \quad \text{False}$$

The ordered pair $(5, 2)$ is *not* a solution of the system since it does not make both equations true.

7. $x + y = 4$
$2x - y = 2$

Graph the line $x + y = 4$ through its intercepts, $(4, 0)$ and $(0, 4)$, and the line $2x - y = 2$ through it intercepts, $(1, 0)$ and $(0, -2)$. The lines appear to intersect at $(2, 2)$.

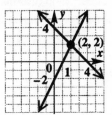

To check, substitute 2 for x and 2 for y in each equation of the system. Since $(2, 2)$ makes both equations true, the solution set of the system is $\{(2, 2)\}$.

9. $2x - 5y = 11 \quad (1)$
$3x + y = 8 \quad (2)$
To eliminate y, multiply equation (2) by 5 and add the result to equation (1).

$$\begin{array}{rcll} 2x - 5y &=& 11 & (1) \\ 15x + 5y &=& 40 & (3) \quad 5 \times (2) \\ \hline 17x &=& 51 & \\ x &=& 3 & \end{array}$$

To find y, substitute 3 for x in equation (2).

$$3x + y = 8 \quad (2)$$
$$3(3) + y = 8$$
$$9 + y = 8$$
$$y = -1$$

The ordered pair $(3, -1)$ satisfies both equations, so it checks.

Solution set: $\{(3, -1)\}$

11. $3x + 4y = -6 \quad (1)$
$5x + 3y = 1 \quad (2)$
To eliminate x, multiply equation (1) by 5 and equation (2) by -3. Then add the results.

$$\begin{array}{rcll} 15x + 20y &=& -30 & (3) \quad 5 \times (1) \\ -15x - 9y &=& -3 & (4) \quad -3 \times (2) \\ \hline 11y &=& -33 & \\ y &=& -3 & \end{array}$$

To find x, substitute -3 for y in equation (2).

$$5x + 3y = 1 \quad (2)$$
$$5x + 3(-3) = 1$$
$$5x - 9 = 1$$
$$5x = 10$$
$$x = 2$$

continued

The ordered pair $(2, -3)$ satisfies both equations, so it checks.

Solution set: $\{(2, -3)\}$

13. $3x + 3y = 0$ (1)
$4x + 2y = 3$ (2)

To eliminate y, multiply equation (1) by 2 and equation (2) by -3. Then add the results.

$$\begin{array}{rcll} 6x + 6y &=& 0 & \text{(3)}\quad 2 \times (1) \\ -12x - 6y &=& -9 & \text{(4)}\quad -3 \times (2) \\ \hline -6x &=& -9 & \\ x &=& \dfrac{-9}{-6} = \dfrac{3}{2} & \end{array}$$

To find y, substitute $\dfrac{3}{2}$ for x in equation (1).

$$3x + 3y = 0 \quad (1)$$
$$3\left(\frac{3}{2}\right) + 3y = 0$$
$$\frac{9}{2} + 3y = 0$$
$$3y = -\frac{9}{2}$$
$$y = \frac{1}{3}\left(-\frac{9}{2}\right) = -\frac{3}{2}$$

The solution $\left(\dfrac{3}{2}, -\dfrac{3}{2}\right)$ checks.

Solution set: $\left\{\left(\dfrac{3}{2}, -\dfrac{3}{2}\right)\right\}$

When you get a solution that has non-integer components, it is sometimes more difficult to check the problem than it was to solve it. A graphing calculator can be very helpful in this case. Just store the values for x and y in their respective memory locations, and then type the expressions as shown in the following screen. The results 0 and 3 (the right sides of the equations) indicate that we have found the correct solution.

```
3/2→X: -3/2→Y
                 -1.5
3X+3Y
                    0
4X+2Y
                    3
```

15. $7x + 2y = 6$ (1)
$-14x - 4y = -12$ (2)

To eliminate y, multiply equation (1) by 2 and add the result to equation (2).

$$\begin{array}{rcll} 14x + 4y &=& 12 & \text{(3)}\quad 2 \times (1) \\ -14x - 4y &=& -12 & \text{(2)} \\ \hline 0 &=& 0 & \textit{True} \end{array}$$

Multiplying equation (1) by -2 gives equation (2). The equations are dependent, and the solution is the set of all points on the line.

Solution set: $\{(x, y) \mid 7x + 2y = 6\}$

17. $\dfrac{x}{2} + \dfrac{y}{3} = -\dfrac{1}{3}$ (1)
$\dfrac{x}{2} + 2y = -7$ (2)

Eliminate the fractions by multiplying equation (1) by -6 and equation (2) by 6. Then add the results to eliminate x.

$$\begin{array}{rcll} -3x - 2y &=& 2 & \text{(3)}\quad -6 \times (1) \\ 3x + 12y &=& -42 & \text{(4)}\quad 6 \times (2) \\ \hline 10y &=& -40 & \\ y &=& -4 & \end{array}$$

To find x, substitute -4 for y in equation (3).

$$-3x - 2y = 2 \quad (3)$$
$$-3x - 2(-4) = 2$$
$$-3x + 8 = 2$$
$$-3x = -6$$
$$x = 2$$

The solution $(2, -4)$ checks.

Solution set: $\{(2, -4)\}$

19. $5x - 5y = 3$ (1)
$x - y = 12$ (2)

To eliminate x, multiply equation (2) by -5 and add the result to equation (1).

$$\begin{array}{rcll} 5x - 5y &=& 3 & \text{(1)} \\ -5x + 5y &=& -60 & \text{(3)}\quad -5 \times (2) \\ \hline 0 &=& -57 & \textit{False} \end{array}$$

The system is inconsistent. Since the graphs of the equations are parallel lines, there are no ordered pairs that satisfy both equations.

Solution set: $\emptyset$

21. **(a)** The graphs of these two linear equations will intersect once.

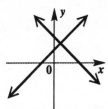

(b) The graphs of these two linear equations will not intersect. They are parallel lines.

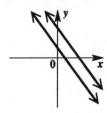

(c) The graphs of these two linear equations will be the same line.

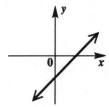

23. $3x + 7y = 4$
$6x + 14y = 3$
Write each equation in slope-intercept form by solving for y.

$$3x + 7y = 4$$
$$7y = -3x + 4$$
$$y = -\frac{3}{7}x + \frac{4}{7}$$

$$6x + 14y = 3$$
$$14y = -6x + 3$$
$$y = -\frac{6}{14}x + \frac{3}{14}$$
$$y = -\frac{3}{7}x + \frac{3}{14}$$

Since the equations have the same slope, $-\frac{3}{7}$, but different y-intercepts, $\frac{4}{7}$ and $\frac{3}{14}$, the lines when graphed are parallel. The system is inconsistent and has no solution.

25. $2x = -3y + 1$
$6x = -9y + 3$
Write each equation in slope-intercept form by solving for y.

$$2x = -3y + 1$$
$$3y = -2x + 1$$
$$y = -\frac{2}{3}x + \frac{1}{3}$$

$$6x = -9y + 3$$
$$9y = -6x + 3$$
$$y = -\frac{6}{9}x + \frac{3}{9}$$
$$y = -\frac{2}{3}x + \frac{1}{3}$$

Since both equations are the same, the solution set is all points on the line $y = -\frac{2}{3}x + \frac{1}{3}$. The system has infinitely many solutions.

27. $4x + y = 6$ (1)
      $y = 2x$ (2)
Since equation (2) is already solved for y, substitute $2x$ for y in equation (1).

$$4x + y = 6 \quad (1)$$
$$4x + 2x = 6 \quad \text{Let } y=2x.$$
$$6x = 6$$
$$x = 1$$

Substitute 1 for x in (2).

$$y = 2(1) = 2$$

The solution $(1, 2)$ checks.
Solution set: $\{(1, 2)\}$

29. $3x - 4y = -22$ (1)
$-3x + y = 0$ (2)
Solve equation (2) for y to get

$$y = 3x.$$

Substitute $3x$ for y in equation (1).

$$3x - 4y = -22 \qquad\qquad (1)$$
$$3x - 4(3x) = -22 \qquad \text{Let } y=3x.$$
$$3x - 12x = -22$$
$$-9x = -22$$
$$x = \frac{-22}{-9} = \frac{22}{9}$$

Substitute $\frac{22}{9}$ for x in $y = 3x$ to get

$$y = 3\left(\frac{22}{9}\right) = \frac{22}{3}.$$

The solution $\left(\frac{22}{9}, \frac{22}{3}\right)$ checks.

Solution set: $\left\{\left(\frac{22}{9}, \frac{22}{3}\right)\right\}$

31.
$$-x - 4y = -14 \quad (1)$$
$$2x = y + 1 \quad (2)$$

Solve equation (2) for y.

$$2x = y + 1 \quad (2)$$
$$2x - 1 = y \quad (3)$$

Substitute $2x - 1$ for y in equation (1) and solve for x.

$$-x - 4y = -14 \quad (1)$$
$$-x - 4(2x - 1) = -14$$
$$-x - 8x + 4 = -14$$
$$-9x = -18$$
$$x = 2$$

Substitute 2 for x in (3)

$$y = 2(2) - 1 = 3$$

The solution $(2, 3)$ checks.

Solution set: $\{(2, 3)\}$

33.
$$5x - 4y = 9 \quad (1)$$
$$3 - 2y = -x \quad (2)$$

Solve equation (2) for x.

$$3 - 2y = -x \quad (2)$$
$$2y - 3 = x \quad (3) \quad -1 \times (2)$$

Substitute $2y - 3$ for x in equation (1).

$$5x - 4y = 9 \quad (1)$$
$$5(2y - 3) - 4y = 9 \quad \text{Let } x = 2y{-}3.$$
$$10y - 15 - 4y = 9$$
$$6y - 15 = 9$$
$$6y = 24$$
$$y = 4$$

Substitute 4 for y in $x = 2y - 3$.

$$x = 2(4) - 3 = 5$$

The solution $(5, 4)$ checks.

Solution set: $\{(5, 4)\}$

35.
$$x = 3y + 5 \quad (1)$$
$$x = \frac{3}{2}y \quad (2)$$

Both equations are given in terms of x. Choose equation (2), and substitute $\frac{3}{2}y$ for x in equation (1).

$$x = 3y + 5 \quad (1)$$
$$\frac{3}{2}y = 3y + 5 \quad \text{Let } x = \frac{3}{2}y.$$
$$3y = 6y + 10 \quad \text{Multiply by 2.}$$
$$-3y = 10$$
$$y = -\frac{10}{3}$$

Since $x = \frac{3}{2}y$ and $y = -\frac{10}{3}$,

$$x = \frac{3}{2}\left(-\frac{10}{3}\right) = -\frac{10}{2} = -5.$$

The solution $\left(-5, -\frac{10}{3}\right)$ checks.

Solution set: $\left\{\left(-5, -\frac{10}{3}\right)\right\}$

37.
$$\frac{1}{2}x + \frac{1}{3}y = 3 \quad (1)$$
$$y = 3x \quad (2)$$

Multiply (1) by its LCD, 6, to eliminate fractions

$$3x + 2y = 18 \quad (3)$$

From (2), substitute $3x$ for y in (3).

$$3x + 2(3x) = 18$$
$$3x + 6x = 18$$
$$9x = 18$$
$$x = 2$$

Substitute 2 for x in $y = 3x$.

$$y = 3(2) = 6$$

The solution $(2, 6)$ checks.

Solution set: $\{(2, 6)\}$

39. Choose one equation and solve for one of the variables in terms of the other. Substitute this expression into the other equation, and solve for the single variable. Then use this value in either of the original equations to solve for the other variable. For example:

$$x + 2y = 5$$
$$2x + 3y = 9$$

Solve the first equation for x to get $x = 5 - 2y$. Then solve $2(5 - 2y) + 3y = 9$ to get $y = 1$. Replace y with 1 in either original equation to get $x = 3$. Check $(3, 1)$ in the original equations. The solution set is $\{(3, 1)\}$.

41. For a system of two linear equations in two variables to have no solution, the graphs of the two equations must be parallel lines. If the two equations are written in $y = mx + b$ form, they will have the same slope but different y-intercepts. Algebraically, a false statement such as $0 = 1$ will occur.

43.
$$\frac{3}{x} + \frac{4}{y} = \frac{5}{2} \quad (1)$$
$$\frac{5}{x} - \frac{3}{y} = \frac{7}{4} \quad (2)$$

If $p = \dfrac{1}{x}$ and $q = \dfrac{1}{y}$, equations (1) and (2) can be written as

$$3p + 4q = \frac{5}{2} \quad (3)$$
$$5p - 3q = \frac{7}{4}. \quad (4)$$

To eliminate q, multiply equation (3) by 3 and equation (4) by 4. Then add the results.

$$
\begin{array}{rll}
9p + 12q &= \dfrac{15}{2} & (5) \quad 3 \times (3) \\
20p - 12q &= 7 & (6) \quad 4 \times (4) \\
\hline
29p &= \dfrac{29}{2} & \\
p &= \dfrac{1}{2} &
\end{array}
$$

To find q, substitute $\dfrac{1}{2}$ for p in equation (3).

$$3p + 4q = \frac{5}{2} \quad (3)$$
$$3\left(\frac{1}{2}\right) + 4q = \frac{5}{2} \quad Let\ p=\frac{1}{2}.$$
$$\frac{3}{2} + 4q = \frac{5}{2}$$
$$4q = 1$$
$$q = \frac{1}{4}$$

Since $p = \dfrac{1}{x}$ and $p = \dfrac{1}{2}$, $\dfrac{1}{2} = \dfrac{1}{x}$ and $x = 2$.

Since $q = \dfrac{1}{y}$ and $q = \dfrac{1}{4}$, $\dfrac{1}{4} = \dfrac{1}{y}$ and $y = 4$.

The solution $(2, 4)$ checks.

Solution set: $\{(2, 4)\}$

45.
$$\frac{2}{x} - \frac{5}{y} = \frac{3}{2} \quad (1)$$
$$\frac{4}{x} + \frac{1}{y} = \frac{4}{5} \quad (2)$$

Let $p = \dfrac{1}{x}$ and $q = \dfrac{1}{y}$. Rewrite equations (1) and (2).

$$2p - 5q = \frac{3}{2} \quad (3)$$
$$4p + q = \frac{4}{5} \quad (4)$$

To clear the fractions multiply equation (3) by 2 and equation (4) by 5.

$$
\begin{array}{rll}
4p - 10q &= 3 & (5) \quad 2 \times (3) \\
20p + 5q &= 4 & (6) \quad 5 \times (4) \\
\hline
\end{array}
$$

To eliminate q, multiply equation (6) by 2. Then add the result to equation (5).

$$
\begin{array}{rll}
4p - 10q &= 3 & (5) \\
40p + 10q &= 8 & (7) \quad 2 \times (6) \\
\hline
44p &= 11 & \\
p &= \dfrac{1}{4} &
\end{array}
$$

Substitute $\dfrac{1}{4}$ for p in equation (4).

$$4p + q = \frac{4}{5} \quad (4)$$
$$4\left(\frac{1}{4}\right) + q = \frac{4}{5} \quad Let\ p=\frac{1}{4}.$$
$$1 + q = \frac{4}{5}$$
$$q = -\frac{1}{5}$$

Since $p = \dfrac{1}{x}$ and $p = \dfrac{1}{4}$, $\dfrac{1}{4} = \dfrac{1}{x}$ and $x = 4$.

Since $q = \dfrac{1}{y}$ and $q = -\dfrac{1}{5}$, $-\dfrac{1}{5} = \dfrac{1}{y}$ and $y = -5$.

The solution $(4, -5)$ checks.

Solution set: $\{(4, -5)\}$

47. Since the coefficients of x, a and $-a$, are opposites, we can simply add the two equations to eliminate x.

$$
\begin{array}{rll}
ax + by &= 2 & (1) \\
-ax + 2by &= 1 & (2) \\
\hline
3by &= 3 & \\
y &= \dfrac{3}{3b} = \dfrac{1}{b} &
\end{array}
$$

Replace y by $\dfrac{1}{b}$ in equation (1).

$$ax + by = 2 \quad (1)$$
$$ax + b\left(\frac{1}{b}\right) = 2$$
$$ax + 1 = 2$$
$$ax = 1$$
$$x = \frac{1}{a}$$

The solution $\left(\dfrac{1}{a}, \dfrac{1}{b}\right)$ checks.

Solution set: $\left\{\left(\dfrac{1}{a}, \dfrac{1}{b}\right)\right\}$

49. $3ax + 2y = 1$ (1)
$-ax + \ y = 2$ (2)

To eliminate x, multiply equation (2) by 3 and add the result to equation (1).

$$
\begin{array}{rll}
3ax \ + \ 2y \ = \ 1 & (1) \\
-3ax \ + \ 3y \ = \ 6 & (3) \quad 3 \times (2) \\
\hline
5y \ = \ 7 & \\
y \ = \ \dfrac{7}{5} &
\end{array}
$$

Replace y by $\dfrac{7}{5}$ in equation (2).

$$-ax + \frac{7}{5} = 2$$

$$-ax = \frac{3}{5}$$

$$x = -\frac{3}{5a}$$

The solution $\left(-\dfrac{3}{5a}, \dfrac{7}{5}\right)$ checks.

Solution set: $\left\{\left(-\dfrac{3}{5a}, \dfrac{7}{5}\right)\right\}$

51. $3x + \ y = 6$ (1)
$-2x + 3y = 7$ (2)

Multiply equation (1) by -3 and add the result to equation (2).

$$
\begin{array}{rll}
-9x \ - \ 3y \ = \ -18 & (3) \quad -3 \times (1) \\
-2x \ + \ 3y \ = \ \ \ \ 7 & (2) \\
\hline
-11x \ \ \ \ \ \ \ \ = \ -11 & \\
x \ = \ 1 &
\end{array}
$$

To find y, substitute 1 for x in equation (1).

$$
\begin{array}{l}
3x + y = 6 \quad (1) \\
3(1) + y = 6 \\
y = 3
\end{array}
$$

The solution $(1, 3)$ checks.

Solution set: $\{(1, 3)\}$

52. $3x + y = 6$
$y = -3x + 6$
Replace y with $f(x)$.

$$f(x) = -3x + 6$$

Since f is in the form $f(x) = mx + b$, it is a linear function.

53. $-2x + 3y = 7$
$3y = 2x + 7$
$$y = \frac{2}{3}x + \frac{7}{3}$$
Replace y with $g(x)$.

$$g(x) = \frac{2}{3}x + \frac{7}{3}$$

Since g is in the form $g(x) = mx + b$, it is a linear function.

54. Because the graphs of f and g are straight lines that are neither parallel nor coincide, they intersect in exactly _one_ point. The coordinates of the point are (_1_, _3_). Using functional notation, this is given by $f(\underline{1}) = \underline{3}$ and $g(\underline{1}) = \underline{3}$.

55. The table shows that when $x = 3$, $Y_1 = -4$ and $Y_2 = -4$. Since no other values of x in the table give the same values for Y_1 and Y_2, and since the functions are linear, the point $(3, -4)$ is the only point of intersection for the two graphs.

57. $y_1 = 3x - 5$
$y_2 = -4x + 2$
Both of the graphs in (b) have negative y-intercepts. But the y-intercept of y_2 is positive, so the graph in (b) is not acceptable. The slope of y_1 is positive, and y_1 has a negative y-intercept. The slope of y_2 is negative, and y_2 has a positive y-intercept. This fits graph (a).

59. **(a)** $\ x \ + \ y \ = \ 10$ (1)
$\ 2x \ - \ y \ = \ \ \ 5$ (2)

$$
\begin{array}{l}
3x \ \ \ \ \ \ \ \ = \ 15 \\
x \ = \ 5
\end{array}
$$

Substitute 5 for x in equation (1).

$$
\begin{array}{l}
x + y = 10 \quad (1) \\
5 + y = 10 \\
y = 5
\end{array}
$$

Solution set: $\{(5, 5)\}$

(b) Solve (1) and (2) for y.

$$
\begin{array}{ll}
x + y = 10 & (1) \\
y = -x + 10 & (3)
\end{array}
$$

$$
\begin{array}{ll}
2x - y = 5 & (2) \\
2x - 5 = y & (4)
\end{array}
$$

Now graph (3) and (4).

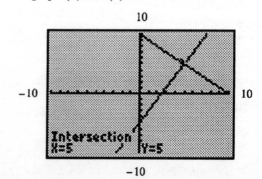

Section 4.2

61. **(a)** $3x - 2y = 4$ (1)
$3x + y = -2$ (2)

Multiply equation (2) by -1 and add the result to equation (1).

$$\begin{array}{rl}
3x - 2y = & 4 \quad (1) \\
-3x - y = & 2 \quad (3) \quad -1 \times (2) \\
\hline
-3y = & 6 \\
y = & -2
\end{array}$$

Substitute -2 for y in equation (1).

$$3x - 2y = 4 \quad (1)$$
$$3x - 2(-2) = 4$$
$$3x + 4 = 4$$
$$3x = 0$$
$$x = 0$$

Solution set: $\{(0, -2)\}$

(b) Solving (1) and (2) for y gives us

$$y = \frac{3}{2}x - 2 \quad (3)$$
$$\text{and} \quad y = -3x - 2. \quad (4)$$

Now graph (3) and (4).

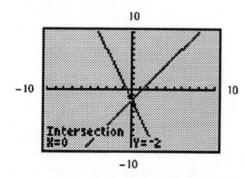

63. **(a)** The growth rates for the United States and Germany intersect at about 1% in the fourth quarter of 1991.

(b) The growth rates for the United States and Japan intersect at about 2.2% in the second quarter of 1992.

(c) The growth rates for Germany and Japan intersect twice in the period shown in the graph. Between those two times, Germany's growth rate was larger than Japan's.

65. **(a)** The graphs for CD production and cassette production intersect at about (1991, 350 million), so in the year 1991 they both were at the level of about 350 million.

(b) The graphs for CD production and LP production intersect at about (1987, 100 million).

1. Substitute 1 for x, 2 for y, and 3 for z in $3x + 2y - z$, which is the left side of each equation.

$$3(1) + 2(2) - (3) = 3 + 4 - 3$$
$$= 7 - 3$$
$$= 4$$

Choice (b) is correct since its right side is 4.

3. $3x + 2y + z = 8$ (1)
$2x - 3y + 2z = -16$ (2)
$x + 4y - z = 20$ (3)

Eliminate z by adding equations (1) and (3).

$$\begin{array}{rl}
3x + 2y + z = & 8 \quad (1) \\
x + 4y - z = & 20 \quad (3) \\
\hline
4x + 6y = & 28 \quad (4)
\end{array}$$

To get another equation without z, multiply equation (3) by 2 and add the result to equation (2).

$$\begin{array}{rl}
2x + 8y - 2z = & 40 \quad 2 \times (3) \\
2x - 3y + 2z = & -16 \quad (2) \\
\hline
4x + 5y = & 24 \quad (5)
\end{array}$$

Use equations (4) and (5) to eliminate x. Multiply equation (4) by -1 and add the result to equation (5).

$$\begin{array}{rl}
-4x - 6y = & -28 \quad -1 \times (4) \\
4x + 5y = & 24 \quad (5) \\
\hline
-y = & -4 \\
y = & 4
\end{array}$$

Substitute 4 for y in equation (5) to find x.

$$4x + 5y = 24 \quad (5)$$
$$4x + 5(4) = 24$$
$$4x + 20 = 24$$
$$4x = 4$$
$$x = 1$$

Substitute 1 for x and 4 for y in equation (3) to find z.

$$x + 4y - z = 20 \quad (3)$$
$$1 + 4(4) - z = 20$$
$$1 + 16 - z = 20$$
$$17 - z = 20$$
$$-z = 3$$
$$z = -3$$

The solution $(1, 4, -3)$ checks in all three of the original equations.

Solution set: $\{(1, 4, -3)\}$

5.
$$2x + 5y + 2z = 0 \quad (1)$$
$$4x - 7y - 3z = 1 \quad (2)$$
$$3x - 8y - 2z = -6 \quad (3)$$

Add equations (1) and (3) to eliminate z.

$$2x + 5y + 2z = 0 \quad (1)$$
$$\underline{3x - 8y - 2z = -6} \quad (3)$$
$$5x - 3y \qquad = -6 \quad (4)$$

Multiply equation (1) by 3 and equation (2) by 2. Then add the results to eliminate z again.

$$6x + 15y + 6z = 0 \qquad 3 \times (1)$$
$$\underline{8x - 14y - 6z = 2} \qquad 2 \times (2)$$
$$14x + y \qquad = 2 \quad (5)$$

Solve the system

$$5x - 3y = -6 \qquad (4)$$
$$14x + y = 2. \qquad (5)$$

Multiply equation (5) by 3 then add this result to (4).

$$5x - 3y = -6 \quad (4)$$
$$\underline{42x + 3y = 6} \qquad 3 \times (5)$$
$$47x \qquad = 0$$
$$x = 0$$

To find y, substitute $x = 0$ into equation (4).

$$5x - 3y = -6 \quad (4)$$
$$5(0) - 3y = -6$$
$$y = 2$$

To find z, substitute $x = 0$ and $y = 2$ in equation (1).

$$2x + 5y + 2z = 0 \qquad (1)$$
$$2(0) + 5(2) + 2z = 0$$
$$10 + 2z = 0$$
$$2z = -10$$
$$z = -5$$

Solution set: $\{(0, 2, -5)\}$

7.
$$x + y - z = -2 \quad (1)$$
$$2x - y + z = -5 \quad (2)$$
$$-x + 2y - 3z = -4 \quad (3)$$

Eliminate y and z by adding equations (1) and (2).

$$x + y - z = -2 \quad (1)$$
$$\underline{2x - y + z = -5} \quad (2)$$
$$3x \qquad = -7$$
$$x = -\frac{7}{3}$$

To get another equation without y, multiply equation (2) by 2 and add the result to equation (3).

$$4x - 2y + 2z = -10 \qquad 2 \times (2)$$
$$\underline{-x + 2y - 3z = -4} \quad (3)$$
$$3x \qquad - z = -14 \quad (4)$$

Substitute $-\dfrac{7}{3}$ for x in equation (4) to find z.

$$3x - z = -14 \quad (4)$$
$$3\left(-\frac{7}{3}\right) - z = -14$$
$$-7 - z = -14$$
$$-z = -7$$
$$z = 7$$

Substitute $-\dfrac{7}{3}$ for x and 7 for z in equation (1) to find y.

$$x + y - z = -2 \quad (1)$$
$$-\frac{7}{3} + y - 7 = -2$$
$$-7 + 3y - 21 = -6 \quad \textit{Multiply by 3.}$$
$$3y - 28 = -6$$
$$3y = 22$$
$$y = \frac{22}{3}$$

Solution set: $\left\{\left(-\dfrac{7}{3}, \dfrac{22}{3}, 7\right)\right\}$

A calculator check reduces the probability of making any arithmetic errors and is highly recommended. The following screen (similar to Exercise 13 solution in Section 4.1) shows the substitution of the solution for x, y, and z along with the three original equations. The evaluation of the three equations, -2, -5, and -4 (the right sides of the three equations), indicates that we have found the correct solution.

```
-7/3→X:22/3→Y:7→
Z:X+Y-Z
                    -2
2X-Y+Z
                    -5
-X+2Y-3Z
                    -4
```

9.
$$
\begin{aligned}
2x - 3y + 2z &= -1 \quad (1)\\
x + 2y + z &= 17 \quad (2)\\
2y - z &= 7 \quad (3)
\end{aligned}
$$

Multiply equation (2) by -2, and add the result to equation (1).

$$
\begin{array}{rl}
2x - 3y + 2z = -1 & (1)\\
-2x - 4y - 2z = -34 & -2 \times (2)\\
\hline
-7y = -35 &\\
y = 5 &
\end{array}
$$

To find z, substitute 5 for y in equation (3).

$$
\begin{aligned}
2y - z &= 7 \quad (3)\\
2(5) - z &= 7\\
10 - z &= 7\\
-z &= -3\\
z &= 3
\end{aligned}
$$

To find x, substitute $y = 5$ and $z = 3$ into equation (1).

$$
\begin{aligned}
2x - 3y + 2z &= -1 \quad (1)\\
2x - 3(5) + 2(3) &= -1\\
2x - 9 &= -1\\
2x &= 8\\
x &= 4
\end{aligned}
$$

Solution set: $\{(4, 5, 3)\}$

11.
$$
\begin{aligned}
4x + 2y - 3z &= 6 \quad (1)\\
x - 4y + z &= -4 \quad (2)\\
-x + 2z &= 2 \quad (3)
\end{aligned}
$$

Equation (3) is missing y. Eliminate y again by multiplying equation (1) by 2 and adding the result to equation (2).

$$
\begin{array}{rl}
8x + 4y - 6z = 12 & 2 \times (1)\\
x - 4y + z = -4 & (2)\\
\hline
9x - 5z = 8 & (4)
\end{array}
$$

Use equations (3) and (4) to eliminate x. Multiply equation (3) by 9 and add the result to equation (4).

$$
\begin{array}{rl}
-9x + 18z = 18 & 9 \times (3)\\
9x - 5z = 8 & (4)\\
\hline
13z = 26 &\\
z = 2 &
\end{array}
$$

Substitute 2 for z in equation (3) to find x.

$$
\begin{aligned}
-x + 2z &= 2 \quad (3)\\
-x + 2(2) &= 2\\
-x + 4 &= 2\\
-x &= -2\\
x &= 2
\end{aligned}
$$

Substitute 2 for x and 2 for z in equation (2) to find y.

$$
\begin{aligned}
x - 4y + z &= -4 \quad (2)\\
2 - 4y + 2 &= -4\\
-4y + 4 &= -4\\
-4y &= -8\\
y &= 2
\end{aligned}
$$

Solution set: $\{(2, 2, 2)\}$

13.
$$
\begin{aligned}
2x + y &= 6 \quad (1)\\
3y - 2z &= -4 \quad (2)\\
3x - 5z &= -7 \quad (3)
\end{aligned}
$$

To eliminate y, multiply equation (1) by -3 and add the result to equation (2).

$$
\begin{array}{rl}
-6x - 3y = -18 & -3 \times (1)\\
3y - 2z = -4 & (2)\\
\hline
-6x - 2z = -22 & (4)
\end{array}
$$

Since equation (3) does not have a y-term, we can multiply equation (3) by 2 and add the result to equation (4) to eliminate x and solve for z.

$$
\begin{array}{rl}
6x - 10z = -14 & 2 \times (3)\\
-6x - 2z = -22 & (4)\\
\hline
-12z = -36 &\\
z = 3 &
\end{array}
$$

To find x, substitute $z = 3$ into equation (3).

$$
\begin{aligned}
3x - 5z &= -7 \quad (3)\\
3x - 5(3) &= -7\\
3x &= 8\\
x &= \frac{8}{3}
\end{aligned}
$$

To find y, substitute $z = 3$ into equation (2).

$$
\begin{aligned}
3y - 2z &= -4 \quad (2)\\
3y - 2(3) &= -4\\
3y &= 2\\
y &= \frac{2}{3}
\end{aligned}
$$

Solution set: $\left\{ \left(\dfrac{8}{3}, \dfrac{2}{3}, 3 \right) \right\}$

15. Answers will vary.

(a) One example is two connecting walls and the ceiling (or floor) of a room, which meet at one point.

(b) One example is two opposite walls and the floor, which have no points in common. Another is the floors of three different levels of an office building.

(c) One example is the plane through the ceiling of a house, the plane through an outside wall, and the plane through one side of a slanted roof, which all meet in one line; therefore, they intersect in infinitely many points. Another example is three pages of this book, which intersect in the spine.

17.

$$
\begin{aligned}
2x + 2y - 6z &= 5 \quad (1)\\
-3x + y - z &= -2 \quad (2)\\
-x - y + 3z &= 4 \quad (3)
\end{aligned}
$$

Multiply equation (3) by 2 and add the result to equation (1).

$$
\begin{array}{ll}
2x + 2y - 6z = 5 & (1)\\
-2x - 2y + 6z = 8 & 2 \times (3)\\
\hline
\qquad\qquad 0 = 13 & \textit{False}
\end{array}
$$

Solution set: $\emptyset$

19.

$$
\begin{aligned}
-5x + 5y - 20z &= -40 \quad (1)\\
x - y + 4z &= 8 \quad (2)\\
3x - 3y + 12z &= 24 \quad (3)
\end{aligned}
$$

Dividing equation (1) by -5 gives equation (2). Dividing equation (3) by 3 also gives equation (2). The resulting equations are the same, so the three equations are dependent. Solution set:

$$\{(x, y, z) \mid x - y + 4z = 8\}$$

21.

$$
\begin{aligned}
2x + y - z &= 6 \quad (1)\\
4x + 2y - 2z &= 12 \quad (2)\\
-x - \tfrac{1}{2}y + \tfrac{1}{2}z &= -3 \quad (3)
\end{aligned}
$$

Multiplying equation (1) by 2 gives equation (2). Multiplying equation (3) by -4 also gives equation (2). The resulting equations are the same, so the three equations are dependent. Solution set:

$$\{(x, y, z) \mid 2x + y - z = 6\}$$

23.

$$
\begin{aligned}
x + y - 2z &= 0 \quad (1)\\
3x - y + z &= 0 \quad (2)\\
4x + 2y - z &= 0 \quad (3)
\end{aligned}
$$

Eliminate z by adding equations (2) and (3).

$$
\begin{array}{ll}
3x - y + z = 0 & (2)\\
4x + 2y - z = 0 & (3)\\
\hline
7x + y \qquad = 0 & (4)
\end{array}
$$

To get another equation without z, multiply equation (2) by 2 and add the result to equation (1).

$$
\begin{array}{ll}
6x - 2y + 2z = 0 & 2 \times (2)\\
x + y - 2z = 0 & (1)\\
\hline
7x - y \qquad = 0 & (5)
\end{array}
$$

Add equations (4) and (5) to find x.

$$
\begin{array}{ll}
7x + y = 0 & (4)\\
7x - y = 0 & (5)\\
\hline
14x \qquad = 0 &\\
x = 0 &
\end{array}
$$

Substitute 0 for x in equation (4) to find y.

$$
\begin{aligned}
7x + y &= 0 \quad (4)\\
7(0) + y &= 0\\
0 + y &= 0\\
y &= 0
\end{aligned}
$$

Substitute 0 for x and 0 for y in equation (1) to find z.

$$
\begin{aligned}
x + y - 2z &= 0 \quad (1)\\
0 + 0 - 2z &= 0\\
-2z &= 0\\
z &= 0
\end{aligned}
$$

Solution set: $\{(0, 0, 0)\}$

25.

$$
\begin{aligned}
x + y + z - w &= 5 \quad (1)\\
2x + y - z + w &= 3 \quad (2)\\
x - 2y + 3z + w &= 18 \quad (3)\\
-x - y + z + 2w &= 8 \quad (4)
\end{aligned}
$$

Eliminate w. Add equations (1) and (2).

$$
\begin{array}{ll}
x + y + z - w = 5 & (1)\\
2x + y - z + w = 3 & (2)\\
\hline
3x + 2y \qquad\qquad = 8 & (5)
\end{array}
$$

Eliminate w again. Add equations (1) and (3).

$$
\begin{array}{ll}
x + y + z - w = 5 & (1)\\
x - 2y + 3z + w = 18 & (3)\\
\hline
2x - y + 4z \qquad = 23 & (6)
\end{array}
$$

Eliminate w again. Multiply equation (2) by -2. Add the result to equation (4).

$$
\begin{array}{ll}
-4x - 2y + 2z - 2w = -6 & -2 \times (2)\\
-x - y + z + 2w = 8 & (4)\\
\hline
-5x - 3y + 3z \qquad = 2 & (7)
\end{array}
$$

Equations (5), (6), and (7) do not contain a w-term. Since (5) does not have a z-term, we will find another equation without a z-term.

Eliminate z. Multiply equation (6) by 3 and equation (7) by -4. Then add the results.

$$6x - 3y + 12z = 69 \qquad 3 \times (6)$$
$$\underline{20x + 12y - 12z = -8 \qquad -4 \times (7)}$$
$$26x + 9y \qquad = 61 \quad (8)$$

Eliminate z again. Multiply equation (5) by 9 and equation (8) by -2. Then add the results.

$$27x + 18y = 72 \qquad 9 \times (5)$$
$$\underline{-52x - 18y = -122 \quad -2 \times (8)}$$
$$-25x \qquad = -50$$
$$x = 2$$

To find y, substitute $x = 2$ into equation (5).

$$3x + 2y = 8 \quad (5)$$
$$3(2) + 2y = 8$$
$$2y = 2$$
$$y = 1$$

To find z, substitute $x = 2$ and $y = 1$ into equation (6).

$$2x - y + 4z = 23 \quad (6)$$
$$2(2) - 1 + 4z = 23$$
$$4z = 20$$
$$z = 5$$

To find w, substitute $x = 2$, $y = 1$, and $z = 5$ into equation (1).

$$x + y + z - w = 5 \quad (1)$$
$$2 + 1 + 5 - w = 5$$
$$-w = -3$$
$$w = 3$$

Solution set: $\{(2, 1, 5, 3)\}$

For Exercises 27–36,

$$f(x) = ax^2 + bx + c \quad (a \neq 0).$$

27. $f(1) = a(1)^2 + b(1) + c$
$$= a + b + c$$
Since $f(1) = 128$, the first equation is
$$a + b + c = 128.$$

28. $f(1.5) = a(1.5)^2 + b(1.5) + c$
$$= 2.25a + 1.5b + c$$
Since $f(1.5) = 140$, the second equation is
$$2.25a + 1.5b + c = 140.$$

29. $f(3) = a(3)^2 + b(3) + c$
$$= 9a + 3b + c$$
Since $f(3) = 80$, the third equation is
$$9a + 3b + c = 80.$$

30. Using the three equations from Exercises 27–29, the system is

$$a + b + c = 128 \quad (1)$$
$$2.25a + 1.5b + c = 140 \quad (2)$$
$$9a + 3b + c = 80. \quad (3)$$

Multiply equation (2) by -1 and add the result to equation (1).

$$a + b + c = 128 \quad (1)$$
$$\underline{-2.25a - 1.5b - c = -140 \quad (4)}$$
$$-1.25a - .5b \qquad = -12 \quad (5)$$

Multiply equation (3) by -1 and add the result to equation (1).

$$a + b + c = 128 \quad (1)$$
$$\underline{-9a - 3b - c = -80 \quad (6)}$$
$$-8a - 2b \qquad = 48 \quad (7)$$

Use equations (5) and (7) to eliminate b. Multiply equation (5) by -4 and add the result to equation (7).

$$5a + 2b = 48 \qquad -4 \times (5)$$
$$\underline{-8a - 2b = 48 \qquad (7)}$$
$$-3a \qquad = 96$$
$$a = -32$$

To find b, substitute $a = -32$ into equation (7).

$$-8a - 2b = 48 \qquad (7)$$
$$-8(-32) - 2b = 48$$
$$256 - 2b = 48$$
$$-2b = -208$$
$$b = 104$$

To find c, substitute $a = -32$ and $b = 104$ into equation (1).

$$a + b + c = 128 \quad (1)$$
$$-32 + 104 + c = 128$$
$$72 + c = 128$$
$$c = 56$$

Solution set: $\{(-32, 104, 56)\}$

31. If $(a, b, c) = (-32, 104, 56)$, then
$$f(x) = -32x^2 + 104x + 56.$$

32. $f(x) = -32x^2 + 104x + 56$
$$f(0) = -32(0)^2 + 104(0) + 56$$
$$= 56$$
The initial height is 56 ft.

33. $f(1.625) = -32(1.625)^2 + 104(1.625) + 56$
$$= -84.5 + 169 + 56$$
$$= 140.5$$
The maximum height is 140.5 ft.

34. $f(3.25) = -32(3.25)^2 + 104(3.25) + 56$

$$= -338 + 338 + 56$$

$$= 56$$

It tells us the projectile is at its original height after 3.25 seconds.

35. $f(x) = ax^2 + bx + c$

$f(1) = a(1)^2 + b(1) + c$

$\quad = a + b + c$

So, $a + b + c = 2$.   (1)

$f(-1) = a(-1)^2 + b(-1) + c$

$\quad = a - b + c$

So, $a - b + c = 0$.   (2)

$f(-2) = a(-2)^2 + b(-2) + c$

$\quad = 4a - 2b + c$

So, $4a - 2b + c = 8$.   (3)

Add equations (1) and (2).

$$
\begin{array}{rcll}
a + b + c &=& 2 & (1) \\
a - b + c &=& 0 & (2) \\
\hline
2a \quad + 2c &=& 2 & (4)
\end{array}
$$

Multiply equation (1) by 2 and add the result to equation (3).

$$
\begin{array}{rcll}
2a + 2b + 2c &=& 4 & 2 \times (1) \\
4a - 2b + c &=& 8 & (3) \\
\hline
6a \quad + 3c &=& 12 & (5)
\end{array}
$$

Multiply equation (4) by -3 and add the result to equation (5).

$$
\begin{array}{rcll}
-6a - 6c &=& -6 & -3 \times (4) \\
6a + 3c &=& 12 & (5) \\
\hline
- 3c &=& 6 & \\
c &=& -2 &
\end{array}
$$

To find a, substitute $c = -2$ into equation (4).

$$
\begin{aligned}
2a + 2c &= 2 \quad (4) \\
2a + 2(-2) &= 2 \\
2a - 4 &= 2 \\
2a &= 6 \\
a &= 3
\end{aligned}
$$

To find b, substitute $a = 3$ and $c = -2$ into equation (1).

$$
\begin{aligned}
a + b + c &= 2 \quad (1) \\
3 + b - 2 &= 2 \\
b + 1 &= 2 \\
b &= 1
\end{aligned}
$$

So, $a = 3$, $b = 1$, and $c = -2$ and the equation is

$$
\begin{aligned}
f(x) &= ax^2 + bx + c \\
f(x) &= 3x^2 + x - 2.
\end{aligned}
$$

36. $Y_1 = ax^2 + bx + c$

Using $(1, 8)$ gives

$a(1)^2 + b(1) + c = 8$

$\quad a + b + c = 8$.   (1)

Using $(2, 15)$ gives

$a(2)^2 + b(2) + c = 15$

$\quad 4a + 2b + c = 15$.   (2)

Using $(3, 24)$ gives

$a(3)^2 + b(3) + c = 24$

$\quad 9a + 3b + c = 24$.   (3)

The system is

$$
\begin{array}{rcll}
a + b + c &=& 8 & (1) \\
4a + 2b + c &=& 15 & (2) \\
9a + 3b + c &=& 24. & (3)
\end{array}
$$

Multiply equation (2) by -1 and add the result to equation (1).

$$
\begin{array}{rcll}
a + b + c &=& 8 & (1) \\
-4a - 2b - c &=& -15 & -1 \times (2) \\
\hline
-3a - b &=& -7 & (4)
\end{array}
$$

Multiply equation (3) by -1 and add the result to equation (1).

$$
\begin{array}{rcll}
a + b + c &=& 8 & (1) \\
-9a - 3b - c &=& -24 & -1 \times (3) \\
\hline
-8a - 2b &=& -16 &
\end{array}
$$

Divide by -2.

$$4a + b = 8 \quad (5)$$

Add equations (4) and (5).

$$
\begin{array}{rcll}
-3a - b &=& -7 & (4) \\
4a + b &=& 8 & (5) \\
\hline
a &=& 1 &
\end{array}
$$

To find b, substitute $a = 1$ into equation (4).

$$
\begin{aligned}
-3a - b &= -7 \quad (4) \\
-3(1) - b &= -7 \\
-3 - b &= -7 \\
-b &= -4 \\
b &= 4
\end{aligned}
$$

To find c, substitute $a = 1$ and $b = 4$ into equation (1).

$$
\begin{aligned}
a + b + c &= 8 \quad (1) \\
1 + 4 + c &= 8 \\
c &= 3
\end{aligned}
$$

So, $a = 1$, $b = 4$, and $c = 3$, and the equation is

$$
\begin{aligned}
Y_1 &= ax^2 + bx + c \\
Y_1 &= x^2 + 4x + 3.
\end{aligned}
$$

37. If one were to eliminate *different* variables in the first two steps, the result would be two equations in three variables, and it would not be possible to solve for a single variable in the next step.

Section 4.3

1. *Step 1*
Let x = the number of games won and y = the number of games lost.

Step 2
Not necessary.

Step 3
They played 82 games, so

$$x + y = 82. \quad (1)$$

They won 56 more games than they lost, so

$$x = 56 + y. \quad (2)$$

Step 4
Substitute $56 + y$ for x in (1).

$$(56 + y) + y = 82$$
$$56 + 2y = 82$$
$$2y = 26$$
$$y = 13$$

Substitute 13 for y in (2),

$$x = 56 + 13 = 69$$

Step 5
The Bulls win-loss record during the 1996–1997 N.B.A. regular season was 69 wins and 13 losses.

Step 6
69 is 56 more than 13 and the sum of 69 and 13 is 82. The solution is correct.

3. *Step 1*
Let x = the number of wins,
y = the number of losses,
and z = the number of ties.

Step 2
Not necessary.

Step 3
They played 82 games, so

$$x + y + z = 82. \quad (1)$$

Their wins and losses totaled 74, so

$$x + y = 74. \quad (2)$$

They tied 18 fewer games than they lost, so

$$z = y - 18. \quad (3)$$

Step 4
Multiply (2) by -1 and add to (1).

$$\begin{array}{rcll} x + y + z &=& 82 & (1) \\ -x - y &=& -74 & -1 \times (2) \\ \hline z &=& 8 & \end{array}$$

Substitute 8 for z in (3).

$$8 = y - 18$$
$$26 = y$$

Substitute 26 for y in (2).

$$x + 26 = 74$$
$$x = 48$$

Step 5
The Stars won 48 games, lost 26 games, and tied 8 games.

Step 6
Adding 48, 26, and 8 gives 82 total games. The wins and losses add up to 74, and there were 18 fewer ties than losses. The solution is correct.

5. Let W = the width of the tennis court and L = the length.

Since the length is 42 ft more than the width,

$$L = W + 42. \quad (1)$$

The perimeter of a rectangle is given by

$$2W + 2L = P.$$

With perimeter $P = 228$ ft,

$$2W + 2L = 228. \quad (2)$$

Substitute $W + 42$ for L in equation (2).

$$2W + 2(W + 42) = 228$$
$$2W + 2W + 84 = 228$$
$$4W = 144$$
$$W = 36$$

Substitute $W = 36$ into equation (1).

$$L = W + 42 \quad (1)$$
$$L = 36 + 42$$
$$= 78$$

The length is 78 ft; the width is 36 ft.

7.   Let $x =$ the length of the original rectangle
and $y =$ the width of the original rectangle.

The length x is 7 ft more than the width y so,

$$x = 7 + y. \quad (1)$$

The perimeter of a rectangle is $P = 2L + 2W$.
Here, $P = 32$, $L = x - 3$, and $W = y + 2$ for the
new rectangle, so

$$32 = 2(x - 3) + 2(y + 2)$$
$$32 = 2x - 6 + 2y + 4$$
$$34 = 2x + 2y$$
or $x + y = 17. \quad (2)$

Solve the system. From (1), substitute $7 + y$ for x
in (2).

$$(7 + y) + y = 17$$
$$7 + 2y = 17$$
$$2y = 10$$
$$y = 5$$

From (1),

$$x = 7 + y = 7 + 5 = 12.$$

The length of the rectangle is 12 ft, and the width
is 5 ft.

9.   From the figure in the text, the angles marked y
and $3x + 10$ are supplementary, so

$$(3x + 10) + y = 180. \quad (1)$$

Also, the angles x and y are complementary, so

$$x + y = 90. \quad (2)$$

Solve equation (2) for y to get

$$y = 90 - x. \quad (3)$$

Substitute $90 - x$ for y in equation (1).

$$(3x + 10) + (90 - x) = 180$$
$$2x + 100 = 180$$
$$2x = 80$$
$$x = 40$$

Substitute $x = 40$ into equation (3) to get

$$y = 90 - x = 90 - 40 = 50.$$

The angles measure 40° and 50°.

11.   Let $x =$ the number of daily newspapers
in Texas
and $y =$ the number of daily newspapers
in Florida.

Since the two states had a total number of 134
dailies,

$$x + y = 134. \quad (1)$$

Since Texas had 52 more daily newspapers than
Florida,

$$x = y + 52. \quad (2)$$

Substitute $y + 52$ for x in equation (1).

$$(y + 52) + y = 134$$
$$2y + 52 = 134$$
$$2y = 82$$
$$y = 41$$

Substitute $y = 41$ into equation (2).

$$x = 41 + 52 = 93$$

Texas had 93 daily newspapers and Florida 41.

13.   Let $x =$ the hockey FCI
and $y =$ the basketball FCI.

The sum is $423.12, so

$$x + y = 423.12. \quad (1)$$

The hockey FCI was $16.36 more than the
basketball FCI, so

$$x = y + 16.36. \quad (2)$$

From (2), substitute $y + 16.36$ for x in (1).

$$(y + 16.36) + y = 423.12$$
$$2y + 16.36 = 423.12$$
$$2y = 406.76$$
$$y = 203.38$$

From (2),

$$x = y + 16.36 = 203.38 + 16.36 = 219.74.$$

The hockey FCI was $219.74 and the basketball
FCI was $203.38.

15.   Let $x =$ the price of a CGA monitor
and $y =$ the price of a VGA monitor.

For the first purchase, the total cost of 4 CGA
monitors and 6 VGA monitors is $4600, so

$$4x + 6y = 4600. \quad (1)$$

For the other purchase, the total cost of 6 CGA
monitors and 4 VGA monitors is $4400, so

$$6x + 4y = 4400. \quad (2)$$

To solve the system, multiply equation (1) by -2
and equation (2) by 3. Then add the results.

$$
\begin{array}{rcl r}
-8x - 12y &=& -9200 & -2 \times (1) \\
18x + 12y &=& 13,200 & 3 \times (2) \\
\hline
10x &=& 4000 \\
x &=& 400
\end{array}
$$

Since $x = 400$,

$$4x + 6y = 4600 \quad (1)$$
$$4(400) + 6y = 4600$$
$$1600 + 6y = 4600$$
$$6y = 3000$$
$$y = 500.$$

A CGA monitor costs $400, and a VGA monitor costs $500.

17. Let x = the number of units of yarn, and y = the number of units of thread.

Make a chart to organize the information in the problem.

	Yarn	Thread	Total Hours
Hours on Machine A	1	1	8
Hours on Machine B	2	1	14

From the chart, write a system of equations.

$$x + y = 8 \quad (1)$$
$$2x + y = 14 \quad (2)$$

Solve the system. Multiply equation (1) by -1 and add the result to equation (2).

$$\begin{array}{r} -x - y = -8 \quad -1 \times (1) \\ 2x + y = 14 \quad (2) \\ \hline x \qquad = 6 \end{array}$$

Substitute $x = 6$ into equation (1) to get $y = 2$. The factory should make 6 units of yarn and 2 units of thread to keep its machines running at capacity.

19. Use the formula (rate of percent) • (base amount) = amount (percentage) of pure acid to compute parts (a) – (d).

(a) $.10(60) = 6$ oz

(b) $.25(60) = 15$ oz

(c) $.40(60) = 24$ oz

(d) $.50(60) = 30$ oz

21. The cost is the price per pound, $.58, times the number of pounds, x, or $.58x$.

23. Let x = the amount of 25% alcohol solution, and y = the amount of 35% alcohol solution.

Make a table. The percent times the amount of solution gives the amount of pure alcohol in the third column.

Kind of Solution	Gallons of Solution	Amount of Pure Alcohol
.25	x	$.25x$
.35	y	$.35y$
.32	20	$.32(20) = 6.4$

The third row gives the total amounts of solution and pure alcohol. From the columns in the table, write a system of equations.

$$x + y = 20 \quad (1)$$
$$.25x + .35y = 6.4 \quad (2)$$

Solve the system. Multiply equation (1) by -25 and equation (2) by 100. Then add the results.

$$\begin{array}{r} -25x - 25y = -500 \quad -25 \times (1) \\ 25x + 35y = 640 \quad 100 \times (2) \\ \hline 10y = 140 \\ y = 14 \end{array}$$

Substitute $y = 14$ into equation (1).

$$x + y = 20 \quad (1)$$
$$x + 14 = 20$$
$$x = 6$$

Mix 6 gal of 25% solution and 14 gal of 35% solution.

25. Let x = the amount of pure acid and y = the amount of 10% acid.

Make a table.

Kind of Solution	Liters of Solution	Amount of Pure Acid
100% = 1	x	$1.00x = x$
.10	y	$.10y$
.20	27	$.20(27) = 5.4$

Solve the following system.

$$x + y = 27 \quad (1)$$
$$x + .10y = 5.4 \quad (2)$$

Multiply equation (2) by 10 to clear the decimals.

$$10x + y = 54 \quad (3)$$

To eliminate y, multiply equation (1) by -1 and add the result to equation (3).

$$\begin{array}{r} -x - y = -27 \quad -1 \times (1) \\ 10x + y = 54 \quad (3) \\ \hline 9x = 27 \\ x = 3 \end{array}$$

Since $x = 3$,

$$x + y = 27 \quad (1)$$
$$3 + y = 27$$
$$y = 24.$$

Use 3 L of pure acid and 24 L of 10% acid.

27. From the "Number of Pounds" column in the text,

$$x + y = 80. \qquad (1)$$

From the "Value of Candy" column in the text,

$$3.60x + 7.20y = 4.95(80). \qquad (2)$$

Solve the system.

$$
\begin{array}{rcll}
-36x - 36y &=& -2880 & -36 \times (1) \\
36x + 72y &=& 3960 & 10 \times (2) \\
\hline
36y &=& 1080 & \\
y &=& 30 &
\end{array}
$$

From (1), $x = 50$.
She should mix 50 lb of \$3.60/lb pecan clusters with 30 lb of \$7.20/lb chocolate truffles.

29. Let $x =$ the number of general admission tickets and $y =$ the number of student tickets.

Make a table.

Ticket	Number	Value of Tickets
General	x	$2.50x = 2.5x$
Student	y	$2.00y = 2y$
Total	184	406

Solve the system.

$$
\begin{array}{rl}
x + y = 184 & (1) \\
2.5x + 2y = 406 & (2)
\end{array}
$$

Multiply equation (2) by 10 to clear the decimal.

$$25x + 20y = 4060 \qquad (3)$$

To eliminate y, multiply equation (1) by -20 and add the result to equation (3).

$$
\begin{array}{rcll}
-20x - 20y &=& -3680 & \\
25x + 20y &=& 4060 & (3) \\
\hline
5x &=& 380 & \\
x &=& 76 &
\end{array}
$$

From (1),

$$76 + y = 184, \text{ so } y = 108.$$

76 general admission tickets and 108 student tickets were sold.

31. Let $x =$ the number of dimes and $y =$ the number of quarters.

Make a table.

Coin	Amount	Value
Dimes (\$.10)	x	$.10x$
Quarters (\$.25)	y	$.25y$
Total	94	19.30

From the table, write a system of equations.

$$
\begin{array}{rl}
x + y = 94 & (1) \\
.10x + .25y = 19.30 & (2)
\end{array}
$$

Solve the system. Multiply equation (1) by -10 and equation (2) by 100. Then add the results.

$$
\begin{array}{rcll}
-10x - 10y &=& -940 & -10 \times (1) \\
10x + 25y &=& 1930 & 100 \times (2) \\
\hline
15y &=& 990 & \\
y &=& 66 &
\end{array}
$$

From (1),

$$x + 66 = 94, \text{ so } x = 28.$$

She has 28 dimes and 66 quarters.

33. From the "Principal" column in the text,

$$x + y = 3000. \qquad (1)$$

From the "Interest" column in the text,

$$.02x + .04y = 100. \qquad (2)$$

Multiply equation (2) by 100 to clear the decimals.

$$2x + 4y = 10,000 \qquad (3)$$

To eliminate x, multiply equation (1) by -2 and add the result to equation (3).

$$
\begin{array}{rcll}
-2x - 2y &=& -6000 & \\
2x + 4y &=& 10,000 & (3) \\
\hline
2y &=& 4000 & \\
y &=& 2000 &
\end{array}
$$

From (1), $x = 1000$.
\$1000 is invested at 2%, and \$2000 is invested at 4%.

35. In the formula $d = rt$, substitute 25 for r (rate or speed) and y for t (time in hours) to get

$$d = 25y \text{ mi}.$$

37. Let $x =$ speed of the freight train and $y =$ the speed of the express train.

Make a table.

	r	t	d
Freight train	x	3	$3x$
Express train	y	3	$3y$

Since $d = rt$ and the trains are 390 km apart,

$$3x + 3y = 390. \qquad (1)$$

The freight train travels 30 km/hr slower than the express train, so

$$x = y - 30. \qquad (2)$$

Substitute $y - 30$ for x in equation (1) and solve for y.

$$3x + 3y = 390 \quad (1)$$
$$3(y - 30) + 3y = 390$$
$$3y - 90 + 3y = 390$$
$$6y = 480$$
$$y = 80$$

From (2), $x = 80 - 30 = 50$.
The freight train travels at 50 km/hr, while the express train travels at 80 km/hr.

39. Let $x =$ the top speed of the snow speeder, and
$y =$ the speed of the wind.

Furthermore,

rate into headwind $= x - y$

and rate with tailwind $= x + y$.

Use these rates and the information in the problem to make a table.

	d	r	t
Into headwind	3600	$x - y$	2
With tailwind	3600	$x + y$	1.5

From the table, use the formula $d = rt$ to write a system of equations.

$$3600 = 2(x - y)$$
$$3600 = 1.5(x + y)$$

Remove parentheses and move the variables to the left side.

$$2x - 2y = 3600 \quad (1)$$
$$1.5x + 1.5y = 3600 \quad (2)$$

Solve the system.

$$
\begin{array}{rcll}
6x & - \ 6y & = \ 10,800 & (1) \times 3 \\
6x & + \ 6y & = \ 14,400 & (2) \times 4 \\
\hline
12x & & = \ 25,200 & \\
x & & = \ 2100 &
\end{array}
$$

Substitute $x = 2100$ into equation (1).

$$2(2100) - 2y = 3600$$
$$4200 - 2y = 3600$$
$$-2y = -600$$
$$y = 300$$

The top speed of the snow speeder is 2100 miles per hour and the speed of the wind is 300 miles per hour.

41. Let $x =$ the number of \$20 fish,
$y =$ the number of \$40 fish,
and $z =$ the number of \$65 fish.

The number of \$40 fish is one less than twice the number of \$20 fish, so

$$y = 2x - 1. \quad (1)$$

The number of fish totals 29, so

$$x + y + z = 29. \quad (2)$$

The fish are worth \$1150, so

$$20x + 40y + 65z = 1150. \quad (3)$$

From (1), substitute $2x - 1$ for y in (2) and (3).

$$
\begin{array}{rl}
x + y + z = 29 & (2) \\
x + (2x - 1) + z = 29 & \\
3x + z = 30 & (4)
\end{array}
$$

$$
\begin{array}{rl}
20x + 40y + 65z = 1150 & (3) \\
20x + 40(2x - 1) + 65z = 1150 & \\
20x + 80x - 40 + 65z = 1150 & \\
100x + 65z = 1190 & (5)
\end{array}
$$

Multiply (4) by -65 and add the result to (5).

$$
\begin{array}{rcll}
-195x & - \ 65z & = \ -1950 & -65 \times (4) \\
100x & + \ 65z & = \ 1190 & (5) \\
\hline
-95x & & = \ -760 & \\
x & & = \ 8 &
\end{array}
$$

From (1), $y = 2(8) - 1 = 15$.
From (2), $z = 29 - x - y$
$$= 29 - 8 - 15 = 6.$$
There are 8 \$20 fish, 15 \$40 fish, and 6 \$65 fish in the collection.

43. Let $x =$ the measure of one angle,
$y =$ the measure of another angle,
and $z =$ the measure of the last angle.

Two equations are given, so

$$z = x + 10$$

or $-x + z = 10 \quad (1)$

and $x + y = 100. \quad (2)$

Since the sum of the measures of the angles of a triangle is 180°, the third equation of the system is

$$x + y + z = 180. \quad (3)$$

Equation (1) is missing y. To eliminate y again, multiply equation (2) by -1 and add the result to equation (3).

$$
\begin{array}{rcll}
-x & - \ y & = \ -100 & -1 \times (2) \\
x & + \ y + z & = \ 180 & (3) \\
\hline
& z & = \ 80 &
\end{array}
$$

Since $z = 80$,

$$
\begin{array}{rl}
-x + z = 10 & (1) \\
-x + 80 = 10 & \\
-x = -70 & \\
x = 70. &
\end{array}
$$

continued

From (2), $y = 30$.
The measures of the angles are 70°, 30°, and 80°.

45. Let $x =$ the measure of the first angle,
$y =$ the measure of the second angle, and
$z =$ the measure of the third angle.

The sum of the angles in a triangle equals 180°, so

$$x + y + z = 180. \quad (1)$$

The measure of the second angle is 10° more than 3 times that of the first angle, so

$$y = 3x + 10. \quad (2)$$

The third angle is equal to the sum of the other two, so

$$z = x + y. \quad (3)$$

Solve the system. Substitute z for $x + y$ in equation (1).

$$(x + y) + z = 180 \quad (1)$$
$$z + z = 180$$
$$2z = 180$$
$$z = 90$$

Substitute $z = 90$ and $3x + 10$ for y in equation (3).

$$z = x + y \qquad (3)$$
$$90 = x + (3x + 10)$$
$$80 = 4x$$
$$20 = x$$

Substitute $x = 20$ and $z = 90$ into equation (3).

$$z = x + y \quad (3)$$
$$90 = 20 + y$$
$$70 = y$$

The three angles have measures of 20°, 70°, and 90°.

47. Let $x =$ the length of the longest side,
$y =$ the length of the middle side,
and $z =$ the length of the shortest side.

Perimeter is the sum of the measures of the sides, so

$$x + y + z = 70. \quad (1)$$

The longest side is 4 cm less than the sum of the other sides, so

$$x = y + z - 4$$
$$\text{or} \quad x - y - z = -4. \qquad (2)$$

Twice the shortest side is 9 cm less than the longest side, so

$$2z = x - 9$$
$$\text{or} \quad -x + 2z = -9 \quad (3)$$

Add equations (1) and (2) to eliminate y and z.

$$\begin{array}{rcrcrcr} x & + & y & + & z & = & 70 \quad (1) \\ x & - & y & - & z & = & -4 \quad (2) \\ \hline 2x & & & & & = & 66 \\ & & & & x & = & 33 \end{array}$$

Since $x = 33$,

$$-x + 2z = -9 \quad (3)$$
$$-33 + 2z = -9$$
$$2z = 24$$
$$z = 12.$$

Since $x = 33$ and $z = 12$,

$$x + y + z = 70 \quad (1)$$
$$33 + y + 12 = 70$$
$$y + 45 = 70$$
$$y = 25.$$

The shortest side is 12 cm long, the middle side is 25 cm long, and the longest side is 33 cm long.

49. Let $x =$ the number of cases sent to wholesaler A,
$y =$ the number of cases sent to wholesaler B, and
$z =$ the number of cases sent to wholesaler C.

The total output is 320 cases per day, so

$$x + y + z = 320. \quad (1)$$

The number of cases to A is three times that sent to B, so

$$x = 3y. \quad (2)$$

Wholesaler C gets 160 cases less than the sum sent to A and B, so

$$z = x + y - 160. \quad (3)$$

Solve equation (3) for $x + y$ to get

$$x + y = z + 160. \quad (4)$$

Substitute $z + 160$ for $x + y$ in equation (1).

$$(x + y) + z = 320 \quad (1)$$
$$(z + 160) + z = 320$$
$$2z = 160$$
$$z = 80$$

Substitute 80 for z and $3y$ for x in equation (3).

$$z = x + y - 160 \quad (3)$$
$$80 = 3y + y - 160$$
$$240 = 4y$$
$$60 = y$$

Substitute $y = 60$ into equation (2).

$$x = 3y = 3(60) = 180$$

She should send 180 cases to Wholesaler A, 60 cases to Wholesaler B, and 80 cases to Wholesaler C.

51. Let $x =$ the amount of jelly beans,
$\quad\quad y =$ the amount of chocolate eggs,
and $z =$ the amount of marshmallow chicks.

The manager plans to make 15 lb of the mixture, so

$$x + y + z = 15. \quad (1)$$

She uses twice as many pounds of jelly beans as eggs and chicks, so

$$x = 2(y + z),$$
$$\text{or} \quad x - 2y - 2z = 0, \quad (2)$$

and five times as many pounds of jelly beans as eggs, so

$$x = 5y. \quad (3)$$

Solve the system of equations (1), (2), and (3). Multiply equation (1) by 2 and add the result to equation (2).

$$
\begin{array}{rcll}
2x + 2y + 2z &=& 30 & 2 \times (1) \\
x - 2y - 2z &=& 0 & (2) \\
\hline
3x \quad\quad\quad &=& 30 & \\
x &=& 10 &
\end{array}
$$

Since $x = 10$,

$$x = 5y \quad (3)$$
$$10 = 5y$$
$$2 = y.$$

Since $x = 10$ and $y = 2$,

$$x + y + z = 15 \quad (1)$$
$$10 + 2 + z = 15$$
$$12 + z = 15$$
$$z = 3.$$

She should use 10 lb of jelly beans, 2 lb of chocolate eggs, and 3 lb of marshmallow chicks.

53. $\quad x^2 + y^2 + ax + by + c = 0$
Let $x = 2$ and $y = 1$.
$$2^2 + 1^2 + a(2) + b(1) + c = 0$$
$$4 + 1 + 2a + b + c = 0$$
$$2a + b + c = -5$$

54. $\quad\quad x^2 + y^2 + ax + by + c = 0$
Let $x = -1$ and $y = 0$.
$$(-1)^2 + 0^2 + a(-1) + b(0) + c = 0$$
$$1 - a + c = 0$$
$$-a + c = -1$$
$$a - c = 1$$

55. $\quad\quad x^2 + y^2 + ax + by + c = 0$
Let $x = 3$ and $y = 3$.
$$3^2 + 3^2 + a(3) + b(3) + c = 0$$
$$9 + 9 + 3a + 3b + c = 0$$
$$3a + 3b + c = -18$$

56. Use the equations from Exercises 53–55 to form a system of equations.

$$
\begin{array}{rcll}
2a + b + c &=& -5 & (1) \\
a \quad\quad - c &=& 1 & (2) \\
3a + 3b + c &=& -18 & (3)
\end{array}
$$

Add equations (1) and (2).

$$
\begin{array}{rcll}
2a + b + c &=& -5 & (1) \\
a \quad\quad - c &=& 1 & (2) \\
\hline
3a + b \quad\quad &=& -4 & (4)
\end{array}
$$

Add equations (2) and (3).

$$
\begin{array}{rcll}
a \quad\quad - c &=& 1 & (2) \\
3a + 3b + c &=& -18 & (3) \\
\hline
4a + 3b \quad\quad &=& -17 & (5)
\end{array}
$$

Multiply equation (4) by -3 and add the result to equation (5).

$$
\begin{array}{rcll}
-9a - 3b &=& 12 & -3 \times (4) \\
4a + 3b &=& -17 & (5) \\
\hline
-5a \quad\quad &=& -5 & \\
a &=& 1 &
\end{array}
$$

Substitute $a = 1$ into equation (4).

$$3a + b = -4 \quad (4)$$
$$3(1) + b = -4$$
$$b = -7$$

Substitute $a = 1$ into equation (2).

$$a - c = 1 \quad (2)$$
$$1 - c = 1$$
$$c = 0$$

Since $a = 1$, $b = -7$, and $c = 0$, the equation of the circle is

$$x^2 + y^2 + ax + by + c = 0$$
$$x^2 + y^2 + x - 7y = 0.$$

57. The graph of the circle is not the graph of a function because it fails the vertical line test.

Section 4.4

1. $\begin{bmatrix} -2 & 3 & 1 \\ 0 & 5 & -3 \\ 1 & 4 & 8 \end{bmatrix}$

   (a) The elements of the second row are 0, 5, and -3.

   (b) The elements of the third column are 1, -3, and 8.

   (c) The matrix is square since the number of rows (three) is the same as the number of columns.

   (d) The matrix obtained by interchanging the first and third rows is

   $$\begin{bmatrix} 1 & 4 & 8 \\ 0 & 5 & -3 \\ -2 & 3 & 1 \end{bmatrix}.$$

   (e) The matrix obtained by multiplying the first row by $-\frac{1}{2}$ is

   $$\begin{bmatrix} -2\left(-\frac{1}{2}\right) & 3\left(-\frac{1}{2}\right) & 1\left(-\frac{1}{2}\right) \\ 0 & 5 & -3 \\ 1 & 4 & 8 \end{bmatrix} = \begin{bmatrix} 1 & -\frac{3}{2} & -\frac{1}{2} \\ 0 & 5 & -3 \\ 1 & 4 & 8 \end{bmatrix}.$$

   (f) The matrix obtained by multiplying the third row by 3 and adding to the first row is

   $$\begin{bmatrix} -2+3(1) & 3+3(4) & 1+3(8) \\ 0 & 5 & -3 \\ 1 & 4 & 8 \end{bmatrix} = \begin{bmatrix} 1 & 15 & 25 \\ 0 & 5 & -3 \\ 1 & 4 & 8 \end{bmatrix}.$$

3. $4x + 8y = 44$
   $2x - y = -3$

   $$\begin{bmatrix} 4 & 8 & | & 44 \\ 2 & -1 & | & -3 \end{bmatrix}$$

   $$\begin{bmatrix} 1 & 2 & | & 11 \\ 2 & -1 & | & -3 \end{bmatrix} \quad \frac{1}{4}R_1$$

   $$\begin{bmatrix} 1 & 2 & | & 11 \\ 0 & -5 & | & -25 \end{bmatrix} \quad -2R_1 + R_2$$

   Note: $\begin{cases} -2(2) + (-1) = -5 \\ -2(11) + (-3) = -25 \end{cases}$

   $$\begin{bmatrix} 1 & 2 & | & 11 \\ 0 & 1 & | & 5 \end{bmatrix} \quad -\frac{1}{5}R_1$$

   This represents the system

   $$x + 2y = 11$$
   $$y = 5.$$

Substitute $y = 5$ in the first equation.

$$x + 2y = 11$$
$$x + 2(5) = 11$$
$$x + 10 = 11$$
$$x = 1$$

Solution set: $\{(1, 5)\}$

5. $x + y = 5$
   $x - y = 3$

   Write the augmented matrix for this system.

   $$\begin{bmatrix} 1 & 1 & | & 5 \\ 1 & -1 & | & 3 \end{bmatrix}$$

   $$\begin{bmatrix} 1 & 1 & | & 5 \\ 0 & -2 & | & -2 \end{bmatrix} \quad -1R_1 + R_2$$

   $$\begin{bmatrix} 1 & 1 & | & 5 \\ 0 & 1 & | & 1 \end{bmatrix} \quad -\frac{1}{2}R_2$$

   This matrix gives the system

   $$x + y = 5$$
   $$y = 1.$$

   Substitute $y = 1$ in the first equation.

   $$x + y = 5$$
   $$x + 1 = 5$$
   $$x = 4$$

   Solution set: $\{(4, 1)\}$

7. $2x + 4y = 6$
   $3x - y = 2$

   Write the augmented matrix.

   $$\begin{bmatrix} 2 & 4 & | & 6 \\ 3 & -1 & | & 2 \end{bmatrix}$$

   The easiest way to get a 1 in the first row, first column position is to multiply the elements in the first row by $\frac{1}{2}$.

   $$\begin{bmatrix} 1 & 2 & | & 3 \\ 3 & -1 & | & 2 \end{bmatrix} \quad \frac{1}{2}R_1$$

   To get a 0 in row two, column 1, we need to subtract 3 from the 3 that is in that position. To do this we will multiply row 1 by -3 and add the result to row 2.

   $$\begin{bmatrix} 1 & 2 & | & 3 \\ 0 & -7 & | & -7 \end{bmatrix} \quad -3R_1 + R_2$$

   $$\begin{bmatrix} 1 & 2 & | & 3 \\ 0 & 1 & | & 1 \end{bmatrix} \quad -\frac{1}{7}R_2$$

   This matrix gives the system

   $$x + 2y = 3$$
   $$y = 1.$$

Substitute $y = 1$ in the first equation.

$$x + 2y = 3$$
$$x + 2(1) = 3$$
$$x = 1$$

Solution set: $\{(1, 1)\}$

9. $3x + 4y = 13$
$2x - 3y = -14$

Write the augmented matrix.

$$\begin{bmatrix} 3 & 4 & | & 13 \\ 2 & -3 & | & -14 \end{bmatrix}$$

$$\begin{bmatrix} 1 & 7 & | & 27 \\ 2 & -3 & | & -14 \end{bmatrix} \quad -1R_2 + R_1$$

$$\begin{bmatrix} 1 & 7 & | & 27 \\ 0 & -17 & | & -68 \end{bmatrix} \quad -2R_1 + R_2$$

$$\begin{bmatrix} 1 & 7 & | & 27 \\ 0 & 1 & | & 4 \end{bmatrix} \quad -\frac{1}{17}R_2$$

This matrix gives the system

$$x + 7y = 27$$
$$y = 4.$$

Substitute $y = 4$ in the first equation.

$$x + 7y = 27$$
$$x + 7(4) = 27$$
$$x + 28 = 27$$
$$x = -1$$

Solution set: $\{(-1, 4)\}$

11. $-4x + 12y = 36$
$x - 3y = 9$

Write the augmented matrix.

$$\begin{bmatrix} -4 & 12 & | & 36 \\ 1 & -3 & | & 9 \end{bmatrix}$$

$$\begin{bmatrix} -1 & 3 & | & 9 \\ 1 & -3 & | & 9 \end{bmatrix} \quad \frac{1}{4}R_1$$

$$\begin{bmatrix} -1 & 3 & | & 9 \\ 0 & 0 & | & 18 \end{bmatrix} \quad R_1 + R_2$$

The corresponding system is

$$-x + 3y = 9$$
$$0 = 18 \quad \textit{False}$$

which is inconsistent and has no solution.

Solution set: $\emptyset$

13. Examples will vary.

(a) A matrix is a rectangular array of numbers.

(b) A horizontal arrangement of elements in a matrix is a row of a matrix.

(c) A vertical arrangement of elements in a matrix is a column of a matrix.

(d) A square matrix contains the same number of rows as columns.

(e) A matrix formed by the coefficients and constants of a linear system is an augmented matrix for the system.

(f) Row operations on a matrix allow it to be transformed into another one in which the solution of the associated system can be found more easily.

15. $x + y - z = -3$
$2x + y + z = 4$
$5x - y + 2z = 23$

$$\begin{bmatrix} 1 & 1 & -1 & | & -3 \\ 2 & 1 & 1 & | & 4 \\ 5 & -1 & 2 & | & 23 \end{bmatrix}$$

$$\begin{bmatrix} 1 & 1 & -1 & | & -3 \\ 0 & -1 & 3 & | & 10 \\ 0 & -6 & 7 & | & 38 \end{bmatrix} \quad \begin{matrix} -2R_1 + R_2 \\ -5R_1 + R_3 \end{matrix}$$

$$\begin{bmatrix} 1 & 1 & -1 & | & -3 \\ 0 & 1 & -3 & | & -10 \\ 0 & -6 & 7 & | & 38 \end{bmatrix} \quad -1R_2$$

$$\begin{bmatrix} 1 & 1 & -1 & | & -3 \\ 0 & 1 & -3 & | & -10 \\ 0 & 0 & -11 & | & -22 \end{bmatrix} \quad 6R_2 + R_3$$

$$\begin{bmatrix} 1 & 1 & -1 & | & -3 \\ 0 & 1 & -3 & | & -10 \\ 0 & 0 & 1 & | & 2 \end{bmatrix} \quad -\frac{1}{11}R_3$$

This matrix gives the system

$$x + y - z = -3$$
$$y - 3z = -10$$
$$z = 2.$$

Substitute $z = 2$ in the second equation.

$$y - 3z = -10$$
$$y - 3(2) = -10$$
$$y - 6 = -10$$
$$y = -4$$

Substitute $y = -4$ and $z = 2$ in the first equation.

$$x + y - z = -3$$
$$x - 4 - 2 = -3$$
$$x - 6 = -3$$
$$x = 3$$

Solution set: $\{(3, -4, 2)\}$

17.
$$x + y - 3z = 1$$
$$2x - y + z = 9$$
$$3x + y - 4z = 8$$

Write the augmented matrix.

$$\begin{bmatrix} 1 & 1 & -3 & | & 1 \\ 2 & -1 & 1 & | & 9 \\ 3 & 1 & -4 & | & 8 \end{bmatrix}$$

$$\begin{bmatrix} 1 & 1 & -3 & | & 1 \\ 0 & -3 & 7 & | & 7 \\ 0 & -2 & 5 & | & 5 \end{bmatrix} \begin{array}{l} -2R_1 + R_2 \\ -3R_1 + R_3 \end{array}$$

$$\begin{bmatrix} 1 & 1 & -3 & | & 1 \\ 0 & 1 & -\frac{7}{3} & | & -\frac{7}{3} \\ 0 & -2 & 5 & | & 5 \end{bmatrix} -\frac{1}{3}R_2$$

$$\begin{bmatrix} 1 & 1 & -3 & | & 1 \\ 0 & 1 & -\frac{7}{3} & | & -\frac{7}{3} \\ 0 & 0 & \frac{1}{3} & | & \frac{1}{3} \end{bmatrix} 2R_2 + R_3$$

$$\begin{bmatrix} 1 & 1 & -3 & | & 1 \\ 0 & 1 & -\frac{7}{3} & | & -\frac{7}{3} \\ 0 & 0 & 1 & | & 1 \end{bmatrix} 3R_3$$

This matrix gives the system

$$x + y - 3z = 1$$
$$y - \frac{7}{3}z = -\frac{7}{3}$$
$$z = 1.$$

Substitute $z = 1$ in the second equation.

$$y - \frac{7}{3}z = -\frac{7}{3}$$
$$y - \frac{7}{3}(1) = -\frac{7}{3}$$
$$y = 0$$

Substitute $y = 0$ and $z = 1$ in the first equation.

$$x + y - 3z = 1$$
$$x + 0 - 3(1) = 1$$
$$x - 3 = 1$$
$$x = 4$$

Solution set: $\{(4, 0, 1)\}$

19.
$$x + y - z = 6$$
$$2x - y + z = -9$$
$$x - 2y + 3z = 1$$

Write the augmented matrix.

$$\begin{bmatrix} 1 & 1 & -1 & | & 6 \\ 2 & -1 & 1 & | & -9 \\ 1 & -2 & 3 & | & 1 \end{bmatrix}$$

$$\begin{bmatrix} 1 & 1 & -1 & | & 6 \\ 0 & -3 & 3 & | & -21 \\ 0 & -3 & 4 & | & -5 \end{bmatrix} \begin{array}{l} -2R_1 + R_2 \\ -1R_1 + R_3 \end{array}$$

$$\begin{bmatrix} 1 & 1 & -1 & | & 6 \\ 0 & 1 & -1 & | & 7 \\ 0 & -3 & 4 & | & -5 \end{bmatrix} -\frac{1}{3}R_2$$

$$\begin{bmatrix} 1 & 1 & -1 & | & 6 \\ 0 & 1 & -1 & | & 7 \\ 0 & 0 & 1 & | & 16 \end{bmatrix} 3R_2 + R_3$$

This matrix gives the system

$$x + y - z = 6$$
$$y - z = 7$$
$$z = 16.$$

Substitute $z = 16$ in the second equation.

$$y - 16 = 7$$
$$y = 23$$

Substitute $y = 23$ and $z = 16$ in the first equation.

$$x + 23 - 16 = 6$$
$$x + 7 = 6$$
$$x = -1$$

Solution set: $\{(-1, 23, 16)\}$

21.
$$x - y = 1$$
$$y - z = 6$$
$$x + z = -1$$

Write the augmented matrix.

$$\begin{bmatrix} 1 & -1 & 0 & | & 1 \\ 0 & 1 & -1 & | & 6 \\ 1 & 0 & 1 & | & -1 \end{bmatrix}$$

$$\begin{bmatrix} 1 & -1 & 0 & | & 1 \\ 0 & 1 & -1 & | & 6 \\ 0 & 1 & 1 & | & -2 \end{bmatrix} -1R_1 + R_3$$

$$\begin{bmatrix} 1 & -1 & 0 & | & 1 \\ 0 & 1 & -1 & | & 6 \\ 0 & 0 & 2 & | & -8 \end{bmatrix} -1R_2 + R_3$$

$$\begin{bmatrix} 1 & -1 & 0 & | & 1 \\ 0 & 1 & -1 & | & 6 \\ 0 & 0 & 1 & | & -4 \end{bmatrix} \frac{1}{2}R_3$$

This matrix gives the system

$$x - y = 1$$
$$y - z = 6$$
$$z = -4.$$

Substitute $z = -4$ in the second equation.

$$y - z = 6$$
$$y - (-4) = 6$$
$$y = 2$$

Substitute $y = 2$ in the first equation.

$$x - y = 1$$
$$x - 2 = 1$$
$$x = 3$$

Solution set: $\{(3, 2, -4)\}$

23.
$$x - 2y + z = 4$$
$$3x - 6y + 3z = 12$$
$$-2x + 4y - 2z = -8$$

Write the augmented matrix.

$$\begin{bmatrix} 1 & -2 & 1 & | & 4 \\ 3 & -6 & 3 & | & 12 \\ -2 & 4 & -2 & | & -8 \end{bmatrix}$$

$$\begin{bmatrix} 1 & -2 & 1 & | & 4 \\ 1 & -2 & 1 & | & 4 \\ -1 & 2 & -1 & | & -4 \end{bmatrix} \begin{matrix} \\ \frac{1}{3}R_2 \\ \frac{1}{2}R_3 \end{matrix}$$

$$\begin{bmatrix} 1 & -2 & 1 & | & 4 \\ 0 & 0 & 0 & | & 0 \\ 0 & 0 & 0 & | & 0 \end{bmatrix} \begin{matrix} \\ -1R_1 + R_2 \\ R_1 + R_3 \end{matrix}$$

This augmented matrix represents a system of dependent equations.

Solution set: $\{(x, y, z) | x - 2y + z = 4\}$

25.
$$4x + y = 5$$
$$2x + y = 3$$

Enter the augmented matrix as [A].

$$\begin{bmatrix} 4 & 1 & 5 \\ 2 & 1 & 3 \end{bmatrix}$$

The TI-83 screen for A follows. (Use MATRX EDIT.)

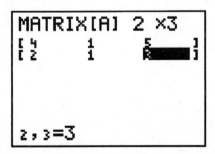

Now use the reduced row echelon form (rref) command to simplify the system. (Use MATRX MATH ALPHA B for rref and MATRX 1 for [A].)

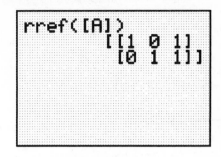

This matrix gives the system

$$1x + 0y = 1$$
$$0x + 1y = 1,$$

or simply, $x = 1$ and $y = 1$.

Solution set: $\{(1, 1)\}$

27.
$$5x + y - 3z = -6$$
$$2x + 3y + z = 5$$
$$-3x - 2y + 4z = 3$$

Enter the augmented matrix as [C].

$$\begin{bmatrix} 5 & 1 & -3 & -6 \\ 2 & 3 & 1 & 5 \\ -3 & -2 & 4 & 3 \end{bmatrix}$$

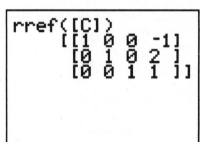

Solution set: $\{(-1, 2, 1)\}$

29.
$$x + z = -3$$
$$y + z = 3$$
$$x + y = 8$$

Enter the augmented matrix as [E].

$$\begin{bmatrix} 1 & 0 & 1 & -3 \\ 0 & 1 & 1 & 3 \\ 1 & 1 & 0 & 8 \end{bmatrix}$$

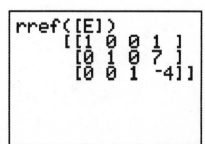

Solution set: $\{(1, 7, -4)\}$

Section 4.5

1. $\begin{vmatrix} -2 & -3 \\ 4 & -6 \end{vmatrix} = -2(-6) - (-3)(4)$

Choice (d) is correct.

3. $\begin{vmatrix} -2 & 5 \\ -1 & 4 \end{vmatrix} = -2(4) - 5(-1)$

$= -8 + 5 = -3$

5. $\begin{vmatrix} 1 & -2 \\ 7 & 0 \end{vmatrix} = 1(0) - (-2)(7)$

$= 0 + 14 = 14$

7. $\begin{vmatrix} 0 & 4 \\ 0 & 4 \end{vmatrix} = 0(4) - 4(0)$

$= 0 - 0 = 0$

9. $\begin{vmatrix} -1 & 2 & 4 \\ -3 & -2 & -3 \\ 2 & -1 & 5 \end{vmatrix}$ Expand by minors about the first column.

$= -1 \begin{vmatrix} -2 & -3 \\ -1 & 5 \end{vmatrix} - (-3) \begin{vmatrix} 2 & 4 \\ -1 & 5 \end{vmatrix}$

$+ 2 \begin{vmatrix} 2 & 4 \\ -2 & -3 \end{vmatrix}$

$= -1[-2(5) - (-3)(-1)]$

$+ 3[2(5) - 4(-1)] + 2[2(-3) - 4(-2)]$

$= -1(-13) + 3(14) + 2(2)$

$= 13 + 42 + 4 = 59$

11. $\begin{vmatrix} 1 & 0 & -2 \\ 0 & 2 & 3 \\ 1 & 0 & 5 \end{vmatrix}$ There are two 0's in column 2. We'll expand about that column since there is only 1 minor to evaluate.

$= -0 \begin{vmatrix} 0 & 3 \\ 1 & 5 \end{vmatrix} + 2 \begin{vmatrix} 1 & -2 \\ 1 & 5 \end{vmatrix} - 0 \begin{vmatrix} 1 & -2 \\ 0 & 3 \end{vmatrix}$

$= 0 + 2[1(5) - (-2)(1)] - 0$

$= 2[5 + 2] = 2(7) = 14$

13. Multiply the upper left and lower right entries. Then multiply the upper right and lower left entries. Subtract the second product from the first to obtain the determinant. For example,

$\begin{vmatrix} 4 & 2 \\ 7 & 1 \end{vmatrix} = 4 \cdot 1 - 2 \cdot 7$

$= 4 - 14 = -10.$

15. $\begin{vmatrix} 4 & 4 & 2 \\ 1 & -1 & -2 \\ 1 & 0 & 2 \end{vmatrix}$ Expand about column 2.

$= -4 \begin{vmatrix} 1 & -2 \\ 1 & 2 \end{vmatrix} + (-1) \begin{vmatrix} 4 & 2 \\ 1 & 2 \end{vmatrix} - 0 \begin{vmatrix} 4 & 2 \\ 1 & -2 \end{vmatrix}$

$= -4[1(2) - (-2)(1)] - 1[4(2) - 2(1)] - 0$

$= -4(4) - 1(6)$

$= -16 - 6 = -22$

17. $\begin{vmatrix} 3 & 5 & -2 \\ 1 & -4 & 1 \\ 3 & 1 & -2 \end{vmatrix}$ Expand about column 1.

$= 3 \begin{vmatrix} -4 & 1 \\ 1 & -2 \end{vmatrix} - 1 \begin{vmatrix} 5 & -2 \\ 1 & -2 \end{vmatrix} + 3 \begin{vmatrix} 5 & -2 \\ -4 & 1 \end{vmatrix}$

$= 3[-4(-2) - 1(1)] - 1[5(-2) - (-2)(1)]$

$+ 3[5(1) - (-2)(-4)]$

$= 3(7) - 1(-8) + 3(-3)$

$= 21 + 8 - 9 = 20$

19. $\begin{vmatrix} 3 & 0 & -2 \\ 1 & -4 & 1 \\ 3 & 1 & -2 \end{vmatrix}$ Expand about row 1.

$= 3 \begin{vmatrix} -4 & 1 \\ 1 & -2 \end{vmatrix} - 0 \begin{vmatrix} 1 & 1 \\ 3 & -2 \end{vmatrix} + (-2) \begin{vmatrix} 1 & -4 \\ 3 & 1 \end{vmatrix}$

$= 3[-4(-2) - 1(1)] - 0$

$+ (-2)[1(1) - (-4)(3)]$

$= 3(7) - 0 - 2(13)$

$= 21 - 26 = -5$

21. By choosing that row or column to expand about, all terms will have a factor of 0, and so the sum of all these terms will be 0.

22. The slope m through the points (x_1, y_1) and (x_2, y_2) is

$$m = \frac{y_2 - y_1}{x_2 - x_1}.$$

23. $y - y_1 = m(x - x_1)$

Let $m = \dfrac{y_2 - y_1}{x_2 - x_1}$ from Exercise 22.

$y - y_1 = \dfrac{y_2 - y_1}{x_2 - x_1}(x - x_1)$

24. $y - y_1 = \dfrac{y_2 - y_1}{x_2 - x_1}(x - x_1)$

Multiply both sides by $x_2 - x_1$.

$(x_2 - x_1)(y - y_1) = (y_2 - y_1)(x - x_1)$

Subtract $(y_2 - y_1)(x - x_1)$ from both sides.

$(x_2 - x_1)(y - y_1) - (y_2 - y_1)(x - x_1) = 0$

Multiply and collect like terms.

$x_2 y - x_2 y_1 - x_1 y + x_1 y_1$

$\qquad - (xy_2 - x_1 y_2 - xy_1 + x_1 y_1) = 0$

$x_2 y - x_2 y_1 - x_1 y + x_1 y_1$

$\qquad - xy_2 + x_1 y_2 + xy_1 - x_1 y_1 = 0$

$x_2 y - x_2 y_1 - x_1 y - xy_2 + x_1 y_2 + xy_1 = 0$

25. $\begin{vmatrix} x & y & 1 \\ x_1 & y_1 & 1 \\ x_2 & y_2 & 1 \end{vmatrix} = 0$ Expand about row 1.

$x \begin{vmatrix} y_1 & 1 \\ y_2 & 1 \end{vmatrix} - y \begin{vmatrix} x_1 & 1 \\ x_2 & 1 \end{vmatrix} + 1 \begin{vmatrix} x_1 & y_1 \\ x_2 & y_2 \end{vmatrix}$

$= x(y_1 - y_2) - y(x_1 - x_2) + 1(x_1 y_2 - x_2 y_1)$

$= xy_1 - xy_2 - x_1 y + x_2 y + x_1 y_2 - x_2 y_1$

Thus,

$xy_1 - xy_2 - x_1 y + x_2 y + x_1 y_2 - x_2 y_1 = 0$

or

$x_2y - x_2y_1 - x_1y - xy_2 + x_1y_2 + xy_1 = 0.$

This is the same equation as that found in Exercise 24.

27. det ([B])

$$= \begin{vmatrix} -1 & 7 & 1 \\ 0 & 2 & 1 \\ -3 & 5 & 4 \end{vmatrix} \quad \text{Expand about column 1.}$$

$$= -1 \begin{vmatrix} 2 & 1 \\ 5 & 4 \end{vmatrix} - 0 \begin{vmatrix} 7 & 1 \\ 5 & 4 \end{vmatrix} - 3 \begin{vmatrix} 7 & 1 \\ 2 & 1 \end{vmatrix}$$

$$= -1[2(4) - 1(5)] - 0 - 3[7(1) - 1(2)]$$

$$= -1(3) - 3(5) = -3 - 15 = -18$$

29. det ([D])

$$= \begin{vmatrix} 0 & 0 & 0 \\ -2 & 3 & 7 \\ 2 & 1 & 0 \end{vmatrix}$$

$= 0$ since row 1 consists of only 0 elements.

31. Enter $\begin{vmatrix} \sqrt{5} & \sqrt{2} & -\sqrt{3} \\ \sqrt{7} & -\sqrt{6} & \sqrt{10} \\ -\sqrt{5} & -\sqrt{2} & \sqrt{17} \end{vmatrix}$ as [A].

Type the keys for det [A].
The result should be -22.04285452.

33.

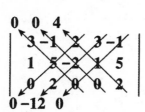

Add the top numbers: $0 + 0 + 4 = 4$
Add the bottom numbers: $0 - 12 + 0 = -12$
Find the difference: $(4) - (-12) = 16$

35.

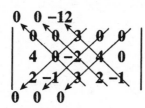

Add the top numbers: $0 + 0 - 12 = -12$
Add the bottom numbers: $0 + 0 + 0 = 0$
Find the difference: $(-12) - (0) = -12$

37.

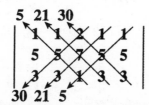

Add the top numbers: $5 + 21 + 30 = 56$
Add the bottom numbers: $30 + 21 + 5 = 56$
Find the difference: $(56) - (56) = 0$

39. $\begin{vmatrix} 5 & 3 \\ x & x \end{vmatrix} = 20$

Evaluate the determinant.

$$\begin{vmatrix} 5 & 3 \\ x & x \end{vmatrix} = 5x - 3x = 2x$$

Solve the equation.

$$2x = 20$$
$$x = 10$$

Solution set: $\{10\}$

41. **(a)** D is the determinant of coefficients—choice IV.

(b) D_x is the determinant with the constants replacing the x-coefficients—choice I.

(c) D_y is the determinant with the constants replacing the y-coefficients—choice III.

(d) D_z is the determinant with the constants replacing the z-coefficients—choice II.

43.
$$3x + 5y = -5$$
$$-2x + 3y = 16$$

$$D = \begin{vmatrix} 3 & 5 \\ -2 & 3 \end{vmatrix} = 3(3) - 5(-2) = 19$$

$$D_x = \begin{vmatrix} -5 & 5 \\ 16 & 3 \end{vmatrix} = -5(3) - 5(16) = -95$$

$$D_y = \begin{vmatrix} 3 & -5 \\ -2 & 16 \end{vmatrix} = 3(16) - (-5)(-2) = 38$$

$$x = \frac{D_x}{D} = \frac{-95}{19} = -5; \; y = \frac{D_y}{D} = \frac{38}{19} = 2$$

Solution set: $\{(-5, 2)\}$

45.
$$8x + 3y = 1$$
$$6x - 5y = 2$$

$$D = \begin{vmatrix} 8 & 3 \\ 6 & -5 \end{vmatrix} = 8(-5) - 3(6) = -58$$

$$D_x = \begin{vmatrix} 1 & 3 \\ 2 & -5 \end{vmatrix} = 1(-5) - 3(2) = -11$$

$$D_y = \begin{vmatrix} 8 & 1 \\ 6 & 2 \end{vmatrix} = 8(2) - 1(6) = 10$$

$$x = \frac{D_x}{D} = \frac{-11}{-58} = \frac{11}{58}; \; y = \frac{D_y}{D} = \frac{10}{-58} = -\frac{5}{29}$$

Solution set: $\left\{ \left(\frac{11}{58}, -\frac{5}{29} \right) \right\}$

47. $2x + 3y = 4$
$5x + 6y = 7$

$$D = \begin{vmatrix} 2 & 3 \\ 5 & 6 \end{vmatrix} = 2(6) - 3(5) = -3$$

$$D_x = \begin{vmatrix} 4 & 3 \\ 7 & 6 \end{vmatrix} = 4(6) - 3(7) = 3$$

$$D_y = \begin{vmatrix} 2 & 4 \\ 5 & 7 \end{vmatrix} = 2(7) - 4(5) = -6$$

$$x = \frac{D_x}{D} = \frac{3}{-3} = -1; \quad y = \frac{D_y}{D} = \frac{-6}{-3} = 2$$

Solution set: $\{(-1, 2)\}$

49. One example is

$$6x + 7y = 8$$
$$9x + 10y = 11.$$

$$D = \begin{vmatrix} 6 & 7 \\ 9 & 10 \end{vmatrix} = 6(10) - 7(9) = -3$$

$$D_x = \begin{vmatrix} 8 & 7 \\ 11 & 10 \end{vmatrix} = 8(10) - 7(11) = 3$$

$$D_y = \begin{vmatrix} 6 & 8 \\ 9 & 11 \end{vmatrix} = 6(11) - 8(9) = -6$$

$$x = \frac{D_x}{D} = \frac{3}{-3} = -1; \quad y = \frac{D_y}{D} = \frac{-6}{-3} = 2$$

As in Exercises 47 and 48, the solution set is $\{(-1, 2)\}$.

51. $2x + 3y + 2z = 15$
$x - y + 2z = 5$
$x + 2y - 6z = -26$

$$D = \begin{vmatrix} 2 & 3 & 2 \\ 1 & -1 & 2 \\ 1 & 2 & -6 \end{vmatrix} \text{ Expand about row 1.}$$

$$= 2\begin{vmatrix} -1 & 2 \\ 2 & -6 \end{vmatrix} - 3\begin{vmatrix} 1 & 2 \\ 1 & -6 \end{vmatrix} + 2\begin{vmatrix} 1 & -1 \\ 1 & 2 \end{vmatrix}$$

$$= 2(2) - 3(-8) + 2(3)$$
$$= 4 + 24 + 6 = 34$$

$$D_x = \begin{vmatrix} 15 & 3 & 2 \\ 5 & -1 & 2 \\ -26 & 2 & -6 \end{vmatrix} \text{ Expand about row 2.}$$

$$= -5\begin{vmatrix} 3 & 2 \\ 2 & -6 \end{vmatrix} + (-1)\begin{vmatrix} 15 & 2 \\ -26 & -6 \end{vmatrix}$$

$$- 2\begin{vmatrix} 15 & 3 \\ -26 & 2 \end{vmatrix}$$

$$= -5(-22) - 1(-38) - 2(108)$$
$$= 110 + 38 - 216 = -68$$

$$D_y = \begin{vmatrix} 2 & 15 & 2 \\ 1 & 5 & 2 \\ 1 & -26 & -6 \end{vmatrix} \text{ Expand about column 1.}$$

$$= 2\begin{vmatrix} 5 & 2 \\ -26 & -6 \end{vmatrix} - 1\begin{vmatrix} 15 & 2 \\ -26 & -6 \end{vmatrix}$$

$$+ 1\begin{vmatrix} 15 & 2 \\ 5 & 2 \end{vmatrix}$$

$$= 2(22) - 1(-38) + 1(20)$$
$$= 44 + 38 + 20 = 102$$

$$D_z = \begin{vmatrix} 2 & 3 & 15 \\ 1 & -1 & 5 \\ 1 & 2 & -26 \end{vmatrix} \text{ Expand about column 1.}$$

$$= 2\begin{vmatrix} -1 & 5 \\ 2 & -26 \end{vmatrix} - 1\begin{vmatrix} 3 & 15 \\ 2 & -26 \end{vmatrix} + 1\begin{vmatrix} 3 & 15 \\ -1 & 5 \end{vmatrix}$$

$$= 2(16) - 1(-108) + 1(30)$$
$$= 32 + 108 + 30 = 170$$

$$x = \frac{D_x}{D} = \frac{-68}{34} = -2; \quad y = \frac{D_y}{D} = \frac{102}{34} = 3$$

$$z = \frac{D_z}{D} = \frac{170}{34} = 5$$

Solution set: $\{(-2, 3, 5)\}$

53. $2x - 3y + 4z = 8$
$6x - 9y + 12z = 24$
$-4x + 6y - 8z = -16$

$$D = \begin{vmatrix} 2 & -3 & 4 \\ 6 & -9 & 12 \\ -4 & 6 & -8 \end{vmatrix} \text{ Expand about row 1.}$$

$$= 2\begin{vmatrix} -9 & 12 \\ 6 & -8 \end{vmatrix} + 3\begin{vmatrix} 6 & 12 \\ -4 & -8 \end{vmatrix} + 4\begin{vmatrix} 6 & -9 \\ -4 & 6 \end{vmatrix}$$

$$= 2(0) + 3(0) + 4(0) = 0$$

Because $D = 0$, Cramer's rule does not apply.

55. $3x + 5z = 0$
$2x + 3y = 1$
$-y + 2z = -11$

$$D = \begin{vmatrix} 3 & 0 & 5 \\ 2 & 3 & 0 \\ 0 & -1 & 2 \end{vmatrix} \text{ Expand about row 1.}$$

$$= 3\begin{vmatrix} 3 & 0 \\ -1 & 2 \end{vmatrix} - 0 + 5\begin{vmatrix} 2 & 3 \\ 0 & -1 \end{vmatrix}$$

$$= 3(6) + 5(-2)$$
$$= 18 - 10 = 8$$

$$D_x = \begin{vmatrix} 0 & 0 & 5 \\ 1 & 3 & 0 \\ -11 & -1 & 2 \end{vmatrix} \text{ Expand about row 1.}$$

$$= 0 - 0 + 5\begin{vmatrix} 1 & 3 \\ -11 & -1 \end{vmatrix}$$

$$= 5(32) = 160$$

$$D_y = \begin{vmatrix} 3 & 0 & 5 \\ 2 & 1 & 0 \\ 0 & -11 & 2 \end{vmatrix} \text{ Expand about row 1.}$$

$$= 3\begin{vmatrix} 1 & 0 \\ -11 & 2 \end{vmatrix} - 0 + 5\begin{vmatrix} 2 & 1 \\ 0 & -11 \end{vmatrix}$$

$$= 3(2) + 5(-22)$$

$$= 6 - 110 = -104$$

$$D_z = \begin{vmatrix} 3 & 0 & 0 \\ 2 & 3 & 1 \\ 0 & -1 & -11 \end{vmatrix} \text{ Expand about row 1.}$$

$$= 3\begin{vmatrix} 3 & 1 \\ -1 & -11 \end{vmatrix} - 0 + 0$$

$$= 3(-32) = -96$$

$$x = \frac{D_x}{D} = \frac{160}{8} = 20; \ y = \frac{D_y}{D} = \frac{-104}{8} = -13$$

$$z = \frac{D_z}{D} = \frac{-96}{8} = -12$$

Solution set: $\{(20, -13, -12)\}$

57. $x \ - \ 3y \qquad \quad = \ 13$
$\qquad \qquad 2y + z = \quad 5$
$\quad -x \qquad \quad + z = -7$

$$D = \begin{vmatrix} 1 & -3 & 0 \\ 0 & 2 & 1 \\ -1 & 0 & 1 \end{vmatrix} \text{ Expand about column 1.}$$

$$= 1\begin{vmatrix} 2 & 1 \\ 0 & 1 \end{vmatrix} - 0 + (-1)\begin{vmatrix} -3 & 0 \\ 2 & 1 \end{vmatrix}$$

$$= 1(2) - 1(-3) = 2 + 3 = 5$$

$$D_x = \begin{vmatrix} 13 & -3 & 0 \\ 5 & 2 & 1 \\ -7 & 0 & 1 \end{vmatrix} \text{ Expand about column 3.}$$

$$= 0 - 1\begin{vmatrix} 13 & -3 \\ -7 & 0 \end{vmatrix} + 1\begin{vmatrix} 13 & -3 \\ 5 & 2 \end{vmatrix}$$

$$= -1(-21) + 1(41)$$

$$= 21 + 41 = 62$$

$$D_y = \begin{vmatrix} 1 & 13 & 0 \\ 0 & 5 & 1 \\ -1 & -7 & 1 \end{vmatrix} \text{ Expand about column 1.}$$

$$= 1\begin{vmatrix} 5 & 1 \\ -7 & 1 \end{vmatrix} - 0 + (-1)\begin{vmatrix} 13 & 0 \\ 5 & 1 \end{vmatrix}$$

$$= 1(12) - 1(13) = 12 - 13 = -1$$

$$D_z = \begin{vmatrix} 1 & -3 & 13 \\ 0 & 2 & 5 \\ -1 & 0 & -7 \end{vmatrix} \text{ Expand about column 1.}$$

$$= 1\begin{vmatrix} 2 & 5 \\ 0 & -7 \end{vmatrix} - 0 - 1\begin{vmatrix} -3 & 13 \\ 2 & 5 \end{vmatrix}$$

$$= 1(-14) - 1(-41) = -14 + 41 = 27$$

$$x = \frac{D_x}{D} = \frac{62}{5}; \ y = \frac{D_y}{D} = -\frac{1}{5}$$

$$z = \frac{D_z}{D} = \frac{27}{5}$$

Solution set: $\left\{\left(\dfrac{62}{5}, -\dfrac{1}{5}, \dfrac{27}{5}\right)\right\}$

59.

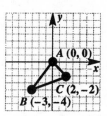

60. Use $(0,0)$ as (x_1, y_1), $(-3, -4)$ as (x_2, y_2), and $(2, -2)$ as (x_3, y_3). Then

$$\frac{1}{2}\begin{vmatrix} x_1 & y_1 & 1 \\ x_2 & y_2 & 1 \\ x_3 & y_3 & 1 \end{vmatrix} = \frac{1}{2}\begin{vmatrix} 0 & 0 & 1 \\ -3 & -4 & 1 \\ 2 & -2 & 1 \end{vmatrix}.$$

61. $\dfrac{1}{2}\begin{vmatrix} 0 & 0 & 1 \\ -3 & -4 & 1 \\ 2 & -2 & 1 \end{vmatrix}$ Expand about row 1.

$$= \frac{1}{2}\left(0 - 0 + 1\begin{vmatrix} -3 & -4 \\ 2 & -2 \end{vmatrix}\right)$$

$$= \frac{1}{2}[1(6 + 8)]$$

$$= \frac{1}{2}(14) = 7$$

The area is 7 square units.

62. $\dfrac{1}{2}\begin{vmatrix} 3 & 8 & 1 \\ -1 & 4 & 1 \\ 0 & 1 & 1 \end{vmatrix}$ Expand about row 3.

$$= \frac{1}{2}\left(0 - 1\begin{vmatrix} 3 & 1 \\ -1 & 1 \end{vmatrix} + 1\begin{vmatrix} 3 & 8 \\ -1 & 4 \end{vmatrix}\right)$$

$$= \frac{1}{2}(-1[3 + 1] + 1[12 + 8])$$

$$= \frac{1}{2}(-4 + 20) = \frac{1}{2}(16) = 8$$

The area is 8 square units.

63. $x \ + \ 2y \ + \quad z = \quad 10$
$\quad 2x \ - \quad y \ - \ 3z = -20$
$\ -x \ + \ 4y \ + \quad z = \quad 18$

Use a calculator to evaluate each determinant.

continued

$$D = \begin{vmatrix} 1 & 2 & 1 \\ 2 & -1 & -3 \\ -1 & 4 & 1 \end{vmatrix} = 20$$

$$D_x = \begin{vmatrix} 10 & 2 & 1 \\ -20 & -1 & -3 \\ 18 & 4 & 1 \end{vmatrix} = -20$$

$$D_y = \begin{vmatrix} 1 & 10 & 1 \\ 2 & -20 & -3 \\ -1 & 18 & 1 \end{vmatrix} = 60$$

$$D_z = \begin{vmatrix} 1 & 2 & 10 \\ 2 & -1 & -20 \\ -1 & 4 & 18 \end{vmatrix} = 100$$

$$x = \frac{D_x}{D} = \frac{-20}{20} = -1;\ y = \frac{D_y}{D} = \frac{60}{20} = 3$$

$$z = \frac{D_z}{D} = \frac{100}{20} = 5$$

Solution set: $\{(-1, 3, 5)\}$

65.
$$\begin{aligned}
-8w + 4x - 2y + z &= -28 \\
-w + x - y + z &= -10 \\
w + x + y + z &= -4 \\
27w + 9x + 3y + z &= 2
\end{aligned}$$

Use a calculator to evaluate each determinant.

$$D = \begin{vmatrix} -8 & 4 & -2 & 1 \\ -1 & 1 & -1 & 1 \\ 1 & 1 & 1 & 1 \\ 27 & 9 & 3 & 1 \end{vmatrix} = 240$$

$$D_w = \begin{vmatrix} -28 & 4 & -2 & 1 \\ -10 & 1 & -1 & 1 \\ -4 & 1 & 1 & 1 \\ 2 & 9 & 3 & 1 \end{vmatrix} = 240$$

$$D_x = \begin{vmatrix} -8 & -28 & -2 & 1 \\ -1 & -10 & -1 & 1 \\ 1 & -4 & 1 & 1 \\ 27 & 2 & 3 & 1 \end{vmatrix} = -720$$

$$D_y = \begin{vmatrix} -8 & 4 & -28 & 1 \\ -1 & 1 & -10 & 1 \\ 1 & 1 & -4 & 1 \\ 27 & 9 & 2 & 1 \end{vmatrix} = 480$$

$$D_z = \begin{vmatrix} -8 & 4 & -2 & -28 \\ -1 & 1 & -1 & -10 \\ 1 & 1 & 1 & -4 \\ 27 & 9 & 3 & 2 \end{vmatrix} = -960$$

$$w = \frac{D_w}{D} = \frac{240}{240} = 1;\ x = \frac{D_x}{D} = \frac{-720}{240} = -3;$$

$$y = \frac{D_y}{D} = \frac{480}{240} = 2;\ z = \frac{D_z}{D} = \frac{-960}{240} = -4$$

Solution set: $\{(1, -3, 2, -4)\}$

Chapter 4 Review Exercises

1. From the graph, the usage rates of core competencies and total quality management both reached about 72% in 1995.

2. The graphs of core competencies and strategic alliances intersect twice. Between these times (about mid-1995 to 1997), core competencies were more popular.

3. **(a)** Since the graphs for supply and demand intersect at (4, 300), supply equals demand at this point. The price is $4.

(b) Supply equals demand for 300 half-gallons.

(c) By reading the graphs when the price equals $2, we see that the supply is 200 half-gallons and the demand is 400 half-gallons.

4.
$$\begin{aligned}
x + 3y &= 8 \quad (1) \\
2x - y &= 2 \quad (2)
\end{aligned}$$

To eliminate y, multiply equation (2) by 3 and add the result to equation (1).

$$\begin{array}{rll}
x + 3y &= 8 & (1) \\
6x - 3y &= 6 & 3 \times (2) \\
\hline
7x &= 14 & \\
x &= 2 &
\end{array}$$

Substitute 2 for x in equation (1) to find y.

$$\begin{aligned}
x + 3y &= 8 \quad (1) \\
2 + 3y &= 8 \\
3y &= 6 \\
y &= 2
\end{aligned}$$

Solution set: $\{(2, 2)\}$

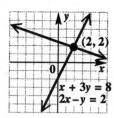

5.
$$\begin{aligned}
x - 4y &= -4 \quad (1) \\
3x + y &= 1 \quad (2)
\end{aligned}$$

To eliminate y, multiply equation (2) by 4 and add the result to equation (1).

$$\begin{array}{rll}
x - 4y &= -4 & (1) \\
12x + 4y &= 4 & 4 \times (2) \\
\hline
13x &= 0 & \\
x &= 0 &
\end{array}$$

Substitute 0 for x in equation (1) to find y.

$$x - 4y = -4 \quad (1)$$
$$0 - 4y = -4$$
$$y = 1$$

Solution set: $\{(0, 1)\}$

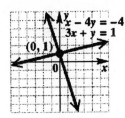

6.
$$6x + 5y = 4 \quad (1)$$
$$-4x + 2y = 8 \quad (2)$$

To eliminate x, multiply equation (1) by 2 and equation (2) by 3. Then add the results.

$$\begin{array}{rcl} 12x + 10y &=& 8 \quad 2 \times (1) \\ -12x + 6y &=& 24 \quad 3 \times (2) \\ \hline 16y &=& 32 \\ y &=& 2 \end{array}$$

Since $y = 2$,

$$-4x + 2y = 8 \quad (2)$$
$$-4x + 2(2) = 8$$
$$-4x + 4 = 8$$
$$-4x = 4$$
$$x = -1.$$

Solution set: $\{(-1, 2)\}$

7.
$$\frac{x}{6} + \frac{y}{6} = -\frac{1}{2} \quad (1)$$
$$x - y = -9 \quad (2)$$

Multiply equation (1) by 6 to clear the fractions. Add the result to equation (2) to eliminate y.

$$\begin{array}{rcl} x + y &=& -3 \quad 6 \times (1) \\ x - y &=& -9 \quad (2) \\ \hline 2x &=& -12 \\ x &=& -6 \end{array}$$

Since $x = -6$,

$$x - y = -9 \quad (2)$$
$$-6 - y = -9$$
$$-y = -3$$
$$y = 3.$$

Solution set: $\{(-6, 3)\}$

8.
$$9x - y = -4 \quad (1)$$
$$y = x + 4$$
$$\text{or} \quad -x + y = 4 \quad (2)$$

To eliminate y, add equations (1) and (2).

$$\begin{array}{rcl} 9x - y &=& -4 \quad (1) \\ -x + y &=& 4 \quad (2) \\ \hline 8x &=& 0 \\ x &=& 0 \end{array}$$

Since $x = 0$, $y = x + 4 = 0 + 4 = 4$.
Solution set: $\{(0, 4)\}$

9.
$$-3x + y = 6 \quad (1)$$
$$y = 6 + 3x \quad (2)$$

Since equation (2) can be rewritten as $-3x + y = 6$, the two equations are the same, and hence, dependent.
Solution set: $\{(x, y) | -3x + y = 6\}$

10.
$$5x - 4y = 2 \quad (1)$$
$$-10x + 8y = 7 \quad (2)$$

Multiply equation (1) by 2 and add the result to equation (2).

$$\begin{array}{rcl} 10x - 8y &=& 4 \quad 2 \times (1) \\ -10x + 8y &=& 7 \quad (2) \\ \hline 0 &=& 11 \; \textit{False} \end{array}$$

Since a false statement results, the system is inconsistent. The solution set is $\emptyset$.

11.
$$3x + y = -4 \quad (1)$$
$$x = \frac{2}{3}y \quad (2)$$

Substitute $\frac{2}{3}y$ for x in equation (1) and solve for y.

$$3x + y = -4 \quad (1)$$
$$3\left(\frac{2}{3}y\right) + y = 4$$
$$2y + y = -4$$
$$3y = -4$$
$$y = -\frac{4}{3}$$

Since $x = \frac{2}{3}y$ and $y = -\frac{4}{3}$,

$$x = \frac{2}{3}\left(-\frac{4}{3}\right) = -\frac{8}{9}.$$

Solution set: $\left\{\left(-\frac{8}{9}, -\frac{4}{3}\right)\right\}$

12. $-5x + 2y = -2$ (1)
   $x + 6y = 26$ (2)

Solve equation (2) for x.

$$x = 26 - 6y$$

Substitute $26 - 6y$ for x in equation (1).

$$-5x + 2y = -2 \quad (1)$$
$$-5(26 - 6y) + 2y = -2$$
$$-130 + 30y + 2y = -2$$
$$-130 + 32y = -2$$
$$32y = 128$$
$$y = 4$$

Since $x = 26 - 6y$ and $y = 4$,

$$x = 26 - 6(4) = 26 - 24 = 2.$$

Solution set: $\{(2, 4)\}$

13. **(a)** The system is inconsistent. Parallel lines do not intersect, so there is no ordered pair that would satisfy both equations.

(b) The equations are equivalent and have the same line as graph. The equations are dependent.

14. $y = 3x + 2$
   $y = 3x - 4$

Since both equations are in slope-intercept form, their slopes and y-intercepts can be easily determined. Since they have the same slope but different y-intercepts, they are parallel lines. Thus, there is no point common to both lines, and the system has $\emptyset$ as its solution set.

15. $2x + 3y - z = -16$ (1)
   $x + 2y + 2z = -3$ (2)
   $-3x + y + z = -5$ (3)

To eliminate z, add equations (1) and (3).

$$\begin{array}{rl} 2x + 3y - z = -16 & (1) \\ -3x + y + z = -5 & (3) \\ \hline -x + 4y = -21 & (4) \end{array}$$

To eliminate z again, multiply equation (1) by 2 and add the result to equation (2).

$$\begin{array}{rl} 4x + 6y - 2z = -32 & 2 \times (1) \\ x + 2y + 2z = -3 & (2) \\ \hline 5x + 8y = -35 & (5) \end{array}$$

Use equations (4) and (5) to eliminate x. Multiply equation (4) by 5 and add the result to equation (5).

$$\begin{array}{rl} -5x + 20y = -105 & 5 \times (4) \\ 5x + 8y = -35 & (5) \\ \hline 28y = -140 \\ y = -5 \end{array}$$

Substitute -5 for y in equation (4) to find x.

$$-x + 4y = -21 \quad (4)$$
$$-x + 4(-5) = -21$$
$$-x - 20 = -21$$
$$-x = -1$$
$$x = 1$$

Substitute 1 for x and -5 for y in equation (2) to find z.

$$x + 2y + 2z = -3 \quad (2)$$
$$1 + 2(-5) + 2z = -3$$
$$1 - 10 + 2z = -3$$
$$2z = 6$$
$$z = 3$$

The solution set is $\{(1, -5, 3)\}$.

16. $4x - y = 2$ (1)
   $3y + z = 9$ (2)
   $x + 2z = 7$ (3)

To eliminate y, multiply equation (1) by 3 and add the result to equation (2).

$$\begin{array}{rl} 12x - 3y = 6 & 3 \times (1) \\ 3y + z = 9 & (2) \\ \hline 12x + z = 15 & (4) \end{array}$$

To eliminate z, multiply equation (4) by -2 and add the result to equation (3).

$$\begin{array}{rl} -24x - 2z = -30 & -2 \times (4) \\ x + 2z = 7 & (3) \\ \hline -23x = -23 \\ x = 1 \end{array}$$

Substitute 1 for x in equation (3) to find z.

$$x + 2z = 7 \quad (3)$$
$$1 + 2z = 7$$
$$2z = 6$$
$$z = 3$$

Substitute 1 for x in equation (1) to find y.

$$4x - y = 2 \quad (1)$$
$$4(1) - y = 2$$
$$4 - y = 2$$
$$-y = -2$$
$$y = 2$$

The solution set is $\{(1, 2, 3)\}$.

17.
$$3x - y - z = -8 \quad (1)$$
$$4x + 2y + 3z = 15 \quad (2)$$
$$-6x + 2y + 2z = 10 \quad (3)$$

To eliminate y, multiply equation (1) by 2 and add the result to equation (3).

$$\begin{array}{rcll} 6x - 2y - 2z &=& -16 & 2 \times (1) \\ -6x + 2y + 2z &=& 10 & (3) \\ \hline 0 &=& -6 & \textit{False} \end{array}$$

Since a false statement results, equations (1) and (3) have no common solution. The system is inconsistent. The solution set is $\emptyset$.

18. Let $L =$ the length of the court
and $W =$ the width of the court.

Twice the width is 6 more than the length, so

$$2W = 6 + L \quad \text{or} \quad L = 2W - 6.$$

The perimeter is 288 ft. Use $P = 2L + 2W$.

$$288 = 2L + 2W$$

Substitute $2W - 6$ for L.

$$288 = 2(2W - 6) + 2W$$
$$288 = 4W - 12 + 2W$$
$$300 = 6W$$
$$W = 50$$
$$L = 2(50) - 6 = 94$$

The length is 94 ft and the width is 50 ft.

19. Let $x =$ the number of touchdowns
and $y =$ the number of interceptions.

40 passes resulted in either a touchdown or an interception, so

$$x + y = 40. \quad (1)$$

The number of touchdowns was 2 less than twice the number of interceptions, so

$$x = 2y - 2. \quad (2)$$

Substitute $2y - 2$ for x in (1).

$$(2y - 2) + y = 40$$
$$3y = 42$$
$$y = 14$$

From (2), $x = 2y - 2 = 2(14) - 2 = 26$. He threw 26 touchdowns and 14 interceptions.

20. Let $x =$ the speed of the plane
and $y =$ the speed of the wind.

Complete the chart.

	r	t	d
With wind	$x + y$	1.75	$1.75(x + y)$
Against wind	$x - y$	2	$2(x - y)$

The distance each way is 560 miles. From the chart,

$$1.75(x + y) = 560.$$

Divide by 1.75.

$$x + y = 320 \quad (1)$$

From the chart,

$$2(x - y) = 560$$
$$x - y = 280. \quad (2)$$

Solve the system by adding equations (1) and (2) to eliminate y.

$$\begin{array}{rcll} x + y &=& 320 & (1) \\ x - y &=& 280 & (2) \\ \hline 2x &=& 600 & \\ x &=& 300 & \end{array}$$

From (1), $x + y = 320$, $y = 20$.
The speed of the plane was 300 mph, and the speed of the wind was 20 mph.

21. Let $x =$ the amount of $2-a-pound nuts
and $y =$ the amount of $1-a-pound chocolate candy.

Complete the chart.

	Pounds	Price per Pound (in Dollars)	Value
Nuts	x	2	$2x$
Chocolate	y	1	$1y = y$
Total	100	1.30	$1.30(100) = 130$

Solve the system.

$$x + y = 100 \quad (1)$$
$$2x + y = 130 \quad (2)$$

Solve equation (1) for y.

$$y = 100 - x$$

Substitute $100 - x$ for y in equation (2).

$$2x + (100 - x) = 130$$
$$x = 30$$

Since $y = 100 - x$ and $x = 30$,

$$y = 100 - 30 = 70.$$

She should use 30 lb of $2-a-pound nuts and 70 lb of $1-a-pound candy.

22. Let x = the measure of the largest angle,

y = the measure of the middle-sized angle,

and z = the measure of the smallest angle.

Since the sum of the measures of the angles of a triangle is 180°,

$$x + y + z = 180. \quad (1)$$

Since the largest angle measures 10° less than the sum of the other two,

$$x = y + z - 10$$

$$\text{or} \quad x - y - z = -10. \quad (2)$$

Since the measure of the middle-sized angle is the average of the other two,

$$y = \frac{x + z}{2}$$

$$2y = x + z$$

$$-x + 2y - z = 0. \quad (3)$$

Solve the system.

$$
\begin{array}{rcr}
x + y + z &=& 180 \quad (1) \\
x - y - z &=& -10 \quad (2) \\
-x + 2y - z &=& 0 \quad (3)
\end{array}
$$

Add equations (1) and (2) to find x.

$$
\begin{array}{rcr}
x + y + z &=& 180 \quad (1) \\
x - y - z &=& -10 \quad (2) \\
\hline
2x &=& 170 \\
x &=& 85
\end{array}
$$

Add equations (1) and (3), to find y.

$$
\begin{array}{rcr}
x + y + z &=& 180 \quad (1) \\
-x + 2y - z &=& 0 \quad (3) \\
\hline
3y &=& 180 \\
y &=& 60
\end{array}
$$

Substitute 85 for x and 60 for y in equation (1) to find z.

$$
\begin{array}{rcl}
x + y + z &=& 180 \quad (1) \\
85 + 60 + z &=& 180 \\
145 + z &=& 180 \\
z &=& 35
\end{array}
$$

The three angles are 85°, 60°, and 35°.

23. Let x = the value of sales at 10%,

y = the value of sales at 6%,

and z = the value of sales at 5%.

Since his total sales were $280,000,

$$x + y + z = 280,000 \quad (1)$$

Since his commissions on the sales totaled $17,000,

$$.10x + .06y + .05z = 17,000.$$

Multiply by 100 to clear the decimals, so

$$10x + 6y + 5z = 1,700,000. \quad (2)$$

Since the 5% sale amounted to the sum of the other two sales,

$$z = x + y. \quad (3)$$

Solve the system.

$$
\begin{array}{rcrl}
x + y + z &=& 280,000 & (1) \\
10x + 6y + 5z &=& 1,700,000 & (2) \\
z &=& x + y & (3)
\end{array}
$$

Since equation (3) is given in terms of z, substitute $x + y$ for z in equations (1) and (2).

$$
\begin{array}{rcl}
x + y + z &=& 280,000 \quad (1) \\
x + y + (x + y) &=& 280,000 \\
2x + 2y &=& 280,000 \\
x + y &=& 140,000 \quad (4)
\end{array}
$$

$$
\begin{array}{rcl}
10x + 6y + 5z &=& 1,700,000 \quad (2) \\
10x + 6y + 5(x + y) &=& 1,700,000 \\
10x + 6y + 5x + 5y &=& 1,700,000 \\
15x + 11y &=& 1,700,000 \quad (5)
\end{array}
$$

To eliminate x, multiply equation (4) by -11 and add the result to equation (5).

$$
\begin{array}{rcrl}
-11x - 11y &=& -1,540,000 & -11 \times (4) \\
15x + 11y &=& 1,700,000 & (5) \\
\hline
4x &=& 160,000 & \\
x &=& 40,000 &
\end{array}
$$

From (4), $y = 100,000$.

From (3), $z = 40,000 + 100,000 = 140,000$.

He sold $40,000 at 10%, $100,000 at 6%, and $140,000 at 5%.

24. Let x = the number of liters of 8% solution,

y = the number of liters of 10% solution,

and z = the number of liters of 20% solution.

Since the amount of the mixture will be 8 L,

$$x + y + z = 8. \quad (1)$$

Since the final solution will be 12.5% hydrogen peroxide,

$$.08x + .10y + .20z = .125(8).$$

Multiply by 100 to clear the decimals.

$$8x + 10y + 20z = 100 \quad (2)$$

Since the amount of 8% solution used must be 2 L more than the amount of 20% solution,

$$x = z + 2. \quad (3)$$

Solve the system.

$$x + y + z = 8 \quad (1)$$
$$8x + 10y + 20z = 100 \quad (2)$$
$$x = z + 2 \quad (3)$$

Since equation (3) is given in terms of x, substitute $z + 2$ for x in equations (1) and (2).

$$x + y + z = 8 \quad (1)$$
$$(z + 2) + y + z = 8$$
$$y + 2z = 6 \quad (4)$$

$$8x + 10y + 20z = 100 \quad (2)$$
$$8(z + 2) + 10y + 20z = 100$$
$$8z + 16 + 10y + 20z = 100$$
$$10y + 28z = 84 \quad (5)$$

To eliminate y, multiply equation (4) by -10 and add the result to equation (5).

$$
\begin{array}{rl}
-10y - 20z = -60 & \quad -10 \times (4) \\
\underline{10y + 28z = 84} & \quad (5) \\
8z = 24 & \\
z = 3 &
\end{array}
$$

From (3), $x = z + 2 = 3 + 2 = 5$.
From (4), $y = 6 - 2z = 6 - 2(3) = 0$.
Mix 5 L of 8% solution, none of 10% solution, and 3 L of 20% solution.

25. Let $x =$ the number of home runs hit by Mantle,

$y =$ the number of home runs hit by Maris,

and $z =$ the number of home runs hit by Blanchard.

They combined for 136 home runs, so

$$x + y + z = 136. \quad (1)$$

Mantle hit 7 fewer than Maris, so

$$x = y - 7. \quad (2)$$

Maris hit 40 more than Blanchard, so

$$y = z + 40 \quad \text{or} \quad z = y - 40. \quad (3)$$

Substitute $y - 7$ for x and $y - 40$ for z in (1).

$$x + y + z = 136 \quad (1)$$
$$(y - 7) + y + (y - 40) = 136$$
$$3y - 47 = 136$$
$$3y = 183$$
$$y = 61$$

From (2), $x = y - 7 = 61 - 7 = 54$.
From (3), $z = y - 40 = 61 - 40 = 21$.
Mantle hit 54 home runs, Maris hit 61 home runs, and Blanchard hit 21 home runs.

26. $2x + 5y = -4$
$4x - y = 14$

Write the augmented matrix.

$$\begin{bmatrix} 2 & 5 & \vline & -4 \\ 4 & -1 & \vline & 14 \end{bmatrix}$$

$$\begin{bmatrix} 2 & 5 & \vline & -4 \\ 0 & -11 & \vline & 22 \end{bmatrix} \quad -2R_1 + R_2$$

$$\begin{bmatrix} 2 & 5 & \vline & -4 \\ 0 & 1 & \vline & -2 \end{bmatrix} \quad -\tfrac{1}{11}R_2$$

This matrix gives the system

$$2x + 5y = -4$$
$$y = -2.$$

Substitute $y = -2$ in the first equation.

$$2x + 5y = -4$$
$$2x + 5(-2) = -4$$
$$2x - 10 = -4$$
$$2x = 6$$
$$x = 3$$

Solution set: $\{(3, -2)\}$

27. $6x + 3y = 9$
$-7x + 2y = 17$

Write the augmented matrix.

$$\begin{bmatrix} 6 & 3 & \vline & 9 \\ -7 & 2 & \vline & 17 \end{bmatrix}$$

$$\begin{bmatrix} 1 & \tfrac{1}{2} & \vline & \tfrac{3}{2} \\ -7 & 2 & \vline & 17 \end{bmatrix} \quad \tfrac{1}{6}R_1$$

$$\begin{bmatrix} 1 & \tfrac{1}{2} & \vline & \tfrac{3}{2} \\ 0 & \tfrac{11}{2} & \vline & \tfrac{55}{2} \end{bmatrix} \quad 7R_1 + R_2$$

$$\begin{bmatrix} 1 & \tfrac{1}{2} & \vline & \tfrac{3}{2} \\ 0 & 1 & \vline & 5 \end{bmatrix} \quad \tfrac{2}{11}R_2$$

This matrix gives the system

$$x + \frac{1}{2}y = \frac{3}{2}$$
$$y = 5.$$

Substitute $y = 5$ in the first equation.

$$x + \frac{1}{2}y = \frac{3}{2}$$
$$x + \frac{1}{2}(5) = \frac{3}{2}$$
$$x + \frac{5}{2} = \frac{3}{2}$$
$$x = -1$$

Solution set: $\{(-1, 5)\}$

28.
$$x + 2y - z = 1$$
$$3x + 4y + 2z = -2$$
$$-2x - y + z = -1$$

$$\begin{bmatrix} 1 & 2 & -1 & | & 1 \\ 3 & 4 & 2 & | & -2 \\ -2 & -1 & 1 & | & -1 \end{bmatrix}$$

$$\begin{bmatrix} 1 & 2 & -1 & | & 1 \\ 0 & -2 & 5 & | & -5 \\ 0 & 3 & -1 & | & 1 \end{bmatrix} \quad \begin{matrix} -3R_1 + R_2 \\ 2R_1 + R_3 \end{matrix}$$

$$\begin{bmatrix} 1 & 2 & -1 & | & 1 \\ 0 & 1 & 4 & | & -4 \\ 0 & 3 & -1 & | & 1 \end{bmatrix} \quad R_3 + R_2$$

$$\begin{bmatrix} 1 & 2 & -1 & | & 1 \\ 0 & 1 & 4 & | & -4 \\ 0 & 0 & -13 & | & 13 \end{bmatrix} \quad -3R_2 + R_3$$

$$\begin{bmatrix} 1 & 2 & -1 & | & 1 \\ 0 & 1 & 4 & | & -4 \\ 0 & 0 & 1 & | & -1 \end{bmatrix} \quad -\tfrac{1}{13}R_3$$

This matrix gives the system

$$x + 2y - z = 1$$
$$y + 4z = -4$$
$$z = -1.$$

Substitute $z = -1$ in the second equation.

$$y + 4z = -4$$
$$y + 4(-1) = -4$$
$$y = 0$$

Substitute $y = 0$ and $z = -1$ in the first equation.

$$x + 2y - z = 1$$
$$x + 2(0) - (-1) = 1$$
$$x + 1 = 1$$
$$x = 0$$

Solution set: $\{(0, 0, -1)\}$

29.
$$x + 3y = 7$$
$$3x + z = 2$$
$$y - 2z = 4$$

$$\begin{bmatrix} 1 & 3 & 0 & | & 7 \\ 3 & 0 & 1 & | & 2 \\ 0 & 1 & -2 & | & 4 \end{bmatrix}$$

$$\begin{bmatrix} 1 & 3 & 0 & | & 7 \\ 0 & -9 & 1 & | & -19 \\ 0 & 1 & -2 & | & 4 \end{bmatrix} \quad -3R_1 + R_2$$

$$\begin{bmatrix} 1 & 3 & 0 & | & 7 \\ 0 & 1 & -2 & | & 4 \\ 0 & -9 & 1 & | & -19 \end{bmatrix} \quad R_2 \leftrightarrow R_3$$

We use $\leftrightarrow$ to represent the interchanging of 2 rows.

$$\begin{bmatrix} 1 & 3 & 0 & | & 7 \\ 0 & 1 & -2 & | & 4 \\ 0 & 0 & -17 & | & 17 \end{bmatrix} \quad 9R_2 + R_3$$

$$\begin{bmatrix} 1 & 3 & 0 & | & 7 \\ 0 & 1 & -2 & | & 4 \\ 0 & 0 & 1 & | & -1 \end{bmatrix} \quad -\tfrac{1}{17}R_3$$

This matrix gives the system

$$x + 3y = 7$$
$$y - 2z = 4$$
$$z = -1.$$

Substitute $z = -1$ in the second equation.

$$y - 2z = 4$$
$$y - 2(-1) = 4$$
$$y + 2 = 4$$
$$y = 2$$

Substitute $y = 2$ in the first equation.

$$x + 3y = 7$$
$$x + 3(2) = 7$$
$$x + 6 = 7$$
$$x = 1$$

Solution set: $\{(1, 2, -1)\}$

30. (a) $\begin{vmatrix} 3 & 2 \\ 2 & 3 \end{vmatrix} = 3(3) - 2(2)$

$$= 9 - 4 = 5$$

(b) $\begin{vmatrix} 4 & 2 \\ -3 & 2 \end{vmatrix} = 4(2) - 2(-3)$

$$= 8 + 6 = 14$$

(c) $\begin{vmatrix} -1 & 1 \\ 8 & 8 \end{vmatrix} = -1(8) - 1(8)$

$$= -8 - 8 = -16$$

(d) $\begin{vmatrix} 1 & 2 \\ 6 & 12 \end{vmatrix} = 1(12) - 2(6)$

$$= 12 - 12 = 0$$

The answer is (d).

31. $\begin{vmatrix} 2 & -9 \\ 8 & 4 \end{vmatrix} = 2(4) - (-9)(8)$

$$= 8 + 72 = 80$$

32. $\begin{vmatrix} 7 & 0 \\ 5 & -3 \end{vmatrix} = 7(-3) - 0(5)$

$$= -21$$

33. $\begin{vmatrix} 2 & 10 & 4 \\ 0 & 1 & 3 \\ 0 & 6 & -1 \end{vmatrix}$ Expand about column 1.

$$= 2\begin{vmatrix} 1 & 3 \\ 6 & -1 \end{vmatrix} - 0 + 0$$

$$= 2[1(-1) - 3(6)]$$

$$= 2(-19) = -38$$

34. $\begin{vmatrix} -1 & 7 & 2 \\ 3 & 0 & 5 \\ -1 & 2 & 6 \end{vmatrix}$ Expand about row 2.

$$-3\begin{vmatrix} 7 & 2 \\ 2 & 6 \end{vmatrix} + 0 - 5\begin{vmatrix} -1 & 7 \\ -1 & 2 \end{vmatrix}$$

$$= -3(42 - 4) - 5(-2 + 7)$$

$$= -114 - 25$$

$$= -139$$

35. If $D = 0$, Cramer's rule does not apply.

36. For three unknowns, three equations are needed.

37. $3x - 4y = -32$
$2x + y = -3$

$$D = \begin{vmatrix} 3 & -4 \\ 2 & 1 \end{vmatrix} = 3(1) - (-4)(2) = 11$$

$$D_x = \begin{vmatrix} -32 & -4 \\ -3 & 1 \end{vmatrix} = -32(1) - (-4)(-3)$$
$$= -32 - 12 = -44$$

$$D_y = \begin{vmatrix} 3 & -32 \\ 2 & -3 \end{vmatrix} = 3(-3) - (-32)(2)$$
$$= -9 + 64 = 55$$

$$x = \frac{D_x}{D} = \frac{-44}{11} = -4; \; y = \frac{D_y}{D} = \frac{55}{11} = 5$$

Solution set: $\{(-4, 5)\}$

38. $-4x + 3y = -12$
$2x + 6y = 15$

$$D = \begin{vmatrix} -4 & 3 \\ 2 & 6 \end{vmatrix} = -4(6) - 3(2)$$
$$= -24 - 6 = -30$$

$$D_x = \begin{vmatrix} -12 & 3 \\ 15 & 6 \end{vmatrix} = -12(6) - 3(15)$$
$$= -72 - 45 = -117$$

$$D_y = \begin{vmatrix} -4 & -12 \\ 2 & 15 \end{vmatrix} = -4(15) - (-12)(2)$$
$$= -60 + 24 = -36$$

$$x = \frac{D_x}{D} = \frac{-117}{-30} = \frac{39}{10}; \; y = \frac{D_y}{D} = \frac{-36}{-30} = \frac{6}{5}$$

Solution set: $\left\{\left(\dfrac{39}{10}, \dfrac{6}{5}\right)\right\}$

39. $4x + y + z = 11$
$x - y - z = 4$
$y + 2z = 0$

Expand about row 3 to find D, D_x, D_y, and D_z.

$$D = \begin{vmatrix} 4 & 1 & 1 \\ 1 & -1 & -1 \\ 0 & 1 & 2 \end{vmatrix}$$

$$= 0 - 1\begin{vmatrix} 4 & 1 \\ 1 & -1 \end{vmatrix} + 2\begin{vmatrix} 4 & 1 \\ 1 & -1 \end{vmatrix}$$

$$= -1(-5) + 2(-5)$$

$$= 5 - 10 = -5$$

$$D_x = \begin{vmatrix} 11 & 1 & 1 \\ 4 & -1 & -1 \\ 0 & 1 & 2 \end{vmatrix}$$

$$= 0 - 1\begin{vmatrix} 11 & 1 \\ 4 & -1 \end{vmatrix} + 2\begin{vmatrix} 11 & 1 \\ 4 & -1 \end{vmatrix}$$

$$= -1(-15) + 2(-15)$$

$$= 15 - 30 = -15$$

$$D_y = \begin{vmatrix} 4 & 11 & 1 \\ 1 & 4 & -1 \\ 0 & 0 & 2 \end{vmatrix} = 0 - 0 + 2\begin{vmatrix} 4 & 11 \\ 1 & 4 \end{vmatrix}$$
$$= 2(5) = 10$$

$$D_z = \begin{vmatrix} 4 & 1 & 11 \\ 1 & -1 & 4 \\ 0 & 1 & 0 \end{vmatrix} = 0 - 1\begin{vmatrix} 4 & 11 \\ 1 & 4 \end{vmatrix} + 0$$
$$= -1(5) = -5$$

$$x = \frac{D_x}{D} = \frac{-15}{-5} = 3; \; y = \frac{D_y}{D} = \frac{10}{-5} = -2$$

$$z = \frac{D_z}{D} = \frac{-5}{-5} = 1$$

Solution set: $\{(3, -2, 1)\}$

40. $-x + 3y - 4z = 4$
$2x + 4y + z = -14$
$3x - y + 2z = -8$

$$D = \begin{vmatrix} -1 & 3 & -4 \\ 2 & 4 & 1 \\ 3 & -1 & 2 \end{vmatrix}$$ Expand about column 1.

$$= -1\begin{vmatrix} 4 & 1 \\ -1 & 2 \end{vmatrix} - 2\begin{vmatrix} 3 & -4 \\ -1 & 2 \end{vmatrix} + 3\begin{vmatrix} 3 & -4 \\ 4 & 1 \end{vmatrix}$$

$$= -1(9) - 2(2) + 3(19)$$

$$= -9 - 4 + 57 = 44$$

$$D_x = \begin{vmatrix} 4 & 3 & -4 \\ -14 & 4 & 1 \\ -8 & -1 & 2 \end{vmatrix}$$ Expand about column 3.

$$= -4\begin{vmatrix} -14 & 4 \\ -8 & -1 \end{vmatrix} - 1\begin{vmatrix} 4 & 3 \\ -8 & -1 \end{vmatrix}$$
$$+ 2\begin{vmatrix} 4 & 3 \\ -14 & 4 \end{vmatrix}$$

$$= -4(46) - 1(20) + 2(58)$$

$$= -184 - 20 + 116 = -88$$

continued

$$D_y = \begin{vmatrix} -1 & 4 & -4 \\ 2 & -14 & 1 \\ 3 & -8 & 2 \end{vmatrix} \quad \text{Expand about column 1.}$$

$$= -1 \begin{vmatrix} -14 & 1 \\ -8 & 2 \end{vmatrix} - 2 \begin{vmatrix} 4 & -4 \\ -8 & 2 \end{vmatrix}$$

$$+ 3 \begin{vmatrix} 4 & -4 \\ -14 & 1 \end{vmatrix}$$

$$= -1(-20) - 2(-24) + 3(-52)$$

$$= 20 + 48 - 156 = -88$$

$$D_z = \begin{vmatrix} -1 & 3 & 4 \\ 2 & 4 & -14 \\ 3 & -1 & -8 \end{vmatrix} \quad \text{Expand about column 1.}$$

$$= -1 \begin{vmatrix} 4 & -14 \\ -1 & -8 \end{vmatrix} - 2 \begin{vmatrix} 3 & 4 \\ -1 & -8 \end{vmatrix}$$

$$+ 3 \begin{vmatrix} 3 & 4 \\ 4 & -14 \end{vmatrix}$$

$$= -1(-46) - 2(-20) + 3(-58)$$

$$= 46 + 40 - 174 = -88$$

Since $D_x = D_y = D_z$,

$$x = y = z = \frac{-88}{44} = -2.$$

Solution set: $\{(-2, -2, -2)\}$

41.
$$\begin{aligned} 2x + y + 3z &= 1 \quad (1) \\ x - 2y + z &= -3 \quad (2) \\ -3x + y - 2z &= -4 \quad (3) \end{aligned}$$

Eliminate x from equations (1) and (2) by multiplying equation (2) by -2. Then add the results.

$$\begin{aligned} 2x + y + 3z &= 1 \quad (1) \\ -2x + 4y - 2z &= 6 \qquad -2 \times (2) \\ \hline 5y + z &= 7 \quad (4) \end{aligned}$$

Eliminate x from equations (2) and (3) by multiplying equation (2) by 3. Then add the results.

$$\begin{aligned} 3x - 6y + 3z &= -9 \qquad 3 \times (2) \\ -3x + y - 2z &= -4 \quad (3) \\ \hline -5y + z &= -13 \quad (5) \end{aligned}$$

The resulting system is

$$\begin{aligned} 5y + z &= 7 \quad (4) \\ -5y + z &= -13. \quad (5) \end{aligned}$$

42.
$$\begin{aligned} 5y + z &= 7 \quad (4) \\ -5y + z &= -13 \quad (5) \\ \hline 2z &= -6 \quad \textit{Add.} \\ z &= -3 \end{aligned}$$

Substitute $z = -3$ into equation (4).

$$\begin{aligned} 5y + z &= 7 \quad (4) \\ 5y - 3 &= 7 \\ 5y &= 10 \\ y &= 2 \end{aligned}$$

Solution set: $\{(y, z) = (2, -3)\}$

43. Substitute $y = 2$ and $z = -3$ into equation (2).

$$\begin{aligned} x - 2y + z &= -3 \quad (2) \\ x - 2(2) + (-3) &= -3 \\ x - 7 &= -3 \\ x &= 4 \end{aligned}$$

Solution set: $\{(4, 2, -3)\}$

44. The augmented matrix is

$$\left[\begin{array}{ccc|c} 2 & 1 & 3 & 1 \\ 1 & -2 & 1 & -3 \\ -3 & 1 & -2 & -4 \end{array} \right].$$

$$\left[\begin{array}{ccc|c} 1 & -2 & 1 & -3 \\ 2 & 1 & 3 & 1 \\ -3 & 1 & -2 & -4 \end{array} \right] \quad R_1 \leftrightarrow R_2$$

$$\left[\begin{array}{ccc|c} 1 & -2 & 1 & -3 \\ 0 & 5 & 1 & 7 \\ 0 & -5 & 1 & -13 \end{array} \right] \quad \begin{array}{l} -2R_1 + R_2 \\ 3R_1 + R_3 \end{array}$$

$$\left[\begin{array}{ccc|c} 1 & -2 & 1 & -3 \\ 0 & 1 & \frac{1}{5} & \frac{7}{5} \\ 0 & -5 & 1 & -13 \end{array} \right] \quad \frac{1}{5}R_2$$

$$\left[\begin{array}{ccc|c} 1 & -2 & 1 & -3 \\ 0 & 1 & \frac{1}{5} & \frac{7}{5} \\ 0 & 0 & 2 & -6 \end{array} \right] \quad 5R_2 + R_3$$

$$\left[\begin{array}{ccc|c} 1 & -2 & 1 & -3 \\ 0 & 1 & \frac{1}{5} & \frac{7}{5} \\ 0 & 0 & 1 & -3 \end{array} \right] \quad \frac{1}{2}R_3$$

This matrix gives the system

$$\begin{aligned} x - 2y + z &= -3 \\ y + \frac{1}{5}z &= \frac{7}{5} \\ z &= -3. \end{aligned}$$

Substitute $z = -3$ in the second equation.

$$y + \frac{1}{5}z = \frac{7}{5}$$

$$y + \frac{1}{5}(-3) = \frac{7}{5}$$

$$y - \frac{3}{5} = \frac{7}{5}$$

$$y = 2$$

Substitute $y = 2$ and $z = -3$ in the first equation.

$$x - 2y + z = -3$$
$$x - 2(2) + (-3) = -3$$
$$x - 7 = -3$$
$$x = 4$$

The solution set is $\{(4, 2, -3)\}$, the same as that found in Exercise 43.

45. (a)

$$D = \begin{vmatrix} 2 & 1 & 3 \\ 1 & -2 & 1 \\ -3 & 1 & -2 \end{vmatrix} \quad \text{Expand about row 1.}$$

$$= 2\begin{vmatrix} -2 & 1 \\ 1 & -2 \end{vmatrix} - 1\begin{vmatrix} 1 & 1 \\ -3 & -2 \end{vmatrix} + 3\begin{vmatrix} 1 & -2 \\ -3 & 1 \end{vmatrix}$$

$$= 2(4 - 1) - 1(-2 + 3) + 3(1 - 6)$$
$$= 2(3) - 1(1) + 3(-5)$$
$$= 6 - 1 - 15 = -10$$

(b)

$$D_x = \begin{vmatrix} 1 & 1 & 3 \\ -3 & -2 & 1 \\ -4 & 1 & -2 \end{vmatrix} \quad \text{Expand about row 1.}$$

$$= 1\begin{vmatrix} -2 & 1 \\ 1 & -2 \end{vmatrix} - 1\begin{vmatrix} -3 & 1 \\ -4 & -2 \end{vmatrix} + 3\begin{vmatrix} -3 & -2 \\ -4 & 1 \end{vmatrix}$$

$$= 1(4 - 1) - 1(6 + 4) + 3(-3 - 8)$$
$$= 1(3) - 1(10) + 3(-11)$$
$$= 3 - 10 - 33 = -40$$

(c)

$$D_y = \begin{vmatrix} 2 & 1 & 3 \\ 1 & -3 & 1 \\ -3 & -4 & -2 \end{vmatrix} \quad \text{Expand about row 1.}$$

$$= 2\begin{vmatrix} -3 & 1 \\ -4 & -2 \end{vmatrix} - 1\begin{vmatrix} 1 & 1 \\ -3 & -2 \end{vmatrix} + 3\begin{vmatrix} 1 & -3 \\ -3 & -4 \end{vmatrix}$$

$$= 2(6 + 4) - 1(-2 + 3) + 3(-4 - 9)$$
$$= 2(10) - 1(1) + 3(-13)$$
$$= 20 - 1 - 39 = -20$$

(d)

$$D_z = \begin{vmatrix} 2 & 1 & 1 \\ 1 & -2 & -3 \\ -3 & 1 & -4 \end{vmatrix} \quad \text{Expand about row 1.}$$

$$= 2\begin{vmatrix} -2 & -3 \\ 1 & -4 \end{vmatrix} - 1\begin{vmatrix} 1 & -3 \\ -3 & -4 \end{vmatrix} + 1\begin{vmatrix} 1 & -2 \\ -3 & 1 \end{vmatrix}$$

$$= 2(8 + 3) - 1(-4 - 9) + 1(1 - 6)$$
$$= 2(11) - 1(-13) + 1(-5)$$
$$= 22 + 13 - 5 = 30$$

46. $x = \dfrac{D_x}{D} = \dfrac{-40}{-10} = 4$

$y = \dfrac{D_y}{D} = \dfrac{-20}{-10} = 2$

$z = \dfrac{D_z}{D} = \dfrac{30}{-10} = -3$

The solution set, $\{(4, 2, -3)\}$, is the same as before.

47. $\dfrac{2}{3}x + \dfrac{1}{6}y = \dfrac{19}{2}$ (1)

$\dfrac{1}{3}x - \dfrac{2}{9}y = 2$ (2)

Multiply equation (1) by 6 and equation (2) by 9 to clear the fractions.

$$4x + y = 57 \quad (3) \ 6 \times (1)$$
$$3x - 2y = 18 \quad (4) \ 9 \times (2)$$

To eliminate y, multiply equation (3) by 2 and add the result to equation (4).

$$\begin{array}{rll} 8x + 2y = & 114 & 2 \times (3) \\ 3x - 2y = & 18 & (4) \\ \hline 11x = & 132 & \\ x = & 12 & \end{array}$$

Substitute 12 for x in equation (3) to find y.

$$4x + y = 57 \quad (3)$$
$$4(12) + y = 57$$
$$48 + y = 57$$
$$y = 9$$

Solution set: $\{(12, 9)\}$

48. $\begin{array}{rll} 2x + 5y - z = & 12 & (1) \\ -x + y - 4z = & -10 & (2) \\ -8x - 20y + 4z = & 31 & (3) \end{array}$

Multiply equation (1) by 4 and add the result to equation (3).

$$\begin{array}{rll} 8x + 20y - 4z = & 48 & 4 \times (1) \\ -8x - 20y + 4z = & 31 & (3) \\ \hline 0 = & 79 & \textit{False} \end{array}$$

Since a false statement results, the system is inconsistent. The solution set is $\emptyset$.

49. $x = 7y + 10$ (1)

$2x + 3y = 3$ (2)

Since equation (1) is given in terms of x, substitute $7y + 10$ for x in equation (2) and solve for y.

$$2(7y + 10) + 3y = 3$$
$$14y + 20 + 3y = 3$$
$$17y = -17$$
$$y = -1$$

Since $x = 7y + 10$ and $y = -1$,

$$x = 7(-1) + 10 = -7 + 10 = 3.$$

Solution set: $\{(3, -1)\}$

50.
$$x + 4y = 17 \quad (1)$$
$$-3x + 2y = -9 \quad (2)$$

To eliminate x, multiply equation (1) by 3 and add the result to equation (2).

$$
\begin{array}{rr}
3x + 12y = 51 & 3 \times (1) \\
-3x + 2y = -9 & (2) \\
\hline
14y = 42 & \\
y = 3 &
\end{array}
$$

Substitute 3 for y in equation (1) to find x.

$$x + 4y = 17 \quad (1)$$
$$x + 4(3) = 17$$
$$x + 12 = 17$$
$$x = 5$$

Solution set: $\{(5, 3)\}$

51.
$$-7x + 3y = 12 \quad (1)$$
$$5x + 2y = 8 \quad (2)$$

To eliminate y, multiply equation (1) by 2 and equation (2) by -3. Then add the results.

$$
\begin{array}{rr}
-14x + 6y = 24 & 2 \times (1) \\
-15x - 6y = -24 & -3 \times (2) \\
\hline
-29x = 0 & \\
x = 0 &
\end{array}
$$

Substitute 0 for x in equation (1) to find y.

$$-7x + 3y = 12 \quad (1)$$
$$-7(0) + 3y = 12$$
$$3y = 12$$
$$y = 4$$

Solution set: $\{(0, 4)\}$

52.
$$2x - 5y = 8 \quad (1)$$
$$3x + 4y = 10 \quad (2)$$

To eliminate y, multiply equation (1) by 4 and equation (2) by 5 and add the results.

$$
\begin{array}{rr}
8x - 20y = 32 & 4 \times (1) \\
15x + 20y = 50 & 5 \times (2) \\
\hline
23x = 82 & \\
x = \dfrac{82}{23} &
\end{array}
$$

Instead of substituting to find y, we'll choose different multipliers and eliminate x from the original system.

$$
\begin{array}{rr}
6x - 15y = 24 & 3 \times (1) \\
-6x - 8y = -20 & -2 \times (2) \\
\hline
-23y = 4 & \\
y = -\dfrac{4}{23} &
\end{array}
$$

Solution set: $\left\{ \left(\dfrac{82}{23}, -\dfrac{4}{23} \right) \right\}$

53. Let $x =$ the price of an AC adaptor and $y =$ the price of a rechargeable flashlight.

Since 7 AC adaptors and 2 rechargeable flashlights cost $86,

$$7x + 2y = 86. \quad (1)$$

Since 3 AC adaptors and 4 rechargeable flashlights cost $84,

$$3x + 4y = 84. \quad (2)$$

Solve the system.

$$7x + 2y = 86 \quad (1)$$
$$3x + 4y = 84 \quad (2)$$

To eliminate y, multiply equation (1) by -2 and add the result to equation (2).

$$
\begin{array}{rr}
-14x - 4y = -172 & -2 \times (1) \\
3x + 4y = 84 & (2) \\
\hline
-11x = -88 & \\
x = 8 &
\end{array}
$$

Substitute 8 for x in equation (1) to find y.

$$7x + 2y = 86 \quad (1)$$
$$7(8) + 2y = 86$$
$$56 + 2y = 86$$
$$2y = 30$$
$$y = 15$$

An AC adaptor costs $8, and a rechargeable flashlight costs $15.

54. Let $x =$ the number of gold medals won by Canada,

$y =$ the number of silver medals won by Canada,

and $z =$ the number of bronze medals won by Canada.

They won a total of 22 medals, so

$$x + y + z = 22. \quad (1)$$

There were 5 fewer gold medals than bronze, so

$$x = z - 5. \quad (2)$$

There were 3 fewer bronze than silver, so

$$z = y - 3 \quad \text{or} \quad y = z + 3. \quad (3)$$

Substitute $z - 5$ for x and $z + 3$ for y in (1).

$$x + y + z = 22 \quad (1)$$
$$(z - 5) + (z + 3) + z = 22$$
$$3z - 2 = 22$$
$$3z = 24$$
$$z = 8$$

From (2), $x = z - 5 = 8 - 5 = 3$.
From (3), $y = z + 3 = 8 + 3 = 11$.
Canada won 3 gold medals, 11 silver medals, and 8 bronze medals.

Chapter 4 Test

1. No; The graph for Babe Ruth lies completely below the graph for Aaron, indicating that Ruth's total was always lower than Aaron's.

2. Aaron had the most and Ruth had the fewest.

3. When each equation of the system

$$x + y = 7$$
$$x - y = 5$$

is graphed, the point of intersection appears to be $(6, 1)$. To check, substitute 6 for x and 1 for y in each of the equations. Since $(6, 1)$ makes both equations true, the solution set of the system is $\{(6, 1)\}$.

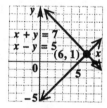

4.
$$3x + y = 12 \quad (1)$$
$$2x - y = 3 \quad (2)$$

To eliminate y, add equations (1) and (2).

$$\begin{array}{rl} 3x + y = 12 & (1) \\ \underline{2x - y = 3} & (2) \\ 5x = 15 & \\ x = 3 & \end{array}$$

Substitute 3 for x in equation (1) to find y.

$$3x + y = 12 \quad (1)$$
$$3(3) + y = 12$$
$$9 + y = 12$$
$$y = 3$$

Solution set: $\{(3, 3)\}$

5.
$$-5x + 2y = -4 \quad (1)$$
$$6x + 3y = -6 \quad (2)$$

To eliminate x, multiply equation (1) by 6 and equation (2) by 5. Then add the results.

$$\begin{array}{rl} -30x + 12y = -24 & 6 \times (1) \\ \underline{30x + 15y = -30} & 5 \times (2) \\ 27y = -54 & \\ y = -2 & \end{array}$$

Substitute -2 for y in equation (1) to find x.

$$-5x + 2y = -4 \quad (1)$$
$$-5x + 2(-2) = -4$$
$$-5x - 4 = -4$$
$$-5x = 0$$
$$x = 0$$

Solution set: $\{(0, -2)\}$

6.
$$\begin{array}{rl} 3x + 4y = 8 & (1) \\ 8y = 7 - 6x & \end{array}$$
or $6x + 8y = 7 \quad (2)$

Multiply equation (1) by -2 and add the result to equation (2).

$$\begin{array}{rl} -6x - 8y = -16 & -2 \times (1) \\ \underline{6x + 8y = 7} & \\ 0 = -9 & \textit{False} \end{array}$$

Since a false statement results, the system is inconsistent. The solution set is $\emptyset$.

7.
$$\begin{array}{rl} 3x + 5y + 3z = 2 & (1) \\ 6x + 5y + z = 0 & (2) \\ 3x + 10y - 2z = 6 & (3) \end{array}$$

To eliminate x, multiply equation (1) by -1 and add the result to equation (3).

$$\begin{array}{rl} -3x - 5y - 3z = -2 & -1 \times (1) \\ \underline{3x + 10y - 2z = 6} & (3) \\ 5y - 5z = 4 & (4) \end{array}$$

To eliminate x again, multiply equation (1) by -2 and add the result to equation (2).

$$\begin{array}{rl} -6x - 10y - 6z = -4 & -2 \times (1) \\ \underline{6x + 5y + z = 0} & (2) \\ -5y - 5z = -4 & (5) \end{array}$$

To eliminate y, add equations (4) and (5).

$$\begin{array}{rl} 5y - 5z = 4 & (4) \\ \underline{-5y - 5z = -4} & (5) \\ -10z = 0 & \\ z = 0 & \end{array}$$

Substitute 0 for z in equation (4) to find y.

continued

$$5y - 5z = 4 \quad (4)$$
$$5y - 5(0) = 4$$
$$5y - 0 = 4$$
$$5y = 4$$
$$y = \frac{4}{5}$$

Substitute $\frac{4}{5}$ for y and 0 for z in equation (1) to find x.

$$3x + 5y + 3z = 2 \quad (1)$$
$$3x + 5\left(\frac{4}{5}\right) + 3(0) = 2$$
$$3x + 4 + 0 = 2$$
$$3x = -2$$
$$x = -\frac{2}{3}$$

Solution set: $\left\{\left(-\frac{2}{3}, \frac{4}{5}, 0\right)\right\}$

8. $2x - 3y = 24 \quad (1)$

$$y = -\frac{2}{3}x \quad (2)$$

Since equation (2) is solved for y, substitute $-\frac{2}{3}x$ for y in equation (1) and solve for x.

$$2x - 3y = 24 \quad (1)$$
$$2x - 3\left(-\frac{2}{3}x\right) = 24$$
$$2x + 2x = 24$$
$$4x = 24$$
$$x = 6$$

Since $y = -\frac{2}{3}x$ and $x = 6$,

$$y = -\frac{2}{3}(6) = -4.$$

Solution set: $\{(6, -4)\}$

9. $12x - 5y = 8 \quad (1)$

$$3x = \frac{5}{4}y + 2$$

or $x = \frac{5}{12}y + \frac{2}{3} \quad (2)$

Substitute $\frac{5}{12}y + \frac{2}{3}$ for x in equation (1) and solve for y.

$$12x - 5y = 8 \quad (1)$$
$$12\left(\frac{5}{12}y + \frac{2}{3}\right) - 5y = 8$$
$$5y + 8 - 5y = 8$$
$$8 = 8 \quad \textit{True}$$

Equations (1) and (2) are dependent.

Solution set: $\{(x, y) | 12x - 5y = 8\}$

10. Let $x =$ the speed of the fast car and $y =$ the speed of the slow car.

Make a table.

	r	t	d
Fast car	x	3.5	$3.5x$
Slow car	y	3.5	$3.5y$

Since the fast car travels 30 mph faster than the slow car,

$$x - y = 30. \quad (1)$$

Since the cars travel a total of 420 miles,

$$3.5x + 3.5y = 420.$$

Multiply by 10 to clear the decimals.

$$35x + 35y = 4200 \quad (2)$$

To eliminate y, multiply equation (1) by 35 and add the result to equation (2).

$$\begin{array}{lll} 35x - 35y = 1050 & \quad 35 \times (1) \\ \underline{35x + 35y = 4200} & \quad (2) \\ 70x \qquad\quad = 5250 \\ \quad\, x \qquad\quad = 75 \end{array}$$

Substitute 75 for x in equation (1) to find y.

$$x - y = 30 \quad (1)$$
$$75 - y = 30$$
$$-y = -45$$
$$y = 45$$

The fast car is traveling at 75 mph, and the slow car is traveling at 45 mph.

11. Let $x =$ the number of liters of 20% solution and $y =$ the number of liters of 50% solution.

Make a table.

Kind of Solution	Liters of Solution	Liters of Pure Alcohol
.20	x	$.20x$
.50	y	$.50y$
.40	12	$.40(12) = 4.8$

Since 12 L of the mixture are needed,

$$x + y = 12. \quad (1)$$

Since the amount of pure alcohol in the 20% solution plus the amount of pure alcohol in the 50% solution must equal the amount of alcohol in the mixture,

$$.2x + .5y = 4.8.$$

Multiply by 10 to clear the decimals.

$$2x + 5y = 48 \quad (2)$$

Multiply equation (1) by -2 and add the result to equation (2).

$$
\begin{array}{rcll}
-2x - 2y &=& -24 & -2 \times (1) \\
2x + 5y &=& 48 & (2) \\
\hline
3y &=& 24 & \\
y &=& 8 &
\end{array}
$$

From (1), $x + y = 12$, $x = 4$.
4 L of 20% solution and 8 L of 50% solution are needed.

12. Let $x =$ the number of points scored by the Bulls and $y =$ the number of points scored by the Jazz.

The total number of points scored is 176, so

$$x + y = 176. \quad (1)$$

The Bulls scored 4 points more than the Jazz, so

$$x = y + 4. \quad (2)$$

Substitute $y + 4$ for x in (1).

$$
\begin{array}{rcl}
x + y &=& 176 \quad (1) \\
(y + 4) + y &=& 176 \\
2y + 4 &=& 176 \\
2y &=& 172 \\
y &=& 86
\end{array}
$$

From (2), $x = y + 4 = 86 + 4 = 90$.
The score of the game was:

Bulls 90, Jazz 86

13. Let $x =$ the cost of a sheet of colored paper and $y =$ the cost of a marker pen.

For the first purchase $8x$ represents the cost of the paper and $3y$ the cost of the pens. The total cost was $6.50, so

$$8x + 3y = 6.50. \quad (1)$$

For the second purchase,

$$2x + 2y = 3.00. \quad (2)$$

To eliminate x, multiply equation (2) by -4 and add the result to equation (1).

$$
\begin{array}{rcll}
8x + 3y &=& 6.50 & (1) \\
-8x - 8y &=& -12.00 & -4 \times (2) \\
\hline
-5y &=& -5.50 & \\
y &=& 1.10 &
\end{array}
$$

Substitute $y = 1.10$ in equation (2).

$$
\begin{array}{rcl}
2x + 2y &=& 3.00 \quad (2) \\
2x + 2(1.10) &=& 3.00 \\
2x + 2.20 &=& 3.00 \\
2x &=& .80 \\
x &=& .40
\end{array}
$$

She paid $1.10 for each marker pen and $.40 for each sheet of colored paper.

14.
$$
\begin{array}{rcl}
3x + 2y &=& 4 \\
5x + 5y &=& 9
\end{array}
$$

Write the augmented matrix.

$$\left[\begin{array}{cc|c} 3 & 2 & 4 \\ 5 & 5 & 9 \end{array}\right]$$

$$\left[\begin{array}{cc|c} 1 & \frac{2}{3} & \frac{4}{3} \\ 5 & 5 & 9 \end{array}\right] \quad \frac{1}{3}R_1$$

$$\left[\begin{array}{cc|c} 1 & \frac{2}{3} & \frac{4}{3} \\ 0 & \frac{5}{3} & \frac{7}{3} \end{array}\right] \quad -5R_1 + R_2$$

$$\left[\begin{array}{cc|c} 1 & \frac{2}{3} & \frac{4}{3} \\ 0 & 1 & \frac{7}{5} \end{array}\right] \quad \frac{3}{5}R_2$$

This matrix gives the system

$$
\begin{aligned}
x + \frac{2}{3}y &= \frac{4}{3} \\
y &= \frac{7}{5}.
\end{aligned}
$$

Substitute $y = \frac{7}{5}$ in the first equation.

$$
\begin{aligned}
x + \frac{2}{3}\left(\frac{7}{5}\right) &= \frac{4}{3} \\
x + \frac{14}{15} &= \frac{4}{3} \\
x &= \frac{20}{15} - \frac{14}{15} = \frac{6}{15} = \frac{2}{5}
\end{aligned}
$$

Solution set: $\left\{\left(\dfrac{2}{5}, \dfrac{7}{5}\right)\right\}$

15.
$$
\begin{array}{rcrcr}
x &+ 3y &+ 2z &=& 11 \\
3x &+ 7y &+ 4z &=& 23 \\
5x &+ 3y &- 5z &=& -14
\end{array}
$$

Write the augmented matrix.

continued

$$\begin{bmatrix} 1 & 3 & 2 & | & 11 \\ 3 & 7 & 4 & | & 23 \\ 5 & 3 & -5 & | & -14 \end{bmatrix}$$

$$\begin{bmatrix} 1 & 3 & 2 & | & 11 \\ 0 & -2 & -2 & | & -10 \\ 0 & -12 & -15 & | & -69 \end{bmatrix} \quad \begin{matrix} -3R_1 + R_2 \\ -5R_1 + R_3 \end{matrix}$$

$$\begin{bmatrix} 1 & 3 & 2 & | & 11 \\ 0 & 1 & 1 & | & 5 \\ 0 & -12 & -15 & | & -69 \end{bmatrix} \quad -\tfrac{1}{2}R_2$$

$$\begin{bmatrix} 1 & 3 & 2 & | & 11 \\ 0 & 1 & 1 & | & 5 \\ 0 & 0 & -3 & | & -9 \end{bmatrix} \quad 12R_2 + R_3$$

$$\begin{bmatrix} 1 & 3 & 2 & | & 11 \\ 0 & 1 & 1 & | & 5 \\ 0 & 0 & 1 & | & 3 \end{bmatrix} \quad -\tfrac{1}{3}R_3$$

This matrix gives the system

$$x + 3y + 2z = 11$$
$$y + z = 5$$
$$z = 3.$$

Substitute $z = 3$ in the second equation.

$$y + z = 5$$
$$y + 3 = 5$$
$$y = 2$$

Substitute $y = 2$ and $z = 3$ in the first equation.

$$x + 3y + 2z = 11$$
$$x + 3(2) + 2(3) = 11$$
$$x + 6 + 6 = 11$$
$$x = -1$$

Solution set: $\{(-1, 2, 3)\}$

16.
$$\begin{aligned} 4x - 2y \quad &= -8 \\ 3y - 5z &= 14 \\ 2x \quad + z &= -10 \end{aligned}$$

Write the augmented matrix.

$$\begin{bmatrix} 4 & -2 & 0 & | & -8 \\ 0 & 3 & -5 & | & 14 \\ 2 & 0 & 1 & | & -10 \end{bmatrix}$$

$$\begin{bmatrix} 1 & -\tfrac{1}{2} & 0 & | & -2 \\ 0 & 3 & -5 & | & 14 \\ 2 & 0 & 1 & | & -10 \end{bmatrix} \quad \tfrac{1}{4}R_1$$

$$\begin{bmatrix} 1 & -\tfrac{1}{2} & 0 & | & -2 \\ 0 & 3 & -5 & | & 14 \\ 0 & 1 & 1 & | & -6 \end{bmatrix} \quad -2R_1 + R_3$$

$$\begin{bmatrix} 1 & -\tfrac{1}{2} & 0 & | & -2 \\ 0 & 1 & -\tfrac{5}{3} & | & \tfrac{14}{3} \\ 0 & 1 & 1 & | & -6 \end{bmatrix} \quad \tfrac{1}{3}R_2$$

$$\begin{bmatrix} 1 & -\tfrac{1}{2} & 0 & | & -2 \\ 0 & 1 & -\tfrac{5}{3} & | & \tfrac{14}{3} \\ 0 & 0 & \tfrac{8}{3} & | & -\tfrac{32}{3} \end{bmatrix} \quad -1R_2 + R_3$$

$$\begin{bmatrix} 1 & -\tfrac{1}{2} & 0 & | & -2 \\ 0 & 1 & -\tfrac{5}{3} & | & \tfrac{14}{3} \\ 0 & 0 & 1 & | & -4 \end{bmatrix} \quad \tfrac{3}{8}R_3$$

This matrix gives the system

$$x - \frac{1}{2}y = -2$$
$$y - \frac{5}{3}z = \frac{14}{3}$$
$$z = -4.$$

Substitute $z = -4$ in the second equation.

$$y - \frac{5}{3}z = \frac{14}{3}$$
$$y - \frac{5}{3}(-4) = \frac{14}{3}$$
$$y + \frac{20}{3} = \frac{14}{3}$$
$$y = -2$$

Substitute $y = -2$ in the first equation.

$$x - \frac{1}{2}y = -2$$
$$x - \frac{1}{2}(-2) = -2$$
$$x + 1 = -2$$
$$x = -3$$

Solution set: $\{(-3, -2, -4)\}$

17. $\begin{vmatrix} 6 & -3 \\ 5 & -2 \end{vmatrix} = 6(-2) - (-3)(5)$

$$= -12 + 15 = 3$$

18. $\begin{vmatrix} 4 & 1 & 0 \\ -2 & 7 & 3 \\ 0 & 5 & 2 \end{vmatrix}$ Expand about row 1.

$$= 4\begin{vmatrix} 7 & 3 \\ 5 & 2 \end{vmatrix} - 1\begin{vmatrix} -2 & 3 \\ 0 & 2 \end{vmatrix} + 0$$
$$= 4[7(2) - 3(5)] - 1[-2(2) - 3(0)]$$
$$= 4(-1) - 1(-4)$$
$$= -4 + 4 = 0$$

19. $3x - y = -8$

$2x + 6y = 3$

$D = \begin{vmatrix} 3 & -1 \\ 2 & 6 \end{vmatrix} = 18 - (-2) = 20$

$D_x = \begin{vmatrix} -8 & -1 \\ 3 & 6 \end{vmatrix} = -48 - (-3) = -45$

$D_y = \begin{vmatrix} 3 & -8 \\ 2 & 3 \end{vmatrix} = 9 - (-16) = 25$

$x = \dfrac{D_x}{D} = \dfrac{-45}{20} = -\dfrac{9}{4}; \; y = \dfrac{D_y}{D} = \dfrac{25}{20} = \dfrac{5}{4}$

Solution set: $\left\{ \left(-\dfrac{9}{4}, \dfrac{5}{4} \right) \right\}$

20. $x + y + z = 6$

$2x - 2y + z = 5$

$-x + 3y + z = 0$

$D = \begin{vmatrix} 1 & 1 & 1 \\ 2 & -2 & 1 \\ -1 & 3 & 1 \end{vmatrix}$ Expand about row 1.

$= 1\begin{vmatrix} -2 & 1 \\ 3 & 1 \end{vmatrix} - 1\begin{vmatrix} 2 & 1 \\ -1 & 1 \end{vmatrix} + 1\begin{vmatrix} 2 & -2 \\ -1 & 3 \end{vmatrix}$

$= 1(-2 - 3) - 1(2 + 1) + 1(6 - 2)$

$= -5 - 3 + 4 = -4$

$D_x = \begin{vmatrix} 6 & 1 & 1 \\ 5 & -2 & 1 \\ 0 & 3 & 1 \end{vmatrix}$ Expand about row 3.

$= 0 - 3\begin{vmatrix} 6 & 1 \\ 5 & 1 \end{vmatrix} + 1\begin{vmatrix} 6 & 1 \\ 5 & -2 \end{vmatrix}$

$= -3(6 - 5) + 1(-12 - 5)$

$= -3 - 17 = -20$

$D_y = \begin{vmatrix} 1 & 6 & 1 \\ 2 & 5 & 1 \\ -1 & 0 & 1 \end{vmatrix}$ Expand about row 3.

$= -1\begin{vmatrix} 6 & 1 \\ 5 & 1 \end{vmatrix} - 0 + 1\begin{vmatrix} 1 & 6 \\ 2 & 5 \end{vmatrix}$

$= -1(6 - 5) + 1(5 - 12)$

$= -1 - 7 = -8$

$D_z = \begin{vmatrix} 1 & 1 & 6 \\ 2 & -2 & 5 \\ -1 & 3 & 0 \end{vmatrix}$ Expand about row 3.

$= -1\begin{vmatrix} 1 & 6 \\ -2 & 5 \end{vmatrix} - 3\begin{vmatrix} 1 & 6 \\ 2 & 5 \end{vmatrix} + 0$

$= -1(5 + 12) - 3(5 - 12)$

$= -17 + 21 = 4$

$x = \dfrac{D_x}{D} = \dfrac{-20}{-4} = 5; \; y = \dfrac{D_y}{D} = \dfrac{-8}{-4} = 2$

$z = \dfrac{D_z}{D} = \dfrac{4}{-4} = -1$

Solution set: $\{(5, 2, -1)\}$

Cumulative Review Exercises Chapters 1–4

1. $(-3)^4 = (-3)(-3)(-3)(-3) = 81$

2. $-3^4 = -(3)(3)(3)(3) = -81$

3. $-(-3)^4 = -(-3)(-3)(-3)(-3) = -81$

4. $\sqrt{.49} = .7$, since $.7$ is positive and $(.7)^2 = .49$.

5. $-\sqrt{.49} = -.7$, since $(.7)^2 = .49$ and the negative sign in front of the radical must be applied.

6. $\sqrt{-.49}$ is not a real number because of the negative sign under the radical. No real number squared is negative.

7. Using the same concept as with $\sqrt{}$, $\sqrt[3]{64} = 4$, since $4^3 = 64$.

8. $\sqrt[3]{-64} = -4$, since $(-4)^3 = -64$.

In Exercises 9–11, let $x = -4$, $y = 3$, and $z = 6$.

9. $|2x| + 3y - z^3$

$= |(2)(-4)| + 3(3) - (6)^3$

$= |-8| + 9 - 216$

$= 8 + 9 - 216$

$= -199$

10. $-5(x^3 - y^3) = -5[(-4)^3 - (3)^3]$

$= -5(-64 - 27)$

$= -5(-91)$

$= 455$

11. The commutative property says that $3 \cdot 6 = 6 \cdot 3$, so that is the property that justifies the given statement.

12. $7(2x + 3) - 4(2x + 1) = 2(x + 1)$

$14x + 21 - 8x - 4 = 2x + 2$

$6x + 17 = 2x + 2$

$4x = -15$

$x = -\dfrac{15}{4}$

Solution set: $\left\{ -\dfrac{15}{4} \right\}$

13. $|6x - 8| = 4$

$6x - 8 = 4$ or $6x - 8 = -4$

$6x = 12$ $6x = 4$

$x = 2$ or $x = \dfrac{4}{6} = \dfrac{2}{3}$

Solution set: $\left\{ \dfrac{2}{3}, 2 \right\}$

14.
$$ax + by = cx + d$$

To solve for x, get all terms with x alone on one side of the equals sign.

$$ax - cx = d - by$$
$$x(a - c) = d - by$$
$$x = \frac{d - by}{a - c}$$

or
$$x = \frac{by - d}{c - a}$$

if the x-terms were put on the right side of the equals sign.

15. $.04x + .06(x - 1) = 1.04$

Multiply both sides by 100 to clear the decimals.
$$4x + 6(x - 1) = 104$$
$$4x + 6x - 6 = 104$$
$$10x - 6 = 104$$
$$10x = 110$$
$$x = 11$$

Solution set: $\{11\}$

16.
$$\frac{2}{3}y + \frac{5}{12}y \leq 20$$

Multiply both sides by 12.
$$12\left(\frac{2}{3}y + \frac{5}{12}y\right) \leq 12(20)$$
$$8y + 5y \leq 240$$
$$13y \leq 240$$
$$y \leq \frac{240}{13}$$

Solution set: $\left(-\infty, \dfrac{240}{13}\right]$

17. $|3x + 2| \leq 4$
$$-4 \leq 3x + 2 \leq 4$$
$$-6 \leq 3x \leq 2 \qquad \textit{Subtract 2.}$$
$$-2 \leq x \leq \frac{2}{3} \qquad \textit{Divide by 3.}$$

Solution set: $\left[-2, \dfrac{2}{3}\right]$

18. $|12t + 7| \geq 0$

The solution set is $(-\infty, \infty)$ since the absolute value of any number is greater than or equal to 0.

19. Let $x = $ the number of "guilty" votes and $x + 10 = $ the number of "not guilty" votes. The total number of votes was 200, so

$$x + (x + 10) = 200$$
$$2x + 10 = 200$$
$$2x = 190$$
$$x = 95,$$

and $x + 10 = 105.$

There were 105 "not guilty" votes and 95 "guilty" votes.

20. Let $h = $ the height of the triangle.

Use the formula $A = \dfrac{1}{2}bh$. Here, $A = 42$ and $b = 14$, so substitute these values in the formula and solve for h.

$$A = \frac{1}{2}bh$$
$$42 = \frac{1}{2}(14)h$$
$$42 = 7h$$
$$6 = h$$

The height is 6 m.

21. Let $x = $ the number of nickels,
$x + 1 = $ the number of dimes, and
$x + 6 = $ the number of pennies.

The total value is $\$4.80$, so
$$.05x + .10(x + 1) + .01(x + 6) = 4.80.$$

Multiply both sides by 100 to clear the decimals.
$$5x + 10(x + 1) + 1(x + 6) = 480$$
$$5x + 10x + 10 + x + 6 = 480$$
$$16x + 16 = 480$$
$$16x = 464$$
$$x = 29$$

Then, $x + 1 = 29 + 1 = 30$,
and $x + 6 = 29 + 6 = 35$.
There are 35 pennies, 29 nickels, and 30 dimes.

22. Let $x = $ the measure of the equal angles and $2x - 4 = $ the measure of the third angle.

The sum of the measures of the angles in a triangle is 180, so

$$x + x + 2x - 4 = 180$$
$$4x - 4 = 180$$
$$4x = 184$$
$$x = 46.$$

So, $2x - 4 = 2(46) - 4 = 92 - 4 = 88.$
The measures of the angles are 46°, 46°, and 88°.

23. A horizontal line through the point (x, k) has equation $y = k$. Since point A has coordinates $(-2, 6)$, $k = 6$. The equation of the horizontal line through A is $y = 6$.

24. A vertical line through the point (k, y) has equation $x = k$. Since point B has coordinates $(4, -2)$, $k = 4$. The equation of the vertical line through B is $x = 4$.

25. Let $(x_1, y_1) = (-2, 6)$ and $(x_2, y_2) = (4, -2)$. Then,

$$m = \frac{y_2 - y_1}{x_2 - x_1} = \frac{-2 - 6}{4 - (-2)} = \frac{-8}{6} = -\frac{4}{3}.$$

The slope is $-\dfrac{4}{3}$.

26. Perpendicular lines have slopes that are negative reciprocals of each other. The slope of line AB is $-\dfrac{4}{3}$ (from Exercise 25). The negative reciprocal of $-\dfrac{4}{3}$ is $\dfrac{3}{4}$, so the slope of a line perpendicular to line AB is $\dfrac{3}{4}$.

27. Let $m = -\dfrac{4}{3}$ and $(x_1, y_1) = (4, -2)$ in the point-slope form.

$$y - y_1 = m(x - x_1)$$
$$y - (-2) = -\frac{4}{3}(x - 4)$$
$$y + 2 = -\frac{4}{3}x + \frac{16}{3}$$

Multiply by 3 to clear the fractions, and then write the equation in standard form, $Ax + By = C$.

$$3y + 6 = -4x + 16$$
$$4x + 3y = 10$$

28. First locate the point $(-1, -3)$ on a graph. Then use the definition of slope to find a second point on the line.

$$m = \frac{\text{change in } y}{\text{change in } x} = \frac{2}{3}$$

From $(-1, -3)$, move 2 units up and 3 units to the right. The line through $(-1, -3)$ and the new point, $(2, -1)$, is the graph.

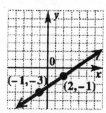

29. $-3x - 2y \leq 6$

Graph the line $-3x - 2y = 6$ as a solid line through its intercepts, $(-2, 0)$ and $(0, -3)$, since the inequality involves $\leq$.

To determine the region that belongs to the graph, test $(0, 0)$.

$$-3x - 2y \leq 6$$
$$-3(0) - 2(0) \leq 6$$
$$0 \leq 6 \quad \textit{True}$$

Since the result is true, shade the region that includes $(0, 0)$.

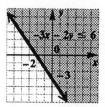

30. $f(x) = x^2 + 3x - 6$

(a) $f(-3) = (-3)^2 + 3(-3) - 6$
$$= 9 - 9 - 6 = -6$$

(b) $f(a) = (a)^2 + 3(a) - 6$
$$= a^2 + 3a - 6$$

31. If y varies directly as x, then

$$y = kx$$

for some constant k. If $y = 5$ when $x = 12$, then

$$5 = k(12),$$
$$\text{so} \quad k = \frac{5}{12}$$
$$\text{and} \quad y = \frac{5}{12}x.$$

When $x = 42$,

$$y = \frac{5}{12}(42)$$
$$= \frac{5 \cdot 6 \cdot 7}{2 \cdot 6} = \frac{35}{2} \text{ or } 17.5.$$

32.
$$-2x + 3y = -15 \quad (1)$$
$$4x - y = 15 \quad (2)$$

To eliminate x, multiply equation (1) by 2 and add the result to equation (2).

$$\begin{array}{rl} -4x + 6y = -30 & 2 \times (1) \\ \underline{4x - y = 15} & (2) \\ 5y = -15 & \\ y = -3 & \end{array}$$

Substitute -3 for y in equation (2) to find x.

$$4x - y = 15 \quad (2)$$
$$4x - (-3) = 15$$
$$4x + 3 = 15$$
$$4x = 12$$
$$x = 3$$

Solution set: $\{(3, -3)\}$

33.
$$x + y + z = 10 \quad (1)$$
$$x - y - z = 0 \quad (2)$$
$$-x + y - z = -4 \quad (3)$$

Add equations (1) and (2) to eliminate y and z. The result is

$$2x = 10$$
$$x = 5.$$

Add equations (2) and (3) to eliminate x and y. The result is

$$-2z = -4$$
$$z = 2.$$

Substitute 5 for x and 2 for z in equation (1) to find y.

$$x + y + z = 10 \quad (1)$$
$$5 + y + 2 = 10$$
$$y + 7 = 10$$
$$y = 3$$

Solution set: $\{(5, 3, 2)\}$

34.
$$\begin{vmatrix} 1 & 2 & 3 \\ 0 & 5 & 1 \\ -1 & 0 & 4 \end{vmatrix} \quad \text{Expand about row 2.}$$

$$= 0 + 5 \begin{vmatrix} 1 & 3 \\ -1 & 4 \end{vmatrix} - 1 \begin{vmatrix} 1 & 2 \\ -1 & 0 \end{vmatrix}$$
$$= 5[4 - (-3)] - 1[0 - (-2)]$$
$$= 5(7) - 1(2) = 35 - 2 = 33$$

35. Let x = the amount of the oranges and y = the amount of the apples.

Since she bought 6 lb of fruit,

$$x + y = 6. \quad (1)$$

The total cost of x lb of oranges at \$.90/lb and y lb of apples at \$.70/lb is \$5.20, so

$$.90x + .70y = 5.20.$$

Multiply by 10 to clear the decimals.

$$9x + 7y = 52 \quad (2)$$

To solve the system, solve equation (1) for x.

$$x + y = 6 \quad (1)$$
$$x = 6 - y$$

Substitute $6 - y$ for x in equation (2) and solve for y.

$$9x + 7y = 52 \quad (2)$$
$$9(6 - y) + 7y = 52$$
$$54 - 9y + 7y = 52$$
$$-2y = -2$$
$$y = 1$$

Since $x = 6 - y$ and $y = 1$,

$$x = 6 - 1 = 5.$$

She bought 5 lb of oranges and 1 lb of apples.

36. Let x = the average retail price of Elmo and y = the average retail price of Kid.

Elmo cost \$8.63 less than Kid, so

$$x = y - 8.63. \quad (1)$$

Together they cost \$63.89, so

$$x + y = 63.89. \quad (2)$$

From (1), substitute $y - 8.63$ for x in equation (2).

$$(y - 8.63) + y = 63.89$$
$$2y - 8.63 = 63.89$$
$$2y = 72.52$$
$$y = 36.26$$

From (1), $x = y - 8.63 = 36.26 - 8.63 = 27.63$. The average retail price of Elmo was \$27.63 and of Kid was \$36.26.

37. Let $x =$ the cost of a small box
and $y =$ the cost of a large box.

The cost of 10 small and 20 large boxes is $65, so

$$10x + 20y = 65. \quad (1)$$

The cost of 6 small and 10 large boxes is $34, so

$$6x + 10y = 34. \quad (2)$$

To eliminate y, multiply equation (2) by -2 and add the result to equation (1).

$$
\begin{array}{rcll}
10x + 20y &=& 65 & (1) \\
-12x - 20y &=& -68 & -2 \times (2) \\
\hline
-2x &=& -3 & \\
x &=& \dfrac{3}{2} & \text{or } 1.5
\end{array}
$$

Substitute $\dfrac{3}{2}$ for x in equation (2) to find y.

$$6x + 10y = 34 \qquad (2)$$
$$6\left(\frac{3}{2}\right) + 10y = 34$$
$$9 + 10y = 34$$
$$10y = 25$$
$$y = \frac{25}{10} \text{ or } \frac{5}{2} \text{ or } 2.5$$

A small box costs $1.50, and a large box costs $2.50.

38. Let $x =$ the cost of a pound of peanuts
and $y =$ the cost of a pound of cashews.

The cost of 6 lb of peanuts and 12 lb of cashews is $60, so

$$6x + 12y = 60. \quad (1)$$

The cost of 3 lb of peanuts and 4 lb of cashews is $22, so

$$3x + 4y = 22. \quad (2)$$

To eliminate x, multiply equation (2) by -2 and add the result to equation (1).

$$
\begin{array}{rcll}
6x + 12y &=& 60 & (1) \\
-6x - 8y &=& -44 & -2 \times (2) \\
\hline
4y &=& 16 & \\
y &=& 4 &
\end{array}
$$

Substitute 4 for y in equation (2) to find x.

$$3x + 4y = 22 \quad (2)$$
$$3x + 4(4) = 22$$
$$3x + 16 = 22$$
$$3x = 6$$
$$x = 2$$

Peanuts cost $2/lb, and cashews cost $4/lb.

39. The lines intersect at $(8, 3000)$, so the cost equals the revenue at $x = 8$ (which is 800 items). The revenue is $3000.

40. On the sale of 1100 parts ($x = 11$), the revenue is about $4100 and the cost is about $3600.

$$
\begin{aligned}
\text{Profit} &= \text{Revenue} - \text{Cost} \\
&\approx 4100 - 3600 \\
&= 500
\end{aligned}
$$

The profit is about $500.

CHAPTER 5   EXPONENTS AND POLYNOMIALS

Section 5.1

1. $(ab)^2 = a^2b^2$ by a power rule. Since $a^2b^2 \neq ab^2$, the expression $(ab)^2 = ab^2$ has been simplified incorrectly. The exponent should apply to both a and b.

3. $\left(\dfrac{4}{a}\right)^3 = \dfrac{4^3}{a^3}$

   Since $\dfrac{4^3}{a^3} \neq \dfrac{4^3}{a}$, the expression

   $$\left(\dfrac{4}{a}\right)^3 = \dfrac{4^3}{a}$$

   has been simplified incorrectly.

5. $x^3 \cdot x^4 = x^{3+4} = x^7$
   This expression has been simplified correctly.

7. The product rule says that when like bases are multiplied, the base stays the same and the exponents are added. For example, $x^5 \cdot x^6 = x^{11}$.

   The quotient rule says that when like bases are divided, the base stays the same and the exponents are subtracted. For example, $\dfrac{x^8}{x^5} = x^3$.

9. $a^5 \cdot a^3 = a^{5+3} = a^8$

11. $y^5 \cdot y^4 \cdot y^{-3} = y^{5+4+(-3)} = y^6$

13. $\left(9x^2y^3\right)\left(-2x^3y^5\right)$
    $= 9(-2)x^2x^3y^3y^5$
    $= -18x^{2+3}y^{3+5}$
    $= -18x^5y^8$

15. $\dfrac{p^{19}}{p^5} = p^{19-5} = p^{14}$

17. $\dfrac{z^{-6}}{z^{-12}} = z^{-6-(-12)} = z^6$

19. $\dfrac{r^{13}r^{-4}r^{-3}}{r^{-2}r^{-5}r^0} = \dfrac{r^{13+(-4)+(-3)}}{r^{-2+(-5)+0}}$
    $= \dfrac{r^6}{r^{-7}}$
    $= r^{6-(-7)} = r^{13}$

21. $7k^2(-2k)\left(4k^{-5}\right)^0 = 7(-2)k^2k \cdot 1$
    $= -14k^{2+1}$
    $= -14k^3$

23. $-4\left(2x^3\right)(3x) = -4(2)(3)x^3x$
    $= -24x^{3+1}$
    $= -24x^4$

25. $\dfrac{(3pq)q^2}{6p^2q^4} = \dfrac{3p^1q^{1+2}}{6p^2q^4} = \dfrac{3pq^3}{6p^2q^4}$
    $= \dfrac{1}{2}p^{1-2}q^{3-4}$
    $= \dfrac{1}{2}p^{-1}q^{-1} = \dfrac{1}{2} \cdot \dfrac{1}{p^1} \cdot \dfrac{1}{q^1}$
    $= \dfrac{1}{2pq}$

27. $\dfrac{6x^{-5}y^{-2}}{(3x^{-3})(2x^{-2}y^{-2})} = \dfrac{6x^{-5}y^{-2}}{6x^{-3-2}y^{-2}}$
    $= \dfrac{6x^{-5}y^{-2}}{6x^{-5}y^{-2}}$
    $= 1$

29. Your friend multiplied the bases which is incorrect. Instead, keep the same base and add the exponents.

    $$4^5 \cdot 4^2 = 4^{5+2} = 4^7$$

31. $\left(\dfrac{2}{3}\right)^2 = \dfrac{2}{3} \cdot \dfrac{2}{3} = \dfrac{4}{9}$

33. $4^{-3} = \dfrac{1}{4^3} = \dfrac{1}{64}$

35. $-4^{-3} = -\dfrac{1}{4^3} = -\dfrac{1}{64}$

37. $(-4)^{-3} = \dfrac{1}{(-4)^3} = \dfrac{1}{-64} = -\dfrac{1}{64}$

39. $\dfrac{1}{3^{-2}} = \dfrac{1}{\dfrac{1}{3^2}} = 1 \cdot \dfrac{3^2}{1} = 9$

41. $\dfrac{-3^{-1}}{4^{-2}} = \dfrac{-\dfrac{1}{3}}{\dfrac{1}{4^2}} = -\dfrac{1}{3}\left(\dfrac{4^2}{1}\right) = -\dfrac{16}{3}$

43. $\left(\dfrac{2}{3}\right)^{-3} = \left(\dfrac{3}{2}\right)^3 = \dfrac{3}{2} \cdot \dfrac{3}{2} \cdot \dfrac{3}{2} = \dfrac{27}{8}$

45. $3^{-1} + 2^{-1} = \dfrac{1}{3} + \dfrac{1}{2} = \dfrac{2}{6} + \dfrac{3}{6} = \dfrac{5}{6}$

47. $6^{-1} - 4^{-1} = \dfrac{1}{6} - \dfrac{1}{4} = \dfrac{2}{12} - \dfrac{3}{12} = -\dfrac{1}{12}$

49. $(6-4)^{-1} = (2)^{-1} = \dfrac{1}{(2)^1} = \dfrac{1}{2}$

51. $\dfrac{3^{-5}}{3^{-2}} = 3^{-5-(-2)} = 3^{-3} = \dfrac{1}{3^3} = \dfrac{1}{27}$

53. $\dfrac{9^{-1}}{-9} = -\dfrac{9^{-1}}{9^1} = -9^{-1-1} = -9^{-2}$

$= -\dfrac{1}{9^2} = -\dfrac{1}{81}$

55. $25^0 = 1$, since $a^0 = 1$ for any nonzero base a.

57. $-7^0 = -(7)^0 = -1$

59. $(-7)^0 = 1$ since -7 is in parentheses.

61. $-4^0 - m^0 = -1 - 1 = -2$

63. The expression $(-3)^{-2}$ is a positive number, not a negative number.

$(-3)^{-2} = \dfrac{1}{(-3)^2} = \dfrac{1}{(-3)(-3)} = \dfrac{1}{9}$

The negative sign is part of the base, -3, of the expression. A negative number squared is a positive number.
If the expression had been written -3^{-2}, then it would have been a negative number because

$-3^{-2} = -\dfrac{1}{3^2} = -\dfrac{1}{9}.$

Here, the base of the expression is 3.

65. $\left(2^{-3} \cdot 5^{-1}\right)^3 = 2^{-3 \cdot 3} \cdot 5^{-1 \cdot 3}$

$= 2^{-9} \cdot 5^{-3}$

$= \dfrac{1}{2^9 \cdot 5^3}$

67. $\left(5^{-4} \cdot 6^{-2}\right)^{-3} = 5^{-4(-3)} \cdot 6^{-2(-3)}$

$= 5^{12} \cdot 6^6$

69. $\left(k^2\right)^{-3} k^4 = k^{2(-3)} k^4$

$= k^{-6} k^4$

$= k^{-6+4}$

$= k^{-2}$

$= \dfrac{1}{k^2}$

71. $-4r^{-2}\left(r^4\right)^2 = -4r^{-2}\left(r^8\right)$

$= -4r^{-2+8}$

$= -4r^6$

73. $\left(5a^{-1}\right)^4\left(a^2\right)^{-3} = 5^4 a^{-1 \cdot 4} a^{2(-3)}$

$= 5^4 a^{-4} a^{-6}$

$= 5^4 a^{-4-6} = 5^4 a^{-10}$

$= \dfrac{5^4}{a^{10}}$ or $\dfrac{625}{a^{10}}$

75. $\left(z^{-4} x^3\right)^{-1} = z^{-4(-1)} x^{3(-1)}$

$= z^4 x^{-3}$

$= \dfrac{z^4}{x^3}$

77. $\dfrac{\left(p^{-2}\right)^0}{5p^{-4}} = \dfrac{1 \cdot p^4}{5} = \dfrac{p^4}{5}$

79. $\dfrac{4a^5\left(a^{-1}\right)^3}{\left(a^{-2}\right)^{-2}} = \dfrac{4a^5 a^{-1 \cdot 3}}{a^{-2(-2)}} = \dfrac{4a^5 a^{-3}}{a^4}$

$= 4a^{5-3-4}$

$= 4a^{-2}$

$= \dfrac{4}{a^2}$

81. The first step may be the most confusing.

$\left(-y^{-4}\right)^2 = \left(-1 \cdot y^{-4}\right)^2$

$= (-1)^2\left(y^{-4}\right)^2$

$= 1 \cdot y^{-4(2)} = y^{-8}$

It is not necessary to include these steps once the concept of squaring a negative is committed to memory.

$\dfrac{\left(-y^{-4}\right)^2}{6\left(y^{-5}\right)^{-1}} = \dfrac{y^{-4(2)}}{6y^{-5(-1)}}$

$= \dfrac{y^{-8}}{6y^5}$

$= \dfrac{1}{6y^5 y^8}$

$= \dfrac{1}{6y^{13}}$

83. $\dfrac{(2k)^2 m^{-5}}{(km)^{-3}} = \dfrac{2^2 k^2 m^{-5}}{k^{-3} m^{-3}}$

$= 2^2 k^{2-(-3)} m^{-5-(-3)}$

$= 2^2 k^5 m^{-2}$

$= \dfrac{2^2 k^5}{m^2}$ or $\dfrac{4k^5}{m^2}$

85. $\dfrac{(2k)^2 k^3}{k^{-1} k^{-5}}\left(5k^{-2}\right)^{-3} = \dfrac{2^2 k^2 k^3}{k^{-1} k^{-5}}\left(5^{-3} k^{-2(-3)}\right)$

$= \dfrac{2^2 k^5}{k^{-6}}\left(5^{-3} k^6\right)$

$= 2^2 k^{11}\left(5^{-3} k^6\right)$

$= 2^2 \cdot 5^{-3} k^{11+6}$

$= \dfrac{2^2 k^{17}}{5^3}$ or $\dfrac{4k^{17}}{125}$

87. $\left(\dfrac{3k^{-2}}{k^4}\right)^{-1} \cdot \dfrac{2}{k} = \dfrac{\left(3k^{-2}\right)^{-1}}{\left(k^4\right)^{-1}} \cdot \dfrac{2}{k}$

$= \dfrac{3^{-1} k^2 \cdot 2}{k^{-4} k^1}$

$= \dfrac{3^{-1} \cdot 2k^2}{k^{-3}}$

$= \dfrac{2k^2 k^3}{3}$

$= \dfrac{2k^5}{3}$

89. $\left(\dfrac{2p}{q^2}\right)^3 \left(\dfrac{3p^4}{q^{-4}}\right)^{-1} = \dfrac{(2p)^3}{(q^2)^3} \cdot \dfrac{(3p^4)^{-1}}{(q^{-4})^{-1}}$

$= \dfrac{2^3 p^3}{q^6} \cdot \dfrac{3^{-1}p^{-4}}{q^4}$

$= \dfrac{2^3 p^{3-4}}{3^1 q^{6+4}}$

$= \dfrac{8p^{-1}}{3q^{10}} = \dfrac{8}{3pq^{10}}$

91. $\dfrac{2^2 y^4 (y^{-3})^{-1}}{2^5 y^{-2}} = \dfrac{y^4 y^3 y^2}{2^{5-2}}$

$= \dfrac{y^{4+3+2}}{2^3}$

$= \dfrac{y^9}{8}$

93. $\dfrac{(2m^2 p^3)^2 (4m^2 p)^{-2}}{(-3mp^4)^{-1}(2m^3 p^4)^3}$

$= \dfrac{2^2 m^4 p^6 4^{-2} m^{-4} p^{-2}}{(-3)^{-1} m^{-1} p^{-4} 2^3 m^9 p^{12}}$ *Power rule*

$= \dfrac{4(-3)^1 (m^{4-4})(p^{6-2})}{4^2 \cdot 2^3 (m^{-1+9})(p^{-4+12})}$ *Product rule*

$= \dfrac{-3m^0 p^4}{4 \cdot 2^3 m^8 p^8}$

$= \dfrac{-3(p^{4-8})}{32m^8}$ *Quotient rule*

$= \dfrac{-3p^{-4}}{32m^8}$

$= -\dfrac{3}{32m^8 p^4}$

95. $\dfrac{(-3y^3 x^3)(-4y^4 x^2)(x^2)^{-4}}{18x^3 y^2 (y^3)^3 (x^3)^{-2}}$

$= \dfrac{(-3y^3 x^3)(-4y^4 x^2)(x^{-8})}{18x^3 y^2 y^9 x^{-6}}$ *Power rule*

$= \dfrac{12x^{3+2-8} y^{3+4}}{18x^{3-6} y^{2+9}}$ *Product rule*

$= \dfrac{2x^{-3} y^7}{3x^{-3} y^{11}}$

$= \dfrac{2x^{-3-(-3)} y^{7-11}}{3}$ *Quotient rule*

$= \dfrac{2y^{-4}}{3}$

$= \dfrac{2}{3y^4}$

97.

$\left(\dfrac{p^2 q^{-1}}{2p^{-2}}\right)^2 \cdot \left(\dfrac{p^3 \cdot 4q^{-2}}{3q^{-5}}\right)^{-1} \cdot \left(\dfrac{pq^{-5}}{q^{-2}}\right)^3$

$= \dfrac{p^4 q^{-2} p^{-3} 4^{-1} q^2 p^3 q^{-15}}{2^2 p^{-4} 3^{-1} q^5 q^{-6}}$ *Power rule*

$= \dfrac{4^{-1} p^{4-3+3} q^{-2+2-15}}{2^2 3^{-1} p^{-4} q^{5-6}}$ *Product rule*

$= \dfrac{3p^4 q^{-15}}{2^2 \cdot 4 p^{-4} q^{-1}}$

$= \dfrac{3p^{4-(-4)} q^{-15-(-1)}}{4 \cdot 4}$ *Quotient rule*

$= \dfrac{3p^8 q^{-14}}{16}$

$= \dfrac{3p^8}{16q^{14}}$

In Exercises 99–102, use the expression

$$\left(\dfrac{a^{-8} b^2}{a^{-5} b^{-4}}\right)^{-2} \ (a, b \neq 0).$$

99. $\left(\dfrac{a^{-8} b^2}{a^{-5} b^{-4}}\right)^{-2} = \left(a^{-8-(-5)} b^{2-(-4)}\right)^{-2}$

$= \left(a^{-3} b^6\right)^{-2}$

$= \left(\dfrac{b^6}{a^3}\right)^{-2}$

$= \dfrac{b^{6(-2)}}{a^{3(-2)}}$

$= \dfrac{b^{-12}}{a^{-6}}$

$= \dfrac{a^6}{b^{12}}$

100. $\left(\dfrac{a^{-8} b^2}{a^{-5} b^{-4}}\right)^{-2} = \dfrac{a^{-8(-2)} b^{2(-2)}}{a^{-5(-2)} b^{-4(-2)}}$

$= \dfrac{a^{16} b^{-4}}{a^{10} b^8}$

$= a^{16-10} b^{-4-8}$

$= a^6 b^{-12}$

$= \dfrac{a^6}{b^{12}}$

101. In each case, the same answer, $\dfrac{a^6}{b^{12}}$, was obtained.

102. "Both methods are correct."

103. $530 = 5{\scriptstyle\wedge}30.$ ← Decimal point

           $\vee\vee$

Count 2 places.

Since the number 5.3 is to be made larger, the exponent on 10 is positive.

$$530 = 5.3 \times 10^2$$

105. $.830 = .8_\wedge 30$

Count 1 place.

Since the number 8.3 is to be made smaller, the exponent on 10 is negative.

$$.830 = 8.3 \times 10^{-1}$$

107. $.00000692 = .000006_\wedge 92$

Count 6 places.

Since the number 6.92 is to be made smaller, the exponent on 10 is negative.

$$.00000692 = 6.92 \times 10^{-6}$$

109. $-38,500 = -3_\wedge 8500.$

Count 4 places.

Since the number 3.85 is to be made larger, the exponent on 10 is positive. Also, affix a negative sign in front of the number.

$$-38,500 = -3.85 \times 10^4$$

111. $7.2 \times 10^4 = 72,000$

Move the decimal point 4 places to the *right* because of the *positive* exponent. Attach extra zeros.

113. $2.54 \times 10^{-3} = .00254$

Since the exponent is *negative*, move the decimal point 3 places to the *left*.

115. $-6 \times 10^4 = -60,000$

Move the decimal point 4 places to the *right* because of the *positive* exponent. Attach extra zeros.

117. $1.2 \times 10^{-5} = .000012$

Since the exponent is *negative*, move the decimal point 5 places to the *left*.

119.
$$\frac{12 \times 10^4}{2 \times 10^6} = \frac{12}{2} \times \frac{10^4}{10^6}$$
$$= 6 \times 10^{4-6}$$
$$= 6 \times 10^{-2}$$
$$= .06$$

121.
$$\frac{3 \times 10^{-2}}{12 \times 10^3} = \frac{3 \times 10^{-2}}{1.2 \times 10^4}$$
$$= \frac{3}{1.2} \times \frac{10^{-2}}{10^4}$$
$$= 2.5 \times 10^{-6}$$
$$= .0000025$$

123.
$$\frac{.05 \times 1600}{.0004} = \frac{5 \times 10^{-2} \times 1.6 \times 10^3}{4 \times 10^{-4}}$$
$$= \frac{5(1.6)}{4} \times \frac{10^{-2} \times 10^3}{10^{-4}}$$
$$= 2 \times 10^{-2+3-(-4)}$$
$$= 2 \times 10^5$$
$$= 200,000$$

125.
$$\frac{20,000 \times .018}{300 \times .0004} = \frac{2 \times 10^4 \times 1.8 \times 10^{-2}}{3 \times 10^2 \times 4 \times 10^{-4}}$$
$$= \frac{2 \times 1.8}{3 \times 4} \times \frac{10^4 \times 10^{-2}}{10^2 \times 10^{-4}}$$
$$= .3 \times 10^{4-2-2-(-4)}$$
$$= .3 \times 10^4$$
$$= 3000$$

127. $382,553,000,000 = 3.82553 \times 10^{11}$

129. If x represents the number of operations per second that the normal desktop computer is capable of, then

$$15,000x = 3.9 \times 10^8.$$
$$x = \frac{3.9 \times 10^8}{1.5 \times 10^4}$$
$$= \frac{3.9}{1.5} \times \frac{10^8}{10^4}$$
$$= 2.6 \times 10^4$$
$$= 26,000$$

131. To solve the problem, divide the number of miles by the number of miles in a parsec.

$$\frac{1.8 \times 10^7 \text{ miles}}{19 \times 10^{12} \text{ miles/parsec}} = \frac{1.8}{19} \times \frac{10^7}{10^{12}}$$
$$\approx .09474 \times 10^{-5}$$
$$= .0000009474$$
$$\text{or } 9.474 \times 10^{-7}$$

The distance from Uranus to the sun is approximately 9.474×10^{-7} parsec.

133. Since $d = rt$, $t = \frac{d}{r}$. Divide the distance traveled by the speed of light.

$$\frac{9 \times 10^{12}}{3 \times 10^{10}} = \frac{9}{3} \times 10^{12-10} = 3 \times 10^2$$

It will take 300 seconds.

135. First find the number of seconds in a year.

$$1 \text{ year} = 365 \text{ days}$$
$$= 365(24) \text{ hr}$$
$$= 8760 \text{ hr}$$
$$= 8760(60) \text{ min}$$
$$= 525,600 \text{ min}$$
$$= 525,600(60) \text{ sec}$$
$$= 31,536,000 \text{ sec}$$
$$= 3.1536 \times 10^7 \text{ sec}$$

Now use $d = rt$ and multiply the rate light travels by the number of seconds in a year.

$$1 \text{ light year} = \left(1.86 \times 10^5 \text{ mi/sec}\right)$$
$$\cdot \left(3.1536 \times 10^7 \text{ sec}\right)$$
$$= 1.86(3.1536) \times 10^{5+7}$$
$$\approx 5.87 \times 10^{12} \text{ mi}$$

There are about 5.87×10^{12} miles in a light year.

137. **(a)** The distance from Mercury to the sun is 3.6×10^7 mi and the distance from Venus to the sun is 6.7×10^7 mi, so the distance between Mercury and Venus in miles is

$$\left(6.7 \times 10^7\right) - \left(3.6 \times 10^7\right)$$
$$= (6.7 - 3.6) \times 10^7$$
$$= 3.1 \times 10^7.$$

Use $d = rt$, or $\dfrac{d}{r} = t$, where $d = 3.1 \times 10^7$ and $r = 1.55 \times 10^3$.

$$\frac{3.1 \times 10^7}{1.55 \times 10^3} = \frac{3.1}{1.55} \times 10^{7-3}$$
$$= 2 \times 10^4$$
$$= 20,000$$

It would take $20,000$ hr.

(b) From part (a), it takes $20,000$ hr for a spacecraft to travel from Venus to Mercury. Convert this to days (24 hours = 1 day).

$$20,000 \text{ hr} = \frac{20,000}{24} \text{ days}$$
$$\approx 833 \text{ days}$$

It would take about 833 days.

139. Males between 29 and 44

$$123,000 = 1.23 \times 10^5$$

141. Males between 45 and 64

$$424,000 = 4.24 \times 10^5$$

143. Males 65 or older

$$440,000 = 4.4 \times 10^5$$

145. $(1.5\,\text{E}\,12)*(5\,\text{E}\,{}^-3)$
$$= \left(1.5 \times 10^{12}\right)\left(5 \times 10^{-3}\right)$$
$$= (1.5 \times 5)\left(10^{12-3}\right)$$
$$= 7.5 \times 10^9$$

147. $(8.4\,\text{E}\,14)/(2.1\,\text{E}\,{}^-3)$
$$= \frac{8.4 \times 10^{14}}{2.1 \times 10^{-3}}$$
$$= \frac{8.4}{2.1} \times \frac{10^{14}}{10^{-3}}$$
$$= 4 \times 10^{14-(-3)}$$
$$= 4 \times 10^{17}$$

Section 5.2

1. $2x^3 + x - 3x^2 + 4$

The polynomial is written in descending powers of the variable if the exponents on the terms of the polynomial decrease from left to right.

$$2x^3 - 3x^2 + x + 4$$

3. $4p^3 - 8p^5 + p^7 = p^7 - 8p^5 + 4p^3$

5. $-m^3 + 5m^2 + 3m^4 + 10$
$$= 3m^4 - m^3 + 5m^2 + 10$$

7. 25 is one term, so it's a *monomial*. 25 is a nonzero constant, so it has degree zero.

9. $7m - 22$ has two terms, so it's a *binomial*. The exponent on m is 1, so $7m - 22$ has degree 1.

11. $2r^3 + 4r^2 + 5r$ has three terms, so it's a *trinomial*. The greatest exponent is 3, so the degree is 3.

13. $-6p^4q - 3p^3q^2 + 2pq^3 - q^4$ has four terms, so it is classified as *none of these*. The greatest sum of exponents on any term is 5, so the polynomial has degree 5.

15. Only choice (a) is a trinomial (it has three terms) with descending powers and having degree 6.

17. In $7z$, the coefficient is 7 and, since $7z = 7z^1$, the degree is 1.

19. In $-15p^2$, the coefficient is -15 and the degree is 2.

21. In x^4, since $x^4 = 1x^4$, the coefficient is 1 and the degree is 4.

23. In $-mn^5$, the coefficient is -1, since $-mn^5 = -1mn^5$, and the degree is 6, since the sum of the exponents on m and n is $1 + 5 = 6$.

25. $5z^4 + 3z^4 = (5+3)z^4 = 8z^4$

27. $-m^3 + 2m^3 + 6m^3 = (-1+2+6)m^3 = 7m^3$

29. $x + x + x + x + x$
$= (1 + 1 + 1 + 1 + 1)x$
$= 5x$

31. $y^2 + 7y - 4y^2 = (y^2 - 4y^2) + 7y$
$= (1 - 4)y^2 + 7y$
$= -3y^2 + 7y$

33. $2k + 3k^2 + 5k^2 - 7$
$= (3k^2 + 5k^2) + 2k - 7$
$= (3 + 5)k^2 + 2k - 7$
$= 8k^2 + 2k - 7$

35. $n^4 - 2n^3 + n^2 - 3n^4 + n^3$
$= n^4 - 3n^4 - 2n^3 + n^3 + n^2$
$= (1 - 3)n^4 + (-2 + 1)n^3 + n^2$
$= -2n^4 - n^3 + n^2$

37. $[4 - (2 + 3m)] + (6m + 9)$
Distribute over innermost grouping symbols first.
$= [4 - 2 - 3m] + (6m + 9)$
$= 2 - 3m + 6m + 9$
$= (-3 + 6)m + (2 + 9)$
$= 3m + 11$

39. $[(6 + 3p) - (2p + 1)] - (2p + 9)$
$= 6 + 3p - 2p - 1 - 2p - 9$
$= (3 - 2 - 2)p + (6 - 1 - 9)$
$= -p - 4$

41. $(3p^2 + 2p - 5) + (7p^2 - 4p^3 + 3p)$
$= -4p^3 + 3p^2 + 7p^2 + 2p + 3p - 5$
$= -4p^3 + (3 + 7)p^2 + (2 + 3)p - 5$
$= -4p^3 + 10p^2 + 5p - 5$

43. $(2x^5 - 2x^4 + x^3 - 1) + (x^4 - 3x^3 + 2)$
$= 2x^5 - 2x^4 + x^4 + x^3 - 3x^3 - 1 + 2$
$= 2x^5 + (-2 + 1)x^4 + (1 - 3)x^3 + (-1 + 2)$
$= 2x^5 - x^4 - 2x^3 + 1$

45. $(9a - 5a) - [2a - (4a + 3)]$
$= 4a - [2a - 4a - 3]$
$= 4a - [-2a - 3]$
$= 4a + 2a + 3$
$= 6a + 3$

47. To add or subtract polynomials, add or subtract the coefficients of the like terms. For example, the sum $(3x^2 + 2x + 1) + (5x^2 - 3x + 9)$ is $(3 + 5)x^2 + (2 - 3)x + (1 + 9) = 8x^2 - x + 10$. The difference $(8y^2 + 9y + 4) - (2y^2 + 7y + 1)$ is $(8 - 2)y^2 + (9 - 7)y + (4 - 1) = 6y^2 + 2y + 3$.

49. $241 = (2 \times 10^2) + (4 \times 10^1) + (1 \times 10^0)$

50. Corresponding place values are not aligned in columns.

51. Align corresponding place values in columns. Then add the numbers vertically, beginning with the ones on the right and moving left place by place.

52. Corresponding powers of the variable are not aligned in columns.

53. Align the polynomials in columns with like terms in the same columns. Then add vertically.

54. While only like terms of polynomials can be added, the commutative property of addition allows us to express the terms of an individual polynomial in any order without changing its value. This is not the case with place value where order determines value. Changing the order of digits in a number will almost always change its value (unless all the digits are the same, for example).

55.
$$\begin{array}{r} 21p - 8 \\ -9p + 4 \\ \hline 12p - 4 \end{array}$$ *Add vertically.*

57.
$$\begin{array}{r} -12p^2 + 4p - 1 \\ 3p^2 + 7p - 8 \\ \hline -9p^2 + 11p - 9 \end{array}$$ *Add vertically.*

59.
$$\begin{array}{r} 6m^2 - 11m + 5 \\ -8m^2 + 2m - 1 \end{array}$$

To subtract, change all the signs in the second polynomial, and add.

$$\begin{array}{r} 6m^2 - 11m + 5 \\ 8m^2 - 2m + 1 \\ \hline 14m^2 - 13m + 6 \end{array}$$

61.
$$\begin{array}{r} 5q^3 - 5q + 2 \\ -3q^3 + 2q - 9 \end{array}$$

To subtract, change all the signs in the second polynomial, and add.

$$\begin{array}{r} 5q^3 - 5q + 2 \\ 3q^3 - 2q + 9 \\ \hline 8q^3 - 7q + 11 \end{array}$$

63. $7y^2 - 6y + 5 - (4y^2 - 2y + 3)$
$= 7y^2 - 6y + 5 - 4y^2 + 2y - 3$
$= 7y^2 - 4y^2 - 6y + 2y + 5 - 3$
$= 3y^2 - 4y + 2$

65. Simplify the expression in brackets first.

$$\left(3m^2 - 5n^2 + 2n\right) + \left(-3m^2\right) + 4n^2$$
$$= 3m^2 - 5n^2 + 2n - 3m^2 + 4n^2$$
$$= 3m^2 - 3m^2 - 5n^2 + 4n^2 + 2n$$
$$= -n^2 + 2n$$

Now perform the subtraction.

$$\left(-4m^2 + 3n^2 - 5n\right) - \left(-n^2 + 2n\right)$$
$$= -4m^2 + 3n^2 - 5n + n^2 - 2n$$
$$= -4m^2 + 3n^2 + n^2 - 5n - 2n$$
$$= -4m^2 + 4n^2 - 7n$$

67. $\left[-\left(y^4 - y^2 + 1\right) - \left(y^4 + 2y^2 + 1\right)\right]$
$$+ \left(3y^4 - 3y^2 - 2\right)$$
$$= -y^4 + y^2 - 1 - y^4 - 2y^2 - 1$$
$$\quad + 3y^4 - 3y^2 - 2$$
$$= -y^4 - y^4 + 3y^4 + y^2 - 2y^2 - 3y^2$$
$$\quad - 1 - 1 - 2$$
$$= y^4 - 4y^2 - 4$$

69. $-\left(3z^2 + 5z - [2z^2 - 6z]\right)$
$$+ \left[\left(8z^2 - [5z - z^2]\right) + 2z^2\right]$$
$$= -\left(3z^2 + 5z - 2z^2 + 6z\right)$$
$$\quad + \left(8z^2 - 5z + z^2 + 2z^2\right)$$
$$= -\left(3z^2 - 2z^2 + 5z + 6z\right)$$
$$\quad + \left(8z^2 + z^2 + 2z^2 - 5z\right)$$
$$= -\left(z^2 + 11z\right) + \left(11z^2 - 5z\right)$$
$$= -z^2 - 11z + 11z^2 - 5z$$
$$= -z^2 + 11z^2 - 11z - 5z$$
$$= 10z^2 - 16z$$

Section 5.3

1. $f(x) = 6x - 4$

  (a) $f(-1) = 6(-1) - 4$
$$= -6 - 4 = -10$$

  (b) $f(2) = 6(2) - 4$
$$= 12 - 4 = 8$$

3. $f(x) = x^2 - 3x + 4$

  (a) $f(-1) = (-1)^2 - 3(-1) + 4$
$$= 1 + 3 + 4 = 8$$

  (b) $f(2) = (2)^2 - 3(2) + 4$
$$= 4 - 6 + 4 = 2$$

5. $f(x) = 5x^4 - 3x^2 + 6$

  (a) $f(-1) = 5(-1)^4 - 3(-1)^2 + 6$
$$= 5 \cdot 1 - 3 \cdot 1 + 6$$
$$= 5 - 3 + 6 = 8$$

  (b) $f(2) = 5(2)^4 - 3(2)^2 + 6$
$$= 5 \cdot 16 - 3 \cdot 4 + 6$$
$$= 80 - 12 + 6 = 74$$

7. $f(x) = -x^2 + 2x^3 - 8$

  (a) $f(-1) = -(-1)^2 + 2(-1)^3 - 8$
$$= -(1) + 2(-1) - 8$$
$$= -1 - 2 - 8 = -11$$

  (b) $f(2) = -(2)^2 + 2(2)^3 - 8$
$$= -(4) + 2 \cdot 8 - 8$$
$$= -4 + 16 - 8 = 4$$

9. $Y_1(x) = -x^2 + 3x - 4$
$$Y_1(-3) = -(-3)^2 + 3(-3) - 4$$
$$= -(9) - 9 - 4$$
$$= -9 - 9 - 4 = -22$$

11. $Y_3(x) = x^3 - 3x^2 + 2x - 4$
$$Y_3(6) = (6)^3 - 3(6)^2 + 2(6) - 4$$
$$= 216 - 3(36) + 12 - 4$$
$$= 216 - 108 + 8 = 116$$

13. $f(x) = 411x^2 + 194x + 24,050$

  (a) 1980 corresponds to $x = 0$

$$f(0) = 411(0)^2 + 194(0) + 24,050$$
$$= 0 + 0 + 24,050 = 24,050$$

Medicaid payments: $24,050 million

  (b) 1985 corresponds to $x = 5$

$$f(5) = 411(5)^2 + 194(5) + 24,050$$
$$= 10,275 + 970 + 24,050$$
$$= 35,295$$

Medicaid payments: $35,295 million

  (c) 1990 corresponds to $x = 10$

$$f(10) = 411(10)^2 + 194(10) + 24,050$$
$$= 41,100 + 1940 + 24,050$$
$$= 67,090$$

Medicaid payments: $67,090 million

  (d) 1995 corresponds to $x = 15$

$$f(15) = 411(15)^2 + 194(15) + 24,050$$
$$= 92,475 + 2910 + 24,050$$
$$= 119,435$$

Medicaid payments: $119,435 million

15. $f(x) = .39x^2 + .71x + 33.0$

(a) 1990 corresponds to $x = 0$

$$f(0) = .39(0)^2 + .71(0) + 33.0$$
$$= 0 + 0 + 33.0 = 33.0$$

The number of people enrolled in HMOs in 1990 was about 33.0 million.

Part	Year	x	$f(x)$ (in millions)
(b)	1991	1	34.1
(c)	1992	2	36.0
(d)	1993	3	38.6
(e)	1994	4	42.1
(f)	1995	5	46.3

17. **(a)** $(f + g)(x) = f(x) + g(x)$
$$= (5x - 10) + (3x + 7)$$
$$= 8x - 3$$

(b) $(f - g)(x) = f(x) - g(x)$
$$= (5x - 10) - (3x + 7)$$
$$= (5x - 10) + (-3x - 7)$$
$$= 2x - 17$$

19. **(a)**
$$(f + g)(x) = f(x) + g(x)$$
$$= (4x^2 + 8x - 3) + (-5x^2 + 4x - 9)$$
$$= -x^2 + 12x - 12$$

(b)
$$(f - g)(x) = f(x) - g(x)$$
$$= (4x^2 + 8x - 3) - (-5x^2 + 4x - 9)$$
$$= (4x^2 + 8x - 3) + (5x^2 - 4x + 9)$$
$$= 9x^2 + 4x + 6$$

21. Answers will vary. Let $f(x) = x^3$ and $g(x) = x^4$.
$$(f - g)(x) = f(x) - g(x)$$
$$= x^3 - x^4$$
$$(g - f)(x) = g(x) - f(x)$$
$$= x^4 - x^3$$

Because the two differences are not equal, subtraction of polynomial functions is not commutative.

23.

x	$f(x) = -2x + 1$
-2	$-2(-2) + 1 = 5$
-1	$-2(-1) + 1 = 3$
0	$-2(0) + 1 = 1$
1	$-2(1) + 1 = -1$
2	$-2(2) + 1 = -3$

This is a linear function, so plot the points and draw a line through them.

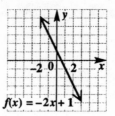

$f(x) = -2x + 1$

Any x-value can be used, so the domain is $(-\infty, \infty)$. From the graph, we see that any y-value can be obtained from the function, so the range is $(-\infty, \infty)$.

25.

x	$f(x) = -3x^2$
-2	$-3(-2)^2 = -12$
-1	$-3(-1)^2 = -3$
0	$-3(0)^2 = 0$
1	$-3(1)^2 = -3$
2	$-3(2)^2 = -12$

Since the greatest exponent is 2, the graph of f is a parabola.

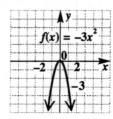

$f(x) = -3x^2$

Any x-value can be used, so the domain is $(-\infty, \infty)$. From the graph, we see that the y-values are at most 0, so the range is $(-\infty, 0]$.

27.

x	$f(x) = x^3 + 1$
-2	$(-2)^3 + 1 = -7$
-1	$(-1)^3 + 1 = 0$
0	$(0)^3 + 1 = 1$
1	$(1)^3 + 1 = 2$
2	$(2)^3 + 1 = 9$

The greatest exponent is 3, so the graph of f is s-shaped.

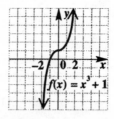

$f(x) = x^3 + 1$

Any x-value can be used, so the domain is $(-\infty, \infty)$. From the graph, we see that any y-value can be obtained from the function, so the range is $(-\infty, \infty)$.

Section 5.4

1. $(2x - 5)(3x + 4)$

$$\quad\text{F}\qquad\quad\text{O}\qquad\quad\text{I}\qquad\quad\text{L}$$
$$= (2x)(3x) + (2x)(4) + (-5)(3x) + (-5)(4)$$
$$= 6x^2 + 8x - 15x - 20$$
$$= 6x^2 - 7x - 20 \quad \text{(Choice C)}$$

3. $(2x - 5)(3x - 4)$

$$\quad\text{F}\qquad\quad\text{O}\qquad\quad\text{I}\qquad\quad\text{L}$$
$$= (2x)(3x) + (2x)(-4) + (-5)(3x) + (-5)(-4)$$
$$= 6x^2 - 8x - 15x + 20$$
$$= 6x^2 - 23x + 20 \quad \text{(Choice D)}$$

5. $(-2x^4)(4x^3) = -2 \cdot 4x^{4+3}$
$$= -8x^7$$

7. $(14x^2y^3)(-2x^5y) = 14(-2)x^{2+5}y^{3+1}$
$$= -28x^7y^4$$

9. $-5(3p^2 + 2p^3)$
$$= -5(3p^2) - 5(2p^3)$$
$$= -15p^2 - 10p^3$$

11. $(3x + 7)(x - 4)$

$$\quad\text{F}\qquad\quad\text{O}\qquad\quad\text{I}\qquad\quad\text{L}$$
$$= (3x)(x) + (3x)(-4) + (7)(x) + (7)(-4)$$
$$= 3x^2 - 12x + 7x - 28$$
$$= 3x^2 - 5x - 28$$

13. $(2t + 3s)(3t - 2s)$

$$\quad\text{F}\qquad\quad\text{O}\qquad\quad\text{I}\qquad\quad\text{L}$$
$$= (2t)(3t) + (2t)(-2s) + (3s)(3t) + (3s)(-2s)$$
$$= 6t^2 - 4st + 9st - 6s^2$$
$$= 6t^2 + 5st - 6s^2$$

15.
$$
\begin{array}{r}
2y + 3 \\
3y - 4 \\
\hline
-8y - 12 \leftarrow -4(2y+3) \\
6y^2 + 9y \qquad\quad \leftarrow 3y(2y+3) \\
\hline
6y^2 + y - 12 \qquad \text{\textit{Combine}}\\
\text{\textit{like terms.}}
\end{array}
$$

17. By the product of the sum and difference of two terms,

$$(5m - 3n)(5m + 3n)$$
$$= (5m)^2 - (3n)^2$$
$$= 25m^2 - 9n^2.$$

19. $m(m + 5)(m - 8)$
$$= m(m^2 - 8m + 5m - 40)$$
$$= m(m^2 - 3m - 40)$$
$$= m(m^2) + m(-3m) + m(-40)$$
$$= m^3 - 3m^2 - 40m$$

21. $4z(2z + 1)(3z - 4)$
$$= 4z(6z^2 - 8z + 3z - 4)$$
$$= 4z(6z^2 - 5z - 4)$$
$$= 4z(6z^2) + 4z(-5z) + 4z(-4)$$
$$= 24z^3 - 20z^2 - 16z$$

23. $x^2(2x + 3)(2x - 3)$
$$= x^2[(2x)^2 - (3)^2]$$
$$= x^2(4x^2 - 9)$$
$$= x^2(4x^2) + x^2(-9)$$
$$= 4x^4 - 9x^2$$

25. $(2m + 3)(3m^2 - 4m - 1)$
$$= 2m(3m^2 - 4m - 1) + 3(3m^2 - 4m - 1)$$
$$= (6m^3 - 8m^2 - 2m) + (9m^2 - 12m - 3)$$
$$= 6m^3 + (-8 + 9)m^2 + (-2 - 12)m - 3$$
$$= 6m^3 + m^2 - 14m - 3$$

27.
$$
\begin{array}{r}
6m^2 + 2m - 1 \\
2m + 3 \\
\hline
18m^2 + 6m - 3 \\
12m^3 + 4m^2 - 2m \qquad\quad \\
\hline
12m^3 + 22m^2 + 4m - 3
\end{array}
$$

29.
$$
\begin{array}{r}
2z^3 - 5z^2 + 8z - 1 \\
4z + 3 \\
\hline
6z^3 - 15z^2 + 24z - 3 \\
8z^4 - 20z^3 + 32z^2 - 4z \qquad\quad \\
\hline
8z^4 - 14z^3 + 17z^2 + 20z - 3
\end{array}
$$

31.
$$
\begin{array}{r}
-x^2 + 8x - 3 \\
2x^3 + 5x \\
\hline
-5x^3 + 40x^2 - 15x \\
-2x^5 + 16x^4 - 6x^3 \qquad\qquad\quad \\
\hline
-2x^5 + 16x^4 - 11x^3 + 40x^2 - 15x
\end{array}
$$

33.
$$
\begin{array}{r}
2p^2 + 3p + 6 \\
3p^2 - 4p - 1 \\
\hline
-2p^2 - 3p - 6 \\
-8p^3 - 12p^2 - 24p \qquad\quad \\
6p^4 + 9p^3 + 18p^2 \qquad\qquad\qquad \\
\hline
6p^4 + p^3 + 4p^2 - 27p - 6
\end{array}
$$

35. We use the FOIL method to multiply two binomials. We multiply the first terms (F), the outside terms (O), the inside terms (I), and the last terms (L). Then we combine like terms.

37. Use the formula for the product of the sum and difference of two terms.

$$(2p - 3)(2p + 3) = (2p)^2 - (3)^2$$
$$= 4p^2 - 9$$

39. $(5m - 1)(5m + 1) = (5m)^2 - (1)^2$
$$= 25m^2 - 1$$

41. $(3a + 2c)(3a - 2c) = (3a)^2 - (2c)^2$
$$= 9a^2 - 4c^2$$

43. $\left(4x - \dfrac{2}{3}\right)\left(4x + \dfrac{2}{3}\right) = (4x)^2 - \left(\dfrac{2}{3}\right)^2$
$$= 16x^2 - \dfrac{4}{9}$$

45. $\left(4m + 7n^2\right)\left(4m - 7n^2\right)$
$$= (4m)^2 - \left(7n^2\right)^2$$
$$= 16m^2 - 49n^4$$

47. $\left(5y^3 + 2\right)\left(5y^3 - 2\right) = \left(5y^3\right)^2 - (2)^2$
$$= 25y^6 - 4$$

49. Use the formula for the square of a binomial.
$$(y - 5)^2 = y^2 - 2(y)(5) + 5^2$$
$$= y^2 - 10y + 25$$

51. $(2p + 7)^2 = (2p)^2 + 2(2p)(7) + 7^2$
$$= 4p^2 + 28p + 49$$

53. $(4n - 3m)^2$
$$= (4n)^2 - 2(4n)(3m) + (3m)^2$$
$$= 16n^2 - 24nm + 9m^2$$

55. $\left(k - \dfrac{5}{7}p\right)^2 = k^2 - 2(k)\left(\dfrac{5}{7}p\right) + \left(\dfrac{5}{7}p\right)^2$
$$= k^2 - \dfrac{10}{7}kp + \dfrac{25}{49}p^2$$

57. $(x + y)^2 = x^2 + 2xy + y^2$. The expression $x^2 + y^2$ is missing the $2xy$ term, so $(x + y)^2 \neq x^2 + y^2$ in general.

59. $[(5x + 1) + 6y]^2$
$$= (5x + 1)^2 + 2(5x + 1)(6y) + (6y)^2$$
Square of a binomial
$$= (5x + 1)^2 + 12y(5x + 1) + 36y^2$$
$$= [(5x)^2 + 2(5x)(1) + 1^2]$$
$$+ 60xy + 12y + 36y^2$$
Square of a binomial
$$= 25x^2 + 10x + 1 + 60xy + 12y + 36y^2$$

61. $[(2a + b) - 3]^2$
$$= (2a + b)^2 - 2(2a + b)(3) + 3^2$$
$$= [(2a)^2 + 2(2a)(b) + b^2]$$
$$- 6(2a + b) + 3^2$$
$$= 4a^2 + 4ab + b^2 - 12a - 6b + 9$$

63. $[(2a + b) - 3][(2a + b) + 3]$
$$= (2a + b)^2 - 3^2$$
*Product of the sum and
difference of two terms*
$$= [(2a)^2 + 2(2a)(b) + b^2] - 9$$
Square of a binomial
$$= 4a^2 + 4ab + b^2 - 9$$

65. $(2a + b)(3a^2 + 2ab + b^2)$
Rewrite vertically and multiply.

$$
\begin{array}{r}
3a^2 + 2ab + b^2 \\
2a + b \\
\hline
3a^2b + 2ab^2 + b^3 \\
6a^3 + 4a^2b + 2ab^2 \\
\hline
6a^3 + 7a^2b + 4ab^2 + b^3
\end{array}
$$

67.
$$
\begin{array}{r}
z^3 - 4z^2x + 2zx^2 - x^3 \\
4z - x \\
\hline
- \quad z^3x + 4z^2x^2 - 2zx^3 + x^4 \\
4z^4 - 16z^3x + 8z^2x^2 - 4zx^3 \\
\hline
4z^4 - 17z^3x + 12z^2x^2 - 6zx^3 + x^4
\end{array}
$$

69.
$$
\begin{array}{r}
m^2 - 2mp + p^2 \\
m^2 + 2mp - p^2 \\
\hline
- \quad m^2p^2 + 2mp^3 - p^4 \\
2m^3p - 4m^2p^2 + 2mp^3 \\
m^4 - 2m^3p + m^2p^2 \\
\hline
m^4 \qquad - 4m^2p^2 + 4mp^3 - p^4
\end{array}
$$

71. $(a + b)(a + 2b)(a - 3b)$
$$= [(a + b)(a + 2b)](a - 3b)$$
$$= [a^2 + 3ab + 2b^2](a - 3b)$$

$$
\begin{array}{r}
a^2 + 3ab + 2b^2 \\
a - 3b \\
\hline
- \quad 3a^2b - 9ab^2 - 6b^3 \\
a^3 + 3a^2b + 2ab^2 \\
\hline
a^3 \qquad - 7ab^2 - 6b^3
\end{array}
$$

Now multiply the last polynomial times ab.
$$ab\left(a^3 - 7ab^2 - 6b^3\right)$$
$$= a^4b - 7a^2b^3 - 6ab^4$$

73. $(y + 2)^3 = (y + 2)^2(y + 2)$
$$= [y^2 + 2(y)(2) + 2^2](y + 2)$$
$$= \left(y^2 + 4y + 4\right)(y + 2)$$

$$
\begin{array}{r}
y^2 + 4y + 4 \\
y + 2 \\
\hline
2y^2 + 8y + 8 \\
y^3 + 4y^2 + 4y \\
\hline
y^3 + 6y^2 + 12y + 8
\end{array}
$$

75. $(q-2)^4 = (q-2)^2(q-2)^2$
$= \left(q^2 - 4q + 4\right)\left(q^2 - 4q + 4\right)$

$$
\begin{array}{r}
q^2 - 4q + 4 \\
q^2 - 4q + 4 \\
\hline
4q^2 - 16q + 16 \\
-4q^3 + 16q^2 - 16q \\
q^4 - 4q^3 + 4q^2 \\
\hline
q^4 - 8q^3 + 24q^2 - 32q + 16
\end{array}
$$

In Exercises 77–80, substitute 3 for x and 4 for y.

77. $(x+y)^2 = (3+4)^2 = 7^2 = 49$
$x^2 + y^2 = 3^2 + 4^2 = 9 + 6 = 25$
Since $49 \neq 25$, $(x+y)^2 \neq x^2 + y^2$.

79. $(x+y)^4 = (3+4)^4 = 7^4 = 2401$
$x^4 + y^4 = 3^4 + 4^4 = 81 + 256 = 337$
Since $2401 \neq 337$, $(x+y)^4 \neq x^4 + y^4$.

81. Although the expressions are equal for this *particular* case, they are not equal *in general*.

83. The formula for the area of a triangle is
$A = \dfrac{1}{2}bh$. Use $b = 3x + 2y$ and $h = 3x - 2y$.

$$
\begin{aligned}
A &= \frac{1}{2}(3x + 2y)(3x - 2y) \\
&= \frac{1}{2}\left(9x^2 - 4y^2\right) \\
&= \frac{9}{2}x^2 - 2y^2
\end{aligned}
$$

85. The formula for the area of a parallelogram is
$A = bh$. Use $b = 5x + 6$ and $h = 3x - 4$.

$$
\begin{aligned}
A &= (5x + 6)(3x - 4) \\
&= 15x^2 - 20x + 18x - 24 \\
&= 15x^2 - 2x - 24
\end{aligned}
$$

87. The length of each side of the entire square is a. To find the length of each side of the blue square, subtract b, the length of a side of the green rectangle, that is, $a - b$.

88. The formula for the area A of a square with side s is $A = s^2$. Since $s = a - b$, the formula for the area of the blue square would be $A = (a - b)^2$.

89. The formula for the area of a rectangle is
$A = LW$. The green rectangle has length $a - b$ and width b, so each green rectangle has an area of $(a - b)b$ or $ab - b^2$.

Since there are two green rectangles, the total area in green is $2(ab - b^2)$ or $2ab - 2b^2$.

90. The length of each side of the yellow square is b, so the yellow square has an area of b^2.

91. The area of the entire colored region is a^2, because each side of the entire colored region has length a.

92. Using the results from Exercises 87–91, the area of the blue square equals

$$a^2 - \left(2ab - 2b^2\right) - b^2 = a^2 - 2ab + b^2.$$

93. Both expressions for the area of the blue square must be equal to each other; that is, $(a - b)^2$ from Exercise 88 and $a^2 - 2ab + b^2$ from Exercise 92.

94. From Exercise 88, the area of the blue square is $(a - b)^2$. From Exercise 92, the area of the blue square is $a^2 - 2ab + b^2$. Since these expressions must be equal

$$(a - b)^2 = a^2 - 2ab + b^2.$$

This reinforces the special product for the square of a binomial difference.

95. $f(x) = 2x$, $g(x) = 5x - 1$

$$
\begin{aligned}
(fg)(x) &= f(x) \cdot g(x) \\
&= 2x(5x - 1) \\
&= 10x^2 - 2x
\end{aligned}
$$

97. $f(x) = x + 1$, $g(x) = 2x - 3$

$$
\begin{aligned}
(fg)(x) &= f(x) \cdot g(x) \\
&= (x + 1)(2x - 3) \\
&= 2x^2 - 3x + 2x - 3 \\
&= 2x^2 - x - 3
\end{aligned}
$$

99. $f(x) = 2x - 3$, $g(x) = 4x^2 + 6x + 9$

$$
\begin{aligned}
(fg)(x) &= f(x) \cdot g(x) \\
&= (2x - 3)(4x^2 + 6x + 9)
\end{aligned}
$$

Multiply vertically.

$$
\begin{array}{r}
4x^2 + 6x + 9 \\
2x - 3 \\
\hline
-12x^2 - 18x - 27 \\
8x^3 + 12x^2 + 18x \\
\hline
8x^3 \qquad\qquad - 27
\end{array}
$$

Section 5.5

1. To factor a polynomial means to find polynomials whose product is the original polynomial. In this section, factoring out the greatest common factor is covered. As an example, x^2y is the GCF in $3x^3y + 2x^2y$, and the factored form of this polynomial is $x^2y(3x + 2)$. The second method covered is factoring by grouping. For example, $x^2 + xy + 3x + 3y$ factors as
$(x^2 + xy) + (3x + 3y) = x(x + y) + 3(x + y)$
$= (x + y)(x + 3)$.

3. The GCF of 7 and 9 is 1. The GCF of z^2 and z^3 is z^2. The GCF of $(m+n)^4$ and $(m+n)^5$ is $(m+n)^4$. The GCF of $7z^2(m+n)^4$ and $9z^3(m+n)^5$ is the product of 1, z^2, and $(m+n)^4$; that is,

$$z^2(m+n)^4.$$

5. $\quad 12m + 60 = 12 \cdot m + 12 \cdot 5$
$$= 12(m + 5)$$

7. $\quad 8k^3 + 24k = 8k \cdot k^2 + 8k \cdot 3$
$$= 8k(k^2 + 3)$$

9. $\quad xy - 5xy^2 = xy \cdot 1 - xy \cdot 5y$
$$= xy(1 - 5y)$$

11. $\quad -4p^3q^4 - 2p^2q^5 = -2p^2q^4 \cdot 2p - 2p^2q^4 \cdot q$
$$= -2p^2q^4(2p + q)$$

13. $\quad 21x^5 + 35x^4 + 14x^3$
$$= 7x^3(3x^2 + 5x + 2)$$

15. $\quad 15a^2c^3 - 25ac^2 + 5ac$
$$= 5ac(3ac^2 - 5c + 1)$$

17. $\quad -27m^3p^5 + 36m^4p^3 - 72m^5p^4$
$$= -9m^3p^3(3p^2 - 4m + 8m^2p)$$

19. $\quad (m-4)(m+2) + (m-4)(m+3)$
The GCF is $(m-4)$.
$$= (m-4)[(m+2) + (m+3)]$$
$$= (m-4)(m+2+m+3)$$
$$= (m-4)(2m+5)$$

21. $\quad (2z-1)(z+6) - (2z-1)(z-5)$
$$= (2z-1)[(z+6) - (z-5)]$$
$$= (2z-1)[z+6-z+5]$$
$$= (2z-1)(11)$$
$$= 11(2z-1)$$

23. $\quad -y^5(r+w) - y^6(z+k)$
The GCF is $-y^5$.
$$= -y^5[(r+w) + y(z+k)]$$
$$= -y^5(r+w+yz+yk)$$

25. $\quad 5(2-x)^2 - (2-x)^3 + 4(2-x)$
$$= (2-x)[5(2-x) - (2-x)^2 + 4]$$
$$= (2-x)[10 - 5x - (4 - 4x + x^2) + 4]$$
$$= (2-x)[10 - 5x - 4 + 4x - x^2 + 4]$$
$$= (2-x)(10 - x - x^2)$$

27. $\quad 4(3-x)^2 - (3-x)^3 + 3(3-x)$
$$= (3-x)[4(3-x) - (3-x)^2 + 3]$$
$$= (3-x)[12 - 4x - (9 - 6x + x^2) + 3]$$
$$= (3-x)[12 - 4x - 9 + 6x - x^2 + 3]$$
$$= (3-x)(6 + 2x - x^2)$$

29. $\quad 15(2z+1)^3 + 10(2z+1)^2 - 25(2z+1)$
The GCF is $5(2z+1)$.
$$= 5(2z+1)$$
$$\quad \cdot [3(2z+1)^2 + 2(2z+1) - 5]$$
$$= 5(2z+1)$$
$$\quad \cdot [3(4z^2 + 4z + 1) + 4z + 2 - 5]$$
$$= 5(2z+1)[12z^2 + 12z + 3 + 4z + 2 - 5]$$
$$= 5(2z+1)[12z^2 + 16z]$$
$$= 5(2z+1)[4z(3z+4)]$$
$$= 20z(2z+1)(3z+4)$$

31. $\quad 5(m+p)^3 - 10(m+p)^2 - 15(m+p)^4$
The GCF is $5(m+p)^2$.
$$= 5(m+p)^2[(m+p) - 2 - 3(m+p)^2]$$
$$= 5(m+p)^2$$
$$\quad \cdot [m+p-2-3(m^2+2mp+p^2)]$$
$$= 5(m+p)^2$$
$$\quad \cdot (m+p-2-3m^2-6mp-3p^2)$$

33. The greatest common factor of $6x^3y^4 - 12x^5y^2 + 24x^4y^8$ is $6x^3y^2$ so (a) has the correct factorization.

35. $\quad -2x^5 + 6x^3 + 4x^2$
Factor out $2x^2$.
$$= 2x^2(-x^3 + 3x + 2)$$
Factor out $-2x^2$.
$$= -2x^2(x^3 - 3x - 2)$$

37. $\quad -32a^4m^5 - 16a^2m^3 - 64a^5m^6$
Factor out $16a^2m^3$.
$$= 16a^2m^3(-2a^2m^2 - 1 - 4a^3m^3)$$
Factor out $-16a^2m^3$.
$$= -16a^2m^3(2a^2m^2 + 1 + 4a^3m^3)$$

39. $\quad 60 = 2 \cdot 30$
$$= 2 \cdot 2 \cdot 15$$
$$= 2 \cdot 2 \cdot 3 \cdot 5$$
$$60 = 2^2 \cdot 3 \cdot 5$$

40. $\quad 420 = 2 \cdot 210$
$$= 2 \cdot 2 \cdot 105$$
$$= 2 \cdot 2 \cdot 3 \cdot 35$$
$$= 2 \cdot 2 \cdot 3 \cdot 5 \cdot 7$$
$$420 = 2^2 \cdot 3 \cdot 5 \cdot 7$$

41. $\quad 420 = 360 + 60$
$$= 2^3 \cdot 3^2 \cdot 5 + 2^2 \cdot 3 \cdot 5$$

42. The GCF of 360 and 60 is
$$2^2 \cdot 3 \cdot 5.$$

43. Factor $2^2 \cdot 3 \cdot 5$ out of each term.

$$2^3 \cdot 3^2 \cdot 5 + 2^2 \cdot 3 \cdot 5$$
$$= 2^2 \cdot 3 \cdot 5(2 \cdot 3 + 1)$$

44. $(2 \cdot 3 + 1) = 6 + 1 = 7$

45. $360 + 60 = 2^2 \cdot 3 \cdot 5(2 \cdot 3 + 1)$
$$= 2^2 \cdot 3 \cdot 5 \cdot 7$$

46. The answer is the same, $2^2 \cdot 3 \cdot 5 \cdot 7$, in each case.

47. $mx + 3qx + my + 3qy$
$$= (mx + 3qx) + (my + 3qy)$$
$$= x(m + 3q) + y(m + 3q)$$
$$= (m + 3q)(x + y)$$

49. $10m + 2n + 5mk + nk$
$$= (10m + 2n) + (5mk + nk)$$
$$= 2(5m + n) + k(5m + n)$$
$$= (5m + n)(2 + k)$$

51. $m^2 - 3m - 15 + 5m$
$$= (m^2 - 3m) + (-15 + 5m)$$
$$= m(m - 3) + 5(-3 + m)$$
$$= m(m - 3) + 5(m - 3)$$
$$= (m - 3)(m + 5)$$

53. $p^2 - 4zq + pq - 4pz$
$$= (p^2 + pq) + (-4zq - 4pz)$$
$$= p(p + q) - 4z(q + p)$$
$$= (p + q)(p - 4z)$$

55. $3a^2 + 15a - 10 - 2a$
$$= (3a^2 + 15a) + (-10 - 2a)$$
$$= 3a(a + 5) - 2(5 + a)$$
$$= 3a(a + 5) - 2(a + 5)$$
$$= (a + 5)(3a - 2)$$

57. $-15p^2 + 5pq - 6pq + 2q^2$
$$= (-15p^2 + 5pq) + (-6pq + 2q^2)$$
$$= 5p(-3p + q) + 2q(-3p + q)$$
$$= (-3p + q)(5p + 2q)$$

59. $-3a^3 - 3ab^2 + 2a^2b + 2b^3$
$$= (-3a^3 - 3ab^2) + (2a^2b + 2b^3)$$
$$= -3a(a^2 + b^2) + 2b(a^2 + b^2)$$
$$= (a^2 + b^2)(-3a + 2b)$$

61. $4 + xy - 2y - 2x$
$$= xy - 2x - 2y + 4$$
$$= (xy - 2x) + (-2y + 4)$$
$$= x(y - 2) - 2(y - 2)$$
$$= (y - 2)(x - 2)$$

63. $8 + 9y^4 - 6y^3 - 12y$
$$= 9y^4 - 6y^3 - 12y + 8$$
$$= (9y^4 - 6y^3) + (-12y + 8)$$
$$= 3y^3(3y - 2) - 4(3y - 2)$$
$$= (3y - 2)(3y^3 - 4)$$

65. $1 - a + ab - b$
$$= 1 - a - b + ab$$
$$= (1 - a) + (-b + ab)$$
$$= 1(1 - a) - b(1 - a)$$
$$= (1 - a)(1 - b)$$

67. **(a)** $(a - 1)(b - 1)$
$$= ab - a - b + 1$$
$$= 1 - a + ab - b$$

(b) $(-a + 1)(-b + 1)$
$$= ab - a - b + 1$$
$$= 1 - a + ab - b$$

(c) $(-1 + a)(-1 + b)$
$$= 1 - b - a + ab$$
$$= 1 - a + ab - b$$

(d) $(1 - a)(b + 1)$
$$= b + 1 - ab - a$$
$$= 1 - a - ab + b$$
$$\neq 1 - a + ab - b$$

The only choice that is not a factored form of $1 - a + ab - b$ is choice (d).

69. $3m^{-5} + m^{-3}$

m^{-5} is the GCF since -5 is the least exponent on m.

$$= m^{-5}(3) + m^{-5}(m^2)$$
$$= m^{-5}(3 + m^2)$$
$$= \frac{3 + m^2}{m^5}$$

71. $3p^{-3} + 2p^{-2}$

-3 is the least exponent on p.

$$= p^{-3}(3) + p^{-3}(2p^1)$$
$$= p^{-3}(3 + 2p)$$
$$= \frac{3 + 2p}{p^3}$$

Section 5.6

1. **(a)** $x^2 + x - 12$
Examine choices A–F. The only choices that would have the term $1x^2$ when multiplied together are B and E. Choice E, $(x + 4)(x - 3)$, is correct since its x-term is $4x - 3x = x$.

(b) $x^2 - x - 12$
Similar to part (a), choice B is correct.

(c) $2x^2 - 3x - 35$
Choices A and D would have $2x^2$ as their x^2-term. Choice A, $(2x + 7)(x - 5)$, is correct since its x-term is $7x - 10x = -3x$.

(d) $2x^2 + 3x - 35$
Similar to part (c), choice D is correct.

(e) $12x^2 + 23x - 24$
Choices C and F would have $12x^2$ as their x^2-term. Choice C, $(3x + 8)(4x - 3)$, is correct since its x-term is $32x - 9x = 23x$.

(f) $12x^2 - 23x - 24$
Similar to part (e), choice F is correct.

3. To factor $a^2 - 2a - 15$, we need two integer factors whose product is -15 (the last term) and whose sum is -2 (coefficient of the middle term). Since $-5 \cdot 3 = -15$ and $-5 + 3 = -2$,
$$a^2 - 2a - 15 = (a - 5)(a + 3).$$

5. $p^2 + 11p + 24$
Two integer factors whose product is 24 and whose sum is 11 are 8 and 3.
$$= (p + 8)(p + 3)$$

7. $r^2 - 15r + 36$
Two integer factors whose product is 36 and whose sum is -15 are -12 and -3.
$$= (r - 12)(r - 3)$$

9. $a^2 - 2ab - 35b^2$
Two integer factors whose product is -35 and whose sum is -2 are 5 and -7.
$$= (a + 5b)(a - 7b)$$

11. $y^2 - 8yq + 15q^2$
Two integer factors whose product is 15 and whose sum is -8 are -5 and -3.
$$= (y - 5q)(y - 3q)$$

13. $x^2y^2 + 12xy + 18$
The positive integer factors of 18 are:

$$1 \text{ and } 18, \ 2 \text{ and } 9, \ 3 \text{ and } 6$$

The sums of these pairs are 19, 11, and 9, but *not* 12. This trinomial is prime.

15. $6m^2 + 13m - 15$
Multiply the first and last coefficients to get $6(-15) = -90$.
Two integer factors whose product is -90 and whose sum is 13 are -5 and 18.
Rewrite the trinomial in a form that can be factored by grouping.

$$6m^2 + (+13m) - 15$$
$$= 6m^2 + (-5m + 18m) - 15$$
$$= (6m^2 - 5m) + (18m - 15)$$
$$= m(6m - 5) + 3(6m - 5)$$
$$= (6m - 5)(m + 3)$$

Note: These exercises can be worked using the alternative method of repeated combinations and FOIL or the grouping method.

17. $10x^2 + 3x - 18$
Two integer factors whose product is $(10)(-18) = -180$ and whose sum is 3 are 15 and -12.
Rewrite the trinomial in a form that can be factored by grouping.
$$10x^2 + 3x - 18$$
$$= 10x^2 + 15x - 12x - 18$$
$$= 5x(2x + 3) - 6(2x + 3)$$
$$= (2x + 3)(5x - 6)$$

19. $20k^2 + 47k + 24$
Two integer factors whose product is $(20)(24) = 480$ and whose sum is 47 are 15 and 32.
Rewrite the trinomial in a form that can be factored by grouping.
$$= 20k^2 + 15k + 32k + 24$$
$$= 5k(4k + 3) + 8(4k + 3)$$
$$= (4k + 3)(5k + 8)$$

21. $15a^2 - 22ab + 8b^2$
Two integer factors whose product is $(15)(8) = 120$ and whose sum is -22 are -10 and -12.
Rewrite the trinomial in a form that can be factored by grouping.
$$= 15a^2 - 10ab - 12ab + 8b^2$$
$$= 5a(3a - 2b) - 4b(3a - 2b)$$
$$= (3a - 2b)(5a - 4b)$$

23. $40x^2 + xy + 6y^2$
There are no integer factors of $(40)(6) = 240$ that add up to 1, so this trinomial is prime.

25. $25r^2 - 90r + 81$
Use the alternative method and write $25r^2$ as $5r \cdot 5r$ and 81 as $9 \cdot 9$. Use these factors in the binomial factors to obtain.

$$25r^2 - 90r + 81 = (5r - 9)(5r - 9)$$
$$= (5r - 9)^2.$$

27. $6x^2z^2 + 5xz - 4$
Two integer factors whose product is $(6)(-4) = -24$ and whose sum is 5 are 8 and -3.
Rewrite the trinomial in a form that can be

factored by grouping.

$$= 6x^2z^2 + 8xz - 3xz - 4$$
$$= 2xz(3xz + 4) - 1(3xz + 4)$$
$$= (3xz + 4)(2xz - 1)$$

29. $24x^2 + 42x + 15$

Always factor out the GCF first.

$$= 3(8x^2 + 14x + 5)$$

Now factor $8x^2 + 14x + 5$ by the alternative method.

$$8x^2 + 14x + 5 = (4x + 5)(2x + 1)$$

The final factored form is

$$3(4x + 5)(2x + 1).$$

31. $15a^2 + 70a - 120$
$$= 5(3a^2 + 14a - 24)$$
$$= 5(a + 6)(3a - 4)$$

33. $4m^3 + 12m^2 - 40m$
$$= 4m(m^2 + 3m - 10)$$
$$= 4m(m + 5)(m - 2)$$

35. $11x^3 - 110x^2 + 264x$
$$= 11x(x^2 - 10x + 24)$$
$$= 11x(x - 6)(x - 4)$$

37. $2x^3y^3 - 48x^2y^4 + 288xy^5$
$$= 2xy^3(x^2 - 24xy + 144y^2)$$
$$= 2xy^3(x - 12y)(x - 12y)$$
or $2xy^3(x - 12y)^2$

39. $18a^2 - 15a - 18$
$$= 3(6a^2 - 5a - 6)$$
$$= 3(3a + 2)(2a - 3)$$

41. $6a^3 + 12a^2 - 90a$
$$= 6a(a^2 + 2a - 15)$$
$$= 6a(a - 3)(a + 5)$$

43. $13y^3 + 39y^2 - 52y$
$$= 13y(y^2 + 3y - 4)$$
$$= 13y(y + 4)(y - 1)$$

45. $12p^3 - 12p^2 + 3p$
$$= 3p(4p^2 - 4p + 1)$$
$$= 3p(2p - 1)(2p - 1)$$
or $3p(2p - 1)^2$

47. $-x^2 + 7x + 18 = -1(x^2 - 7x - 18)$
$$= -(x - 9)(x + 2)$$

49. $-18a^2 + 17a + 15$
$$= -1(18a^2 - 17a - 15)$$

Two integer factors whose product is $(18)(-15) = -270$ and whose sum is -17 are 10

and -27.
$$= -(18a^2 + 10a - 27a - 15)$$
$$= -[2a(9a + 5) - 3(9a + 5)]$$
$$= -(9a + 5)(2a - 3)$$

51. $-14r^3 + 19r^2 + 3r$
$$= -r(14r^2 - 19r - 3)$$
$$= -r(7r + 1)(2r - 3)$$

53. There is a GCF of 2. She did not factor the polynomial *completely*. The factor $(4x + 10)$ can be factored further as $2(2x + 5)$, giving the final form as $2(2x + 5)(x - 2)$.

55. They are both correct: in each case the product is $-4x^2 - 29x + 24$.

57. No, there is no number that, when multiplied by 2, equals 45, so 2 is not a factor of 45.

58. The positive integer factors of 45 are 1, 3, 5, 9, 15, and 45. 2 is not a factor of any of these factors either.

59. No, 3 is not a factor of 20.

60. The positive integer factors of 20 are 1, 2, 4, 5, 10, and 20. 3 is not a factor of any of these factors either.

61. No, 5 is not a factor of $10x^2 + 29x + 10$ because it is not a factor of each term.

62. To factor $10x^2 + 29x + 10$, look for two integers whose product is $10(10) = 100$ and whose sum is 29. The numbers are 25 and 4.

$10x^2 + 29x + 10$
$$= 10x^2 + 25x + 4x + 10$$
$$= 5x(2x + 5) + 2(2x + 5)$$
$$= (2x + 5)(5x + 2)$$

5 is not a factor of either of these factors.

63. Since k is odd, 2 is not a factor of $2x^2 + kx + 8$, and because 2 is a factor of $2x + 4$, the binomial $2x + 4$ cannot be a factor of $2x^2 + kx + 8$.

64. 3 is a factor of $3y + 15$, but it is not a factor of $12y^2 - 11y - 15$. Therefore, $3y + 15$ cannot be a factor of $12y^2 - 11y - 15$.

65. In $p^4 - 10p^2 + 16$, let $x = p^2$ to obtain

$$x^2 - 10x + 16 = (x - 8)(x - 2).$$

Replace x with p^2.

$$p^4 - 10p^2 + 16 = (p^2 - 8)(p^2 - 2)$$

67. In $2x^4 - 9x^2 - 18$, let $y = x^2$ to obtain

$$2y^2 - 9y - 18 = (2y + 3)(y - 6).$$

Replace y with x^2.

$$2x^4 - 9x^2 - 18 = (2x^2 + 3)(x^2 - 6)$$

69. In $16x^4 + 16x^2 + 3$, let $m = x^2$ to obtain

$$16m^2 + 16m + 3 = (4m + 3)(4m + 1).$$

Replace m with x^2.

$$16x^4 + 16x^2 + 3 = (4x^2 + 3)(4x^2 + 1)$$

71. In $12p^6 - 32p^3r + 5r^2$, let $x = p^3$ to obtain

$$12x^2 - 32xr + 5r^2 = (6x - r)(2x - 5r).$$

Replace x with p^3.

$$12p^6 - 32p^3r + 5r^2 = (6p^3 - r)(2p^3 - 5r)$$

73. $10(k + 1)^2 - 7(k + 1) + 1$

Let $x = k + 1$ to obtain
$$10x^2 - 7x + 1 = (5x - 1)(2x - 1).$$
Replace x with $k + 1$.
$$10(k + 1)^2 - 7(k + 1) + 1$$
$$= [5(k + 1) - 1][2(k + 1) - 1]$$
$$= (5k + 5 - 1)(2k + 2 - 1)$$
$$= (5k + 4)(2k + 1)$$

75. $3(m + p)^2 - 7(m + p) - 20$

Let $x = m + p$ to obtain
$$3x^2 - 7x - 20 = (3x + 5)(x - 4).$$
Replace x with $m + p$.
$$3(m + p)^2 - 7(m + p) - 20$$
$$= [3(m + p) + 5][(m + p) - 4]$$
$$= (3m + 3p + 5)(m + p - 4)$$

77. $a^2(a + b)^2 - ab(a + b)^2 - 6b^2(a + b)^2$

Factor out the GCF, $(a + b)^2$.
$$= (a + b)^2(a^2 - ab - 6b^2)$$
Factor the trinomial.
$$= (a + b)^2(a - 3b)(a + 2b)$$

79. $p^2(p + q) + 4pq(p + q) + 3q^2(p + q)$

Factor out the GCF, $p + q$.
$$= (p + q)(p^2 + 4pq + 3q^2)$$
Factor the trinomial.
$$= (p + q)(p + q)(p + 3q)$$
$$= (p + q)^2(p + 3q)$$

81. $z^2(z - x) - zx(x - z) - 2x^2(z - x)$

Factor out -1 from the middle term:
$$x - z = -1(z - x).$$
$$= z^2(z - x) + zx(z - x) - 2x^2(z - x)$$

Factor out the GCF, $z - x$.
$$= (z - x)(z^2 + zx - 2x^2)$$
Factor the trinomial.
$$= (z - x)(z + 2x)(z - x)$$
$$= (z - x)^2(z + 2x)$$

83. The first kind of trinomial has 1 as the coefficient of the squared term. An example is $x^2 + 5x + 6$.

Incorrect factored form: $(x + 6)(x + 1)$
This is incorrect because the middle term of this product is $7x$, not $5x$.
Correct factored form: $(x + 3)(x + 2)$

The second kind of trinomial has a number other than 1 as the coefficient of the squared term. An example is $2x^2 + 7x + 3$.

Incorrect factored form: $(2x + 3)(x + 1)$
This is incorrect because the middle term of this product is $5x$, not $7x$.
Correct factored form: $(2x + 1)(x + 3)$

Section 5.7

1. **(a)** Use the formula for the difference of two squares.

$$x^2 - y^2 = (x + y)(x - y)$$

Choice B is equivalent.

(b) $y^2 - x^2 = (y + x)(y - x)$
Choice D is equivalent.

(c) Use the formula for the difference of two cubes.

$$x^3 - y^3 = (x - y)(x^2 + xy + y^2)$$

Choice C is correct.

(d) Use the formula for the sum of two cubes.

$$x^3 + y^3 = (x + y)(x^2 - xy + y^2)$$

Choice A is correct.

3. **(a)** $64 + y^3$ is the sum of two cubes, $4^3 + y^3$.

(b) $125 - p^6$ is the difference of two cubes, $5^3 - (p^2)^3$.

(c) $9x^3 + 125$ is neither the sum nor the difference of cubes.

(d) $(x + y)^3 - 1$ is the difference of two cubes $(x + y)^3 - 1^3$.

Choices (a), (b), and (d) are sums or differences of cubes.

5. $p^2 - 16 = p^2 - 4^2$
$$= (p + 4)(p - 4)$$

7. $25x^2 - 4 = (5x)^2 - 2^2$
$$= (5x + 2)(5x - 2)$$

9. $9a^2 - 49b^2 = (3a)^2 - (7b)^2$
$$= (3a + 7b)(3a - 7b)$$

11. $64m^4 - 4y^4$
$$= 4(16m^4 - y^4)$$
$$= 4\left[(4m^2)^2 - (y^2)^2\right]$$
Factor the difference of squares.
$$= 4(4m^2 + y^2)(4m^2 - y^2)$$
$$= 4(4m^2 + y^2)[(2m)^2 - y^2]$$
Factor the difference of squares again.
$$= 4(4m^2 + y^2)(2m + y)(2m - y)$$

13. $(y + z)^2 - 81$
$$= (y + z)^2 - 9^2$$
$$= [(y + z) + 9][(y + z) - 9]$$
$$= (y + z + 9)(y + z - 9)$$

15. $16 - (x + 3y)^2$

$$= 4^2 - z^2 \quad \text{Let } z = (x + 3y).$$
$$= (4 + z)(4 - z)$$
Substitute $x + 3y$ for z.
$$= [4 + (x + 3y)][4 - (x + 3y)]$$
$$= (4 + x + 3y)(4 - x - 3y)$$

17. $(p + q)^2 - (p - q)^2$
$$= [(p + q) + (p - q)]$$
$$\quad \cdot [(p + q) - (p - q)]$$
$$= (p + q + p - q)(p + q - p + q)$$
$$= 2p(2q)$$
$$= 4pq$$

19. $k^2 - 6k + 9 = (k)^2 - 2(k)(3) + 3^2$
$$= (k - 3)^2$$

21. $4z^2 + 4zw + w^2$
$$= (2z)^2 + 2(2z)(w) + w^2$$
$$= (2z + w)^2$$

23. $16m^2 - 8m + 1 - n^2$
Group the first three terms.
$$= (16m^2 - 8m + 1) - n^2$$
$$= [(4m)^2 - 2(4m)(1) + 1^2] - n^2$$
$$= (4m - 1)^2 - n^2$$
$$= [(4m - 1) + n][(4m - 1) - n]$$
$$= (4m - 1 + n)(4m - 1 - n)$$

25. $4r^2 - 12r + 9 - s^2$
Group the first three terms.
$$= (4r^2 - 12r + 9) - s^2$$
$$= [(2r)^2 - 2(2r)(3) + 3^2] - s^2$$
$$= (2r - 3)^2 - s^2$$
$$= [(2r - 3) + s][(2r - 3) - s]$$
$$= (2r - 3 + s)(2r - 3 - s)$$

27. $x^2 - y^2 + 2y - 1$
Group the last three terms.
$$= x^2 - (y^2 - 2y + 1)$$
$$= x^2 - (y - 1)^2$$
$$= [x + (y - 1)][x - (y - 1)]$$
$$= (x + y - 1)(x - y + 1)$$

29. $98m^2 + 84mn + 18n^2$
$$= 2(49m^2 + 42mn + 9n^2)$$
$$= 2[(7m)^2 + 2(7m)(3n) + (3n)^2]$$
$$= 2(7m + 3n)^2$$

31. $(p + q)^2 + 2(p + q) + 1$
$$= x^2 + 2x + 1 \qquad \text{Let } x = p + q.$$
$$= (x + 1)^2$$
$$= (p + q + 1)^2 \qquad \text{Resubstitute.}$$

33. $(a - b)^2 + 8(a - b) + 16$
$$= (a - b)^2 + 2(a - b)(4) + 4^2$$
$$= [(a - b) + 4]^2$$
$$= (a - b + 4)^2$$

35. $8x^3 - y^3$
$$= (2x)^3 - y^3$$
$$= [2x - y][(2x)^2 + (2x)(y) + y^2]$$
$$= (2x - y)(4x^2 + 2xy + y^2)$$

37. $64g^3 + 27h^3$
$$= (4g)^3 + (3h)^3$$
$$= (4g + 3h)[(4g)^2 - (4g)(3h) + (3h)^2]$$
$$= (4g + 3h)(16g^2 - 12gh + 9h^2)$$

39. $24n^3 + 81p^3$
$$= 3(8n^3 + 27p^3)$$
$$= 3[(2n)^3 + (3p)^3]$$
$$= 3[2n + 3p][(2n)^2 - (2n)(3p) + (3p)^2]$$
$$= 3(2n + 3p)(4n^2 - 6np + 9p^2)$$

41. $(y + z)^3 - 64$
$$= (y + z)^3 - 4^3$$
$$= [(y + z) - 4][(y + z)^2 + (y + z)(4) + 4^2]$$
$$= (y + z - 4)(y^2 + 2yz + z^2 + 4y + 4z + 16)$$

43. $64y^6 + 1$

$= \left(4y^2\right)^3 + 1^3$

$= \left(4y^2 + 1\right)\left[\left(4y^2\right)^2 - \left(4y^2\right)(1) + 1^2\right]$

$= \left(4y^2 + 1\right)\left(16y^4 - 4y^2 + 1\right)$

45. $1000x^9 - 27$

$= \left(10x^3\right)^3 - 3^3$

$= \left(10x^3 - 3\right)\left[\left(10x^3\right)^2 + \left(10x^3\right)(3) + 3^2\right]$

$= \left(10x^3 - 3\right)\left(100x^6 + 30x^3 + 9\right)$

47. $512t^6 - p^3$

$= \left(8t^2\right)^3 - p^3$

$= \left(8t^2 - p\right)\left[\left(8t^2\right)^2 + \left(8t^2\right)(p) + p^2\right]$

$= \left(8t^2 - p\right)\left(64t^4 + 8t^2p + p^2\right)$

49. $x^6 - y^6$

$= \left(x^3\right)^2 - \left(y^3\right)^2$

$= \left(x^3 + y^3\right)\left(x^3 - y^3\right)$

$= \left[(x + y)\left(x^2 - xy + y^2\right)\right]$

$\quad \cdot \left[(x - y)\left(x^2 + xy + y^2\right)\right]$

$= (x + y)\left(x^2 - xy + y^2\right)(x - y)$

$\quad \cdot \left(x^2 + xy + y^2\right)$

50. $x^6 - y^6$

$= (x - y)(x + y)$

$\quad \cdot \underline{\left(x^2 + xy + y^2\right)\left(x^2 - xy + y^2\right)}$

51. $x^6 - y^6$

$= \left(x^2\right)^3 - \left(y^2\right)^3$

$= \left(x^2 - y^2\right)\left(x^4 + x^2y^2 + y^4\right)$

$= (x + y)(x - y)\left(x^4 + x^2y^2 + y^4\right)$

52. $x^6 - y^6 = (x - y)(x + y)$

$\quad \cdot \underline{\left(x^4 + x^2y^2 + y^4\right)}$

53. The product written on the blank in Exercise 50 must equal the product written on the blank in Exercise 52. To verify this, multiply the two factors written in Exercise 50.

$\left(x^2 + xy + y^2\right)\left(x^2 - xy + y^2\right)$

$= x^2\left(x^2 - xy + y^2\right)$

$\quad + xy\left(x^2 - xy + y^2\right)$

$\quad + y^2\left(x^2 - xy + y^2\right)$

$= x^4 - x^3y + x^2y^2 + x^3y - x^2y^2$

$\quad + xy^3 + x^2y^2 - xy^3 + y^4$

$= x^4 + x^2y^2 + y^4$

They are equal.

54. Start by factoring as the difference of two squares since doing so resulted in the complete factorization more directly.

55. $p^2 + 8p + c$ factored as a perfect square trinomial is

$$p^2 + 2px + x^2 = (p + x)^2.$$

The middle term, $2px$, equals $8p$, so

$$2px = 8p$$
$$\frac{2px}{2p} = \frac{8p}{2p}$$
$$x = 4.$$

Substitute 4 for x in $p^2 + 2px + x^2$.

$$p^2 + 2px + x^2 = p^2 + 2p(4) + 4^2$$
$$= p^2 + 8p + 16$$

Therefore, the value of c is 16.

57. $9z^2 + 30z + c$ factored as a perfect square trinomial is

$$(3z)^2 + 2(3z)x + x^2 = (3z + x)^2.$$

The middle term, $6zx$, equals $30z$, so

$$6zx = 30z$$
$$\frac{6zx}{6z} = \frac{30z}{6z}$$
$$x = 5.$$

Substitute 5 for x in $(3z)^2 + 2(3z)x + x^2$.

$$(3z)^2 + 2(3z)x + x^2 = (3z)^2 + 2(3z)(5) + 5^2$$
$$= 9z^2 + 30z + 25$$

Therefore, the value of c is 25.

59. $16x^2 + bx + 49$ factored as a perfect square trinomial is

$$(4x)^2 + 2(4x)(7) + 7^2 = (4x + 7)^2$$

or

$$(4x)^2 - 2(4x)(7) + 7^2 = (4x - 7)^2.$$

The middle term, $2(4x)(7)$ or $-2(4x)(7)$, must equal bx, so

$$2(4x)(7) = bx \quad \text{or} \quad -2(4x)(7) = bx$$
$$56x = bx \qquad\qquad -56x = bx$$
$$56 = b \quad \text{or} \qquad -56 = b.$$

Therefore, the value of b is 56 or -56.

61. $27x^3 + 9x^2 + y^3 - y^2$

$= \left(27x^3 + y^3\right) + \left(9x^2 - y^2\right)$

$= \left[(3x)^3 + y^3\right] + \left[(3x)^2 - y^2\right]$

Factor within groups.

$= \left[(3x + y)\left(9x^2 - 3xy + y^2\right)\right]$

$\quad + \left[(3x + y)(3x - y)\right]$

Factor out the GCF, $3x + y$.
$$= (3x + y)$$
$$\quad \bullet \left[\left(9x^2 - 3xy + y^2 \right) + (3x - y) \right]$$
$$= (3x + y)$$
$$\quad \bullet \left(9x^2 - 3xy + y^2 + 3x - y \right)$$

63. $1000k^3 + 20k - m^3 - 2m$
$$= \left(1000k^3 - m^3 \right) + (20k - 2m)$$
$$= \left[(10k)^3 - m^3 \right] + (20k - 2m)$$
Factor within groups.
$$= \left[(10k - m)\left(100k^2 + 10km + m^2 \right) \right]$$
$$\quad + 2(10k - m)$$
Factor out the GCF, $10k - m$.
$$= (10k - m)\left[\left(100k^2 + 10km + m^2 \right) + 2 \right]$$
$$= (10k - m)\left(100k^2 + 10km + m^2 + 2 \right)$$

65. $y^4 + y^3 + y + 1$
$$= \left(y^4 + y^3 \right) + (y + 1)$$
$$= y^3(y + 1) + 1(y + 1)$$
Factor out the GCF, $y + 1$.
$$= (y + 1)\left(y^3 + 1 \right)$$
Factor the sum of two cubes.
$$= (y + 1)(y + 1)\left(y^2 - y + 1 \right)$$
$$= (y + 1)^2 \left(y^2 - y + 1 \right)$$

67. $10x^2 + 5x^3 - 10y^2 + 5y^3$
$$= \left(10x^2 - 10y^2 \right) + \left(5x^3 + 5y^3 \right)$$
$$= 10\left(x^2 - y^2 \right) + 5\left(x^3 + y^3 \right)$$
Factor within groups.
$$= 10(x + y)(x - y)$$
$$\quad + 5(x + y)\left(x^2 - xy + y^2 \right)$$
Factor out the GCF, $5(x + y)$.
$$= 5(x + y)\left[2(x - y) + \left(x^2 - xy + y^2 \right) \right]$$
$$= 5(x + y)\left(2x - 2y + x^2 - xy + y^2 \right)$$

Section 5.8

1. $100a^2 - 9b^2$
$$= (10a)^2 - (3b)^2 \qquad \textit{Difference of two squares}$$
$$= (10a + 3b)(10a - 3b)$$

3. $3p^4 - 3p^3 - 90p^2$
$$= 3p^2 \left(p^2 - p - 30 \right)$$
$$= 3p^2(p - 6)(p + 5)$$

5. $3a^2pq + 3abpq - 90b^2pq$
$$= 3pq\left(a^2 + ab - 30b^2 \right)$$
$$= 3pq(a + 6b)(a - 5b)$$

7. $225p^2 + 256$ cannot be factored. The binomial is prime.

9. $6b^2 - 17b - 3$
Two integer factors whose product is $(6)(-3) = -18$ and whose sum is -17 are -18 and 1.
$$= 6b^2 - 18b + b - 3$$
$$= 6b(b - 3) + 1(b - 3)$$
$$= (b - 3)(6b + 1).$$

11. $18m^3n + 3m^2n^2 - 6mn^3$
$$= 3mn\left(6m^2 + mn - 2n^2 \right)$$
Factor the trinomial.
$$= 3mn(3m + 2n)(2m - n)$$

13. $2p^2 + 11pq + 15q^2$
Two integer factors whose product is $(2)(15) = 30$ and whose sum is 11 are 5 and 6.
$$= 2p^2 + 5pq + 6pq + 15q^2$$
$$= p(2p + 5q) + 3q(2p + 5q)$$
$$= (2p + 5q)(p + 3q)$$

15. $9m^2 - 45m + 18m^3$
Factor out the GCF, $9m$.
$$= 9m\left(m - 5 + 2m^2 \right) \text{ or } 9m\left(2m^2 + m - 5 \right)$$
There is no pair of integers with a product of -10 and a sum of 1, so this cannot be factored further.

17. $54m^3 - 2000$
Factor out the GCF, 2.
$$2\left(27m^3 - 1000 \right)$$
$$= 2[(3m)^3 - 10^3] \qquad \textit{Difference of two cubes}$$
$$= 2(3m - 10)$$
$$\quad \bullet [(3m)^2 + (3m)(10) + 10^2]$$
$$= 2(3m - 10)\left(9m^2 + 30m + 100 \right)$$

19. $2a^2 - 7a - 4$
Two integer factors whose product is $(2)(-4) = -8$ and whose sum is -7 are -8 and 1.
$$= 2a^2 - 8a + a - 4$$
$$= 2a(a - 4) + 1(a - 4)$$
$$= (a - 4)(2a + 1)$$

21. $kq - 9q + kr - 9r$
$$= q(k - 9) + r(k - 9)$$
$$= (k - 9)(q + r)$$

23. $9r^2 + 100 = (3r)^2 + 10^2$
The sum of two squares cannot be factored. The binomial is prime.

25. $9x^2 + 36y^2 = 9\left(x^2 + 4y^2 \right)$; A sum of two squares can be factored when there is a GCF to be factored out. Two other examples of sums of two squares that can be factored are as follows.
$$16x^2 + 64y^2 = 16\left(x^2 + 4y^2 \right)$$
$$100x^2 + 400y^2 = 100\left(x^2 + 4y^2 \right)$$

27. $x^4 - 625$

$= (x^2)^2 - 25^2$ *Difference of two squares*

$= (x^2 + 25)(x^2 - 25)$

$= (x^2 + 25)(x + 5)(x - 5)$ *Difference of two squares again*

29. $p^3 + 64$

$= p^3 + 4^3$ *Sum of two cubes*

$= (p + 4)(p^2 - 4p + 4^2)$

$= (p + 4)(p^2 - 4p + 16)$

31. $64m^2 - 625$

$= (8m)^2 - 25^2$ *Difference of two squares*

$= (8m + 25)(8m - 25)$

33. $12z^3 - 6z^2 + 18z$

Factor out the GCF, $6z$.

$$6z(2z^2 - z + 3)$$

Further factoring is not possible. There is no pair of integers whose product is 6 and whose sum is -1.

35. $256b^2 - 400c^2$

$= 16(16b^2 - 25c^2)$

$= 16[(4b)^2 - (5c)^2]$ *Difference of two squares*

$= 16(4b + 5c)(4b - 5c)$

37. $1000z^3 + 512$

$= 8(125z^3 + 64)$

$= 8[(5z)^3 + 4^3]$ *Sum of two cubes*

$= 8[5z + 4][(5z)^2 - (5z)(4) + 4^2]$

$= 8(5z + 4)(25z^2 - 20z + 16)$

39. $10r^2 + 23rs - 5s^2$

Two integer factors whose product is $(10)(-5) = -50$ and whose sum is 23 are 25 and -2.

$= 10r^2 + 25rs - 2rs - 5s^2$

$= 5r(2r + 5s) - s(2r + 5s)$

$= (2r + 5s)(5r - s)$

41. $24p^3q + 52p^2q^2 + 20pq^3$

$= 4pq(6p^2 + 13pq + 5q^2)$

Two integer factors whose product is $(6)(5) = 30$ and whose sum is 13 are 10 and 3.

$= 4pq(6p^2 + 10pq + 3pq + 5q^2)$

$= 4pq[2p(3p + 5q) + q(3p + 5q)]$

$= 4pq(3p + 5q)(2p + q)$

43. $48k^4 - 243$

$= 3(16k^4 - 81)$

$= 3[(4k^2)^2 - 9^2]$

$= 3(4k^2 + 9)(4k^2 - 9)$

$= 3(4k^2 + 9)[(2k)^2 - 3^2]$

$= 3(4k^2 + 9)(2k + 3)(2k - 3)$

45. $m^3 + m^2 - n^3 - n^2$

$= (m^3 - n^3) + (m^2 - n^2)$

Difference of   Difference of
two cubes;   two squares

$= (m - n)(m^2 + mn + n^2)$
$\quad + (m + n)(m - n)$

$= (m - n)[(m^2 + mn + n^2) + (m + n)]$

$= (m - n)(m^2 + mn + n^2 + m + n)$

47. $x^2 - 4m^2 - 4mn - n^2$

$= x^2 - (4m^2 + 4mn + n^2)$

$= x^2 - [(2m)^2 + 2(2m)n + n^2]$

     Perfect square trinomial

$= x^2 - (2m + n)^2$

     Difference of two squares

$= [x + (2m + n)][x - (2m + n)]$

$= (x + 2m + n)(x - 2m - n)$

49. Use the formula $A = LW$, and substitute $2W^2 + 9W$ for A. Solve for L.

$$A = LW$$
$$2W^2 + 9W = LW$$
$$(2W + 9)W = LW$$

Thus, we must have $L = 2W + 9$.

51. $2x^2 - 2x - 40 = 2(x^2 - x - 20)$
$$= 2(x + 4)(x - 5)$$

53. $(2m + n)^2 - (2m - n)^2$

          Difference of two squares

$= [(2m + n) + (2m - n)]$
$\quad \bullet [(2m + n) - (2m - n)]$

$= (2m + n + 2m - n)(2m + n - 2m + n)$

$= 4m(2n)$

$= 8mn$

55. $50p^2 - 162$

$= 2(25p^2 - 81)$

$= 2[(5p)^2 - 9^2]$

$= 2(5p + 9)(5p - 9)$

57. $12m^2rx + 4mnrx + 40n^2rx$

Factor out the GCF, $4rx$.

$= 4rx(3m^2 + mn + 10n^2)$

59. $21a^2 - 5ab - 4b^2$

Two integer factors whose product is $(21)(-4) = -84$ and whose sum is -5 are -12 and 7.

$$= 21a^2 - 12ab + 7ab - 4b^2$$
$$= 3a(7a - 4b) + b(7a - 4b)$$
$$= (7a - 4b)(3a + b)$$

61. $x^2 - y^2 - 4$ cannot be factored. The polynomial is prime.

63. $(p + 8q)^2 - 10(p + 8q) + 25$

$$= x^2 - 10x + 25 \qquad \text{Let } x = p + 8q.$$
$$= (x - 5)^2$$
$$= [(p + 8q) - 5]^2 \qquad \textit{Resubstitute.}$$
$$= (p + 8q - 5)^2$$

65. $21m^4 - 32m^2 - 5$

$$= 21x^2 - 32x - 5 \quad \text{Let } x = m^2.$$

Two integer factors whose product is $(21)(-5) = -105$ and whose sum is -32 are -35 and 3.

$$= 21x^2 - 35x + 3x - 5$$
$$= 7x(3x - 5) + 1(3x - 5)$$
$$= (3x - 5)(7x + 1)$$
$$= (3m^2 - 5)(7m^2 + 1) \qquad \textit{Resubstitute.}$$

67. $(r + 2t)^3 + (r - 3t)^3$

Let $x = r + 2t$ and $y = r - 3t$.

$$= x^3 + y^3 \qquad \textit{Sum of two cubes.}$$
$$= (x + y)(x^2 - xy + y^2)$$

Substitute $r + 2t$ for x and $r - 3t$ for y.

$$= [(r + 2t) + (r - 3t)]$$
$$\quad \cdot [(r + 2t)^2 - (r + 2t)(r - 3t)$$
$$\quad + (r - 3t)^2]$$
$$= (2r - t)$$
$$\quad \cdot (r^2 + 4rt + 4t^2 - r^2 + rt + 6t^2$$
$$\quad + r^2 - 6rt + 9t^2)$$
$$= (2r - t)(r^2 - rt + 19t^2)$$

69. $x^5 + 3x^4 - x - 3$

$$= x^4(x + 3) - 1(x + 3)$$
$$= (x + 3)(x^4 - 1)$$
$$= (x + 3)\left[(x^2)^2 - 1^2\right]$$
$$= (x + 3)(x^2 + 1)(x^2 - 1)$$
$$= (x + 3)(x^2 + 1)[x^2 - 1^2]$$
$$= (x + 3)(x^2 + 1)(x + 1)(x - 1)$$

71. $m^2 - 4m + 4 - n^2 + 6n - 9$

$$= (m^2 - 4m + 4) - (n^2 - 6n + 9)$$
$$= (m^2 - 2(2)m + 2^2)$$
$$\quad - (n^2 - 2(3)n + 3^2)$$

Perfect square trinomials

$$= (m - 2)^2 - (n - 3)^2$$
$$= [(m - 2) + (n - 3)]$$
$$\quad \cdot [(m - 2) - (n - 3)]$$

Difference of two squares

$$= (m - 2 + n - 3)(m - 2 - n + 3)$$
$$= (m + n - 5)(m - n + 1)$$

73. $a + ay = a \cdot 1 + a \cdot y$
$$= a(1 + y)$$

74. The 1 must appear so that when the distributive property is used to check, the term $a \cdot 1 = a$ will appear in the original binomial.

75. $x^2 + 4 + x^2 y + 4y$

$$= (x^2 + 4) + (x^2 y + 4y)$$
$$= 1(x^2 + 4) + y(x^2 + 4)$$
$$= (x^2 + 4)(1 + y)$$

76. If 1 did not appear, the terms x^2 and 4 would not appear in the product when the factoring was checked.

Section 5.9

1. One side of the equation must be 0. By factoring the other side, the zero-factor property can be used. Set each factor equal to 0, and solve each of the resulting equations.

3. $3x^2 - 8x = 0$
Factor $3x^2 - 8x$ as $x(3x - 8)$.

5. $x^2 - 4x = 12$
Subtract 12 to get $x^2 - 4x - 12$ on the left, and then factor as $(x - 6)(x + 2)$.

7. $(x + 2)(2x - 9) = 0$
The equation is ready to solve as it is given.

In the exercises in this section, check all solutions to the equations by substituting them back in the original equations.

9. $(x - 5)(x + 10) = 0$

$$x - 5 = 0 \quad \text{or} \quad x + 10 = 0$$
$$x = 5 \quad \text{or} \qquad x = -10$$

Check $x = 5$: $\qquad (0)(15) = 0 \quad \textit{True}$

Check $x = -10$: $\quad (-15)(0) = 0 \quad \textit{True}$

Solution set: $\{5, -10\}$

11. $(2k - 5)(3k + 8) = 0$

$2k - 5 = 0$ or $3k + 8 = 0$

$2k = 5$ $3k = -8$

$k = \dfrac{5}{2}$ or $k = -\dfrac{8}{3}$

Solution set: $\left\{\dfrac{5}{2}, -\dfrac{8}{3}\right\}$

13. $(m + 6)(4m - 3)(m - 1) = 0$

$m + 6 = 0$ or $4m - 3 = 0$ or $m - 1 = 0$

$m = -6$ or $m = \dfrac{3}{4}$ or $m = 1$

Solution set: $\left\{-6, \dfrac{3}{4}, 1\right\}$

15. $r(r - 4)(2r + 5) = 0$

$r = 0$ or $r - 4 = 0$ or $2r + 5 = 0$

$r = 4$ $2r = -5$

$r = -\dfrac{5}{2}$

Solution set: $\left\{0, 4, -\dfrac{5}{2}\right\}$

17. $m^2 - 3m - 10 = 0$

$(m + 2)(m - 5) = 0$

Use the zero-factor property.

$m + 2 = 0$ or $m - 5 = 0$

$m = -2$ or $m = 5$

Solution set: $\{-2, 5\}$

19. $z^2 + 9z + 18 = 0$

$(z + 6)(z + 3) = 0$

$z + 6 = 0$ or $z + 3 = 0$

$z = -6$ or $z = -3$

Solution set: $\{-6, -3\}$

21. $2x^2 = 7x + 4$

Get 0 on one side.

$2x^2 - 7x - 4 = 0$

$(2x + 1)(x - 4) = 0$

$2x + 1 = 0$ or $x - 4 = 0$

$2x = -1$ $x = 4$

$x = -\dfrac{1}{2}$

Solution set: $\left\{-\dfrac{1}{2}, 4\right\}$

23. $15k^2 - 7k = 4$

$15k^2 - 7k - 4 = 0$

$(3k + 1)(5k - 4) = 0$

$3k + 1 = 0$ or $5k - 4 = 0$

$3k = -1$ $5k = 4$

$k = -\dfrac{1}{3}$ or $k = \dfrac{4}{5}$

Solution set: $\left\{-\dfrac{1}{3}, \dfrac{4}{5}\right\}$

25. $2y^2 - 12 - 4y = y^2 - 3y$

$y^2 - y - 12 = 0$

$(y + 3)(y - 4) = 0$

$y + 3 = 0$ or $y - 4 = 0$

$y = -3$ or $y = 4$

Solution set: $\{-3, 4\}$

27. $8m^2 - 72 = 0$

Factor out the GCF, 8.

$8(m^2 - 9) = 0$

$8(m + 3)(m - 3) = 0$

$m + 3 = 0$ or $m - 3 = 0$

$m = -3$ or $m = 3$

Solution set: $\{-3, 3\}$

29. $5k^2 + 3k = 0$

$k(5k + 3) = 0$

$k = 0$ or $5k + 3 = 0$

$5k = -3$

$k = -\dfrac{3}{5}$

Solution set: $\left\{-\dfrac{3}{5}, 0\right\}$

31. $16x^2 + 24x + 9 = 0$

$(4x + 3)^2 = 0$

$4x + 3 = 0$

$4x = -3$

$x = -\dfrac{3}{4}$

Solution set: $\left\{-\dfrac{3}{4}\right\}$

33.
$$4x^2 = 9$$
$$4x^2 - 9 = 0$$
$$(2x + 3)(2x - 3) = 0$$

$$2x + 3 = 0 \quad \text{or} \quad 2x - 3 = 0$$
$$2x = -3 \qquad\qquad 2x = 3$$
$$x = -\frac{3}{2} \quad \text{or} \qquad x = \frac{3}{2}$$

Solution set: $\left\{-\dfrac{3}{2}, \dfrac{3}{2}\right\}$

35.
$$-3m^2 + 27 = 0$$
$$-3(m^2 - 9) = 0$$
$$-3(m + 3)(m - 3) = 0$$

$$m + 3 = 0 \quad \text{or} \quad m - 3 = 0$$
$$m = -3 \quad \text{or} \qquad m = 3$$

Solution set: $\{-3, 3\}$

37. $(x - 3)(x + 5) = -7$

Multiply the factors, and then add 7 on both sides of the equation to get 0 on the right.
$$x^2 + 5x - 3x - 15 = -7$$
$$x^2 + 2x - 8 = 0$$
Now factor the polynomial.
$$(x + 4)(x - 2) = 0$$

$$x + 4 = 0 \quad \text{or} \quad x - 2 = 0$$
$$x = -4 \quad \text{or} \qquad x = 2$$

Solution set: $\{-4, 2\}$

39.
$$(2x + 1)(x - 3) = 6x + 3$$
$$2x^2 - 6x + x - 3 = 6x + 3$$
$$2x^2 - 5x - 3 = 6x + 3$$
$$2x^2 - 11x - 6 = 0$$
$$(2x + 1)(x - 6) = 0$$

$$2x + 1 = 0 \quad \text{or} \quad x - 6 = 0$$
$$2x = -1 \qquad\qquad x = 6$$
$$x = -\frac{1}{2}$$

Solution set: $\left\{-\dfrac{1}{2}, 6\right\}$

41. $(x + 3)(x - 6) = (2x + 2)(x - 6)$
$$x^2 - 3x - 18 = 2x^2 - 10x - 12$$
$$0 = x^2 - 7x + 6$$
$$0 = (x - 1)(x - 6)$$

$$x - 1 = 0 \quad \text{or} \quad x - 6 = 0$$
$$x = 1 \quad \text{or} \qquad x = 6$$

Solution set: $\{1, 6\}$

43. The left side factors as $x(ax + b)$. When the factor x is set equal to 0, the solution 0 is apparent.

45.
$$2x^3 - 9x^2 - 5x = 0$$
$$x(2x^2 - 9x - 5) = 0$$
$$x(2x + 1)(x - 5) = 0$$

$$x = 0 \quad \text{or} \quad 2x + 1 = 0 \quad \text{or} \quad x - 5 = 0$$
$$2x = -1 \qquad\qquad x = 5$$
$$x = -\frac{1}{2}$$

Solution set: $\left\{-\dfrac{1}{2}, 0, 5\right\}$

47.
$$9t^3 = 16t$$
$$9t^3 - 16t = 0$$
$$t(9t^2 - 16) = 0$$
$$t(3t + 4)(3t - 4) = 0$$

$$t = 0 \quad \text{or} \quad 3t + 4 = 0 \quad \text{or} \quad 3t - 4 = 0$$
$$3t = -4 \qquad\qquad 3t = 4$$
$$t = -\frac{4}{3} \quad \text{or} \qquad t = \frac{4}{3}$$

Solution set: $\left\{-\dfrac{4}{3}, 0, \dfrac{4}{3}\right\}$

49.
$$2r^3 + 5r^2 - 2r - 5 = 0$$
$$(2r^3 - 2r) + (5r^2 - 5) = 0$$
$$2r(r^2 - 1) + 5(r^2 - 1) = 0$$
$$(r^2 - 1)(2r + 5) = 0$$
$$(r + 1)(r - 1)(2r + 5) = 0$$

$$r + 1 = 0 \quad \text{or} \quad r - 1 = 0 \quad \text{or} \quad 2r + 5 = 0$$
$$r = -1 \quad \text{or} \qquad r = 1 \quad \text{or} \qquad r = -\frac{5}{2}$$

Solution set: $\left\{-\dfrac{5}{2}, -1, 1\right\}$

51. By dividing both sides by a variable expression, she "lost" the solution, 0.

53.
$$2x^2 - 7x - 4 = 0$$
$$(2x + 1)(x - 4) = 0$$

$$2x + 1 = 0 \quad \text{or} \quad x - 4 = 0$$
$$2x = -1 \qquad\qquad x = 4$$
$$x = -\frac{1}{2}$$

These values correspond to the zeros shown on the screens.

Solution set: $\{-.5, 4\}$

55. $-x^2 + 3x = -10$

$$0 = x^2 - 3x - 10$$
$$0 = (x + 2)(x - 5)$$

$x + 2 = 0$ or $x - 5 = 0$
$\quad x = -2 \qquad\qquad x = 5$

These values correspond to the zeros shown on the screens.

Solution set: $\{-2, 5\}$

57. $2(x - 1)^2 - 7(x - 1) - 15 = 0$

Let $y = x - 1$.

$$2y^2 - 7y - 15 = 0$$
$$(2y + 3)(y - 5) = 0$$

$2y + 3 = 0$ or $y - 5 = 0$
$\quad 2y = -3 \qquad\qquad y = 5$
$\quad y = -\dfrac{3}{2}$

Substitute $x - 1$ for y.

$x - 1 = -\dfrac{3}{2}$ or $x - 1 = 5$

$\quad x = -\dfrac{1}{2}$ or $\quad x = 6$

Solution set: $\left\{-\dfrac{1}{2}, 6\right\}$

59. $5(3a - 1)^2 + 3 = -16(3a - 1)$

Let $x = 3a - 1$.

$$5x^2 + 3 = -16x$$
$$5x^2 + 16x + 3 = 0$$
$$(x + 3)(5x + 1) = 0$$

$x + 3 = 0$ or $5x + 1 = 0$
$\quad x = -3 \qquad\qquad 5x = -1$
$\qquad\qquad\qquad\qquad x = -\dfrac{1}{5}$

Substitute $3a - 1$ for x.

$3a - 1 = -3$ or $3a - 1 = -\dfrac{1}{5}$

$\quad 3a = -2 \qquad\qquad 3a = \dfrac{4}{5}$

$\quad a = -\dfrac{2}{3} \qquad\qquad a = \dfrac{4}{15}$

Solution set: $\left\{-\dfrac{2}{3}, \dfrac{4}{15}\right\}$

61. $(x - 1)^2 - (2x - 5)^2 = 0$

$\left(x^2 - 2x + 1\right) - \left(4x^2 - 20x + 25\right) = 0$
$$-3x^2 + 18x - 24 = 0$$
$$x^2 - 6x + 8 = 0 \qquad \textit{Divide by } -3.$$
$$(x - 2)(x - 4) = 0$$

$x - 2 = 0$ or $x - 4 = 0$
$\quad x = 2$ or $\qquad x = 4$

Solution set: $\{2, 4\}$

63. $\qquad (2k - 3)^2 = 16k^2$

$$4k^2 - 12k + 9 = 16k^2$$
$$-12k^2 - 12k + 9 = 0$$
$$4k^2 + 4k - 3 = 0 \qquad \textit{Divide by } -3.$$
$$(2k + 3)(2k - 1) = 0$$

$2k + 3 = 0$ or $2k - 1 = 0$
$\quad 2k = -3 \qquad\qquad 2k = 1$
$\quad k = -\dfrac{3}{2}$ or $\quad k = \dfrac{1}{2}$

Solution set: $\left\{-\dfrac{3}{2}, \dfrac{1}{2}\right\}$

65. Let $x =$ the width of the garden.

$x + 4 =$ the length of the garden.

The area of the rectangular-shaped garden is 320 ft^2, so use the formula $A = LW$ and substitute 320 for A, $x + 4$ for L, and x for W.

$$A = LW$$
$$320 = (x + 4)x$$
$$320 = x^2 + 4x$$
$$0 = x^2 + 4x - 320$$
$$0 = (x - 16)(x + 20)$$

$x - 16 = 0$ or $x + 20 = 0$
$\quad x = 16$ or $\qquad x = -20$

A rectangle cannot have a width that is a negative measure, so reject -20 as a solution. The only possible solution is 16.

The width of the garden is 16 ft, and the length is $16 + 4 = 20$ ft.

67. Let $h =$ the length of the height.

$h - 3 =$ the length of the base.

The area of the triangle is 44 m^2, so use the formula $A = \dfrac{1}{2}bh$ and substitute 44 for A and $h - 3$ for b.

$$A = \dfrac{1}{2}bh$$
$$44 = \dfrac{1}{2}(h - 3)h$$
$$88 = (h - 3)h \qquad \textit{Multiply by 2.}$$
$$88 = h^2 - 3h$$
$$0 = h^2 - 3h - 88$$
$$0 = (h - 11)(h + 8)$$

$h - 11 = 0$ or $h + 8 = 0$
$\quad h = 11$ or $\qquad h = -8$

A triangle cannot have a height that is negative, so reject -8 as a solution. The only possible solution is 11.

The height of the triangle is 11 m, and the base is $11 - 3 = 8$ m.

69. Let $L =$ the length.

   $W =$ the width.

Use the formula for perimeter, $P = 2L + 2W$, and solve for W in terms of L. The perimeter is 300 ft.

$$300 = 2L + 2W$$
$$300 - 2L = 2W$$
$$150 - L = W$$

Now use the formula for area, $A = LW$, substitute 5000 for A, and solve for L.

$$5000 = L(150 - L)$$
$$5000 = 150L - L^2$$
$$L^2 - 150L + 5000 = 0$$
$$(L - 50)(L - 100) = 0$$

$$L - 50 = 0 \quad \text{or} \quad L - 100 = 0$$
$$L = 50 \quad \text{or} \quad L = 100$$

When $L = 50$, $W = 150 - 50 = 100$.
When $L = 100$, $W = 150 - 100 = 50$.
The dimensions should be 50 ft by 100 ft.

71. Let $x =$ one integer.

   $x + 1 =$ the next consecutive integer.

The sum of their squares is 61, so

$$x^2 + (x + 1)^2 = 61.$$
$$x^2 + x^2 + 2x + 1 = 61$$
$$2x^2 + 2x - 60 = 0 \quad \textit{Divide by 2.}$$
$$x^2 + x - 30 = 0$$
$$(x + 6)(x - 5) = 0$$

$$x + 6 = 0 \quad \text{or} \quad x - 5 = 0$$
$$x = -6 \quad \text{or} \quad x = 5$$

The two possible pairs of consecutive integers are -6 and -5 or 5 and 6.

73. Let $w =$ the width of the cardboard.

   $w + 6 =$ the length of the cardboard.

If squares that measure 2 inches are cut from each corner of the cardboard, then the width becomes $w - 4$ and the length becomes $(w + 6) - 4 = w + 2$. Use the formula $V = LWH$ and substitute 110 for V, $w + 2$ for L, $w - 4$ for W, and 2 for H.

$$V = LWH$$
$$110 = (w + 2)(w - 4)2$$
$$110 = (w^2 - 2w - 8)2$$
$$55 = w^2 - 2w - 8$$
$$0 = w^2 - 2w - 63$$
$$0 = (w - 9)(w + 7)$$

$$w - 9 = 0 \quad \text{or} \quad w + 7 = 0$$
$$w = 9 \quad \text{or} \quad w = -7$$

A box cannot have a negative width, so reject -7 as a solution. The only possible solution is 9. The piece of cardboard has width 9 inches and length $9 + 6 = 15$ inches.

75. From Example 7,

$$h(t) = -16t^2 + 128t.$$
$$240 = -16t^2 + 128t \quad \textit{Let h(t)=240.}$$
$$16t^2 - 128t + 240 = 0 \quad \textit{Standard form}$$
$$t^2 - 8t + 15 = 0 \quad \textit{Divide by 16.}$$
$$(t - 3)(t - 5) = 0$$

$$t - 3 = 0 \quad \text{or} \quad t - 5 = 0$$
$$t = 3 \quad \text{or} \quad t = 5$$

The height of the rocket will be 240 feet after 3 seconds (on the way up) and after 5 seconds (on the way down).

$$112 = -16t^2 + 128t \quad \textit{Let h(t)=112.}$$
$$16t^2 - 128t + 112 = 0 \quad \textit{Standard form}$$
$$t^2 - 8t + 7 = 0 \quad \textit{Divide by 16.}$$
$$(t - 1)(t - 7) = 0$$

$$t - 1 = 0 \quad \text{or} \quad t - 7 = 0$$
$$t = 1 \quad \text{or} \quad t = 7$$

The height of the rocket will be 112 feet after 1 second (on the way up) and after 7 seconds (on the way down).

77. Use $f(t) = -16t^2 + 625$ with $f(t) = 0$.

$$0 = -16t^2 + 625$$
$$0 = 16t^2 - 625 \quad \textit{Multiply by -1.}$$
$$0 = (4t + 25)(4t - 25)$$

$$4t + 25 = 0 \quad \text{or} \quad 4t - 25 = 0$$
$$4t = -25 \qquad\qquad 4t = 25$$
$$t = -\frac{25}{4} \quad \text{or} \quad t = \frac{25}{4}$$

Time cannot be negative, so reject $-\dfrac{25}{4}$ as a solution. The only possible solution is $\dfrac{25}{4}$ or $6\dfrac{1}{4}$.

The ball will hit the ground after $6\dfrac{1}{4}$ seconds.

79. $f(x) = 4x^4 - 4x^2$

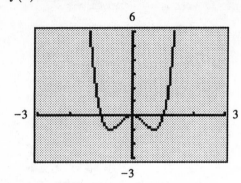

The exact zeros of f are $-1, 0,$ and 1.

Solution set: $\{-1, 0, 1\}$

81. $f(x) = -2.47x^3 - 6.58x^2 - 3.33x + .14$

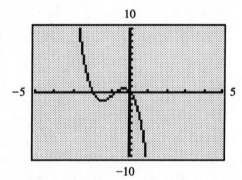

The approximate zeros (to the nearest hundredth) of f are $-1.96, -.74,$ and $.04$.

Solution set: $\{-1.96, -.74, .04\}$

Chapter 5 Review Exercises

1. $(-3x^4y^3)(4x^{-2}y^5) = -3(4)x^4x^{-2}y^3y^5$
$$= -12x^{4-2}y^{3+5}$$
$$= -12x^2y^8$$

2. $\dfrac{6m^{-4}n^3}{-3mn^2} = -2m^{-4-1}n^{3-2}$
$$= -2m^{-5}n^1$$
$$= -\dfrac{2n}{m^5} \text{ or } \dfrac{-2n}{m^5}$$

3. $\dfrac{(5p^{-2}q)(4p^5q^{-3})}{2p^{-5}q^5} = \dfrac{20p^{-2+5}q^{1-3}}{2p^{-5}q^5}$
$$= \dfrac{10p^3q^{-2}}{p^{-5}q^5}$$
$$= 10p^{3-(-5)}q^{-2-5}$$
$$= 10p^8q^{-7}$$
$$= \dfrac{10p^8}{q^7}$$

4. In $(-6)^0$, the base is -6 and the expression simplifies to 1. In -6^0, the base is 6 and the expression simplifies to -1.

5. $4^3 = 4 \cdot 4 \cdot 4 = 64$

6. $\left(\dfrac{1}{3}\right)^4 = \dfrac{1}{3} \cdot \dfrac{1}{3} \cdot \dfrac{1}{3} \cdot \dfrac{1}{3} = \dfrac{1}{81}$

7. $(-5)^3 = (-5)(-5)(-5) = -125$

8. $\dfrac{2}{(-3)^{-2}} = \dfrac{2}{\dfrac{1}{(-3)^2}}$
$$= 2 \cdot (-3)^2$$
$$= 2 \cdot (-3)(-3)$$
$$= 18$$

9. $\left(\dfrac{2}{3}\right)^{-4} = \left(\dfrac{3}{2}\right)^4$
$$= \dfrac{3}{2} \cdot \dfrac{3}{2} \cdot \dfrac{3}{2} \cdot \dfrac{3}{2}$$
$$= \dfrac{81}{16}$$

10. $\left(\dfrac{5}{4}\right)^{-2} = \left(\dfrac{4}{5}\right)^2 = \dfrac{4}{5} \cdot \dfrac{4}{5} = \dfrac{16}{25}$

11. $5^{-1} + 6^{-1} = \dfrac{1}{5} + \dfrac{1}{6} = \dfrac{6}{30} + \dfrac{5}{30} = \dfrac{11}{30}$

12. $(5 + 6)^{-1} = 11^{-1} = \dfrac{1}{11}$

13. $-3^0 + 3^0 = -1 + 1 = 0$

14. For example, if $a = 4$,
$$(2a)^{-3} = (2 \cdot 4)^{-3} = 8^{-3} = \dfrac{1}{512}, \text{ while}$$
$$\dfrac{2}{a^3} = \dfrac{2}{4^3} = \dfrac{2}{64} = \dfrac{1}{32}. \quad \dfrac{1}{512} \neq \dfrac{1}{32}.$$

15. $\left(3^{-4}\right)^2 = 3^{(-4) \cdot 2} = 3^{-8} = \dfrac{1}{3^8}$

16. $\left(x^{-4}\right)^{-2} = x^{-4(-2)} = x^8$

17. $\left(xy^{-3}\right)^{-2} = x^{1(-2)}y^{(-3)(-2)}$
$$= x^{-2}y^6$$
$$= \dfrac{1}{x^2} \cdot y^6 = \dfrac{y^6}{x^2}$$

18. $\left(z^{-3}\right)^3 z^{-6} = z^{-9}z^{-6}$
$$= z^{-9+(-6)}$$
$$= z^{-15} = \dfrac{1}{z^{15}}$$

19. $\left(5m^{-3}\right)^2\left(m^4\right)^{-3} = 5^2m^{-6}m^{-12}$
$$= 25m^{-6-12}$$
$$= 25m^{-18} = \dfrac{25}{m^{18}}$$

20. $\dfrac{(3r)^2 r^4}{r^{-2}r^{-3}}(9r^{-3})^{-2} = \dfrac{3^2 r^2 r^4 9^{-2} r^6}{r^{-2}r^{-3}}$

$= \dfrac{9^1 9^{-2} r^{2+4+6}}{r^{-2-3}}$

$= \dfrac{9^{-1} r^{12}}{r^{-5}}$

$= \dfrac{r^{12-(-5)}}{9} = \dfrac{r^{17}}{9}$

21. $\left(\dfrac{5z^{-3}}{z^{-1}}\right)\dfrac{5}{z^2} = \dfrac{25z^{-3}}{z^{-1+2}}$

$= \dfrac{25}{z^3 z^1}$

$= \dfrac{25}{z^{3+1}} = \dfrac{25}{z^4}$

22. $\left(\dfrac{6m^{-4}}{m^{-9}}\right)^{-1}\left(\dfrac{m^{-2}}{16}\right) = \dfrac{6^{-1}m^4}{m^9} \cdot \dfrac{m^{-2}}{16}$

$= \dfrac{1}{6 \cdot 16}m^{4+(-2)-9}$

$= \dfrac{1}{96}m^{-7} = \dfrac{1}{96m^7}$

23. $\left(\dfrac{3r^5}{5r^{-3}}\right)^{-2}\left(\dfrac{9r^{-1}}{2r^{-5}}\right)^3$

$= \dfrac{3^{-2}r^{-10}}{5^{-2}r^6} \cdot \dfrac{9^3 r^{-3}}{2^3 r^{-15}}$

$= \dfrac{5^2}{3^2}r^{-10-6} \cdot \dfrac{9^3}{2^3}r^{-3-(-15)}$

$= \dfrac{25}{9}r^{-16} \cdot \dfrac{729}{8}r^{12}$

$= \dfrac{(25)(729)}{(9)(8)}r^{-16+12}$

$= \dfrac{2025}{8}r^{-4} = \dfrac{2025}{8r^4}$

24. $\left(\dfrac{a^{-2}b^{-1}}{3a^2}\right)^{-2}\left(\dfrac{b^{-2} \cdot 3a^4}{2b^{-3}}\right)^{-2}\left(\dfrac{a^{-4}b^5}{a^3}\right)^{-2}$

$= \dfrac{a^4 b^2 b^4 3^{-2} a^{-8} b^{-10}}{3^{-2}a^{-4}2^{-2}b^6 a^{-6}}$

$= \dfrac{4a^{4-8+8}b^{2+4-10}}{a^{-4-6}b^6}$

$= \dfrac{4a^4 b^{-4}}{a^{-10}b^6}$

$= 4a^{4-(-10)}b^{-4-6}$

$= 4a^{14}b^{-10} = \dfrac{4a^{14}}{b^{10}}$

25. Yes, $\left(\dfrac{a}{b}\right)^{-1} = \dfrac{a^{-1}}{b^{-1}}$ for all $a, b \neq 0$.

26. No, $(ab)^{-1} \neq ab^{-1}$ for all $a, b \neq 0$. For example, let $a = 3$ and $b = 4$. Then $(ab)^{-1} = (3 \cdot 4)^{-1} = 12^{-1} = \dfrac{1}{12}$, while $ab^{-1} = 3 \cdot 4^{-1} = 3 \cdot \dfrac{1}{4} = \dfrac{3}{4}$. $\dfrac{1}{12} \neq \dfrac{3}{4}$.

27. Let $x = 2$ and $y = 3$. Then

$$\left(x^2 + y^2\right)^2 = \left(2^2 + 3^2\right)^2$$
$$= (4 + 9)^2$$
$$= 13^2 = 169,$$
and $\quad x^4 + y^4 = 2^4 + 3^4$
$$= 16 + 81 = 97.$$

Since $169 \neq 97$,

$$\left(x^2 + y^2\right)^2 \neq x^4 + y^4.$$

28. $13,450 = 1\wedge 3\,4\,5\,0.$

Place a caret to the right of the first nonzero digit. Count 4 places.

Since the number 1.345 is to be made larger, the exponent on 10 is positive.

$$13,450 = 1.345 \times 10^4$$

29. $.0000000765 = .0\,0\,0\,0\,0\,0\,7\wedge 65$

Count 8 places

Since the number 7.65 is to be made smaller, the exponent on 10 is negative.

$$.0000000765 = 7.65 \times 10^{-8}$$

30. $.138 = .1\wedge 38$

Count 1 place.

Since the number 1.38 is to be made smaller, the exponent on 10 is negative.

$$.138 = 1.38 \times 10^{-1}$$

31. $1.21 \times 10^6 = 1,210,000$

Move the decimal point 6 places to the right because the exponent is positive. Attach extra zeros.

32. $5.8 \times 10^{-3} = .0058$

Move the decimal point 3 places to the left because the exponent is negative.

33. $568,000 = 5\text{\textasciicircum}68000.$

Place a caret to the right of the first nonzero digit. Count 5 places.

Since the number 5.68 is to be made larger, the exponent on 10 is positive.

$$\$568,000 = \$5.68 \times 10^5$$

34. $\dfrac{16 \times 10^4}{8 \times 10^8} = \dfrac{16}{8} \times 10^{4-8}$

$$= 2 \times 10^{-4} \text{ or } .0002$$

35. $\dfrac{6 \times 10^{-2}}{4 \times 10^{-5}} = \dfrac{6}{4} \times 10^{-2-(-5)}$

$$= 1.5 \times 10^3 \text{ or } 1500$$

36. $\dfrac{.0000000164}{.0004} = \dfrac{1.64 \times 10^{-8}}{4 \times 10^{-4}}$

$$= \dfrac{1.64}{4} \times 10^{-8-(-4)}$$

$$= .41 \times 10^{-4}$$

$$= 4.1 \times 10^{-5} \text{ or } .000041$$

37. $\dfrac{.0009 \times 12,000,000}{400,000}$

$$= \dfrac{9 \times 10^{-4} \times 1.2 \times 10^7}{4 \times 10^5}$$

$$= \dfrac{9 \times 1.2}{4} \times \dfrac{10^{-4} \times 10^7}{10^5}$$

$$= \dfrac{10.8}{4} \times 10^{-4+7-5}$$

$$= 2.7 \times 10^{-2} \text{ or } .027$$

38. The density D is the population P divided by the area A.

$$D = \dfrac{P}{A}$$

We want to find the area, so solve the formula for A.

$$DA = P$$

$$A = \dfrac{P}{D}$$

$$= \dfrac{3.92 \times 10^5}{400}$$

$$= \dfrac{3.92 \times 10^5}{4 \times 10^2}$$

$$= .98 \times 10^{5-2}$$

$$= .98 \times 10^3 = 980$$

The area is 980 mi^2.

39. $5449 = 5\text{\textasciicircum}449.$

Count 3 places.

$$= 5.449 \times 10^3$$

40. As in Exercise 38, use $A = \dfrac{P}{D}$.

$$\dfrac{3.45 \times 10^5}{5.449 \times 10^3} = \dfrac{3.45}{5.449} \times \dfrac{10^5}{10^3}$$

$$\approx .63 \times 10^{5-3}$$

$$= .63 \times 10^2$$

$$= 63$$

The area is approximately 63 mi^2.

41. The coefficient of $14p^5$ is 14.

42. The coefficient of $-z$ is -1.

43. The coefficient of $504p^3r^5$ is 504.

44. $9k + 11k^3 - 3k^2$

(a) In descending powers of k, the polynomial is

$$11k^3 - 3k^2 + 9k.$$

(b) The polynomial is a trinomial since it has three terms.

(c) The degree of the polynomial is 3 since the highest power of k is 3.

45. $14m^6 + 9m^7$

(a) In descending powers of m, the polynomial is

$$9m^7 + 14m^6.$$

(b) The polynomial is a binomial since it has two terms.

(c) The degree of the polynomial is 7 since the highest power of m is 7.

46. $-5y^4 + 3y^3 + 7y^2 - 2y$

(a) The polynomial is already written in descending powers of y.

(b) The polynomial has four terms, so it is none of these choices.

(c) The degree of the polynomial is 4 since the highest power of y is 4.

47. $-7q^5r^3$

(a) The polynomial is already written in descending powers.

(b) The polynomial is a monomial since it has just one term.

(c) The degree is $5 + 3 = 8$, the sum of the exponents of this term.

48. One example of a polynomial in the variable x that has degree 5, is lacking a third-degree term, and is written in descending powers of the variable is

$$x^5 + 2x^4 - x^2 + x + 2.$$

49. Add by columns.

$$
\begin{array}{r}
3x^2 - 5x + 6 \\
-4x^2 + 2x - 5 \\
\hline
-1x^2 - 3x + 1
\end{array}
\text{ or } -x^2 - 3x + 1
$$

50. Subtract.

$$
\begin{array}{r}
-5y^3 \quad\;\; + 8y - 3 \\
4y^2 + 2y + 9 \\
\hline
\end{array}
$$

Change the signs in the second polynomial and add.

$$
\begin{array}{r}
-5y^3 \quad\;\; + 8y - 3 \\
- 4y^2 - 2y - 9 \\
\hline
-5y^3 - 4y^2 + 6y - 12
\end{array}
$$

51. $\left(4a^3 - 9a + 15\right) - \left(-2a^3 + 4a^2 + 7a\right)$

$= 4a^3 - 9a + 15 + 2a^3 - 4a^2 - 7a$

$= 4a^3 + 2a^3 - 4a^2 - 9a - 7a + 15$

$= 6a^3 - 4a^2 - 16a + 15$

52. $\left(3y^2 + 2y - 1\right) + \left(5y^2 - 11y + 6\right)$

$= 3y^2 + 5y^2 + 2y - 11y - 1 + 6$

$= 8y^2 - 9y + 5$

53. To find the perimeter, add the measures of the three sides.

$\left(4x^2 + 2\right) + \left(6x^2 + 5x + 2\right) + \left(2x^2 + 3x + 1\right)$

$= 4x^2 + 6x^2 + 2x^2 + 5x + 3x + 2 + 2 + 1$

$= 12x^2 + 8x + 5$

The perimeter is $12x^2 + 8x + 5$.

54. $f(x) = -2x^2 + 5x + 7$

(a) $f(-2) = -2(-2)^2 + 5(-2) + 7$

$= -2(4) - 10 + 7$

$= -8 - 3 = -11$

(b) $f(3) = -2(3)^2 + 5(3) + 7$

$= -2(9) + 15 + 7$

$= -18 + 22 = 4$

55. **(a)** $(f + g)(x) = f(x) + g(x)$

$= (2x + 3) + \left(5x^2 - 3x + 2\right)$

$= 5x^2 + 2x - 3x + 3 + 2$

$= 5x^2 - x + 5$

(b) $(f - g)(x) = f(x) - g(x)$

$= (2x + 3) - \left(5x^2 - 3x + 2\right)$

$= 2x + 3 - 5x^2 + 3x - 2$

$= -5x^2 + 5x + 1$

56. $f(x) = 22x^2 + 243x + 8992$

$f(x)$ represents the number of people employed in health service industries (in thousands).

Year	x	$f(x)$
1994	0	8992
1995	1	9257
1996	2	9566

57. **(a)**

x	$f(x) = -2x + 5$
-2	$-2(-2) + 5 = 9$
-1	$-2(-1) + 5 = 7$
0	$-2(0) + 5 = 5$
1	$-2(1) + 5 = 3$
2	$-2(2) + 5 = 1$

This is a linear function, so plot the points and draw a line through them.

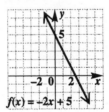

Any x-value can be used, so the domain is $(-\infty, \infty)$. From the graph, we see that any y-value can be obtained from the function, so the range is $(-\infty, \infty)$.

(b)

x	$f(x) = x^2 - 6$
-2	$(-2)^2 - 6 = -2$
-1	$(-1)^2 - 6 = -5$
0	$(0)^2 - 6 = -6$
1	$(1)^2 - 6 = -5$
2	$(2)^2 - 6 = -2$

Since the greatest exponent is 2, the graph of f is a parabola.

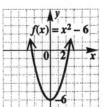

Any x-value can be used, so the domain is $(-\infty, \infty)$. From the graph, we see that the y-values are at least -6, so the range is $[-6, \infty)$.

(c)

x	$f(x) = -x^3 + 1$
-2	$-(-2)^3 + 1 = 9$
-1	$-(-1)^3 + 1 = 2$
0	$-(0)^3 + 1 = 1$
1	$-(1)^3 + 1 = 0$
2	$-(2)^3 + 1 = -7$

The greatest exponent is 3, so the graph of f is s-shaped.

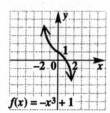

$f(x) = -x^3 + 1$

Any x-value can be used, so the domain is $(-\infty, \infty)$. From the graph, we see that any y-value can be obtained from the function, so the range is $(-\infty, \infty)$.

58. $-6k(2k^2 + 7) = -6k(2k^2) - 6k(7)$
$$= -12k^3 - 42k$$

59. $(3m - 2)(5m + 1)$
$$= 15m^2 + 3m - 10m - 2$$
$$= 15m^2 - 7m - 2$$

60. $(7y - 8)(2y + 3)$
$$= 14y^2 + 21y - 16y - 24$$
$$= 14y^2 + 5y - 24$$

61. $(3w - 2t)(2w - 3t)$
$$= 6w^2 - 9wt - 4wt + 6t^2$$
$$= 6w^2 - 13wt + 6t^2$$

62. $(2p^2 + 6p)(5p^2 - 4)$
$$= 10p^4 - 8p^2 + 30p^3 - 24p$$
$$= 10p^4 + 30p^3 - 8p^2 - 24p$$

63. $(3q^2 + 2q - 4)(q - 5)$
$$= 3q^3 - 15q^2 + 2q^2 - 10q - 4q + 20$$
$$= 3q^3 - 13q^2 - 14q + 20$$

64. $(3z^3 - 2z^2 + 4z - 1)(3z - 2)$
$$= 9z^4 - 6z^3 - 6z^3 + 4z^2 + 12z^2$$
$$- 8z - 3z + 2$$
$$= 9z^4 - 12z^3 + 16z^2 - 11z + 2$$

65. $(6r^2 - 1)(6r^2 + 1) = (6r^2)^2 - 1^2$
$$= 36r^4 - 1$$

66. $\left(z + \dfrac{3}{5}\right)\left(z - \dfrac{3}{5}\right) = z^2 - \left(\dfrac{3}{5}\right)^2$
$$= z^2 - \dfrac{9}{25}$$

67. $(4m + 3)^2 = (4m)^2 + 2(4m)(3) + 3^2$
$$= 16m^2 + 24m + 9$$

68. $(2n - 10)^2$
$$= (2n)^2 - 2(2n)(10) + 10^2$$
$$= 4n^2 - 40n + 100$$

69. $12p^2 - 6p = 6p(2p - 1)$

70. $21y^2 + 35y = 7y(3y + 5)$

71. $12q^2b + 8qb^2 - 20q^3b^2$
$$= 4qb(3q + 2b - 5q^2b)$$

72. $6r^3t - 30r^2t^2 + 18rt^3$
$$= 6rt(r^2 - 5rt + 3t^2)$$

73. $(x + 3)(4x - 1) - (x + 3)(3x + 2)$

The GCF is $(x + 3)$.
$$= (x + 3)[(4x - 1) - (3x + 2)]$$
$$= (x + 3)(4x - 1 - 3x - 2)$$
$$= (x + 3)(x - 3)$$

74. $(z + 1)(z - 4) + (z + 1)(2z + 3)$
$$= (z + 1)[(z - 4) + (2z + 3)]$$
$$= (z + 1)(3z - 1)$$

75. $4m + nq + mn + 4q$

Rearrange the terms.
$$= 4m + 4q + mn + nq$$
$$= 4(m + q) + n(m + q)$$
$$= (m + q)(4 + n)$$

76. $x^2 + 5y + 5x + xy$
$$= x^2 + xy + 5x + 5y$$
$$= x(x + y) + 5(x + y)$$
$$= (x + y)(x + 5)$$

77. $2m + 6 - am - 3a$
$$= 2(m + 3) - a(m + 3)$$
$$= (m + 3)(2 - a)$$

78. $2am - 2bm - ap + bp$
$$= 2m(a - b) - p(a - b)$$
$$= (a - b)(2m - p)$$

79. $3p^2 - p - 4$

Two integer factors whose product is $(3)(-4) = -12$ and whose sum is -1 are -4 and 3.
$$= 3p^2 - 4p + 3p - 4$$
$$= p(3p - 4) + 1(3p - 4)$$
$$= (3p - 4)(p + 1)$$

80. $6k^2 + 11k - 10$

Two integer factors whose product is $(6)(-10) = -60$ and whose sum is 11 are 15 and -4.

$$= 6k^2 + 15k - 4k - 10$$
$$= 3k(2k + 5) - 2(2k + 5)$$
$$= (2k + 5)(3k - 2)$$

81. $12r^2 - 5r - 3$

Two integer factors whose product is $(12)(-3) = -36$ and whose sum is -5 are -9 and 4.

$$= 12r^2 - 9r + 4r - 3$$
$$= 3r(4r - 3) + 1(4r - 3)$$
$$= (4r - 3)(3r + 1)$$

82. $10m^2 + 37m + 30$

Two integer factors whose product is $(10)(30) = 300$ and whose sum is 37 are 12 and 25.

$$= 10m^2 + 12m + 25m + 30$$
$$= 2m(5m + 6) + 5(5m + 6)$$
$$= (5m + 6)(2m + 5)$$

83. $10k^2 - 11kh + 3h^2$

Two integer factors whose product is $(10)(3) = 30$ and whose sum is -11 are -6 and -5.

$$= 10k^2 - 6kh - 5kh + 3h^2$$
$$= 2k(5k - 3h) - h(5k - 3h)$$
$$= (5k - 3h)(2k - h)$$

84. $9x^2 + 4xy - 2y^2$

There are no integers that have a product of $9(-2) = -18$ and a sum of 4. Therefore, the trinomial cannot be factored and is prime.

85. $24x - 2x^2 - 2x^3$

$$= 2x(12 - x - x^2)$$
$$= 2x(4 + x)(3 - x)$$

86. $6b^3 - 9b^2 - 15b$

$$= 3b(2b^2 - 3b - 5)$$
$$= 3b(2b - 5)(b + 1)$$

87. $y^4 + 2y^2 - 8$

$$= (y^2)^2 + 2y^2 - 8$$
$$= (y^2 + 4)(y^2 - 2)$$

88. $2k^4 - 5k^2 - 3$

$$= 2(k^2)^2 - 5k^2 - 3$$
$$= (2k^2 + 1)(k^2 - 3)$$

89. $p^2(p + 2)^2 + p(p + 2)^2 - 6(p + 2)^2$

Factor out $(p + 2)^2$.

$$= (p + 2)^2(p^2 + p - 6)$$
$$= (p + 2)^2(p + 3)(p - 2)$$

90. $3(r + 5)^2 - 11(r + 5) - 4$

$$= 3x^2 - 11x - 4 \qquad \textit{Let } x = r + 5.$$
$$= (3x + 1)(x - 4)$$
$$= [3(r + 5) + 1][(r + 5) - 4] \quad \textit{Resubstitute.}$$
$$= (3r + 15 + 1)(r + 1)$$
$$= (3r + 16)(r + 1)$$

91. The student's answer is

$$x^2y^2 - 6x^2 + 5y^2 - 30$$
$$= x^2(y^2 - 6) + 5(y^2 - 6).$$

This is incorrect because the polynomial still has two terms, so it is not factored. The correct answer is

$$(y^2 - 6)(x^2 + 5).$$

92. Since area equals length times width, factor the polynomial given for the area.

$$4p^2 + 3p - 1 = (4p - 1)(p + 1)$$

Since the length is given as $4p - 1$, the width is $p + 1$.

93. $16x^2 - 25 = (4x)^2 - 5^2 \qquad$ *Difference of two squares*

$$= (4x + 5)(4x - 5)$$

94. $9t^2 - 49 = (3t)^2 - 7^2$

$$= (3t + 7)(3t - 7)$$

95. $x^2 + 14x + 49 = x^2 + 2(x)(7) + 7^2 \qquad$ *Perfect square trinomial*

$$= (x + 7)^2$$

96. $9k^2 - 12k + 4 = (3k)^2 - 2(3k)(2) + 2^2$

$$= (3k - 2)^2$$

97. $r^3 + 27 = r^3 + 3^3 \qquad$ *Sum of two cubes*

$$= (r + 3)(r^2 - 3r + 9)$$

98. $125x^3 - 1 = (5x)^3 - 1^3 \qquad$ *Difference of two cubes*

$$= (5x - 1)(25x^2 + 5x + 1)$$

99. $m^6 - 1 = \left(m^3\right)^2 - 1^2$

  Difference of two squares

  $= \left(m^3 + 1\right)\left(m^3 - 1\right)$

  $= \left(m^3 + 1^3\right)\left(m^3 - 1^3\right)$

  Sum of   Difference of

  two cubes;   two cubes

  $= (m + 1)\left(m^2 - m + 1\right)$

  $\quad \cdot (m - 1)\left(m^2 + m + 1\right)$

100. $x^8 - 1 = \left(x^4\right)^2 - 1^2$

  Difference of two squares

  $= \left(x^4 + 1\right)\left(x^4 - 1\right)$

  $= \left(x^4 + 1\right)\left[\left(x^2\right)^2 - 1^2\right]$

  Difference of two squares again

  $= \left(x^4 + 1\right)\left(x^2 + 1\right)\left(x^2 - 1\right)$

  Difference of two squares again

  $= \left(x^4 + 1\right)\left(x^2 + 1\right)(x + 1)(x - 1)$

101. $x^2 + 6x + 9 - 25y^2$

  $= \left(x^2 + 6x + 9\right) - 25y^2$

  $= \left[x^2 + 2(x)(3) + 3^2\right] - 25y^2$

  Perfect square trinomial

  $= (x + 3)^2 - (5y)^2$

  Difference of two squares

  $= [(x + 3) + 5y][(x + 3) - 5y]$

  $= (x + 3 + 5y)(x + 3 - 5y)$

102. $(a + b)^3 - (a - b)^3$

  Difference of two cubes

  $= [(a + b) - (a - b)]$

  $\quad \cdot [(a + b)^2 + (a + b)(a - b)$

  $\quad + (a - b)^2]$

  $= [2b]\left(a^2 + 2ab + b^2 + a^2 - b^2\right.$

  $\quad \left. + a^2 - 2ab + b^2\right)$

  $= 2b\left(3a^2 + b^2\right)$

103. $x^5 - x^3 - 8x^2 + 8$

  $= x^3\left(x^2 - 1\right) - 8\left(x^2 - 1\right)$

  $= \left(x^2 - 1\right)\left(x^3 - 8\right)$

  $= \left(x^2 - 1^2\right)\left(x^3 - 2^3\right)$

  Difference of   Difference of

  two squares;   two cubes

  $= (x + 1)(x - 1)(x - 2)\left(x^2 + 2x + 4\right)$

104. Choice (a) is still written as a sum. Before the zero-factor property can be applied, it must be written as the product of factors.

105. (a) $(2x + 9)(x - 3) + (4x + 7)(x - 3) = 0$

  $(x - 3)[(2x + 9) + (4x + 7)] = 0$

  $(x - 3)(6x + 16) = 0,$

  or factor out 2 from the second factor to get

  $2(x - 3)(3x + 8) = 0.$

Now the zero-factor property can be directly applied.

(b) $2(x - 3)(3x + 8) = 0$

  $x - 3 = 0 \quad$ or $\quad 3x + 8 = 0$

  $x = 3 \qquad\qquad 3x = -8$

  $\qquad\qquad\qquad x = -\dfrac{8}{3}$

Solution set: $\left\{3, -\dfrac{8}{3}\right\}$

106. $(5x + 2)(x + 1) = 0$

  $5x + 2 = 0 \quad$ or $\quad x + 1 = 0$

  $5x = -2 \qquad\qquad x = -1$

  $x = -\dfrac{2}{5}$

Solution set: $\left\{-1, -\dfrac{2}{5}\right\}$

107. $p^2 - 5p + 6 = 0$

  $(p - 2)(p - 3) = 0$

  $p - 2 = 0 \quad$ or $\quad p - 3 = 0$

  $p = 2 \quad$ or $\qquad p = 3$

  Solution set: $\{2, 3\}$

108. $q^2 + 2q = 8$

  $q^2 + 2q - 8 = 0$

  $(q + 4)(q - 2) = 0$

  $q + 4 = 0 \quad$ or $\quad q - 2 = 0$

  $q = -4 \quad$ or $\qquad q = 2$

  Solution set: $\{-4, 2\}$

109. $6z^2 = 5z + 50$

  $6z^2 - 5z - 50 = 0$

  $(3z - 10)(2z + 5) = 0$

  $3z - 10 = 0 \quad$ or $\quad 2z + 5 = 0$

  $3z = 10 \qquad\qquad 2z = -5$

  $z = \dfrac{10}{3} \quad$ or $\qquad z = -\dfrac{5}{2}$

  Solution set: $\left\{-\dfrac{5}{2}, \dfrac{10}{3}\right\}$

110.
$$6r^2 + 7r = 3$$
$$6r^2 + 7r - 3 = 0$$
$$(2r + 3)(3r - 1) = 0$$

$2r + 3 = 0$ or $3r - 1 = 0$

$\quad 2r = -3 \qquad\qquad 3r = 1$

$\quad r = -\dfrac{3}{2}$ or $r = \dfrac{1}{3}$

Solution set: $\left\{-\dfrac{3}{2}, \dfrac{1}{3}\right\}$

111.
$$8k^2 + 14k + 3 = 0$$
$$(2k + 3)(4k + 1) = 0$$

$2k + 3 = 0$ or $4k + 1 = 0$

$\quad 2k = -3 \qquad\qquad 4k = -1$

$\quad k = -\dfrac{3}{2}$ or $k = -\dfrac{1}{4}$

Solution set: $\left\{-\dfrac{3}{2}, -\dfrac{1}{4}\right\}$

112.
$$-4m^2 + 36 = 0$$
$$m^2 - 9 = 0 \quad \textit{Divide by } -4.$$
$$(m + 3)(m - 3) = 0$$

$m + 3 = 0$ or $m - 3 = 0$

$\quad m = -3$ or $m = 3$

Solution set: $\{-3, 3\}$

113.
$$6y^2 + 9y = 0$$
$$3y(2y + 3) = 0$$

$3y = 0$ or $2y + 3 = 0$

$\quad y = 0 \qquad\qquad 2y = -3$

$$y = -\dfrac{3}{2}$$

Solution set: $\left\{-\dfrac{3}{2}, 0\right\}$

114. $(2x + 1)(x - 2) = -3$
$$2x^2 - 3x - 2 = -3$$
$$2x^2 - 3x + 1 = 0$$
$$(2x - 1)(x - 1) = 0$$

$2x - 1 = 0$ or $x - 1 = 0$

$\quad 2x = 1 \qquad\qquad x = 1$

$\quad x = \dfrac{1}{2}$

Solution set: $\left\{\dfrac{1}{2}, 1\right\}$

115. $(r + 2)(r - 2) = (r - 2)(r + 3) - 2$
$$r^2 - 4 = r^2 + r - 6 - 2$$
$$-4 = r - 8$$
$$4 = r$$

Solution set: $\{4\}$

116.
$$2x^3 - x^2 - 28x = 0$$
$$x(2x^2 - x - 28) = 0$$
$$x(2x + 7)(x - 4) = 0$$

$x = 0$ or $2x + 7 = 0$ or $x - 4 = 0$

$\qquad\qquad\quad 2x = -7 \qquad\qquad x = 4$

$$x = -\dfrac{7}{2}$$

Solution set: $\left\{-\dfrac{7}{2}, 0, 4\right\}$

117. $-t^3 - 3t^2 + 4t + 12 = 0$
$$t^3 + 3t^2 - 4t - 12 = 0 \quad \textit{Multiply by } -1.$$
$$t^2(t + 3) - 4(t + 3) = 0$$
$$(t + 3)(t^2 - 4) = 0$$
$$(t + 3)(t + 2)(t - 2) = 0$$

$t + 3 = 0$ or $t + 2 = 0$ or $t - 2 = 0$

$\quad t = -3$ or $t = -2$ or $t = 2$

Solution set: $\{-3, -2, 2\}$

118. $(r + 2)(5r^2 - 9r - 18) = 0$
$$(r + 2)(5r + 6)(r - 3) = 0$$

$r + 2 = 0$ or $5r + 6 = 0$ or $r - 3 = 0$

$\quad r = -2 \qquad\qquad 5r = -6 \qquad\qquad r = 3$

$$r = -\dfrac{6}{5}$$

Solution set: $\left\{-2, -\dfrac{6}{5}, 3\right\}$

119. Let x be the length of the shorter side. Then, the length of the longer side will be $2x + 1$. The area is 10.5 ft^2. Use the formula for area of a triangle, $A = \dfrac{1}{2}bh$.

$$\dfrac{1}{2}(2x + 1)(x) = 10.5$$
$$x(2x + 1) = 21 \quad \textit{Multiply by 2.}$$
$$2x^2 + x = 21$$
$$2x^2 + x - 21 = 0$$
$$(2x + 7)(x - 3) = 0$$

$2x + 7 = 0$ or $x - 3 = 0$

$\quad 2x = -7 \qquad\qquad x = 3$

$$x = -\dfrac{7}{2}$$

The side cannot have a negative length, so reject $x = -\dfrac{7}{2}$.

The length of the shorter side is 3 ft.

120. Let w be the width of the lot. Then $w + 20$ will be the length of the lot. The area is 2400 ft^2. Use the formula $LW = A$.

$$(w + 20)w = 2400$$
$$w^2 + 20w = 2400$$
$$w^2 + 20w - 2400 = 0$$
$$(w - 40)(w + 60) = 0$$
$$w - 40 = 0 \quad \text{or} \quad w + 60 = 0$$
$$w = 40 \quad \text{or} \quad w = -60$$

The lot cannot have a negative width, so reject $w = -60$.
The width of the lot is 40 ft, and the length is $40 + 20 = 60$ ft.

121. The height is 0 when the rock returns to the ground.

$$f(t) = -16t^2 + 256t$$
$$0 = -16t^2 + 256t$$
$$0 = -16t(t - 16)$$
$$-16t = 0 \quad \text{or} \quad t - 16 = 0$$
$$t = 0 \quad \text{or} \quad t = 16$$

The rock is on the ground when $t = 0$. It will return to the ground again after 16 seconds.

122. $f(t) = -16t^2 + 256t$
$$240 = -16t^2 + 256t \qquad \textit{Let f(t)=240.}$$
$$0 = -16t^2 + 256t - 240$$
$$0 = -16(t^2 - 16t + 15)$$
$$0 = -16(t - 15)(t - 1)$$
$$t - 15 = 0 \quad \text{or} \quad t - 1 = 0$$
$$t = 15 \quad \text{or} \quad t = 1$$

The rock will be 240 ft above the ground after 1 second and again after 15 seconds.

123. The question in Exercise 122 has two answers because the rock will be 240 ft above the ground after 1 second on the way up and again after 15 seconds on the way back down.

124. $(4x + 1)(2x - 3)$
$$= 8x^2 - 12x + 2x - 3$$
$$= 8x^2 - 10x - 3$$

125. $\dfrac{6^{-1}y^3(y^2)^{-2}}{6y^{-4}(y^{-1})} = \dfrac{y^3 y^{-4}}{6^1 \cdot 6y^{-4}y^{-1}}$

$$= \dfrac{y^{3-4-(-4-1)}}{36}$$

$$= \dfrac{y^{-1+5}}{36} = \dfrac{y^4}{36}$$

126. $5^{-3} = \dfrac{1}{5^3} = \dfrac{1}{5 \cdot 5 \cdot 5} = \dfrac{1}{125}$

127. $\left(y^6\right)^{-5}\left(2y^{-3}\right)^{-4}$
$$= y^{-30}(2)^{-4}y^{12}$$
$$= \dfrac{y^{-30+12}}{2^4}$$
$$= \dfrac{y^{-18}}{2^4} = \dfrac{1}{16y^{18}}$$

128. $(-5 + 11w) + (6 + 5w) + (-15 - 8w^2)$
$$= -5 + 6 - 15 + 11w + 5w - 8w^2$$
$$= -14 + 16w - 8w^2$$

129. $7p^5\left(3p^4 + p^3 + 2p^2\right)$
$$= 7p^5\left(3p^4\right) + 7p^5\left(p^3\right) + 7p^5\left(2p^2\right)$$
$$= 21p^9 + 7p^8 + 14p^7$$

130. $\dfrac{\left(-z^{-2}\right)^3}{5\left(z^{-3}\right)^{-1}} = \dfrac{(-1)^3 z^{-2(3)}}{5z^{-3(-1)}}$

$$= \dfrac{-z^{-6}}{5z^3}$$

$$= \dfrac{-z^{-6-3}}{5}$$

$$= \dfrac{-z^{-9}}{5} = -\dfrac{1}{5z^9}$$

131. $-(-3)^2 = -(9) = -9$

132. $\dfrac{\left(5z^2x^3\right)^2\left(2zx^2\right)^{-1}}{\left(-10zx^{-3}\right)^{-2}\left(3z^{-1}x^{-4}\right)^2}$

$$= \dfrac{5^2 z^4 x^6 2^{-1} z^{-1} x^{-2}}{(-10)^{-2} z^{-2} x^6 3^2 z^{-2} x^{-8}}$$

$$= \dfrac{25(-10)^2 z^3 x^4}{2 \cdot 9 z^{-4} x^{-2}}$$

$$= \dfrac{25(100) z^7 x^6}{2 \cdot 9}$$

$$= \dfrac{1250 z^7 x^6}{9}$$

133. $(2k - 1) - \left(3k^2 - 2k + 6\right)$
$$= 2k - 1 - 3k^2 + 2k - 6$$
$$= -3k^2 + 2k + 2k - 1 - 6$$
$$= -3k^2 + 4k - 7$$

134. $30a + am - am^2$
$$= a\left(30 + m - m^2\right)$$
$$= a(6 - m)(5 + m)$$

135. $11k + 12k^2 = k(11 + 12k)$

136. $8 - a^3$
$$= 2^3 - a^3$$
$$= (2 - a)\left(2^2 + 2a + a^2\right)$$
$$= (2 - a)\left(4 + 2a + a^2\right)$$

137. $9x^2 + 13xy - 3y^2$ is prime since it cannot be factored further.

138. $15y^3 + 20y^2 = 5y^2(3y + 4)$

139. $25z^2 - 30zm + 9m^2$

$= (5z)^2 - 2(5z)(3m) + (3m)^2$

$= (5z - 3m)^2$

140. $5x^2 - 17x - 12 = 0$

$(5x + 3)(x - 4) = 0$

$5x + 3 = 0 \quad$ or $\quad x - 4 = 0$

$5x = -3 \qquad\qquad x = 4$

$x = -\dfrac{3}{5}$

Solution set: $\left\{-\dfrac{3}{5}, 4\right\}$

141. $x^3 - x = 0$

$x(x^2 - 1) = 0$

$x(x + 1)(x - 1) = 0$

$x = 0 \quad$ or $\quad x + 1 = 0 \quad$ or $\quad x - 1 = 0$

$x = -1 \qquad\qquad x = 1$

Solution set: $\{-1, 0, 1\}$

142. Let x be the width of the frame. Then $x + 2$ will be the length of the frame. The area is 48 in^2. Use the formula $LW = A$.

$$(x + 2)x = 48$$

$$x^2 + 2x = 48$$

$$x^2 + 2x - 48 = 0$$

$$(x + 8)(x - 6) = 0$$

$x + 8 = 0 \quad$ or $\quad x - 6 = 0$

$x = -8 \quad$ or $\quad x = 6$

The frame cannot have a negative width, so reject $x = -8$.

The width of the frame is 6 in.

Exercises 143–148 are steps in the factorization of

$$x^{14} - x^2 - 4x^{13} + 4x + 4x^{12} - 4.$$

143. $x^{14} - x^2 - 4x^{13} + 4x + 4x^{12} - 4$

$= (x^{14} - x^2) - (4x^{13} - 4x) + (4x^{12} - 4)$

144. $= x^2(x^{12} - 1) - 4x(x^{12} - 1) + 4(x^{12} - 1)$

145. $= (x^{12} - 1)(x^2 - 4x + 4)$

146. $= \left[(x^6)^2 - 1^2\right][x^2 - 2(x)(2) + 2^2]$

$= (x^6 + 1)(x^6 - 1)(x - 2)^2$

147. $= \left[(x^2)^3 + 1^3\right]\left[(x^3)^2 - 1^2\right](x - 2)^2$

$= (x^2 + 1)(x^4 - x^2 + 1)(x^3 + 1)$

$\cdot (x^3 - 1)(x - 2)^2$

148. $= (x^2 + 1)(x^4 - x^2 + 1)(x + 1)$

$\cdot (x^2 - x + 1)(x - 1)$

$\cdot (x^2 + x + 1)(x - 2)^2$

Chapter 5 Test

1. **(a)** $7^{-2} = \dfrac{1}{7^2} = \dfrac{1}{49}$ (C)

(b) $7^0 = 1$ (A)

(c) $-7^0 = -(1) = -1$ (D)

(d) $(-7)^0 = 1$ (A)

(e) $-7^2 = -49$ (E)

(f) $7^{-1} + 2^{-1} = \dfrac{1}{7} + \dfrac{1}{2}$

$= \dfrac{2}{14} + \dfrac{7}{14} = \dfrac{9}{14}$ (F)

(g) $(7 + 2)^{-1} = 9^{-1} = \dfrac{1}{9}$ (B)

(h) $\dfrac{7^{-1}}{2^{-1}} = \dfrac{2^1}{7^1} = \dfrac{2}{7}$ (G)

(i) $(-7)^{-2} = \dfrac{1}{(-7)^2} = \dfrac{1}{49}$ (C)

2. $(3x^{-2}y^3)^{-2}(4x^3y^{-4})$

$= 3^{-2}x^{-2(-2)}y^{3(-2)}4x^3y^{-4}$

$= 3^{-2}x^4y^{-6}4x^3y^{-4}$

$= \dfrac{4x^{4+3}y^{-6-4}}{3^2}$

$= \dfrac{4x^7y^{-10}}{9} = \dfrac{4x^7}{9y^{10}}$

3. $\dfrac{36r^{-4}(r^2)^{-3}}{6r^4} = \dfrac{36r^{-4}r^{2(-3)}}{6r^4}$

$= \dfrac{6r^{-4}r^{-6}}{r^4}$

$= \dfrac{6r^{-10}}{r^4} = \dfrac{6}{r^{14}}$

4. $\left(\dfrac{4p^2}{q^4}\right)^3\left(\dfrac{6p^8}{q^{-8}}\right)^{-2}$

$= \dfrac{4^3p^6}{q^{12}} \cdot \dfrac{6^{-2}p^{-16}}{q^{16}}$

$= \dfrac{4^3p^{-10}}{6^2q^{28}}$

$= \dfrac{64}{36p^{10}q^{28}} = \dfrac{16}{9p^{10}q^{28}}$

5. $(-2x^4y^{-3})^0(-4x^{-3}y^{-8})^2$

$= 1(-4)^2x^{-6}y^{-16}$

$= \dfrac{16}{x^6y^{16}}$

6.   **(a)** $9.1 \times 10^{-7} = .00000091$

Move the decimal point 7 places to the left because the exponent is negative.

(b) $\dfrac{(2,500,000)(.00003)}{(.05)(5,000,000)}$

$= \dfrac{(2.5 \times 10^6)(3 \times 10^{-5})}{(5 \times 10^{-2})(5 \times 10^6)}$

$= \dfrac{7.5 \times 10^1}{25 \times 10^4}$

$= .3 \times 10^{1-4}$

$= .3 \times 10^{-3}$

$= 3 \times 10^{-4}$ or $.0003$

7.   **(a)** $f(x) = -2x^2 + 5x - 6$

$f(4) = -2(4)^2 + 5(4) - 6$

$\quad = -2 \cdot 16 + 20 - 6$

$\quad = -32 + 14 = -18$

(b) $(f + g)(x) = f(x) + g(x)$

$= \left(-2x^2 + 5x - 6\right) + \left(7x - 3\right)$

$= -2x^2 + 12x - 9$

(c) $(f - g)(x) = f(x) - g(x)$

$= \left(-2x^2 + 5x - 6\right) - \left(7x - 3\right)$

$= -2x^2 + 5x - 6 - 7x + 3$

$= -2x^2 - 2x - 3$

8.

x	$f(x) = -2x^2 + 3$
-2	$-2(-2)^2 + 3 = -5$
-1	$-2(-1)^2 + 3 = 1$
0	$-2(0)^2 + 3 = 3$
1	$-2(1)^2 + 3 = 1$
2	$-2(2)^2 + 3 = -5$

Since the greatest exponent is 2, the graph of f is a parabola.

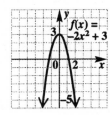

9. $f(x) = .5x^2 + 68.9x + 1852$

1996 corresponds to $x = 6$

$f(6) = .5(6)^2 + 68.9(6) + 1852$

$\quad = 18 + 413.4 + 1852$

$\quad = 2283.4$

There were about 2283 first-year students in 1996.

10. $\left(4x^3 - 3x^2 + 2x - 5\right)$

$\quad - \left(3x^3 + 11x + 8\right) + \left(x^2 - x\right)$

$= 4x^3 - 3x^2 + 2x - 5 - 3x^3 - 11x$

$\quad - 8 + x^2 - x$

$= x^3 - 2x^2 - 10x - 13$

11. $(5x - 3)(2x + 1)$

$= 10x^2 + 5x - 6x - 3$

$= 10x^2 - x - 3$

12. $(2m - 5)\left(3m^2 + 4m - 5\right)$

$= 6m^3 + 8m^2 - 10m$

$\quad - 15m^2 - 20m + 25$

$= 6m^3 - 7m^2 - 30m + 25$

13. $(6x + y)(6x - y) = (6x)^2 - y^2$

$= 36x^2 - y^2$

14. $(3k + q)^2 = (3k)^2 + 2(3k)(q) + q^2$

$= 9k^2 + 6kq + q^2$

15. $[2y + (3z - x)][2y - (3z - x)]$

$= (2y)^2 - (3z - x)^2$

$= 4y^2 - \left(9z^2 - 6zx + x^2\right)$

$= 4y^2 - 9z^2 + 6zx - x^2$

16.   It is not in factored form because there are two terms: $(x^2 + 2y)p$ and $3(x^2 + 2y)$. The common factor is $x^2 + 2y$, and the factored form is $(x^2 + 2y)(p + 3)$.

17. $11z^2 - 44z = 11z(z - 4)$

18. $3x + by + bx + 3y$

$= 3x + 3y + bx + by$

$= 3(x + y) + b(x + y)$

$= (x + y)(3 + b)$

19. $4p^2 + 3pq - q^2$

Two integer factors whose product is $(4)(-1) = -4$ and whose sum is 3 are 4 and -1.

$= 4p^2 + 4pq - pq - q^2$

$= 4p(p + q) - q(p + q)$

$= (p + q)(4p - q)$

20. $16a^2 + 40ab + 25b^2$

$= (4a)^2 + 2(4a)(5b) + (5b)^2$

$= (4a + 5b)^2$

21. $y^3 - 216 = y^3 - 6^3$

$= (y - 6)\left(y^2 + 6y + 6^2\right)$

$= (y - 6)\left(y^2 + 6y + 36\right)$

22. $9k^2 - 121j^2 = (3k)^2 - (11j)^2$

$= (3k + 11j)(3k - 11j)$

23. $6k^4 - k^2 - 35 = 6(k^2)^2 - k^2 - 35$

Two integer factors whose product is
$(6)(-35) = -210$ and whose sum is -1 are -15
and 14.

$$= 6(k^2)^2 - 15k^2 + 14k^2 - 35$$
$$= 3k^2(2k^2 - 5) + 7(2k^2 - 5)$$
$$= (2k^2 - 5)(3k^2 + 7)$$

24. $27x^6 + 1$
$$= (3x^2)^3 + (1)^3$$
$$= (3x^2 + 1)[(3x^2)^2 - (3x^2)(1) + 1^2]$$
$$= (3x^2 + 1)(9x^4 - 3x^2 + 1)$$

25. **(a)** $(3 - x)(x + 4)$
$$= 3x + 12 - x^2 - 4x$$
$$= -x^2 - x + 12$$

   (b) $-(x - 3)(x + 4)$
$$= -(x^2 + 4x - 3x - 12)$$
$$= -(x^2 + x - 12)$$
$$= -x^2 - x + 12$$

   (c) $(-x + 3)(x + 4)$
$$= -x^2 - 4x + 3x + 12$$
$$= -x^2 - x + 12$$

   (d) $(x - 3)(-x + 4)$
$$= -x^2 + 4x + 3x - 12$$
$$= -x^2 + 7x - 12$$

Therefore, only (d) is not a factored form of
$-x^2 - x + 12$.

26. $3x^2 + 8x + 4 = 0$
$(x + 2)(3x + 2) = 0$

$x + 2 = 0$ or $3x + 2 = 0$
$x = -2$ $3x = -2$
$$x = -\frac{2}{3}$$

Solution set: $\left\{-2, -\frac{2}{3}\right\}$

27. $10x^2 = 17x - 3$
$10x^2 - 17x + 3 = 0$
$(5x - 1)(2x - 3) = 0$

$5x - 1 = 0$ or $2x - 3 = 0$
$5x = 1$ $2x = 3$
$$x = \frac{1}{5}$$ or $$x = \frac{3}{2}$$

Solution set: $\left\{\frac{1}{5}, \frac{3}{2}\right\}$

28. $5m(m - 1) = 2(1 - m)$
$5m^2 - 5m = 2 - 2m$
$5m^2 - 3m - 2 = 0$
$(5m + 2)(m - 1) = 0$

$5m + 2 = 0$ or $m - 1 = 0$
$5m = -2$ $m = 1$
$$m = -\frac{2}{5}$$

Solution set: $\left\{-\frac{2}{5}, 1\right\}$

29. Using $A = LW$, substitute 40 for A, $x + 7$ for L,
and $2x + 3$ for W.

$$A = LW$$
$$40 = (x + 7)(2x + 3)$$
$$40 = 2x^2 + 3x + 14x + 21$$
$$40 = 2x^2 + 17x + 21$$
$$0 = 2x^2 + 17x - 19$$
$$0 = (2x + 19)(x - 1)$$

$2x + 19 = 0$ or $x - 1 = 0$
$2x = -19$ $x = 1$
$$x = -\frac{19}{2}$$

Length and width will be negative if $x = -\frac{19}{2}$,
so reject it as a possible solution. If $x = 1$, then

$$x + 7 = 1 + 7 = 8$$

and

$$2x + 3 = 2(1) + 3 = 5.$$

The length is 8 inches, and the width is 5 inches.

30. Substitute 128 for $f(t)$ in the equation.
$$f(t) = -16t^2 + 96t$$
$$128 = -16t^2 + 96t$$
$$16t^2 - 96t + 128 = 0$$
$$16(t^2 - 6t + 8) = 0$$
$$16(t - 4)(t - 2) = 0$$

$t - 4 = 0$ or $t - 2 = 0$
$t = 4$ or $t = 2$

The ball is 128 ft high at 2 seconds (on the way
up) and again at 4 seconds (on the way down).

Cumulative Review Exercises Chapters 1–5

1. $-2(m - 3) = -2(m) - 2(-3) = -2m + 6$

2. $-(-4m + 3) = -(-4m) - (3) = 4m - 3$

3. $3x^2 - 4x + 4 + 9x - x^2$

$\quad = 3x^2 - x^2 - 4x + 9x + 4$

$\quad = 2x^2 + 5x + 4$

For Exercises 4–7, let $p = -4, q = -2$, and $r = 5$.

4. $-3(2q - 3p) = -3[2(-2) - 3(-4)]$

$\qquad\qquad\qquad = -3(-4 + 12)$

$\qquad\qquad\qquad = -3(8) = -24$

5. $8r^2 + q^2 = 8(5)^2 + (-2)^2$

$\qquad\qquad = 8(25) + 4$

$\qquad\qquad = 200 + 4 = 204$

6. $\dfrac{\sqrt{r}}{-p + 2q} = \dfrac{\sqrt{5}}{-(-4) + 2(-2)}$

$\qquad\qquad = \dfrac{\sqrt{5}}{4 - 4} = \dfrac{\sqrt{5}}{0}$

This is undefined since the denominator is zero.

7. $\dfrac{5p + 6r^2}{p^2 + q - 1} = \dfrac{5(-4) + 6(5)^2}{(-4)^2 + (-2) - 1}$

$\qquad\qquad\qquad = \dfrac{-20 + 6(25)}{16 - 2 - 1}$

$\qquad\qquad\qquad = \dfrac{-20 + 150}{13}$

$\qquad\qquad\qquad = \dfrac{130}{13} = 10$

8. $2z - 5 + 3z = 4 - (z + 2)$

$\quad\; 5z - 5 = 4 - z - 2$

$\quad\; 5z - 5 = 2 - z$

$\qquad\quad 6z = 7$

$\qquad\quad\; z = \dfrac{7}{6}$

Solution set: $\left\{\dfrac{7}{6}\right\}$

9. $\dfrac{3a - 1}{5} + \dfrac{a + 2}{2} = -\dfrac{3}{10}$

$2(3a - 1) + 5(a + 2) = -3$ *Multiply by 10.*

$\quad\; 6a - 2 + 5a + 10 = -3$

$\qquad\qquad\quad 11a + 8 = -3$

$\qquad\qquad\qquad 11a = -11$

$\qquad\qquad\qquad\; a = -1$

Solution set: $\{-1\}$

10. $\qquad -\dfrac{4}{3}d \geq -5$

$-\dfrac{3}{4}\left(-\dfrac{4}{3}d\right) \leq -\dfrac{3}{4}(-5)$

$\qquad\qquad d \leq \dfrac{15}{4}$

Solution set: $\left(-\infty, \dfrac{15}{4}\right]$

11. $3 - 2(m + 3) < 4m$

$\quad\; 3 - 2m - 6 < 4m$

$\qquad\; -2m - 3 < 4m$

$\qquad\qquad -6m < 3$

$\qquad\qquad\quad m > -\dfrac{3}{6} \text{ or } -\dfrac{1}{2}$

Solution set: $\left(-\dfrac{1}{2}, \infty\right)$

12. $2k + 4 < 10 \quad$ and $\quad 3k - 1 > 5$

$\quad 2k < 6 \qquad\qquad\qquad 3k > 6$

$\quad\; k < 3 \quad$ and $\qquad\quad k > 2$

The overlap of these inequalities is the set of all numbers between 2 and 3.

Solution set: $(2, 3)$

13. $2k + 4 > 10 \quad$ or $\quad 3k - 1 < 5$

$\quad 2k > 6 \qquad\qquad\qquad 3k < 6$

$\quad\; k > 3 \quad$ or $\qquad\quad\; k < 2$

The solution set is the set of numbers that are either greater than 3 or less than 2.

Solution set: $(-\infty, 2) \cup (3, \infty)$

14. $|5x + 3| - 10 = 3$

$\qquad\; |5x + 3| = 13$

$5x + 3 = 13 \quad$ or $\quad 5x + 3 = -13$

$\quad\; 5x = 10 \qquad\qquad\quad 5x = -16$

$\qquad x = 2 \quad$ or $\qquad\quad x = -\dfrac{16}{5}$

Solution set: $\left\{-\dfrac{16}{5}, 2\right\}$

15. $|x + 2| < 9$

$\quad -9 < x + 2 < 9$

$\; -11 < x < 7$

Solution set: $(-11, 7)$

16. $|2y - 5| \geq 9$

$2y - 5 \geq 9 \quad$ or $\quad 2y - 5 \leq -9$

$\quad 2y \geq 14 \qquad\qquad\quad 2y \leq -4$

$\quad\; y \geq 7 \quad$ or $\qquad\quad\; y \leq -2$

Solution set: $(-\infty, -2] \cup [7, \infty)$

17. Solve $V = lwh$ for h.

$\quad \dfrac{V}{lw} = \dfrac{lwh}{lw}$

$\quad \dfrac{V}{lw} = h$

18. Let x be the time it takes for the planes to be 2100 mi apart. Use the formula $d = rt$ to compete the table.

Plane	r	t	d
Eastbound	550	x	$550x$
Westbound	500	x	$500x$

The total distance is 2100 mi.

$$550x + 500x = 2100$$
$$1050x = 2100$$
$$x = 2$$

It will take 2 hours for the planes to be 2100 miles apart.

19. $4x + 2y = -8$

Draw a line through the x- and y-intercepts, $(-2, 0)$ and $(0, -4)$, respectively.

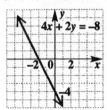

20. The slope m of the line through $(-4, 8)$ and $(-2, 6)$ is

$$m = \frac{6 - 8}{-2 - (-4)} = \frac{-2}{2} = -1.$$

21. $y = -3$ is the equation of a horizontal line. Its slope is 0.

22. $f(x) = 2x + 7$
$f(-4) = 2(-4) + 7$
$= -8 + 7 = -1$

23. To find the x-intercept of the graph of $f(x) = 2x + 7$, let $f(x) = 0$ (which is the same as letting $y = 0$) and solve for x.

$$0 = 2x + 7$$
$$-7 = 2x$$
$$-\frac{7}{2} = x$$

The x-intercept is $\left(-\frac{7}{2}, 0\right)$.

24. $3x - 2y = -7$ (1)
$2x + 3y = 17$ (2)
To eliminate y, multiply (1) by 3 and (2) by 2, and then add the resulting equations.

$$\begin{array}{rl} 9x - 6y = -21 & 3 \times (1) \\ 4x + 6y = 34 & 2 \times (2) \\ \hline 13x = 13 & \\ x = 1 & \end{array}$$

Substitute 1 for x in (1).

$$3x - 2y = -7 \quad (1)$$
$$3(1) - 2y = -7$$
$$-2y = -10$$
$$y = 5$$

Solution set: $\{(1, 5)\}$

25. $2x + 3y - 6z = 5$ (1)
$8x - y + 3z = 7$ (2)
$3x + 4y - 3z = 7$ (3)

To eliminate z, add (2) and (3).

$$\begin{array}{rl} 8x - y + 3z = 7 & (2) \\ 3x + 4y - 3z = 7 & (3) \\ \hline 11x + 3y = 14 & (4) \end{array}$$

To eliminate z again, multiply (2) by 2 and add the result to (1).

$$\begin{array}{rl} 2x + 3y - 6z = 5 & (1) \\ 16x - 2y + 6z = 14 & 2 \times (2) \\ \hline 18x + y = 19 & (5) \end{array}$$

To eliminate y, multiply (5) by -3 and add the result to (4).

$$\begin{array}{rl} 11x + 3y = 14 & (4) \\ -54x - 3y = -57 & -3 \times (5) \\ \hline -43x = -43 & \\ x = 1 & \end{array}$$

From (5), $18(1) + y = 19$, so $y = 1$.
To find z, let $x = 1$ and $y = 1$ in (1).

$$2x + 3y - 6z = 5 \quad (1)$$
$$2(1) + 3(1) - 6z = 5$$
$$5 - 6z = 5$$
$$-6z = 0$$
$$z = 0$$

Solution set: $\{(1, 1, 0)\}$

26. $\begin{vmatrix} -3 & 2 \\ -1 & 8 \end{vmatrix} = -3(8) - 2(-1)$
$= -24 + 2 = -22$

27. $(3x^2y^{-1})^{-2}(2x^{-3}y)^{-1}$
$= 3^{-2}x^{-4}y^2 \cdot 2^{-1}x^3y^{-1}$
$= 3^{-2}2^{-1}x^{-1}y$
$= \frac{y}{3^2 \cdot 2x} = \frac{y}{18x}$

28. $\frac{5m^{-2}y^3}{3m^{-3}y^{-1}} = \frac{5}{3} \cdot \frac{m^{-2}}{m^{-3}} \cdot \frac{y^3}{y^{-1}}$
$= \frac{5}{3}m^{-2-(-3)}y^{3-(-1)}$
$= \frac{5}{3}m^1y^4 \text{ or } \frac{5my^4}{3}$

29. $\left(3x^3 + 4x^2 - 7\right) - \left(2x^3 - 8x^2 + 3x\right)$

$\quad = 3x^3 + 4x^2 - 7 - 2x^3 + 8x^2 - 3x$

$\quad = x^3 + 12x^2 - 3x - 7$

30. $(7x + 3y)^2 = (7x)^2 + 2(7x)(3y) + (3y)^2$

$\quad\quad\quad\quad = 49x^2 + 42xy + 9y^2$

31. $(2p + 3)\left(5p^2 - 4p - 8\right)$

$\quad = 10p^3 - 8p^2 - 16p + 15p^2 - 12p - 24$

$\quad = 10p^3 + 7p^2 - 28p - 24$

32. $16w^2 + 50wz - 21z^2$

Two integer factors whose product is
$(16)(-21) = -336$ and whose sum is 50 are 56
and -6.

$\quad = 16w^2 + 56wz - 6wz - 21z^2$

$\quad = 8w(2w + 7z) - 3z(2w + 7z)$

$\quad = (2w + 7z)(8w - 3z)$

33. $4x^2 - 4x + 1 - y^2$

Group the first three terms.

$\quad = \left(4x^2 - 4x + 1\right) - y^2$

$\quad = (2x - 1)^2 - y^2$

$\quad = [(2x - 1) + y][(2x - 1) - y]$

$\quad = (2x - 1 + y)(2x - 1 - y)$

34. $4y^2 - 36y + 81$

$\quad = (2y)^2 - 2(2y)(9) + 9^2$

$\quad = (2y - 9)^2$

35. $100x^4 - 81 = \left(10x^2\right)^2 - 9^2$

$\quad\quad\quad\quad\quad = \left(10x^2 + 9\right)\left(10x^2 - 9\right)$

36. $8p^3 + 27$

$\quad = (2p)^3 + 3^3$

$\quad = (2p + 3)[(2p)^2 - (2p)(3) + 3^2]$

$\quad = (2p + 3)\left(4p^2 - 6p + 9\right)$

37. $(p - 1)(2p + 3)(p + 4) = 0$

$p - 1 = 0 \quad \text{or} \quad 2p + 3 = 0 \quad \text{or} \quad p + 4 = 0$

$\quad p = 1 \quad\quad\quad 2p = -3 \quad\quad\quad p = -4$

$\quad\quad\quad\quad\quad\quad p = -\dfrac{3}{2}$

Solution set: $\left\{-4, -\dfrac{3}{2}, 1\right\}$

38.
$$9q^2 = 6q - 1$$
$$9q^2 - 6q + 1 = 0$$
$$(3q - 1)^2 = 0$$
$$3q - 1 = 0$$
$$3q = 1$$
$$q = \frac{1}{3}$$

Solution set: $\left\{\dfrac{1}{3}\right\}$

39. Let x be the length of the base. Then $x + 3$ will be
the height. The area is 14 square feet. Use the

formula $A = \dfrac{1}{2}bh$, and substitute 14 for A, x for

b, and $x + 3$ for h.

$$\frac{1}{2}bh = A$$

$$\frac{1}{2}(x)(x + 3) = 14$$

$$x(x + 3) = 28 \quad \textit{Multiply by 2.}$$

$$x^2 + 3x = 28$$

$$x^2 + 3x - 28 = 0$$

$$(x + 7)(x - 4) = 0$$

$$x + 7 = 0 \quad \text{or} \quad x - 4 = 0$$

$$x = -7 \quad \text{or} \quad x = 4$$

The length cannot be negative, so reject -7 as a
solution. The only possible solution is 4. The base
is 4 feet long.

40. Let x be the distance between the longer sides.
(This is actually the width.) Then $x + 2$ will be
the length of the longer side. The area of the
rectangle is 288 in^2. Use the formula $LW = A$.
Substitute 288 for A, $x + 2$ for L, and x for W.

$$(x + 2)x = 288$$

$$x^2 + 2x = 288$$

$$x^2 + 2x - 288 = 0$$

$$(x + 18)(x - 16) = 0$$

$$x + 18 = 0 \quad \text{or} \quad x - 16 = 0$$

$$x = -18 \quad \text{or} \quad x = 16$$

The distance cannot be negative, so reject -18 as
a solution. The only possible solution is 16.
The distance between the longer sides is 16
inches, and the length of the longer sides is
$16 + 2 = 18$ inches.

CHAPTER 6   RATIONAL EXPRESSIONS

Section 6.1

1. $\dfrac{x-3}{x+4} = \dfrac{(-1)(x-3)}{(-1)(x+4)}$

$\phantom{\dfrac{x-3}{x+4}} = \dfrac{-x+3}{-x-4}$ or $\dfrac{3-x}{-x-4}$ **(C)**

3. $\dfrac{x-3}{x-4} = \dfrac{(-1)(x-3)}{(-1)(x-4)}$

$\phantom{\dfrac{x-3}{x-4}} = \dfrac{-x+3}{-x+4}$ **(D)**

5. $\dfrac{3-x}{x+4} = \dfrac{(-1)(3-x)}{(-1)(x+4)}$

$\phantom{\dfrac{3-x}{x+4}} = \dfrac{-3+x}{-x-4}$ or $\dfrac{x-3}{-x-4}$ **(E)**

7. Replacing k with 2 makes the denominator of the fraction 0 and the value of the fraction undefined. To find the values excluded from the domain, set the denominator equal to 0 and solve the equation. All solutions of the equation are excluded from the domain.

9. $f(z) = \dfrac{z}{z-7}$

Set the denominator equal to zero, and solve the equation.

$$z - 7 = 0$$
$$z = 7$$

The number 7 makes the rational expression undefined, so 7 is not in the domain of the function.

11. $f(p) = \dfrac{6p-5}{7p+1}$

Set the denominator equal to zero, and solve the equation.

$$7p + 1 = 0$$
$$7p = -1$$
$$p = -\frac{1}{7}$$

The number $-\dfrac{1}{7}$ makes the rational expression undefined, so $-\dfrac{1}{7}$ is not in the domain of the function.

13. $f(x) = \dfrac{12x+3}{x}$

Set the denominator equal to zero and solve.

$$x = 0$$

The number 0 makes the rational expression undefined, so 0 is not in the domain of the function.

15. $f(x) = \dfrac{3x+1}{2x^2+x-6}$

Set the denominator equal to zero and solve.

$$2x^2 + x - 6 = 0$$
$$(x+2)(2x-3) = 0$$

$x + 2 = 0$ or $2x - 3 = 0$

$ x = -2 \qquad 2x = 3$

$$x = \frac{3}{2}$$

The numbers -2 and $\dfrac{3}{2}$ are not in the domain of the function.

17. $f(x) = \dfrac{x+2}{14}$

The denominator is never zero, so all numbers are in the domain of the function.

19. $f(x) = \dfrac{2x^2-3x+4}{3x^2+8}$

Set the denominator equal to zero and solve.

$$3x^2 + 8 = 0$$
$$3x^2 = -8$$
$$x^2 = -\frac{8}{3}$$

The square of any real number x is positive or zero. There are no real numbers which make this rational expression undefined, so all numbers are in the domain of the function.

21. **(a)** $\dfrac{x^2+4x}{x+4}$

The two terms in the numerator are x^2 and $4x$. The two terms in the denominator are x and 4.

(b) To express the rational expression in lowest terms, factor the numerator and denominator and replace the quotient of common factors with 1.

$$\frac{x^2+4x}{x+4} = \frac{x(x+4)}{x+4}$$
$$= x \cdot 1 = x$$

23. From the graph, the ratio of domestic sales to total sales is

$$\frac{6.3}{8.2} \approx .77$$

for 1992 and

$$\frac{7.2}{8.5} \approx .85$$

for 1996. Since $.85 > .77$, the domestic sales were a larger fraction of total sales in 1996.

25. Factor out the GCF, $6xy^4$. (Here 1 is the least exponent on x, and 4 the least exponent on y.)

$$\frac{24x^2y^4}{18xy^5} = \frac{4x \cdot 6xy^4}{3y \cdot 6xy^4} = \frac{4x}{3y} \cdot 1 = \frac{4x}{3y}$$

27. Divide numerator and denominator by $x + 4$.

$$\frac{(x+4)(x-3)}{(x+5)(x+4)} = \frac{x-3}{x+5}$$

29. $$\frac{4x(x+3)}{8x^2(x-3)} = \frac{(x+3) \cdot 4x}{2x(x-3) \cdot 4x}$$
$$= \frac{x+3}{2x(x-3)}$$

31. $\dfrac{3x+7}{3}$ Since the numerator and denominator have no common factors, the expression is already in lowest terms.

33. $$\frac{6m+18}{7m+21} = \frac{6(m+3)}{7(m+3)} = \frac{6}{7}$$

35. $$\frac{3z^2+z}{18z+6} = \frac{z(3z+1)}{6(3z+1)} = \frac{z}{6}$$

37. $$\frac{2t+6}{t^2-9} = \frac{2(t+3)}{(t-3)(t+3)}$$
$$= \frac{2}{t-3}$$

39. $$\frac{x^2+2x-15}{x^2+6x+5} = \frac{(x+5)(x-3)}{(x+5)(x+1)}$$
$$= \frac{x-3}{x+1}$$

41. $$\frac{8x^2-10x-3}{8x^2-6x-9} = \frac{(4x+1)(2x-3)}{(4x+3)(2x-3)}$$
$$= \frac{4x+1}{4x+3}$$

43. $$\frac{a^3+b^3}{a+b} = \frac{(a+b)(a^2-ab+b^2)}{a+b}$$
$$= a^2-ab+b^2$$

45. $$\frac{2c^2+2cd-60d^2}{2c^2-12cd+10d^2}$$
$$= \frac{2(c^2+cd-30d^2)}{2(c^2-6cd+5d^2)}$$
$$= \frac{2(c+6d)(c-5d)}{2(c-d)(c-5d)}$$
$$= \frac{c+6d}{c-d}$$

47. $$\frac{ac-ad+bc-bd}{ac-ad-bc+bd}$$
$$= \frac{a(c-d)+b(c-d)}{a(c-d)-b(c-d)}$$
$$= \frac{(c-d)(a+b)}{(c-d)(a-b)} \qquad \textit{Factor by}$$
$$\qquad\qquad\qquad\qquad \textit{grouping}$$
$$= \frac{a+b}{a-b}$$

49. (a) $$\frac{3-x}{x-4} = \frac{-1(x-3)}{-1(4-x)} = \frac{x-3}{4-x}$$

(b) $\dfrac{x+3}{4+x}$ cannot be transformed to equal $\dfrac{x-3}{4-x}$.

(c) $$-\frac{3-x}{4-x} = \frac{-(3-x)}{4-x} = \frac{x-3}{4-x}$$

(d) $$-\frac{x-3}{x-4} = \frac{x-3}{-(x-4)} = \frac{x-3}{4-x}$$

Only the expression in (b) is not equivalent to $\dfrac{x-3}{4-x}$.

51. $$\frac{7-b}{b-7} = \frac{-1(b-7)}{b-7} = -1$$

In Exercises 53–56, there are several other acceptable ways to express the answers.

53. $$\frac{x^2-y^2}{y-x} = \frac{(x-y)(x+y)}{y-x}$$
$$= \frac{-1(y-x)(x+y)}{y-x}$$
$$= -(x+y)$$

55. $$\frac{(a-3)(x+y)}{(3-a)(x-y)} = \frac{(a-3)(x+y)}{-1(a-3)(x-y)}$$
$$= \frac{x+y}{-1(x-y)}$$
$$= -\frac{x+y}{x-y}$$

57. $$\frac{5k-10}{20-10k} = \frac{-5(2-k)}{10(2-k)} = \frac{-5}{10} = -\frac{1}{2}$$

59. $$\frac{a^2-b^2}{a^2+b^2} = \frac{(a+b)(a-b)}{a^2+b^2}$$

The numerator and denominator have no common factors except 1, so the original expression is already in lowest terms.

61. Multiply the numerators, multiply the denominators, factor each numerator and denominator (this step can be performed first). Divide the numerator and denominator by any common factors to reduce to lowest terms. For example,

$$\frac{6r-5s}{3r+2s} \cdot \frac{6r+4s}{5s-6r} =$$
$$\frac{(6r-5s)(6r+4s)}{(3r+2s)(5s-6r)} =$$
$$\frac{(6r-5s)2(3r+2s)}{(3r+2s)(-1)(6r-5s)} = \frac{2}{-1} = -2.$$

63. $\dfrac{x^3}{3y} \cdot \dfrac{9y^2}{x^5} = \dfrac{3y \cdot 3x^3 y}{x^2 \cdot 3x^3 y} = \dfrac{3y}{x^2} \cdot 1 = \dfrac{3y}{x^2}$

65. $\dfrac{5a^4 b^2}{16a^2 b} \div \dfrac{25a^2 b}{60a^3 b^2}$

Multiply by the reciprocal of the divisor.

$$= \frac{5a^4 b^2}{16a^2 b} \cdot \frac{60a^3 b^2}{25a^2 b}$$
$$= \frac{300a^7 b^4}{400a^4 b^2}$$
$$= \frac{100a^4 b^2 \cdot 3a^3 b^2}{100a^4 b^2 \cdot 4}$$
$$= \frac{3a^3 b^2}{4}$$

67. $\dfrac{(5pq^2)^2}{60p^3 q^6} \div \dfrac{5p^2 q^2}{16p^2 q^3}$

$$= \frac{5^2 p^2 q^4}{60p^3 q^6} \cdot \frac{16p^2 q^3}{5p^2 q^2}$$
$$= \frac{5^2 4^2 p^4 q^7}{4 \cdot 3 \cdot 5 \cdot 5p^5 q^8}$$
$$= \frac{4}{3pq}$$

69. $\dfrac{4x}{8x+4} \cdot \dfrac{14x+7}{6} = \dfrac{4x \cdot 7(2x+1)}{4(2x+1) \cdot 6}$
$$= \frac{7x}{6}$$

For Exercises 69–72, there are several other ways to express the answer.

71. $\dfrac{p^2-25}{4p} \cdot \dfrac{2}{5-p} = \dfrac{(p+5)(p-5)2}{2 \cdot 2p(-1)(p-5)}$
$$= \frac{p+5}{(2p)(-1)} = -\frac{p+5}{2p}$$

73. $\dfrac{m^2-49}{m+1} \div \dfrac{7-m}{m}$
$$= \frac{(m-7)(m+7)}{m+1} \cdot \frac{m}{7-m}$$
$$= \frac{(-1)(7-m)(m+7)}{m+1} \cdot \frac{m}{7-m}$$
$$= \frac{-m(m+7)}{m+1}$$

75. $\dfrac{12x-10y}{3x+2y} \cdot \dfrac{6x+4y}{10y-12x}$
$$= \frac{2(6x-5y) \cdot 2(3x+2y)}{(3x+2y) \cdot 2(5y-6x)}$$
$$= \frac{2(-1)(5y-6x)}{(5y-6x)} = -2$$

77. $\dfrac{x^2-25}{x^2+x-20} \cdot \dfrac{x^2+7x+12}{x^2-2x-15}$
$$= \frac{(x-5)(x+5)}{(x+5)(x-4)} \cdot \frac{(x+3)(x+4)}{(x-5)(x+3)}$$
$$= \frac{x+4}{x-4}$$

79. $\dfrac{6x^2+5xy-6y^2}{12x^2-11xy+2y^2} \div \dfrac{4x^2-12xy+9y^2}{8x^2-14xy+3y^2}$
$$= \frac{(3x-2y)(2x+3y)}{(3x-2y)(4x-y)} \cdot \frac{(4x-y)(2x-3y)}{(2x-3y)(2x-3y)}$$
$$= \frac{2x+3y}{2x-3y}$$

81. $\dfrac{3k^2+17kp+10p^2}{6k^2+13kp-5p^2} \div \dfrac{6k^2+kp-2p^2}{6k^2-5kp+p^2}$
$$= \frac{(3k+2p)(k+5p)}{(3k-p)(2k+5p)} \cdot \frac{(3k-p)(2k-p)}{(3k+2p)(2k-p)}$$
$$= \frac{k+5p}{2k+5p}$$

83. $\left(\dfrac{6k^2-13k-5}{k^2+7k} \div \dfrac{2k-5}{k^3+6k^2-7k} \right)$
$$\cdot \frac{k^2-5k+6}{3k^2-8k-3}$$

Factor k from the denominator of the divisor; multiply by the reciprocal.

$$= \left[\frac{6k^2-13k-5}{k^2+7k} \cdot \frac{k(k^2+6k-7)}{2k-5} \right]$$
$$\cdot \frac{k^2-5k+6}{3k^2-8k-3}$$
$$= \left[\frac{(3k+1)(2k-5)}{k(k+7)} \cdot \frac{k(k+7)(k-1)}{2k-5} \right]$$
$$\cdot \frac{(k-2)(k-3)}{(3k+1)(k-3)}$$
$$= (k-1)(k-2)$$

85. $\dfrac{a^2(2a+b)+6a(2a+b)+5(2a+b)}{3a^2(a+2b)-2a(a+2b)-(a+2b)} \div \dfrac{a+1}{a-1}$
$$= \frac{(2a+b)(a^2+6a+5)}{(a+2b)(3a^2-2a-1)} \cdot \frac{a-1}{a+1}$$
$$= \frac{(2a+b)(a+5)(a+1)(a-1)}{(a+2b)(3a+1)(a-1)(a+1)}$$
$$= \frac{(a+5)(2a+b)}{(3a+1)(a+2b)}$$

87. From the table, Y_1 is undefined when $x = 2$, so $x - 2$ is a factor of the numerator and denominator of Y_1. Since $Y_2 = x - 3$,
$$Y_1 = \frac{(x-3)(x-2)}{x-2} = \frac{x^2 - 5x + 6}{x-2}.$$

89. From the table, Y_1 is undefined when $x = 2$, so $x - 2$ is a factor of the numerator and denominator of Y_1. Since $Y_2 = x + 5$,
$$Y_1 = \frac{(x+5)(x-2)}{x-2} = \frac{x^2 + 3x - 10}{x-2}.$$

Section 6.2

1. $\dfrac{1}{x} + \dfrac{1}{y} = \dfrac{1}{4} + \dfrac{1}{2}$ *Let x=4 and y=2.*
$$= \frac{1}{4} + \frac{2}{4} = \frac{3}{4}$$

2. $\dfrac{1}{x+y} = \dfrac{1}{4+2}$ *Let x=4 and y=2.*
$$= \frac{1}{6}$$

3. The answers for Exercises 1 and 2 are not the same. We conclude that we cannot find the sum $\dfrac{1}{x} + \dfrac{1}{y}$ by adding the denominators and keeping the common numerator.

4. $\dfrac{1}{x} - \dfrac{1}{y} = \dfrac{1}{3} - \dfrac{1}{5}$ *Let x=3 and y=5.*
$$= \frac{5}{15} - \frac{3}{15} = \frac{2}{15}$$

5. $\dfrac{1}{x-y} = \dfrac{1}{3-5}$ *Let x=3 and y=5.*
$$= \frac{1}{-2} = -\frac{1}{2}$$

6. The answers for Exercises 4 and 5 are not the same. We conclude that we cannot find the difference $\dfrac{1}{x} - \dfrac{1}{y}$ by subtracting the denominators and keeping the common numerator.

7. $\dfrac{7}{t} + \dfrac{2}{t} = \dfrac{7+2}{t} = \dfrac{9}{t}$

9. $\dfrac{11}{5x} - \dfrac{1}{5x} = \dfrac{11-1}{5x} = \dfrac{10}{5x} = \dfrac{2}{x}$

11. $\dfrac{5x+4}{6x+5} + \dfrac{x+1}{6x+5}$
$$= \frac{5x+4+x+1}{6x+5}$$
$$= \frac{6x+5}{6x+5} = 1$$

13. $\dfrac{x^2}{x+5} - \dfrac{25}{x+5} = \dfrac{x^2 - 25}{x+5}$
$$= \frac{(x+5)(x-5)}{x+5}$$
$$= x - 5$$

15. $\dfrac{-3p+7}{p^2 + 7p + 12} + \dfrac{8p+13}{p^2 + 7p + 12}$
$$= \frac{-3p+7+8p+13}{p^2 + 7p + 12}$$
$$= \frac{5p+20}{(p+3)(p+4)}$$
$$= \frac{5(p+4)}{(p+3)(p+4)} = \frac{5}{p+3}$$

17. $\dfrac{a^3}{a^2 + ab + b^2} - \dfrac{b^3}{a^2 + ab + b^2}$
$$= \frac{a^3 - b^3}{a^2 + ab + b^2}$$
$$= \frac{(a-b)(a^2 + ab + b^2)}{a^2 + ab + b^2}$$
$$= a - b$$

19. First add or subtract the numerators. Then place the result over the common denominator. Write the answer in lowest terms. We give one example:
$$\frac{5}{x} - \frac{3x+1}{x} = \frac{5 - (3x+1)}{x}$$
$$= \frac{5 - 3x - 1}{x} = \frac{4 - 3x}{x}.$$

21. $18x^2y^3,\ 24x^4y^5$

Factor each denominator.
$$18x^2y^3 = 2 \cdot 3 \cdot 3 \cdot x^2 \cdot y^3$$
$$= 2 \cdot 3^2 \cdot x^2 \cdot y^3$$
$$24x^4y^5 = 2 \cdot 2 \cdot 2 \cdot 3 \cdot x^4 \cdot y^5$$
$$= 2^3 \cdot 3 \cdot x^4 \cdot y^5$$

The least common denominator (LCD) is the product of all the different factors, with each factor raised to the greatest power in any denominator.
$$\text{LCD} = 2^3 \cdot 3^2 \cdot x^4 \cdot y^5$$
$$= 8 \cdot 9 \cdot x^4 \cdot y^5$$
$$= 72x^4y^5$$

The LCD is $72x^4y^5$.

23. $z - 2,\ z$

Both $z - 2$ and z have only 1 and themselves for factors.
$$\text{LCD} = z(z - 2)$$

25. $2y + 8,\ y + 4$

Factor each denominator.

$$2y + 8 = 2(y + 4)$$

The second denominator, $y + 4$, is already factored. The LCD is

$$2(y + 4).$$

27. $6x + 18,\ 5x + 15$

Factor each denominator.

$$6x + 18 = 6(x + 3) = 2 \cdot 3(x + 3)$$
$$5x + 15 = 5(x + 3)$$

The LCD is $2 \cdot 3 \cdot 5(x + 3) = 30(x + 3)$.

29. $m + n,\ m - n$

Both $m + n$ and $m - n$ have only 1 and themselves for factors.

$$\text{LCD} = (m + n)(m - n)$$

31. $\dfrac{x + 8}{x^2 - 3x - 4},\ \dfrac{-9}{x + x^2}$

Factor each denominator.

$$x^2 - 3x - 4 = (x - 4)(x + 1)$$
$$x + x^2 = x(1 + x) = x(x + 1)$$

The LCD is $x(x - 4)(x + 1)$.

33. $\dfrac{t}{2t^2 + 7t - 15},\ \dfrac{t}{t^2 + 3t - 10}$

Factor each denominator.

$$2t^2 + 7t - 15 = (2t - 3)(t + 5)$$
$$t^2 + 3t - 10 = (t + 5)(t - 2)$$

The LCD is $(2t - 3)(t + 5)(t - 2)$.

35. $\dfrac{y}{2y + 6},\ \dfrac{3}{y^2 - 9},\ \dfrac{6}{y}$

Factor each denominator.

$$2y + 6 = 2(y + 3)$$
$$y^2 - 9 = (y + 3)(y - 3)$$

Remember the factor y from the third denominator. The LCD is

$$2y(y + 3)(y - 3).$$

37. $\dfrac{5}{6x},\ \dfrac{3}{x^2},\ \dfrac{7}{x + 1}$

The LCD is $6x^2(x + 1)$.

39. Yes, they are both correct, because the expressions are equivalent. Multiplying $\dfrac{3}{5 - y}$ by $\dfrac{-1}{-1}$ gives

$$\dfrac{-3}{y - 5}.$$

41. $\dfrac{8}{t} + \dfrac{7}{3t}$ The LCD is $3t$.

$$\dfrac{8}{t} + \dfrac{7}{3t} = \dfrac{8 \cdot 3}{t \cdot 3} + \dfrac{7}{3t}$$
$$= \dfrac{24 + 7}{3t} = \dfrac{31}{3t}$$

43. $\dfrac{5}{12x^2y} - \dfrac{11}{6xy}$ The LCD is $12x^2y$.

$$\dfrac{5}{12x^2y} - \dfrac{11}{6xy} = \dfrac{5}{12x^2y} - \dfrac{11 \cdot 2x}{6xy \cdot 2x}$$
$$= \dfrac{5}{12x^2y} - \dfrac{22x}{12x^2y}$$
$$= \dfrac{5 - 22x}{12x^2y}$$

45. $\dfrac{1}{x - 1} - \dfrac{1}{x}$ LCD $= x(x - 1)$

$$\dfrac{1}{x - 1} - \dfrac{1}{x} = \dfrac{1 \cdot x}{(x - 1)x} - \dfrac{1 \cdot (x - 1)}{x(x - 1)}$$
$$= \dfrac{x - (x - 1)}{x(x - 1)}$$
$$= \dfrac{x - x + 1}{x(x - 1)}$$
$$= \dfrac{1}{x(x - 1)}$$

47. $\dfrac{3a}{a + 1} + \dfrac{2a}{a - 3}$ LCD $= (a + 1)(a - 3)$

$$= \dfrac{3a(a - 3)}{(a + 1)(a - 3)} + \dfrac{2a(a + 1)}{(a - 3)(a + 1)}$$
$$= \dfrac{3a(a - 3) + 2a(a + 1)}{(a + 1)(a - 3)}$$
$$= \dfrac{3a^2 - 9a + 2a^2 + 2a}{(a + 1)(a - 3)}$$
$$= \dfrac{5a^2 - 7a}{(a + 1)(a - 3)}$$

49. $\dfrac{3x+2}{4-x} + \dfrac{5-3x}{x-4}$

To get a common denominator of $x-4$, multiply both the numerator and denominator of the first expression by -1.

$$= \dfrac{(3x+2)(-1)}{(4-x)(-1)} + \dfrac{5-3x}{x-4}$$

$$= \dfrac{-3x-2}{x-4} + \dfrac{5-3x}{x-4}$$

$$= \dfrac{-3x-2+5-3x}{x-4}$$

$$= \dfrac{-6x+3}{x-4}$$

If you chose $4-x$ for the LCD, then you should have obtained the equivalent answer, $\dfrac{6x-3}{4-x}$.

51. $\dfrac{-3w+2z}{w-z} - \dfrac{4w-z}{z-w}$

$w-z$ and $z-w$ are opposites, so factor out -1 from $z-w$ to get a common denominator.

$$= \dfrac{-3w+2z}{w-z} - \dfrac{4w-z}{-1(w-z)}$$

$$= \dfrac{-3w+2z}{w-z} + \dfrac{4w-z}{w-z}$$

$$= \dfrac{-3w+2z+4w-z}{w-z}$$

$$= \dfrac{w+z}{w-z} \text{ or } \dfrac{-w-z}{z-w}$$

53. $\dfrac{4x}{x-1} - \dfrac{2}{x+1} - \dfrac{4}{x^2-1}$

$x^2-1 = (x+1)(x-1)$, the LCD.

$\dfrac{4x}{x-1} - \dfrac{2}{x+1} - \dfrac{4}{x^2-1}$

$$= \dfrac{4x(x+1)}{(x-1)(x+1)} - \dfrac{2(x-1)}{(x+1)(x-1)}$$
$$- \dfrac{4}{(x+1)(x-1)}$$

$$= \dfrac{4x(x+1) - 2(x-1) - 4}{(x+1)(x-1)}$$

$$= \dfrac{4x^2 + 4x - 2x + 2 - 4}{(x-1)(x+1)}$$

$$= \dfrac{4x^2 + 2x - 2}{(x-1)(x+1)}$$

$$= \dfrac{2(2x^2 + x - 1)}{(x-1)(x+1)}$$

$$= \dfrac{2(2x-1)(x+1)}{(x-1)(x+1)}$$

$$= \dfrac{2(2x-1)}{x-1}$$

55. $\dfrac{5}{x-2} + \dfrac{1}{x} + \dfrac{2}{x^2-2x}$

$x^2-2x = x(x-2)$, the LCD.

$\dfrac{5}{x-2} + \dfrac{1}{x} + \dfrac{2}{x^2-2x}$

$$= \dfrac{5x}{(x-2)x} + \dfrac{1(x-2)}{x(x-2)} + \dfrac{2}{x(x-2)}$$

$$= \dfrac{5x+x-2+2}{x(x-2)}$$

$$= \dfrac{6x}{x(x-2)} = \dfrac{6}{x-2}$$

57. $\dfrac{3x}{x+1} + \dfrac{4}{x-1} - \dfrac{6}{x^2-1}$

$x^2-1 = (x+1)(x-1)$, the LCD.

$$= \dfrac{3x(x-1)}{(x+1)(x-1)} + \dfrac{4(x+1)}{(x-1)(x+1)}$$

$$- \dfrac{6}{(x+1)(x-1)}$$

$$= \dfrac{3x(x-1) + 4(x+1) - 6}{(x+1)(x-1)}$$

$$= \dfrac{3x^2 - 3x + 4x + 4 - 6}{(x+1)(x-1)}$$

$$= \dfrac{3x^2 + x - 2}{(x+1)(x-1)}$$

$$= \dfrac{(3x-2)(x+1)}{(x+1)(x-1)}$$

$$= \dfrac{3x-2}{x-1}$$

59. $\dfrac{4}{x+1} + \dfrac{1}{x^2-x+1} - \dfrac{12}{x^3+1}$

$x^3+1 = (x+1)(x^2-x+1)$, the LCD.

$$= \dfrac{4(x^2-x+1)}{(x+1)(x^2-x+1)}$$

$$+ \dfrac{1\cdot(x+1)}{(x^2-x+1)(x+1)}$$

$$- \dfrac{12}{(x+1)(x^2-x+1)}$$

$$= \dfrac{4(x^2-x+1) + (x+1) - 12}{(x+1)(x^2-x+1)}$$

$$= \dfrac{4x^2 - 4x + 4 + x + 1 - 12}{(x+1)(x^2-x+1)}$$

$$= \dfrac{4x^2 - 3x - 7}{(x+1)(x^2-x+1)}$$

$$= \dfrac{(4x-7)(x+1)}{(x+1)(x^2-x+1)}$$

$$= \dfrac{4x-7}{x^2-x+1}$$

61. $\dfrac{2x+4}{x+3} + \dfrac{3}{x} - \dfrac{6}{x^2+3x}$

$x^2 + 3x = x(x+3)$, the LCD.

$= \dfrac{(2x+4)x}{(x+3)x} + \dfrac{3(x+3)}{x(x+3)} - \dfrac{6}{x(x+3)}$

$= \dfrac{(2x+4)x + 3(x+3) - 6}{x(x+3)}$

$= \dfrac{2x^2 + 4x + 3x + 9 - 6}{x(x+3)}$

$= \dfrac{x^2 + 7x + 3}{x(x+3)}$

$= \dfrac{(2x+1)(x+3)}{x(x+3)} = \dfrac{2x+1}{x}$

63. $\dfrac{2}{m+1} + \dfrac{5}{m}$

(a) Use the common denominator, $m^2(m+1)$.

$\dfrac{2}{m+1} + \dfrac{5}{m} = \dfrac{2m^2}{(m+1)m^2} + \dfrac{5 \cdot m(m+1)}{m \cdot m(m+1)}$

$= \dfrac{2m^2 + 5m^2 + 5m}{m^2(m+1)}$

$= \dfrac{7m^2 + 5m}{m^2(m+1)}$

$= \dfrac{m(7m+5)}{m^2(m+1)}$

$= \dfrac{7m+5}{m(m+1)}$

(b) Use the common denominator, $m(m+1)$.

$\dfrac{2}{m+1} + \dfrac{5}{m} = \dfrac{2m}{m(m+1)} + \dfrac{5(m+1)}{m(m+1)}$

$= \dfrac{2m + 5m + 5}{m(m+1)}$

$= \dfrac{7m+5}{m(m+1)}$

(c) Use the common denominator, $m(m+1)^2$.

$\dfrac{2}{m+1} + \dfrac{5}{m}$

$= \dfrac{2 \cdot m(m+1)}{(m+1) \cdot m(m+1)} + \dfrac{5(m+1)^2}{m(m+1)^2}$

$= \dfrac{2m(m+1)}{m(m+1)^2} + \dfrac{5(m^2+2m+1)}{m(m+1)^2}$

$= \dfrac{2m^2 + 2m}{m(m+1)^2} + \dfrac{5m^2 + 10m + 5}{m(m+1)^2}$

$= \dfrac{7m^2 + 12m + 5}{m(m+1)^2}$

$= \dfrac{(7m+5)(m+1)}{m(m+1)^2}$

$= \dfrac{7m+5}{m(m+1)}$

(d) Use the common denominator, $m^2(m+1)^2$.

$\dfrac{2}{m+1} + \dfrac{5}{m}$

$= \dfrac{2}{m+1} \cdot \dfrac{m^2(m+1)}{m^2(m+1)} + \dfrac{5}{m} \cdot \dfrac{m(m+1)^2}{m(m+1)^2}$

$= \dfrac{2m^2(m+1)}{m^2(m+1)^2} + \dfrac{5m(m+1)^2}{m^2(m+1)^2}$

$= \dfrac{2m^3 + 2m^2}{m^2(m+1)^2} + \dfrac{5m(m^2+2m+1)}{m^2(m+1)^2}$

$= \dfrac{2m^3 + 2m^2}{m^2(m+1)^2} + \dfrac{5m^3 + 10m^2 + 5m}{m^2(m+1)^2}$

$= \dfrac{7m^3 + 12m^2 + 5m}{m^2(m+1)^2}$

$= \dfrac{m(7m^2 + 12m + 5)}{m^2(m+1)^2}$

$= \dfrac{m(m+1)(7m+5)}{m^2(m+1)^2}$

$= \dfrac{7m+5}{m(m+1)}$

65. $\dfrac{3}{(p-2)^2} - \dfrac{5}{p-2} + 4$

$= \dfrac{3}{(p-2)^2} - \dfrac{5(p-2)}{(p-2)^2} + \dfrac{4(p-2)^2}{(p-2)^2}$

$\qquad\qquad LCD=(p-2)^2$

$= \dfrac{3 - 5(p-2) + 4(p^2 - 4p + 4)}{(p-2)^2}$

$= \dfrac{3 - 5p + 10 + 4p^2 - 16p + 16}{(p-2)^2}$

$= \dfrac{4p^2 - 21p + 29}{(p-2)^2}$

67. $\dfrac{3}{x^2 - 5x + 6} - \dfrac{2}{x^2 - x - 2}$

$= \dfrac{3}{(x-2)(x-3)} - \dfrac{2}{(x-2)(x+1)}$

$= \dfrac{3(x+1)}{(x-2)(x-3)(x+1)}$

$\quad - \dfrac{2(x-3)}{(x-2)(x+1)(x-3)}$

$\qquad\qquad LCD=(x-2)(x-3)(x+1)$

$= \dfrac{3x + 3 - 2x + 6}{(x-2)(x-3)(x+1)}$

$= \dfrac{x+9}{(x-2)(x-3)(x+1)}$

69. $\dfrac{5x - y}{x^2 + xy - 2y^2} - \dfrac{3x + 2y}{x^2 + 5xy - 6y^2}$

Factor each denominator.

$$x^2 + xy - 2y^2 = (x + 2y)(x - y)$$
$$x^2 + 5xy - 6y^2 = (x + 6y)(x - y)$$

The LCD is $(x + 2y)(x - y)(x + 6y)$.

$$\dfrac{5x - y}{(x + 2y)(x - y)} - \dfrac{3x + 2y}{(x + 6y)(x - y)}$$

$$= \dfrac{(5x - y)(x + 6y)}{(x + 2y)(x - y)(x + 6y)}$$

$$- \dfrac{(3x + 2y)(x + 2y)}{(x + 6y)(x - y)(x + 2y)}$$

$$= \dfrac{(5x - y)(x + 6y) - (3x + 2y)(x + 2y)}{(x + 6y)(x - y)(x + 2y)}$$

$$= \dfrac{5x^2 + 29xy - 6y^2 - (3x^2 + 8xy + 4y^2)}{(x + 2y)(x - y)(x + 6y)}$$

$$= \dfrac{2x^2 + 21xy - 10y^2}{(x + 2y)(x - y)(x + 6y)}$$

71. $\dfrac{r + s}{3r^2 + 2rs - s^2} - \dfrac{s - r}{6r^2 - 5rs + s^2}$

Factor each denominator.

$$3r^2 + 2rs - s^2 = (3r - s)(r + s)$$
$$6r^2 - 5rs + s^2 = (3r - s)(2r - s)$$

The LCD is $(3r - s)(r + s)(2r - s)$.

$$\dfrac{r + s}{3r^2 + 2rs - s^2} - \dfrac{s - r}{6r^2 - 5rs + s^2}$$

$$= \dfrac{r + s}{(3r - s)(r + s)} - \dfrac{s - r}{(3r - s)(2r - s)}$$

$$= \dfrac{(r + s)(2r - s)}{(3r - s)(r + s)(2r - s)}$$

$$- \dfrac{(s - r)(r + s)}{(3r - s)(2r - s)(r + s)}$$

$$= \dfrac{(r + s)(2r - s) - (s - r)(s + r)}{(3r - s)(r + s)(2r - s)}$$

$$= \dfrac{2r^2 + rs - s^2 - (s^2 - r^2)}{(3r - s)(r + s)(2r - s)}$$

$$= \dfrac{2r^2 + rs - s^2 - s^2 + r^2}{(3r - s)(r + s)(2r - s)}$$

$$= \dfrac{3r^2 + rs - 2s^2}{(3r - s)(r + s)(2r - s)}$$

$$= \dfrac{(3r - 2s)(r + s)}{(3r - s)(r + s)(2r - s)}$$

$$= \dfrac{3r - 2s}{(3r - s)(2r - s)}$$

73. (a) $\dfrac{-x}{4xy + 3y^2} + \dfrac{8x + 6y}{16x^2 - 9y^2}$

Factor the denominators to get

$$= \dfrac{-x}{y(4x + 3y)} + \dfrac{8x + 6y}{(4x + 3y)(4x - 3y)}.$$

(b) $\dfrac{-x}{4xy + 3y^2} \cdot \dfrac{8x + 6y}{16x^2 - 9y^2}$

Factor all numerators and denominators to get

$$= \dfrac{-x}{y(4x + 3y)} \cdot \dfrac{2(4x + 3y)}{(4x + 3y)(4x - 3y)}.$$

74. (a) $\dfrac{-x}{y(4x + 3y)} + \dfrac{8x + 6y}{(4x + 3y)(4x - 3y)}$

The LCD is $y(4x + 3y)(4x - 3y)$.

$$= \dfrac{-x(4x - 3y)}{y(4x + 3y)(4x - 3y)}$$

$$+ \dfrac{y(8x + 6y)}{y(4x + 3y)(4x - 3y)}$$

(b) $\dfrac{-x}{y(4x + 3y)} \cdot \dfrac{2(4x + 3y)}{(4x + 3y)(4x - 3y)}$

$$= \dfrac{-2x(4x + 3y)}{y(4x + 3y)^2(4x - 3y)}$$

75. (a) $\dfrac{-x(4x - 3y)}{y(4x + 3y)(4x - 3y)}$

$$+ \dfrac{y(8x + 6y)}{y(4x + 3y)(4x - 3y)}$$

$$= \dfrac{-x(4x - 3y) + y(8x + 6y)}{y(4x + 3y)(4x - 3y)}$$

$$= \dfrac{-4x^2 + 3xy + 8xy + 6y^2}{y(4x + 3y)(4x - 3y)}$$

$$= \dfrac{-4x^2 + 11xy + 6y^2}{y(4x + 3y)(4x - 3y)}$$

(b) $\dfrac{-2x(4x + 3y)}{y(4x + 3y)^2(4x - 3y)}$

$$= \dfrac{-2x}{y(4x + 3y)(4x - 3y)}$$

76. When adding the rational expressions, we had to factor the denominators to find the least common denominator. Only after each expression was written with this LCD could we add the numerators and simplify the result.

Multiplying the two rational expressions was a bit easier process. After factoring both numerators and denominators, we multiplied them directly and simplified.

77. **(a)** $c(x) = \dfrac{1010}{49(101-x)} - \dfrac{10}{49}$

$= \dfrac{1010}{49(101-x)} - \dfrac{10(101-x)}{49(101-x)}$

$= \dfrac{1010 - 1010 + 10x}{49(101-x)}$

$= \dfrac{10x}{49(101-x)}$

(b) $c(95) = \dfrac{10(95)}{49(101-95)}$

$= \dfrac{950}{294} \approx 3.23$

It would cost approximately 3.23 thousand dollars to win 95 points.

Section 6.3

1. *Method 1*: Begin by simplifying the numerator to a single fraction. Then simplify the denominator to a single fraction. Write as a division problem, and multiply by the reciprocal of the denominator. *Method 2*: Find the LCD of all fractions in the complex fraction. Multiply the numerator and denominator of the complex fraction by this LCD. Simplify the result, if possible.

3. $\dfrac{\dfrac{12}{x-1}}{\dfrac{6}{x}} = \dfrac{12}{x-1} \div \dfrac{6}{x}$

Multiply by the reciprocal of the divisor.

$= \dfrac{12}{x-1} \cdot \dfrac{x}{6}$

$= \dfrac{2x}{x-1}$

5. $\dfrac{\dfrac{k+1}{2k}}{\dfrac{3k-1}{4k}} = \dfrac{k+1}{2k} \cdot \dfrac{4k}{3k-1}$

$= \dfrac{4k(k+1)}{2k(3k-1)}$

$= \dfrac{2(k+1)}{3k-1}$

7. $\dfrac{\dfrac{4z^2x^4}{9}}{\dfrac{12x^2z^5}{15}} = \dfrac{\dfrac{4z^2x^4}{9}}{\dfrac{4x^2z^5}{5}}$

$= \dfrac{4z^2x^4}{9} \div \dfrac{4x^2z^5}{5}$

$= \dfrac{4z^2x^4}{9} \cdot \dfrac{5}{4x^2z^5}$

$= \dfrac{5z^2x^4}{9x^2z^5} = \dfrac{5x^2}{9z^3}$

9. $\dfrac{\dfrac{1}{x}+1}{-\dfrac{1}{x}+1}$

Multiply the numerator and denominator by x, the LCD of all the fractions.

$= \dfrac{x\left(\dfrac{1}{x}+1\right)}{x\left(-\dfrac{1}{x}+1\right)}$

$= \dfrac{x \cdot \dfrac{1}{x} + x \cdot 1}{x\left(-\dfrac{1}{x}\right) + x \cdot 1}$

$= \dfrac{1+x}{-1+x}$

11. $\dfrac{\dfrac{3}{x}+\dfrac{3}{y}}{\dfrac{3}{x}-\dfrac{3}{y}}$

Multiply the numerator and denominator by xy, the LCD of all the fractions.

$= \dfrac{\left(\dfrac{3}{x}+\dfrac{3}{y}\right)xy}{\left(\dfrac{3}{x}-\dfrac{3}{y}\right)xy}$

$= \dfrac{\dfrac{3}{x} \cdot xy + \dfrac{3}{y} \cdot xy}{\dfrac{3}{x} \cdot xy - \dfrac{3}{y} \cdot xy}$

$= \dfrac{3y+3x}{3y-3x}$

$= \dfrac{3(y+x)}{3(y-x)}$

$= \dfrac{y+x}{y-x}$

13. $\dfrac{\dfrac{8x - 24y}{10}}{\dfrac{x - 3y}{5x}} = \dfrac{8x - 24y}{10} \cdot \dfrac{5x}{x - 3y}$

$= \dfrac{8(x - 3y)5x}{10(x - 3y)}$

$= \dfrac{40x}{10} = 4x$

15. $\dfrac{\dfrac{x^2 - 16y^2}{xy}}{\dfrac{1}{y} - \dfrac{4}{x}}$

Multiply the numerator and denominator by xy, the LCD of all the fractions.

$= \dfrac{\left(\dfrac{x^2 - 16y^2}{xy}\right)xy}{\left(\dfrac{1}{y} - \dfrac{4}{x}\right)xy}$

$= \dfrac{x^2 - 16y^2}{\dfrac{1}{y} \cdot xy - \dfrac{4}{x} \cdot xy}$

$= \dfrac{x^2 - 16y^2}{x - 4y}$

$= \dfrac{(x + 4y)(x - 4y)}{x - 4y}$

$= x + 4y$

17. $\dfrac{y - \dfrac{y - 3}{3}}{\dfrac{4}{9} + \dfrac{2}{3y}}$

Multiply the numerator and denominator by $9y$, the LCD of all the fractions.

$= \dfrac{9y\left(y - \dfrac{y - 3}{3}\right)}{9y\left(\dfrac{4}{9} + \dfrac{2}{3y}\right)}$

$= \dfrac{9y^2 - 3y(y - 3)}{4y + 6}$

$= \dfrac{9y^2 - 3y^2 + 9y}{4y + 6}$

$= \dfrac{6y^2 + 9y}{4y + 6}$

$= \dfrac{3y(2y + 3)}{2(2y + 3)} = \dfrac{3y}{2}$

19. $\dfrac{\dfrac{x + 2}{x} + \dfrac{1}{x + 2}}{\dfrac{5}{x} + \dfrac{x}{x + 2}}$

Multiply the numerator and denominator by $x(x + 2)$, the LCD of all the fractions.

$= \dfrac{x(x + 2)\left(\dfrac{x + 2}{x} + \dfrac{1}{x + 2}\right)}{x(x + 2)\left(\dfrac{5}{x} + \dfrac{x}{x + 2}\right)}$

$= \dfrac{x(x + 2)\left(\dfrac{x + 2}{x}\right) + x(x + 2)\left(\dfrac{1}{x + 2}\right)}{x(x + 2)\left(\dfrac{5}{x}\right) + x(x + 2)\left(\dfrac{x}{x + 2}\right)}$

$= \dfrac{(x + 2)(x + 2) + x}{5(x + 2) + x^2}$

$= \dfrac{x^2 + 4x + 4 + x}{5x + 10 + x^2}$

$= \dfrac{x^2 + 5x + 4}{x^2 + 5x + 10}$

21. $\dfrac{1}{x^{-2} + y^{-2}} = \dfrac{1}{\dfrac{1}{x^2} + \dfrac{1}{y^2}}$

$= \dfrac{x^2 y^2(1)}{x^2 y^2\left(\dfrac{1}{x^2} + \dfrac{1}{y^2}\right)}$ $LCD = x^2 y^2$

$= \dfrac{x^2 y^2}{y^2 + x^2}$

23. $\dfrac{x^{-2} + y^{-2}}{x^{-1} + y^{-1}}$

$= \dfrac{\dfrac{1}{x^2} + \dfrac{1}{y^2}}{\dfrac{1}{x} + \dfrac{1}{y}}$

Multiply the numerator and denominator by $x^2 y^2$, the LCD of all the fractions.

$= \dfrac{x^2 y^2\left(\dfrac{1}{x^2} + \dfrac{1}{y^2}\right)}{x^2 y^2\left(\dfrac{1}{x} + \dfrac{1}{y}\right)}$

$$= \frac{x^2 y^2 \cdot \dfrac{1}{x^2} + x^2 y^2 \cdot \dfrac{1}{y^2}}{x^2 y^2 \cdot \dfrac{1}{x} + x^2 y^2 \cdot \dfrac{1}{y}}$$

$$= \frac{y^2 + x^2}{xy^2 + x^2 y} \text{ or } \frac{y^2 + x^2}{xy(y + x)}$$

25. $\left(r^{-1} + s^{-1}\right)^{-1} = \left(\dfrac{1}{r} + \dfrac{1}{s}\right)^{-1}$

$$= \left(\frac{s + r}{rs}\right)^{-1}$$

$$= \left(\frac{rs}{s + r}\right)^{1}$$

$$= \frac{rs}{s + r}$$

27. (a) $\dfrac{\dfrac{3}{mp} - \dfrac{4}{p} + \dfrac{8}{m}}{2m^{-1} - 3p^{-1}}$

$$= \frac{\dfrac{3}{mp} - \dfrac{4}{p} + \dfrac{8}{m}}{\dfrac{2}{m} - \dfrac{3}{p}}$$

(b) $2m^{-1} = \dfrac{2}{m}$, not $\dfrac{1}{2m}$, since the exponent

applies only to m, not 2. Likewise, $3p^{-1} = \dfrac{3}{p}$,

not $\dfrac{1}{3p}$.

(c) $\dfrac{\dfrac{3}{mp} - \dfrac{4}{p} + \dfrac{8}{m}}{2m^{-1} - 3p^{-1}}$

Multiply the numerator and denominator by mp, the LCD of all the fractions.

$$= \frac{mp\left(\dfrac{3}{mp} - \dfrac{4}{p} + \dfrac{8}{m}\right)}{mp\left(\dfrac{2}{m} - \dfrac{3}{p}\right)}$$

$$= \frac{mp \cdot \dfrac{3}{mp} - mp \cdot \dfrac{4}{p} + mp \cdot \dfrac{8}{m}}{mp \cdot \dfrac{2}{m} - mp \cdot \dfrac{3}{p}}$$

$$= \frac{3 - 4m + 8p}{2p - 3m}$$

29. $1 - \dfrac{3}{3 - \dfrac{1}{2y}}$

$$= 1 - \frac{3}{\left(3 - \dfrac{1}{2y}\right)} \cdot \frac{2y}{2y}$$

$$= 1 - \frac{6y}{6y - 1}$$

$$= \frac{6y - 1}{6y - 1} - \frac{6y}{6y - 1}$$

$$= \frac{6y - 1 - 6y}{6y - 1}$$

$$= \frac{-1}{6y - 1} \text{ or } \frac{1}{1 - 6y}$$

31. $\dfrac{1}{p + \dfrac{1}{p + \dfrac{1}{1 + p}}}$

Simplify the least complex fraction first.

$$= \frac{1}{p + \dfrac{1}{\dfrac{p(1 + p)}{1 + p} + \dfrac{1}{1 + p}}}$$

$$= \frac{1}{p + \dfrac{1}{\dfrac{p + p^2 + 1}{1 + p}}}$$

$$= \frac{1}{p + \dfrac{1 + p}{p + p^2 + 1}}$$

$$= \frac{1}{\dfrac{p(p + p^2 + 1)}{p + p^2 + 1} + \dfrac{1 + p}{p + p^2 + 1}}$$

$$= \frac{1}{\dfrac{p^2 + p^3 + p + 1 + p}{p + p^2 + 1}}$$

$$= \frac{p + p^2 + 1}{p^3 + p^2 + 2p + 1}$$

In Exercises 33–38, use the complex fraction

$$\dfrac{\dfrac{4}{m} + \dfrac{m+2}{m-1}}{\dfrac{m+2}{m} - \dfrac{2}{m-1}}.$$

33. To add the fractions in the numerator, use the LCD $m(m-1)$.

$$\dfrac{4}{m} + \dfrac{m+2}{m-1} = \dfrac{4(m-1)}{m(m-1)} + \dfrac{m(m+2)}{m(m-1)}$$
$$= \dfrac{4m - 4 + m^2 + 2m}{m(m-1)}$$
$$= \dfrac{m^2 + 6m - 4}{m(m-1)}$$

34. To subtract the fractions in the denominator, use the same LCD, $m(m-1)$.

$$\dfrac{m+2}{m} - \dfrac{2}{m-1}$$
$$= \dfrac{(m+2)(m-1)}{m(m-1)} - \dfrac{m \cdot 2}{m(m-1)}$$
$$= \dfrac{m^2 + m - 2 - 2m}{m(m-1)}$$
$$= \dfrac{m^2 - m - 2}{m(m-1)}$$

35.

Exercise 33	Exercise 34
answer	answer
$\downarrow$	$\downarrow$

$$\dfrac{m^2 + 6m - 4}{m(m-1)} \div \dfrac{m^2 - m - 2}{m(m-1)}$$

Multiply by the reciprocal.

$$= \dfrac{m^2 + 6m - 4}{m(m-1)} \cdot \dfrac{m(m-1)}{m^2 - m - 2}$$
$$= \dfrac{m^2 + 6m - 4}{m^2 - m - 2}$$

36. The LCD of all the denominators in the complex fraction is $m(m-1)$.

37.

$$\dfrac{\left(\dfrac{4}{m} + \dfrac{m+2}{m-1}\right) \cdot m(m-1)}{\left(\dfrac{m+2}{m} - \dfrac{2}{m-1}\right) \cdot m(m-1)}$$
$$= \dfrac{4(m-1) + m(m+2)}{(m+2)(m-1) - 2m}$$
$$= \dfrac{4m - 4 + m^2 + 2m}{m^2 + m - 2 - 2m}$$
$$= \dfrac{m^2 + 6m - 4}{m^2 - m - 2}$$

38. Answers will vary. Because of the complicated nature of the numerator and denominator of the complex fraction, Method 1 takes much longer to simplify the complex fraction. Method 2 is a simpler, more direct means of simplifying and is most likely the preferred method.

Section 6.4

1. We find the quotient of two monomials by using the _quotient_ rule for _exponents_.

3. When dividing polynomials that are not monomials, first write them in _descending powers_.

5.
$$\dfrac{9y^2 + 12y - 15}{3y} = \dfrac{9y^2}{3y} + \dfrac{12y}{3y} - \dfrac{15}{3y}$$
$$= 3y + 4 - \dfrac{5}{y}$$

7.
$$\dfrac{15m^3 + 25m^2 + 30m}{5m^2}$$
$$= \dfrac{15m^3}{5m^2} + \dfrac{25m^2}{5m^2} + \dfrac{30m}{5m^2}$$
$$= 3m + 5 + \dfrac{6}{m}$$

9.
$$\dfrac{14m^2n^2 - 21mn^3 + 28m^2n}{14m^2n}$$
$$= \dfrac{14m^2n^2}{14m^2n} - \dfrac{21mn^3}{14m^2n} + \dfrac{28m^2n}{14m^2n}$$
$$= n - \dfrac{3n^2}{2m} + 2$$

11.
$$\dfrac{8wxy^2 + 3wx^2y + 12w^2xy}{4wx^2y}$$
$$= \dfrac{8wxy^2}{4wx^2y} + \dfrac{3wx^2y}{4wx^2y} + \dfrac{12w^2xy}{4wx^2y}$$
$$= \dfrac{2y}{x} + \dfrac{3}{4} + \dfrac{3w}{x}$$

13.

$$\begin{array}{r}
r^2 \phantom{{}+{}} - 7r \phantom{{}+{}} + 6 \\
3r - 1 \overline{\smash{)}\, 3r^3 \phantom{{}+{}} - 22r^2 \phantom{{}+{}} + 25r \phantom{{}+{}} - 6} \\
\underline{3r^3 \phantom{{}+{}} - r^2 \phantom{{}+{}}} \\
-21r^2 \phantom{{}+{}} + 25r \\
\underline{-21r^2 \phantom{{}+{}} + 7r} \\
18r \phantom{{}+{}} - 6 \\
\underline{18r \phantom{{}+{}} - 6} \\
0
\end{array}$$

Answer: $r^2 - 7r + 6$

15.

$$
\begin{array}{r}
y \;-\; 3 \\
y + 6\overline{\smash{\big)}\,y^2 \;+\; 3y \;-\; 18} \\
\underline{y^2 \;+\; 6y} \\
-3y \;-\; 18 \\
\underline{-3y \;-\; 18} \\
0
\end{array}
$$

Answer: $y - 3$

17.

$$
\begin{array}{r}
t \;+\; 5 \\
3t + 2\overline{\smash{\big)}\,3t^2 \;+\; 17t \;+\; 10} \\
\underline{3t^2 \;+\; 2t} \\
15t \;+\; 10 \\
\underline{15t \;+\; 10} \\
0
\end{array}
$$

Answer: $t + 5$

19.

$$
\begin{array}{r}
z^2 \;+\; 3 \\
2z - 5\overline{\smash{\big)}\,2z^3 \;-\; 5z^2 \;+\; 6z \;-\; 15} \\
\underline{2z^3 \;-\; 5z^2} \\
6z \;-\; 15 \\
\underline{6z \;-\; 15} \\
0
\end{array}
$$

Answer: $z^2 + 3$

21.

$$
\begin{array}{r}
x^2 \;+\; 2x \;-\; 3 \\
4x + 1\overline{\smash{\big)}\,4x^3 \;+\; 9x^2 \;-\; 10x \;+\; 3} \\
\underline{4x^3 \;+\; x^2} \\
8x^2 \;-\; 10x \\
\underline{8x^2 \;+\; 2x} \\
-12x \;+\; 3 \\
\underline{-12x \;-\; 3} \\
6
\end{array}
$$

Remainder

Answer: $x^2 + 2x - 3 + \dfrac{6}{4x+1}$

23.

$$
\begin{array}{r}
2x \;-\; 5 \\
3x^2 - 2x + 4\overline{\smash{\big)}\,6x^3 \;-\; 19x^2 \;+\; 14x \;-\; 15} \\
\underline{6x^3 \;-\; 4x^2 \;+\; 8x} \\
-15x^2 \;+\; 6x \;-\; 15 \\
\underline{-15x^2 \;+\; 10x \;-\; 20} \\
-4x \;+\; 5
\end{array}
$$

Remainder

Answer: $2x - 5 + \dfrac{-4x+5}{3x^2-2x+4}$

25.

$$
\begin{array}{r}
2k^2 \;+\; 3k \;-\; 1 \\
2k^2 + 1\overline{\smash{\big)}\,4k^4 \;+\; 6k^3 \;+\; 0k^2 \;+\; 3k \;-\; 1} \\
\underline{4k^4 \;+\; 2k^2} \\
6k^3 \;-\; 2k^2 \;+\; 3k \\
\underline{6k^3 \;+\; 3k} \\
-2k^2 \;-\; 1 \\
\underline{-2k^2 \;-\; 1} \\
0
\end{array}
$$

Answer: $2k^2 + 3k - 1$

27.

$$
\begin{array}{r}
9z^2 \;-\; 4z \;+\; 1 \\
z^2 - z + 2\overline{\smash{\big)}\,9z^4 \;-\; 13z^3 \;+\; 23z^2 \;-\; 10z \;+\; 8} \\
\underline{9z^4 \;-\; 9z^3 \;+\; 18z^2} \\
-4z^3 \;+\; 5z^2 \;-\; 10z \\
\underline{-4z^3 \;+\; 4z^2 \;-\; 8z} \\
z^2 \;-\; 2z \;+\; 8 \\
\underline{z^2 \;-\; z \;+\; 2} \\
-z \;+\; 6
\end{array}
$$

Remainder

Answer: $9z^2 - 4z + 1 + \dfrac{-z+6}{z^2-z+2}$

29.

$$
\begin{array}{r}
p^2 \;+\; p \;+\; 1 \\
p - 1\overline{\smash{\big)}\,p^3 \;+\; 0p^2 \;+\; 0p \;-\; 1} \\
\underline{p^3 \;-\; p^2} \\
p^2 \;+\; 0p \\
\underline{p^2 \;-\; p} \\
p \;-\; 1 \\
\underline{p \;-\; 1} \\
0
\end{array}
$$

Answer: $p^2 + p + 1$

31. $\dfrac{P}{Q}(x) = \dfrac{P(x)}{Q(x)}$

$$
\begin{array}{r}
2x \;+\; 7 \\
x^2 + 4x\overline{\smash{\big)}\,2x^3 \;+\; 15x^2 \;+\; 28x \;+\; 0} \\
\underline{2x^3 \;+\; 8x^2} \\
7x^2 \;+\; 28x \\
\underline{7x^2 \;+\; 28x} \\
0
\end{array}
$$

Answer: $2x + 7$

33. $\left(4a^3 + 5a^2 + 4a + 5\right) \div (4a + 5)$

$= a^2 + a + 1$

To check this division problem, multiply the quotient by the divisor.

$\left(a^2 + a + 1\right)(4a + 5)$

$= a^2(4a + 5) + a(4a + 5) + 1(4a + 5)$

$= 4a^3 + 5a^2 + 4a^2 + 5a + 4a + 5$

$= 4a^3 + 9a^2 + 9a + 5$

Since we did not get the dividend, $4a^3 + 5a^2 + 4a + 5$, the quotient is incorrect. Do the division problem.

$$
\begin{array}{r}
a^2 \qquad\ + 1 \\
4a + 5 \overline{\smash{\big)}\ 4a^3 + 5a^2 + 4a + 5} \\
\underline{4a^3 + 5a^2} \qquad\qquad \\
4a + 5 \\
\underline{4a + 5} \\
0
\end{array}
$$

The quotient is $a^2 + 1$. Multiply to check.

$\left(a^2 + 1\right)(4a + 5) = 4a^3 + 5a^2 + 4a + 5$

This is the dividend, so the correct quotient is $a^2 + 1$.

35. To start: $\dfrac{2x^2}{3x} = \dfrac{2}{3}x$

$$
\begin{array}{r}
\frac{2}{3}x - 1 \\
3x + 1 \overline{\smash{\big)}\ 2x^2 - \frac{7}{3}x - 1} \\
\underline{2x^2 + \frac{2}{3}x} \qquad\quad \\
-3x - 1 \\
\underline{-3x - 1} \\
0
\end{array}
$$

Answer: $\dfrac{2}{3}x - 1$

37.

$$
\begin{array}{r}
\frac{3}{4}a - 2 \\
4a + 3 \overline{\smash{\big)}\ 3a^2 - \frac{23}{4}a - 5} \\
\underline{3a^2 + \frac{9}{4}a} \qquad\quad \\
-8a - 5 \\
\underline{-8a - 6} \\
1
\end{array}
$$

Remainder

Answer: $\dfrac{3}{4}a - 2 + \dfrac{1}{4a + 3}$

39. The volume of a box is the product of the height, length, and width. Use the formula $V = LWH$.

$$V = LWH$$

$$\frac{V}{LH} = W$$

Here,

$L \cdot H = (p + 4)p = p^2 + 4p$, so

$$W = \frac{V}{LH} = \frac{2p^3 + 15p^2 + 28p}{p^2 + 4p}.$$

$$
\begin{array}{r}
2p + 7 \\
p^2 + 4p \overline{\smash{\big)}\ 2p^3 + 15p^2 + 28p} \\
\underline{2p^3 + 8p^2} \qquad\qquad \\
7p^2 + 28p \\
\underline{7p^2 + 28p} \\
0
\end{array}
$$

The width is $2p + 7$.

41. $P(x) = x^3 - 4x^2 + 3x - 5$

$P(-1) = (-1)^3 - 4(-1)^2 + 3(-1) - 5$

$\qquad = -1 - 4 - 3 - 5 = -13$

Now divide the given polynomial by $x + 1$.

$$
\begin{array}{r}
x^2 - 5x + 8 \\
x + 1 \overline{\smash{\big)}\ x^3 - 4x^2 + 3x - 5} \\
\underline{x^3 + x^2} \qquad\qquad\quad \\
-5x^2 + 3x \\
\underline{-5x^2 - 5x} \\
8x - 5 \\
\underline{8x + 8} \\
-13
\end{array}
$$

Remainder

The remainder in the division is the same as $P(-1)$, -13. This suggests that if a polynomial is divided by $x - r$, in this case $x - (-1)$ or $x + 1$, then the remainder is equal to $P(r)$, in this case $P(-1)$.

43. $\left(\dfrac{f}{g}\right)(x) = \dfrac{f(x)}{g(x)} = \dfrac{10x^2 - 2x}{2x}$

$\qquad\qquad\quad = \dfrac{10x^2}{2x} - \dfrac{2x}{2x}$

$\qquad\qquad\quad = 5x - 1$

The x-values that are not in the domain of the quotient function g are found by solving $g(x) = 0$.

$$2x = 0$$

$$x = 0$$

45.

$$
\begin{array}{r}
2x - 3 \\
x + 1 \overline{\smash{\big)}\ 2x^2 - x - 3} \\
\underline{2x^2 + 2x} \\
-3x - 3 \\
\underline{-3x - 3} \\
0
\end{array}
$$

Quotient: $2x - 3$

$$
g(x) = 0
$$
$$
x + 1 = 0
$$
$$
x = -1
$$

47.

$$
\begin{array}{r}
4x^2 + 6x + 9 \\
2x - 3 \overline{\smash{\big)}\ 8x^3 + 0x^2 + 0x - 27} \\
\underline{8x^3 - 12x^2} \\
12x^2 \\
\underline{12x^2 - 18x} \\
18x - 27 \\
\underline{18x - 27} \\
0
\end{array}
$$

Quotient: $4x^2 + 6x + 9$

$$
g(x) = 0
$$
$$
2x - 3 = 0
$$
$$
2x = 3
$$
$$
x = \frac{3}{2}
$$

Section 6.5

1. Synthetic division provides a quick, easy way to divide a polynomial by a binomial of the form $x - k$.

3. $\dfrac{x^2 - 6x + 5}{x - 1}$

$$
\begin{array}{r|rrr}
1 & 1 & -6 & 5 \\
  &   & 1 & -5 \\
\hline
  & 1 & -5 & 0
\end{array}
$$
$\leftarrow$ *Coefficients of numerator*

Write the answer from the bottom row.

$$
\begin{array}{cc}
\downarrow & \downarrow \\
x & -5
\end{array}
$$

Answer: $x - 5$

5. $\dfrac{4m^2 + 19m - 5}{m + 5}$

$m + 5 = m - (-5)$, so use -5.

$$
\begin{array}{r|rrr}
-5 & 4 & 19 & -5 \\
   &   & -20 & 5 \\
\hline
   & 4 & -1 & 0
\end{array}
$$

Answer: $4m - 1$

7. $\dfrac{2a^2 + 8a + 13}{a + 2}$

$a + 2 = a - (-2)$, so use -2.

$$
\begin{array}{r|rrr}
-2 & 2 & 8 & 13 \\
   &   & -4 & -8 \\
\hline
   & 2 & 4 & 5
\end{array}
$$
$\leftarrow$ *Remainder*

Answer: $2a + 4 + \dfrac{5}{a + 2}$

9. $(p^2 - 3p + 5) \div (p + 1)$

$$
\begin{array}{r|rrr}
-1 & 1 & -3 & 5 \\
   &   & -1 & 4 \\
\hline
   & 1 & -4 & 9
\end{array}
$$

Answer: $p - 4 + \dfrac{9}{p + 1}$

11. $\dfrac{4a^3 - 3a^2 + 2a - 3}{a - 1}$

$$
\begin{array}{r|rrrr}
1 & 4 & -3 & 2 & -3 \\
  &   & 4 & 1 & 3 \\
\hline
  & 4 & 1 & 3 & 0
\end{array}
$$

Answer: $4a^2 + a + 3$

13. $(x^5 - 2x^3 + 3x^2 - 4x - 2) \div (x - 2)$

Insert 0 for the missing x^4-term.

$$
\begin{array}{r|rrrrrr}
2 & 1 & 0 & -2 & 3 & -4 & -2 \\
  &   & 2 & 4 & 4 & 14 & 20 \\
\hline
  & 1 & 2 & 2 & 7 & 10 & 18
\end{array}
$$
$\leftarrow$ *Remainder*

Answer: $x^4 + 2x^3 + 2x^2 + 7x + 10 + \dfrac{18}{x - 2}$

15. $(-4r^6 - 3r^5 - 3r^4 + 5r^3 - 6r^2 + 3r + 3) \div (r - 1)$

$$
\begin{array}{r|rrrrrrr}
1 & -4 & -3 & -3 & 5 & -6 & 3 & 3 \\
  &   & -4 & -7 & -10 & -5 & -11 & -8 \\
\hline
  & -4 & -7 & -10 & -5 & -11 & -8 & -5
\end{array}
$$
$\leftarrow$ *Remainder*

Answer:

$-4r^5 - 7r^4 - 10r^3 - 5r^2 - 11r - 8 + \dfrac{-5}{r - 1}$

17. $(-3y^5 + 2y^4 - 5y^3 - 6y^2 - 1) \div (y + 2)$

Insert 0 for the missing y-term.

$$
\begin{array}{r|rrrrrr}
-2 & -3 & 2 & -5 & -6 & 0 & -1 \\
   &   & 6 & -16 & 42 & -72 & 144 \\
\hline
   & -3 & 8 & -21 & 36 & -72 & 143
\end{array}
$$
$\leftarrow$ *Remainder*

Answer:

$-3y^4 + 8y^3 - 21y^2 + 36y - 72 + \dfrac{143}{y + 2}$

19. $\dfrac{y^3 + 1}{y - 1} = \dfrac{y^3 + 0y^2 + 0y + 1}{y - 1}$

$$
\begin{array}{r|rrrr}
1 & 1 & 0 & 0 & 1 \\
  &   & 1 & 1 & 1 \\
\hline
  & 1 & 1 & 1 & 2
\end{array}
\quad \leftarrow \;\; Remainder
$$

Answer: $y^2 + y + 1 + \dfrac{2}{y - 1}$

21. $P(x) = 2x^3 - 4x^2 + 5x - 3; \; k = 2$

To find $P(2)$, divide the polynomial by $x - 2$. $P(2)$ will be the remainder.

$$
\begin{array}{r|rrrr}
2 & 2 & -4 & 5 & -3 \\
  &   & 4 & 0 & 10 \\
\hline
  & 2 & 0 & 5 & 7
\end{array}
\quad \leftarrow \;\; Remainder
$$

By the remainder theorem, $P(2) = 7$.

23. $P(r) = -r^3 - 5r^2 - 4r - 2; \; k = -4$

Divide by $r + 4$. The remainder is equal to $P(-4)$.

$$
\begin{array}{r|rrrr}
-4 & -1 & -5 & -4 & -2 \\
   &   & 4 & 4 & 0 \\
\hline
   & -1 & -1 & 0 & -2
\end{array}
\quad \leftarrow \;\; Remainder
$$

By the remainder theorem, $P(-4) = -2$.

25. $P(y) = 2y^3 - 4y^2 + 5y - 33; \; k = 3$

Divide by $y - 3$. The remainder is equal to $P(3)$.

$$
\begin{array}{r|rrrr}
3 & 2 & -4 & 5 & -33 \\
  &   & 6 & 6 & 33 \\
\hline
  & 2 & 2 & 11 & 0
\end{array}
\quad \leftarrow \;\; Remainder
$$

By the remainder theorem, $P(3) = 0$.

27. By the remainder theorem, a zero remainder means that $P(k) = 0$; that is, k is a number that makes $P(x) = 0$.

29. Is $x = -2$ a solution of

$$x^3 - 2x^2 - 3x + 10 = 0?$$

To decide whether -2 is a solution to the given equation, divide the polynomial by $x + 2$.

$$
\begin{array}{r|rrrr}
-2 & 1 & -2 & -3 & 10 \\
   &   & -2 & 8 & -10 \\
\hline
   & 1 & -4 & 5 & 0
\end{array}
\quad \leftarrow \;\; Remainder
$$

Since the remainder is 0, -2 is a solution of the equation.

31. Is $m = -2$ a solution of

$$m^4 + 2m^3 - 3m^2 + 8m - 8 = 0?$$

To decide whether -2 is a solution to the given equation, divide the polynomial by $m + 2$.

$$
\begin{array}{r|rrrrr}
-2 & 1 & 2 & -3 & 8 & -8 \\
   &   & -2 & 0 & 6 & -28 \\
\hline
   & 1 & 0 & -3 & 14 & -36
\end{array}
\quad \leftarrow \;\; Remainder
$$

Since the remainder is not 0, -2 is not a solution of the equation.

33. Is $a = -2$ a solution of

$$3a^3 + 2a^2 - 2a + 11 = 0?$$

$$
\begin{array}{r|rrrr}
-2 & 3 & 2 & -2 & 11 \\
   &   & -6 & 8 & -12 \\
\hline
   & 3 & -4 & 6 & -1
\end{array}
\quad \leftarrow \;\; Remainder
$$

Since the remainder is not 0, -2 is not a solution of the equation.

35. Is $x = -3$ a solution of

$$2x^3 - x^2 - 13x + 24 = 0?$$

$$
\begin{array}{r|rrrr}
-3 & 2 & -1 & -13 & 24 \\
   &   & -6 & 21 & -24 \\
\hline
   & 2 & -7 & 8 & 0
\end{array}
\quad \leftarrow \;\; Remainder
$$

Since the remainder is 0, -3 is a solution of the equation.

In Exercises 37–41,

$$P(x) = 2x^2 + 5x - 12.$$

37. Factor $P(x)$.

$$2x^2 + 5x - 12 = (2x - 3)(x + 4)$$

38. Solve $P(x) = 0$.

$$2x^2 + 5x - 12 = 0$$
$$(2x - 3)(x + 4) = 0$$

$2x - 3 = 0 \quad$ or $\quad x + 4 = 0$

$2x = 3 \qquad\qquad\quad x = -4$

$x = \dfrac{3}{2}$

Solution set: $\left\{ \dfrac{3}{2}, -4 \right\}$

39. $P(-4) = 2(-4)^2 + 5(-4) - 12$
$$= 2(16) - 20 - 12$$
$$= 32 - 20 - 12 = 0$$

$$P\left(\frac{3}{2}\right) = 2\left(\frac{3}{2}\right)^2 + 5\left(\frac{3}{2}\right) - 12$$
$$= 2\left(\frac{9}{4}\right) + \frac{15}{2} - 12$$
$$= \frac{9}{2} + \frac{15}{2} - \frac{24}{2} = 0$$

40. If $P(a) = 0$, then $x - \underline{a}$ is a factor of $P(x)$.

41. $Q(x) = 3x^3 - 4x^2 - 17x + 6$
$Q(3) = 3(3)^3 - 4(3)^2 - 17(3) + 6$
$$= 81 - 36 - 51 + 6 = 0$$

Since $Q(3) = 0$, $x - 3$ is a factor of $Q(x)$. To check, use synthetic division to see if 3 is a solution of the equation.

$$\begin{array}{r|rrrr} 3 & 3 & -4 & -17 & 6 \\ & & 9 & 15 & -6 \\ \hline & 3 & 5 & -2 & 0 \end{array}$$

Therefore, $x - 3$ is a factor of the polynomial and

$$3x^3 - 4x^2 - 17x + 6 = (x - 3)(3x^2 + 5x - 2)$$
$$Q(x) = (x - 3)(3x - 1)(x + 2).$$

43. From the graph, it appears that $x = 3$ is a solution of the equation

$$x^3 - x^2 - 21x + 45 = 0.$$

Check this with synthetic division.

$$\begin{array}{r|rrrr} 3 & 1 & -1 & -21 & 45 \\ & & 3 & 6 & -45 \\ \hline & 1 & 2 & -15 & 0 \end{array} \leftarrow \textit{Remainder}$$

Since the remainder is 0, 3 is a solution of the equation and $x - 3$ is a factor of the polynomial. From the last row, we see that the other factor is $x^2 + 2x - 15$, so

$$x^3 - x^2 - 21x + 45$$
$$= (x - 3)(x^2 + 2x - 15)$$
$$= (x - 3)(x - 3)(x + 5)$$
$$= (x - 3)^2(x + 5).$$

45. From the graph, it appears that $x = -1$ is a solution of the equation

$$x^3 + 3x^2 - 13x - 15 = 0.$$

Check this with synthetic division.

$$\begin{array}{r|rrrr} -1 & 1 & 3 & -13 & -15 \\ & & -1 & -2 & 15 \\ \hline & 1 & 2 & -15 & 0 \end{array} \leftarrow \textit{Remainder}$$

Since the remainder is 0, -1 is a solution of the equation and $x + 1$ is a factor of the polynomial. From the last row, we see that the other factor is $x^2 + 2x - 15$, so

$$x^3 + 3x^2 - 13x - 15$$
$$= (x + 1)(x^2 + 2x - 15)$$
$$= (x + 1)(x - 3)(x + 5).$$

Section 6.6

1. $\dfrac{4}{x} + \dfrac{5}{x - 1} = 10$

The expression we are looking for is the LCD. In this case, the LCD is

$$x(x - 1).$$

3. $\dfrac{m}{m + 1} - \dfrac{2}{m + 2} = \dfrac{m - 1}{m + 1}$

LCD $= (m + 1)(m + 2)$

5. Because -1 makes the denominators of $m + 1$ equal 0, it is not in the domain of the equation. The other number excluded from the domain is -2.

7. $\dfrac{1}{x + 1} - \dfrac{1}{x - 2} = 0$

Set each denominator equal to 0 and solve.

$$\begin{array}{ccc} x + 1 = 0 & \text{or} & x - 2 = 0 \\ x = -1 & \text{or} & x = 2 \end{array}$$

Solutions of -1 and 2 would be rejected since these values would make a denominator of the original equation equal to 0.

9. $\dfrac{5}{3x + 5} - \dfrac{1}{x} = \dfrac{1}{2x + 3}$

Set each denominator equal to 0 and solve.

$$\begin{array}{ccccc} 3x + 5 = 0 & \text{or} & x = 0 & \text{or} & 2x + 3 = 0 \\ 3x = -5 & & & & 2x = -3 \\ x = -\dfrac{5}{3} & & & & x = -\dfrac{3}{2} \end{array}$$

So $-\dfrac{5}{3}$, 0, and $-\dfrac{3}{2}$ would have to be rejected as potential solutions.

11. $\dfrac{3x+1}{x-4} = \dfrac{6x+5}{2x-7}$

Set each denominator equal to 0 and solve.

$$x - 4 = 0 \quad \text{or} \quad 2x - 7 = 0$$
$$x = 4 \qquad\qquad 2x = 7$$
$$x = \frac{7}{2}$$

So 4 and $\dfrac{7}{2}$ would have to be rejected as potential solutions.

In Exercises 13–33, check each potential solution in the original equation.

13. $\dfrac{4}{x} - \dfrac{6}{x} = \dfrac{2}{3}$

Multiply by the LCD, $3x$. Note that $x \neq 0$.

$$3x\left(\frac{4}{x} - \frac{6}{x}\right) = 3x\left(\frac{2}{3}\right)$$
$$3 \cdot 4 - 3 \cdot 6 = x \cdot 2$$
$$12 - 18 = 2x$$
$$-6 = 2x$$
$$-3 = x$$

Check $x = -3$: $\quad -\dfrac{4}{3} + 2 = \dfrac{2}{3}$ *True*

Solution set: $\{-3\}$

15. $\dfrac{x+8}{5} = \dfrac{6+x}{3}$

Multiply by the LCD, 15.

$$15\left(\frac{x+8}{5}\right) = 15\left(\frac{6+x}{3}\right)$$
$$3(x+8) = 5(6+x)$$
$$3x + 24 = 30 + 5x$$
$$-2x = 6$$
$$x = -3$$

Check $x = -3$: $1 = 1$ *True*

Solution set: $\{-3\}$

17. $\dfrac{3x+1}{x-4} = \dfrac{6x+5}{2x-7}$

Multiply by the LCD, $(x-4)(2x-7)$.

Note that $x \neq 4$ and $x \neq \dfrac{7}{2}$.

$$(x-4)(2x-7)\left(\frac{3x+1}{x-4}\right)$$
$$= (x-4)(2x-7)\left(\frac{6x+5}{2x-7}\right)$$
$$(2x-7)(3x+1) = (x-4)(6x+5)$$
$$6x^2 - 19x - 7 = 6x^2 - 19x - 20$$
$$-7 = -20 \quad \textit{False}$$

The false statement indicates that the original

equation has no solution.

Solution set: $\emptyset$

19. $\dfrac{-5}{2x} + \dfrac{3}{4x} = \dfrac{-7}{4}$

Multiply by the LCD, $4x$. $(x \neq 0)$

$$4x\left(\frac{-5}{2x} + \frac{3}{4x}\right) = 4x\left(\frac{-7}{4}\right)$$
$$2(-5) + 1(3) = x(-7)$$
$$-10 + 3 = -7x$$
$$-7 = -7x$$
$$1 = x$$

Check $x = 1$: $\quad -\dfrac{5}{2} + \dfrac{3}{4} = -\dfrac{7}{4}$ *True*

Solution set: $\{1\}$

21. $x - \dfrac{24}{x} = -2$

Multiply by the LCD, x. $(x \neq 0)$

$$x\left(x - \frac{24}{x}\right) = -2 \cdot x$$
$$x^2 - 24 = -2x$$
$$x^2 + 2x - 24 = 0$$
$$(x+6)(x-4) = 0$$

$$x + 6 = 0 \quad \text{or} \quad x - 4 = 0$$
$$x = -6 \quad \text{or} \qquad x = 4$$

Check $x = -6$: $-6 + 4 = -2$ *True*

Check $x = 4$: $4 - 6 = -2$ *True*

Solution set: $\{-6, 4\}$

23. $\dfrac{1}{y-1} + \dfrac{5}{12} = \dfrac{-4}{3y-3}$

$$\frac{1}{y-1} + \frac{5}{12} = \frac{-4}{3(y-1)}$$

Multiply by the LCD, $12(y-1)$. $(y \neq 1)$

$$12(y-1)\left(\frac{1}{y-1} + \frac{5}{12}\right)$$
$$= 12(y-1)\left(\frac{-4}{3(y-1)}\right)$$

$$12 + 5(y-1) = -16$$
$$12 + 5y - 5 = -16$$
$$5y + 7 = -16$$
$$5y = -23$$
$$y = -\frac{23}{5}$$

A calculator check is suggested.

Check $y = -\dfrac{23}{5}$: $\dfrac{5}{21} = \dfrac{5}{21}$ *True*

Solution set: $\left\{-\dfrac{23}{5}\right\}$

25.

$$\dfrac{3}{k+2} - \dfrac{2}{k^2-4} = \dfrac{1}{k-2}$$

$$\dfrac{3}{k+2} - \dfrac{2}{(k+2)(k-2)} = \dfrac{1}{k-2}$$

Multiply by the LCD, $(k+2)(k-2)$.

$(k \neq -2, 2)$

$$(k+2)(k-2)\left(\dfrac{3}{k+2} - \dfrac{2}{(k+2)(k-2)}\right)$$

$$= (k+2)(k-2)\left(\dfrac{1}{k-2}\right)$$

$$3(k-2) - 2 = k+2$$

$$3k - 6 - 2 = k+2$$

$$3k - 8 = k+2$$

$$2k = 10$$

$$k = 5$$

Check $k = 5$: $\dfrac{1}{3} = \dfrac{1}{3}$ *True*

Solution set: $\{5\}$

27.

$$\dfrac{1}{t+3} + \dfrac{4}{t+5} = \dfrac{2}{t^2+8t+15}$$

$$\dfrac{1}{t+3} + \dfrac{4}{t+5} = \dfrac{2}{(t+5)(t+3)}$$

Multiply by the LCD, $(t+5)(t+3)$.

$(t \neq -5, -3)$

$$1(t+5) + 4(t+3) = 2$$

$$t + 5 + 4t + 12 = 2$$

$$5t + 17 = 2$$

$$5t = -15$$

$$t = -3$$

But t cannot equal -3 because that would make the denominator $t+3$ equal to 0. Since division by 0 is undefined, the equation has no solution.

Solution set: $\emptyset$

29.

$$\dfrac{2x}{x-3} + \dfrac{4}{x+3} = \dfrac{-24}{x^2-9}$$

$$\dfrac{2x}{x-3} + \dfrac{4}{x+3} = \dfrac{-24}{(x+3)(x-3)}$$

Multiply by the LCD, $(x+3)(x-3)$.

$(x \neq -3, 3)$

$$2x(x+3) + 4(x-3) = -24$$

$$2x^2 + 6x + 4x - 12 = -24$$

$$2x^2 + 10x - 12 = -24$$

$$2x^2 + 10x + 12 = 0$$

$$x^2 + 5x + 6 = 0$$

$$(x+3)(x+2) = 0$$

$$x + 3 = 0 \quad \text{or} \quad x + 2 = 0$$

$$x = -3 \quad \text{or} \quad x = -2$$

But x cannot equal -3, so we only need to check -2.

Check $x = -2$: $\dfrac{24}{5} = \dfrac{24}{5}$ *True*

Solution set: $\{-2\}$

31.

$$\dfrac{7}{x-4} + \dfrac{3}{x} = \dfrac{-12}{x^2-4x}$$

$$\dfrac{7}{x-4} + \dfrac{3}{x} = \dfrac{-12}{x(x-4)}$$

Multiply by the LCD, $x(x-4)$. $(x \neq 0, 4)$

$$7(x) + 3(x-4) = -12$$

$$7x + 3x - 12 = -12$$

$$10x = 0$$

$$x = 0$$

But $x \neq 0$, so the there is no solution.

Solution set: $\emptyset$

33.

$$\dfrac{4x-7}{4x^2-9} = \dfrac{-2x^2+5x-4}{4x^2-9} + \dfrac{x+1}{2x+3}$$

$$\dfrac{4x+7}{(2x+3)(2x-3)}$$

$$= \dfrac{-2x^2+5x-4}{(2x+3)(2x-3)} + \dfrac{x+1}{2x+3}$$

Multiply by the LCD, $(2x+3)(2x-3)$.

$$\left(x \neq -\dfrac{3}{2}, \dfrac{3}{2}\right)$$

$$4x - 7 = -2x^2 + 5x - 4 + (2x-3)(x+1)$$

$$4x - 7 = -2x^2 + 5x - 4 + 2x^2 - x - 3$$

$$4x - 7 = 4x - 7 \quad \text{*True*}$$

This equation is true for every real number value of x, but we have already determined that

$x \neq -\dfrac{3}{2}$ or $x \neq \dfrac{3}{2}$. So every real number except

$-\dfrac{3}{2}$ and $\dfrac{3}{2}$ is a solution.

Solution set:

$$\left(-\infty, -\dfrac{3}{2}\right) \cup \left(-\dfrac{3}{2}, \dfrac{3}{2}\right) \cup \left(\dfrac{3}{2}, \infty\right)$$

35. $\dfrac{x}{2} + \dfrac{x}{3} = -5$ and $\dfrac{x}{2} + \dfrac{x}{3}$

In each problem, the LCD is 6.

36. (a) $6\left(\dfrac{x}{2} + \dfrac{x}{3}\right) = 6(-5)$ (b) $\dfrac{x}{2} \cdot \dfrac{3}{3} + \dfrac{x}{3} \cdot \dfrac{2}{2}$

$\qquad\qquad 3x + 2x = -30 \qquad\qquad = \dfrac{3x}{6} + \dfrac{2x}{6}$

37. (a) $5x = -30$ (b) $= \dfrac{3x + 2x}{6}$

$\qquad\quad x = -6 \qquad\qquad = \dfrac{5x}{6}$

Solution set: $\{-6\}$

38. The answer to (a), $\{-6\}$, is a solution that results in a true statement in the original equation. The answer to (b), $\dfrac{5x}{6}$, is a simplified form of the given expression.

39. Simplifying the indicated expression requires adding and subtracting the rational expressions. The result is a single rational expression. However, solving an equation requires finding a set of numbers that satisfy the equation, and the result is a solution set.

40. The word "Solve" refers to finding the solution set of an equation. What appears here is not an equation, but an expression. "Solve" should be replaced by "Simplify" or "Add".

41. (a)

$$f(x) = \dfrac{125,000 - 25x}{125 + 2x}$$

$$300 = \dfrac{125,000 - 25x}{125 + 2x} \quad \textit{Let f(x)=300.}$$

$$300(125 + 2x) = 125,000 - 25x$$

$$\textit{Multiply by the LCD, 125+2x.}$$

$$37,500 + 600x = 125,000 - 25x$$

$$625x = 87,500$$

$$x = \dfrac{87,500}{625} = 140$$

Since x is measured in hundreds of gallons per day, the amount of gasoline produced is $140(100) = 14,000$ gallons per day.

(b) There are only so many resources available. If the amount of resources allocated to the production of heating oil increases, the amount of resources remaining for the production of gasoline must decrease. The following table (with $0 \leq x \leq 300$ and $Y_1 = f(x)$) illustrates this relationship and supports the answer in part (a).

X	Y₁	
0	1000	
50	550	
100	376.92	
150	285.29	
200	228.57	
250	190	
300	162.07	

$Y_1 \boxminus (125000 - 25X)...$

43. $w(x) = \dfrac{x^2}{2(1 - x)}$

(a) $w(.1) = \dfrac{(.1)^2}{2(1 - .1)}$

$\qquad\quad = \dfrac{.01}{2(.9)} \approx .006$

To the nearest tenth, $w(.1)$ is 0.

(b) $w(.8) = \dfrac{(.8)^2}{2(1 - .8)}$

$\qquad\quad = \dfrac{.64}{2(.2)} = 1.6$

(c) $w(.9) = \dfrac{(.9)^2}{2(1 - .9)}$

$\qquad\quad = \dfrac{.81}{2(.1)} = 4.05 \approx 4.1$

(d) Based on the answers in (a), (b), and (c), we see that as the traffic intensity increases, the waiting time also increases.

45. The number of solutions of the equation $f(x) = 0$ is the same as the number of x-intercepts, four.

47. The x-intercepts are -2, 0, and 3, so the solution set is $\{-2, 0, 3\}$.

Summary: Exercises on Operations and Equations with Rational Expressions

1. $\dfrac{2}{x} - \dfrac{4}{3x} = 5$

There is an equals sign, so this is an *equation*.

$3x\left(\dfrac{2}{x} - \dfrac{4}{3x}\right) = 3x(5) \quad \textit{Multiply by 3x.}$

$\qquad\qquad 6 - 4 = 15x$

$\qquad\qquad\quad 2 = 15x$

$\qquad\qquad \dfrac{2}{15} = x$

Check $x = \dfrac{2}{15}$: $\ 15 - 10 = 5 \quad \textit{True}$

Solution set: $\left\{\dfrac{2}{15}\right\}$

3. No equals sign appears so this is an *expression*.

$$\frac{4x - 20}{x^2 - 25} \cdot \frac{(x + 5)^2}{10}$$

$$= \frac{4(x - 5)}{(x + 5)(x - 5)} \cdot \frac{(x + 5)(x + 5)}{10}$$

$$= \frac{2(x + 5)}{5}$$

5. No equals sign appears so this is an *expression*.

$$\frac{\dfrac{1}{x} + \dfrac{1}{y}}{\dfrac{1}{x} - \dfrac{1}{y}}$$

Multiply the numerator and denominator by the LCD of all the fractions, xy.

$$= \frac{xy\left(\dfrac{1}{x} + \dfrac{1}{y}\right)}{xy\left(\dfrac{1}{x} - \dfrac{1}{y}\right)} = \frac{y + x}{y - x}$$

7.
$$\frac{x - 5}{3} + \frac{1}{3} = \frac{x - 2}{5}$$

There is an equals sign, so this is an *equation*.
Multiply by the LCD, 15.

$$15\left(\frac{x - 5}{3} + \frac{1}{3}\right) = 15\left(\frac{x - 2}{5}\right)$$

$$5(x - 5) + 5 = 3(x - 2)$$

$$5x - 25 + 5 = 3x - 6$$

$$5x - 20 = 3x - 6$$

$$2x = 14$$

$$x = 7$$

Check $x = 7$: $\dfrac{2}{3} + \dfrac{1}{3} = 1$ *True*

Solution set: $\{7\}$

9.
$$\frac{4}{x} - \frac{8}{x + 1} = 0$$

There is an equals sign, so this is an *equation*.
Multiply by the LCD, $x(x + 1)$. $(x \neq -1, 0)$

$$x(x + 1)\left(\frac{4}{x} - \frac{8}{x + 1}\right) = x(x + 1) \cdot 0$$

$$4(x + 1) - 8x = 0$$

$$4x + 4 - 8x = 0$$

$$-4x = -4$$

$$x = 1$$

Check $x = 1$: $4 - 4 = 0$ *True*

Solution set: $\{1\}$

11. No equals sign appears so this is an *expression*.

$$\frac{8}{r + 2} - \frac{7}{4r + 8}$$

$$= \frac{8}{r + 2} - \frac{7}{4(r + 2)}$$

$$= \frac{4(8)}{4(r + 2)} - \frac{7}{4(r + 2)} \quad LCD = 4(r + 2)$$

$$= \frac{32}{4(r + 2)} - \frac{7}{4(r + 2)}$$

$$= \frac{25}{4(r + 2)}$$

13. No equals sign appears so this is an *expression*.

$$\frac{3p^2 - 6p}{p + 5} \div \frac{p^2 - 4}{8p + 40}$$

$$= \frac{3p(p - 2)}{p + 5} \cdot \frac{8(p + 5)}{(p + 2)(p - 2)}$$

$$= \frac{24p(p - 2)(p + 5)}{(p + 2)(p - 2)(p + 5)}$$

$$= \frac{24p}{p + 2}$$

15.
$$\frac{a - 4}{3} + \frac{11}{6} = \frac{a + 1}{2}$$

There is an equals sign, so this is an *equation*.
Multiply by the LCD, 6.

$$6\left(\frac{a - 4}{3} + \frac{11}{6}\right) = 6\left(\frac{a + 1}{2}\right)$$

$$2(a - 4) + 11 = 3(a + 1)$$

$$2a - 8 + 11 = 3a + 3$$

$$2a + 3 = 3a + 3$$

$$0 = a$$

Check $a = 0$: $-\dfrac{4}{3} + \dfrac{11}{6} = \dfrac{1}{2}$ *True*

Solution set: $\{0\}$

17. No equals sign appears so this is an *expression*.

$$\frac{10z^2 - 5z}{3z^3 - 6z^2} \div \frac{2z^2 + 5z - 3}{z^2 + z - 6}$$

$$= \frac{5z(2z - 1)}{3z^2(z - 2)} \cdot \frac{(z + 3)(z - 2)}{(2z - 1)(z + 3)}$$

$$= \frac{5z(2z - 1)(z + 3)(z - 2)}{3z^2(2z - 1)(z + 3)(z - 2)}$$

$$= \frac{5}{3z}$$

19. $\dfrac{6}{t+1} + \dfrac{4}{5t+5} = \dfrac{34}{15}$

$\dfrac{6}{t+1} + \dfrac{4}{5(t+1)} = \dfrac{34}{15}$

There is an equals sign, so this is an *equation*.

Multiply by the LCD, $15(t+1)$. $(t \neq -1)$

$15(t+1)\left(\dfrac{6}{t+1} + \dfrac{4}{5(t+1)}\right)$

$= 15(t+1)\left(\dfrac{34}{15}\right)$

$90 + 12 = 34(t+1)$

$102 = 34t + 34$

$68 = 34t$

$2 = t$

Check $t = 2$: $2 + \dfrac{4}{15} = \dfrac{34}{15}$

Solution set: $\{2\}$

21. No equals sign appears so this is an *expression*.

$\dfrac{\dfrac{5}{x} - \dfrac{3}{y}}{\dfrac{9x^2 - 25y^2}{x^2 y}}$

Multiply the numerator and denominator by the LCD of all the fractions, $x^2 y$.

$= \dfrac{x^2 y\left(\dfrac{5}{x} - \dfrac{3}{y}\right)}{x^2 y\left(\dfrac{9x^2 - 25y^2}{x^2 y}\right)}$

$= \dfrac{5xy - 3x^2}{9x^2 - 25y^2}$

$= \dfrac{-x(3x - 5y)}{(3x + 5y)(3x - 5y)}$

$= \dfrac{-x}{(3x + 5y)}$

23. No equals sign appears so this is an *expression*.

$\dfrac{\dfrac{2r^{-1} + 5s^{-1}}{4s^2 - 25r^2}}{3rs}$

$= \dfrac{\dfrac{\dfrac{2}{r} + \dfrac{5}{s}}{4s^2 - 25r^2}}{3rs}$

Multiply the numerator and denominator by the LCD of all the fractions, $3rs$.

$= \dfrac{3rs\left(\dfrac{2}{r} + \dfrac{5}{s}\right)}{3rs\left(\dfrac{4s^2 - 25r^2}{3rs}\right)}$

$= \dfrac{6s + 15r}{4s^2 - 25r^2}$

$= \dfrac{3(2s + 5r)}{(2s - 5r)(2s + 5r)}$

$= \dfrac{3}{2s - 5r}$

25. $\dfrac{8}{3k+9} - \dfrac{8}{15} = \dfrac{2}{5k+15}$

$\dfrac{8}{3(k+3)} - \dfrac{8}{15} = \dfrac{2}{5(k+3)}$

There is an equals sign, so this is an *equation*.

Multiply by the LCD, $15(k+3)$. $(k \neq -3)$

$5(8) - 8(k+3) = 3(2)$

$40 - 8k - 24 = 6$

$-8k + 16 = 6$

$-8k = -10$

$k = \dfrac{-10}{-8} = \dfrac{5}{4}$

Check $k = \dfrac{5}{4}$: $\dfrac{32}{51} - \dfrac{8}{15} = \dfrac{8}{85}$ *True*

Solution set: $\left\{\dfrac{5}{4}\right\}$

27. No equals sign appears so this is an *expression*.

$\dfrac{6z^2 - 5z - 6}{6z^2 + 5z - 6} \cdot \dfrac{12z^2 - 17z + 6}{12z^2 - z - 6}$

$= \dfrac{(2z - 3)(3z + 2)}{(2z + 3)(3z - 2)} \cdot \dfrac{(3z - 2)(4z - 3)}{(3z + 2)(4z - 3)}$

$= \dfrac{2z - 3}{2z + 3}$

29. No equals sign appears so this is an *expression*.

$\dfrac{\dfrac{t}{4} - \dfrac{1}{t}}{1 + \dfrac{t+4}{t}}$

Multiply the numerator and denominator by the LCD of all the fractions, $4t$.

$$= \frac{4t\left(\dfrac{t}{4} - \dfrac{1}{t}\right)}{4t\left(1 + \dfrac{t+4}{t}\right)} = \frac{t^2 - 4}{4t + 4(t+4)}$$

$$= \frac{t^2 - 4}{4t + 4t + 16}$$

$$= \frac{t^2 - 4}{8t + 16}$$

$$= \frac{(t+2)(t-2)}{8(t+2)} = \frac{t-2}{8}$$

31. No equals sign appears so this is an *expression*.

$$\frac{7}{2x^2 - 8x} + \frac{3}{x^2 - 16}$$

$$= \frac{7}{2x(x-4)} + \frac{3}{(x-4)(x+4)}$$

$$= \frac{7(x+4)}{2x(x-4)(x+4)} + \frac{(3)2x}{(x-4)(x+4)2x}$$

$$LCD = 2x(x-4)(x+4)$$

$$= \frac{7(x+4)}{2x(x-4)(x+4)} + \frac{6x}{2x(x-4)(x+4)}$$

$$= \frac{7x + 28 + 6x}{2x(x-4)(x+4)}$$

$$= \frac{13x + 28}{2x(x-4)(x+4)}$$

33. No equals sign appears so this is an *expression*.

$$\frac{2k + \dfrac{5}{k-1}}{3k - \dfrac{2}{k}}$$

Multiply the numerator and denominator by the LCD of all the fractions, $k(k-1)$.

$$= \frac{k(k-1)\left(2k + \dfrac{5}{k-1}\right)}{k(k-1)\left(3k - \dfrac{2}{k}\right)}$$

$$= \frac{2k^2(k-1) + 5k}{3k^2(k-1) - 2(k-1)}$$

$$= \frac{k[2k(k-1) + 5]}{(k-1)(3k^2 - 2)} \quad \text{Factor out } k \text{ and } k\text{--}1.$$

$$= \frac{k(2k^2 - 2k + 5)}{(k-1)(3k^2 - 2)}$$

Section 6.7

1. Let $x =$ the number of girls in the class. Write and solve a proportion.

$$\frac{3}{4} = \frac{x}{20}$$

Multiply by the LCD, 20.

$$20\left(\frac{3}{4}\right) = 20\left(\frac{x}{20}\right)$$

$$15 = x$$

There are 15 girls and $20 - 15 = 5$ boys in the class.

3. Marin's rate

$$= \frac{1 \text{ job}}{\text{time to complete 1 job}}$$

$$= \frac{1 \text{ job}}{2 \text{ hours}}$$

$$= \frac{1}{2} \text{ job per hour}$$

5. **(a)** $b = \dfrac{p}{r}$ is the same as $p = br$.

(b) $r = \dfrac{b}{p}$ is the same as $b = pr$.

(c) $b = \dfrac{r}{p}$ is the same as $r = bp$.

(d) $p = \dfrac{r}{b}$ is the same as $r = bp$.

Choice (a) is correct.

7. **(a)** $a = mF$ is the same as $m = \dfrac{a}{F}$.

(b) $F = \dfrac{m}{a}$ is the same as $m = Fa$.

(c) $F = \dfrac{a}{m}$ is the same as $Fm = a$,

which is the same as $m = \dfrac{a}{F}$.

(d) $F = ma$ is the same as $m = \dfrac{F}{a}$.

Choice (d) is correct.

9. Solve $F = \dfrac{GMm}{d^2}$ for M.

$Fd^2 = GMm$ *Multiply by d^2.*

$\dfrac{Fd^2}{Gm} = M$ *Divide by Gm.*

Substitute 10 for F, 6.67×10^{-11} for G, 1 for m, and 3×10^{-6} for d.

$$M = \dfrac{Fd^2}{Gm}$$

$$M = \dfrac{(10)(3 \times 10^{-6})^2}{(6.67 \times 10^{-11})(1)}$$

$$\approx 1.349$$

11. $\dfrac{1}{a} = \dfrac{1}{b} + \dfrac{1}{c}$

Let $a = 8$ and $c = 12$.

$\dfrac{1}{8} = \dfrac{1}{b} + \dfrac{1}{12}$

Multiply by the LCD, $24b$.

$24b\left(\dfrac{1}{8}\right) = 24b\left(\dfrac{1}{b} + \dfrac{1}{12}\right)$

$3b = 24 + 2b$

$b = 24$

13. Solve $F = \dfrac{GMm}{d^2}$ for G.

$Fd^2 = GMm$ *Multiply by d^2.*

$\dfrac{Fd^2}{Mm} = G$ *Divide by Mm.*

15. Solve $\dfrac{1}{a} = \dfrac{1}{b} + \dfrac{1}{c}$ for a.

Multiply by the LCD, abc.

$abc\left(\dfrac{1}{a}\right) = abc\left(\dfrac{1}{b} + \dfrac{1}{c}\right)$

$bc = ac + ab$

$bc = a(c + b)$ *Factor out a.*

$\dfrac{bc}{c + b} = a$ *Divide by $c + b$.*

17. Solve $\dfrac{PV}{T} = \dfrac{pv}{t}$ for v.

$\dfrac{PVt}{T} = pv$ *Multiply by t.*

$\dfrac{PVt}{pT} = v$ *Divide by p.*

19. Solve $I = \dfrac{nE}{R + nr}$ for r.

$I(R + nr) = nE$

$IR + Inr = nE$

$Inr = nE - IR$

$r = \dfrac{nE - IR}{In}$

21. Solve $A = \dfrac{1}{2}h(B + b)$ for b.

$\dfrac{2}{h}(A) = \dfrac{2}{h}\left[\dfrac{1}{2}h(B + b)\right]$ *Multiply by $\dfrac{2}{h}$.*

$\dfrac{2A}{h} = B + b$

$\dfrac{2A}{h} - B = b$ or $b = \dfrac{2A - Bh}{h}$

23. $\dfrac{E}{e} = \dfrac{R + r}{r}$ for r

$Er = e(R + r)$ *Multiply by er.*

$Er = eR + er$

$Er - er = eR$ *Subtract eR.*

$r(E - e) = eR$

$r = \dfrac{eR}{E - e}$ *Divide by $E - e$.*

25. To solve the equation $m = \dfrac{ab}{a - b}$ for a, the first step is to multiply both sides of the equation by the LCD, $a - b$.

27. In 1991, light truck sales were about 4 million (rounded to the nearest .5 million) and new car sales were about 8.5 million. The ratio is

$$\dfrac{4}{8.5} \text{ or } \dfrac{8}{17}.$$

29. The ratio of light truck sales in 1991 to light truck sales in 1995 is

$$\dfrac{4}{6} \text{ or } \dfrac{2}{3}.$$

If you used 6.5, the ratio is

$$\dfrac{4}{6.5} \text{ or } \dfrac{8}{13}.$$

31. Let $x =$ the amount that 75 shares would earn.

Write a proportion.

$$\dfrac{\text{unknown earnings}}{\text{shares}} = \dfrac{\text{earnings}}{\text{shares}}$$

$$\dfrac{x}{75} = \dfrac{191.50}{50}$$

$$x = \dfrac{75(191.50)}{50}$$

$$= 287.25$$

The difference is $\$287.25 - \$191.50 = \$95.75$.

33. Let x = the number of fish in the lake.
Write and solve a proportion.

$$\frac{\text{total in lake}}{\text{tagged in lake}} = \frac{\text{total in sample}}{\text{tagged in sample}}$$

$$\frac{x}{500} = \frac{400}{8}$$

$$x = 500(50)$$

$$= 25,000$$

There are approximately $25,000$ fish in the lake.

35. *Step 1*
Find the distance from Tulsa to Detroit.
Let x represent that distance.

Step 2
Complete the table.

	d	r	t
Actual trip	x	50	$\dfrac{x}{50}$
Alternative trip	x	60	$\dfrac{x}{60}$

Step 3
At 60 mph, his time at 50 mph would be decreased 3 hr.

$$\frac{x}{60} = \frac{x}{50} - 3$$

Step 4
Multiply by 300.

$$5x = 6x - 900$$

$$900 = x$$

Step 5
The distance from Tulsa to Detroit is 900 miles.

Step 6
Check: 900 miles at 50 mph takes $\dfrac{900}{50}$ or
18 hours; 900 miles at 60 mph takes $\dfrac{900}{60}$ or
15 hours; $15 = 18 - 3$ as required.

37. *Step 1*
Let x = the distance to the fishing hole.

Step 2

	d	r	t
Old highway	x	30	$\dfrac{x}{30}$
Interstate	x	50	$\dfrac{x}{50}$

Step 3
The time on the interstate is 2 hr less than the time on the old highway.

$$\frac{x}{50} = \frac{x}{30} - 2$$

Step 4
Multiply by 150.

$$3x = 5x - 300$$

$$300 = 2x$$

$$150 = x$$

Step 5
The distance to the fishing hole is 150 miles.

Step 6
Check: 150 miles at 30 mph takes $\dfrac{150}{30}$ or 5
hours; 150 miles at 50 mph takes $\dfrac{150}{50}$ or 3 hours.
The interstate time is 2 less than the highway time, as required.

39. *Step 1*
Let x and y represent the calm water speeds of the container ship and the FastShip, respectively.

Step 2
Complete the table.

	Distance	Rate	Time
Container ship	2400	$x - 6$	218
FastShip	2400	$y - .02y$	72

Step 3
The rate for the FastShip is $y - .02y = .98y$ (which makes sense—losing 2% is the same as retaining 98%). Since the speed of a FastShip in calm water is twice the speed of the container ship, $y = 2x$. Thus, the rate of the FastShip is

$$.98y = .98(2x) = 1.96x.$$

The distance each ship traveled is the same, so

$$\text{Distance}_{\text{container ship}} = \text{Distance}_{\text{FastShip}}.$$

Using $D = RT$, we obtain

$$(x - 6)(218) = (1.96x)(72).$$

Step 4

$$218x - 1308 = 141.12x$$

$$76.88x = 1308$$

$$x = \frac{1308}{76.88} \approx 17$$

Step 5
To the nearest knot, the speed of the container ship in calm water is 17 knots and the speed of the FastShip is 34 knots.

Step 6
The container ship travels at $17 - 6 = 11$ knots for 218 hours. Its distance is
$11(218) = 2398 \approx 2400$. The FastShip travels at
$34 - 2\%(34) = 33.32$ knots for 72 hours. Its distance is $33.32(72) = 2399.04 \approx 2400$.

41. To solve problems about distance, rate, and time, we use the formula $d = rt$. To solve problems about work, we use the similar formula $A = rt$, where A represents the part of the job completed. $A = 1$ if 1 job is completed.

43. Let $x = $ the time it would take them working together.

Complete the table.

Worker	Rate	Time Together	Part of the Job Done
Lou	$\dfrac{1}{8}$	x	$\dfrac{1}{8}x$
Janet	$\dfrac{1}{5}$	x	$\dfrac{1}{5}x$

$$\begin{array}{ccccc} \text{Part done} & & \text{part done} & & \text{1 whole} \\ \text{by Butch} & + & \text{by Peggy} & = & \text{job.} \\ \dfrac{1}{8}x & + & \dfrac{1}{5}x & = & 1 \end{array}$$

Multiply by the LCD, 40.

$$40\left(\frac{1}{8}x + \frac{1}{5}x\right) = 40 \cdot 1$$
$$5x + 8x = 40$$
$$13x = 40$$
$$x = \frac{40}{13} = 3\frac{1}{13}$$

Together they could groom the dogs in $\dfrac{40}{13}$ or $3\dfrac{1}{13}$ hr.

45. Let $x = $ the time for Carolyn to do the job alone. Make a table.

Worker	Rate	Time Together	Part of Job Done
Bernard	$\dfrac{1}{7}$	5	$\dfrac{1}{7}(5) = \dfrac{5}{7}$
Carolyn	$\dfrac{1}{x}$	5	$\dfrac{1}{x}(5) = \dfrac{5}{x}$

$$\begin{array}{ccccc} \text{Part done} & & \text{part done} & & \text{1 whole} \\ \text{by Bernard} & + & \text{by Carolyn} & = & \text{job.} \\ \dfrac{5}{7} & + & \dfrac{5}{x} & = & 1 \end{array}$$

Multiply by the LCD, $7x$.

$$7x\left(\frac{5}{7} + \frac{5}{x}\right) = 7x \cdot 1$$
$$5x + 35 = 7x$$
$$-2x = -35$$
$$x = \frac{35}{2} = 17\frac{1}{2}$$

Carolyn can do the job alone in $\dfrac{35}{2}$ or $17\dfrac{1}{2}$ hours.

Another solution: Bernard will do $\dfrac{5}{7}$ of the job, leaving $\dfrac{2}{7}$ of the job for Carolyn.

$$A = rt$$
$$\frac{2}{7} = \frac{1}{x} \cdot 5$$
$$x = 5 \cdot \frac{7}{2} = \frac{35}{2}$$

47. Let $x = $ the time it will take to fill the barrel if both pipes are open.

	Rate	Time to Fill Together	Fractional Part of Job Done
Inlet Pipe	$\dfrac{1}{9}$	x	$\dfrac{1}{9}x$
Outlet Pipe	$-\dfrac{1}{12}$	x	$-\dfrac{1}{12}x$

Notice that the rate of the outlet pipe is negative because it will empty the barrel, not fill it.

$$\begin{array}{ccccc} \text{Part done} & & \text{Part done} & & \\ \text{with the} & & \text{with the} & & \text{1 whole} \\ \text{inlet pipe} & + & \text{outlet pipe} & = & \text{job.} \\ \text{open} & & \text{open} & & \\ \dfrac{1}{9}x & + & \left(-\dfrac{1}{12}x\right) & = & 1 \end{array}$$

Multiply by the LCD, 36.

$$36\left(\frac{1}{9}x - \frac{1}{12}x\right) = 36 \cdot 1$$
$$4x - 3x = 36$$
$$x = 36$$

It will take 36 hours to fill the barrel.

49. Let $x = $ the time it will take to fill the pond if both pipes are open.

	Rate	Time to Fill Together	Fractional Part of Job Done
Inlet Pipe	$\dfrac{1}{60}$	x	$\dfrac{1}{60}x$
Outlet Pipe	$-\dfrac{1}{80}$	x	$-\dfrac{1}{80}x$

Notice that the rate of the outlet pipe is negative because it will empty the pond, not fill it.

| Part done with the inlet pipe open | + | Part done with the outlet pipe open | = | 1 whole job. |

$$\frac{1}{60}x \quad + \quad \left(-\frac{1}{80}x\right) \quad = \quad 1$$

$$\frac{1}{60}x - \frac{1}{80}x = 1$$

Multiply by the LCD, 240.

$$240\left(\frac{1}{60}x - \frac{1}{80}x\right) = 240 \cdot 1$$

$$4x - 3x = 240$$

$$x = 240$$

It will take 240 minutes (4 hours) to fill the pond.

51. Since $\frac{2}{3} = \frac{4}{6} = \frac{6}{9}$, use the proportion

$$\frac{2}{3} = \frac{2x+1}{2x+5}.$$

$$2(2x+5) = 3(2x+1)$$

$$4x + 10 = 6x + 3$$

$$7 = 2x$$

$$\frac{7}{2} = x$$

If $x = \frac{7}{2}$, then

$$AC = 2x + 1 = 2\left(\frac{7}{2}\right) + 1 = 8$$

and

$$DF = 2x + 5 = 2\left(\frac{7}{2}\right) + 5 = 12.$$

Chapter 6 Review Exercises

1. A fraction is any quotient indicated by a division bar, such as $\frac{21}{4}$ and $\frac{x}{2x-1}$. A rational expression is a fraction that is the quotient of two polynomials: $\frac{x}{2x-1}$ is an example, as is $\frac{x^2+3}{(x-1)^2}$. A complex fraction is a fraction in which either the numerator or denominator (or both) contains a fraction. $\frac{\frac{1}{a}}{2a+3}$, $\frac{\frac{x}{y+1}}{\frac{y}{x-2}}$, and

$$\frac{1 + \frac{1}{m}}{1 - \frac{1}{m}}$$ are examples.

2. $g(t) = \dfrac{-7}{3t + 18}$

Set the denominator equal to zero and solve.

$$3t + 18 = 0$$
$$3t = -18$$
$$t = -6$$

The number -6 makes the expression undefined, so it is excluded from the domain.

3. $f(r) = \dfrac{5r + 17}{r^2 - 7r + 10}$

Set the denominator equal to zero and solve.

$$r^2 - 7r + 10 = 0$$
$$(r - 5)(r - 2) = 0$$
$$r - 5 = 0 \quad \text{or} \quad r - 2 = 0$$
$$r = 5 \quad \text{or} \qquad r = 2$$

The numbers 2 and 5 make the expression undefined, so they are excluded from the domain.

4. To write $\dfrac{x^2 + 3x}{5x + 15}$ in lowest terms:

Step 1 Factor the numerator and the denominator to find their GCF.

Step 2 Simplify using the fundamental principle.

$$\frac{x^2 + 3x}{5x + 15} = \frac{x(x + 3)}{5(x + 3)} = \frac{x}{5}$$

5. $\dfrac{55m^4n^3}{10m^5n} = \dfrac{5m^4n}{5m^4n} \cdot \dfrac{11n^2}{2m} = \dfrac{11n^2}{2m}$

6. $\dfrac{12x^2 + 6x}{24x + 12} = \dfrac{6x(2x + 1)}{12(2x + 1)} = \dfrac{x}{2}$

7. $\dfrac{25m^2 - n^2}{25m^2 - 10mn + n^2} = \dfrac{(5m + n)(5m - n)}{(5m - n)(5m - n)}$
$$= \dfrac{5m + n}{5m - n}$$

8. $\dfrac{r - 2}{4 - r^2} = \dfrac{r - 2}{(2 + r)(2 - r)}$
$$= \dfrac{(-1)(2 - r)}{(2 + r)(2 - r)}$$
$$= \dfrac{-1}{2 + r}$$

9. The reciprocal of a rational expression is another rational expression, such that the two rational expressions have a product of 1.

10. $\dfrac{25p^3q^2}{8p^4q} \div \dfrac{15pq^2}{16p^5}$

Multiply by the reciprocal.

$= \dfrac{25p^3q^2}{8p^4q} \cdot \dfrac{16p^5}{15pq^2}$

$= \dfrac{5 \cdot 5 \cdot 4 \cdot 2 \cdot 2p^8q^2}{2 \cdot 4 \cdot 3 \cdot 5p^5q^3}$

$= \dfrac{5 \cdot 4 \cdot 2p^5q^2 \cdot 5 \cdot 2p^3}{2 \cdot 4 \cdot 5p^5q^2 \cdot 3q}$

$= \dfrac{10p^3}{3q}$

11. $\dfrac{w^2 - 16}{w} \cdot \dfrac{3}{4 - w}$

$= \dfrac{(w - 4)(w + 4)}{w} \cdot \dfrac{3}{4 - w}$

$= \dfrac{(-1)(4 - w)(w + 4)}{w} \cdot \dfrac{3}{4 - w}$

$= \dfrac{-3(w + 4)}{w}$

12. $\dfrac{z^2 - z - 6}{z - 6} \cdot \dfrac{z^2 - 6z}{z^2 + 2z - 15}$

$= \dfrac{(z - 3)(z + 2)}{z - 6} \cdot \dfrac{z(z - 6)}{(z - 3)(z + 5)}$

$= \dfrac{z(z + 2)}{z + 5}$

13. $\dfrac{m^3 - n^3}{m^2 - n^2} \div \dfrac{m^2 + mn + n^2}{m + n}$

Multiply by the reciprocal.

$= \dfrac{m^3 - n^3}{m^2 - n^2} \cdot \dfrac{m + n}{m^2 + mn + n^2}$

$= \dfrac{(m - n)(m^2 + mn + n^2)}{(m - n)(m + n)} \cdot \dfrac{m + n}{m^2 + mn + n^2}$

$= 1$

14. The terms x and 5 in the denominator cannot be separated into denominators of two fractions.

$$\dfrac{x^2 + 5x}{x + 5} \neq \dfrac{x^2}{x} + \dfrac{5x}{x}$$

Instead, $\dfrac{x^2 + 5x}{x + 5} = \dfrac{x(x + 5)}{x + 5} = x.$

15. To find the least common denominator, factor each denominator. Then take the product of all the different factors in the denominators, with each factor raised to the greatest power that occurs in any denominator.

16. $\dfrac{5a}{32b^3}, \dfrac{31}{24b^5}$

Factor each denominator.

$32b^3 = 2 \cdot 2 \cdot 2 \cdot 2 \cdot 2 \cdot b^3 = 2^5 \cdot b^3$

$24b^5 = 2 \cdot 2 \cdot 2 \cdot 3 \cdot b^5 = 2^3 \cdot 3 \cdot b^5$

$LCD = 2^5 \cdot 3 \cdot b^5 = 96b^5$

The LCD is $96b^5$.

17. $\dfrac{17}{9r^2}, \dfrac{5r - 3}{3r + 1}$

Factor each denominator.

$$9r^2 = 3^2 \cdot r^2$$

The second denominator is already in factored form. The LCD is

$$3^2 \cdot r^2 \cdot (3r + 1) \text{ or } 9r^2(3r + 1).$$

18. $\dfrac{4x - 9}{6x^2 + 13x - 5}, \dfrac{x + 15}{9x^2 + 9x - 4}$

Factor each denominator.

$$6x^2 + 13x - 5 = (3x - 1)(2x + 5)$$
$$9x^2 + 9x - 4 = (3x - 1)(3x + 4)$$

The LCD is

$$(3x - 1)(2x + 5)(3x + 4).$$

19. $\dfrac{8}{z} - \dfrac{3}{2z^2}$

The LCD is $2z^2$.

$= \dfrac{8 \cdot 2z}{z \cdot 2z} - \dfrac{3}{2z^2}$

$= \dfrac{16z}{2z^2} - \dfrac{3}{2z^2}$

$= \dfrac{16z - 3}{2z^2}$

20. $\dfrac{5y + 13}{y + 1} - \dfrac{1 - 7y}{y + 1}$

$= \dfrac{5y + 13 - (1 - 7y)}{y + 1}$

$= \dfrac{5y + 13 - 1 + 7y}{y + 1}$

$= \dfrac{12y + 12}{y + 1}$

$= \dfrac{12(y + 1)}{y + 1} = 12$

21. $\dfrac{6}{5a+10} + \dfrac{7}{6a+12}$

$= \dfrac{6}{5(a+2)} + \dfrac{7}{6(a+2)}$

The LCD is $30(a+2)$.

$= \dfrac{6 \cdot 6}{5(a+2) \cdot 6} + \dfrac{7 \cdot 5}{6(a+2) \cdot 5}$

$= \dfrac{36}{30(a+2)} + \dfrac{35}{30(a+2)}$

$= \dfrac{36+35}{30(a+2)} = \dfrac{71}{30(a+2)}$

22. $\dfrac{3r}{10r^2 - 3rs - s^2} + \dfrac{2r}{2r^2 + rs - s^2}$

$= \dfrac{3r}{(5r+s)(2r-s)} + \dfrac{2r}{(2r-s)(r+s)}$

The LCD is $(5r+s)(2r-s)(r+s)$.

$= \dfrac{3r(r+s)}{(5r+s)(2r-s)(r+s)}$

$\quad + \dfrac{2r(5r+s)}{(2r-s)(r+s)(5r+s)}$

$= \dfrac{3r^2 + 3rs + 10r^2 + 2rs}{(5r+s)(2r-s)(r+s)}$

$= \dfrac{13r^2 + 5rs}{(5r+s)(2r-s)(r+s)}$

23. $\dfrac{\dfrac{3}{t} + 2}{\dfrac{4}{t} - 7}$

Multiply the numerator and denominator by the LCD of all the fractions, t.

$= \dfrac{t\left(\dfrac{3}{t} + 2\right)}{t\left(\dfrac{4}{t} - 7\right)} = \dfrac{3+2t}{4-7t}$

24. $\dfrac{\dfrac{4m^5 n^6}{mn}}{\dfrac{8m^7 n^3}{m^4 n^2}} = \dfrac{4m^5 n^6}{mn} \div \dfrac{8m^7 n^3}{m^4 n^2}$

Multiply by the reciprocal.

$= \dfrac{4m^5 n^6}{mn} \cdot \dfrac{m^4 n^2}{8m^7 n^3}$

$= \dfrac{4m^9 n^8}{8m^8 n^4} = \dfrac{mn^4}{2}$

25. $\dfrac{\dfrac{3}{p} - \dfrac{2}{q}}{\dfrac{9q^2 - 4p^2}{qp}}$

Multiply the numerator and denominator by the LCD of all the fractions, qp.

$= \dfrac{qp\left(\dfrac{3}{p} - \dfrac{2}{q}\right)}{qp\left(\dfrac{9q^2 - 4p^2}{qp}\right)}$

$= \dfrac{3q - 2p}{9q^2 - 4p^2}$

$= \dfrac{3q - 2p}{(3q + 2p)(3q - 2p)}$

$= \dfrac{1}{3q + 2p}$

26. $\dfrac{4y^3 - 12y^2 + 5y}{4y}$

$= \dfrac{4y^3}{4y} - \dfrac{12y^2}{4y} + \dfrac{5y}{4y}$

$= y^2 - 3y + \dfrac{5}{4}$

27. $\dfrac{2p^3 + 9p^2 + 27}{2p - 3}$

$$
\begin{array}{r}
p^2 \;+\; 6p \;+\; 9 \\
2p - 3 \overline{\smash{\big)}\, 2p^3 \;+\; 9p^2 \;+\; 0p \;+\; 27} \\
\underline{2p^3 \;-\; 3p^2 } \\
12p^2 \;+\; 0p \\
\underline{12p^2 \;-\; 18p } \\
18p \;+\; 27 \\
\underline{18p \;-\; 27} \\
54
\end{array}
$$

Remainder

Answer: $p^2 + 6p + 9 + \dfrac{54}{2p - 3}$

28. $\dfrac{5p^4 + 15p^3 - 33p^2 - 9p + 18}{5p^2 - 3}$

$$
\begin{array}{r}
p^2 \;+\; 3p \;-\; 6 \\
5p^2 - 3 \overline{\smash{\big)}\, 5p^4 \;+\; 15p^3 \;-\; 33p^2 \;-\; 9p \;+\; 18} \\
\underline{5p^4 -\; 3p^2 } \\
15p^3 \;-\; 30p^2 \;-\; 9p \\
\underline{15p^3 -\; 9p } \\
-30p^2 +\; 18 \\
\underline{-30p^2 +\; 18} \\
0
\end{array}
$$

Answer: $p^2 + 3p - 6$

29. $\dfrac{3p^2 - p - 2}{p - 1}$

$$1 \,\big|\; \begin{array}{rrr} 3 & -1 & -2 \\ & 3 & 2 \\ \hline 3 & 2 & 0 \end{array}$$

Answer: $3p + 2$

30. $\left(2k^3 - 5k^2 + 12\right) \div (k - 3)$

Insert 0 for the missing k-term.

$$3 \,\big|\; \begin{array}{rrrr} 2 & -5 & 0 & 12 \\ & 6 & 3 & 9 \\ \hline 2 & 1 & 3 & 21 \end{array} \;\; \leftarrow \;\; Remainder$$

Answer: $2k^2 + k + 3 + \dfrac{21}{k - 3}$

31. $2w^3 + 8w^2 - 14w - 20 = 0$

Divide the polynomial by $w + 5$.

$$-5 \,\big|\; \begin{array}{rrrr} 2 & 8 & -14 & -20 \\ & -10 & 10 & 20 \\ \hline 2 & -2 & -4 & 0 \end{array}$$

Since the remainder is 0, -5 is a solution of the equation.

32. $-3q^4 + 2q^3 + 5q^2 - 9q + 1 = 0$

Divide the polynomial by $q + 5$.

$$-5 \,\big|\; \begin{array}{rrrrr} -3 & 2 & 5 & -9 & 1 \\ & 15 & -85 & 400 & -1955 \\ \hline -3 & 17 & -80 & 391 & -1954 \end{array}$$

Since the remainder is not 0, -5 is not a solution of the equation.

33. $P(x) = 3x^3 - 5x^2 + 4x - 1; \; k = -1$

Divide the polynomial by $x + 1$.

$$-1 \,\big|\; \begin{array}{rrrr} 3 & -5 & 4 & -1 \\ & -3 & 8 & -12 \\ \hline 3 & -8 & 12 & -13 \end{array}$$

By the remainder theorem, $P(-1) = -13$.

34. $P(z) = z^4 - 2z^3 - 9z - 5; \; k = 3$

Divide the polynomial by $z - 3$. Insert 0 for the missing z^2-term.

$$3 \,\big|\; \begin{array}{rrrrr} 1 & -2 & 0 & -9 & -5 \\ & 3 & 3 & 9 & 0 \\ \hline 1 & 1 & 3 & 0 & -5 \end{array}$$

By the remainder theorem, $P(3) = -5$.

35. The graph in choice (c) has an asymptote, so it is the graph of a rational function.

36.
$$\dfrac{1}{t + 4} + \dfrac{1}{2} = \dfrac{3}{2t + 8}$$
$$\dfrac{1}{t + 4} + \dfrac{1}{2} = \dfrac{3}{2(t + 4)}$$

Multiply by the LCD, $2(t + 4)$. $(t \neq -4)$

$$2(t + 4)\left(\dfrac{1}{t + 4} + \dfrac{1}{2}\right) = 2(t + 4)\left(\dfrac{3}{2(t + 4)}\right)$$
$$2 + (t + 4) = 3$$
$$t + 6 = 3$$
$$t = -3$$

Check $t = -3$: $1 + \dfrac{1}{2} = \dfrac{3}{2}$ *True*

Solution set: $\{-3\}$

37.
$$\dfrac{-5m}{m + 1} + \dfrac{m}{3m + 3} = \dfrac{56}{6m + 6}$$
$$\dfrac{-5m}{m + 1} + \dfrac{m}{3(m + 1)} = \dfrac{56}{6(m + 1)}$$
$$\dfrac{-5m}{m + 1} + \dfrac{m}{3(m + 1)} = \dfrac{28}{3(m + 1)}$$

Multiply by the LCD, $3(m + 1)$. $(m \neq -1)$

$$3(m + 1)\left(\dfrac{-5m}{m + 1} + \dfrac{m}{3(m + 1)}\right)$$
$$= 3(m + 1)\left(\dfrac{28}{3(m + 1)}\right)$$
$$-15m + m = 28$$
$$-14m = 28$$
$$m = -2$$

Check $m = -2$: $-10 + \dfrac{2}{3} = -\dfrac{56}{6}$ *True*

Solution set: $\{-2\}$

38.
$$\dfrac{2}{k - 1} - \dfrac{4k + 1}{k^2 - 1} = \dfrac{-1}{k + 1}$$
$$\dfrac{2}{k - 1} - \dfrac{4k + 1}{(k + 1)(k - 1)} = \dfrac{-1}{k + 1}$$

Multiply by the LCD, $(k + 1)(k - 1)$.

$$(k + 1)(k - 1)\left(\dfrac{2}{k - 1} - \dfrac{4k + 1}{(k + 1)(k - 1)}\right)$$
$$= (k + 1)(k - 1)\left(\dfrac{-1}{k + 1}\right)$$
$$2(k + 1) - (4k + 1) = -1(k - 1)$$
$$2k + 2 - 4k - 1 = -k + 1$$
$$-2k + 1 = -k + 1$$
$$0 = k$$

Check $k = 0$: $-2 + 1 = -1$ *True*

Solution set: $\{0\}$

39.
$$\frac{5}{x+2}+\frac{3}{x+3}=\frac{x}{x^2+5x+6}$$
$$\frac{5}{x+2}+\frac{3}{x+3}=\frac{x}{(x+2)(x+3)}$$
Multiply by the LCD, $(x+2)(x+3)$.
$(x \neq -3, -2)$
$$(x+2)(x+3)\left(\frac{5}{x+2}+\frac{3}{x+3}\right)$$
$$=(x+2)(x+3)\left(\frac{x}{(x+2)(x+3)}\right)$$
$$5(x+3)+3(x+2)=x$$
$$5x+15+3x+6=x$$
$$8x+21=x$$
$$7x=-21$$
$$x=-3$$

Substituting -3 in the original equation results in division by 0, so -3 is not a solution.
Solution set: $\emptyset$

40. The student is wrong, even though she did not make an error in her steps. The number 3 cannot be a solution because it causes division by 0 when substituted in the equation.

41. Simplifying the expression means to write the problem as a single rational expression. Solving the equation means to find the values of the variable that make the equation true.

42.
$$\frac{1}{A}=\frac{1}{B}+\frac{1}{C}$$
Let $B=30$ and $C=10$.
$$\frac{1}{A}=\frac{1}{30}+\frac{1}{10}$$
To solve for A, multiply both sides by the LCD, $30A$.
$$30A\left(\frac{1}{A}\right)=30A\left(\frac{1}{30}+\frac{1}{10}\right)$$
$$30=A+3A$$
$$30=4A$$
$$A=\frac{30}{4}=\frac{15}{2}$$

43. Solve $F=\frac{GMm}{d^2}$ for m.
$$Fd^2=GMm \qquad \textit{Multiply by } d^2.$$
$$\frac{Fd^2}{GM}=m \qquad \textit{Divide by } GM.$$

44. Solve $\mu=\frac{Mv}{M+m}$ for M.
$$\mu(M+m)=Mv \qquad \textit{Multiply by } M+m.$$
$$\mu M+\mu m=Mv$$
$$\mu m=Mv-\mu M$$
$$m\mu=M(v-\mu)$$
$$M=\frac{m\mu}{v-\mu}$$

45. The ratio of the deer population in 1980 to the population in 1998 is
$$\frac{.48}{1.0}=\frac{48}{100}=\frac{12}{25}.$$

46. The ratio of the bear population in 1998 to the population in 1986 is
$$\frac{14.0}{4.3}=\frac{140}{43}.$$

47. The change in the deer population over the 18 year period from 1980 to 1998 was $1.0-.48=.52$ million. Thus, the average annual rate of change of the deer population for that period was
$$\frac{.52}{18}\approx .03$$
million deer per year.

48. The change in the bear population over the 12 year period from 1986 to 1998 was $14.0-4.3=9.7$ thousand. Thus, the average annual rate of change of the bear population for that period was
$$\frac{9.7}{12}\approx .81$$
thousand bears per year.

49. Since the deer population barely doubled in 18 years and the bear population more than tripled in 12 years, the bear population had the higher growth rate.

50. Let $x=$ the number of passenger-kilometers per day provided by high-speed trains.
Write a proportion.
$$\frac{x}{15,000}=\frac{23,200}{58,000}$$
$$x=\frac{23,200(15,000)}{58,000}$$
$$=6000$$

51. Let $x=$ the speed of the boat in still water.
Use $d=rt$, or $t=\frac{d}{r}$, to make a table.

	Distance	Rate	Time
Downstream	40	$x+4$	$\frac{40}{x+4}$
Upstream	24	$x-4$	$\frac{24}{x-4}$

Because the times are equal,

continued

$$\frac{40}{x+4} = \frac{24}{x-4}.$$

Multiply by the LCD, $(x+4)(x-4)$. $(x \neq -4, 4)$

$$(x+4)(x-4)\left(\frac{40}{x+4}\right) = (x+4)(x-4)\left(\frac{24}{x-4}\right)$$

$$40(x-4) = 24(x+4)$$
$$40x - 160 = 24x + 96$$
$$16x = 256$$
$$x = 16$$

The speed of the boat in still water is 16 km/hr.

52.   Let $x =$ the time it takes to fill the sink with both taps open.

Make a table.

	Rate	Time Together	Part of Job Done
Cold	$\frac{1}{8}$	x	$\frac{x}{8}$
Hot	$\frac{1}{12}$	x	$\frac{x}{12}$

Part done by hot	plus	part done by cold	equals	1 whole job.
$\frac{x}{8}$	$+$	$\frac{x}{12}$	$=$	1

Multiply by the LCD, 24.

$$24\left(\frac{x}{8} + \frac{x}{12}\right) = 24 \cdot 1$$
$$3x + 2x = 24$$
$$5x = 24$$
$$x = \frac{24}{5} \text{ or } 4\frac{4}{5}$$

The sink will be filled in $\frac{24}{5}$ or $4\frac{4}{5}$ minutes.

53.   Let $x =$ the time to do the job working together.
Make a table.

Worker	Rate	Time Together	Part of Job Done
Jane	$\frac{1}{9}$	x	$\frac{x}{9}$
Jason	$\frac{1}{6}$	x	$\frac{x}{6}$

Part done by Jane	plus	part done by Jason	equals	1 whole job.
$\frac{x}{9}$	$+$	$\frac{x}{6}$	$=$	1

Multiply by the LCD, 36.

$$36\left(\frac{x}{9} + \frac{x}{6}\right) = 36 \cdot 1$$
$$4x + 6x = 36$$
$$10x = 36$$
$$x = \frac{36}{10}$$
$$x = \frac{18}{5} \text{ or } 3\frac{3}{5}$$

Working together, they can do the job in $\frac{18}{5}$ or $3\frac{3}{5}$ hours.

54. $\dfrac{14}{6k+3}, \dfrac{7k^2 + 2k + 1}{10k + 5}, \dfrac{-11k}{18k + 9}$

Factor each denominator.

$$6k + 3 = 3(2k + 1)$$
$$10k + 5 = 5(2k + 1)$$
$$18k + 9 = 9(2k + 1)$$
$$= 3^2 \cdot (2k + 1)$$

LCD $= 5 \cdot 3^2 \cdot (2k + 1) = 45(2k + 1)$
The LCD is $45(2k + 1)$.

55. $\dfrac{2}{m} + \dfrac{5}{3m^2}$
The LCD is $3m^2$.

$$= \frac{2 \cdot 3m}{m \cdot 3m} + \frac{5}{3m^2}$$
$$= \frac{6m}{3m^2} + \frac{5}{3m^2} = \frac{6m + 5}{3m^2}$$

56. $\dfrac{k^2 - 6k + 9}{1 - 216k^3} \cdot \dfrac{6k^2 + 17k - 3}{9 - k^2}$

Factor $1 - 216k^3$ as the difference of cubes, $1^3 - (6k)^3$.

$$= \frac{(k - 3)(k - 3)}{(1 - 6k)(1 + 6k + 36k^2)}$$
$$\cdot \frac{(6k - 1)(k + 3)}{(3 - k)(3 + k)}$$
$$= \frac{(k - 3)(k - 3)}{(-1)(6k - 1)(1 + 6k + 36k^2)}$$
$$\cdot \frac{(6k - 1)(k + 3)}{(-1)(k - 3)(k + 3)}$$
$$= \frac{k - 3}{1 + 6k + 36k^2} \text{ or } \frac{k - 3}{36k^2 + 6k + 1}$$

57. $\dfrac{\dfrac{-3}{x} + \dfrac{x}{2}}{1 + \dfrac{x+1}{x}}$

Multiply the numerator and denominator by the LCD of all the fractions, $2x$.

$= \dfrac{2x\left(\dfrac{-3}{x} + \dfrac{x}{2}\right)}{2x\left(1 + \dfrac{x+1}{x}\right)}$

$= \dfrac{-6 + x^2}{2x + 2(x+1)}$

$= \dfrac{x^2 - 6}{2x + 2x + 2}$

$= \dfrac{x^2 - 6}{4x + 2} = \dfrac{x^2 - 6}{2(2x+1)}$

58. $\dfrac{9x^2 + 46x + 5}{3x^2 - 2x - 1} \div \dfrac{x^2 + 11x + 30}{x^3 + 5x^2 - 6x}$

Multiply by the reciprocal.

$= \dfrac{9x^2 + 46x + 5}{3x^2 - 2x - 1} \cdot \dfrac{x(x^2 + 5x - 6)}{x^2 + 11x + 30}$

$= \dfrac{(9x+1)(x+5)}{(3x+1)(x-1)} \cdot \dfrac{x(x+6)(x-1)}{(x+6)(x+5)}$

$= \dfrac{x(9x+1)}{3x+1}$

59. $\dfrac{3x^{-1} - 5}{6 + x^{-1}} = \dfrac{\dfrac{3}{x} - 5}{6 + \dfrac{1}{x}}$

Multiply the numerator and denominator by the LCD of all the fractions, x.

$= \dfrac{x\left(\dfrac{3}{x} - 5\right)}{x\left(6 + \dfrac{1}{x}\right)}$

$= \dfrac{3 - 5x}{6x + 1}$

60. $\dfrac{9}{3-x} - \dfrac{2}{x-3}$

$= \dfrac{9}{3-x} - \dfrac{2(-1)}{(x-3)(-1)}$

$= \dfrac{9}{3-x} - \dfrac{-2}{3-x}$

$= \dfrac{9 - (-2)}{3-x}$

$= \dfrac{11}{3-x}$ or $\dfrac{-11}{x-3}$

61. $\dfrac{4y+16}{30} \div \dfrac{2y+8}{5}$

Multiply by the reciprocal.

$= \dfrac{4y+16}{30} \cdot \dfrac{5}{2y+8}$

$= \dfrac{4(y+4)}{30} \cdot \dfrac{5}{2(y+4)}$

$= \dfrac{4 \cdot 5}{2 \cdot 30} = \dfrac{2}{6} = \dfrac{1}{3}$

62. $\dfrac{t^{-2} + s^{-2}}{t^{-1} - s^{-1}} = \dfrac{\dfrac{1}{t^2} + \dfrac{1}{s^2}}{\dfrac{1}{t} - \dfrac{1}{s}}$

Multiply the numerator and denominator by the LCD of all the fractions, $t^2 s^2$.

$= \dfrac{t^2 s^2\left(\dfrac{1}{t^2} + \dfrac{1}{s^2}\right)}{t^2 s^2\left(\dfrac{1}{t} - \dfrac{1}{s}\right)}$

$= \dfrac{s^2 + t^2}{ts^2 - t^2 s}$

$= \dfrac{s^2 + t^2}{st(s - t)}$

63. $\dfrac{4a}{a^2 - ab - 2b^2} - \dfrac{6b - a}{a^2 + 4ab + 3b^2}$

$= \dfrac{4a}{(a - 2b)(a + b)} - \dfrac{6b - a}{(a + 3b)(a + b)}$

The LCD is $(a + 3b)(a - 2b)(a + b)$.

$= \dfrac{4a(a + 3b)}{(a - 2b)(a + b)(a + 3b)}$

$\quad - \dfrac{(6b - a)(a - 2b)}{(a + 3b)(a + b)(a - 2b)}$

$= \dfrac{4a(a + 3b) - (6b - a)(a - 2b)}{(a + 3b)(a + b)(a - 2b)}$

$= \dfrac{4a^2 + 12ab - (6ab - 12b^2 - a^2 + 2ab)}{(a + 3b)(a + b)(a - 2b)}$

$= \dfrac{4a^2 + 12ab - 6ab + 12b^2 + a^2 - 2ab}{(a + 3b)(a + b)(a - 2b)}$

$= \dfrac{5a^2 + 4ab + 12b^2}{(a + 3b)(a + b)(a - 2b)}$

64. $(-a^4 + 19a^2 + 18a + 15) \div (a + 4)$

Insert 0 for the missing a^3-term.

$$\begin{array}{r|rrrrr} -4 & -1 & 0 & 19 & 18 & 15 \\ & & 4 & -16 & -12 & -24 \\ \hline & -1 & 4 & 3 & 6 & -9 \end{array} \leftarrow \textit{Remainder}$$

Answer: $-a^3 + 4a^2 + 3a + 6 + \dfrac{-9}{a + 4}$

65. $\dfrac{12y^4 + 7y^2 - 2y + 1}{3y^2 + 1}$

$$
\begin{array}{r}
4y^2 \qquad\quad + 1 \\
3y^2 + 1 \,\overline{\smash{\big)}\, 12y^4 + 7y^2 - 2y + 1} \\
\underline{12y^4 + 4y^2} \qquad\qquad \\
3y^2 - 2y + 1 \\
\underline{3y^2 \qquad + 1} \\
-2y \\
\text{Remainder}
\end{array}
$$

Answer: $4y^2 + 1 + \dfrac{-2y}{3y^2 + 1}$

66. $7z^3 - z^2 + 5z - 3 = 0$

Divide the polynomial by $z - 3$.

$$
\begin{array}{r|rrrr}
3 & 7 & -1 & 5 & -3 \\
  &   & 21 & 60 & 195 \\
\hline
  & 7 & 20 & 65 & 192
\end{array}
$$

Since the remainder is not 0, 3 is not a solution of the equation.

67. $\dfrac{x + 3}{x^2 - 5x + 4} - \dfrac{1}{x} = \dfrac{2}{x^2 - 4x}$

$\dfrac{x + 3}{(x - 4)(x - 1)} - \dfrac{1}{x} = \dfrac{2}{x(x - 4)}$

Multiply by the LCD, $x(x - 4)(x - 1)$.
$(x \neq 0, 1, 4)$

$x(x - 4)(x - 1)\left(\dfrac{x + 3}{(x - 4)(x - 1)} - \dfrac{1}{x}\right)$

$= x(x - 4)(x - 1) \cdot \left(\dfrac{2}{x(x - 4)}\right)$

$x(x + 3) - (x - 4)(x - 1) = 2(x - 1)$

$x^2 + 3x - (x^2 - 5x + 4) = 2x - 2$

$x^2 + 3x - x^2 + 5x - 4 = 2x - 2$

$8x - 4 = 2x - 2$

$6x = 2$

$x = \dfrac{1}{3}$

Check $x = \dfrac{1}{3}:\ \dfrac{15}{11} - 3 = -\dfrac{18}{11}$ *True*

Solution set: $\left\{\dfrac{1}{3}\right\}$

68. Solve $A = \dfrac{Rr}{R + r}$ for r.

$A(R + r) = Rr$

$AR + Ar = Rr$

$AR = Rr - Ar$

$AR = (R - A)r$

$\dfrac{AR}{R - A} = r$ or $r = \dfrac{-AR}{A - R}$

69. $1 - \dfrac{5}{r} = \dfrac{-4}{r^2}$

Multiply by the LCD, r^2. $(r \neq 0)$

$r^2\left(1 - \dfrac{5}{r}\right) = r^2\left(\dfrac{-4}{r^2}\right)$

$r^2 - 5r = -4$

$r^2 - 5r + 4 = 0$

$(r - 4)(r - 1) = 0$

$r - 4 = 0$ or $r - 1 = 0$

$r = 4$ or $r = 1$

Check $r = 1:\ 1 - 5 = -4$ *True*

Check $r = 4:\ 1 - \dfrac{5}{4} = -\dfrac{1}{4}$ *True*

Solution set: $\{1, 4\}$

70. $\dfrac{3x}{x - 4} + \dfrac{2}{x} = \dfrac{48}{x^2 - 4x}$

$\dfrac{3x}{x - 4} + \dfrac{2}{x} = \dfrac{48}{x(x - 4)}$

Multiply by the LCD, $x(x - 4)$. $(x \neq 0, 4)$

$x(x - 4)\left(\dfrac{3x}{x - 4} + \dfrac{2}{x}\right) = x(x - 4)\left(\dfrac{48}{x(x - 4)}\right)$

$3x^2 + 2(x - 4) = 48$

$3x^2 + 2x - 8 = 48$

$3x^2 + 2x - 56 = 0$

$(3x + 14)(x - 4) = 0$

$3x + 14 = 0$ or $x - 4 = 0$

$3x = -14$

$x = -\dfrac{14}{3}$ or $x = 4$

The number 4 is not allowed as a solution because substituting it in the original equation results in division by 0.

Check $x = -\dfrac{14}{3}:\ \dfrac{21}{13} - \dfrac{3}{7} = \dfrac{108}{91}$ *True*

Solution set: $\left\{-\dfrac{14}{3}\right\}$

71. $a = \dfrac{337}{d}$

(a) $a = \dfrac{337}{40.5}$ *Let d=40.5*

≈ 8.32

(b) $7.51 = \dfrac{337}{d}$ *Let a=7.51*

$7.51d = 337$

$d = \dfrac{337}{7.51} \approx 44.9$

72. Let $x =$ the average speed of the automobile. Complete the table.

	d	r	t
Automobile	7	x	$\dfrac{7}{x}$
Airplane	100	$x + 558$	$\dfrac{100}{x + 558}$

The times are equal.

$$\frac{7}{x} = \frac{100}{x + 558}$$
$$7(x + 558) = 100x$$
$$7x + 3906 = 100x$$
$$3906 = 93x$$
$$x = \frac{3906}{93} = 42$$

The speed of the automobile is 42 kilometers per hour and the speed of the airplane is $42 + 558 = 600$ kilometers per hour.

73. Let $x =$ the time to fill the tub working together. Make a table.

	Rate	Time Together	Part of Job Done
Hot	$\dfrac{1}{20}$	x	$\dfrac{x}{20}$
Cold	$\dfrac{1}{15}$	x	$\dfrac{x}{15}$

Part done by hot	plus	part done by cold	equals	1 whole job.
$\dfrac{x}{20}$	$+$	$\dfrac{x}{15}$	$=$	1

Multiply by the LCD, 60.

$$60\left(\frac{x}{20} + \frac{x}{15}\right) = 60 \cdot 1$$
$$3x + 4x = 60$$
$$7x = 60$$
$$x = \frac{60}{7} \text{ or } 8\frac{4}{7}$$

Working together, the tub can be filled in $\dfrac{60}{7}$ or $8\dfrac{4}{7}$ minutes.

74. Let $x =$ the cost of 13 gallons. Write a proportion.

$$\frac{x}{13} = \frac{4.86}{3}$$
$$x = \frac{13(4.86)}{3}$$
$$= 21.06$$

The cost of 13 gallons of unleaded gasoline is $21.06.

Chapter 6 Test

1. $f(k) = \dfrac{2k - 1}{3k^2 + 2k - 8}$

Set the denominator equal to zero and solve.

$$3k^2 + 2k - 8 = 0$$
$$(3k - 4)(k + 2) = 0$$

$$3k - 4 = 0 \quad \text{or} \quad k + 2 = 0$$
$$3k = 4$$
$$k = \frac{4}{3} \quad \text{or} \quad k = -2$$

The numbers -2 and $\dfrac{4}{3}$ make the rational expression undefined and are excluded from the domain of f.

2. $\dfrac{6x^2 - 13x - 5}{9x^3 - x} = \dfrac{(3x + 1)(2x - 5)}{x(9x^2 - 1)}$

$$= \frac{(3x + 1)(2x - 5)}{x(3x + 1)(3x - 1)}$$
$$= \frac{2x - 5}{x(3x - 1)}$$

3. $\dfrac{4x^2y^5}{7xy^8} \div \dfrac{8xy^6}{21xy}$

Multiply by the reciprocal.

$$= \frac{4x^2y^5}{7xy^8} \cdot \frac{21xy}{8xy^6}$$
$$= \frac{4 \cdot 3 \cdot 7x^3y^6}{7 \cdot 2 \cdot 4x^2y^{14}} = \frac{3x}{2y^8}$$

4. $\dfrac{y^2 - 16}{y^2 - 25} \cdot \dfrac{y^2 + 2y - 15}{y^2 - 7y + 12}$

$$= \frac{(y + 4)(y - 4)}{(y + 5)(y - 5)} \cdot \frac{(y + 5)(y - 3)}{(y - 4)(y - 3)}$$
$$= \frac{y + 4}{y - 5}$$

5. $\dfrac{x^2 - 9}{x^3 + 3x^2} \div \dfrac{x^2 + x - 12}{x^3 + 9x^2 + 20x}$

Multiply by the reciprocal.

$$= \frac{x^2 - 9}{x^3 + 3x^2} \cdot \frac{x(x^2 + 9x + 20)}{x^2 + x - 12}$$
$$= \frac{(x + 3)(x - 3)}{x^2(x + 3)} \cdot \frac{x(x + 5)(x + 4)}{(x + 4)(x - 3)}$$
$$= \frac{x + 5}{x}$$

6. $t^2 + t - 6, t^2 + 3t, t^2$

Factor each denominator.

$$t^2 + t - 6 = (t + 3)(t - 2)$$
$$t^2 + 3t = t(t + 3)$$

The third denominator is already in factored form. The LCD is

$$t^2(t + 3)(t - 2).$$

7. $\dfrac{7}{6t^2} - \dfrac{1}{3t}$

The LCD is $6t^2$.

$$= \frac{7}{6t^2} - \frac{1 \cdot 2t}{3t \cdot 2t}$$

$$= \frac{7}{6t^2} - \frac{2t}{6t^2} = \frac{7 - 2t}{6t^2}$$

8. $\dfrac{9}{x-7} + \dfrac{4}{x+7}$

The LCD is $(x-7)(x+7)$.

$$= \frac{9(x+7)}{(x-7)(x+7)} + \frac{4(x-7)}{(x+7)(x-7)}$$

$$= \frac{9(x+7) + 4(x-7)}{(x-7)(x+7)}$$

$$= \frac{9x + 63 + 4x - 28}{(x-7)(x+7)}$$

$$= \frac{13x + 35}{(x-7)(x+7)}$$

9. $\dfrac{6}{x+4} + \dfrac{1}{x+2} - \dfrac{3x}{x^2 + 6x + 8}$

$$= \frac{6}{x+4} + \frac{1}{x+2} - \frac{3x}{(x+4)(x+2)}$$

The LCD is $(x+4)(x+2)$.

$$= \frac{6(x+2)}{(x+4)(x+2)} + \frac{1(x+4)}{(x+2)(x+4)}$$

$$- \frac{3x}{(x+4)(x+2)}$$

$$= \frac{6(x+2) + x + 4 - 3x}{(x+4)(x+2)}$$

$$= \frac{6x + 12 + x + 4 - 3x}{(x+4)(x+2)}$$

$$= \frac{4x + 16}{(x+4)(x+2)}$$

$$= \frac{4(x+4)}{(x+4)(x+2)} = \frac{4}{x+2}$$

10. $\dfrac{\dfrac{12}{r+4}}{\dfrac{11}{6r+24}} = \dfrac{12}{r+4} \div \dfrac{11}{6r+24}$

Multiply by the reciprocal.

$$= \frac{12}{r+4} \cdot \frac{6r + 24}{11}$$

$$= \frac{12}{r+4} \cdot \frac{6(r+4)}{11} = \frac{72}{11}$$

11. $\dfrac{\dfrac{1}{a} - \dfrac{1}{b}}{\dfrac{a}{b} - \dfrac{b}{a}}$

Multiply the numerator and the denominator by the LCD of all the fractions, ab.

$$= \frac{ab\left(\dfrac{1}{a} - \dfrac{1}{b}\right)}{ab\left(\dfrac{a}{b} - \dfrac{b}{a}\right)} = \frac{b - a}{a^2 - b^2}$$

$$= \frac{b - a}{(a-b)(a+b)} = \frac{(-1)(a-b)}{(a-b)(a+b)}$$

$$= \frac{-1}{a+b}$$

12. $\dfrac{16p^3 - 32p^2 + 24p}{4p^2}$

$$= \frac{16p^3}{4p^2} - \frac{32p^2}{4p^2} + \frac{24p}{4p^2}$$

$$= 4p - 8 + \frac{6}{p}$$

13. $\dfrac{9q^4 - 18q^3 + 11q^2 + 10q - 10}{3q - 2}$

$$
\require{enclose}
\begin{array}{r}
3q^3 - 4q^2 + q + 4 \\
3q - 2 \enclose{longdiv}{9q^4 - 18q^3 + 11q^2 + 10q - 10} \\
\underline{9q^4 - 6q^3} \\
-12q^3 + 11q^2 \\
\underline{-12q^3 + 8q^2} \\
3q^2 + 10q \\
\underline{3q^2 - 2q} \\
12q - 10 \\
\underline{12q - 8} \\
-2
\end{array}
$$

Remainder

Answer: $3q^3 - 4q^2 + q + 4 + \dfrac{-2}{3q - 2}$

14. $\dfrac{6y^4 - 4y^3 + 5y^2 + 6y - 9}{2y^2 + 3}$

$$
\require{enclose}
\begin{array}{r}
3y^2 - 2y - 2 \\
2y^2 + 3 \enclose{longdiv}{6y^4 - 4y^3 + 5y^2 + 6y - 9} \\
\underline{6y^4 + 9y^2} \\
-4y^3 - 4y^2 + 6y \\
\underline{-4y^3 - 6y} \\
-4y^2 + 12y - 9 \\
\underline{-4y^2 - 6} \\
12y - 3
\end{array}
$$

Remainder

Answer: $3y^2 - 2y - 2 + \dfrac{12y - 3}{2y^2 + 3}$

15. $x^4 - 8x^3 + 21x^2 - 14x - 24 = 0$

Divide the polynomial by $x - 4$.

$$
\begin{array}{r|rrrrr}
4 & 1 & -8 & 21 & -14 & -24 \\
  &   & 4 & -16 & 20 & 24 \\
\hline
  & 1 & -4 & 5 & 6 & 0
\end{array}
$$

Since the remainder is 0, 4 is a solution of the equation.

16. $\left(9x^5 + 40x^4 - 23x^3 + 8x^2 - 6x + 22\right) \div (x + 5)$

$$
\begin{array}{r|rrrrrr}
-5 & 9 & 40 & -23 & 8 & -6 & 22 \\
   &   & -45 & 25 & -10 & 10 & -20 \\
\hline
   & 9 & -5 & 2 & -2 & 4 & 2
\end{array}
$$

Remainder

Answer: $9x^4 - 5x^3 + 2x^2 - 2x + 4 + \dfrac{2}{x + 5}$

17. (a) Simplify this expression.

$$\frac{2x}{3} + \frac{x}{4} - \frac{11}{2}$$

The LCD is 12.

$$
\begin{aligned}
&= \frac{2x \cdot 4}{3 \cdot 4} + \frac{x \cdot 3}{4 \cdot 3} - \frac{11 \cdot 6}{2 \cdot 6} \\
&= \frac{8x}{12} + \frac{3x}{12} - \frac{66}{12} \\
&= \frac{8x + 3x - 66}{12} \\
&= \frac{11x - 66}{12} = \frac{11(x - 6)}{12}
\end{aligned}
$$

(b) Solve this equation.

$$\frac{2x}{3} + \frac{x}{4} = \frac{11}{2}$$

Multiply by the LCD, 12.

$$
\begin{aligned}
12\left(\frac{2x}{3} + \frac{x}{4}\right) &= 12\left(\frac{11}{2}\right) \\
8x + 3x &= 66 \\
11x &= 66 \\
x &= 6
\end{aligned}
$$

Check $x = 6$: $4 + \dfrac{3}{2} = \dfrac{11}{2}$ *True*

Solution set: $\{6\}$

18.

$$\frac{1}{x} - \frac{4}{3x} = \frac{1}{x - 2}$$

Multiply by the LCD, $3x(x - 2)$. $(x \neq 0, 2)$

$$
\begin{aligned}
3x(x - 2)\left(\frac{1}{x} - \frac{4}{3x}\right) &= 3x(x - 2)\left(\frac{1}{x - 2}\right) \\
3(x - 2) - 4(x - 2) &= 3x \\
3x - 6 - 4x + 8 &= 3x \\
-x + 2 &= 3x \\
-4x &= -2 \\
x &= \frac{1}{2}
\end{aligned}
$$

Check $x = \dfrac{1}{2}$: $2 - \dfrac{8}{3} = -\dfrac{2}{3}$ *True*

Solution set: $\left\{\dfrac{1}{2}\right\}$

19.

$$\frac{y}{y + 2} - \frac{1}{y - 2} = \frac{8}{y^2 - 4}$$

$$\frac{y}{y + 2} - \frac{1}{y - 2} = \frac{8}{(y + 2)(y - 2)}$$

Multiply by the LCD, $(y + 2)(y - 2)$.

$(y \neq -2, 2)$

$$
\begin{aligned}
(y + 2)(y - 2)&\left(\frac{y}{y + 2} - \frac{1}{y - 2}\right) \\
&= (y + 2)(y - 2)\left(\frac{8}{(y + 2)(y - 2)}\right) \\
y(y - 2) - 1(y + 2) &= 8 \\
y^2 - 2y - y - 2 &= 8 \\
y^2 - 3y - 10 &= 0 \\
(y - 5)(y + 2) &= 0
\end{aligned}
$$

$$
\begin{array}{ccc}
y - 5 = 0 & \text{or} & y + 2 = 0 \\
y = 5 & \text{or} & y = -2
\end{array}
$$

The number -2 is not allowed as a solution because substituting it in the original equation results in division by 0.

Check $y = 5$: $\dfrac{5}{7} - \dfrac{1}{3} = \dfrac{8}{21}$ *True*

Solution set: $\{5\}$

20. A solution cannot make a denominator 0. This is illustrated in Exercise 19 where -2 is not allowed as a solution for this reason.

21. Solve $S = \dfrac{n}{2}(a + l)$ for l.

To solve for l, multiply both sides by $\dfrac{2}{n}$.

$$
\begin{aligned}
\frac{2}{n} \cdot S &= \frac{2}{n} \cdot \frac{n}{2}(a + l) \\
\frac{2S}{n} &= a + l \\
\frac{2S}{n} - a = l \quad &\text{or} \quad l = \frac{2S - na}{n}
\end{aligned}
$$

22. Let $x =$ the time to do the job working together. Make a table.

Worker	Rate	Time Together	Part of Job Done
Wayne	$\dfrac{1}{9}$	x	$\dfrac{x}{9}$
Susan	$\dfrac{1}{5}$	x	$\dfrac{x}{5}$

Part done by Wayne	plus	part done by Susan	equals	1 whole job.
$\dfrac{x}{9}$	$+$	$\dfrac{x}{5}$	$=$	1

Multiply by the LCD, 45.

$$45\left(\frac{x}{9} + \frac{x}{5}\right) = 45 \cdot 1$$
$$5x + 9x = 45$$
$$14x = 45$$
$$x = \frac{45}{14} \text{ or } 3\frac{3}{14}$$

Working together, they can do the job in $\dfrac{45}{14}$ or $3\dfrac{3}{14}$ hours.

23. Let $x =$ the speed of the boat in still water.

Use $d = rt$, or $t = \dfrac{d}{r}$, to complete the table.

	d	r	t
Downstream	36	$x+3$	$\dfrac{36}{x+3}$
Upstream	24	$x-3$	$\dfrac{24}{x-3}$

Because the times are equal,

$$\frac{36}{x+3} = \frac{24}{x-3}.$$

Multiply by the LCD, $(x+3)(x-3)$. $(x \neq -3, 3)$

$$(x+3)(x-3)\left(\frac{36}{x+3}\right) = (x+3)(x-3)\left(\frac{24}{x-3}\right)$$
$$36(x-3) = 24(x+3)$$
$$36x - 108 = 24x + 72$$
$$12x = 180$$
$$x = 15$$

The speed of the boat in still water is 15 mph.

24. Let $x =$ the number of fish in Lake Linda. Write a proportion.

$$\frac{x}{600} = \frac{800}{10}$$

Multiply by the LCD, 600.

$$600\left(\frac{x}{600}\right) = 600\left(\frac{800}{10}\right)$$
$$x = 60 \cdot 800$$
$$x = 48,000$$

There are about 48,000 fish in the lake.

25. $g(x) = \dfrac{5x}{2+x}$

(a)
$$3 = \frac{5x}{2+x} \qquad \text{Let } g(x)=3$$
$$3(2+x) = 5x$$
$$6 + 3x = 5x$$
$$6 = 2x$$
$$x = 3 \text{ units}$$

(b) If no food is available, then $x = 0$.

$$g(0) = \frac{5(0)}{2+0}$$
$$= \frac{0}{2} = 0$$

The growth rate is 0.

Cumulative Review Exercises Chapters 1–6

In Exercises 1–2, $x = -4$, $y = 3$, and $z = 6$.

1. $|2x| + 3y - z^3$
$$= |(2)(-4)| + 3(3) - (6)^3$$
$$= |-8| + 9 - 216$$
$$= 8 - 207 = -199$$

2. $\dfrac{x(2x-1)}{3y-z} = \dfrac{-4[2(-4)-1]}{3(3)-6}$
$$= \frac{-4[-8-1]}{9-6}$$
$$= \frac{-4(-9)}{3} = 12$$

3. $7(2x+3) - 4(2x+1) = 2(x+1)$
$$14x + 21 - 8x - 4 = 2x + 2$$
$$6x + 17 = 2x + 2$$
$$4x = -15$$
$$x = -\frac{15}{4}$$

Solution set: $\left\{-\dfrac{15}{4}\right\}$

4. $|6x - 8| - 4 = 0$
$$|6x - 8| = 4$$

$$6x - 8 = 4 \qquad \text{or} \qquad 6x - 8 = -4$$
$$6x = 12 \qquad \qquad \qquad 6x = 4$$
$$x = 2 \qquad \text{or} \qquad x = \frac{4}{6} = \frac{2}{3}$$

Solution set: $\left\{\dfrac{2}{3}, 2\right\}$

5. Solve $ax + by = cx + d$ for x.
Get the x-terms on one side.

$$ax - cx = d - by$$
$$x(a - c) = d - by$$
$$x = \frac{d - by}{a - c}$$
or
$$x = \frac{by - d}{c - a}$$

6.
$$\frac{2}{3}y + \frac{5}{12}y \le 20$$
Multiply both sides by 12.

$$12\left(\frac{2}{3}y + \frac{5}{12}y\right) \le 12(20)$$
$$8y + 5y \le 240$$
$$13y \le 240$$
$$y \le \frac{240}{13}$$

Solution set: $\left(-\infty, \dfrac{240}{13}\right]$

7. $|3x + 2| \ge 4$

$$3x + 2 \ge 4 \quad \text{or} \quad 3x + 2 \le -4$$
$$3x \ge 2 \quad \text{or} \quad 3x \le -6$$
$$x \ge \frac{2}{3} \quad \text{or} \quad x \le -2$$

Solution set: $\left(-\infty, -2\right] \cup \left[\dfrac{2}{3}, \infty\right)$

8. Let $x =$ the number of votes
for the Republicans;
$x + 8.2 =$ the number of votes
for the Democrats.

The total number of votes cast was 96.2 million,
but we must exclude the 9.7 million votes for "All
other candidates."

$$96.2 - 9.7 = 86.5$$

Thus, the total number of votes for Republicans
and Democrats was 86.5 million.

$$x + (x + 8.2) = 86.5$$
$$2x = 78.3$$
$$x = 39.15$$

The Republicans received 39.15 million votes and
the Democrats received $39.15 + 8.2 = 47.35$
million votes.

9. Let $h =$ the height of the triangle.
Use the formula $A = \dfrac{1}{2}bh$. Here, $A = 42$ and
$b = 14$.

$$A = \frac{1}{2}bh$$
$$42 = \frac{1}{2}(14)h$$
$$42 = 7h$$
$$6 = h$$

The height is 6 meters.

10. $-4x + 2y = 8$
To find the x-intercept, let $y = 0$.

$$-4x + 2(0) = 8$$
$$-4x = 8$$
$$x = -2$$

The x-intercept is $(-2, 0)$.
To find the y-intercept, let $x = 0$.

$$-4(0) + 2y = 8$$
$$2y = 8$$
$$y = 4$$

The y-intercept is $(0, 4)$. Plot the intercepts, and
draw the line through them.

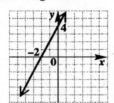

11. Through $(-5, 8)$ and $(-1, 2)$
Let $(x_1, y_1) = (-5, 8)$ and $(x_2, y_2) = (-1, 2)$.
Then,

$$m = \frac{y_2 - y_1}{x_2 - x_1} = \frac{2 - 8}{-1 - (-5)} = \frac{-6}{4} = -\frac{3}{2}.$$

The slope is $-\dfrac{3}{2}$.

12. Perpendicular to $4x - 3y = 12$

Solve for y to write the equation in slope-intercept form and find the slope.

$$4x - 3y = 12$$
$$-3y = -4x + 12$$
$$y = \frac{4}{3}x - 4$$

The slope is $\frac{4}{3}$. Perpendicular lines have slopes that are negative reciprocals of each other. The negative reciprocal of $\frac{4}{3}$ is $-\frac{3}{4}$. The slope of a line perpendicular to the given line is $-\frac{3}{4}$.

13. Use $(x_1, y_1) = (-5, 8)$ and $m = -\frac{3}{2}$ in the point-slope form.

$$y - y_1 = m(x - x_1)$$
$$y - 8 = -\frac{3}{2}[x - (-5)]$$
$$y - 8 = -\frac{3}{2}(x + 5)$$
$$y - 8 = -\frac{3}{2}x - \frac{15}{2}$$
$$y = -\frac{3}{2}x + \frac{1}{2}$$

14. $2x + 5y > 10$

Graph the line $2x + 5y = 10$ by drawing a dashed line (since the inequality involves $>$) through the intercepts $(5, 0)$ and $(0, 2)$.

Test a point not on this line, such as $(0, 0)$.

$$2x + 5y > 10$$
$$2(0) + 5(0) > 10$$
$$0 > 10 \quad \textit{False}$$

Shade the side of the line not containing $(0, 0)$.

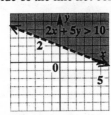

15. $x - y \geq 3$ and $3x + 4y \leq 12$

Graph the solid line $x - y = 3$ through $(3, 0)$ and $(0, -3)$. The inequality $x - y \geq 3$ can be written as $y \leq x - 3$, so shade the region below the boundary line.

Graph the solid line $3x + 4y = 12$ through $(4, 0)$ and $(0, 3)$. The inequality $3x + 4y \leq 12$ can be written as $y \leq -\frac{3}{4}x + 3$, so shade the region below the boundary line.

The required graph is the common shaded area as well as the portions of the lines that bound it.

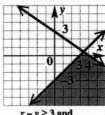

$$x - y \geq 3 \text{ and}$$
$$3x + 4y \leq 12$$

16. The relation is a function since each country is assigned one amount. The domain is the set of countries, {Venezuela, Canada, Saudi Arabia, Mexico}. The range is the set of amounts in thousand barrels per day, $\{1657, 1415, 1363, 1240\}$.

17. The relation is not a function since a vertical line may intersect the graph in more than one point. The domain is the set of x-values, $[-2, \infty)$. The range is the set of y-values, $(-\infty, \infty)$.

18. $y = -\sqrt{x + 2}$

The relation is a function since each input value corresponds to exactly one output value. The radicand must be nonnegative; that is, $x + 2 \geq 0$, or $x \geq -2$. The domain is $[-2, \infty)$. The values of $\sqrt{x + 2}$ are nonnegative, so the values of $-\sqrt{x + 2}$ are nonpositive. Thus, the range is $(-\infty, 0]$.

19. (a) Solve the equation for y.

$$5x - 3y = 8$$
$$5x - 8 = 3y$$
$$\frac{5x - 8}{3} = y$$

So $f(x) = \frac{5x - 8}{3}$ or $f(x) = \frac{5}{3}x - \frac{8}{3}$.

(b) $f(1) = \frac{5(1) - 8}{3} = \frac{-3}{3} = -1$

20. Let $V =$ the volume of a sphere and $d =$ its diameter.

V varies directly as the cube of d, so

$$V = kd^3$$

for some constant k. $V = 1.4 \times 10^{15}$ when $d = 1.4 \times 10^5$.

$$1.4 \times 10^{15} = k(1.4 \times 10^5)^3$$
$$k = \frac{1.4 \times 10^{15}}{(1.4 \times 10^5)^3} \approx .51$$

Thus, $V = .51d^3$. When d is 1.3×10^4,

$$V = .51(1.3 \times 10^4)^3$$
$$\approx 1.1 \times 10^{12}$$

The volume of Earth is about 1.1×10^{12} km^3.

21. **(a)** $4x - y = -7$ (1)
        $5x + 2y = 1$ (2)

To eliminate y, multiply equation (1) by 2 and add the result to (2).

$$\begin{array}{rcll} 8x - 2y &=& -14 & 2 \times (1) \\ 5x + 2y &=& 1 & (2) \\ \hline 13x &=& -13 & \\ x &=& -1 & \end{array}$$

To find y, substitute -1 for x in (1).

$$\begin{aligned} 4x - y &= -7 \quad (1) \\ 4(-1) - y &= -7 \\ -4 - y &= -7 \\ -y &= -3 \\ y &= 3 \end{aligned}$$

Solution set: $\{(-1, 3)\}$

(b) Write the augmented matrix.

$$\begin{bmatrix} 4 & -1 & -7 \\ 5 & 2 & 1 \end{bmatrix}$$

$$\begin{bmatrix} 1 & 3 & 8 \\ 5 & 2 & 1 \end{bmatrix} \quad R_2 - R_1$$

$$\begin{bmatrix} 1 & 3 & 8 \\ 0 & -13 & -39 \end{bmatrix} \quad -5R_1 + R_2$$

$$\begin{bmatrix} 1 & 3 & 8 \\ 0 & 1 & 3 \end{bmatrix} \quad -\frac{1}{13}R_2$$

From the last row, $y = 3$. Substitute 3 for y in the equation given by the first row.

$$\begin{aligned} x + 3y &= 8 \\ x + 3(3) &= 8 \\ x &= -1 \end{aligned}$$

Solution set: $\{(-1, 3)\}$

22. We'll use matrix row operations.

$$\begin{bmatrix} 1 & 1 & -2 & -1 \\ 2 & -1 & 1 & -6 \\ 3 & 2 & -3 & -3 \end{bmatrix}$$

$$\begin{bmatrix} 1 & 1 & -2 & -1 \\ 0 & -3 & 5 & -4 \\ 0 & -1 & 3 & 0 \end{bmatrix} \begin{array}{l} -2R_1 + R_2 \\ -3R_1 + R_3 \end{array}$$

$$\begin{bmatrix} 1 & 1 & -2 & -1 \\ 0 & 1 & -3 & 0 \\ 0 & -3 & 5 & -4 \end{bmatrix} \quad -R_3 \leftrightarrow R_2$$

$$\begin{bmatrix} 1 & 1 & -2 & -1 \\ 0 & 1 & -3 & 0 \\ 0 & 0 & -4 & -4 \end{bmatrix} \quad 3R_2 + R_3$$

$$\begin{bmatrix} 1 & 1 & -2 & -1 \\ 0 & 1 & -3 & 0 \\ 0 & 0 & 1 & 1 \end{bmatrix} \quad -\frac{1}{4}R_3$$

This matrix gives the system

$$\begin{aligned} x + y - 2z &= -1 \\ y - 3z &= 0 \\ z &= 1 \end{aligned}$$

Substitute 1 for z in the second equation.

$$\begin{aligned} y - 3(1) &= 0 \\ y &= 3 \end{aligned}$$

Substitute 1 for z and 3 for y in the first equation.

$$\begin{aligned} x + 3 - 2(1) &= -1 \\ x + 1 &= -1 \\ x &= -2 \end{aligned}$$

Solution set: $\{(-2, 3, 1)\}$

23. Let $x =$ the value of shipments of all other tobacco products;
$y =$ the value of shipments of cigarettes.

The total value of shipments was \$33 million, so

$$x + y = 33. \quad (1)$$

The value of shipments of cigarettes was \$16.8 million more than twice the value of shipments of all other tobacco products, so

$$x = 2y + 16.8. \quad (2)$$

Substitute $2y + 16.8$ for x in (1).

$$\begin{aligned} (2y + 16.8) + y &= 33 \\ 3y + 16.8 &= 33 \\ 3y &= 16.2 \\ y &= 5.4 \end{aligned}$$

From (2), $x = 2(5.4) + 16.8 = 27.6$.
The value of shipments of cigarettes was \$27.6 million and the value of all other tobacco product shipments was \$5.4 million.

24. $\begin{vmatrix} 2 & 0 & -3 \\ 5 & 1 & 0 \\ -2 & 0 & 8 \end{vmatrix}$ Expand along row 1.

$$= 2\begin{vmatrix} 1 & 0 \\ 0 & 8 \end{vmatrix} - 0 + (-3)\begin{vmatrix} 5 & 1 \\ -2 & 0 \end{vmatrix}$$

$$= 2(8 - 0) - 3(0 + 2)$$
$$= 16 - 6 = 10$$

25. $\left(\dfrac{a^{-3}b^4}{a^2b^{-1}}\right)^{-2} = \left(\dfrac{b^4b^1}{a^2a^3}\right)^{-2} = \left(\dfrac{b^5}{a^5}\right)^{-2}$

$\qquad = \left(\dfrac{a^5}{b^5}\right)^2 = \dfrac{a^{10}}{b^{10}}$

26. $\left(\dfrac{m^{-4}n^2}{m^2n^{-3}}\right) \cdot \left(\dfrac{m^5n^{-1}}{m^{-2}n^5}\right)$

$\qquad = \left(\dfrac{n^2n^3}{m^2m^4}\right) \cdot \left(\dfrac{m^5m^2}{n^1n^5}\right)$

$\qquad = \left(\dfrac{n^5}{m^6}\right) \cdot \left(\dfrac{m^7}{n^6}\right)$

$\qquad = \dfrac{n^5m^7}{n^6m^6} = \dfrac{m}{n}$

27. $(3y^2 - 2y + 6) - (-y^2 + 5y + 12)$

$\qquad = 3y^2 - 2y + 6 + y^2 - 5y - 12$

$\qquad = 4y^2 - 7y - 6$

28. $-6x^4(x^2 - 3x + 2)$

$\qquad = -6x^6 + 18x^5 - 12x^4$

29. $(4f + 3)(3f - 1) = 12f^2 - 4f + 9f - 3$

$\qquad = 12f^2 + 5f - 3$

30. $(7t^3 + 8)(7t^3 - 8)$

This is the product of the sum and difference of two terms.

$\qquad = (7t^3)^2 - 8^2$

$\qquad = 49t^6 - 64$

31. $\left(\dfrac{1}{4}x + 5\right)^2$

Use the formula for the square of a binomial,
$(a + b)^2 = a^2 + 2ab + b^2$.

$\qquad = \left(\dfrac{1}{4}x\right)^2 + 2\left(\dfrac{1}{4}x\right)(5) + 5^2$

$\qquad = \dfrac{1}{16}x^2 + \dfrac{5}{2}x + 25$

32. $(3x^3 + 13x^2 - 17x - 7) \div (3x + 1)$

$$\begin{array}{r}
x^2 + 4x - 7 \\
3x + 1{\overline{\smash{\big)}\,3x^3 + 13x^2 - 17x - 7}} \\
\underline{3x^3 + x^2} \\
12x^2 - 17x \\
\underline{12x^2 + 4x} \\
-21x - 7 \\
\underline{-21x - 7} \\
0
\end{array}$$

Answer: $x^2 + 4x - 7$

33. $(2x^4 + 3x^3 - 8x^2 + x + 2) \div (x - 1)$

$$\begin{array}{r|rrrrr}
1 & 2 & 3 & -8 & 1 & 2 \\
  &   & 2 & 5 & -3 & -2 \\
\hline
  & 2 & 5 & -3 & -2 & 0
\end{array}$$

Answer: $2x^3 + 5x^2 - 3x - 2$

34. $2x^2 - 13x - 45 = (2x + 5)(x - 9)$

35. $100t^4 - 25 = 25(4t^4 - 1)$

$\qquad = 25\left[(2t^2)^2 - 1^2\right]$

$\qquad = 25(2t^2 + 1)(2t^2 - 1)$

36. Use the sum of two cubes formula,

$\qquad a^3 + b^3 = (a + b)(a^2 - ab + b^2)$.

$\qquad 8p^3 + 125 = (2p)^3 + 5^3$

$\qquad = (2p + 5)(4p^2 - 10p + 25)$

37. $\qquad 3x^2 + 4x = 7$

$\qquad 3x^2 + 4x - 7 = 0$

$\qquad (3x + 7)(x - 1) = 0$

$\qquad 3x + 7 = 0 \qquad \text{or} \quad x - 1 = 0$

$\qquad 3x = -7$

$\qquad x = -\dfrac{7}{3} \quad \text{or} \qquad x = 1$

Solution set: $\left\{-\dfrac{7}{3}, 1\right\}$

38. $\dfrac{y^2 - 16}{y^2 - 8y + 16} = \dfrac{(y + 4)(y - 4)}{(y - 4)(y - 4)}$

$\qquad = \dfrac{y + 4}{y - 4}$

39. $\dfrac{8x^2 - 18}{8x^2 + 4x - 12} = \dfrac{2(4x^2 - 9)}{4(2x^2 + x - 3)}$

$\qquad = \dfrac{2(2x + 3)(2x - 3)}{4(2x + 3)(x - 1)}$

$\qquad = \dfrac{2x - 3}{2(x - 1)}$

40. $\dfrac{2a^2}{a + b} \cdot \dfrac{a - b}{4a} = \dfrac{2a^2(a - b)}{4a(a + b)}$

$\qquad = \dfrac{a(a - b)}{2(a + b)}$

41. $\dfrac{x + 4}{x - 2} + \dfrac{2x - 10}{x - 2} = \dfrac{x + 4 + 2x - 10}{x - 2}$

$\qquad = \dfrac{3x - 6}{x - 2}$

$\qquad = \dfrac{3(x - 2)}{x - 2} = 3$

42. $\dfrac{2x}{2x-1} + \dfrac{4}{2x+1} + \dfrac{8}{4x^2-1}$

$= \dfrac{2x}{2x-1} + \dfrac{4}{2x+1} + \dfrac{8}{(2x+1)(2x-1)}$

The LCD is $(2x+1)(2x-1)$.

$= \dfrac{2x(2x+1)}{(2x-1)(2x+1)} + \dfrac{4(2x-1)}{(2x+1)(2x-1)}$

$\quad + \dfrac{8}{(2x+1)(2x-1)}$

$= \dfrac{2x(2x+1) + 4(2x-1) + 8}{(2x+1)(2x-1)}$

$= \dfrac{4x^2 + 2x + 8x - 4 + 8}{(2x+1)(2x-1)}$

$= \dfrac{4x^2 + 10x + 4}{(2x+1)(2x-1)}$

$= \dfrac{2(2x^2 + 5x + 2)}{(2x+1)(2x-1)}$

$= \dfrac{2(2x+1)(x+2)}{(2x+1)(2x-1)}$

$= \dfrac{2(x+2)}{2x-1}$

43. $\dfrac{-3x}{x+1} + \dfrac{4x+1}{x} = \dfrac{-3}{x^2+x}$

$\dfrac{-3x}{x+1} + \dfrac{4x+1}{x} = \dfrac{-3}{x(x+1)}$

Multiply by the LCD, $x(x+1)$. $(x \neq -1, 0)$

$x(x+1)\left(\dfrac{-3x}{x+1} + \dfrac{4x+1}{x}\right)$

$\qquad\qquad = x(x+1)\left(\dfrac{-3}{x(x+1)}\right)$

$x(-3x) + (x+1)(4x+1) = -3$

$-3x^2 + 4x^2 + x + 4x + 1 = -3$

$x^2 + 5x + 4 = 0$

$(x+4)(x+1) = 0$

$x+4 = 0 \quad$ or $\quad x+1 = 0$

$x = -4 \quad$ or $\qquad x = -1$

The number -1 is not allowed as a solution because substituting it in the original equation results in division by 0.

Check $x = -4$: $\quad -4 + \dfrac{15}{4} = -\dfrac{1}{4} \quad$ *True*

Solution set: $\{-4\}$

44. Solve $\dfrac{1}{f} = \dfrac{1}{p} + \dfrac{1}{q}$ for q.

Multiply by the LCD, fpq.

$fpq\left(\dfrac{1}{f}\right) = fpq\left(\dfrac{1}{p} + \dfrac{1}{q}\right)$

$pq = fq + fp$

Get all terms with q on one side.

$pq - fq = fp$

$q(p - f) = fp$

$q = \dfrac{fp}{p - f}$

or $\qquad q = \dfrac{-fp}{f - p}$

45. Let $x =$ the speed of the plane in still air.

Use $d = rt$, or $t = \dfrac{d}{r}$, to make a table.

	d	r	t
Against the Wind	200	$x - 30$	$\dfrac{200}{x-30}$
With the Wind	300	$x + 30$	$\dfrac{300}{x+30}$

The times are the same, so

$$\dfrac{200}{x - 30} = \dfrac{300}{x + 30}.$$

Multiply by the LCD, $(x - 30)(x + 30)$.

$(x - 30)(x + 30)\left(\dfrac{200}{x - 30}\right)$

$\qquad = (x - 30)(x + 30)\left(\dfrac{300}{x + 30}\right)$

$200(x + 30) = 300(x - 30)$

$200x + 6000 = 300x - 9000$

$-100x = -15{,}000$

$x = 150$

The speed of the plane in still air is 150 mph.

46.   Let $x =$ the time to complete the job
                   working together.

Make a table.

	Rate	Time Together	Part of Job Done
Machine A	$\frac{1}{2}$	x	$\frac{x}{2}$
Machine B	$\frac{1}{3}$	x	$\frac{x}{3}$

Part done by A	plus	part done by B	equals	1 whole job.
$\frac{x}{2}$	$+$	$\frac{x}{3}$	$=$	1

Multiply by the LCD, 6.

$$6\left(\frac{x}{2} + \frac{x}{3}\right) = 6 \cdot 1$$
$$3x + 2x = 6$$
$$5x = 6$$
$$x = \frac{6}{5} \text{ or } 1\frac{1}{5}$$

Working together, the machines can complete the job in $\frac{6}{5}$ or $1\frac{1}{5}$ hours.

CHAPTER 7   ROOTS AND RADICALS

Section 7.1

1. $2^{1/2} = \sqrt[2]{2^1} = \sqrt{2}$ (C)

3. $-\sqrt{16} = -(4) = -4$ (A)

5. $(-32)^{1/5} = [(-2)^5]^{1/5} = (-2)^1 = -2$ (H)

7. $4^{3/2} = [(2^2)^{1/2}]^3 = 2^3 = 8$ (B)

9. $-6^{1/2} = -(6^{1/2}) = -\sqrt{6}$ (D)

11. $169^{1/2} = (13^2)^{1/2} = 13$

13. $729^{1/3} = (9^3)^{1/3} = 9$

15. $16^{1/4} = (2^4)^{1/4} = 2$

17. $\left(\dfrac{64}{81}\right)^{1/2} = \left[\left(\dfrac{8}{9}\right)^2\right]^{1/2} = \dfrac{8}{9}$

19. $(-27)^{1/3} = [(-3)^3]^{1/3} = -3$

21. $100^{3/2} = (100^{1/2})^3$
$= 10^3 = 1000$

23. $-4^{5/2} = -(4^{1/2})^5 = -(2)^5 = -32$

25. $(-144)^{1/2}$ is not a real number because no real number squared equals -144.

27. $64^{-3/2} = \dfrac{1}{64^{3/2}} = \dfrac{1}{(64^{1/2})^3} = \dfrac{1}{8^3} = \dfrac{1}{512}$

29. $\left(-\dfrac{8}{27}\right)^{-2/3} = \left(-\dfrac{27}{8}\right)^{2/3}$
$= \left(\left[\left(-\dfrac{3}{2}\right)^3\right]^{1/3}\right)^2$
$= \left(-\dfrac{3}{2}\right)^2 = \dfrac{9}{4}$

31. $(-64)^{1/2}$ is an even root of a negative number. No real number squared will give -64. On the other hand, $-64^{1/2} = -\sqrt{64} = -8$, which is a real number. ($-64^{1/2}$ is the opposite of $64^{1/2}$.)

33. $\sqrt{36} = 6$, since $6^2 = 36$. Notice that -6 is not an answer because the symbol $\sqrt{36}$ means the *nonnegative* square root of 36.

35. $\sqrt{\dfrac{64}{81}} = \dfrac{8}{9}$, since $\left(\dfrac{8}{9}\right)^2 = \dfrac{64}{81}$.

37. $-\sqrt{-169}$ is not a real number since no real number squared equals -169.

39. $\sqrt[3]{216} = 6$, since $6^3 = 216$.

41. $\sqrt[3]{-64} = -4$, since $(-4)^3 = -64$.

43. $-\sqrt[3]{512} = -(8) = -8$, since $8^3 = 512$.

45. $-\sqrt[4]{81} = -(3) = -3$, since $3^4 = 81$.

47. $\sqrt[4]{-16}$ is not a real number since no real number to the fourth power equals -16. Any real number raised to the fourth power is 0 or positive.

49. Since 6 is an even positive integer, $\sqrt[6]{a^6} = |a|$, so $\sqrt[6]{(-2)^6} = |-2| = 2$.

51. Since 5 is odd, $\sqrt[5]{a^5} = a$, so
$$\sqrt[5]{(-9)^5} = -9.$$

53. Since the index is even, $\sqrt{x^2} = |x|$.

55. Since the index is odd, $\sqrt[3]{x^3} = x$.

57. $\sqrt[3]{x^{15}} = \sqrt[3]{(x^5)^3} = x^5$ (3 is odd)

59. $12^{1/2} = \sqrt{12}$

61. $8^{3/4} = (8^{1/4})^3 = (\sqrt[4]{8})^3$

63. $(9q)^{5/8} - (2x)^{2/3}$
$= [(9q)^{1/8}]^5 - [(2x)^{1/3}]^2$
$= (\sqrt[8]{9q})^5 - (\sqrt[3]{2x})^2$

65. $(2m)^{-3/2} = [(2m)^{1/2}]^{-3}$
$= (\sqrt{2m})^{-3}$
$= \dfrac{1}{(\sqrt{2m})^3}$

67. $(2y+x)^{2/3} = [(2y+x)^{1/3}]^2$
$= (\sqrt[3]{2y+x})^2$

69. $(3m^4 + 2k^2)^{-2/3}$
$= \dfrac{1}{(3m^4 + 2k^2)^{2/3}}$
$= \dfrac{1}{[(3m^4 + 2k^2)^{1/3}]^2}$
$= \dfrac{1}{(\sqrt[3]{3m^4 + 2k^2})^2}$

71. We are to show that, in general,

$\sqrt{a^2 + b^2} \neq a + b$.

When $a = 3$ and $b = 4$,

$$\begin{aligned}
\sqrt{a^2 + b^2} &= \sqrt{3^2 + 4^2} \\
&= \sqrt{9 + 16} \\
&= \sqrt{25} = 5,
\end{aligned}$$

but

$$a + b = 3 + 4 = 7.$$

Since $5 \neq 7$, $\sqrt{a^2 + b^2} \neq a + b$.

73. $\sqrt{2^{12}} = 2^{12/2} = 2^6 = 64$

75. $\sqrt[3]{4^9} = 4^{9/3} = 4^3 = 64$

77. $\sqrt{x^{20}} = x^{20/2} = x^{10}$

79. $\sqrt[3]{x} \cdot \sqrt{x} = x^{1/3} \cdot x^{1/2} = x^{1/3+1/2} = x^{2/6+3/6}$
$\qquad = x^{5/6} = \sqrt[6]{x^5}$

81. $\dfrac{\sqrt[3]{t^4}}{\sqrt[5]{t^4}} = \dfrac{t^{4/3}}{t^{4/5}} = \dfrac{t^{20/15}}{t^{12/15}} = t^{20/15 - 12/15}$
$\qquad = t^{8/15} = \sqrt[15]{t^8}$

83. $3^{1/2} \cdot 3^{3/2} = 3^{1/2+3/2} = 3^{4/2} = 3^2 = 9$

85. $\dfrac{64^{5/3}}{64^{4/3}} = 64^{5/3 - 4/3} = 64^{1/3} = \sqrt[3]{64} = 4$

87. $y^{7/3} \cdot y^{-4/3} = y^{7/3 + (-4/3)} = y^{3/3} = y$

89. $\dfrac{k^{1/3}}{k^{2/3} \cdot k^{-1}} = \dfrac{k^{1/3}}{k^{-1/3}} = k^{1/3 - (-1/3)} = k^{2/3}$

91. $\left(27x^{12}y^{15}\right)^{2/3}$
$\qquad = 27^{2/3}x^{12(2/3)}y^{15(2/3)}$
$\qquad = \left(27^{1/3}\right)^2 x^8 y^{10}$
$\qquad = 3^2 x^8 y^{10} = 9x^8 y^{10}$

93. $\dfrac{\left(x^{2/3}\right)^2}{\left(x^2\right)^{7/3}} = \dfrac{x^{4/3}}{x^{14/3}} = x^{4/3 - 14/3}$

$\qquad = x^{-10/3} = \dfrac{1}{x^{10/3}}$

95. $\dfrac{m^{3/4}n^{-1/4}}{\left(m^2 n\right)^{1/2}} = \dfrac{m^{3/4}n^{-1/4}}{m^1 n^{1/2}}$

$\qquad = m^{3/4 - 1}n^{-1/4 - 1/2}$

$\qquad = m^{3/4 - 4/4}n^{-1/4 - 2/4}$

$\qquad = m^{-1/4}n^{-3/4}$

$\qquad = \dfrac{1}{m^{1/4}n^{3/4}}$

97. $\dfrac{p^{1/5}p^{7/10}p^{1/2}}{\left(p^3\right)^{-1/5}} = \dfrac{p^{2/10 + 7/10 + 5/10}}{p^{-3/5}}$

$\qquad = \dfrac{p^{14/10}}{p^{-6/10}}$

$\qquad = p^{14/10 - (-6/10)}$

$\qquad = p^{20/10} = p^2$

99. $\left(\dfrac{b^{-3/2}}{c^{-5/3}}\right)^2 \left(b^{-1/4}c^{-1/3}\right)^{-1}$

$\qquad = \left(\dfrac{c^{5/3}}{b^{3/2}}\right)^2 \left(b^{1/4}c^{1/3}\right)$

$\qquad = \dfrac{c^{10/3}}{b^3}\left(b^{1/4}c^{1/3}\right)$

$\qquad = \dfrac{c^{10/3}b^{1/4}c^{1/3}}{b^3}$

$\qquad = c^{10/3 + 1/3}b^{1/4 - 3}$

$\qquad = c^{11/3}b^{-11/4}$

$\qquad = \dfrac{c^{11/3}}{b^{11/4}}$

101. $\left(\dfrac{p^{-1/4}q^{-3/2}}{3^{-1}p^{-2}q^{-2/3}}\right)^{-2}$

$\qquad = \dfrac{p^{1/2}q^3}{3^2 p^4 q^{4/3}}$

$\qquad = \dfrac{q^{3 - 4/3}}{9p^{4 - 1/2}}$

$\qquad = \dfrac{q^{5/3}}{9p^{7/2}}$

103. $p^{2/3}\left(p^{1/3} + 2p^{4/3}\right)$
$\qquad = p^{2/3}p^{1/3} + p^{2/3}\left(2p^{4/3}\right)$
$\qquad = p^{2/3 + 1/3} + 2p^{2/3 + 4/3}$
$\qquad = p^{3/3} + 2p^{6/3}$
$\qquad = p^1 + 2p^2$
$\qquad = p + 2p^2$

105. $k^{1/4}\left(k^{3/2} - k^{1/2}\right)$
$\qquad = k^{1/4 + 3/2} - k^{1/4 + 1/2}$
$\qquad = k^{1/4 + 6/4} - k^{1/4 + 2/4}$
$\qquad = k^{7/4} - k^{3/4}$

107. $6a^{7/4}\left(a^{-7/4} + 3a^{-3/4}\right)$
$\qquad = 6a^{7/4 + (-7/4)} + 18a^{7/4 + (-3/4)}$
$\qquad = 6a^0 + 18a^{4/4}$
$\qquad = 6(1) + 18a^1 = 6 + 18a$

109. $\sqrt[5]{x^3} \cdot \sqrt[4]{x} = x^{3/5} \cdot x^{1/4}$
$\qquad = x^{3/5 + 1/4}$
$\qquad = x^{12/20 + 5/20}$
$\qquad = x^{17/20}$

111. $\dfrac{\sqrt{x^5}}{x^4} = \dfrac{\left(x^5\right)^{1/2}}{x^4}$

$= x^{5/2-4}$

$= x^{-3/2} = \dfrac{1}{x^{3/2}}$

113. $\sqrt{y} \cdot \sqrt[3]{yz} = y^{1/2} \cdot (yz)^{1/3}$

$= y^{1/2}y^{1/3}z^{1/3}$

$= y^{1/2+1/3}z^{1/3}$

$= y^{3/6+2/6}z^{1/3}$

$= y^{5/6}z^{1/3}$

115. $\sqrt[4]{\sqrt[3]{m}} = \sqrt[4]{m^{1/3}} = \left(m^{1/3}\right)^{1/4} = m^{1/12}$

117. The length $\sqrt{98}$ is closer to $\sqrt{100} = 10$ than to $\sqrt{81} = 9$. The width $\sqrt{26}$ is closer to $\sqrt{25} = 5$ than to $\sqrt{36} = 6$. Use the estimates $L = 10$ and $W = 5$ in $A = LW$ to find an estimate of the area.

$$A \approx 10 \cdot 5 = 50$$

Choice (c) is the best estimate.

In Exercises 119–136, use a calculator and round to three decimal places

119. $\sqrt{9483} \approx 97.381$

121. $\sqrt{284.361} \approx 16.863$

123. $\sqrt{7} \approx 2.646$

125. $-\sqrt{82} \approx -9.055$

127. $\sqrt[3]{423} \approx 7.507$

129. $\sqrt[4]{100} \approx 3.162$

131. $\sqrt[5]{23.8} \approx 1.885$

133. $59^{2/3} \approx 15.155$

135. $26^{-2/5} \approx .272$

137. $f = \dfrac{1}{2\pi\sqrt{LC}}$

$= \dfrac{1}{2\pi\sqrt{(7.237 \times 10^{-5})(2.5 \times 10^{-10})}}$

$\approx 1{,}183{,}235$

or about $1{,}183{,}000$ cycles per second.

139. Since $H = 44 + 6 = 50$ ft, substitute 50 for H in the formula.

$$D = \sqrt{2H} = \sqrt{2 \cdot 50} = \sqrt{100} = 10$$

She will be able to see about 10 miles.

141. Use $T = .07D^{3/2}$ with $D = 16$.

$T = .07(16)^{3/2} = .07\left(16^{1/2}\right)^3$

$= .07(4)^3 = .07(64) = 4.48$

To the nearest tenth of an hour, time T is 4.5 hours.

143. Let $a = b = 246.75$ and $c = 438.14$.

$s = \dfrac{1}{2}(a + b + c)$

$= \dfrac{1}{2}(246.75 + 246.75 + 438.14)$

$= \dfrac{931.64}{2} = 465.82$

$A = \sqrt{s(s-a)(s-b)(s-c)}$

$= \sqrt{465.82(219.07)^2(27.68)}$

$\approx 24{,}875.7$

The area of this enclosure is about $24{,}900$ square feet.

Section 7.2

1. Does $2\sqrt{12} = \sqrt{48}$?

$2\sqrt{12} = 2\sqrt{4 \cdot 3} = 2\sqrt{4} \cdot \sqrt{3} = 2 \cdot 2 \cdot \sqrt{3}$

$= 4\sqrt{3}$

$\sqrt{48} = \sqrt{16 \cdot 3} = \sqrt{16} \cdot \sqrt{3} = 4\sqrt{3}$

The calculator approximation for each expression is 6.92820323. The statement is true.

3. Does $3\sqrt{8} = 2\sqrt{18}$?

$3\sqrt{8} = 3\sqrt{4 \cdot 2} = 3\sqrt{4}\sqrt{2} = 3 \cdot 2\sqrt{2} = 6\sqrt{2}$

$2\sqrt{18} = 2\sqrt{9 \cdot 2} = 2\sqrt{9}\sqrt{2} = 2 \cdot 3\sqrt{2} = 6\sqrt{2}$

The calculator approximation for each expression is 8.485281374. The statement is true.

5. **(a)** $.5 = \dfrac{1}{2}$ so $\sqrt{.5} = \sqrt{\dfrac{1}{2}}$

(b) $\dfrac{2}{4} = \dfrac{1}{2}$ so $\sqrt{\dfrac{2}{4}} = \sqrt{\dfrac{1}{2}}$

(c) $\dfrac{3}{6} = \dfrac{1}{2}$ so $\sqrt{\dfrac{3}{6}} = \sqrt{\dfrac{1}{2}}$

(d) $\dfrac{\sqrt{4}}{\sqrt{16}} = \sqrt{\dfrac{4}{16}} = \sqrt{\dfrac{1}{4}} \neq \sqrt{\dfrac{1}{2}}$

Choice (d) is not equal to $\sqrt{\dfrac{1}{2}}$.

7. $\sqrt{5} \cdot \sqrt{6} = \sqrt{5 \cdot 6} = \sqrt{30}$

9. $\sqrt[3]{7x} \cdot \sqrt[3]{2y} = \sqrt[3]{7x \cdot 2y} = \sqrt[3]{14xy}$

11. $\sqrt[4]{11} \cdot \sqrt[4]{3} = \sqrt[4]{11 \cdot 3} = \sqrt[4]{33}$

13. $\sqrt[3]{7} \cdot \sqrt[4]{3}$ cannot be multiplied using the product rule, because the indexes (3 and 4) are different.

15. To multiply two radical expressions with the same index, multiply the radicands and keep the same index. For example, $\sqrt[3]{3} \cdot \sqrt[3]{5} = \sqrt[3]{15}$.

17. $\sqrt{\dfrac{64}{121}} = \dfrac{\sqrt{64}}{\sqrt{121}} = \dfrac{8}{11}$

19. $\sqrt{\dfrac{3}{25}} = \dfrac{\sqrt{3}}{\sqrt{25}} = \dfrac{\sqrt{3}}{5}$

21. $\sqrt{\dfrac{x}{25}} = \dfrac{\sqrt{x}}{\sqrt{25}} = \dfrac{\sqrt{x}}{5}$

23. $\sqrt{\dfrac{p^6}{81}} = \dfrac{\sqrt{p^6}}{\sqrt{81}} = \dfrac{\sqrt{(p^3)^2}}{9} = \dfrac{p^3}{9}$

25. $\sqrt[3]{\dfrac{27}{64}} = \dfrac{\sqrt[3]{27}}{\sqrt[3]{64}} = \dfrac{3}{4}$

27. $\sqrt[3]{-\dfrac{r^2}{8}} = \dfrac{\sqrt[3]{r^2}}{\sqrt[3]{-8}} = \dfrac{\sqrt[3]{r^2}}{-2} = -\dfrac{\sqrt[3]{r^2}}{2}$

29. $-\sqrt[4]{\dfrac{81}{x^4}} = -\dfrac{\sqrt[4]{3^4}}{\sqrt[4]{x^4}} = -\dfrac{3}{x}$

31. $\sqrt[5]{\dfrac{1}{x^{15}}} = \dfrac{\sqrt[5]{1}}{\sqrt[5]{(x^3)^5}} = \dfrac{1}{x^3}$

33. $\sqrt{28} = \sqrt{4 \cdot 7} = \sqrt{4} \cdot \sqrt{7} = 2\sqrt{7}$

35. $-\sqrt{32} = -\sqrt{16 \cdot 2} = -\sqrt{16} \cdot \sqrt{2} = -4\sqrt{2}$

37. $\sqrt{300} = \sqrt{100 \cdot 3} = \sqrt{100} \cdot \sqrt{3} = 10\sqrt{3}$

39. $\sqrt[3]{128} = \sqrt[3]{64 \cdot 2} = \sqrt[3]{64} \cdot \sqrt[3]{2} = 4\sqrt[3]{2}$

41. $\sqrt[3]{-16} = \sqrt[3]{-8 \cdot 2} = \sqrt[3]{-8} \cdot \sqrt[3]{2} = -2\sqrt[3]{2}$

43. $\sqrt[3]{40} = \sqrt[3]{8 \cdot 5} = \sqrt[3]{8} \cdot \sqrt[3]{5} = 2\sqrt[3]{5}$

45. $-\sqrt[4]{512} = -\sqrt[4]{256 \cdot 2} = -\sqrt[4]{4^4} \cdot \sqrt[4]{2} = -4\sqrt[4]{2}$

47. $\sqrt[5]{64} = \sqrt[5]{32 \cdot 2} = \sqrt[5]{2^5} \cdot \sqrt[5]{2} = 2\sqrt[5]{2}$

49. His reasoning was incorrect. The radicand 14 must be written as a product of two factors (not a sum of two terms) where one of the two factors is a perfect cube.

51. $\sqrt{72k^2} = \sqrt{36k^2 \cdot 2} = \sqrt{36k^2} \cdot \sqrt{2} = 6k\sqrt{2}$

53. $\sqrt[3]{\dfrac{81}{64}} = \dfrac{\sqrt[3]{81}}{\sqrt[3]{4^3}} = \dfrac{\sqrt[3]{27 \cdot 3}}{4} = \dfrac{3\sqrt[3]{3}}{4}$

55. $\sqrt{121x^6} = \sqrt{(11x^3)^2} = 11x^3$

57. $-\sqrt[3]{27t^{12}} = -\sqrt[3]{(3t^4)^3} = -3t^4$

59. $-\sqrt{100m^8z^4} = -\sqrt{(10m^4z^2)^2} = -10m^4z^2$

61. $-\sqrt[3]{-125a^6b^9c^{12}} = -\sqrt[3]{(-5a^2b^3c^4)^3}$
$= -(-5a^2b^3c^4)$
$= 5a^2b^3c^4$

63. $\sqrt[4]{\dfrac{1}{16}r^8t^{20}} = \sqrt[4]{\left(\dfrac{1}{2}r^2t^5\right)^4} = \dfrac{1}{2}r^2t^5$

65. $-\sqrt{13x^7y^8} = -\sqrt{(x^6y^8)(13x)}$
$= -\sqrt{x^6y^8} \cdot \sqrt{13x}$
$= -x^3y^4\sqrt{13x}$

67. $\sqrt[3]{8z^6w^9} = \sqrt[3]{(2z^2w^3)^3} = 2z^2w^3$

69. $\sqrt[3]{-16z^5t^7} = \sqrt[3]{(-2^3z^3t^6)(2z^2t)}$
$= -2zt^2\sqrt[3]{2z^2t}$

71. $\sqrt[4]{81x^{12}y^{16}} = \sqrt[4]{(3x^3y^4)^4} = 3x^3y^4$

73. $-\sqrt[4]{162r^{15}s^9} = -\sqrt[4]{81r^{12}s^8(2r^3s)}$
$= -\sqrt[4]{81r^{12}s^8} \cdot \sqrt[4]{2r^3s}$
$= -3r^3s^2\sqrt[4]{2r^3s}$

75. $\sqrt{\dfrac{y^{11}}{36}} = \dfrac{\sqrt{y^{11}}}{\sqrt{36}} = \dfrac{\sqrt{y^{10} \cdot y}}{6} = \dfrac{y^5\sqrt{y}}{6}$

77. $\sqrt[3]{\dfrac{x^{16}}{27}} = \dfrac{\sqrt[3]{x^{15} \cdot x}}{\sqrt[3]{27}} = \dfrac{x^5\sqrt[3]{x}}{3}$

79. $\sqrt[4]{48^2} = 48^{2/4} = 48^{1/2} = \sqrt{48} = \sqrt{16 \cdot 3}$
$= \sqrt{16} \cdot \sqrt{3} = 4\sqrt{3}$

81. $\sqrt[4]{25} = 25^{1/4} = \left(5^2\right)^{1/4} = 5^{2/4} = 5^{1/2} = \sqrt{5}$

83. $\sqrt[10]{x^{25}} = x^{25/10} = x^{5/2} = \sqrt{x^5}$
$= \sqrt{x^4 \cdot x} = x^2\sqrt{x}$

85. $\sqrt[3]{4} \cdot \sqrt{3}$
The least common index of 3 and 2 is 6. Write each radical as a sixth root.

$$\sqrt[3]{4} = 4^{1/3} = 4^{2/6} = \sqrt[6]{4^2} = \sqrt[6]{16}$$
$$\sqrt{3} = 3^{1/2} = 3^{3/6} = \sqrt[6]{3^3} = \sqrt[6]{27}$$

Therefore,

$$\sqrt[3]{4} \cdot \sqrt{3} = \sqrt[6]{16} \cdot \sqrt[6]{27}$$
$$= \sqrt[6]{16 \cdot 27} = \sqrt[6]{432}.$$

87. $\sqrt[4]{3} \cdot \sqrt[3]{4}$

The least common index of 4 and 3 is 12. Write each radical as a twelfth root.

$$\sqrt[4]{3} = 3^{1/4} = 3^{3/12} = \sqrt[12]{3^3} = \sqrt[12]{27}$$
$$\sqrt[3]{4} = 4^{1/3} = 4^{4/12} = \sqrt[12]{4^4} = \sqrt[12]{256}$$

Therefore,

$$\sqrt[4]{3} \cdot \sqrt[3]{4} = \sqrt[12]{27} \cdot \sqrt[12]{256}$$
$$= \sqrt[12]{27 \cdot 256} = \sqrt[12]{6912}.$$

89. $\sqrt{x} = x^{1/2} = x^{3/6} = \sqrt[6]{x^3}$

$\sqrt[3]{x} = x^{1/3} = x^{2/6} = \sqrt[6]{x^2}$

So $\sqrt{x} \cdot \sqrt[3]{x} = \sqrt[6]{x^3} \cdot \sqrt[6]{x^2}$
$$= \sqrt[6]{x^3 \cdot x^2} = \sqrt[6]{x^5}.$$

91. $\sqrt{48} = \sqrt{16 \cdot 3} = \sqrt{16}\sqrt{3} = 4\sqrt{3}$

The statement is true so the calculator should return a 1.

93. $\sqrt{5} = 5^{1/2} = 5^{3/6} = \sqrt[6]{5^3} = \sqrt[6]{125}$

$\sqrt[3]{4} = 4^{1/3} = 4^{2/6} = \sqrt[6]{4^2} = \sqrt[6]{16}$

So $\sqrt{5} \cdot \sqrt[3]{4} = \sqrt[6]{125 \cdot 16} = \sqrt[6]{2000}.$
The statement is true so the calculator should return a 1.

95. Substitute 3 for a and 4 for b in the Pythagorean formula to find c.

$$c^2 = a^2 + b^2$$
$$c = \sqrt{a^2 + b^2} = \sqrt{3^2 + 4^2}$$
$$= \sqrt{9 + 16} = \sqrt{25} = 5$$

The length of the hypotenuse is 5.

97. Substitute 12 for c and 4 for a in the Pythagorean formula to find b.

$$a^2 + b^2 = c^2$$
$$b = \sqrt{c^2 - a^2} = \sqrt{12^2 - 4^2}$$
$$= \sqrt{144 - 16} = \sqrt{128}$$
$$= \sqrt{64}\sqrt{2} = 8\sqrt{2}$$

The length of the unknown leg is $8\sqrt{2}$.

99. Substitute 27 for c and 16 for a in the Pythagorean formula to find b.

$$a^2 + b^2 = c^2$$
$$b = \sqrt{c^2 - a^2} = \sqrt{27^2 - 16^2}$$
$$= \sqrt{729 - 256} = \sqrt{473} \approx 21.7$$

The width of the screen is about 21.7 inches.

101. Substitute 282 for E, 100 for R, 264 for L, and 120π for ω in the formula to find I.

$$I = \frac{E}{\sqrt{R^2 + \omega^2 L^2}}$$
$$= \frac{282}{\sqrt{100^2 + (120\pi)^2(264)^2}}$$
$$\approx \frac{282}{99,526} \approx .003$$

103. Substitute 640 for k and 2 for I in the formula to find d.

$$d = \sqrt{\frac{k}{I}} = \sqrt{\frac{640}{2}} = \sqrt{320} = \sqrt{64 \cdot 5}$$
$$= \sqrt{64} \cdot \sqrt{5} = 8\sqrt{5} \approx 17.9$$

The illumination will be 2 foot-candles at $8\sqrt{5}$ feet or about 17.9 feet from the light source.

In Exercises 105–112, use the distance formula

$$d = \sqrt{(x_2 - x_1)^2 + (y_2 - y_1)^2}.$$

105. $(-8, 2)$ and $(-4, 1)$

$$d = \sqrt{[-8 - (-4)]^2 + (2 - 1)^2}$$
$$= \sqrt{(-4)^2 + 1^2}$$
$$= \sqrt{16 + 1} = \sqrt{17}$$

107. $(-1, 4)$ and $(5, 3)$

$$d = \sqrt{(-1 - 5)^2 + (4 - 3)^2}$$
$$= \sqrt{(-6)^2 + 1^2} = \sqrt{36 + 1} = \sqrt{37}$$

109. $(4.7, 2.3)$ and $(1.7, -1.7)$

$$d = \sqrt{(4.7 - 1.7)^2 + [2.3 - (-1.7)]^2}$$
$$= \sqrt{3^2 + 4^2} = \sqrt{9 + 16}$$
$$= \sqrt{25} = 5$$

111. $(x + y, y)$ and $(x - y, x)$

$$d = \sqrt{[(x + y) - (x - y)]^2 + (y - x)^2}$$
$$= \sqrt{(2y)^2 + (y - x)^2}$$
$$= \sqrt{4y^2 + y^2 - 2xy + x^2}$$
$$= \sqrt{5y^2 - 2xy + x^2}$$

113. Since $\sqrt{a} = a^{1/2}$, the distance formula

$$d = \sqrt{(x_2 - x_1)^2 + (y_2 - y_1)^2}$$

may be expressed as

$$d = \left[(x_2 - x_1)^2 + (y_2 - y_1)^2\right]^{1/2}.$$

115. To find the lengths of the three sides of the triangle, use the distance formula to find the distance between each pair of points. Then add the distances to find the perimeter.

$$P = \sqrt{(-3-2)^2 + (-3-6)^2}$$
$$+ \sqrt{(2-6)^2 + (6-2)^2}$$
$$+ \sqrt{[6-(-3)]^2 + [2-(-3)]^2}$$
$$= \sqrt{(-5)^2 + (-9)^2} + \sqrt{(-4)^2 + 4^2}$$
$$+ \sqrt{9^2 + 5^2}$$
$$= \sqrt{25+81} + \sqrt{16+16} + \sqrt{81+25}$$
$$= \sqrt{106} + \sqrt{32} + \sqrt{106}$$
$$= 2\sqrt{106} + \sqrt{16}\cdot\sqrt{2}$$
$$= 2\sqrt{106} + 4\sqrt{2}$$

Section 7.3

1. Only choice (b) has like radical factors, so it can be simplified without first simplifying the individual radical expressions.

$$3\sqrt{6} + 9\sqrt{6} = 12\sqrt{6}$$

3. $\sqrt{64} + \sqrt[3]{125} + \sqrt[4]{16} = \sqrt{8^2} + \sqrt[3]{5^3} + \sqrt[4]{2^4}$
$$= 8 + 5 + 2 = 15$$

This sum can be found easily since each radicand has a whole number power corresponding to the index of the radical. In other words, each term is a perfect root.

5. Simplify each radical and subtract.

$$\sqrt{36} - \sqrt{100} = 6 - 10 = -4$$

7. $-2\sqrt{48} + 3\sqrt{75}$
$$= -2\sqrt{16\cdot3} + 3\sqrt{25\cdot3}$$
$$= -2\cdot4\sqrt{3} + 3\cdot5\sqrt{3}$$
$$= -8\sqrt{3} + 15\sqrt{3} = 7\sqrt{3}$$

9. $\sqrt[3]{16} + 4\sqrt[3]{54}$
$$= \sqrt[3]{8\cdot2} + 4\sqrt[3]{27\cdot2}$$
$$= \sqrt[3]{8}\sqrt[3]{2} + 4\sqrt[3]{27}\sqrt[3]{2}$$
$$= 2\sqrt[3]{2} + 4\cdot3\sqrt[3]{2}$$
$$= 2\sqrt[3]{2} + 12\sqrt[3]{2} = 14\sqrt[3]{2}$$

11. $\sqrt[4]{32} + 3\sqrt[4]{2}$
$$= \sqrt[4]{16\cdot2} + 3\sqrt[4]{2}$$
$$= \sqrt[4]{16}\sqrt[4]{2} + 3\sqrt[4]{2}$$
$$= 2\sqrt[4]{2} + 3\sqrt[4]{2} = 5\sqrt[4]{2}$$

13. $6\sqrt{18} - \sqrt{32} + 2\sqrt{50}$
$$= 6\sqrt{9\cdot2} - \sqrt{16\cdot2} + 2\sqrt{25\cdot2}$$
$$= 6\cdot3\sqrt{2} - 4\sqrt{2} + 2\cdot5\sqrt{2}$$
$$= 18\sqrt{2} - 4\sqrt{2} + 10\sqrt{2}$$
$$= 24\sqrt{2}$$

15. $-2\sqrt{63} + 2\sqrt{28} + 2\sqrt{7}$
$$= -2\sqrt{9\cdot7} + 2\sqrt{4\cdot7} + 2\sqrt{7}$$
$$= -2\cdot3\sqrt{7} + 2\cdot2\sqrt{7} + 2\sqrt{7}$$
$$= -6\sqrt{7} + 4\sqrt{7} + 2\sqrt{7} = 0$$

17. $2\sqrt{5} + 3\sqrt{20} + 4\sqrt{45}$
$$= 2\sqrt{5} + 3\sqrt{4\cdot5} + 4\sqrt{9\cdot5}$$
$$= 2\sqrt{5} + 3\cdot2\sqrt{5} + 4\cdot3\sqrt{5}$$
$$= 2\sqrt{5} + 6\sqrt{5} + 12\sqrt{5}$$
$$= 20\sqrt{5}$$

19. $8\sqrt{2x} - \sqrt{8x} + \sqrt{72x}$
$$= 8\sqrt{2x} - \sqrt{4\cdot2x} + \sqrt{36\cdot2x}$$
$$= 8\sqrt{2x} - 2\sqrt{2x} + 6\sqrt{2x}$$
$$= 12\sqrt{2x}$$

21. $3\sqrt{72m^2} - 5\sqrt{32m^2} - 3\sqrt{18m^2}$
$$= 3\sqrt{36m^2\cdot2} - 5\sqrt{16m^2\cdot2} - 3\sqrt{9m^2\cdot2}$$
$$= 3\cdot6m\sqrt{2} - 5\cdot4m\sqrt{2} - 3\cdot3m\sqrt{2}$$
$$= 18m\sqrt{2} - 20m\sqrt{2} - 9m\sqrt{2}$$
$$= (18m - 20m - 9m)\sqrt{2} = -11m\sqrt{2}$$

23. $-\sqrt[3]{54} + 2\sqrt[3]{16} = -\sqrt[3]{27\cdot2} + 2\sqrt[3]{8\cdot2}$
$$= -3\sqrt[3]{2} + 2\cdot2\sqrt[3]{2}$$
$$= -3\sqrt[3]{2} + 4\sqrt[3]{2}$$
$$= \sqrt[3]{2}$$

25. $2\sqrt[3]{27x} - 2\sqrt[3]{8x}$
$$= 2\sqrt[3]{27\cdot x} - 2\sqrt[3]{8\cdot x}$$
$$= 2\cdot3\sqrt[3]{x} - 2\cdot2\sqrt[3]{x}$$
$$= 6\sqrt[3]{x} - 4\sqrt[3]{x}$$
$$= 2\sqrt[3]{x}$$

27. $5\sqrt[4]{32} + 3\sqrt[4]{162} = 5\sqrt[4]{16\cdot2} + 3\sqrt[4]{81\cdot2}$
$$= 5\cdot2\sqrt[4]{2} + 3\cdot3\sqrt[4]{2}$$
$$= 10\sqrt[4]{2} + 9\sqrt[4]{2}$$
$$= 19\sqrt[4]{2}$$

29. $3\sqrt[4]{x^5y} - 2x\sqrt[4]{xy}$
$$= 3\sqrt[4]{x^4\cdot xy} - 2x\sqrt[4]{xy}$$
$$= 3x\sqrt[4]{xy} - 2x\sqrt[4]{xy}$$
$$= (3x - 2x)\sqrt[4]{xy} = x\sqrt[4]{xy}$$

31. $\sqrt[3]{64xy^2} + \sqrt[3]{27x^4y^5}$

$= \sqrt[3]{64 \cdot xy^2} + \sqrt[3]{27x^3y^3 \cdot xy^2}$

$= 4\sqrt[3]{xy^2} + 3xy\sqrt[3]{xy^2}$

$= (4 + 3xy)\sqrt[3]{xy^2}$

33. $\sqrt{\dfrac{8}{9}} + \sqrt{\dfrac{18}{36}} = \dfrac{\sqrt{8}}{\sqrt{9}} + \dfrac{\sqrt{18}}{\sqrt{36}}$

$= \dfrac{\sqrt{4}\sqrt{2}}{3} + \dfrac{\sqrt{9}\sqrt{2}}{6}$

$= \dfrac{2\sqrt{2}}{3} + \dfrac{3\sqrt{2}}{6}$

$= \dfrac{4\sqrt{2}}{6} + \dfrac{3\sqrt{2}}{6}$

$= \dfrac{4\sqrt{2} + 3\sqrt{2}}{6} = \dfrac{7\sqrt{2}}{6}$

35. $\dfrac{\sqrt{32}}{3} + \dfrac{2\sqrt{2}}{3} - \dfrac{\sqrt{2}}{\sqrt{9}}$

$= \dfrac{\sqrt{16}\sqrt{2}}{3} + \dfrac{2\sqrt{2}}{3} - \dfrac{\sqrt{2}}{3}$

$= \dfrac{4\sqrt{2} + 2\sqrt{2} - \sqrt{2}}{3} = \dfrac{5\sqrt{2}}{3}$

37. $3\sqrt[3]{\dfrac{m^5}{27}} - 2m\sqrt[3]{\dfrac{m^2}{64}}$

$= \dfrac{3\sqrt[3]{m^5}}{\sqrt[3]{27}} - \dfrac{2m\sqrt[3]{m^2}}{\sqrt[3]{64}}$

$= \dfrac{3\sqrt[3]{m^3}\sqrt[3]{m^2}}{3} - \dfrac{2m\sqrt[3]{m^2}}{4}$

$= \dfrac{m\sqrt[3]{m^2}}{1} - \dfrac{m\sqrt[3]{m^2}}{2}$

$= \dfrac{2m\sqrt[3]{m^2} - m\sqrt[3]{m^2}}{2} = \dfrac{m\sqrt[3]{m^2}}{2}$

39. $3\sqrt{32} - 2\sqrt{8} \approx 11.3137085$

$\qquad 8\sqrt{2} \approx 11.3137085$

Both calculator approximations are the same, supporting (but not proving) the truth of the statement.

41. $2\sqrt{40} + 6\sqrt{90} - 3\sqrt{160} \approx 31.6227766$

$\qquad 10\sqrt{10} \approx 31.6227766$

Both calculator approximations are the same, supporting (but not proving) the truth of the statement.

43. Let $L = \sqrt{192} \approx 14$ and $W = \sqrt{48} \approx 7$.
The best estimate is choice (a).

45. The perimeter, P, of a triangle is the sum of the measures of the sides.

$$P = 3\sqrt{20} + 2\sqrt{45} + \sqrt{75}$$

$= 3\sqrt{4 \cdot 5} + 2\sqrt{9 \cdot 5} + \sqrt{25 \cdot 3}$

$= 3 \cdot 2\sqrt{5} + 2 \cdot 3\sqrt{5} + 5\sqrt{3}$

$= 6\sqrt{5} + 6\sqrt{5} + 5\sqrt{3}$

$= 12\sqrt{5} + 5\sqrt{3}$

The perimeter is $12\sqrt{5} + 5\sqrt{3}$ inches.

47. To find the perimeter, add the lengths of the sides.

$4\sqrt{18} + \sqrt{108} + 2\sqrt{72} + 3\sqrt{12}$

$= 4\sqrt{9}\sqrt{2} + \sqrt{36}\sqrt{3} + 2\sqrt{36}\sqrt{2} + 3\sqrt{4}\sqrt{3}$

$= 4 \cdot 3\sqrt{2} + 6\sqrt{3} + 2 \cdot 6\sqrt{2} + 3 \cdot 2\sqrt{3}$

$= 12\sqrt{2} + 6\sqrt{3} + 12\sqrt{2} + 6\sqrt{3}$

$= 24\sqrt{2} + 12\sqrt{3}$

The perimeter is $24\sqrt{2} + 12\sqrt{3}$ inches.

Section 7.4

1. $(x + \sqrt{y})(x - \sqrt{y})$

$= x^2 - x\sqrt{y} + x\sqrt{y} - (\sqrt{y})^2$

$= x^2 - y$ (E)

3. $(\sqrt{x} + \sqrt{y})(\sqrt{x} - \sqrt{y})$

$= (\sqrt{x})^2 - (\sqrt{y})^2$

$= x - y$ (A)

5. $(\sqrt{x} - \sqrt{y})^2$

$= (\sqrt{x})^2 - 2\sqrt{x}\sqrt{y} + (\sqrt{y})^2$

$= x - 2\sqrt{xy} + y$ (D)

7. $\sqrt[3]{x} \cdot \sqrt[3]{x}$ is not equal to x because the product rule leads to $\sqrt[3]{x^2}$, not $\sqrt[3]{x^3}$.

9. $\sqrt{3}\left(\sqrt{12} - 4\right) = \sqrt{3 \cdot 12} - 4\sqrt{3}$

$= \sqrt{36} - 4\sqrt{3}$

$= 6 - 4\sqrt{3}$

11. $\sqrt{2}\left(\sqrt{18} - \sqrt{3}\right) = \sqrt{2 \cdot 18} - \sqrt{2 \cdot 3}$

$= \sqrt{36} - \sqrt{6}$

$= 6 - \sqrt{6}$

13. $\left(\sqrt{6} + 2\right)\left(\sqrt{6} - 2\right) = \left(\sqrt{6}\right)^2 - 2^2$

$= 6 - 4 = 2$

15. $\left(\sqrt{12} - \sqrt{3}\right)\left(\sqrt{12} + \sqrt{3}\right)$

$= \left(\sqrt{12}\right)^2 - \left(\sqrt{3}\right)^2$

$= 12 - 3 = 9$

17. $\left(\sqrt{3}+2\right)\left(\sqrt{6}-5\right)$

$= \sqrt{3}\cdot\sqrt{6}-5\sqrt{3}+2\sqrt{6}-10$

$= \sqrt{18}-5\sqrt{3}+2\sqrt{6}-10$

$= \sqrt{9\cdot2}-5\sqrt{3}+2\sqrt{6}-10$

$= 3\sqrt{2}-5\sqrt{3}+2\sqrt{6}-10$

19. $\left(\sqrt{3x}+2\right)\left(\sqrt{3x}-2\right) = \left(\sqrt{3x}\right)^2-2^2$

$= 3x-4$

21. $\left(2\sqrt{x}+\sqrt{y}\right)\left(2\sqrt{x}-\sqrt{y}\right)$

$= \left(2\sqrt{x}\right)^2-\left(\sqrt{y}\right)^2$

$= 2^2\left(\sqrt{x}\right)^2-y$

$= 4x-y$

23. $\left(4\sqrt{x}+3\right)^2 = \left(4\sqrt{x}\right)^2+2\left(4\sqrt{x}\right)(3)+(3)^2$

$= 16x+24\sqrt{x}+9$

25. $\left(9-\sqrt[3]{2}\right)\left(9+\sqrt[3]{2}\right)$

$= 9^2-\left(\sqrt[3]{2}\right)^2$

$= 81-\sqrt[3]{2^2}$

$= 81-\sqrt[3]{4}$

27. $\left[\left(\sqrt{2}+\sqrt{3}\right)-\sqrt{6}\right]\left[\left(\sqrt{2}+\sqrt{3}\right)+\sqrt{6}\right]$

$= \left(\sqrt{2}+\sqrt{3}\right)^2-\left(\sqrt{6}\right)^2$

$= \left[\left(\sqrt{2}\right)^2+2\sqrt{2}\sqrt{3}+\left(\sqrt{3}\right)^2\right]-6$

$= \left(2+2\sqrt{6}+3\right)-6$

$= 2\sqrt{6}-1$

29. $\left[\left(\sqrt{2}+\sqrt{3}\right)+\sqrt{5}\right]^2$

$= \left(\sqrt{2}+\sqrt{3}\right)^2+2\left(\sqrt{2}+\sqrt{3}\right)\sqrt{5}$

$\quad +\left(\sqrt{5}\right)^2$

$= \left(\sqrt{2}\right)^2+2\sqrt{2}\sqrt{3}+\left(\sqrt{3}\right)^2$

$\quad +2\sqrt{2}\sqrt{5}+2\sqrt{3}\sqrt{5}+5$

$= 2+2\sqrt{6}+3+2\sqrt{10}+2\sqrt{15}+5$

$= 10+2\sqrt{6}+2\sqrt{10}+2\sqrt{15}$

31. $6-4\sqrt{3}$ cannot be simplified further. Only radical expressions with the same index and the same radicand may be combined. For example,

$$6\sqrt{3}-4\sqrt{3}=2\sqrt{3}.$$

33. $\dfrac{7}{\sqrt{7}} = \dfrac{7\cdot\sqrt{7}}{\sqrt{7}\cdot\sqrt{7}} = \dfrac{7\sqrt{7}}{7} = \sqrt{7}$

35. $\dfrac{15}{\sqrt{3}} = \dfrac{15\cdot\sqrt{3}}{\sqrt{3}\cdot\sqrt{3}} = \dfrac{15\sqrt{3}}{3} = 5\sqrt{3}$

37. $\dfrac{\sqrt{3}}{\sqrt{2}} = \dfrac{\sqrt{3}\cdot\sqrt{2}}{\sqrt{2}\cdot\sqrt{2}} = \dfrac{\sqrt{6}}{2}$

39. $\dfrac{9\sqrt{3}}{\sqrt{5}} = \dfrac{9\sqrt{3}\cdot\sqrt{5}}{\sqrt{5}\cdot\sqrt{5}} = \dfrac{9\sqrt{15}}{5}$

41. $\dfrac{-6}{\sqrt{18}} = \dfrac{-6}{\sqrt{9\cdot2}} = \dfrac{-6}{3\sqrt{2}} = \dfrac{-2}{\sqrt{2}} = \dfrac{-2\cdot\sqrt{2}}{\sqrt{2}\cdot\sqrt{2}}$

$= \dfrac{-2\sqrt{2}}{2} = -\sqrt{2}$

43. $\sqrt{\dfrac{7}{2}} = \dfrac{\sqrt{7}}{\sqrt{2}} = \dfrac{\sqrt{7}\cdot\sqrt{2}}{\sqrt{2}\cdot\sqrt{2}} = \dfrac{\sqrt{14}}{2}$

45. $-\sqrt{\dfrac{7}{50}} = -\dfrac{\sqrt{7}}{\sqrt{25\cdot2}} = -\dfrac{\sqrt{7}}{5\sqrt{2}}$

$= -\dfrac{\sqrt{7}\cdot\sqrt{2}}{5\sqrt{2}\cdot\sqrt{2}} = -\dfrac{\sqrt{14}}{5\cdot2} = -\dfrac{\sqrt{14}}{10}$

47. $\sqrt{\dfrac{24}{x}} = \dfrac{\sqrt{24}}{\sqrt{x}} = \dfrac{\sqrt{4\cdot6}}{\sqrt{x}} = \dfrac{2\sqrt{6}}{\sqrt{x}}$

$= \dfrac{2\sqrt{6}\cdot\sqrt{x}}{\sqrt{x}\cdot\sqrt{x}} = \dfrac{2\sqrt{6x}}{x}$

49. $\dfrac{-8\sqrt{3}}{\sqrt{k}} = \dfrac{-8\sqrt{3}\cdot\sqrt{k}}{\sqrt{k}\cdot\sqrt{k}} = \dfrac{-8\sqrt{3k}}{k}$

51. $-\sqrt{\dfrac{150m^5}{n^3}} = \dfrac{-\sqrt{150m^5}}{\sqrt{n^3}}$

$= \dfrac{-\sqrt{25m^4\cdot6m}}{\sqrt{n^2\cdot n}} = \dfrac{-5m^2\sqrt{6m}}{n\sqrt{n}}$

$= \dfrac{-5m^2\sqrt{6m}\cdot\sqrt{n}}{n\sqrt{n}\cdot\sqrt{n}}$

$= \dfrac{-5m^2\sqrt{6mn}}{n\cdot n}$

$= \dfrac{-5m^2\sqrt{6mn}}{n^2}$

53. $\sqrt{\dfrac{288x^7}{y^9}} = \dfrac{\sqrt{288x^7}}{\sqrt{y^9}} = \dfrac{\sqrt{144x^6\cdot2x}}{\sqrt{y^8\cdot y}}$

$= \dfrac{12x^3\sqrt{2x}}{y^4\sqrt{y}} = \dfrac{12x^3\sqrt{2x}\cdot\sqrt{y}}{y^4\sqrt{y}\cdot\sqrt{y}}$

$= \dfrac{12x^3\sqrt{2xy}}{y^4\cdot y} = \dfrac{12x^3\sqrt{2xy}}{y^5}$

55. $\sqrt{\dfrac{52}{y}} = \dfrac{\sqrt{52}}{\sqrt{y}} = \dfrac{\sqrt{4 \cdot 13}}{\sqrt{y}} = \dfrac{2\sqrt{13}}{\sqrt{y}}$

$= \dfrac{2\sqrt{13} \cdot \sqrt{y}}{\sqrt{y} \cdot \sqrt{y}} = \dfrac{2\sqrt{13y}}{y}$

$\sqrt{\dfrac{52}{y}} = \dfrac{\sqrt{52}}{\sqrt{y}} = \dfrac{\sqrt{4 \cdot 13}}{\sqrt{y}} = \dfrac{2\sqrt{13}}{\sqrt{y}}$

$= \dfrac{2\sqrt{13} \cdot \sqrt{y^3}}{\sqrt{y} \cdot \sqrt{y^3}} = \dfrac{2\sqrt{13y^3}}{\sqrt{y^4}}$

$= \dfrac{2\sqrt{y^2 \cdot 13y}}{\sqrt{(y^2)^2}} = \dfrac{2y\sqrt{13y}}{y^2} = \dfrac{2\sqrt{13y}}{y}$

The first method is preferable since the answer is obtained more directly with less simplifying involved.

57. $\sqrt[3]{\dfrac{2}{3}} = \dfrac{\sqrt[3]{2} \cdot \sqrt[3]{9}}{\sqrt[3]{3} \cdot \sqrt[3]{9}} = \dfrac{\sqrt[3]{18}}{\sqrt[3]{27}} = \dfrac{\sqrt[3]{18}}{3}$

59. $\sqrt[3]{\dfrac{4}{9}} = \dfrac{\sqrt[3]{4}}{\sqrt[3]{9}} = \dfrac{\sqrt[3]{4}}{\sqrt[3]{3^2}} = \dfrac{\sqrt[3]{4} \cdot \sqrt[3]{3}}{\sqrt[3]{3^2} \cdot \sqrt[3]{3}}$

$= \dfrac{\sqrt[3]{12}}{\sqrt[3]{3^3}} = \dfrac{\sqrt[3]{12}}{3}$

61. $-\sqrt[3]{\dfrac{2p}{r^2}} = -\dfrac{\sqrt[3]{2p}}{\sqrt[3]{r^2}} = -\dfrac{\sqrt[3]{2p} \cdot \sqrt[3]{r}}{\sqrt[3]{r^2} \cdot \sqrt[3]{r}}$

$= -\dfrac{\sqrt[3]{2pr}}{\sqrt[3]{r^3}} = -\dfrac{\sqrt[3]{2pr}}{r}$

63. $\sqrt[4]{\dfrac{16}{x}} = \dfrac{\sqrt[4]{16}}{\sqrt[4]{x}} = \dfrac{2}{\sqrt[4]{x}} = \dfrac{2 \cdot \sqrt[4]{x^3}}{\sqrt[4]{x} \cdot \sqrt[4]{x^3}}$

$= \dfrac{2\sqrt[4]{x^3}}{\sqrt[4]{x^4}} = \dfrac{2\sqrt[4]{x^3}}{x}$

65. To rationalize the denominator in $\dfrac{2}{4 + \sqrt{3}}$,

multiply both the numerator and the denominator by the conjugate of the denominator, $4 - \sqrt{3}$.

Multiply both numerator and denominator by $4 + \sqrt{3}$:

$\dfrac{2}{4 + \sqrt{3}} = \dfrac{2(4 + \sqrt{3})}{(4 + \sqrt{3})(4 + \sqrt{3})}$

$= \dfrac{2(4 + \sqrt{3})}{16 + 8\sqrt{3} + 3}$

$= \dfrac{2(4 + \sqrt{3})}{19 + 8\sqrt{3}}.$

The denominator is not rational because we did not multiply it by an expression that would eliminate the radical.

67. $\dfrac{6}{5 + \sqrt{2}}$

Multiply by the conjugate of the denominator, $5 - \sqrt{2}$.

$= \dfrac{6(5 - \sqrt{2})}{(5 + \sqrt{2})(5 - \sqrt{2})}$

$= \dfrac{6(5 - \sqrt{2})}{25 - 2} = \dfrac{6(5 - \sqrt{2})}{23}$

69. $\dfrac{12}{\sqrt{6} + \sqrt{3}}$

Multiply by the conjugate of the denominator, $\sqrt{6} - \sqrt{3}$.

$= \dfrac{12(\sqrt{6} - \sqrt{3})}{(\sqrt{6} + \sqrt{3})(\sqrt{6} - \sqrt{3})}$

$= \dfrac{12(\sqrt{6} - \sqrt{3})}{6 - 3} = \dfrac{12(\sqrt{6} - \sqrt{3})}{3}$

$= 4(\sqrt{6} - \sqrt{3})$

71. $\dfrac{-3}{\sqrt{2} + \sqrt{5}}$

$= \dfrac{-3(\sqrt{2} - \sqrt{5})}{(\sqrt{2} + \sqrt{5})(\sqrt{2} - \sqrt{5})}$

$= \dfrac{-3(\sqrt{2} - \sqrt{5})}{2 - 5}$

$= \dfrac{-3(\sqrt{2} - \sqrt{5})}{-3} = \sqrt{2} - \sqrt{5}$

73. $\dfrac{-1 - \sqrt{3}}{\sqrt{6} + \sqrt{5}} = \dfrac{(-1 - \sqrt{3})(\sqrt{6} - \sqrt{5})}{(\sqrt{6} + \sqrt{5})(\sqrt{6} - \sqrt{5})}$

$= \dfrac{-\sqrt{6} + \sqrt{5} - \sqrt{18} + \sqrt{15}}{6 - 5}$

$= -\sqrt{6} + \sqrt{5} - 3\sqrt{2} + \sqrt{15}$

75. $\dfrac{5\sqrt{r}}{3\sqrt{r} + \sqrt{s}}$

Multiply by the conjugate of the denominator, $3\sqrt{r} - \sqrt{s}$.

$= \dfrac{5\sqrt{r}(3\sqrt{r} - \sqrt{s})}{(3\sqrt{r} + \sqrt{s})(3\sqrt{r} - \sqrt{s})}$

$= \dfrac{5\sqrt{r}(3\sqrt{r} - \sqrt{s})}{9r - s}$

77.
$$\frac{\sqrt{a}+\sqrt{b}}{\sqrt{a}-\sqrt{b}} = \frac{\left(\sqrt{a}+\sqrt{b}\right)\left(\sqrt{a}+\sqrt{b}\right)}{\left(\sqrt{a}-\sqrt{b}\right)\left(\sqrt{a}+\sqrt{b}\right)}$$

$$= \frac{\left(\sqrt{a}\right)^2 + 2\sqrt{a}\sqrt{b} + \left(\sqrt{b}\right)^2}{\left(\sqrt{a}\right)^2 - \left(\sqrt{b}\right)^2}$$

$$= \frac{a + 2\sqrt{ab} + b}{a - b}$$

79.
$$\frac{25 + 10\sqrt{6}}{20} = \frac{5\left(5 + 2\sqrt{6}\right)}{5 \cdot 4} = \frac{5 + 2\sqrt{6}}{4}$$

81.
$$\frac{16 + 4\sqrt{8}}{12} = \frac{16 + 4\left(2\sqrt{2}\right)}{12} = \frac{16 + 8\sqrt{2}}{12}$$

$$= \frac{4\left(4 + 2\sqrt{2}\right)}{4 \cdot 3} = \frac{4 + 2\sqrt{2}}{3}$$

83.
$$\frac{6x + \sqrt{24x^3}}{3x}$$

$$= \frac{6x + \sqrt{4x^2 \cdot 6x}}{3x} = \frac{6x + 2x\sqrt{6x}}{3x}$$

$$= \frac{x\left(6 + 2\sqrt{6x}\right)}{3x} = \frac{6 + 2\sqrt{6x}}{3}$$

85.
$$\frac{1}{\sqrt{2}} \cdot \frac{\sqrt{3}}{2} - \frac{1}{\sqrt{2}} \cdot \frac{1}{2} = \frac{\sqrt{3}}{2\sqrt{2}} - \frac{1}{2\sqrt{2}}$$

$$= \frac{\sqrt{3} - 1}{2\sqrt{2}} = \frac{\left(\sqrt{3} - 1\right)\sqrt{2}}{\left(2\sqrt{2}\right)\sqrt{2}}$$

$$= \frac{\sqrt{6} - \sqrt{2}}{2 \cdot 2} = \frac{\sqrt{6} - \sqrt{2}}{4}$$

Using a calculator,

$$\frac{1}{\sqrt{2}} \cdot \frac{\sqrt{3}}{2} - \frac{1}{\sqrt{2}} \cdot \frac{1}{2} \approx .2588190451 \text{ and}$$

$$\frac{\sqrt{6} - \sqrt{2}}{4} \approx .2588190451.$$

87. $x - 7 = \left(\sqrt{x}\right)^2 - \left(\sqrt{7}\right)^2$

$$= \left(\sqrt{x} + \sqrt{7}\right)\left(\sqrt{x} - \sqrt{7}\right)$$

88. $x - 7 = \left(\sqrt[3]{x}\right)^3 - \left(\sqrt[3]{7}\right)^3$

$$= \left(\sqrt[3]{x} - \sqrt[3]{7}\right)$$

$$\cdot \left[\left(\sqrt[3]{x}\right)^2 + \sqrt[3]{x} \cdot \sqrt[3]{7} + \left(\sqrt[3]{7}\right)^2\right]$$

$$= \left(\sqrt[3]{x} - \sqrt[3]{7}\right)\left(\sqrt[3]{x^2} + \sqrt[3]{7x} + \sqrt[3]{49}\right)$$

89. $x + 7 = \left(\sqrt[3]{x}\right)^3 + \left(\sqrt[3]{7}\right)^3$

$$= \left(\sqrt[3]{x} + \sqrt[3]{7}\right)$$

$$\cdot \left[\left(\sqrt[3]{x}\right)^2 - \sqrt[3]{x} \cdot \sqrt[3]{7} + \left(\sqrt[3]{7}\right)^2\right]$$

$$= \left(\sqrt[3]{x} + \sqrt[3]{7}\right)\left(\sqrt[3]{x^2} - \sqrt[3]{7x} + \sqrt[3]{49}\right)$$

90.
$$\frac{x + 3}{\sqrt{x} - \sqrt{7}} = \frac{(x + 3)\left(\sqrt{x} + \sqrt{7}\right)}{\left(\sqrt{x} - \sqrt{7}\right)\left(\sqrt{x} + \sqrt{7}\right)}$$

$$= \frac{(x + 3)\left(\sqrt{x} + \sqrt{7}\right)}{x - 7}$$

91.
$$\frac{x + 3}{\sqrt[3]{x} - \sqrt[3]{7}}$$

$$= \frac{(x + 3)\left(\sqrt[3]{x^2} + \sqrt[3]{7x} + \sqrt[3]{49}\right)}{\left(\sqrt[3]{x} - \sqrt[3]{7}\right)\left(\sqrt[3]{x^2} + \sqrt[3]{7x} + \sqrt[3]{49}\right)}$$

$$= \frac{(x + 3)\left(\sqrt[3]{x^2} + \sqrt[3]{7x} + \sqrt[3]{49}\right)}{x - 7}$$

92.
$$\frac{x + 3}{\sqrt[3]{x^2} - \sqrt[3]{7x} + \sqrt[3]{49}}$$

$$= \frac{(x + 3)\left(\sqrt[3]{x} + \sqrt[3]{7}\right)}{\left(\sqrt[3]{x^2} - \sqrt[3]{7x} + \sqrt[3]{49}\right)\left(\sqrt[3]{x} + \sqrt[3]{7}\right)}$$

$$= \frac{(x + 3)\left(\sqrt[3]{x} + \sqrt[3]{7}\right)}{x + 7}$$

93. $2 = 5 - 3 = \left(\sqrt[3]{5}\right)^3 - \left(\sqrt[3]{3}\right)^3$

$$= \left(\sqrt[3]{5} - \sqrt[3]{3}\right)$$

$$\cdot \left[\left(\sqrt[3]{5}\right)^2 + \sqrt[3]{5} \cdot \sqrt[3]{3} + \left(\sqrt[3]{3}\right)^2\right]$$

$$= \left(\sqrt[3]{5} - \sqrt[3]{3}\right)\left(\sqrt[3]{25} + \sqrt[3]{15} + \sqrt[3]{9}\right)$$

94.
$$\frac{2}{\sqrt[3]{5} - \sqrt[3]{3}}$$

$$= \frac{2\left(\sqrt[3]{25} + \sqrt[3]{15} + \sqrt[3]{9}\right)}{\left(\sqrt[3]{5} - \sqrt[3]{3}\right)\left(\sqrt[3]{25} + \sqrt[3]{15} + \sqrt[3]{9}\right)}$$

$$= \frac{2\left(\sqrt[3]{25} + \sqrt[3]{15} + \sqrt[3]{9}\right)}{5 - 3}$$

$$= \sqrt[3]{25} + \sqrt[3]{15} + \sqrt[3]{9}$$

Section 7.5

1.　$f(x) = \sqrt{x + 3}$

For the radicand to be nonnegative, we must have

$$x + 3 \geq 0 \quad \text{or} \quad x \geq -3.$$

Thus, the domain is $[-3, \infty)$.
The function values are positive or zero (the result of the radical), so the range is $[0, \infty)$.

x	$f(x) = \sqrt{x + 3}$
-3	$\sqrt{-3 + 3} = 0$
-2	$\sqrt{-2 + 3} = 1$
1	$\sqrt{1 + 3} = 2$

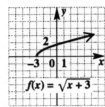

3.　$f(x) = \sqrt{x} - 2$

For the radicand to be nonnegative, we must have

$$x \geq 0.$$

Note that the "-2" does not affect the domain, which is $[0, \infty)$.
The result of the radical is positive or zero, but the function values are 2 less than those values, so the range is $[-2, \infty)$.

x	$f(x) = \sqrt{x} - 2$
0	$\sqrt{0} - 2 = -2$
1	$\sqrt{1} - 2 = -1$
4	$\sqrt{4} - 2 = 0$

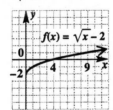

5.　$f(x) = \sqrt[3]{x} - 3$

Since we can take the cube root of any real number, the domain is $(-\infty, \infty)$.
The result of a cube root can be any real number, so the range is $(-\infty, \infty)$. (The "-3" does not affect the range.)

x	$f(x) = \sqrt[3]{x} - 3$
-8	$\sqrt[3]{-8} - 3 = -5$
-1	$\sqrt[3]{-1} - 3 = -4$
0	$\sqrt[3]{0} - 3 = -3$
1	$\sqrt[3]{1} - 3 = -2$
8	$\sqrt[3]{8} - 3 = -1$

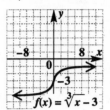

7.　$f(x) = \sqrt[3]{x - 3}$

Both the domain and range are $(-\infty, \infty)$.

x	$f(x) = \sqrt[3]{x - 3}$
-5	$\sqrt[3]{-5 - 3} = -2$
2	$\sqrt[3]{2 - 3} = -1$
3	$\sqrt[3]{3 - 3} = 0$
4	$\sqrt[3]{4 - 3} = 1$
11	$\sqrt[3]{11 - 3} = 2$

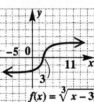

9.　$\sqrt{3x + 18} = x$

(a) Check $x = 6$.

$$\sqrt{3(6) + 18} \overset{?}{=} 6$$
$$\sqrt{18 + 18} \overset{?}{=} 6$$
$$\sqrt{36} \overset{?}{=} 6$$
$$6 = 6 \quad \textit{True}$$

The number 6 is a solution.

(b) Check $x = -3$.

$\sqrt{3(-3) + 18} = -3$ is a false statement since the square root of a number is nonnegative. The number -3 is not a solution.

11. $\sqrt{x+2} = \sqrt{9x-2} - 2\sqrt{x-1}$

  (a) Check $x = 2$.

$$\sqrt{2+2} \overset{?}{=} \sqrt{9(2)-2} - 2\sqrt{2-1}$$
$$\sqrt{4} \overset{?}{=} \sqrt{16} - 2\sqrt{1}$$
$$2 \overset{?}{=} 4 - 2$$
$$2 = 2 \quad \textit{True}$$

The number 2 is a solution.

  (b) Check $x = 7$.

$$\sqrt{7+2} \overset{?}{=} \sqrt{9(7)-2} - 2\sqrt{7-1}$$
$$\sqrt{9} \overset{?}{=} \sqrt{61} - 2\sqrt{6}$$
$$3 = \sqrt{61} - 2\sqrt{6} \quad \textit{False}$$

The number 7 is not a solution.

13. $\sqrt{9} = 3$, not -3. There is no solution of $\sqrt{x} = -3$ since the result of a radical cannot equal a negative number.

In Exercises 15–26, check each solution in the original equation.

15. $\sqrt{x-3} = 4$

Square both sides.

$$\left(\sqrt{x-3}\right)^2 = 4^2$$
$$x - 3 = 16$$
$$x = 19$$

Check the potential solution, 19.

Check $x = 19$: $\sqrt{16} \overset{?}{=} 4$ *True*

Solution set: $\{19\}$

17. $\sqrt{3k-2} - 8 = -2$

Isolate the radical by adding 8 to both sides.

$$\sqrt{3k-2} = 6$$

Square both sides.

$$\left(\sqrt{3k-2}\right)^2 = (6)^2$$
$$3k - 2 = 36$$
$$3k = 38$$
$$k = \frac{38}{3}$$

Check $k = \frac{38}{3}$: $\sqrt{36} - 8 \overset{?}{=} -2$ *True*

Solution set: $\left\{\dfrac{38}{3}\right\}$

19. $\sqrt{x} + 9 = 0$

$$\sqrt{x} = -9$$

This equation has no solution because $\sqrt{x}$ cannot be negative.

Solution set: $\emptyset$

21. $\sqrt{3x-6} - 3 = 0$

$$\sqrt{3x-6} = 3 \quad \textit{Isolate}$$
$$\left(\sqrt{3x-6}\right)^2 = 3^2 \quad \textit{Square}$$
$$3x - 6 = 9$$
$$3x = 15$$
$$x = 5$$

Check $x = 5$: $\sqrt{9} - 3 \overset{?}{=} 0$

Solution set: $\{5\}$

23. $\sqrt{6x+2} - \sqrt{5x+3} = 0$

Get one radical on each side of the equals sign.

$$\sqrt{6x+2} = \sqrt{5x+3}$$

Square both sides.

$$\left(\sqrt{6x+2}\right)^2 = \left(\sqrt{5x+3}\right)^2$$
$$6x + 2 = 5x + 3$$
$$x = 1$$

Check $x = 1$: $\sqrt{8} - \sqrt{8} \overset{?}{=} 0$ *True*

Solution set: $\{1\}$

25. $3\sqrt{x} = \sqrt{8x+9}$

$$\left(3\sqrt{x}\right)^2 = \left(\sqrt{8x+9}\right)^2 \quad \textit{Square}$$
$$9x = 8x + 9$$
$$x = 9$$

Check $x = 9$: $3\sqrt{9} \overset{?}{=} \sqrt{81}$ *True*

Solution set: $\{9\}$

27. $x = 3$

Solution set: $\{3\}$

28. $x = -3$

Solution set: $\{-3\}$

29. $x^2 = 9$

$$3^2 = 9 \quad \text{or} \quad (-3)^2 = 9$$

Solution set: $\{-3, 3\}$

30. $x^3 = 27$

$$3^3 = 27$$

Solution set: $\{3\}$

31. $x^4 = 81$

$$3^4 = 81 \quad \text{or} \quad (-3)^4 = 81$$

Solution set: $\{-3, 3\}$

32. Suppose both sides of $x = k$ are raised to the nth power.

  (a) If n is even, the number of solutions of the new equation is _more than_ the number of solutions of the original equation.

(b) If n is odd, the number of solutions of the new equation, is *the same as* the number of solutions of the original equation.

33. $\sqrt{3x+4} = 8 - x$

$(8-x)^2$ equals $64 - 16x + x^2$, not $64 + x^2$. The first step should be

$$3x + 4 = 64 - 16x + x^2$$
$$0 = x^2 - 19x + 60$$
$$0 = (x-4)(x-15)$$

$$x - 4 = 0 \quad \text{or} \quad x - 15 = 0$$
$$x = 4 \quad \text{or} \quad x = 15$$

Check $x = 4$: $\sqrt{16} \overset{?}{=} 8 - 4$ *True*

Check $x = 15$: $\sqrt{49} \overset{?}{=} 8 - 15$ *False*

Solution set: $\{4\}$

35. See the solution in Exercise 33.

37. $\sqrt{13 + 4t} = t + 4$

$$\left(\sqrt{13 + 4t}\right)^2 = (t+4)^2$$
$$13 + 4t = t^2 + 8t + 16$$
$$0 = t^2 + 4t + 3$$
$$0 = (t+3)(t+1)$$

$$t + 3 = 0 \quad \text{or} \quad t + 1 = 0$$
$$t = -3 \quad \text{or} \quad t = -1$$

Check $t = -3$: $\sqrt{1} \overset{?}{=} 1$ *True*

Check $t = -1$: $\sqrt{9} \overset{?}{=} 3$ *True*

Solution set: $\{-3, -1\}$

39. $\sqrt{r^2 - 15r + 15} + 5 = r$

$$\sqrt{r^2 - 15r + 15} = r - 5 \qquad \textit{Isolate}$$
$$\left(\sqrt{r^2 - 15r + 15}\right)^2 = (r-5)^2$$
$$r^2 - 15r + 15 = r^2 - 10r + 25$$
$$-5r = 10$$
$$r = -2$$

Check $r = -2$: $\sqrt{49} + 5 \overset{?}{=} -2$ *False*

Solution set: $\emptyset$

41. $\sqrt{3x + 7} - 3x = 5$

$$\sqrt{3x + 7} = 3x + 5 \qquad \textit{Isolate}$$
$$\left(\sqrt{3x + 7}\right)^2 = (3x + 5)^2$$
$$3x + 7 = 9x^2 + 30x + 25$$
$$0 = 9x^2 + 27x + 18$$
$$0 = x^2 + 3x + 2$$
$$0 = (x + 1)(x + 2)$$

$$x + 1 = 0 \quad \text{or} \quad x + 2 = 0$$
$$x = -1 \quad \text{or} \quad x = -2$$

Check $x = -1$: $\sqrt{4} + 3 \overset{?}{=} 5$ *True*

Check $x = -2$: $\sqrt{1} + 6 \overset{?}{=} 5$ *False*

Solution set: $\{-1\}$

43. $\sqrt{4x + 2} - 4x = 0$

$$\sqrt{4x + 2} = 4x$$
$$\left(\sqrt{4x + 2}\right)^2 = (4x)^2$$
$$4x + 2 = 16x^2$$
$$0 = 16x^2 - 4x - 2$$
$$0 = 8x^2 - 2x - 1$$
$$0 = (4x + 1)(2x - 1)$$

$$4x + 1 = 0 \quad \text{or} \quad 2x - 1 = 0$$
$$x = -\frac{1}{4} \quad \text{or} \quad x = \frac{1}{2}$$

Check $x = -\frac{1}{4}$: $\sqrt{1} + 1 \overset{?}{=} 0$ *False*

Check $x = \frac{1}{2}$: $\sqrt{4} - 2 \overset{?}{=} 0$ *True*

Solution set: $\left\{\dfrac{1}{2}\right\}$

45. $\sqrt{r + 4} - \sqrt{r - 4} = 2$

Get one radical on each side of the equals sign.

$$\sqrt{r + 4} = \sqrt{r - 4} + 2$$

Square both sides.

$$\left(\sqrt{r + 4}\right)^2 = \left(\sqrt{r - 4} + 2\right)^2$$
$$r + 4 = r - 4 + 4\sqrt{r - 4} + 4$$

Isolate the remaining radical.

$$r + 4 = r + 4\sqrt{r - 4}$$
$$4 = 4\sqrt{r - 4}$$
$$1 = \sqrt{r - 4}$$

Square both sides again.

$$1^2 = \left(\sqrt{r - 4}\right)^2$$
$$1 = r - 4$$
$$5 = r$$

Check $r = 5$: $\sqrt{9} - \sqrt{1} \overset{?}{=} 2$ *True*

Solution set: $\{5\}$

47.
$$\sqrt{11 + 2q} + 1 = \sqrt{5q + 1}$$
Square both sides.
$$\left(\sqrt{11 + 2q} + 1\right)^2 = \left(\sqrt{5q + 1}\right)^2$$
$$\left(\sqrt{11 + 2q}\right)^2 + 2 \cdot \sqrt{11 + 2q} + 1$$
$$= 5q + 1$$
$$11 + 2q + 2\sqrt{11 + 2q} + 1 = 5q + 1$$
Isolate the remaining radical.
$$2\sqrt{11 + 2q} = 3q - 11$$
Square both sides again.
$$\left(2\sqrt{11 + 2q}\right)^2 = (3q - 11)^2$$
$$4(11 + 2q) = 9q^2 - 66q + 121$$
$$44 + 8q = 9q^2 - 66q + 121$$
$$0 = 9q^2 - 74q + 77$$
$$0 = (9q - 11)(q - 7)$$

$9q - 11 = 0$ or $q - 7 = 0$

$q = \dfrac{11}{9}$ or $q = 7$

Check $q = \dfrac{11}{9}$: $\sqrt{\dfrac{121}{9}} + 1 \overset{?}{=} \sqrt{\dfrac{64}{9}}$ *False*

Check $q = 7$: $\sqrt{25} + 1 \overset{?}{=} \sqrt{36}$ *True*

Solution set: $\{7\}$

49. $\sqrt{3 - 3p} - \sqrt{3p + 2} = 3$

Get one radical on each side.
$$\sqrt{3 - 3p} = \sqrt{3p + 2} + 3$$
Square both sides.
$$\left(\sqrt{3 - 3p}\right)^2 = \left(\sqrt{3p + 2} + 3\right)^2$$
$$3 - 3p = 3p + 2 + 6\sqrt{3p + 2} + 9$$
Isolate the remaining radical.
$$-6p - 8 = 6\sqrt{3p + 2}$$
$$-3p - 4 = 3\sqrt{3p + 2}$$
Square both sides again.
$$(-3p - 4)^2 = \left(3\sqrt{3p + 2}\right)^2$$
$$9p^2 + 24p + 16 = 9(3p + 2)$$
$$9p^2 + 24p + 16 = 27p + 18$$
$$9p^2 - 3p - 2 = 0$$
$$(3p + 1)(3p - 2) = 0$$

$3p + 1 = 0$ or $3p - 2 = 0$

$p = -\dfrac{1}{3}$ or $p = \dfrac{2}{3}$

Check $p = -\dfrac{1}{3}$: $\sqrt{4} - \sqrt{1} \overset{?}{=} 3$ *False*

Check $p = \dfrac{2}{3}$: $\sqrt{1} - \sqrt{4} \overset{?}{=} 3$ *False*

Solution set: $\emptyset$

51.
$$\sqrt{5x - 6} = 2 + \sqrt{3x - 6}$$
$$\left(\sqrt{5x - 6}\right)^2 = \left(2 + \sqrt{3x - 6}\right)^2$$
$$5x - 6 = 4 + 4\sqrt{3x - 6} + 3x - 6$$
$$2x - 4 = 4\sqrt{3x - 6}$$
$$x - 2 = 2\sqrt{3x - 6}$$
$$(x - 2)^2 = \left(2\sqrt{3x - 6}\right)^2$$
$$x^2 - 4x + 4 = 4(3x - 6)$$
$$x^2 - 4x + 4 = 12x - 24$$
$$x^2 - 16x + 28 = 0$$
$$(x - 2)(x - 14) = 0$$

$x = 2$ or $x = 14$

Check $x = 2$: $\sqrt{4} \overset{?}{=} 2 + \sqrt{0}$ *True*

Check $x = 14$: $\sqrt{64} \overset{?}{=} 2 + \sqrt{36}$ *True*

Solution set: $\{2, 14\}$

53. Rewrite $\sqrt[3]{x + 3} = \sqrt[3]{5 + 4x}$ as
$(x + 3)^{1/3} = (5 + 4x)^{1/3}$. To eliminate the 1/3 power, raise both sides to the power of 3 to get

$$\left[(x + 3)^{1/3}\right]^3 = \left[(5 + 4x)^{1/3}\right]^3,$$

so $x + 3 = 5 + 4x$ by the power rule. Thus, 3 is the smallest power to which you can raise both sides to eliminate the radicals.

55. $\sqrt[3]{2x^2 + 3x - 7} = \sqrt[3]{2x^2 + 4x + 6}$
Cube both sides.
$$\left(\sqrt[3]{2x^2 + 3x - 7}\right)^3 = \left(\sqrt[3]{2x^2 + 4x + 6}\right)^3$$
$$2x^2 + 3x - 7 = 2x^2 + 4x + 6$$
$$-13 = x$$
Check $x = -13$: $\sqrt[3]{292} \overset{?}{=} \sqrt[3]{292}$ *True*
Solution set: $\{-13\}$

57. $\sqrt[3]{1 - 2k} - \sqrt[3]{-k - 13} = 0$
$$\sqrt[3]{1 - 2k} = \sqrt[3]{-k - 13}$$
Cube both sides.
$$\left(\sqrt[3]{1 - 2k}\right)^3 = \left(\sqrt[3]{-k - 13}\right)^3$$
$$1 - 2k = -k - 13$$
$$14 = k$$
Check $k = 14$: $\sqrt[3]{-27} - \sqrt[3]{-27} \overset{?}{=} 0$ *True*
Solution set: $\{14\}$

59. $\sqrt[4]{x - 1} + 2 = 0$
$$\sqrt[4]{x - 1} = -2$$
This equation has no solution because $\sqrt[4]{x - 1}$ cannot be negative.
Solution set: $\emptyset$

61. $\sqrt[4]{x+7} = \sqrt[4]{2x}$

Raise each side to the fourth power.

$$\left(\sqrt[4]{x+7}\right)^4 = \left(\sqrt[4]{2x}\right)^4$$
$$x + 7 = 2x$$
$$7 = x$$

Check $x = 7$: $\sqrt[4]{14} \overset{?}{=} \sqrt[4]{14}$ *True*

Solution set: $\{7\}$

63. $(2x - 9)^{1/2} = 2 + (x - 8)^{1/2}$
$$\sqrt{2x - 9} = 2 + \sqrt{x - 8}$$
$$\left(\sqrt{2x - 9}\right)^2 = \left(2 + \sqrt{x - 8}\right)^2$$
$$2x - 9 = 4 + 4\sqrt{x - 8} + x - 8$$
$$x - 5 = 4\sqrt{x - 8}$$
$$(x - 5)^2 = \left(4\sqrt{x - 8}\right)^2$$
$$x^2 - 10x + 25 = 16(x - 8)$$
$$x^2 - 10x + 25 = 16x - 128$$
$$x^2 - 26x + 153 = 0$$
$$(x - 9)(x - 17) = 0$$

Check $x = 9$: $9^{1/2} \overset{?}{=} 2 + 1^{1/2}$

   $3 \overset{?}{=} 2 + 1$ *True*

Check $x = 17$: $25^{1/2} \overset{?}{=} 2 + 9^{1/2}$

   $5 \overset{?}{=} 2 + 3$ *True*

Solution set: $\{9, 17\}$

65. $(2w - 1)^{2/3} - w^{1/3} = 0$
$$\sqrt[3]{(2w - 1)^2} = \sqrt[3]{w}$$
$$\left[\sqrt[3]{(2w - 1)^2}\right]^3 = \left(\sqrt[3]{w}\right)^3$$
$$(2w - 1)^2 = w$$
$$4w^2 - 4w + 1 = w$$
$$4w^2 - 5w + 1 = 0$$
$$(4w - 1)(w - 1) = 0$$

$4w - 1 = 0$ or $w - 1 = 0$

$w = \dfrac{1}{4}$ or $w = 1$

Check $w = \dfrac{1}{4}$: $\left(-\dfrac{1}{2}\right)^{2/3} - \left(\dfrac{1}{4}\right)^{1/3} \overset{?}{=} 0$

   $\left(\dfrac{1}{4}\right)^{1/3} - \left(\dfrac{1}{4}\right)^{1/3} \overset{?}{=} 0$ *True*

Check $w = 1$: $1 - 1 \overset{?}{=} 0$ *True*

Solution set: $\left\{\dfrac{1}{4}, 1\right\}$

67.   Solve $V = \sqrt{\dfrac{2K}{m}}$ for K.

$$(V)^2 = \left(\sqrt{\dfrac{2K}{m}}\right)^2$$ *Square*

$$V^2 = \dfrac{2K}{m}$$ *Multiply by $\dfrac{m}{2}$.*

$$\dfrac{V^2 m}{2} = K$$

69.   Solve $f = \dfrac{1}{2\pi\sqrt{LC}}$ for L.

$$2\pi f\sqrt{LC} = 1$$
$$\left(2\pi f\sqrt{LC}\right)^2 = 1^2$$
$$4\pi^2 f^2 LC = 1$$
$$L = \dfrac{1}{4\pi^2 f^2 C}$$

71.   **(a)** Solve $N = \dfrac{1}{2\pi}\sqrt{\dfrac{a}{r}}$ for r.

$$2\pi N = \sqrt{\dfrac{a}{r}}$$
$$(2\pi N)^2 = \left(\sqrt{\dfrac{a}{r}}\right)^2$$
$$4\pi^2 N^2 = \dfrac{a}{r}$$
$$4\pi^2 N^2 r = a$$
$$r = \dfrac{a}{4\pi^2 N^2}$$

(b) $r = \dfrac{9.8}{4\pi^2(.063)^2}$

   ≈ 62.5 meters

(c) $r = \dfrac{9.8}{4\pi^2(.04)^2}$

   ≈ 155.1 meters

73.   Use the given equation

$$y = x^{.7}.$$

Substitute the values for x.
For 1920, $x = 20$ and

$$y = 20^{.7} \approx 8 \text{ billion ft}^3.$$

For 1952, $x = 52$ and

$$y = 52^{.7} \approx 16 \text{ billion ft}^3.$$

For 1976, $x = 76$ and

$$y = 76^{.7} \approx 21 \text{ billion ft}^3.$$

For 1991, $x = 91$ and

$$y = 91^{.7} \approx 24 \text{ billion ft}^3.$$

75. The values obtained from the equations are fairly close to the estimates obtained from the figure. The approximation is best in 1976 where the equation yielded 21 billion ft^3 and the estimate was 22.5 billion ft^3.

77. Use the given equation

$$y = x^{.62},$$

and substitute the values for x.
For 1920, $x = 20$ and

$$y = 20^{.62} \approx 6 \text{ billion ft}^3.$$

For 1952, $x = 52$ and

$$y = 52^{.62} \approx 12 \text{ billion ft}^3.$$

For 1976, $x = 76$ and

$$y = 76^{.62} \approx 15 \text{ billion ft}^3.$$

For 1991, $x = 91$ and

$$y = 91^{.62} \approx 16 \text{ billion ft}^3.$$

79. $(-8)^{2/3} = \left[(-8)^{1/3} \right]^2$

Using a graphing calculator, $\left[(-8)^{1/3} \right]^2 = 4$. The power rule of exponents applies here.

81. Graph the functions

$$Y_1 = \sqrt{13 + 4x} \quad \text{and} \quad Y_2 = x + 4.$$

Find the points of intersection of the graphs.

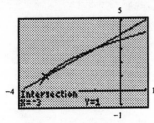

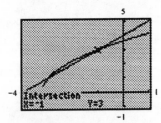

The x-values of the points of intersection are the solutions of the equation. Note that the y-values of the points of intersection are the "check" values obtained in Exercise 37.
Solution set: $\{-3, -1\}$

83. Graph the functions

$$Y_1 = \sqrt{3 - 3x} \quad \text{and} \quad Y_2 = 3 + \sqrt{3x + 2}.$$

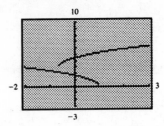

The graphs do not intersect, so the solution set is $\emptyset$.
To find the domain of

$$y = \sqrt{3 - 3x} - 3 - \sqrt{3x + 2}$$

we must have

$$3 - 3x \geq 0 \quad \text{and} \quad 3x + 2 \geq 0$$
$$-3x \geq -3 \qquad\qquad 3x \geq -2$$
$$x \leq 1 \quad \text{and} \quad x \geq -\frac{2}{3}.$$

The domain is $\left[-\dfrac{2}{3}, 1 \right]$. This is the region where

the graphs of Y_1 and Y_2 vertically overlap.

Section 7.6

1. $\sqrt{-1} = i$

3. $i^2 = -1$

5. $\dfrac{1}{i} = \dfrac{1}{i} \cdot \dfrac{-i}{-i} \quad -i$ *is the conjugate of* i

$$= \frac{-i}{-i^2} = \frac{i}{i^2} = \frac{i}{-1} = -i$$

7. $\sqrt{-169} = i\sqrt{169} = 13i$

9. $-\sqrt{-144} = -i\sqrt{144} = -12i$

11. $\sqrt{-5} = i\sqrt{5}$

13. $\sqrt{-48} = i\sqrt{48} = i\sqrt{16 \cdot 3} = 4i\sqrt{3}$

15. $\sqrt{-15} \cdot \sqrt{-15} = i\sqrt{15} \cdot i\sqrt{15} = i^2 \left(\sqrt{15} \right)^2$
$$= -1(15) = -15$$

17. $\sqrt{-4} \cdot \sqrt{-25} = i\sqrt{4} \cdot i\sqrt{25} = 2i \cdot 5i = 10i^2$
$$= 10(-1) = -10$$

19. $\dfrac{\sqrt{-300}}{\sqrt{-100}} = \dfrac{i\sqrt{300}}{i\sqrt{100}} = \sqrt{\dfrac{300}{100}} = \sqrt{3}$

21. $\dfrac{\sqrt{-75}}{\sqrt{3}} = \dfrac{i\sqrt{75}}{\sqrt{3}} = i\sqrt{\dfrac{75}{3}} = i\sqrt{25} = 5i$

23. **(a)** Since any real number can be written as $a + bi$, where $b = 0$, every real number is also a complex number.

(b) Not every complex number is a real number. For example, any number $a + bi$, $b \neq 0$, such as $3 + 7i$, is an imaginary number, which is a complex number that is not a real number.

25. $(6 + 2i) + (4 + 6i)$
$$= (6 + 4) + (2 + 6)i$$
$$= 10 + 8i$$

27. $(3 + 2i) + (-4 + 5i)$
$$= [(3 + (-4)] + (2 + 5)i$$
$$= -1 + 7i$$

29. $(5 - i) + (-5 + i)$
$$= (5 - 5) + (-1 + 1)i$$
$$= 0$$

31. $(4 + i) - (-3 - 2i)$
$$= [(4 - (-3)] + [(1 - (-2)]i$$
$$= 7 + 3i$$

33. $(-3 - 4i) - (-1 - 4i)$
$$= [-3 - (-1)] + [-4 - (-4)]i$$
$$= -2$$

35. $(-4 + 11i) + (-2 - 4i) + (7 + 6i)$
$$= (-4 - 2 + 7) + (11 - 4 + 6)i$$
$$= 1 + 13i$$

37. $[(7 + 3i) - (4 - 2i)] + (3 + i)$
Work inside the brackets first.
$$= [(7 - 4) + (3 + 2)i] + (3 + i)$$
$$= (3 + 5i) + (3 + i)$$
$$= (3 + 3) + (5 + 1)i$$
$$= 6 + 6i$$

39. If $a - c = b$, then $b + c = a$.
So, $(4 + 2i) - (3 + i) = 1 + i$ implies that
$$(1 + i) + (3 + i) = \underline{4 + 2i}.$$

41. $(3i)(27i) = 81i^2 = 81(-1) = -81$

43. $(-8i)(-2i) = 16i^2 = 16(-1) = -16$

45. $5i(-6 + 2i) = (5i)(-6) + (5i)(2i)$
$$= -30i + 10i^2$$
$$= -30i + 10(-1)$$
$$= -10 - 30i$$

47. $(4 + 3i)(1 - 2i)$

$$\quad\mathbf{F}\qquad\mathbf{O}\qquad\mathbf{I}\qquad\mathbf{L}$$
$$= (4)(1) + 4(-2i) + (3i)(1) + (3i)(-2i)$$
$$= 4 - 8i + 3i - 6i^2$$
$$= 4 - 5i - 6(-1)$$
$$= 4 - 5i + 6 = 10 - 5i$$

49. $(4 + 5i)^2 = 4^2 + 2(4)(5i) + (5i)^2$
$$= 16 + 40i + 25i^2$$
$$= 16 + 40i + 25(-1)$$
$$= 16 + 40i - 25$$
$$= -9 + 40i$$

51. $2i(-4 - i)^2$
$$= 2i[(-4)^2 - 2(-4)(i) + i^2]$$
$$= 2i(16 + 8i - 1)$$
$$= 2i(15 + 8i) = 30i + 16i^2$$
$$= 30i + 16(-1) = -16 + 30i$$

53. $(12 + 3i)(12 - 3i)$
$$= 12^2 - (3i)^2 = 144 - 9i^2$$
$$= 144 - 9(-1) = 144 + 9 = 153$$

55. $(4 + 9i)(4 - 9i)$
$$= 4^2 - (9i)^2 = 16 - 81i^2$$
$$= 16 - 81(-1) = 16 + 81 = 97$$

57. $\dfrac{2}{1 - i}$

Multiply the numerator and the denominator by the conjugate of the denominator, $1 + i$.

$$= \frac{2(1 + i)}{(1 - i)(1 + i)} = \frac{2(1 + i)}{1^2 - i^2}$$
$$= \frac{2(1 + i)}{1 - (-1)} = \frac{2(1 + i)}{2} = 1 + i$$

59. $\dfrac{-7 + 4i}{3 + 2i}$

Multiply the numerator and the denominator by the conjugate of the denominator, $3 - 2i$.

$$= \frac{(-7 + 4i)(3 - 2i)}{(3 + 2i)(3 - 2i)}$$

In the denominator, we make use of the fact that

$$(a + bi)(a - bi) = a^2 + b^2.$$

$$= \frac{-21 + 14i + 12i + 8}{3^2 + 2^2}$$
$$= \frac{-13 + 26i}{13} = \frac{13(-1 + 2i)}{13} = -1 + 2i$$

61. $\dfrac{8i}{2+2i}$

Write in lowest terms.

$= \dfrac{2 \cdot 4i}{2(1+i)} = \dfrac{4i}{1+i}$

Multiply the numerator and the denominator by the conjugate of the denominator, $1-i$.

$= \dfrac{4i(1-i)}{(1+i)(1-i)} = \dfrac{4(i-i^2)}{1^2+1^2}$

$= \dfrac{4(i+1)}{2} = 2(i+1) = 2+2i$

63. $\dfrac{2-3i}{2+3i}$

Multiply the numerator and the denominator by the conjugate of the denominator, $2-3i$.

$= \dfrac{(2-3i)(2-3i)}{(2+3i)(2-3i)} = \dfrac{2^2-2(2)(3i)+(3i)^2}{2^2+3^2}$

$= \dfrac{4-12i+9i^2}{4+9} = \dfrac{4-12i-9}{13}$

$= \dfrac{-5-12i}{13} = -\dfrac{5}{13} - \dfrac{12}{13}i$

65. $(8-5i)+2(11+6i)-(1+i)^2$

$= 8-5i+22+12i-(1+2i-1)$

$= 30+7i-(2i)$

$= 30+5i$

67. $(26+32i)/(2+4i)-(2i)(1+i)^{-1}$

$= \dfrac{26+32i}{2+4i} - \dfrac{2i}{1+i}$

$= \dfrac{2(13+16i)}{2(1+2i)} - \dfrac{2i}{1+i}$

$= \dfrac{(13+16i)(1-2i)}{(1+2i)(1-2i)} - \dfrac{2i(1-i)}{(1+i)(1-i)}$

$= \dfrac{13-26i+16i-32i^2}{1^2+2^2} - \dfrac{2i-2i^2}{1^2+1^2}$

$= \dfrac{13-10i+32}{5} - \dfrac{2i+2}{2}$

$= \dfrac{45-10i}{5} - \dfrac{2+2i}{2}$

$= \dfrac{5(9-2i)}{5} - \dfrac{2(1+i)}{2}$

$= (9-2i)-(1+i) = 8-3i$

69. **(a)** $(x+2)+(3x-1)$

$= (1+3)x+(2-1)$

$= 4x+1$

(b) $(1+2i)+(3-i)$

$= (1+3)+(2-1)i$

$= 4+i$

70. **(a)** $(x+2)-(3x-1)$

$= (x-3x)+[2-(-1)]$

$= (1-3)x+(2+1)$

$= -2x+3$

(b) $(1+2i)-(3-i)$

$= (1-3)+[2-(-1)]i$

$= -2+3i$

71. **(a)** $(x+2)(3x-1)$

$= 3x^2-x+6x-2$

$= 3x^2+5x-2$

(b) $(1+2i)(3-i)$

$= 3-i+6i-2i^2$

$= 3+5i-2(-1)$

$= 3+5i+2$

$= 5+5i$

72. **(a)** $\dfrac{\sqrt{3}-1}{1+\sqrt{2}} = \dfrac{\left(\sqrt{3}-1\right)\left(1-\sqrt{2}\right)}{\left(1+\sqrt{2}\right)\left(1-\sqrt{2}\right)}$

$= \dfrac{\sqrt{3}-\sqrt{6}-1+\sqrt{2}}{1^2-\left(\sqrt{2}\right)^2}$

$= \dfrac{\sqrt{3}-\sqrt{6}-1+\sqrt{2}}{1-2}$

$= \dfrac{\sqrt{3}-\sqrt{6}-1+\sqrt{2}}{-1}$

$= -\left(\sqrt{3}-\sqrt{6}-1+\sqrt{2}\right)$

$= -\sqrt{3}+\sqrt{6}+1-\sqrt{2}$

(b) $\dfrac{3-i}{1+2i} = \dfrac{(3-i)(1-2i)}{(1+2i)(1-2i)}$

$= \dfrac{3-6i-i+2i^2}{1-(2i)^2}$

$= \dfrac{3-7i+2(-1)}{1-4(-1)}$

$= \dfrac{3-7i-2}{1+4}$

$= \dfrac{1-7i}{5}$

$= \dfrac{1}{5} - \dfrac{7}{5}i$

73. In parts (a) and (b) of Exercises 69 and 70, real and imaginary parts are added, just like coefficients of similar terms in the binomials, and the answers correspond. In Exercise 71, introducing $i^2 = -1$ when a product is found leads to answers that do not correspond.

74. In parts (a) and (b) of Exercises 69 and 70, real and imaginary parts are added, just like coefficients of similar terms in the binomials, and the answers correspond. In Exercise 72, introducing $i^2 = -1$ when performing the division leads to answers that do not correspond.

75. The reciprocal of $5 - 4i$ is $\dfrac{1}{5-4i}$.

$$\frac{1}{5-4i} = \frac{1 \cdot (5+4i)}{(5-4i)(5+4i)}$$
$$= \frac{5+4i}{25 - 16i^2} = \frac{5+4i}{25 - 16(-1)}$$
$$= \frac{5+4i}{25+16} = \frac{5+4i}{41} = \frac{5}{41} + \frac{4}{41}i$$

77. $i^{18} = i^{16} \cdot i^2 = \left(i^4\right)^4 \cdot i^2$
$$= 1^4 \cdot (-1) = 1 \cdot (-1) = -1$$

79. $i^{89} = i^{88} \cdot i = \left(i^4\right)^{22} \cdot i = 1^{22} \cdot i$
$$= 1 \cdot i = i$$

81. $i^{-5} = \dfrac{1}{i^5} = \dfrac{1}{i^4 \cdot i} = \dfrac{1}{1 \cdot i} = \dfrac{1}{i}$

From Exercise 5, $\dfrac{1}{i} = -i$.

83. Because $i^{20} = \left(i^4\right)^5 = 1^5 = 1$, multiplying by i^{20} is an application of the identity property for multiplication.

85. $I = \dfrac{E}{R + (X_L - X_c)i}$

Substitute $2 + 3i$ for E, 5 for R, 4 for X_L, and 3 for X_c.

$$I = \frac{2+3i}{5 + (4-3)i} = \frac{2+3i}{5+i}$$
$$= \frac{(2+3i)(5-i)}{(5+i)(5-i)} = \frac{10 - 2i + 15i - 3i^2}{5^2 + 1^2}$$
$$= \frac{10 + 3 + 13i}{25 + 1} = \frac{13 + 13i}{26}$$
$$= \frac{13(1+i)}{13 \cdot 2} = \frac{1+i}{2} = \frac{1}{2} + \frac{1}{2}i$$

87. To check that $1 + 5i$ is a solution of the equation, substitute $1 + 5i$ for x.

$$x^2 - 2x + 26 = 0$$
$$(1+5i)^2 - 2(1+5i) + 26 \overset{?}{=} 0$$
$$\left(1 + 10i + 25i^2\right) - 2 - 10i + 26 \overset{?}{=} 0$$
$$1 + 10i - 25 - 2 - 10i + 26 \overset{?}{=} 0$$
$$(1 - 25 - 2 + 26) + (10 - 10)i \overset{?}{=} 0$$
$$0 \overset{?}{=} 0$$
True

Thus, $1 + 5i$ is a solution of

$$x^2 - 2x + 26 = 0.$$

Now substitute $1 - 5i$ for x.

$$(1-5i)^2 - 2(1-5i) + 26 \overset{?}{=} 0$$
$$\left(1 - 10i + 25i^2\right) - 2 + 10i + 26 \overset{?}{=} 0$$
$$1 - 10i - 25 - 2 + 10i + 26 \overset{?}{=} 0$$
$$(1 - 25 - 2 + 26) + (-10 + 10)i \overset{?}{=} 0$$
$$0 \overset{?}{=} 0$$
True

Thus, $1 - 5i$ is also a solution of the given equation.

Chapter 7 Review Exercises

1. One way to evaluate $8^{2/3}$ is to first find the _cube (or third)_ root of _8_, which is _2_. Then raise that result to the _second_ power, to get an answer of _4_. Therefore, $8^{2/3} =$ _4_.

2. **(a)** $(-27)^{2/3} = \left[(-27)^{1/3}\right]^2$

This number is a square, so it is a positive number.

(b) $(-64)^{5/3} = \left[(-64)^{1/3}\right]^5$

This number is the odd power of an odd root of a negative number, so it is a negative number.

(c) $(-100)^{1/2}$

This number is the square root of a negative number, so it is not a real number.

(d) $(-32)^{1/5}$

This number is an odd root of a negative number, so it is a negative number.

The only positive number is Choice (a).

3. $a^{m/n} = \sqrt[n]{a^m}$

Since n is odd, $\sqrt[n]{a^m}$ is positive if a^m is positive and negative if a^m is negative. Since a is negative, a^m is positive if m is even and negative if m is odd.

(a) If a is negative and n is odd, then $a^{m/n}$ is positive if m is even.

(b) If a is negative and n is odd, then $a^{m/n}$ is negative if m is odd.

4. If a is negative and n is even, then $a^{1/n}$ is not a real number. An example is $(-4)^{1/2}$, not a real number.

5. $49^{1/2} = \left(7^2\right)^{1/2} = 7$

6. $-121^{1/2} = -\left(11^2\right)^{1/2} = -11$

7. $16^{5/4} = \left[\left(2^4\right)^{1/4}\right]^5 = 2^5 = 32$

8. $-8^{2/3} = -\left[\left(2^3\right)^{1/3}\right]^2 = -2^2 = -4$

9. $-\left(\dfrac{36}{25}\right)^{3/2} = -\left(\left[\left(\dfrac{6}{5}\right)^2\right]^{1/2}\right)^3$

$= -\left(\dfrac{6}{5}\right)^3 = -\dfrac{216}{125}$

10. $\left(-\dfrac{1}{8}\right)^{-5/3} = (-8)^{5/3} = \left(\left[(-2)^3\right]^{1/3}\right)^5$

$= (-2)^5 = -32$

11. $\left(\dfrac{81}{10,000}\right)^{-3/4} = \left(\dfrac{10,000}{81}\right)^{3/4}$

$= \left(\left[\left(\dfrac{10}{3}\right)^4\right]^{1/4}\right)^3$

$= \left(\dfrac{10}{3}\right)^3 = \dfrac{1000}{27}$

12.   Solve $a^2 + b^2 = c^2$ for b. $(b > 0)$

$b^2 = c^2 - a^2$

$b = \sqrt{c^2 - a^2}$

13.   The expression with fractional exponents, $a^{m/n}$, is equivalent to the radical expression, $\sqrt[n]{a^m}$. The denominator of the exponent is the index of the radical. For example, $\sqrt[3]{8^2} = \sqrt[3]{64} = 4$, and $8^{2/3} = \left(8^{1/3}\right)^2 = 2^2 = 4$.

14. $\sqrt{1764} = 42$, since $42^2 = 1764$.

15. $-\sqrt{289} = -(17) = -17$, since $17^2 = 289$.

16. $\sqrt[3]{216} = 6$, since $6^3 = 216$.

17. $\sqrt[3]{-125} = -5$, since $(-5)^3 = -125$.

18. $-\sqrt[3]{27z^{12}} = -\sqrt[3]{(3z^4)^3} = -3z^4$

19. $\sqrt[5]{-32} = -2$, since $(-2)^5 = -32$.

20. $\sqrt[n]{a}$ is not a real number if n is even and a is negative.

21.   (a) $\sqrt{x^2} = |x|$

(b) $-\sqrt{x^2} = -|x|$

(c) $\sqrt[3]{x^3} = x$

22. $(m + 3n)^{1/2} = \sqrt{m + 3n}$

23. $(3a + b)^{-5/3} = \dfrac{1}{(3a + b)^{5/3}}$

$= \dfrac{1}{\left((3a + b)^{1/3}\right)^5}$

$= \dfrac{1}{\left(\sqrt[3]{3a + b}\right)^5}$

or $\dfrac{1}{\sqrt[3]{(3a + b)^5}}$

24. $\sqrt{7^9} = (7^9)^{1/2} = 7^{9/2}$

25. $\sqrt[5]{p^4} = (p^4)^{1/5} = p^{4/5}$

26. $5^{1/4} \cdot 5^{7/4} = 5^{1/4+7/4} = 5^{8/4}$

$= 5^2$ or 25

27. $\dfrac{96^{2/3}}{96^{-1/3}} = 96^{2/3-(-1/3)} = 96^1 = 96$

28. $\dfrac{\left(a^{1/3}\right)^4}{a^{2/3}} = \dfrac{a^{4/3}}{a^{2/3}} = a^{4/3-2/3} = a^{2/3}$

29. $\dfrac{y^{-1/3} \cdot y^{5/6}}{y} = \dfrac{y^{-2/6}y^{5/6}}{y^{6/6}}$

$= y^{-2/6+5/6-6/6}$

$= y^{-3/6} = y^{-1/2} = \dfrac{1}{y^{1/2}}$

30. $\left(\dfrac{z^{-1}x^{-3/5}}{2^{-2}z^{-1/2}x}\right)^{-1} = \dfrac{z^1 x^{3/5}}{2^2 z^{1/2}x^{-1}}$

$= \dfrac{1}{4}z^{1-1/2}x^{3/5-(-1)}$

$= \dfrac{z^{1/2}x^{8/5}}{4}$

31. $r^{-1/2}\left(r + r^{3/2}\right)$

$= r^{-1/2}(r) + r^{-1/2}\left(r^{3/2}\right)$

$= r^{-1/2+1} + r^{-1/2+3/2}$

$= r^{1/2} + r^{2/2}$

$= r^{1/2} + r$

32. $\sqrt[8]{s^4} = (s^4)^{1/8} = s^{4/8} = s^{1/2}$

33. $\sqrt[6]{r^9} = (r^9)^{1/6} = r^{9/6} = r^{3/2}$

34. $\dfrac{\sqrt{p^5}}{p^2} = \dfrac{p^{5/2}}{p^2} = p^{5/2-2}$

$= p^{5/2-4/2}$ or $p^{1/2}$

35. $\sqrt[4]{k^3} \cdot \sqrt{k^3} = \left(k^3\right)^{1/4}\left(k^3\right)^{1/2}$

$= k^{3/4}k^{3/2} = k^{3/4+3/2}$

$= k^{3/4+6/4} = k^{9/4}$

36. $\sqrt[3]{m^5} \cdot \sqrt[3]{m^8} = \left(m^5\right)^{1/3}\left(m^8\right)^{1/3}$

$= m^{5/3}m^{8/3} = m^{5/3+8/3} = m^{13/3}$

37. $\sqrt[4]{\sqrt[3]{z}} = \sqrt[4]{z^{1/3}} = \left(z^{1/3}\right)^{1/4} = z^{1/12}$

38. $\sqrt{\sqrt{\sqrt{x}}} = \sqrt{\sqrt{x^{1/2}}} = \sqrt{\left(x^{1/2}\right)^{1/2}}$

$= \sqrt{x^{1/4}} = \left(x^{1/4}\right)^{1/2} = x^{1/8}$

39. $-\sqrt{47} \approx -6.856$

40. $\sqrt[3]{-129} \approx -5.053$

41. $\sqrt[4]{605} \approx 4.960$

42. $500^{-3/4} \approx .009$

43. $-500^{4/3} \approx -3968.503$

44. $-28^{-1/2} \approx -.189$

45. The product rule for exponents applies only if the bases are the same.

46. The base $\sqrt{38}$ is closest to $\sqrt{36} = 6$. The height $\sqrt{99}$ is closest to $\sqrt{100} = 10$. Use the estimates $b = 6$ and $h = 10$ in $A = \frac{1}{2}bh$ to find an estimate of the area.

$$A \approx \frac{1}{2}(6)(10) = 30$$

Choice (b) is the best estimate.

47. $\sqrt{6} \cdot \sqrt{11} = \sqrt{6 \cdot 11} = \sqrt{66}$

48. $\sqrt{5} \cdot \sqrt{r} = \sqrt{5 \cdot r} = \sqrt{5r}$

49. $\sqrt[3]{6} \cdot \sqrt[3]{5} = \sqrt[3]{6 \cdot 5} = \sqrt[3]{30}$

50. $\sqrt[4]{7} \cdot \sqrt[4]{3} = \sqrt[4]{7 \cdot 3} = \sqrt[4]{21}$

51. $\sqrt{20} = \sqrt{4 \cdot 5} = \sqrt{4}\sqrt{5} = 2\sqrt{5}$

52. $\sqrt{75} = \sqrt{25 \cdot 3} = \sqrt{25}\sqrt{3} = 5\sqrt{3}$

53. $-\sqrt{125} = -\sqrt{25 \cdot 5} = -5\sqrt{5}$

54. $\sqrt[3]{-108} = \sqrt[3]{-27 \cdot 4} = -3\sqrt[3]{4}$

55. $\sqrt{100y^7} = \sqrt{100y^6 \cdot y} = 10y^3\sqrt{y}$

56. $\sqrt[3]{64p^4q^6} = \sqrt[3]{64p^3q^6 \cdot p} = 4pq^2\sqrt[3]{p}$

57. $\sqrt[3]{108a^8b^5} = \sqrt[3]{27a^6b^3 \cdot 4a^2b^2}$

$= 3a^2b\sqrt[3]{4a^2b^2}$

58. $\sqrt[3]{632r^8t^4} = \sqrt[3]{8r^6t^3 \cdot 79r^2t}$

$= 2r^2t\sqrt[3]{79r^2t}$

59. $\sqrt{\dfrac{y^3}{144}} = \dfrac{\sqrt{y^3}}{\sqrt{144}} = \dfrac{\sqrt{y^2 \cdot y}}{12} = \dfrac{y\sqrt{y}}{12}$

60. $\sqrt[3]{\dfrac{m^{15}}{27}} = \dfrac{\sqrt[3]{m^{15}}}{\sqrt[3]{27}} = \dfrac{\sqrt[3]{(m^5)^3}}{\sqrt[3]{3^3}} = \dfrac{m^5}{3}$

61. $\sqrt[3]{\dfrac{r^2}{8}} = \dfrac{\sqrt[3]{r^2}}{\sqrt[3]{8}} = \dfrac{\sqrt[3]{r^2}}{2}$

62. $\sqrt[4]{\dfrac{a^9}{81}} = \dfrac{\sqrt[4]{a^9}}{\sqrt[4]{81}} = \dfrac{\sqrt[4]{a^8 \cdot a}}{3} = \dfrac{a^2\sqrt[4]{a}}{3}$

63. $\sqrt[6]{15^3} = 15^{3/6} = 15^{1/2} = \sqrt{15}$

64. $\sqrt[4]{p^6} = \left(p^6\right)^{1/4} = p^{6/4} = p^{3/2}$

$= p^{2/2}p^{1/2} = p\sqrt{p}$

65. $\sqrt[3]{2} \cdot \sqrt[4]{5} = 2^{1/3} \cdot 5^{1/4}$

$= 2^{4/12} \cdot 5^{3/12}$

$= \left(2^4 \cdot 5^3\right)^{1/12}$

$= \sqrt[12]{16 \cdot 125} = \sqrt[12]{2000}$

66. $\sqrt{x} \cdot \sqrt[5]{x} = x^{1/2} \cdot x^{1/5}$

$= x^{5/10} \cdot x^{2/10}$

$= x^{7/10} = \sqrt[10]{x^7}$

67. Substitute 8 for a and 6 for b in the Pythagorean formula to find the hypotenuse, c.

$$c^2 = a^2 + b^2$$

$$c = \sqrt{a^2 + b^2} = \sqrt{8^2 + 6^2}$$

$$= \sqrt{64 + 36} = \sqrt{100} = 10$$

The length of the hypotenuse is 10.

68. $(-4, 7)$ and $(10, 6)$

$$d = \sqrt{(x_2 - x_1)^2 + (y_2 - y_1)^2}$$

$$= \sqrt{[10 - (-4)]^2 + (6 - 7)^2}$$

$$= \sqrt{14^2 + (-1)^2}$$

$$= \sqrt{196 + 1} = \sqrt{197}$$

69. $2\sqrt{8} - 3\sqrt{50} = 2\sqrt{4 \cdot 2} - 3\sqrt{25 \cdot 2}$

$= 2 \cdot 2\sqrt{2} - 3 \cdot 5\sqrt{2}$

$= 4\sqrt{2} - 15\sqrt{2} = -11\sqrt{2}$

70. $8\sqrt{80} - 3\sqrt{45} = 8\sqrt{16 \cdot 5} - 3\sqrt{9 \cdot 5}$

$= 8 \cdot 4\sqrt{5} - 3 \cdot 3\sqrt{5}$

$= 32\sqrt{5} - 9\sqrt{5} = 23\sqrt{5}$

71. $-\sqrt{27y} + 2\sqrt{75y} = -\sqrt{9 \cdot 3y} + 2\sqrt{25 \cdot 3y}$
$$= -3\sqrt{3y} + 2 \cdot 5\sqrt{3y}$$
$$= -3\sqrt{3y} + 10\sqrt{3y}$$
$$= 7\sqrt{3y}$$

72. $2\sqrt{54m^3} + 5\sqrt{96m^3}$
$$= 2\sqrt{9m^2 \cdot 6m} + 5\sqrt{16m^2 \cdot 6m}$$
$$= 2 \cdot 3m\sqrt{6m} + 5 \cdot 4m\sqrt{6m}$$
$$= 6m\sqrt{6m} + 20m\sqrt{6m} = 26m\sqrt{6m}$$

73. $3\sqrt[3]{54} + 5\sqrt[3]{16} = 3\sqrt[3]{27 \cdot 2} + 5\sqrt[3]{8 \cdot 2}$
$$= 3 \cdot 3\sqrt[3]{2} + 5 \cdot 2\sqrt[3]{2}$$
$$= 9\sqrt[3]{2} + 10\sqrt[3]{2} = 19\sqrt[3]{2}$$

74. $-6\sqrt[4]{32} + \sqrt[4]{512} = -6\sqrt[4]{16 \cdot 2} + \sqrt[4]{256 \cdot 2}$
$$= -6 \cdot 2\sqrt[4]{2} + 4\sqrt[4]{2}$$
$$= -12\sqrt[4]{2} + 4\sqrt[4]{2} = -8\sqrt[4]{2}$$

75. $\dfrac{3}{\sqrt{16}} - \dfrac{\sqrt{5}}{2} = \dfrac{3}{4} - \dfrac{2\sqrt{5}}{2 \cdot 2} = \dfrac{3 - 2\sqrt{5}}{4}$

76. $\dfrac{4}{\sqrt{25}} + \dfrac{\sqrt{5}}{4} = \dfrac{4}{5} + \dfrac{\sqrt{5}}{4}$
$$= \dfrac{16}{20} + \dfrac{5\sqrt{5}}{20} = \dfrac{16 + 5\sqrt{5}}{20}$$

77. Add the measures of the sides.
$$P = a + b + c + d$$
$$P = 4\sqrt{8} + 6\sqrt{12} + 8\sqrt{2} + 3\sqrt{48}$$
$$= 4\sqrt{4 \cdot 2} + 6\sqrt{4 \cdot 3} + 8\sqrt{2} + 3\sqrt{16 \cdot 3}$$
$$= 4 \cdot 2\sqrt{2} + 6 \cdot 2\sqrt{3} + 8\sqrt{2} + 3 \cdot 4\sqrt{3}$$
$$= 8\sqrt{2} + 12\sqrt{3} + 8\sqrt{2} + 12\sqrt{3}$$
$$= 16\sqrt{2} + 24\sqrt{3}$$

The perimeter is $16\sqrt{2} + 24\sqrt{3}$ feet.

78. Add the measures of the sides.
$$P = a + b + c$$
$$P = 2\sqrt{27} + \sqrt{108} + \sqrt{50}$$
$$= 2\sqrt{9 \cdot 3} + \sqrt{36 \cdot 3} + \sqrt{25 \cdot 2}$$
$$= 2 \cdot 3\sqrt{3} + 6\sqrt{3} + 5\sqrt{2}$$
$$= 6\sqrt{3} + 6\sqrt{3} + 5\sqrt{2}$$
$$= 12\sqrt{3} + 5\sqrt{2}$$

The perimeter is $12\sqrt{3} + 5\sqrt{2}$ feet.

79. $\left(\sqrt{3} + 1\right)\left(\sqrt{3} - 2\right) = 3 - 2\sqrt{3} + \sqrt{3} - 2$
$$= 1 - \sqrt{3}$$

80. $\left(\sqrt{7} + \sqrt{5}\right)\left(\sqrt{7} - \sqrt{5}\right) = \left(\sqrt{7}\right)^2 - \left(\sqrt{5}\right)^2$
$$= 7 - 5 = 2$$

81. $\left(3\sqrt{2} + 1\right)\left(2\sqrt{2} - 3\right)$
$$= 6 \cdot 2 - 9\sqrt{2} + 2\sqrt{2} - 3$$
$$= 12 - 7\sqrt{2} - 3 = 9 - 7\sqrt{2}$$

82. $\left(\sqrt{13} - \sqrt{2}\right)^2$
$$= \left(\sqrt{13}\right)^2 - 2 \cdot \sqrt{13} \cdot \sqrt{2} + \left(\sqrt{2}\right)^2$$
$$= 13 - 2\sqrt{26} + 2 = 15 - 2\sqrt{26}$$

83. $\left(\sqrt[3]{2} + 3\right)\left(\sqrt[3]{4} - 3\sqrt[3]{2} + 9\right)$
$$= \sqrt[3]{2} \cdot \sqrt[3]{4} - \sqrt[3]{2} \cdot 3\sqrt[3]{2} + 9\sqrt[3]{2}$$
$$\quad + 3\sqrt[3]{4} - 3 \cdot 3\sqrt[3]{2} + 27$$
$$= \sqrt[3]{8} - 3\sqrt[3]{4} + 9\sqrt[3]{2} + 3\sqrt[3]{4} - 9\sqrt[3]{2} + 27$$
$$= 2 + 27 = 29$$

84. $\left(\sqrt[3]{4y} - 1\right)\left(\sqrt[3]{4y} + 3\right)$
$$= \sqrt[3]{16y^2} + 3\sqrt[3]{4y} - \sqrt[3]{4y} - 3$$
$$= \sqrt[3]{8 \cdot 2y^2} + 2\sqrt[3]{4y} - 3$$
$$= 2\sqrt[3]{2y^2} + 2\sqrt[3]{4y} - 3$$

85. Show that $15 - 2\sqrt{26} \neq 13\sqrt{26}$.

Find a calculator approximation of each term.

$$4.801960973 \neq 66.28725368$$

Therefore, $15 - 2\sqrt{26} \neq 13\sqrt{26}$.

86. Multiplying by $\sqrt[3]{6}$ still results in an expression with a radical in the denominator.

$$\dfrac{5\sqrt[3]{6}}{\sqrt[3]{6} \cdot \sqrt[3]{6}} = \dfrac{5\sqrt[3]{6}}{\sqrt[3]{36}}$$

To rationalize the denominator, multiply by $\sqrt[3]{6^2}$ or $\sqrt[3]{36}$.

$$\dfrac{5\sqrt[3]{6^2}}{\sqrt[3]{6} \cdot \sqrt[3]{6^2}} = \dfrac{5\sqrt[3]{36}}{\sqrt[3]{6^3}} = \dfrac{5\sqrt[3]{36}}{6}$$

87. $\dfrac{\sqrt{6}}{\sqrt{5}} = \dfrac{\sqrt{6} \cdot \sqrt{5}}{\sqrt{5} \cdot \sqrt{5}} = \dfrac{\sqrt{30}}{5}$

88. $\dfrac{-6\sqrt{3}}{\sqrt{2}} = \dfrac{-6\sqrt{3} \cdot \sqrt{2}}{\sqrt{2} \cdot \sqrt{2}} = \dfrac{-6\sqrt{6}}{2} = -3\sqrt{6}$

89. $\dfrac{3\sqrt{7p}}{\sqrt{y}} = \dfrac{3\sqrt{7p} \cdot \sqrt{y}}{\sqrt{y} \cdot \sqrt{y}} = \dfrac{3\sqrt{7py}}{y}$

90. $\sqrt{\dfrac{11}{8}} = \dfrac{\sqrt{11}}{\sqrt{8}} = \dfrac{\sqrt{11}}{\sqrt{4 \cdot 2}} = \dfrac{\sqrt{11}}{2\sqrt{2}} = \dfrac{\sqrt{11} \cdot \sqrt{2}}{2\sqrt{2} \cdot \sqrt{2}}$
$$= \dfrac{\sqrt{22}}{2 \cdot 2} = \dfrac{\sqrt{22}}{4}$$

91. $-\sqrt[3]{\dfrac{9}{25}} = -\dfrac{\sqrt[3]{9}}{\sqrt[3]{5^2}} = -\dfrac{\sqrt[3]{9} \cdot \sqrt[3]{5}}{\sqrt[3]{5^2} \cdot \sqrt[3]{5}}$

$\phantom{-\sqrt[3]{\dfrac{9}{25}}} = -\dfrac{\sqrt[3]{45}}{\sqrt[3]{5^3}} = -\dfrac{\sqrt[3]{45}}{5}$

92. $\sqrt[3]{\dfrac{108m^3}{n^5}} = \dfrac{\sqrt[3]{108m^3}}{\sqrt[3]{n^5}} = \dfrac{\sqrt[3]{27m^3 \cdot 4}}{\sqrt[3]{n^3 \cdot n^2}}$

$\phantom{\sqrt[3]{\dfrac{108m^3}{n^5}}} = \dfrac{3m\sqrt[3]{4}}{n\sqrt[3]{n^2}} = \dfrac{3m\sqrt[3]{4} \cdot \sqrt[3]{n}}{n\sqrt[3]{n^2} \cdot \sqrt[3]{n}}$

$\phantom{\sqrt[3]{\dfrac{108m^3}{n^5}}} = \dfrac{3m\sqrt[3]{4n}}{n \cdot n} = \dfrac{3m\sqrt[3]{4n}}{n^2}$

93. $\dfrac{1}{\sqrt{2} + \sqrt{7}}$

Multiply by the conjugate of the denominator, $\sqrt{2} - \sqrt{7}$.

$= \dfrac{1\left(\sqrt{2} - \sqrt{7}\right)}{\left(\sqrt{2} + \sqrt{7}\right)\left(\sqrt{2} - \sqrt{7}\right)}$

$= \dfrac{\sqrt{2} - \sqrt{7}}{2 - 7} = \dfrac{\sqrt{2} - \sqrt{7}}{-5}$

94. $\dfrac{-5}{\sqrt{6} - 3}$

Multiply by the conjugate of the denominator, $\sqrt{6} + 3$.

$= \dfrac{-5\left(\sqrt{6} + 3\right)}{\left(\sqrt{6} - 3\right)\left(\sqrt{6} + 3\right)} = \dfrac{-5\left(\sqrt{6} + 3\right)}{6 - 9}$

$= \dfrac{-5\left(\sqrt{6} + 3\right)}{-3} = \dfrac{5\left(\sqrt{6} + 3\right)}{3}$

95. $\dfrac{2 - 2\sqrt{5}}{8} = \dfrac{2\left(1 - \sqrt{5}\right)}{2 \cdot 4} = \dfrac{1 - \sqrt{5}}{4}$

96. $\dfrac{4 - 8\sqrt{8}}{12} = \dfrac{4\left(1 - 2\sqrt{8}\right)}{3 \cdot 4} = \dfrac{1 - 2\sqrt{8}}{3}$

$\phantom{\dfrac{4 - 8\sqrt{8}}{12}} = \dfrac{1 - 2\sqrt{4 \cdot 2}}{3} = \dfrac{1 - 4\sqrt{2}}{3}$

97. $f(x) = \sqrt{x - 1}$

For the radicand to be nonnegative, we must have

$$x - 1 \geq 0 \quad \text{or} \quad x \geq 1.$$

Thus, the domain is $[1, \infty)$.
The function values are positive or zero (the result of the radical), so the range is $[0, \infty)$.

x	$f(x) = \sqrt{x - 1}$
1	$\sqrt{1 - 1} = 0$
2	$\sqrt{2 - 1} = 1$
5	$\sqrt{5 - 1} = 2$

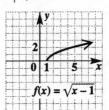

98. $f(x) = \sqrt[3]{x} + 4$

Since we can take the cube root of any real number, the domain is $(-\infty, \infty)$.
The result of a cube root can be any real number, so the range is $(-\infty, \infty)$. (The "+4" does not affect that range.)

x	$f(x) = \sqrt[3]{x} + 4$
-8	$\sqrt[3]{-8} + 4 = 2$
-1	$\sqrt[3]{-1} + 4 = 3$
0	$\sqrt[3]{0} + 4 = 4$
1	$\sqrt[3]{1} + 4 = 5$
8	$\sqrt[3]{8} + 4 = 6$

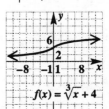

99. $\sqrt{8y + 9} = 5$

$\left(\sqrt{8y + 9}\right)^2 = 5^2 \quad Square$

$8y + 9 = 25$

$8y = 16$

$y = 2$

Check $y = 2$: $\sqrt{25} \overset{?}{=} 5$ *True*

Solution set: $\{2\}$

100. $\sqrt{2z - 3} - 3 = 0$

$\sqrt{2z - 3} = 3 \quad Isolate$

$\left(\sqrt{2z - 3}\right)^2 = 3^2 \quad Square$

$2z - 3 = 9$

$2z = 12$

$z = 6$

Check $z = 6$: $\sqrt{9} - 3 \overset{?}{=} 0$ *True*

Solution set: $\{6\}$

101. $\sqrt{3m+1} - 2 = -3$

$\qquad \sqrt{3m+1} = -1$ *Isolate*

This equation has no solution because $\sqrt{3m+1}$ cannot be negative.

Solution set: $\emptyset$

102. $\sqrt{7z+1} = z+1$

$\left(\sqrt{7z+1}\right)^2 = (z+1)^2$ *Square*

$\qquad 7z+1 = z^2 + 2z + 1$

$\qquad 0 = z^2 - 5z$

$\qquad 0 = z(z-5)$

$z = 0 \quad$ or $\quad z = 5$

Check $z = 0$: $\sqrt{1} \overset{?}{=} 1$ *True*

Check $z = 5$: $\sqrt{36} \overset{?}{=} 6$ *True*

Solution set: $\{0, 5\}$

103. $3\sqrt{m} = \sqrt{10m-9}$

$\left(3\sqrt{m}\right)^2 = \left(\sqrt{10m-9}\right)^2$ *Square*

$\qquad 9m = 10m - 9$

$\qquad 9 = m$

Check $m = 9$: $3\sqrt{9} \overset{?}{=} \sqrt{81}$ *True*

Solution set: $\{9\}$

104. $\sqrt{p^2 + 3p + 7} = p + 2$

$\left(\sqrt{p^2 + 3p + 7}\right)^2 = (p+2)^2$ *Square*

$\qquad p^2 + 3p + 7 = p^2 + 4p + 4$

$\qquad 3 = p$

Check $p = 3$: $\sqrt{25} \overset{?}{=} 5$ *True*

Solution set: $\{3\}$

105. $\sqrt{a+2} - \sqrt{a-3} = 1$

Get one radical on each side of the equals sign.

$\qquad \sqrt{a+2} = 1 + \sqrt{a-3}$

Square both sides.

$\left(\sqrt{a+2}\right)^2 = \left(1 + \sqrt{a-3}\right)^2$

$\qquad a + 2 = 1 + 2\sqrt{a-3} + a - 3$

$\qquad 4 = 2\sqrt{a-3}$

$\qquad 2 = \sqrt{a-3}$

Square both sides again.

$\qquad 2^2 = \left(\sqrt{a-3}\right)^2$

$\qquad 4 = a - 3$

$\qquad 7 = a$

Check $a = 7$: $\sqrt{9} - \sqrt{4} \overset{?}{=} 1$ *True*

Solution set: $\{7\}$

106. $\sqrt[3]{5m-1} = \sqrt[3]{3m-2}$

Cube both sides.

$\left(\sqrt[3]{5m-1}\right)^3 = \left(\sqrt[3]{3m-2}\right)^3$

$\qquad 5m - 1 = 3m - 2$

$\qquad 2m = -1$

$\qquad m = -\dfrac{1}{2}$

Check $m = -\dfrac{1}{2}$: $\sqrt[3]{-\dfrac{7}{2}} \overset{?}{=} \sqrt[3]{-\dfrac{7}{2}}$ *True*

Solution set: $\left\{-\dfrac{1}{2}\right\}$

107. When the potential solution, 0, is substituted into the original equation, it does not check. Therefore, it is not a solution of the equation. The solution set is $\emptyset$.

108. (a) Solve $L = \sqrt{H^2 + W^2}$ for H.

$\qquad L^2 = H^2 + W^2$

$\qquad L^2 - W^2 = H^2$

$\qquad \sqrt{L^2 - W^2} = H$

(b) Substitute 12 for L and 9 for W in $H = \sqrt{L^2 - W^2}$.

$\qquad H = \sqrt{12^2 - 9^2}$

$\qquad = \sqrt{144 - 81} = \sqrt{63}$

To the nearest tenth of a foot, the height is approximately 7.9 feet.

109. $\sqrt{-25} = i\sqrt{25} = 5i$

110. $\sqrt{-200} = i\sqrt{100 \cdot 2} = 10i\sqrt{2}$

111. If a is a positive real number, then $-a$ is negative. So, $\sqrt{-a}$ is not a real number. Therefore, $-\sqrt{-a}$ is not a real number either.

112. $(-2 + 5i) + (-8 - 7i)$

$\qquad = [-2 + (-8)] + [5 + (-7)]i$

$\qquad = -10 - 2i$

113. $(5 + 4i) - (-9 - 3i)$

$\qquad = [(5 - (-9)] + [(4 - (-3)]i$

$\qquad = 14 + 7i$

114. $\sqrt{-5} \cdot \sqrt{-7} = i\sqrt{5} \cdot i\sqrt{7} = i^2\sqrt{35}$

$\qquad = -1\left(\sqrt{35}\right) = -\sqrt{35}$

115. $\sqrt{-25} \cdot \sqrt{-81} = 5i \cdot 9i = 45i^2$

$\qquad = 45(-1) = -45$

116. $\dfrac{\sqrt{-72}}{\sqrt{-8}} = \dfrac{i\sqrt{72}}{i\sqrt{8}} = \sqrt{\dfrac{72}{8}} = \sqrt{9} = 3$

117. $(2+3i)(1-i) = 2 - 2i + 3i - 3i^2$
$$= 2 + i - 3(-1)$$
$$= 2 + i + 3$$
$$= 5 + i$$

118. $(6-2i)^2 = 6^2 - 2 \cdot 6 \cdot 2i + (2i)^2$
$$= 36 - 24i + 4i^2$$
$$= 36 - 24i + 4(-1)$$
$$= 36 - 24i - 4$$
$$= 32 - 24i$$

119. $\dfrac{3-i}{2+i}$

Multiply by the conjugate of the denominator, $2 - i$.
$$= \frac{(3-i)(2-i)}{(2+i)(2-i)}$$
$$= \frac{6 - 3i - 2i + i^2}{4 - i^2}$$
$$= \frac{6 - 5i - 1}{4 - (-1)} = \frac{5 - 5i}{5}$$
$$= \frac{5(1-i)}{5} = 1 - i$$

120. $\dfrac{5+14i}{2+3i}$

Multiply by the conjugate of the denominator, $2 - 3i$.
$$= \frac{(5+14i)(2-3i)}{(2+3i)(2-3i)}$$
$$= \frac{10 - 15i + 28i - 42i^2}{4 - 9i^2}$$
$$= \frac{10 + 13i - 42(-1)}{4 - 9(-1)} = \frac{52 + 13i}{13}$$
$$= \frac{13(4+i)}{13} = 4 + i$$

121. $i^{11} = i^8 \cdot i^3 = \left(i^4\right)^2 \cdot i^2 \cdot i$
$$= 1^2 \cdot (-1) \cdot i = -i$$

122. $i^{-10} = \dfrac{1}{i^{10}} = \dfrac{1}{i^8 \cdot i^2} = \dfrac{1}{\left(i^4\right)^2 \cdot (-1)}$
$$= \frac{1}{1^2 \cdot (-1)} = \frac{1}{-1} = -1$$

Another method:
$$i^{-10} = i^{-10} \cdot i^{12} = i^2 = -1$$

Note that $i^{12} = \left(i^4\right)^3 = 1^3 = 1$.

123. $-\sqrt[4]{256} = -\left(\sqrt[4]{4^4}\right) = -(4) = -4$

124. $1000^{-2/3} = \dfrac{1}{1000^{2/3}} = \dfrac{1}{\left[(10^3)^{1/3}\right]^2}$
$$= \frac{1}{10^2} = \frac{1}{100}$$

125. $\dfrac{z^{-1/5} \cdot z^{3/10}}{z^{7/10}} = z^{-2/10 + 3/10 - 7/10}$
$$= z^{-6/10} = \frac{1}{z^{6/10}} = \frac{1}{z^{3/5}}$$

126. $\sqrt[4]{k^{24}} = k^{24/4} = k^6$

127. $\sqrt[3]{54z^9 t^8} = \sqrt[3]{27 z^9 t^6 \cdot 2t^2} = 3z^3 t^2 \sqrt[3]{2t^2}$

128. $-5\sqrt{18} + 12\sqrt{72} = -5\sqrt{9 \cdot 2} + 12\sqrt{36 \cdot 2}$
$$= -5 \cdot 3\sqrt{2} + 12 \cdot 6\sqrt{2}$$
$$= -15\sqrt{2} + 72\sqrt{2} = 57\sqrt{2}$$

129. $8\sqrt[3]{x^3 y^2} - 2x\sqrt[3]{y^2} = 8x\sqrt[3]{y^2} - 2x\sqrt[3]{y^2}$
$$= 6x\sqrt[3]{y^2}$$

130. $\left(\sqrt{5} - \sqrt{3}\right)\left(\sqrt{7} + \sqrt{3}\right)$
$$= \sqrt{35} + \sqrt{15} - \sqrt{21} - 3$$

131. $\dfrac{-1}{\sqrt{12}} = \dfrac{-1}{\sqrt{4 \cdot 3}} = \dfrac{-1}{2\sqrt{3}} = \dfrac{-1 \cdot \sqrt{3}}{2\sqrt{3} \cdot \sqrt{3}}$
$$= \frac{-\sqrt{3}}{2 \cdot 3} = \frac{-\sqrt{3}}{6}$$

132. $\sqrt[3]{\dfrac{12}{25}} = \dfrac{\sqrt[3]{12}}{\sqrt[3]{25}} = \dfrac{\sqrt[3]{12}}{\sqrt[3]{5^2}}$
$$= \frac{\sqrt[3]{12} \cdot \sqrt[3]{5}}{\sqrt[3]{5^2} \cdot \sqrt[3]{5}} = \frac{\sqrt[3]{60}}{\sqrt[3]{5^3}} = \frac{\sqrt[3]{60}}{5}$$

133. $i^{-1000} = \dfrac{1}{i^{1000}} = \dfrac{1}{\left(i^4\right)^{250}}$
$$= \frac{1}{1^{250}} = \frac{1}{1} = 1$$

134. $\sqrt{-49} = i\sqrt{49} = 7i$

135. $(4 - 9i) + (-1 + 2i)$
$$= (4 - 1) + (-9 + 2)i$$
$$= 3 - 7i$$

136. $\dfrac{\sqrt{50}}{\sqrt{-2}} = \dfrac{\sqrt{25 \cdot 2}}{i\sqrt{2}} = \dfrac{5\sqrt{2}}{i\sqrt{2}} = \dfrac{5}{i}$

The conjugate of i is $-i$.
$$\frac{5(-i)}{i(-i)} = \frac{-5i}{-i^2} = \frac{-5i}{-1(-1)} = -5i$$

137. $\dfrac{3 + \sqrt{54}}{6} = \dfrac{3 + \sqrt{9 \cdot 6}}{6} = \dfrac{3 + 3\sqrt{6}}{6}$
$$= \frac{3\left(1 + \sqrt{6}\right)}{2 \cdot 3} = \frac{1 + \sqrt{6}}{2}$$

138. $\sqrt{x+4} = x - 2$

$\left(\sqrt{x+4}\right)^2 = (x-2)^2$ *Square*

$x + 4 = x^2 - 4x + 4$

$0 = x^2 - 5x$

$0 = x(x - 5)$

$x = 0$ or $x = 5$

Check $x = 0$: $\sqrt{4} \overset{?}{=} -2$ *False*

Check $x = 5$: $\sqrt{9} \overset{?}{=} 3$ *True*

Solution set: $\{5\}$

139. $\sqrt[3]{2x-9} = \sqrt[3]{5x+3}$

Cube both sides.

$\left(\sqrt[3]{2x-9}\right)^3 = \left(\sqrt[3]{5x+3}\right)^3$

$2x - 9 = 5x + 3$

$-3x = 12$

$x = -4$

Check $x = -4$: $\sqrt[3]{-17} \overset{?}{=} \sqrt[3]{-17}$ *True*

Solution set: $\{-4\}$

140. $\sqrt{6+2y} - 1 = \sqrt{7-2y}$

Square both sides.

$\left(\sqrt{6+2y} - 1\right)^2 = \left(\sqrt{7-2y}\right)^2$

$6 + 2y - 2\sqrt{6+2y} + 1 = 7 - 2y$

$4y = 2\sqrt{6+2y}$

$2y = \sqrt{6+2y}$

Square both sides again.

$(2y)^2 = \left(\sqrt{6+2y}\right)^2$

$4y^2 = 6 + 2y$

$4y^2 - 2y - 6 = 0$

$2y^2 - y - 3 = 0$

$(2y - 3)(y + 1) = 0$

$2y - 3 = 0$ or $y + 1 = 0$

$y = \dfrac{3}{2}$ or $y = -1$

Check $y = \dfrac{3}{2}$: $\sqrt{9} - 1 \overset{?}{=} \sqrt{4}$ *True*

Check $y = -1$: $\sqrt{4} - 1 \overset{?}{=} \sqrt{9}$ *False*

Solution set: $\left\{\dfrac{3}{2}\right\}$

Chapter 7 Test

1. $\left(\dfrac{16}{25}\right)^{-3/2} = \left(\dfrac{25}{16}\right)^{3/2} = \dfrac{25^{3/2}}{16^{3/2}}$

$= \dfrac{\left[(5^2)^{1/2}\right]^3}{\left[(4^2)^{1/2}\right]^3} = \dfrac{5^3}{4^3} = \dfrac{125}{64}$

2. $(-64)^{-4/3} = \dfrac{1}{(-64)^{4/3}} = \dfrac{1}{\left(\left[(-4)^3\right]^{1/3}\right)^4}$

$= \dfrac{1}{(-4)^4} = \dfrac{1}{256}$

3. $\dfrac{3^{2/5}x^{-1/4}y^{2/5}}{3^{-8/5}x^{7/4}y^{1/10}}$

$= 3^{2/5-(-8/5)}x^{-1/4-7/4}y^{2/5-1/10}$

$= 3^{10/5}x^{-8/4}y^{4/10-1/10}$

$= 3^2 x^{-2} y^{3/10} = \dfrac{9y^{3/10}}{x^2}$

4. $7^{3/4} \cdot 7^{-1/4} = 7^{3/4+(-1/4)}$

$= 7^{1/2}$ or $\sqrt{7}$

5. $\sqrt[3]{a^4} \cdot \sqrt[3]{a^7} = a^{4/3} \cdot a^{7/3} = a^{4/3+7/3} = a^{11/3}$

$= \sqrt[3]{a^{11}} = \sqrt[3]{(a^3)^3 \cdot a^2} = a^3\sqrt[3]{a^2}$

6. (a) $\sqrt{478} \approx 21.863$

(b) $\sqrt[3]{-832} \approx -9.405$

(c) $34^{1/4} \approx 2.415$

7. $a^2 + b^2 = c^2$

$12^2 + b^2 = 17^2$ *Let a=12, c=17*

$144 + b^2 = 289$

$b^2 = 145$

$b = \sqrt{145}$

8. $(-4, 2)$ and $(2, 10)$

$d = \sqrt{(x_2 - x_1)^2 + (y_2 - y_1)^2}$

$= \sqrt{[2 - (-4)]^2 + (10 - 2)^2}$

$= \sqrt{6^2 + 8^2} = \sqrt{36 + 64} = \sqrt{100} = 10$

9. $\sqrt{54x^5y^6} = \sqrt{9x^4y^6 \cdot 6x} = 3x^2y^3\sqrt{6x}$

10. $\sqrt[4]{32a^7b^{13}} = \sqrt[4]{16a^4b^{12} \cdot 2a^3b}$

$= 2ab^3\sqrt[4]{2a^3b}$

11. $\sqrt{2} \cdot \sqrt[3]{5} = 2^{1/2} \cdot 5^{1/3} = 2^{3/6} \cdot 5^{2/6}$

$= \left(2^3 \cdot 5^2\right)^{1/6} = \sqrt[6]{2^3 \cdot 5^2}$

$= \sqrt[6]{8 \cdot 25} = \sqrt[6]{200}$

12. $3\sqrt{20} - 5\sqrt{80} + 4\sqrt{500}$

$= 3\sqrt{4 \cdot 5} - 5\sqrt{16 \cdot 5} + 4\sqrt{100 \cdot 5}$

$= 3 \cdot 2\sqrt{5} - 5 \cdot 4\sqrt{5} + 4 \cdot 10\sqrt{5}$

$= 6\sqrt{5} - 20\sqrt{5} + 40\sqrt{5} = 26\sqrt{5}$

13. $\left(7\sqrt{5}+4\right)\left(2\sqrt{5}-1\right)$

$= 14 \cdot 5 - 7\sqrt{5} + 8\sqrt{5} - 4$

$= 70 + \sqrt{5} - 4 = 66 + \sqrt{5}$

14. $\left(\sqrt{3} - 2\sqrt{5}\right)^2$

$= \left(\sqrt{3}\right)^2 - 2 \cdot \sqrt{3} \cdot 2\sqrt{5} + \left(2\sqrt{5}\right)^2$

$= 3 - 4\sqrt{15} + 4 \cdot 5$

$= 3 - 4\sqrt{15} + 20 = 23 - 4\sqrt{15}$

15. $\dfrac{-5}{\sqrt{40}} = \dfrac{-5}{\sqrt{4 \cdot 10}} = \dfrac{-5}{2\sqrt{10}} = \dfrac{-5 \cdot \sqrt{10}}{2\sqrt{10} \cdot \sqrt{10}}$

$= \dfrac{-5\sqrt{10}}{2 \cdot 10} = \dfrac{-5\sqrt{10}}{20} = -\dfrac{\sqrt{10}}{4}$

16. $\dfrac{2}{\sqrt[3]{5}} = \dfrac{2 \cdot \sqrt[3]{5^2}}{\sqrt[3]{5}\sqrt[3]{5^2}} = \dfrac{2\sqrt[3]{25}}{5}$

17. $\dfrac{-4}{\sqrt{7}+\sqrt{5}}$

Multiply by the conjugate of the denominator, $\sqrt{7} - \sqrt{5}$.

$= \dfrac{-4\left(\sqrt{7}-\sqrt{5}\right)}{\left(\sqrt{7}+\sqrt{5}\right)\left(\sqrt{7}-\sqrt{5}\right)}$

$= \dfrac{-4\left(\sqrt{7}-\sqrt{5}\right)}{7-5}$

$= \dfrac{-4\left(\sqrt{7}-\sqrt{5}\right)}{2} = -2\left(\sqrt{7}-\sqrt{5}\right).$

18. $\dfrac{6+\sqrt{24}}{2} = \dfrac{6+\sqrt{4 \cdot 6}}{2} = \dfrac{6+2\sqrt{6}}{2}$

$= \dfrac{2\left(3+\sqrt{6}\right)}{2} = 3+\sqrt{6}$

19. (a) Substitute 50 for V_0, .01 for k, and 30 for T in the formula.

$V = \dfrac{V_0}{\sqrt{1-kT}}$

$V = \dfrac{50}{\sqrt{1-(.01)(30)}}$

$= \dfrac{50}{\sqrt{1-.3}} = \dfrac{50}{\sqrt{.7}} \approx 59.8$

The velocity is about 59.8.

(b) $\qquad V = \dfrac{V_0}{\sqrt{1-kT}}$

$V^2 = \dfrac{V_0^2}{1-kT} \qquad \textit{Square}$

$V^2(1-kT) = V_0^2$

$V^2 - V^2kT = V_0^2$

$V^2 - V_0^2 = V^2kT$

$T = \dfrac{V^2 - V_0^2}{V^2k}$

or $\qquad T = \dfrac{V_0^2 - V^2}{-V^2k}$

20. $f(x) = \sqrt{x+6}$

For the radicand to be nonnegative, we must have

$x+6 \geq 0 \quad$ or $\quad x \geq -6.$

Thus, the domain is $[-6, \infty)$.
The function values are positive or zero (the result of the radical), so the range is $[0, \infty)$.

x	$f(x) = \sqrt{x+6}$
-6	$\sqrt{-6+6} = 0$
-5	$\sqrt{-5+6} = 1$
-2	$\sqrt{-2+6} = 2$

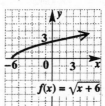

21. $\sqrt[3]{5x} = \sqrt[3]{2x-3}$

$\left(\sqrt[3]{5x}\right)^3 = \left(\sqrt[3]{2x-3}\right)^3$

$5x = 2x - 3$

$3x = -3$

$x = -1$

Check $x = -1$: $\sqrt[3]{-5} \overset{?}{=} \sqrt[3]{-5}$ *True*

Solution set: $\{-1\}$

22. $\sqrt{7-x} + 5 = x$

$\sqrt{7-x} = x - 5$

$\left(\sqrt{7-x}\right)^2 = (x-5)^2$

$7 - x = x^2 - 10x + 25$

$0 = x^2 - 9x + 18$

$0 = (x-3)(x-6)$

$x = 3 \quad$ or $\quad x = 6$

Check $x = 3$: $\sqrt{4} + 5 = 3$ *False*

Check $x = 6$: $\sqrt{1} + 5 = 6$ *True*

Solution set: $\{6\}$

23. (a) $(-2 + 5i) - (3 + 6i) - 7i$
$$= (-2 - 3) + (5 - 6 - 7)i$$
$$= -5 - 8i$$

(b) $(1 + 5i)(3 + i)$
$$= 3 + i + 15i + 5i^2$$
$$= 3 + 16i + 5(-1)$$
$$= -2 + 16i$$

(c) $\dfrac{7 + i}{1 - i}$

Multiply by the conjugate of the denominator, $1 + i$.
$$= \frac{(7 + i)(1 + i)}{(1 - i)(1 + i)}$$
$$= \frac{7 + 7i + i + i^2}{1 - i^2}$$
$$= \frac{7 + 8i - 1}{1 - (-1)}$$
$$= \frac{6 + 8i}{2} = \frac{2(3 + 4i)}{2} = 3 + 4i$$

24. $i^{37} = i^{36} \cdot i = \left(i^4\right)^9 \cdot i = 1^9 \cdot i = 1 \cdot i = i$

25. (a) $i^2 = -1$ is *true*.

(b) $i = \sqrt{-1}$ is *true*.

(c) $i = -1$ is *false*; $i = \sqrt{-1}$.

(d) $\sqrt{-3} = i\sqrt{3}$ is *true*.

Cumulative Review Exercises Chapters 1–7

In Exercises 1 and 2, $a = -3$, $b = 5$, and $c = -4$.

1. $|2a^2 - 3b + c|$
$$= |2(-3)^2 - 3(5) + (-4)|$$
$$= |2(9) - 15 - 4|$$
$$= |18 - 19| = |-1| = 1$$

2. $\dfrac{(a + b)(a + c)}{3b - 6}$
$$= \frac{(-3 + 5)[(-3 + (-4)]}{3(5) - 6}$$
$$= \frac{(2)(-7)}{15 - 6} = -\frac{14}{9}$$

3. $3(x + 2) - 4(2x + 3) = -3x + 2$
$$3x + 6 - 8x - 12 = -3x + 2$$
$$-5x - 6 = -3x + 2$$
$$-2x = 8$$
$$x = -4$$

Check $x = -4$: $-6 + 20 \overset{?}{=} 14$ *True*

Solution set: $\{-4\}$

4. $\dfrac{1}{3}x + \dfrac{1}{4}(x + 8) = x + 7$

Multiply by the LCD, 12.
$$12\left[\frac{1}{3}x + \frac{1}{4}(x + 8)\right] = 12(x + 7)$$
$$4x + 3(x + 8) = 12x + 84$$
$$4x + 3x + 24 = 12x + 84$$
$$7x + 24 = 12x + 84$$
$$-5x = 60$$
$$x = -12$$

Check $x = -12$: $-4 - 1 \overset{?}{=} -5$ *True*

Solution set: $\{-12\}$

5. $.04y + .06(100 - y) = 5.88$

Multiply both sides by 100 to clear decimals.
$$4y + 6(100 - y) = 588$$
$$4y + 600 - 6y = 588$$
$$-2y = -12$$
$$y = 6$$

Check $y = 6$: $.24 + 5.64 \overset{?}{=} 5.88$ *True*

Solution set: $\{6\}$

6. $|6x + 7| = 13$

$6x + 7 = 13$ or $6x + 7 = -13$
$\quad 6x = 6$ $\qquad\qquad 6x = -20$
$\quad\ x = 1$ or $x = -\dfrac{20}{6} = -\dfrac{10}{3}$

Check $x = 1$: $13 \overset{?}{=} 13$ *True*

Check $x = -\dfrac{10}{3}$: $13 \overset{?}{=} 13$ *True*

Solution set: $\left\{-\dfrac{10}{3}, 1\right\}$

7. $|-2x + 4| = |-2x - 3|$
$-2x + 4 = -2x - 3$ or $-2x + 4 = -(-2x - 3)$
$\qquad 4 = -3$ $\qquad\qquad -2x + 4 = 2x + 3$
$\qquad$ *False* $\qquad\qquad\qquad -4x = -1$
$$x = \frac{1}{4}$$

Check $x = \dfrac{1}{4}$: $\dfrac{7}{2} \overset{?}{=} \dfrac{7}{2}$ *True*

Solution set: $\left\{\dfrac{1}{4}\right\}$

8. $-5 - 3(m - 2) < 11 - 2(m + 2)$
$$-5 - 3m + 6 < 11 - 2m - 4$$
$$1 - 3m < 7 - 2m$$
$$-m < 6$$

Multiply by -1; reverse the inequality.
$$m > -6$$

Solution set: $(-6, \infty)$

9. The two angles have the same measure, so

$$10x - 70 = 7x - 25$$
$$3x = 45$$
$$x = 15.$$

Then,

$$10x - 70 = 10(15) - 70$$
$$= 150 - 70 = 80.$$

Each angle measures $80°$.

10. Let x = the number of nickels and $50 - x$ = the number of quarters.

Value of nickels $+$ value of quarters $=$ total value.

$$.05x + .25(50 - x) = 8.90$$

Multiply by 100 to clear the decimals.

$$5x + 25(50 - x) = 890$$
$$5x + 1250 - 25x = 890$$
$$-20x = -360$$
$$x = 18$$

Then,

$$50 - x = 50 - 18 = 32.$$

There are 18 nickels and 32 quarters.

11. Let x = the amount of pure alcohol. Make a table.

Solution Strength	Amount	Pure Alcohol
100%	x	$1 \cdot x = x$
18%	40	$.18(40) = 7.2$
22%	$x + 40$	$.22(x + 40)$

From the last column:

$$x + 7.2 = .22x + 8.8$$
$$.78x = 1.6$$
$$x = \frac{1.6}{.78} = \frac{160}{78} = \frac{80}{39} \text{ or } 2\frac{2}{39}$$

The required amount is $\frac{80}{39}$ or $2\frac{2}{39}$ L of pure alcohol.

12. $4x - 3y = 12$

Let $x = 0$ to find the y-intercept, $(0, -4)$. Let $y = 0$ to find the x-intercept, $(3, 0)$. Draw a line through the intercepts.

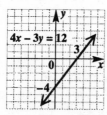

13. $(-4, 6)$ and $(2, -3)$

$$m = \frac{-3 - 6}{2 - (-4)} = \frac{-9}{6} = -\frac{3}{2}$$

Use the point-slope form:

$$y - 6 = -\frac{3}{2}[x - (-4)]$$
$$y - 6 = -\frac{3}{2}(x + 4)$$
$$y - 6 = -\frac{3}{2}x - 6$$
$$y = -\frac{3}{2}x$$

14. $f(x) = 3x - 7$
$$f(-10) = 3(-10) - 7 = -30 - 7 = -37$$

15. $3x - y = 23$ (1)
$2x + 3y = 8$ (2)

To eliminate y, multiply equation (1) by 3 and add the result to equation (2).

$$
\begin{array}{rcll}
9x - 3y & = & 69 & 3 \times (1) \\
2x + 3y & = & 8 & (2) \\
\hline
11x & = & 77 & \\
x & = & 7 &
\end{array}
$$

Substitute 7 for x in (2),

$$2(7) + 3y = 8$$
$$14 + 3y = 8$$
$$3y = -6$$
$$y = -2$$

Solution set: $\{(7, -2)\}$

16. $x + y + z = 1$
$x - y - z = -3$
$x + y - z = -1$

$$D = \begin{vmatrix} 1 & 1 & 1 \\ 1 & -1 & -1 \\ 1 & 1 & -1 \end{vmatrix} \text{ Expand about row 1.}$$

$$= 1\begin{vmatrix} -1 & -1 \\ 1 & -1 \end{vmatrix} - 1\begin{vmatrix} 1 & -1 \\ 1 & -1 \end{vmatrix} + 1\begin{vmatrix} 1 & -1 \\ 1 & 1 \end{vmatrix}$$

$$= 1(2) - 1(0) + 1(2) = 4$$

$$D_x = \begin{vmatrix} 1 & 1 & 1 \\ -3 & -1 & -1 \\ -1 & 1 & -1 \end{vmatrix} \text{ Expand about row 1.}$$

$$= 1\begin{vmatrix} -1 & -1 \\ 1 & -1 \end{vmatrix} - 1\begin{vmatrix} -3 & -1 \\ -1 & -1 \end{vmatrix} + 1\begin{vmatrix} -3 & -1 \\ -1 & 1 \end{vmatrix}$$

$$= 1(2) - 1(2) + 1(-4) = -4$$

continued

$$D_y = \begin{vmatrix} 1 & 1 & 1 \\ 1 & -3 & -1 \\ 1 & -1 & -1 \end{vmatrix} \quad \text{Expand about row 1.}$$

$$= 1\begin{vmatrix} -3 & -1 \\ -1 & -1 \end{vmatrix} - 1\begin{vmatrix} 1 & -1 \\ 1 & -1 \end{vmatrix} + 1\begin{vmatrix} 1 & -3 \\ 1 & -1 \end{vmatrix}$$

$$= 1(2) - 1(0) + 1(2) = 4$$

$$D_z = \begin{vmatrix} 1 & 1 & 1 \\ 1 & -1 & -3 \\ 1 & 1 & -1 \end{vmatrix} \quad \text{Expand about row 1.}$$

$$= 1\begin{vmatrix} -1 & -3 \\ 1 & -1 \end{vmatrix} - 1\begin{vmatrix} 1 & -3 \\ 1 & -1 \end{vmatrix} + 1\begin{vmatrix} 1 & -1 \\ 1 & 1 \end{vmatrix}$$

$$= 1(4) - 1(2) + 1(2) = 4$$

$$x = \frac{D_x}{D} = \frac{-4}{4} = -1; \quad y = \frac{D_y}{D} = \frac{4}{4} = 1$$

$$z = \frac{D_z}{D} = \frac{4}{4} = 1$$

Solution set: $\{(-1, 1, 1)\}$

17. Let $x =$ the number of 2-ounce letters and
$\quad\; y =$ the number of 3-ounce letters.

$$5x + 3y = 5.09 \qquad (1)$$
$$3x + 5y = 5.55 \qquad (2)$$

To eliminate x, multiply (1) by -3 and (2) by 5 and add the results.

$$\begin{array}{rcll} -15x - 9y &=& -15.27 & -3 \times (1) \\ 15x + 25y &=& 27.75 & 5 \times (2) \\ \hline 16y &=& 12.48 & \\ y &=& .78 & \end{array}$$

Substitute $y = .78$ in (1).

$$5x + 3y = 5.09 \quad (1)$$
$$5x + 3(.78) = 5.09$$
$$5x + 2.34 = 5.09$$
$$5x = 2.75$$
$$x = .55$$

The 1997 postage rate for a 2-ounce letter was $.55 and for a 3-ounce letter, $.78.

18. $(3k^2 - 5k^2 + 8k - 2) - (4k^3 + 11k + 7)$
$\quad + (2k^2 - 5k)$
$$= 3k^3 - 4k^3 - 5k^2 + 2k^2$$
$$\quad + 8k - 11k - 5k - 2 - 7$$
$$= -k^3 - 3k^2 - 8k - 9$$

19. $(8x - 7)(x + 3)$
$$= 8x^2 + 24x - 7x - 21$$
$$= 8x^2 + 17x - 21$$

20. $\dfrac{8z^3 - 16z^2 + 24z}{8z^2} = \dfrac{8z^3}{8z^2} - \dfrac{16z^2}{8z^2} + \dfrac{24z}{8z^2}$

$$= z - 2 + \frac{3}{z}$$

21.

$$\begin{array}{r} 3y^3 - 3y^2 + 4y + 1 \\ 2y + 1 \overline{\smash{\big)}\ 6y^4 - 3y^3 + 5y^2 + 6y - 9} \\ \underline{6y^4 + 3y^3} \\ -6y^3 + 5y^2 \\ \underline{-6y^3 - 3y^2} \\ 8y^2 + 6y \\ \underline{8y^2 + 4y} \\ 2y - 9 \\ \underline{2y + 1} \\ -10 \end{array}$$

The answer is

$$3y^3 - 3y^2 + 4y + 1 + \frac{-10}{2y + 1}.$$

22. $2p^2 - 5pq + 3q^2 = (2p - 3q)(p - q)$

23. $3k^4 + k^2 - 4 = (3k^2 + 4)(k^2 - 1)$
$$= (3k^2 + 4)(k + 1)(k - 1)$$

24. $x^3 + 512 = x^3 + 8^3$
$$= (x + 8)(x^2 - 8x + 64)$$

25. $2x^2 + 11x + 15 = 0$
$$(x + 3)(2x + 5) = 0$$

$$x + 3 = 0 \quad \text{or} \quad 2x + 5 = 0$$

$$x = -3 \quad \text{or} \qquad x = -\frac{5}{2}$$

Solution set: $\left\{-3, -\dfrac{5}{2}\right\}$

26. $\qquad 5t(t - 1) = 2(1 - t)$
$$5t^2 - 5t = 2 - 2t$$
$$5t^2 - 3t - 2 = 0$$
$$(5t + 2)(t - 1) = 0$$

$$5t + 2 = 0 \quad \text{or} \quad t - 1 = 0$$

$$t = -\frac{2}{5} \quad \text{or} \qquad t = 1$$

Solution set: $\left\{-\dfrac{2}{5}, 1\right\}$

27. $f(x) = \dfrac{2}{x^2 - 9} = \dfrac{2}{(x + 3)(x - 3)}$

The numbers -3 and 3 make the denominator 0 so they must be excluded from the set of all real numbers for the domain of f.
Domain: $\{x | x \neq \pm 3\}$

28. $\dfrac{y^2 + y - 12}{y^3 + 9y^2 + 20y} \div \dfrac{y^2 - 9}{y^3 + 3y^2}$

$$= \frac{y^2 + y - 12}{y(y^2 + 9y + 20)} \cdot \frac{y^3 + 3y^2}{y^2 - 9}$$

$$= \frac{(y + 4)(y - 3)}{y(y + 4)(y + 5)} \cdot \frac{y^2(y + 3)}{(y + 3)(y - 3)}$$

$$= \frac{y}{y + 5}$$

29. $\dfrac{1}{x+y}+\dfrac{3}{x-y}$ *The LCD is* $(x+y)(x-y)$.

$$= \dfrac{1(x-y)}{(x+y)(x-y)}+\dfrac{3(x+y)}{(x-y)(x+y)}$$

$$= \dfrac{(x-y)+3(x+y)}{(x+y)(x-y)}$$

$$= \dfrac{x-y+3x+3y}{(x+y)(x-y)}$$

$$= \dfrac{4x+2y}{(x+y)(x-y)}$$

30. $\dfrac{\dfrac{-6}{x-2}}{\dfrac{8}{3x-6}} = \dfrac{-6}{x-2} \div \dfrac{8}{3x-6}$

$$= \dfrac{-6}{x-2} \cdot \dfrac{3x-6}{8}$$

$$= \dfrac{-6}{x-2} \cdot \dfrac{3(x-2)}{8}$$

$$= \dfrac{-2\cdot 3\cdot 3}{2\cdot 4} = -\dfrac{9}{4}$$

31. $\dfrac{\dfrac{1}{a}-\dfrac{1}{b}}{\dfrac{a}{b}-\dfrac{b}{a}}$ *The LCD in both numerator and denominator is ab.*

$$= \dfrac{\dfrac{b-a}{ab}}{\dfrac{a^2-b^2}{ab}}$$

$$= \dfrac{b-a}{ab} \div \dfrac{a^2-b^2}{ab}$$

$$= \dfrac{b-a}{ab} \cdot \dfrac{ab}{a^2-b^2}$$

$$= \dfrac{b-a}{a^2-b^2}$$

$$= \dfrac{-(a-b)}{(a-b)(a+b)} = \dfrac{-1}{a+b}$$

32. $\dfrac{x^{-1}}{y-x^{-1}} = \dfrac{\dfrac{1}{x}}{y-\dfrac{1}{x}}$

Multiply the numerator and denominator by x.

$$= \dfrac{x\left(\dfrac{1}{x}\right)}{x\left(y-\dfrac{1}{x}\right)} = \dfrac{1}{xy-1}$$

33. Let $x =$ Chuck's speed and $x+4 =$ Natalie's speed.

Use $d = rt$, or $t = \dfrac{d}{r}$, to make a table.

	Distance	Rate	Time
Chuck	24	x	$\dfrac{24}{x}$
Natalie	48	$x+4$	$\dfrac{48}{x+4}$

Since the times are the same,

$$\dfrac{24}{x} = \dfrac{48}{x+4}.$$

Multiply by the LCD, $x(x+4)$.

$$x(x+4)\left(\dfrac{24}{x}\right) = x(x+4)\left(\dfrac{48}{x+4}\right)$$

$$24(x+4) = 48x$$

$$24x + 96 = 48x$$

$$-24x = -96$$

$$x = 4$$

Then, $x + 4 = 4 + 4 = 8$.
Chuck's speed is 4 mph; Natalie's speed is 8 mph.

34. $\dfrac{p+1}{p-3} = \dfrac{4}{p-3}+6$

Multiply by the LCD, $p - 3$.

$$(p-3)\left(\dfrac{p+1}{p-3}\right) = (p-3)\left(\dfrac{4}{p-3}+6\right)$$

$$p+1 = 4 + 6(p-3)$$

$$p+1 = 4 + 6p - 18$$

$$p+1 = -14 + 6p$$

$$-5p = -15$$

$$p = 3$$

Substituting 3 in the original equation results in division by zero, so 3 cannot be a solution. The solution set is $\emptyset$.

35. $27^{-2/3} = \dfrac{1}{27^{2/3}} = \dfrac{1}{\left[(3^3)^{1/3}\right]^2} = \dfrac{1}{3^2} = \dfrac{1}{9}$

36. $\sqrt{200x^4} = \sqrt{100x^4 \cdot 2}$

$$= \sqrt{10^2(x^2)^2 \cdot 2} = 10x^2\sqrt{2}$$

37. $\sqrt[3]{48x^5y^2} = \sqrt[3]{8x^3 \cdot 6x^2y^2}$

$$= \sqrt[3]{2^3x^3 \cdot 6x^2y^2} = 2x\sqrt[3]{6x^2y^2}$$

38. $\sqrt{50} + \sqrt{8} = \sqrt{25\cdot 2} + \sqrt{4\cdot 2}$

$$= 5\sqrt{2} + 2\sqrt{2} = 7\sqrt{2}$$

39.
$$\frac{1}{\sqrt{10} - \sqrt{8}}$$

Multiply by the conjugate of the denominator, $\sqrt{10} + \sqrt{8}$.

$$= \frac{1\left(\sqrt{10} + \sqrt{8}\right)}{\left(\sqrt{10} - \sqrt{8}\right)\left(\sqrt{10} + \sqrt{8}\right)}$$

$$= \frac{\sqrt{10} + \sqrt{8}}{10 - 8} = \frac{\sqrt{10} + \sqrt{4 \cdot 2}}{2}$$

$$= \frac{\sqrt{10} + 2\sqrt{2}}{2}$$

40. $\left(2\sqrt{x} + \sqrt{y}\right)\left(-3\sqrt{x} - 4\sqrt{y}\right)$

$$= -6x - 8\sqrt{xy} - 3\sqrt{xy} - 4y$$

$$= -6x - 11\sqrt{xy} - 4y$$

41. $\sqrt{3r - 8} = r - 2$

Square both sides.

$$\left(\sqrt{3r - 8}\right)^2 = (r - 2)^2$$

$$3r - 8 = r^2 - 4r + 4$$

$$0 = r^2 - 7r + 12$$

$$0 = (r - 3)(r - 4)$$

$r - 3 = 0$ or $r - 4 = 0$

$r = 0$ or $r = 4$

Check $r = 3$: $\sqrt{1} = 1$ *True*

Check $r = 4$: $\sqrt{4} = 2$ *True*

Solution set: $\{3, 4\}$

42. Substitute 32 for D and 5 for h in the given formula.

$$S = \frac{2.74D}{\sqrt{h}} = \frac{2.74(32)}{\sqrt{5}} = \frac{87.68}{\sqrt{5}} \approx 39.2$$

The fall speed is about 39.2 mph.

43. $(5 + 7i) - (3 - 2i)$

$$= (5 - 3) + [7 - (-2)]i$$

$$= 2 + 9i$$

44. $\dfrac{6 - 2i}{1 - i}$

Multiply by the conjugate of the denominator, $1 + i$.

$$= \frac{(6 - 2i)(1 + i)}{(1 - i)(1 + i)}$$

$$= \frac{6 + 6i - 2i - 2i^2}{1^2 + 1^2}$$

$$= \frac{6 + 4i + 2}{1 + 1}$$

$$= \frac{8 + 4i}{2} = \frac{2(4 + 2i)}{2} = 4 + 2i$$

CHAPTER 8 QUADRATIC EQUATIONS AND INEQUALITIES

Section 8.1

1. To solve $2x^2 + 8x = 9$ by completing the square, the first step would be to divide both sides by 2 so the coefficient of x^2 would be 1.

3. To solve $(2x + 1)^2 = 5$, we use the square root property and find that $2x + 1 = \sqrt{5}$ or $2x + 1 = -\sqrt{5}$. Then we solve for x in both equations.
   To solve $x^2 + 4x = 12$, we find that
   $$x^2 + 4x + 4 = 12 + 4$$
   $$(x + 2)^2 = 16$$
   by completing the square.

5. We need to add the square of half the coefficient of x to get a perfect square trinomial.
   $$\frac{1}{2}(8) = 4 \quad \text{and} \quad 4^2 = 16$$
   Add 16 to $x^2 + 8x$ to get a perfect square trinomial.

7. $\frac{1}{2}(-9) = -\frac{9}{2}$ and $\left(-\frac{9}{2}\right)^2 = \frac{81}{4}$

9. $x^2 = 81$
   $x = 9$ or $x = -9$
   Solution set: $\{-9, 9\}$

11. $m^2 = 32$
    $m = \sqrt{32}$ or $m = -\sqrt{32}$
    $m = 4\sqrt{2}$ or $m = -4\sqrt{2}$
    Solution set: $\left\{-4\sqrt{2}, 4\sqrt{2}\right\}$

13. $(x + 2)^2 = 25$
    $x + 2 = \sqrt{25}$ or $x + 2 = -\sqrt{25}$
    $x + 2 = 5$ $\qquad x + 2 = -5$
    $x = 3$ or $x = -7$
    Solution set: $\{-7, 3\}$

15. $(1 - 3k)^2 = 7$
    $1 - 3k = \sqrt{7}$ or $1 - 3k = -\sqrt{7}$
    $-3k = -1 + \sqrt{7}$ $\qquad -3k = -1 - \sqrt{7}$
    $k = \frac{-1 + \sqrt{7}}{-3}$ $\qquad k = \frac{-1 - \sqrt{7}}{-3}$
    $k = \frac{1 - \sqrt{7}}{3}$ or $k = \frac{1 + \sqrt{7}}{3}$
    Solution set: $\left\{\frac{1 + \sqrt{7}}{3}, \frac{1 - \sqrt{7}}{3}\right\}$

17. $(4p + 1)^2 = 24$
    $4p + 1 = \sqrt{24}$ or $4p + 1 = -\sqrt{24}$
    $4p + 1 = 2\sqrt{6}$ $\qquad 4p + 1 = -2\sqrt{6}$
    $4p = -1 + 2\sqrt{6}$ $\qquad 4p = -1 - 2\sqrt{6}$
    $p = \frac{-1 + 2\sqrt{6}}{4}$ or $p = \frac{-1 - 2\sqrt{6}}{4}$
    Solution set: $\left\{\frac{-1 + 2\sqrt{6}}{4}, \frac{-1 - 2\sqrt{6}}{4}\right\}$

19. $x^2 = -12$
    $x = \sqrt{-12}$ or $x = -\sqrt{-12}$
    $x = i\sqrt{12}$ $\qquad x = -i\sqrt{12}$
    $x = 2i\sqrt{3}$ or $x = -2i\sqrt{3}$
    Solution set: $\left\{-2i\sqrt{3}, 2i\sqrt{3}\right\}$

21. $(r - 5)^2 = -3$
    $r - 5 = \sqrt{-3}$ or $r - 5 = -\sqrt{-3}$
    $r = 5 + \sqrt{-3}$ $\qquad r = 5 - \sqrt{-3}$
    $r = 5 + i\sqrt{3}$ or $r = 5 - i\sqrt{3}$
    Solution set: $\left\{5 + i\sqrt{3}, 5 - i\sqrt{3}\right\}$

23. $(6k - 1)^2 = -8$
    $6k - 1 = \sqrt{-8}$ or $6k - 1 = -\sqrt{-8}$
    $6k - 1 = i\sqrt{8}$ $\qquad 6k - 1 = -i\sqrt{8}$
    $6k - 1 = 2i\sqrt{2}$ $\qquad 6k - 1 = -2i\sqrt{2}$
    $6k = 1 + 2i\sqrt{2}$ $\qquad 6k = 1 - 2i\sqrt{2}$
    $k = \frac{1 + 2i\sqrt{2}}{6}$ or $k = \frac{1 - 2i\sqrt{2}}{6}$
    Solution set: $\left\{\frac{1 + 2i\sqrt{2}}{6}, \frac{1 - 2i\sqrt{2}}{6}\right\}$

25. $x^2 - 2x - 24 = 0$
    Get the variable terms alone on the left side.
    $$x^2 - 2x = 24$$
    Complete the square by taking half of -2, the coefficient of x, and squaring the result.
    $$\left[\frac{1}{2}(-2)\right]^2 = (-1)^2 = 1$$
    Add 1 to each side.
    $$x^2 - 2x + 1 = 24 + 1$$
    Factor the left side.
    $$(x - 1)^2 = 25$$
    Use the square root property.

continued

$$x - 1 = \sqrt{25} \quad \text{or} \quad x - 1 = -\sqrt{25}$$
$$x - 1 = 5 \quad \text{or} \quad x - 1 = -5$$
$$x = 6 \quad \text{or} \quad x = -4$$

Solution set: $\{-4, 6\}$

27.
$$3y^2 + y = 24$$

Divide by 3.
$$y^2 + \frac{1}{3}y = 8$$

Complete the square by taking half of $\frac{1}{3}$, the coefficient of y, and squaring the result.
$$\left(\frac{1}{2} \cdot \frac{1}{3}\right)^2 = \left(\frac{1}{6}\right)^2 = \frac{1}{36}$$

Add $\frac{1}{36}$ to both sides.
$$y^2 + \frac{1}{3}y + \frac{1}{36} = 8 + \frac{1}{36}$$
$$\left(y + \frac{1}{6}\right)^2 = \frac{288}{36} + \frac{1}{36}$$
$$\left(y + \frac{1}{6}\right)^2 = \frac{289}{36}$$
$$y + \frac{1}{6} = \sqrt{\frac{289}{36}} \quad \text{or} \quad y + \frac{1}{6} = -\sqrt{\frac{289}{36}}$$
$$y = -\frac{1}{6} + \frac{\sqrt{289}}{\sqrt{36}} \qquad y = -\frac{1}{6} - \frac{\sqrt{289}}{\sqrt{36}}$$
$$y = -\frac{1}{6} + \frac{17}{6} \qquad y = -\frac{1}{6} - \frac{17}{6}$$
$$y = \frac{16}{6} \qquad y = -\frac{18}{6}$$
$$y = \frac{8}{3} \quad \text{or} \quad y = -3$$

Solution set: $\left\{-3, \frac{8}{3}\right\}$

29.
$$2k^2 + 5k - 2 = 0$$
$$2k^2 + 5k = 2$$
$$k^2 + \frac{5}{2}k = 1$$

Complete the square.
$$\left(\frac{1}{2} \cdot \frac{5}{2}\right)^2 = \left(\frac{5}{4}\right)^2 = \frac{25}{16}$$

Add $\frac{25}{16}$ to each side.
$$k^2 + \frac{5}{2}k + \frac{25}{16} = 1 + \frac{25}{16}$$
$$\left(k + \frac{5}{4}\right)^2 = \frac{41}{16}$$

$$k + \frac{5}{4} = \sqrt{\frac{41}{16}} \quad \text{or} \quad k + \frac{5}{4} = -\sqrt{\frac{41}{16}}$$
$$k = -\frac{5}{4} + \frac{\sqrt{41}}{4} \qquad k = -\frac{5}{4} - \frac{\sqrt{41}}{4}$$
$$k = \frac{-5 + \sqrt{41}}{4} \quad \text{or} \quad k = \frac{-5 - \sqrt{41}}{4}$$

Solution set: $\left\{\frac{-5 + \sqrt{41}}{4}, \frac{-5 - \sqrt{41}}{4}\right\}$

31.
$$9x^2 - 24x = -13$$
$$x^2 - \frac{24}{9}x = \frac{-13}{9}$$
$$x^2 - \frac{8}{3}x = \frac{-13}{9}$$

Complete the square.
$$\left(\frac{1}{2} \cdot -\frac{8}{3}\right)^2 = \left(-\frac{4}{3}\right)^2 = \frac{16}{9}$$

Add $\frac{16}{9}$ to each side.
$$x^2 - \frac{8}{3}x + \frac{16}{9} = \frac{-13}{9} + \frac{16}{9}$$
$$\left(x - \frac{4}{3}\right)^2 = \frac{3}{9}$$

$$x - \frac{4}{3} = \sqrt{\frac{3}{9}} \quad \text{or} \quad x - \frac{4}{3} = -\sqrt{\frac{3}{9}}$$
$$x = \frac{4}{3} + \frac{\sqrt{3}}{3} \qquad x = \frac{4}{3} - \frac{\sqrt{3}}{3}$$
$$x = \frac{4 + \sqrt{3}}{3} \quad \text{or} \quad x = \frac{4 - \sqrt{3}}{3}$$

Solution set: $\left\{\frac{4 + \sqrt{3}}{3}, \frac{4 - \sqrt{3}}{3}\right\}$

33.
$$m^2 + 4m + 13 = 0$$
$$m^2 + 4m = -13$$

Complete the square.
$$\left(\frac{1}{2} \cdot 4\right)^2 = 2^2 = 4$$

Add 4 to both sides.
$$m^2 + 4m + 4 = -13 + 4$$
$$(m + 2)^2 = -9$$

$$m + 2 = \sqrt{-9} \quad \text{or} \quad m + 2 = -\sqrt{-9}$$
$$m = -2 + 3i \quad \text{or} \quad m = -2 - 3i$$

Solution set: $\{-2 + 3i, -2 - 3i\}$

35. $\quad z^2 - \dfrac{4}{3}z = -\dfrac{1}{9}$

Complete the square.

$$\left[\frac{1}{2}\left(-\frac{4}{3}\right)\right]^2 = \left(-\frac{2}{3}\right)^2 = \frac{4}{9}$$

Add $\dfrac{4}{9}$ to both sides.

$$z^2 - \frac{4}{3}z + \frac{4}{9} = -\frac{1}{9} + \frac{4}{9}$$

$$\left(z - \frac{2}{3}\right)^2 = \frac{3}{9}$$

$$z - \frac{2}{3} = \sqrt{\frac{3}{9}} \qquad \text{or} \qquad z - \frac{2}{3} = -\sqrt{\frac{3}{9}}$$

$$z = \frac{2}{3} + \frac{\sqrt{3}}{3} \qquad\qquad z = \frac{2}{3} - \frac{\sqrt{3}}{3}$$

$$z = \frac{2 + \sqrt{3}}{3} \qquad \text{or} \qquad z = \frac{2 - \sqrt{3}}{3}$$

Solution set: $\left\{\dfrac{2 + \sqrt{3}}{3}, \dfrac{2 - \sqrt{3}}{3}\right\}$

37. $\quad 3r^2 + 4r + 4 = 0$

$$3r^2 + 4r = -4$$

$$r^2 + \frac{4}{3}r = \frac{-4}{3}$$

Complete the square.

$$\left(\frac{1}{2} \cdot \frac{4}{3}\right)^2 = \left(\frac{2}{3}\right)^2 = \frac{4}{9}$$

Add $\dfrac{4}{9}$ to each side.

$$r^2 + \frac{4}{3}r + \frac{4}{9} = \frac{-4}{3} + \frac{4}{9}$$

$$\left(r + \frac{2}{3}\right)^2 = \frac{-8}{9}$$

$$r + \frac{2}{3} = \frac{\sqrt{-8}}{\sqrt{9}} \qquad \text{or } r + \frac{2}{3} = -\frac{\sqrt{-8}}{\sqrt{9}}$$

$$r = -\frac{2}{3} + \frac{2i\sqrt{2}}{3} \qquad\qquad r = -\frac{2}{3} - \frac{2i\sqrt{2}}{3}$$

$$r = \frac{-2 + 2i\sqrt{2}}{3} \quad \text{or} \quad r = \frac{-2 - 2i\sqrt{2}}{3}$$

Solution set: $\left\{\dfrac{-2 + 2i\sqrt{2}}{3}, \dfrac{-2 - 2i\sqrt{2}}{3}\right\}$

39. $\quad .1x^2 - .2x - .1 = 0$

Multiply both sides by 10 to clear the decimals.

$$x^2 - 2x - 1 = 0$$

$$x^2 - 2x = 1$$

Complete the square.

$$\left[\frac{1}{2}(-2)\right]^2 = (-1)^2 = 1$$

Add 1 to both sides.

$$x^2 - 2x + 1 = 1 + 1$$

$$(x - 1)^2 = 2$$

$$x - 1 = \sqrt{2} \qquad \text{or} \quad x - 1 = -\sqrt{2}$$

$$x = 1 + \sqrt{2} \quad \text{or} \qquad x = 1 - \sqrt{2}$$

Solution set: $\left\{1 + \sqrt{2}, 1 - \sqrt{2}\right\}$

41. $\quad -m^2 - 6m - 12 = 0$

Multiply each side by -1.

$$m^2 + 6m + 12 = 0$$

$$m^2 + 6m = -12$$

Complete the square.

$$\left(\frac{1}{2} \cdot 6\right)^2 = 3^2 = 9$$

Add 9 to each side.

$$m^2 + 6m + 9 = -12 + 9$$

$$(m + 3)^2 = -3$$

$$m + 3 = \sqrt{-3} \qquad \text{or} \quad m + 3 = -\sqrt{-3}$$

$$m = -3 + i\sqrt{3} \quad \text{or} \qquad m = -3 - i\sqrt{3}$$

Solution set: $\left\{-3 + i\sqrt{3}, -3 - i\sqrt{3}\right\}$

43. The zero-factor property cannot be applied because the product of the two factors is 5, not 0.

45. Some quadratic polynomials cannot be factored, so all quadratic equations cannot be solved by factoring. An example is $x^2 + x - 3 = 0$. Therefore, the zero-factor property cannot be used.

47. $\quad x^2 - b = 0$

$$x^2 = b$$

$$x = \sqrt{b} \quad \text{or} \quad x = -\sqrt{b}$$

Solution set: $\left\{-\sqrt{b}, \sqrt{b}\right\}$

49. $\quad 4x^2 = b^2 + 16$

$$x^2 = \frac{b^2 + 16}{4}$$

$$x = \sqrt{\frac{b^2 + 16}{4}} \quad \text{or} \quad x = -\sqrt{\frac{b^2 + 16}{4}}$$

$$x = \frac{\sqrt{b^2 + 16}}{2} \quad \text{or} \quad x = -\frac{\sqrt{b^2 + 16}}{2}$$

Solution set: $\left\{-\dfrac{\sqrt{b^2 + 16}}{2}, \dfrac{\sqrt{b^2 + 16}}{2}\right\}$

51. $(5x - 2b)^2 = 3a$

$$5x - 2b = \sqrt{3a} \qquad \text{or} \quad 5x - 2b = -\sqrt{3a}$$
$$5x = 2b + \sqrt{3a} \qquad\qquad 5x = 2b - \sqrt{3a}$$
$$x = \frac{2b + \sqrt{3a}}{5} \quad \text{or} \qquad x = \frac{2b - \sqrt{3a}}{5}$$

Solution set: $\left\{ \dfrac{2b + \sqrt{3a}}{5}, \dfrac{2b - \sqrt{3a}}{5} \right\}$

53. $P^2 = 140.608$

$$P = \sqrt{140.608} \quad \text{or} \quad P = -\sqrt{140.608}$$
$$P \approx 11.9 \qquad\qquad P \text{ cannot be negative}$$

The sidereal period of Jupiter is about 11.9 years.

55. **(a)** The radius of the orbit of an HEAO satellite is the sum of the altitude and the radius of Earth.

$$4.3 \times 10^5 + 6.37 \times 10^6$$
$$= .43 \times 10^6 + 6.37 \times 10^6$$
$$= 6.8 \times 10^6$$

The radius is about 6.8×10^6 meters.

(b) $v^2 = \dfrac{GM}{r}$

$$v^2 = \frac{3.99 \times 10^{11}}{6.8 \times 10^6}$$
$$\approx 58,676.47$$
$$v \approx \sqrt{58,676.47} \approx 242$$

The velocity of an HEAO satellite is approximately 242 meters per second.

57. The area of the original square is $x \cdot x$, or x^2.

58. Each rectangular strip has length x and width 1, so each strip has an area of $x \cdot 1$, or x.

59. From Exercise 58, the area of a rectangular strip is x. The area of 6 rectangular strips is $6x$.

60. These are 1 by 1 squares, so each has an area of $1 \cdot 1$, or 1.

61. There are 9 small squares, each with area 1 (from Exercise 60), so the total area is $9 \cdot 1$, or 9.

62. The area of the larger square is $(x + 3)^2$. Using the results from Exercises 57–61,

$$(x + 3)^2 = x^2 + 6x + 9.$$

63. As in the example, we have $x^2 = 17$ or $x = \pm\sqrt{17}$.

Section 8.2

1. Multiplying both sides of either equation,

$$x^2 + 3x - 4 = 0 \quad \text{or} \quad -x^2 - 3x + 4 = 0,$$

gives us the other equation. So the equations have the same solution set and the statement is *true*.

3. The quadratic formula *can* be used to solve $4m^2 + 3m = 0$ ($a = 4$, $b = 3$, and $c = 0$), so the statement is *true*.

5.
$$x^2 + kx = 0$$
$$x(x + k) = 0$$

The equation has two real solutions, 0 and $-k$, so the statement is *true*.

7. $m^2 - 8m + 15 = 0$

Here $a = 1$, $b = -8$, and $c = 15$.

$$m = \frac{-b \pm \sqrt{b^2 - 4ac}}{2a}$$
$$m = \frac{-(-8) \pm \sqrt{(-8)^2 - 4(1)(15)}}{2(1)}$$
$$= \frac{8 \pm \sqrt{64 - 60}}{2}$$
$$= \frac{8 \pm \sqrt{4}}{2} = \frac{8 \pm 2}{2}$$
$$m = \frac{8 + 2}{2} = \frac{10}{2} = 5 \text{ or}$$
$$m = \frac{8 - 2}{2} = \frac{6}{2} = 3$$

Solution set: $\{3, 5\}$

9. $2k^2 + 4k + 1 = 0$

Here $a = 2$, $b = 4$, and $c = 1$.

$$k = \frac{-b \pm \sqrt{b^2 - 4ac}}{2a}$$
$$k = \frac{-4 \pm \sqrt{4^2 - 4(2)(1)}}{2(2)}$$
$$= \frac{-4 \pm \sqrt{16 - 8}}{4}$$
$$= \frac{-4 \pm \sqrt{8}}{4} = \frac{-4 \pm 2\sqrt{2}}{4}$$
$$= \frac{2\left(-2 \pm \sqrt{2}\right)}{2 \cdot 2} = \frac{-2 \pm \sqrt{2}}{2}$$

Solution set: $\left\{ \dfrac{-2 + \sqrt{2}}{2}, \dfrac{-2 - \sqrt{2}}{2} \right\}$

11.
$$2x^2 - 2x = 1$$
$$2x^2 - 2x - 1 = 0$$
Here $a = 2$, $b = -2$, and $c = -1$.
$$x = \frac{-b \pm \sqrt{b^2 - 4ac}}{2a}$$
$$x = \frac{-(-2) \pm \sqrt{(-2)^2 - 4(2)(-1)}}{2(2)}$$
$$= \frac{2 \pm \sqrt{4 + 8}}{4} = \frac{2 \pm \sqrt{12}}{4} = \frac{2 \pm 2\sqrt{3}}{4}$$
$$= \frac{2\left(1 \pm \sqrt{3}\right)}{2 \cdot 2} = \frac{1 \pm \sqrt{3}}{2}$$
Solution set: $\left\{ \dfrac{1 + \sqrt{3}}{2}, \dfrac{1 - \sqrt{3}}{2} \right\}$

13.
$$x^2 + 18 = 10x$$
$$x^2 - 10x + 18 = 0$$
Here $a = 1$, $b = -10$, and $c = 18$.
$$x = \frac{-b \pm \sqrt{b^2 - 4ac}}{2a}$$
$$x = \frac{-(-10) \pm \sqrt{(-10)^2 - 4(1)(18)}}{2(1)}$$
$$= \frac{10 \pm \sqrt{100 - 72}}{2} = \frac{10 \pm \sqrt{28}}{2}$$
$$= \frac{10 \pm 2\sqrt{7}}{2} = \frac{2\left(5 \pm \sqrt{7}\right)}{2} = 5 \pm \sqrt{7}$$
Solution set: $\left\{ 5 + \sqrt{7}, 5 - \sqrt{7} \right\}$

15.
$$-2t(t + 2) = -3$$
$$-2t^2 - 4t = -3$$
$$-2t^2 - 4t + 3 = 0$$
Here $a = -2$, $b = -4$, and $c = 3$.
$$t = \frac{-b \pm \sqrt{b^2 - 4ac}}{2a}$$
$$t = \frac{-(-4) \pm \sqrt{(-4)^2 - 4(-2)(3)}}{2(-2)}$$
$$= \frac{4 \pm \sqrt{16 + 24}}{-4} = \frac{4 \pm \sqrt{40}}{-4}$$
$$= \frac{4 \pm 2\sqrt{10}}{-4} = \frac{2\left(2 \pm \sqrt{10}\right)}{-2 \cdot 2}$$
$$= \frac{2 \pm \sqrt{10}}{-2} \cdot \frac{-1}{-1} = \frac{-2 \mp \sqrt{10}}{2}$$
$$= \frac{-2 \pm \sqrt{10}}{2}$$
Solution set: $\left\{ \dfrac{-2 + \sqrt{10}}{2}, \dfrac{-2 - \sqrt{10}}{2} \right\}$

17.
$$(r - 3)(r + 5) = 2$$
$$r^2 + 2r - 15 = 2$$
$$r^2 + 2r - 17 = 0$$
Here $a = 1$, $b = 2$, and $c = -17$.
$$r = \frac{-b \pm \sqrt{b^2 - 4ac}}{2a}$$
$$r = \frac{-2 \pm \sqrt{2^2 - 4(1)(-17)}}{2(1)}$$
$$= \frac{-2 \pm \sqrt{4 + 68}}{2} = \frac{-2 \pm \sqrt{72}}{2}$$
$$= \frac{-2 \pm 6\sqrt{2}}{2} = \frac{-2\left(1 \pm 3\sqrt{2}\right)}{2}$$
$$= -1\left(1 \pm 3\sqrt{2}\right) = -1 \mp 3\sqrt{2} = -1 \pm 3\sqrt{2}$$
Solution set: $\left\{ -1 + 3\sqrt{2}, -1 - 3\sqrt{2} \right\}$

19.
$$p^2 + \frac{p}{3} = \frac{2}{3}$$
$$p^2 + \frac{p}{3} - \frac{2}{3} = 0$$
$$3p^2 + p - 2 = 0 \quad \textit{Multiply by 3.}$$
Here $a = 3$, $b = 1$, and $c = -2$.
$$p = \frac{-b \pm \sqrt{b^2 - 4ac}}{2a}$$
$$p = \frac{-1 \pm \sqrt{1^2 - 4(3)(-2)}}{2(3)}$$
$$= \frac{-1 \pm \sqrt{1 + 24}}{6}$$
$$= \frac{-1 \pm \sqrt{25}}{6} = \frac{-1 \pm 5}{6}$$
$$p = \frac{-1 + 5}{6} = \frac{4}{6} = \frac{2}{3} \text{ or}$$
$$p = \frac{-1 - 5}{6} = \frac{-6}{6} = -1$$
Solution set: $\left\{ -1, \dfrac{2}{3} \right\}$

21.
$$4k(k+1) = 1$$
$$4k^2 + 4k = 1$$
$$4k^2 + 4k - 1 = 0$$
Here $a = 4$, $b = 4$, and $c = -1$.
$$k = \frac{-b \pm \sqrt{b^2 - 4ac}}{2a}$$
$$k = \frac{-4 \pm \sqrt{4^2 - 4(4)(-1)}}{2(4)}$$
$$= \frac{-4 \pm \sqrt{16 + 16}}{8} = \frac{-4 \pm \sqrt{32}}{8}$$
$$= \frac{-4 \pm 4\sqrt{2}}{8} = \frac{4\left(-1 \pm \sqrt{2}\right)}{2 \cdot 4} = \frac{-1 \pm \sqrt{2}}{2}$$
Solution set: $\left\{ \dfrac{-1 + \sqrt{2}}{2}, \dfrac{-1 - \sqrt{2}}{2} \right\}$

23.
$$(g+2)(g-3) = 1$$
$$g^2 - g - 6 = 1$$
$$g^2 - g - 7 = 0$$
Here $a = 1$, $b = -1$, and $c = -7$.
$$g = \frac{-b \pm \sqrt{b^2 - 4ac}}{2a}$$
$$g = \frac{-(-1) \pm \sqrt{(-1)^2 - 4(1)(-7)}}{2(1)}$$
$$= \frac{1 \pm \sqrt{1 + 28}}{2} = \frac{1 \pm \sqrt{29}}{2}$$
Solution set: $\left\{ \dfrac{1 + \sqrt{29}}{2}, \dfrac{1 - \sqrt{29}}{2} \right\}$

25.
$$3x^2 + 2x = 2$$
$$3x^2 + 2x - 2 = 0$$
Here $a = 3$, $b = 2$, and $c = -2$.
$$x = \frac{-b \pm \sqrt{b^2 - 4ac}}{2a}$$
$$x = \frac{-2 \pm \sqrt{2^2 - 4(3)(-2)}}{2(3)}$$
$$= \frac{-2 \pm \sqrt{4 + 24}}{6} = \frac{-2 \pm \sqrt{28}}{6}$$
$$= \frac{-2 \pm 2\sqrt{7}}{6} = \frac{2\left(-1 \pm \sqrt{7}\right)}{2 \cdot 3} = \frac{-1 \pm \sqrt{7}}{3}$$
Solution set: $\left\{ \dfrac{-1 + \sqrt{7}}{3}, \dfrac{-1 - \sqrt{7}}{3} \right\}$

27.
$$y = \frac{5(5-y)}{3(y+1)}$$
$$3y(y+1) = 5(5-y)$$
$$3y^2 + 3y = 25 - 5y$$
$$3y^2 + 8y - 25 = 0$$
Here $a = 3$, $b = 8$, and $c = -25$.
$$y = \frac{-b \pm \sqrt{b^2 - 4ac}}{2a}$$
$$y = \frac{-8 \pm \sqrt{8^2 - 4(3)(-25)}}{2(3)}$$
$$= \frac{-8 \pm \sqrt{64 + 300}}{6} = \frac{-8 \pm \sqrt{364}}{6}$$
$$= \frac{-8 \pm 2\sqrt{91}}{6} = \frac{2\left(-4 \pm \sqrt{91}\right)}{2 \cdot 3} = \frac{-4 \pm \sqrt{91}}{3}$$
Solution set: $\left\{ \dfrac{-4 + \sqrt{91}}{3}, \dfrac{-4 - \sqrt{91}}{3} \right\}$

29. The last step is wrong. Because 5 is not a common factor in the numerator, the fraction cannot be reduced. The solutions are $\dfrac{5 \pm \sqrt{5}}{10}$.

31. $k^2 + 47 = 0$
Here $a = 1$, $b = 0$, and $c = 47$.
$$k = \frac{-b \pm \sqrt{b^2 - 4ac}}{2a}$$
$$k = \frac{-(0) \pm \sqrt{(0)^2 - 4(1)(47)}}{2(1)}$$
$$= \frac{0 \pm \sqrt{0 - 4 \cdot 47}}{2} = \frac{\pm \sqrt{-4 \cdot 47}}{2}$$
$$= \frac{\pm 2i\sqrt{47}}{2} = \pm i\sqrt{47}$$
Solution set: $\left\{ -i\sqrt{47}, i\sqrt{47} \right\}$

33. $r^2 - 6r + 14 = 0$
Here $a = 1$, $b = -6$, and $c = 14$.
$$r = \frac{-b \pm \sqrt{b^2 - 4ac}}{2a}$$
$$r = \frac{-(-6) \pm \sqrt{(-6)^2 - 4(1)(14)}}{2(1)}$$
$$= \frac{6 \pm \sqrt{36 - 56}}{2}$$
$$= \frac{6 \pm \sqrt{-20}}{2} = \frac{6 \pm 2i\sqrt{5}}{2}$$
$$= \frac{2\left(3 \pm i\sqrt{5}\right)}{2} = 3 \pm i\sqrt{5}$$
Solution set: $\left\{ 3 + i\sqrt{5}, 3 - i\sqrt{5} \right\}$

35.
$$4x^2 - 4x = -7$$
$$4x^2 - 4x + 7 = 0$$
Here $a = 4$, $b = -4$, and $c = 7$.
$$x = \frac{-b \pm \sqrt{b^2 - 4ac}}{2a}$$
$$x = \frac{-(-4) \pm \sqrt{(-4)^2 - 4(4)(7)}}{2(4)}$$
$$= \frac{4 \pm \sqrt{16 - 112}}{8} = \frac{4 \pm \sqrt{-96}}{8}$$
$$= \frac{4 \pm 4i\sqrt{6}}{8}$$
$$= \frac{4\left(1 \pm i\sqrt{6}\right)}{2 \cdot 4} = \frac{1 \pm i\sqrt{6}}{2}$$
Solution set: $\left\{ \dfrac{1 + i\sqrt{6}}{2}, \dfrac{1 - i\sqrt{6}}{2} \right\}$

37.
$$x(3x + 4) = -2$$
$$3x^2 + 4x = -2$$
$$3x^2 + 4x + 2 = 0$$
Here $a = 3$, $b = 4$, and $c = 2$.
$$x = \frac{-b \pm \sqrt{b^2 - 4ac}}{2a}$$
$$x = \frac{-4 \pm \sqrt{4^2 - 4(3)(2)}}{2(3)}$$
$$= \frac{-4 \pm \sqrt{16 - 24}}{6} = \frac{-4 \pm \sqrt{-8}}{6}$$
$$= \frac{-4 \pm 2i\sqrt{2}}{6} = \frac{2\left(-2 \pm i\sqrt{2}\right)}{2 \cdot 3} = \frac{-2 \pm i\sqrt{2}}{3}$$
Solution set: $\left\{ \dfrac{-2 + i\sqrt{2}}{3}, \dfrac{-2 - i\sqrt{2}}{3} \right\}$

39.
$$\frac{x + 5}{2x - 1} = \frac{x - 4}{x - 6}$$
Multiply by the LCD, $(2x - 1)(x - 6)$.
$$(2x - 1)(x - 6)\left(\frac{x + 5}{2x - 1}\right)$$
$$= (2x - 1)(x - 6)\left(\frac{x - 4}{x - 6}\right)$$
$$(x - 6)(x + 5) = (2x - 1)(x - 4)$$
$$x^2 - x - 30 = 2x^2 - 9x + 4$$
$$0 = x^2 - 8x + 34$$
Here $a = 1$, $b = -8$, and $c = 34$.

$$x = \frac{-b \pm \sqrt{b^2 - 4ac}}{2a}$$
$$x = \frac{-(-8) \pm \sqrt{(-8)^2 - 4(1)(34)}}{2(1)}$$
$$= \frac{8 \pm \sqrt{64 - 136}}{2} = \frac{8 \pm \sqrt{-72}}{2}$$
$$= \frac{8 \pm 6i\sqrt{2}}{2} = \frac{2\left(4 \pm 3i\sqrt{2}\right)}{2} = 4 \pm 3i\sqrt{2}$$
Solution set: $\left\{ 4 + 3i\sqrt{2}, 4 - 3i\sqrt{2} \right\}$

41.
$$\frac{1}{x^2} + 1 = -\frac{1}{x}$$
Multiply by the LCD, x^2.
$$x^2\left(\frac{1}{x^2} + 1\right) = x^2\left(-\frac{1}{x}\right)$$
$$1 + x^2 = -x$$
$$x^2 + x + 1 = 0$$
Here $a = 1$, $b = 1$, and $c = 1$.
$$x = \frac{-b \pm \sqrt{b^2 - 4ac}}{2a}$$
$$x = \frac{-1 \pm \sqrt{1^2 - 4(1)(1)}}{2(1)} = \frac{-1 \pm \sqrt{1 - 4}}{2}$$
$$= \frac{-1 \pm \sqrt{-3}}{2} = \frac{-1 \pm i\sqrt{3}}{2}$$
Solution set: $\left\{ \dfrac{-1 + i\sqrt{3}}{2}, \dfrac{-1 - i\sqrt{3}}{2} \right\}$

43. Let x be the time in hours required for the faster person to cut the lawn. Then the slower person requires $x + 1$ hours.
Complete the chart.

Worker	Rate	Time Working Together	Fractional Part of the Job Done
Faster person	$\dfrac{1}{x}$	2	$\dfrac{2}{x}$
Slower person	$\dfrac{1}{x + 1}$	2	$\dfrac{2}{x + 1}$

$$\begin{array}{ccccc} \text{Part done by} & & \text{Part done by} & & \text{one whole} \\ \text{faster person} & + & \text{slower person} & = & \text{job.} \\ \dfrac{2}{x} & + & \dfrac{2}{x + 1} & = & 1 \end{array}$$

Multiply both sides by the LCD, $x(x + 1)$.
$$x(x + 1)\left(\frac{2}{x} + \frac{2}{x + 1}\right) = x(x + 1) \cdot 1$$
$$2(x + 1) + 2x = x^2 + x$$
$$2x + 2 + 2x = x^2 + x$$
$$0 = x^2 - 3x - 2$$

continued

Solve for x using the quadratic formula with $a = 1$, $b = -3$, and $c = -2$.

$$x = \frac{-(-3) \pm \sqrt{(-3)^2 - 4(1)(-2)}}{2(1)}$$

$$= \frac{3 \pm \sqrt{9 + 8}}{2} = \frac{3 \pm \sqrt{17}}{2}$$

$$x = \frac{3 + \sqrt{17}}{2} \quad \text{or} \quad x = \frac{3 - \sqrt{17}}{2}$$

$$x \approx 3.6 \quad \text{or} \quad x \approx -.6$$

Discard $-.6$ as a solution since time cannot be negative.
It would take the faster person approximately 3.6 hours.

45. Let x represent the time in hours it takes Nancy to plant the flowers. Then $x + 2$ is the time it takes Rusty.
Organize the information in a chart.

Worker	Rate	Time Working Together	Fractional Part of the Job Done
Nancy	$\dfrac{1}{x}$	12	$\dfrac{12}{x}$
Rusty	$\dfrac{1}{x + 2}$	12	$\dfrac{12}{x + 2}$

$$\begin{array}{ccc} \text{Part done} & \text{part done} & \text{one whole} \\ \text{by Nancy} & + \quad \text{by Rusty} & = \quad \text{job.} \end{array}$$

$$\frac{12}{x} + \frac{12}{x + 2} = 1$$

Multiply both sides by the LCD, $x(x + 2)$.

$$x(x + 2)\left(\frac{12}{x} + \frac{12}{x + 2}\right) = x(x + 2) \cdot 1$$

$$12(x + 2) + 12x = x^2 + 2x$$

$$12x + 24 + 12x = x^2 + 2x$$

$$0 = x^2 - 22x - 24$$

Solve for x using the quadratic formula with $a = 1$, $b = -22$, and $c = -24$.

$$x = \frac{-(-22) \pm \sqrt{(-22)^2 - 4(1)(-24)}}{2(1)}$$

$$= \frac{22 \pm \sqrt{580}}{2}$$

$$x = \frac{22 + \sqrt{580}}{2} \approx 23.0 \quad \text{or}$$

$$x = \frac{22 - \sqrt{580}}{2} \approx -1.0$$

Since x represents time, discard the negative solution.
Nancy takes about 23.0 hours planting flowers alone while Rusty takes about 25.0 hours planting alone.

47. Let x be the time in hours it will take to fill the tank with the faster pipe. Then the slower pipe requires $x + .5$ hours.
Organize the information in a chart.

Pipe	Rate	Time to Fill Together	Fractional Part of the Job Done
Faster	$\dfrac{1}{x}$	4	$\dfrac{4}{x}$
Slower	$\dfrac{1}{x + .5}$	4	$\dfrac{4}{x + .5}$

$$\begin{array}{ccc} \text{Part done by} & \text{part done by} & \text{one whole} \\ \text{the faster pipe} & + \quad \text{the slower pipe} & = \quad \text{job.} \end{array}$$

$$\frac{4}{x} + \frac{4}{x + .5} = 1$$

Multiply both sides by the LCD, $x(x + .5)$.

$$4(x + .5) + 4(x) = x(x + .5)$$

$$4x + 2 + 4x = x^2 + .5x$$

$$0 = x^2 - 7.5x - 2$$

$$0 = 2x^2 - 15x - 4 \quad \textit{Multiply by 2.}$$

Solve for x using the quadratic formula with $a = 2$, $b = -15$, and $c = -4$.

$$x = \frac{-(-15) \pm \sqrt{(-15)^2 - 4(2)(-4)}}{2(2)}$$

$$= \frac{15 \pm \sqrt{257}}{4}$$

$$x = \frac{15 + \sqrt{257}}{4} \approx 7.8 \quad \text{or} \quad x = \frac{15 - \sqrt{257}}{4} \approx -.3$$

Discard the negative solution. The faster pipe takes about 7.8 hours to fill the tank by itself. The slower pipe takes about $7.8 + .5 = 8.3$ hours to fill the tank by itself.

49. Let $s = 213$ in the equation.

$$s = -16t^2 + 128t$$

$$213 = -16t^2 + 128t$$

$$0 = -16t^2 + 128t - 213$$

Here $a = -16$, $b = 128$, and $c = -213$.

$$t = \frac{-b \pm \sqrt{b^2 - 4ac}}{2a}$$

$$t = \frac{-128 \pm \sqrt{128^2 - 4(-16)(-213)}}{2(-16)}$$

$$= \frac{-128 \pm \sqrt{16,384 - 13,632}}{-32}$$

$$= \frac{-128 \pm \sqrt{2752}}{-32}$$

$$t = \frac{-128 + \sqrt{2752}}{-32} \approx 2.4 \text{ or}$$

$$t = \frac{-128 - \sqrt{2752}}{-32} \approx 5.6$$

The ball will be 213 feet from the ground after 2.4 seconds and again after 5.6 seconds.

51. When the rock is 400 ft from ground level,

$$s = -16t^2 + 160t$$

can be written as

$$400 = -16t^2 + 160t.$$

Solve the equation for time, t.

$$16t^2 - 160t + 400 = 0$$
$$t^2 - 10t + 25 = 0$$
$$a = 1, b = -10, c = 25$$
$$t = \frac{-b \pm \sqrt{b^2 - 4ac}}{2a}$$
$$t = \frac{-(-10) \pm \sqrt{(-10)^2 - 4(1)(25)}}{2(1)}$$
$$= \frac{10 \pm \sqrt{100 - 100}}{2}$$
$$= \frac{10 \pm \sqrt{0}}{2} = \frac{10}{2} = 5$$

The rock reaches a height of 400 feet after 5 seconds. This is its maximum height since this is the only time it reaches 400 feet.

53. $f(x) = -88x^2 + 17,108x - 813,360$
$$16,000 = -88x^2 + 17,108x - 813,360$$
$$0 = -88x^2 + 17,108x - 829,360$$
$$0 = 22x^2 - 4277x + 207,340$$
$$x = \frac{-(-4277) \pm \sqrt{(-4277)^2 - 4(22)(207,340)}}{2(22)}$$
$$x = \frac{4277 \pm \sqrt{46,809}}{44}$$
$$\approx 102 \text{ or } 92$$

$x = 92$ represents 1992 and $x = 102$ represents 2002, but 2002 is too far from the given data to estimate. Therefore, the sales reached 16,000 thousand in 1992.

55. $2x^2 - x + 1 = 0$
$a = 2, b = -1, c = 1$
$$b^2 - 4ac = (-1)^2 - 4(2)(1)$$
$$= 1 - 8 = -7$$

Since the discriminant is negative, the solutions are (d) two different imaginary numbers.

57. $6m^2 + 7m - 3 = 0$
$a = 6, b = 7, c = -3$
$$b^2 - 4ac = 7^2 - 4(6)(-3)$$
$$= 49 + 72 = 121$$

Since $11^2 = 121$, the discriminant is positive and the square of an integer. The solutions are (a) two different rational numbers.

59. $x^2 + 4x = -4$
$$x^2 + 4x + 4 = 0$$
$a = 1, b = 4, c = 4$
$$b^2 - 4ac = 4^2 - 4(1)(4)$$
$$= 16 - 16 = 0$$

Since the discriminant is zero, the solution is (b) exactly one rational number.

61. $$9t^2 = 30t - 15$$
$$9t^2 - 30t + 15 = 0$$
$a = 9, b = -30, c = 15$
$$b^2 - 4ac = (-30)^2 - 4(9)(15)$$
$$= 900 - 540 = 360$$

Since the discriminant is positive but not the square of an integer, the solutions are (c) two different irrational numbers.

63. $p^2 + bp + 25 = 0$
For there to be only one rational solution, $b^2 - 4ac$ must equal zero.
Since $a = 1$ and $c = 25$,
$$b^2 - 4(1)(25) = 0$$
$$b^2 - 100 = 0$$
$$b^2 = 100$$
$$b = \pm\sqrt{100}$$
$$b = 10 \text{ or } b = -10.$$

65. $am^2 + 8m + 1 = 0$
For there to be only one rational solution, $b^2 - 4ac$ must equal zero.
Since $b = 8$ and $c = 1$,
$$8^2 - 4(a)(1) = 0$$
$$64 - 4a = 0$$
$$-4a = -64$$
$$a = 16.$$

67. $9y^2 - 30y + c = 0$
Again, $b^2 - 4ac$ must equal zero.
Since $a = 9$ and $b = -30$,
$$(-30)^2 - 4(9)(c) = 0$$
$$900 - 36c = 0$$
$$-36c = -900$$
$$c = 25.$$

69. No, because an irrational solution occurs only if the discriminant is positive, but not the square of an integer. In that case, there will be two irrational solutions.

71. For $24x^2 - 34x - 45 = 0$, the discriminant is

$$b^2 - 4ac = (-34)^2 - 4(24)(-45)$$
$$= 1156 + 4320$$
$$= 5476 \text{ or } 74^2.$$

Since the discriminant is positive and the square of an integer, the polynomial can be factored.

$$24x^2 - 34x - 45 = (6x + 5)(4x - 9)$$

73. For $36x^2 + 21x - 24 = 0$, the discriminant is

$$b^2 - 4ac = 21^2 - 4(36)(-24)$$
$$= 441 + 3456$$
$$= 3897.$$

Since the discriminant is positive, but not the square of an integer, the polynomial $36x^2 + 21x - 24$ cannot be factored.

75. For $12x^2 - 83x - 7 = 0$, the discriminant is

$$b^2 - 4ac = (-83)^2 - 4(12)(-7)$$
$$= 6889 + 336$$
$$= 7225 \text{ or } 85^2.$$

Since the discriminant is positive and the square of an integer, the polynomial can be factored.

$$12x^2 - 83x - 7 = (12x + 1)(x - 7)$$

77. Substitute $1 + 5i$ for r and $1 - 5i$ for s in the equation.

$$(x - r)(x - s) = 0$$
$$[x - (1 + 5i)][x - (1 - 5i)] = 0$$

78. $(x - 1 - 5i)(x - 1 + 5i) = 0$
$$[(x - 1) - 5i][(x - 1) + 5i] = 0$$

79. $(x - 1)^2 - (5i)^2 = 0$
$$(x^2 - 2x + 1) - 25i^2 = 0$$
$$x^2 - 2x + 1 - 25(-1) = 0$$
$$x^2 - 2x + 1 + 25 = 0$$
$$x^2 - 2x + 26 = 0$$

80. $x^2 - 2x + 26 = 0$
$$a = 1, b = -2, c = 26$$
$$x = \frac{-b \pm \sqrt{b^2 - 4ac}}{2a}$$
$$= \frac{-(-2) \pm \sqrt{(-2)^2 - 4(1)(26)}}{2(1)}$$
$$= \frac{2 \pm \sqrt{4 - 104}}{2} = \frac{2 \pm \sqrt{-100}}{2}$$

$$= \frac{2 \pm 10i}{2} = \frac{2(1 \pm 5i)}{2} = 1 \pm 5i$$

81. For $4x^2 + bx - 3 = 0$ and the solution $x_1 = -\frac{5}{2}$, the product of the solutions is $\frac{c}{a}$ or $-\frac{3}{4}$.

Therefore,

$$\left(-\frac{5}{2}\right)x_2 = -\frac{3}{4}$$
$$x_2 = -\frac{2}{5}\left(-\frac{3}{4}\right) = \frac{3}{10}.$$

Since the sum of the solutions is $-\frac{b}{a}$ or $-\frac{b}{4}$,

$$-\frac{5}{2} + \frac{3}{10} = -\frac{b}{4}.$$

Multiply by 20.

$$20\left(-\frac{5}{2}\right) + 20\left(\frac{3}{10}\right) = 20\left(-\frac{b}{4}\right)$$
$$-50 + 6 = -5b$$
$$-44 = -5b$$
$$\frac{44}{5} = b$$

Section 8.3

1. $\dfrac{14}{x} = x - 5$

This is a rational equation, so multiply both sides by the LCD, x.

3. $\left(r^2 + r\right)^2 - 8\left(r^2 + r\right) + 12 = 0$

This is quadratic in form, so substitute a variable for $r^2 + r$.

5. Many answers are possible. See the chart in the text for possible reasons for agreeing or not.

7. $\dfrac{14}{x} = x - 5$

To clear the fraction, multiply each term by the LCD, x.

$$x\left(\frac{14}{x}\right) = x(x) - 5x$$
$$14 = x^2 - 5x$$
$$-x^2 + 5x + 14 = 0$$
$$x^2 - 5x - 14 = 0$$
$$(x + 2)(x - 7) = 0$$

$$x + 2 = 0 \quad \text{or} \quad x - 7 = 0$$
$$x = -2 \quad \text{or} \qquad x = 7$$

Check $x = -2$: $\quad -7 \overset{?}{=} -7 \quad$ *True*

Check $x = 7$: $\quad\quad 2 \overset{?}{=} 2 \quad$ *True*

Solution set: $\{-2, 7\}$

9.
$$1 - \frac{3}{x} - \frac{28}{x^2} = 0$$

Multiply by the LCD, x^2.

$$x^2(1) - x^2\left(\frac{3}{x}\right) - x^2\left(\frac{28}{x^2}\right) = x^2 \cdot 0$$
$$x^2 - 3x - 28 = 0$$
$$(x+4)(x-7) = 0$$

$x + 4 = 0$ or $x - 7 = 0$

$x = -4$ or $x = 7$

Check $x = -4$: $1 + \frac{3}{4} - \frac{7}{4} \overset{?}{=} 0$ *True*

Check $x = 7$: $1 - \frac{3}{7} - \frac{4}{7} \overset{?}{=} 0$ *True*

Solution set: $\{-4, 7\}$

11.
$$\frac{1}{x} + \frac{2}{x+2} = \frac{17}{35}$$

Multiply by the LCD, $35x(x+2)$.

$$35x(x+2)\left(\frac{1}{x}\right) + 35x(x+2)\left(\frac{2}{x+2}\right)$$
$$= 35x(x+2)\left(\frac{17}{35}\right)$$
$$35(x+2) + 35x(2) = 17x(x+2)$$
$$35x + 70 + 70x = 17x^2 + 34x$$
$$70 + 105x = 17x^2 + 34x$$
$$0 = 17x^2 - 71x - 70$$
$$0 = (17x + 14)(x - 5)$$

$17x + 14 = 0$ or $x - 5 = 0$

$x = -\dfrac{14}{17}$ or $x = 5$

Check $x = -\dfrac{14}{17}$: $-\dfrac{17}{14} + \dfrac{17}{10} \overset{?}{=} \dfrac{17}{35}$ *True*

Check $x = 5$: $\dfrac{1}{5} + \dfrac{2}{7} \overset{?}{=} \dfrac{17}{35}$ *True*

Solution set: $\left\{-\dfrac{14}{17}, 5\right\}$

13.
$$\frac{2}{x+1} + \frac{3}{x+2} = \frac{7}{2}$$

Multiply by the LCD, $2(x+1)(x+2)$.

$$2(x+1)(x+2)\left(\frac{2}{x+1} + \frac{3}{x+2}\right)$$
$$= 2(x+1)(x+2)\left(\frac{7}{2}\right)$$
$$2(x+2)(2) + 2(x+1)(3)$$
$$= (x+1)(x+2)(7)$$
$$4x + 8 + 6x + 6 = (x^2 + 3x + 2)(7)$$
$$10x + 14 = 7x^2 + 21x + 14$$
$$7x^2 + 11x = 0$$
$$x(7x + 11) = 0$$

$x = 0$ or $7x + 11 = 0$

$$x = -\frac{11}{7}$$

Check $x = -\dfrac{11}{7}$: $-\dfrac{7}{2} + 7 \overset{?}{=} \dfrac{7}{2}$ *True*

Check $x = 0$: $2 + \dfrac{3}{2} \overset{?}{=} \dfrac{7}{2}$ *True*

Solution set: $\left\{-\dfrac{11}{7}, 0\right\}$

15.
$$\frac{3}{2x} - \frac{1}{2(x+2)} = 1$$

Multiply by the LCD, $2x(x+2)$.

$$2x(x+2)\left(\frac{3}{2x} - \frac{1}{2(x+2)}\right)$$
$$= 2x(x+2) \cdot 1$$
$$3(x+2) - x(1) = 2x(x+2)$$
$$3x + 6 - x = 2x^2 + 4x$$
$$0 = 2x^2 + 2x - 6$$
$$0 = x^2 + x - 3$$

Use $a = 1$, $b = 1$, $c = -3$ in the quadratic formula.

$$x = \frac{-b \pm \sqrt{b^2 - 4ac}}{2a}$$

$$x = \frac{-1 \pm \sqrt{1^2 - 4(1)(-3)}}{2(1)}$$

$$= \frac{-1 \pm \sqrt{1 + 12}}{2} = \frac{-1 \pm \sqrt{13}}{2}$$

Use a calculator to check both potential solutions. Both solutions check.

Solution set: $\left\{\dfrac{-1 + \sqrt{13}}{2}, \dfrac{-1 - \sqrt{13}}{2}\right\}$

17.   The grader's rate $= \dfrac{1 \text{ job}}{\text{time to finish}}$

$$= \frac{1}{m} \text{ job per hour.}$$

19.   Let $x =$ rate of the boat in still water.
With the speed of the current at 15 mph, then
$x - 15 =$ rate going upstream and
$x + 15 =$ rate going downstream.
Complete a table using the information in the problem, the rates given above, and the formula
$d = rt$ or $t = \dfrac{d}{r}$.

	d	r	t
Upstream	4	$x - 15$	$\dfrac{4}{x-15}$
Downstream	16	$x + 15$	$\dfrac{16}{x+15}$

The time, 48 min, is written as $\dfrac{48}{60} = \dfrac{4}{5}$ hr. The time upstream plus the time downstream equals

continued

$\frac{4}{5}$. So, from the table, the equation is written as

$$\frac{4}{x-15} + \frac{16}{x+15} = \frac{4}{5}.$$

Multiply by the LCD, $5(x-15)(x+15)$.

$$5(x-15)(x+15)\left(\frac{4}{x-15} + \frac{16}{x+15}\right)$$
$$= 5(x-15)(x+15)\cdot\frac{4}{5}$$
$$20(x+15) + 80(x-15)$$
$$= 4(x-15)(x+15)$$
$$20x + 300 + 80x - 1200$$
$$= 4(x^2 - 225)$$
$$100x - 900 = 4x^2 - 900$$
$$0 = 4x^2 - 100x$$
$$0 = 4x(x-25)$$

$$4x = 0 \quad \text{or} \quad x - 25 = 0$$
$$x = 0 \quad \text{or} \quad x = 25$$

Reject $x = 0$ mph as a possible boat speed. Yoshiaki's boat had a top speed of 25 mph.

21. Let x = Harry's average speed.

$x - 20$ = Karen's average speed.

	d	r	t
Harry	300	x	$\frac{300}{x}$
Karen	300	$x-20$	$\frac{300}{x-20}$

It takes Harry $1\frac{1}{4}$ or $\frac{5}{4}$ hours less time than Karen.

$$\frac{300}{x} = \frac{300}{x-20} - \frac{5}{4}$$

Multiply by the LCD, $4x(x-20)$.

$$4x(x-20)\left(\frac{300}{x}\right) = 4x(x-20)\left(\frac{300}{x-20} - \frac{5}{4}\right)$$
$$1200(x-20) = 4x(300) - x(x-20)\cdot 5$$
$$1200x - 24,000 = 1200x - 5x^2 + 100x$$
$$5x^2 - 100x - 24,000 = 0$$
$$x^2 - 20x - 4800 = 0$$
$$(x-80)(x+60) = 0$$

$$x - 80 = 0 \quad \text{or} \quad x + 60 = 0$$
$$x = 80 \quad \text{or} \quad x = -60$$

Reject $x = -60$. Harry's average speed is 80 km/hr.

23. Let x = the number of minutes it takes for the cold water tap alone to fill the washer.

$x + 9$ = the number of minutes it takes for the hot water tap alone to fill the washer.

Working together, both taps can fill the washer in 6 minutes. Complete a chart using the above information.

Tap	Rate	Time	Fractional Part of Washer Filled
Cold	$\frac{1}{x}$	6	$\frac{1}{x}(6)$
Hot	$\frac{1}{x+9}$	6	$\frac{1}{x+9}(6)$

Since together the hot and cold taps fill one washer, the sum of their fractional parts is 1; that is,

$$\frac{6}{x} + \frac{6}{x+9} = 1.$$

Multiply by the LCD, $x(x+9)$.

$$x(x+9)\left(\frac{6}{x} + \frac{6}{x+9}\right) = x(x+9)\cdot 1$$
$$6(x+9) + 6x = x(x+9)$$
$$6x + 54 + 6x = x^2 + 9x$$
$$0 = x^2 - 3x - 54$$
$$0 = (x-9)(x+6)$$

$$x - 9 = 0 \quad \text{or} \quad x + 6 = 0$$
$$x = 9 \quad \text{or} \quad x = -6$$

Reject -6 as a possible time. The cold water tap can fill the washer in 9 minutes.

25.
$$2x = \sqrt{11x+3}$$
$$(2x)^2 = \left(\sqrt{11x+3}\right)^2$$
$$4x^2 = 11x + 3$$
$$4x^2 - 11x - 3 = 0$$
$$(4x+1)(x-3) = 0$$

$$4x + 1 = 0 \quad \text{or} \quad x - 3 = 0$$
$$x = -\frac{1}{4} \quad \text{or} \quad x = 3$$

Check $x = -\frac{1}{4}$: $-\frac{1}{2} \overset{?}{=} \sqrt{\frac{1}{4}}$ *False*

Check $x = 3$: $6 \overset{?}{=} \sqrt{36}$ *True*

Solution set: $\{3\}$

27.
$$3y = (16 - 10y)^{1/2}$$
$$(3y)^2 = \left[(16 - 10y)^{1/2}\right]^2$$
$$9y^2 = 16 - 10y$$
$$9y^2 + 10y - 16 = 0$$
$$(9y - 8)(y + 2) = 0$$

$$9y - 8 = 0 \quad \text{or} \quad y + 2 = 0$$
$$y = \frac{8}{9} \quad \text{or} \quad y = -2$$

Check $y = \dfrac{8}{9}$: $\dfrac{8}{3} \overset{?}{=} \sqrt{\dfrac{64}{9}}$ *True*

Check $y = -2$: $-6 \overset{?}{=} \sqrt{36}$ *False*

Solution set: $\left\{ \dfrac{8}{9} \right\}$

29.
$$p - 2\sqrt{p} = 8$$
$$p - 8 = 2\sqrt{p}$$
$$(p - 8)^2 = \left(2\sqrt{p}\right)^2$$
$$p^2 - 16p + 64 = 4p$$
$$p^2 - 20p + 64 = 0$$
$$(p - 4)(p - 16) = 0$$

$$p - 4 = 0 \quad \text{or} \quad p - 16 = 0$$
$$p = 4 \quad \text{or} \quad p = 16$$

Check $p = 4$: $4 - 4 \overset{?}{=} 8$ *False*

Check $p = 16$: $16 - 8 \overset{?}{=} 8$ *True*

Solution set: $\{16\}$

31.
$$m = \sqrt{\frac{6 - 13m}{5}}$$
$$m^2 = \frac{6 - 13m}{5}$$
$$5m^2 = 6 - 13m$$
$$5m^2 + 13m - 6 = 0$$
$$(5m - 2)(m + 3) = 0$$

$$5m - 2 = 0 \quad \text{or} \quad m + 3 = 0$$
$$m = \frac{2}{5} \quad \text{or} \quad m = -3$$

Check $m = \dfrac{2}{5}$: $\dfrac{2}{5} \overset{?}{=} \sqrt{\dfrac{4}{25}}$ *True*

Check $m = -3$: $-3 \overset{?}{=} \sqrt{9}$ *False*

Solution set: $\left\{ \dfrac{2}{5} \right\}$

33.
$$.126\pi r = \sqrt{4.7r}$$
$$(.126\pi r)^2 = \left(\sqrt{4.7r}\right)^2$$
$$(.126\pi)^2 r^2 = 4.7r$$
$$(.126\pi)^2 r^2 - 4.7r = 0$$
$$r\left[(.126\pi)^2 r - 4.7\right] = 0$$

$$r = 0 \quad \text{or} \quad r = \frac{4.7}{(.126\pi)^2} \approx 30$$

Reject $r = 0$, the distance is about 30 meters.

35. $t^4 - 18t^2 + 81 = 0$

Let $u = t^2$, so $u^2 = t^4$. The equation becomes
$$u^2 - 18u + 81 = 0.$$
$$(u - 9)^2 = 0$$
$$u - 9 = 0$$
$$u = 9$$

To find t, substitute t^2 for u.
$$t^2 = 9$$
$$t = 3 \quad \text{or} \quad t = -3$$

Check $t = \pm 3$: $81 - 162 + 81 \overset{?}{=} 0$ *True*

Solution set: $\{-3, 3\}$

37. $4k^4 - 13k^2 + 9 = 0$

Let $u = k^2$ and $u^2 = k^4$ to get
$$4u^2 - 13u + 9 = 0$$
$$(4u - 9)(u - 1) = 0$$

$$4u - 9 = 0 \quad \text{or} \quad u - 1 = 0$$
$$u = \frac{9}{4} \quad \text{or} \quad u = 1.$$

To find k, substitute k^2 for u.

$$k^2 = \frac{9}{4} \quad \text{or} \quad k^2 = 1$$

$$k = \pm \frac{3}{2} \quad \text{or} \quad k = \pm 1$$

Check $k = \pm \dfrac{3}{2}$: $\dfrac{81}{4} - \dfrac{117}{4} + 9 \overset{?}{=} 0$ *True*

Check $k = \pm 1$: $4 - 13 + 9 \overset{?}{=} 0$ *True*

Solution set: $\left\{ -\dfrac{3}{2}, -1, 1, \dfrac{3}{2} \right\}$

39. $(x+3)^2 + 5(x+3) + 6 = 0$

Let $u = x + 3$, so $u^2 = (x+3)^2$.

$$u^2 + 5u + 6 = 0$$
$$(u+3)(u+2) = 0$$

$u + 3 = 0 \quad$ or $\quad u + 2 = 0$

$u = -3 \quad$ or $\qquad u = -2$

To find x, substitute $x + 3$ for u.

$x + 3 = -3 \quad$ or $\quad x + 3 = -2$

$x = -6 \quad$ or $\qquad x = -5$

Check $x = -6$: $\quad 9 - 15 + 6 \overset{?}{=} 0 \quad$ *True*

Check $x = -5$: $\quad 4 - 10 + 6 \overset{?}{=} 0 \quad$ *True*

Solution set: $\{-6, -5\}$

41. $(t+5)^2 + 6 = 7(t+5)$

Let $u = t + 5$, so $u^2 = (t+5)^2$.

$$u^2 + 6 = 7u$$
$$u^2 - 7u + 6 = 0$$
$$(u-6)(u-1) = 0$$

$u - 6 = 0 \quad$ or $\quad u - 1 = 0$

$u = 6 \qquad\qquad u = 1$

To find t, substitute $t + 5$ for u.

$t + 5 = 6 \quad$ or $\quad t + 5 = 1$

$t = 1 \qquad\qquad t = -4$

Check $t = 1$: $\quad 36 + 6 \overset{?}{=} 42 \quad$ *True*

Check $t = -4$: $\quad 1 + 6 \overset{?}{=} 7 \quad$ *True*

Solution set: $\{-4, 1\}$

43. $2 + \dfrac{5}{3k-1} = \dfrac{-2}{(3k-1)^2}$

Let $u = 3k - 1$, so $u^2 = (3k-1)^2$.

$$2 + \frac{5}{u} = -\frac{2}{u^2}$$

Multiply by the LCD, u^2.

$$u^2\left(2 + \frac{5}{u}\right) = u^2\left(-\frac{2}{u^2}\right)$$
$$2u^2 + 5u = -2$$
$$2u^2 + 5u + 2 = 0$$
$$(2u+1)(u+2) = 0$$

$2u + 1 = 0 \quad$ or $\quad u + 2 = 0$

$u = -\dfrac{1}{2} \qquad\qquad u = -2$

To find k, substitute $3k - 1$ for u.

$3k - 1 = -\dfrac{1}{2} \quad$ or $\quad 3k - 1 = -2$

$3k = \dfrac{1}{2} \qquad\qquad 3k = -1$

$k = \dfrac{1}{6} \quad$ or $\qquad k = -\dfrac{1}{3}$

Check $k = \dfrac{1}{6}$: $\quad 2 - 10 \overset{?}{=} -8 \quad$ *True*

Check $k = -\dfrac{1}{3}$: $\quad 2 - \dfrac{5}{2} \overset{?}{=} -\dfrac{1}{2} \quad$ *True*

Solution set: $\left\{-\dfrac{1}{3}, \dfrac{1}{6}\right\}$

45. $2 - 6(m-1)^{-2} = (m-1)^{-1}$

Let $u = m - 1$ to get

$$2 - 6u^{-2} = u^{-1}$$

or $\qquad 2 - \dfrac{6}{u^2} = \dfrac{1}{u}$.

Multiply by the LCD, u^2.

$$2u^2 - 6 = u$$
$$2u^2 - u - 6 = 0$$
$$(2u+3)(u-2) = 0$$

$2u + 3 = 0 \quad$ or $\quad u - 2 = 0$

$u = -\dfrac{3}{2} \quad$ or $\qquad u = 2$

To find m, substitute $m - 1$ for u.

$m - 1 = -\dfrac{3}{2} \quad$ or $\quad m - 1 = 2$

$m = -\dfrac{1}{2} \quad$ or $\qquad m = 3$

Check $m = -\dfrac{1}{2}$: $\quad 2 - \dfrac{8}{3} \overset{?}{=} -\dfrac{2}{3} \quad$ *True*

Check $m = 3$: $\quad 2 - \dfrac{3}{2} \overset{?}{=} \dfrac{1}{2} \quad$ *True*

Solution set: $\left\{-\dfrac{1}{2}, 3\right\}$

47. $4k^{4/3} - 13k^{2/3} + 9 = 0$

Let $x = k^{2/3}$, so $x^2 = k^{4/3}$.

$$4x^2 - 13x + 9 = 0$$
$$(4x-9)(x-1) = 0$$

$4x - 9 = 0 \quad$ or $\quad x - 1 = 0$

$x = \dfrac{9}{4} \quad$ or $\qquad x = 1$

$k^{2/3} = \dfrac{9}{4} \quad$ or $\quad k^{2/3} = 1$

$\left(k^{2/3}\right)^3 = \left(\dfrac{9}{4}\right)^3 \quad$ or $\quad \left(k^{2/3}\right)^3 = 1^3$

$k^2 = \dfrac{729}{64} \quad$ or $\quad k^2 = 1$

$k = \pm\sqrt{\dfrac{729}{64}} \quad$ or $\quad k = \pm\sqrt{1}$

$k = \pm\dfrac{27}{8} \quad$ or $\quad k = \pm 1$

Check $k = \pm \dfrac{27}{8}$: $\dfrac{81}{4} - \dfrac{117}{4} + 9 \stackrel{?}{=} 0$ *True*

Check $k = \pm 1$: $4 - 13 + 9 \stackrel{?}{=} 0$ *True*

Solution set: $\left\{ -\dfrac{27}{8}, -1, 1, \dfrac{27}{8} \right\}$

49. $x^{2/3} + x^{1/3} - 2 = 0$

Let $u = x^{1/3}$, so $u^2 = x^{2/3}$.

$u^2 + u - 2 = 0$

$(u + 2)(u - 1) = 0$

$u + 2 = 0$ or $u - 1 = 0$

$u = -2$ or $u = 1$

To find x, substitute $x^{1/3}$ for u.

$x^{1/3} = -2$ or $x^{1/3} = 1$

Cube both sides of each equation.

$\left(x^{1/3} \right)^3 = (-2)^3$ $\left(x^{1/3} \right)^3 = 1^3$

$x = -8$ or $x = 1$

Check $x = -8$: $4 - 2 - 2 \stackrel{?}{=} 0$ *True*

Check $x = 1$: $1 + 1 - 2 \stackrel{?}{=} 0$ *True*

Solution set: $\{ -8, 1 \}$

51. $2\left(1 + \sqrt{y} \right)^2 = 13 \left(1 + \sqrt{y} \right) - 6$

Let $u = 1 + \sqrt{y}$.

$2u^2 = 13u - 6$

$2u^2 - 13u + 6 = 0$

$(2u - 1)(u - 6) = 0$

$2u - 1 = 0$ or $u - 6 = 0$

$u = \dfrac{1}{2}$ or $u = 6$

Replace u with $1 + \sqrt{y}$.

$1 + \sqrt{y} = \dfrac{1}{2}$ or $1 + \sqrt{y} = 6$

$\sqrt{y} = -\dfrac{1}{2}$ $\sqrt{y} = 5$

Not possible, $y = 25$

since $\sqrt{y} \geq 0$.

Check $y = 25$: $72 \stackrel{?}{=} 78 - 6$ *True*

Solution set: $\{ 25 \}$

53. $2x^4 + x^2 - 3 = 0$

Let $m = x^2$, so $m^2 = x^4$.

$2m^2 + m - 3 = 0$

$(2m + 3)(m - 1) = 0$

$2m + 3 = 0$ or $m - 1 = 0$

$m = -\dfrac{3}{2}$ or $m = 1$

To find x, substitute x^2 for m.

$x^2 = -\dfrac{3}{2}$ or $x^2 = 1$

$x = \pm \sqrt{-\dfrac{3}{2}}$ $x = \pm \sqrt{1}$

$x = \pm \dfrac{\sqrt{3}}{\sqrt{2}} \cdot \dfrac{\sqrt{2}}{\sqrt{2}} i$ $x = \pm 1$

$x = \pm \dfrac{\sqrt{6}}{2} i$

Check $x = \pm \dfrac{\sqrt{6}}{2} i$: $\dfrac{9}{2} - \dfrac{3}{2} - 3 \stackrel{?}{=} 0$ *True*

Check $x = \pm 1$: $2 + 1 - 3 \stackrel{?}{=} 0$ *True*

Solution set: $\left\{ -1, 1, -\dfrac{\sqrt{6}}{2} i, \dfrac{\sqrt{6}}{2} i \right\}$

55. The solutions stated are the values of u, $u = \dfrac{1}{2}$ or $u = 1$. We must give the values of the original variable, m. Substitute $m - 1$ for u.

$u = \dfrac{1}{2}$ or $u = 1$

$m - 1 = \dfrac{1}{2}$ $m - 1 = 1$

$m = \dfrac{3}{2}$ or $m = 2$

Solution set: $\left\{ \dfrac{3}{2}, 2 \right\}$

57. $x^4 - 16x^2 + 48 = 0$

Let $m = x^2$, so $m^2 = x^4$.

$m^2 - 16m + 48 = 0$

$(m - 4)(m - 12) = 0$

$m - 4 = 0$ or $m - 12 = 0$

$m = 4$ or $m = 12$

To find x, substitute x^2 for m.

$x^2 = 4$ or $x^2 = 12$

$x = \pm \sqrt{4}$ $x = \pm \sqrt{12}$

$x = \pm 2$ or $x = \pm 2\sqrt{3}$

Check $x = \pm 2$: $16 - 64 + 48 \stackrel{?}{=} 0$ *True*

Check $x = \pm 2\sqrt{3}$: $144 - 192 + 48 \stackrel{?}{=} 0$ *True*

Solution set: $\left\{ -2\sqrt{3}, -2, 2, 2\sqrt{3} \right\}$

59.
$$\sqrt{2x+3} = 2 + \sqrt{x-2}$$
Square both sides.
$$\left(\sqrt{2x+3}\right)^2 = \left(2+\sqrt{x-2}\right)^2$$
$$2x+3 = 4 + 4\sqrt{x-2} + (x-2)$$
$$2x+3 = x + 2 + 4\sqrt{x-2}$$
Isolate the radical term on one side.
$$x + 1 = 4\sqrt{x-2}$$
Square both sides again.
$$(x+1)^2 = \left(4\sqrt{x-2}\right)^2$$
$$x^2 + 2x + 1 = 16(x-2)$$
$$x^2 + 2x + 1 = 16x - 32$$
$$x^2 - 14x + 33 = 0$$
$$(x-11)(x-3) = 0$$
$$x - 11 = 0 \quad \text{or} \quad x - 3 = 0$$
$$x = 11 \quad \text{or} \quad x = 3$$
Check $x = 11$: $\sqrt{25} \overset{?}{=} 2 + \sqrt{9}$ *True*

Check $x = 3$: $\sqrt{9} \overset{?}{=} 2 + \sqrt{1}$ *True*

Solution set: $\{3, 11\}$

61. $2m^6 + 11m^3 + 5 = 0$
Let $y = m^3$, so $y^2 = m^6$.
$$2y^2 + 11y + 5 = 0$$
$$(2y+1)(y+5) = 0$$
$$2y + 1 = 0 \quad \text{or} \quad y + 5 = 0$$
$$y = -\frac{1}{2} \quad \text{or} \quad y = -5$$
To find m, substitute m^3 for y.
$$m^3 = -\frac{1}{2} \quad \text{or} \quad m^3 = -5$$
Take the cube root of both sides of each equation.
$$m = \sqrt[3]{-\frac{1}{2}} \qquad \text{or} \quad m = \sqrt[3]{-5}$$
$$m = -\sqrt[3]{\frac{1}{2}} \qquad\qquad m = -\sqrt[3]{5}$$
$$= -\frac{\sqrt[3]{1}}{\sqrt[3]{2}} \cdot \frac{\sqrt[3]{2^2}}{\sqrt[3]{2^2}}$$
$$= -\frac{\sqrt[3]{4}}{2}$$

Check $m = -\dfrac{\sqrt[3]{4}}{2}$: $\dfrac{1}{2} - \dfrac{11}{2} + 5 \overset{?}{=} 0$ *True*

Check $m = -\sqrt[3]{5}$: $50 - 55 + 5 \overset{?}{=} 0$ *True*

Solution set: $\left\{ -\sqrt[3]{5}, -\dfrac{\sqrt[3]{4}}{2} \right\}$

63. $2 - (y-1)^{-1} = 6(y-1)^{-2}$
Let $m = (y-1)^{-1}$, so $m^2 = (y-1)^{-2}$.
$$2 - m = 6m^2$$
$$0 = 6m^2 + m - 2$$
$$0 = (3m+2)(2m-1)$$
$$3m + 2 = 0 \quad \text{or} \quad 2m - 1 = 0$$
$$m = -\frac{2}{3} \quad \text{or} \qquad m = \frac{1}{2}$$
To find y, substitute $(y-1)^{-1}$ for m.
$$(y-1)^{-1} = -\frac{2}{3} \quad \text{or} \quad (y-1)^{-1} = \frac{1}{2}$$
$$\frac{1}{y-1} = -\frac{2}{3} \qquad\qquad \frac{1}{y-1} = \frac{1}{2}$$
$$3 = -2y + 2 \qquad\qquad 2 = y - 1$$
$$2y = -1 \qquad\qquad\qquad y = 3$$
$$y = -\frac{1}{2}$$

Check $y = -\dfrac{1}{2}$: $2 + \dfrac{2}{3} \overset{?}{=} \dfrac{8}{3}$ *True*

Check $y = 3$: $2 - \dfrac{1}{2} \overset{?}{=} \dfrac{3}{2}$ *True*

Solution set: $\left\{ -\dfrac{1}{2}, 3 \right\}$

For Exercises 65–69, use the equation
$$\frac{x^2}{(x-3)^2} + \frac{3x}{x-3} - 4 = 0.$$

65. Substituting 3 for x would cause both denominators to be 0, and division by 0 is undefined.

66. $(x-3)^2\left(\dfrac{x^2}{(x-3)^2}\right) + (x-3)^2\left(\dfrac{3x}{x-3}\right)$
$$- (x-3)^2 \cdot 4 = (x-3)^2 \cdot 0$$
$$x^2 + 3x(x-3) - 4(x^2 - 6x + 9) = 0$$
$$x^2 + 3x^2 - 9x - 4x^2 + 24x - 36 = 0$$
$$15x - 36 = 0$$
$$15x = 36$$
$$x = \frac{36}{15} = \frac{12}{5}$$
The solution is $\dfrac{12}{5}$.

67.
$$\frac{x^2}{(x-3)^2} + \frac{3x}{x-3} - 4 = 0$$
$$\left(\frac{x}{x-3}\right)^2 + 3\left(\frac{x}{x-3}\right) - 4 = 0$$

68. If a fraction is equal to 1, then the numerator must be equal to the denominator. But the numerator can never equal the denominator, since the denominator is 3 less than the numerator.

69. From Exercise 67,

$$\left(\frac{x}{x-3}\right)^2 + 3\left(\frac{x}{x-3}\right) - 4 = 0.$$

Let $t = \dfrac{x}{x-3}$, so $t^2 = \left(\dfrac{x}{x-3}\right)^2$.

$$t^2 + 3t - 4 = 0$$
$$(t-1)(t+4) = 0$$

$$t - 1 = 0 \quad \text{or} \quad t + 4 = 0$$
$$t = 1 \quad \text{or} \quad t = -4$$

To find x, substitute $\dfrac{x}{x-3}$ for t.

$$\frac{x}{x-3} = 1 \quad \text{or} \quad \frac{x}{x-3} = -4$$

The equation $\dfrac{x}{x-3} = 1$ has no solution since there is no value of x for which $x = x - 3$. (See Exercise 68.) Therefore, $t = 1$ is impossible.

$$\frac{x}{x-3} = -4$$
$$x = -4(x-3)$$
$$x = -4x + 12$$
$$5x = 12$$
$$x = \frac{12}{5}$$

Solution set: $\left\{\dfrac{12}{5}\right\}$

70. $x^2(x-3)^{-2} + 3x(x-3)^{-1} - 4 = 0$

Let $s = (x-3)^{-1}$, so $s^2 = (x-3)^{-2}$.

$$x^2 s^2 + 3xs - 4 = 0$$
$$(xs+4)(xs-1) = 0$$

$$xs + 4 = 0 \qquad \text{or} \quad xs - 1 = 0$$
$$s = -\frac{4}{x} \qquad \text{or} \qquad s = \frac{1}{x}$$

To find x, substitute $\dfrac{1}{x-3}$ for s.

$$\frac{1}{x-3} = -\frac{4}{x} \quad \text{or} \quad \frac{1}{x-3} = \frac{1}{x}$$
$$x = -4(x-3)$$
$$x = -4x + 12 \qquad\qquad x = x - 3$$
$$5x = 12 \qquad\qquad\quad 0 = -3 \quad \textit{False}$$
$$x = \frac{12}{5} \qquad \text{Thus, } s = \frac{1}{x} \text{ is}$$
$$\text{impossible.}$$

Solution set: $\left\{\dfrac{12}{5}\right\}$

Section 8.4

1. The first step in solving a formula that has the specified variable in the denominator is to multiply both sides by the LCD to clear the equation of fractions.

3. We must recognize that a formula like

$$gw^2 = kw + 24$$

is quadratic in w. So the first step is to write the formula in standard form (with 0 on one side, in decreasing powers of w). This allows us to apply the quadratic formula to solve for w.

5. No. There is only one time after the rock starts moving when it hits the ground.

7. Since the triangle is a right triangle, use the Pythagorean formula with legs m and n and hypotenuse p.

$$m^2 + n^2 = p^2$$
$$m^2 = p^2 - n^2$$
$$m = \sqrt{p^2 - n^2}$$

Only the positive square root is given since m represents the side of a triangle.

9. Solve $d = kt^2$ for t.

$$kt^2 = d$$
$$t^2 = \frac{d}{k} \qquad\qquad \textit{Divide by } k.$$
$$t = \pm\sqrt{\frac{d}{k}} \qquad \textit{Use square root property.}$$
$$= \frac{\pm\sqrt{d}}{\sqrt{k}} \cdot \frac{\sqrt{k}}{\sqrt{k}} \qquad \textit{Rationalize denominator.}$$
$$t = \frac{\pm\sqrt{dk}}{k} \qquad\quad \textit{Simplify}$$

11. Solve $F = \dfrac{kA}{v^2}$ for v.

$$v^2 F = kA \qquad\qquad \textit{Multiply by } v^2.$$
$$v^2 = \frac{kA}{F} \qquad\qquad \textit{Divide by } F.$$
$$v = \pm\sqrt{\frac{kA}{F}} \qquad \textit{Use square root property.}$$
$$= \frac{\pm\sqrt{kA}}{\sqrt{F}} \cdot \frac{\sqrt{F}}{\sqrt{F}} \qquad \textit{Rationalize denominator.}$$
$$v = \frac{\pm\sqrt{kAF}}{F} \qquad\quad \textit{Simplify}$$

13. Solve $V = \frac{1}{3}\pi r^2 h$ for r.

$3V = \pi r^2 h$ *Multiply by 3.*

$\dfrac{3V}{\pi h} = r^2$ *Divide by πh.*

$r = \pm\sqrt{\dfrac{3V}{\pi h}}$ *Use square root property.*

$= \dfrac{\pm\sqrt{3V} \cdot \sqrt{\pi h}}{\sqrt{\pi h} \cdot \sqrt{\pi h}}$ *Rationalize denominator.*

$r = \dfrac{\pm\sqrt{3\pi V h}}{\pi h}$ *Simplify*

15. Solve $At^2 + Bt = -C$ for t.

$At^2 + Bt + C = 0$

Use the quadratic formula.

$$t = \frac{-B \pm \sqrt{B^2 - 4AC}}{2A}$$

17. Solve $D = \sqrt{kh}$ for h.

$D^2 = kh$ *Square both sides.*

$\dfrac{D^2}{k} = h$ *Divide by k.*

19. Solve $p = \sqrt{\dfrac{kl}{g}}$ for l.

$p^2 = \dfrac{kl}{g}$ *Square both sides.*

$p^2 g = kl$ *Multiply by g.*

$\dfrac{p^2 g}{k} = l$ *Divide by k.*

21. If g is positive, the only way to have a real value for p is to have kl positive, since the quotient of two positive numbers is positive. If k and l have different signs, their product is negative, leading to a negative radicand.

23. $s = -16t^2 + 45t + 400$

$200 = -16t^2 + 45t + 400$ *Let s=200.*

$0 = -16t^2 + 45t + 200$

$0 = 16t^2 - 45t - 200$ *Multiply by –1.*

$a = 16,\ b = -45,\ c = -200$

$t = \dfrac{-b \pm \sqrt{b^2 - 4ac}}{2a}$

$t = \dfrac{-(-45) \pm \sqrt{(-45)^2 - 4(16)(-200)}}{2(16)}$

$= \dfrac{45 \pm \sqrt{2025 + 12,800}}{32}$

$= \dfrac{45 \pm \sqrt{14,825}}{32}$

$t = \dfrac{45 + \sqrt{14,825}}{32} \approx 5.2$ or

$t = \dfrac{45 - \sqrt{14,825}}{32} \approx -2.4$

Reject the negative solution since time cannot be negative. The ball will reach a height of 200 feet above the ground after 5.2 seconds.

25. The height of the building could be determined by finding s when $t = 0$.

27. $s = 144t - 16t^2$

(a) Substitute 128 for s.

$128 = 144t - 16t^2$

$0 = -16t^2 + 144t - 128$

Divide by -16.

$0 = t^2 - 9t + 8$

$0 = (t - 8)(t - 1)$

$t - 8 = 0$ or $t - 1 = 0$

$t = 8$ or $t = 1$

The object will be 128 feet above the ground at two times, going up and coming down, or at 1 second and at 8 seconds.

(b) Substitute 0 for s since the ground is at 0 height.

$0 = 144t - 16t^2$

Divide by 16.

$0 = 9t - t^2$

$0 = t(9 - t)$

$t = 0$ or $9 - t = 0$

$9 = t$

The object will be on the ground at the starting point ($t = 0$). Then, it will hit the ground again at 9 seconds.

29. **(a)** From Example 4,

$$y = x - \frac{g}{1922}x^2.$$

For the moon, $g = 3320$, so

$$y = x - \frac{3320}{1922}x^2 \approx x - 1.727x^2.$$

(b) $y = .25 - 1.727(.25)^2$ *Let x=.25*

$\approx .1421$ foot

$y = .5 - 1.727(.5)^2$ *Let x=.5*

$\approx .06825$ foot

(c) $0 = x - 1.727x^2$ *Let y=0.*

$0 = (1 - 1.727x)x$

$1 - 1.727x = 0$ or $x = 0$

$x = \dfrac{1}{1.727} \approx .5790$

The vehicle is on the surface of the moon at 0 seconds and .5790 second.

(d) One time represents the time before the vehicle leaves the surface. The other time represents the time it returns to the surface.

31. Apply the Pythagorean formula.

$$(5m)^2 = (2m)^2 + (2m+3)^2$$
$$25m^2 = 4m^2 + 4m^2 + 12m + 9$$
$$17m^2 - 12m - 9 = 0$$
$$a = 17,\ b = -12,\ c = -9$$
$$m = \frac{-(-12) \pm \sqrt{(-12)^2 - 4(17)(-9)}}{2(17)}$$
$$= \frac{12 \pm \sqrt{144 + 612}}{34} = \frac{12 \pm \sqrt{756}}{34}$$
$$m = \frac{12 + \sqrt{756}}{34} \approx 1.16 \text{ or}$$
$$m = \frac{12 - \sqrt{756}}{34} \approx -.46$$

Reject the negative solution.
If $m = 1.16$, then

$$5m = 5(1.16) = 5.80,$$
$$2m = 2(1.16) = 2.32, \text{ and}$$
$$2m + 3 = 2(1.16) + 3 = 5.32.$$

The lengths of the sides of the triangle are approximately 2.3, 5.3, and 5.8.

33. Let $x =$ the length of the wire.
Use the Pythagorean formula.

$$x^2 = 100^2 + 400^2 \quad Height=400$$
$$= 10,000 + 160,000$$
$$= 170,000$$
$$x = \pm\sqrt{170,000} \approx \pm 412.3$$

Reject the negative solution. The length of the wire is about 412.3 feet.

35. Let $x =$ the distance traveled by the eastbound ship and
$x + 70 =$ the distance traveled by the southbound ship.

Since the ships are traveling at right angles to one another, the distance d between them can be found using the Pythagorean formula.

$$c^2 = a^2 + b^2$$
$$d^2 = x^2 + (x+70)^2$$

Let $d = 170$, and solve for x.

$$170^2 = x^2 + (x+70)^2$$
$$28,900 = x^2 + x^2 + 140x + 4900$$
$$0 = 2x^2 + 140x - 24,000$$
$$0 = x^2 + 70x - 12,000$$
$$0 = (x+150)(x-80)$$

$$x + 150 = 0 \quad \text{or} \quad x - 80 = 0$$
$$x = -150 \quad \text{or} \quad x = 80$$

Distance cannot be negative, so reject -150. If $x = 80$, then $x + 70 = 150$. The eastbound ship traveled 80 miles, and the southbound ship traveled 150 miles.

37. Let $x =$ length of the shorter leg;
$2x + 2 =$ length of the longer leg;
$2x + 2 + 1 =$ length of the hypotenuse.

Use the Pythagorean formula.

$$x^2 + (2x+2)^2 = (2x+3)^2$$
$$x^2 + 4x^2 + 8x + 4 = 4x^2 + 12x + 9$$
$$5x^2 + 8x + 4 = 4x^2 + 12x + 9$$
$$x^2 - 4x - 5 = 0$$
$$(x-5)(x+1) = 0$$

$$x - 5 = 0 \quad \text{or} \quad x + 1 = 0$$
$$x = 5 \qquad\qquad x = -1$$

Since x represents length, discard -1 as a solution. If $x = 5$, then

$$2x + 2 = 2(5) + 2 = 12 \text{ and}$$
$$2x + 2 + 1 = 2(5) + 3 = 13.$$

The lengths should be 5 cm, 12 cm, and 13 cm.

39. Let $x =$ the width of the rectangle;
$2x - 1 =$ the length of the rectangle.

Use the Pythagorean formula.

$$x^2 + (2x-1)^2 = (2.5)^2$$
$$x^2 + 4x^2 - 4x + 1 = 6.25$$
$$5x^2 - 4x - 5.25 = 0$$

Multiply by 4.
$$20x^2 - 16x - 21 = 0$$
$$(2x-3)(10x+7) = 0$$

$$2x - 3 = 0 \quad \text{or} \quad 10x + 7 = 0$$
$$x = \frac{3}{2} \quad \text{or} \quad x = -\frac{7}{10}$$

Discard the negative solution.
If $x = \frac{3}{2}$, then

$$2x - 1 = 2\left(\frac{3}{2}\right) - 1 = 2.$$

The width of the rectangle is 1.5 cm, and the length is 2 cm.

41. Let $x =$ the width of the uncovered strip of flooring.

From the problem,
(length of the rug) • (width of the rug) = 234.

The rug is centered in the room a distance x from the walls (width of the strip x), so the length of the rug

= length of the room − 2 • (width of the strip)
= 20 − 2x,

and the width of the rug

= width of the room − 2 • (width of the strip)
= 15 − 2x.

The equation (length of rug) • (width of rug) = 234 becomes

$$(20 - 2x)(15 - 2x) = 234.$$
$$300 - 70x + 4x^2 = 234$$
$$4x^2 - 70x + 66 = 0$$

Divide by 2.
$$2x^2 - 35x + 33 = 0$$
$$(2x - 33)(x - 1) = 0$$

$2x - 33 = 0$ or $x - 1 = 0$

$x = \dfrac{33}{2}$ or $x = 1$

Reject $\dfrac{33}{2} = 16\dfrac{1}{2}$ since $16\dfrac{1}{2}$ is wider than the room itself. The width of the uncovered strip is 1 foot.

43. Let x be the width of grass.

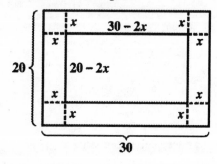

The area of the flower garden is $(20 - 2x)(30 - 2x)$.

Area of back yard	less	area of flower garden	is	area of grass.
20 • 30	−	$(20 - 2x)(30 - 2x)$	=	184

$$600 - (600 - 100x + 4x^2) = 184$$
$$600 - 600 + 100x - 4x^2 = 184$$
$$100x - 4x^2 = 184$$
$$4x^2 - 100x + 184 = 0$$

Divide by 4.
$$x^2 - 25x + 46 = 0$$
$$(x - 23)(x - 2) = 0$$

$x - 23 = 0$ or $x - 2 = 0$

$x = 23$ or $x = 2$

Discard $x = 23$ since the width of the garden cannot be

$$20 - 2x = 20 - 2(23) = -26.$$
Thus, if $x = 2$, then
$$30 - 2x = 30 - 2(2) = 26,$$
and
$$20 - 2x = 20 - 2(2) = 16.$$

The flower garden is 26 m by 16 m.

45. Let $r =$ the interest rate.

$$A = P(1 + r)^2$$

Let $A = 2142.25$ and $P = 2000$. Solve for r.

$$2142.25 = 2000(1 + r)^2$$
$$1.071125 = (1 + r)^2$$

Use the square root property.

$1 + r \approx 1.035$ or $1 + r \approx -1.035$

$r \approx .035$ or $r \approx -2.035$

Since the interest rate cannot be negative, reject -2.035. The interest rate is about 3.5%.

47. Supply and demand are equal when

$$3p - 200 = \frac{3200}{p}.$$

Solve for p.
$$3p^2 - 200p = 3200$$
$$3p^2 - 200p - 3200 = 0$$

Use the quadratic formula with $a = 3$, $b = -200$, and $c = -3200$.

$$p = \frac{-(-200) \pm \sqrt{(-200)^2 - 4(3)(-3200)}}{2(3)}$$

$$= \frac{200 \pm \sqrt{40,000 + 38,400}}{6}$$

$$= \frac{200 \pm \sqrt{78,400}}{6} = \frac{200 \pm 280}{6}$$

$p = \dfrac{480}{6} = 80$ or $p = \dfrac{-80}{6} = -\dfrac{40}{3}$

continued

Discard the negative solution. The supply and demand are equal when the price is 80 cents or $.80.

49. Let F denote the Froude number. Solve

$$F = \frac{v^2}{gl}$$

for v.

$$v^2 = Fgl$$
$$v = \pm\sqrt{Fgl}$$

v is positive, so

$$V = \sqrt{Fgl}.$$

For the rhinoceros, $l = 1.2$ and $F = 2.57$.

$$V = \sqrt{(2.57)(9.8)(1.2)} \approx 5.5$$

or 5.5 meters per second.

51. Write a proportion.

$$\frac{x-4}{3x-19} = \frac{4}{x-3}$$

Multiply by the LCD, $(3x-19)(x-3)$.

$$(3x-19)(x-3)\left(\frac{x-4}{3x-19}\right)$$
$$= (3x-19)(x-3)\left(\frac{4}{x-3}\right)$$

$$(x-3)(x-4) = (3x-19)4$$
$$x^2 - 7x + 12 = 12x - 76$$
$$x^2 - 19x + 88 = 0$$
$$(x-8)(x-11) = 0$$

$$x - 8 = 0 \quad\text{or}\quad x - 11 = 0$$
$$x = 8 \quad\text{or}\quad x = 11$$

If $x = 8$, then

$$3x - 19 = 3(8) - 19 = 5.$$

If $x = 11$, then

$$3x - 19 = 3(11) - 19 = 14.$$

Thus, $AC = 5$ or $AC = 14$.

53. Solve $p = \dfrac{E^2R}{(r+R)^2}$ $(E > 0)$ for R.

$$p(r+R)^2 = E^2R$$
$$p(r^2 + 2rR + R^2) = E^2R$$
$$pr^2 + 2prR + pR^2 = E^2R$$
$$pR^2 + 2prR - E^2R + pr^2 = 0$$
$$pR^2 + (2pr - E^2)R + pr^2 = 0$$

$$a = p,\, b = 2pr - E^2,\, c = pr^2$$

$$R = \frac{-(2pr - E^2) \pm \sqrt{(2pr - E^2)^2 - 4p \cdot pr^2}}{2p}$$

$$= \frac{E^2 - 2pr \pm \sqrt{4p^2r^2 - 4prE^2 + E^4 - 4p^2r^2}}{2p}$$

$$= \frac{E^2 - 2pr \pm \sqrt{E^4 - 4prE^2}}{2p}$$

$$= \frac{E^2 - 2pr \pm \sqrt{E^2(E^2 - 4pr)}}{2p}$$

$$R = \frac{E^2 - 2pr \pm E\sqrt{E^2 - 4pr}}{2p}$$

55. Solve $10p^2c^2 + 7pcr = 12r^2$ for r.

$$0 = 12r^2 - 7pcr - 10p^2c^2$$
$$a = 12,\, b = -7pc,\, c = -10p^2c^2$$

$$r = \frac{-(-7pc) \pm \sqrt{(-7pc)^2 - 4(12)(-10p^2c^2)}}{2(12)}$$

$$= \frac{7pc \pm \sqrt{49p^2c^2 + 480p^2c^2}}{24}$$

$$= \frac{7pc \pm \sqrt{529p^2c^2}}{24} = \frac{7pc \pm 23pc}{24}$$

$$r = \frac{7pc + 23pc}{24} = \frac{30pc}{24} = \frac{5pc}{4} \quad\text{or}$$

$$r = \frac{7pc - 23pc}{24} = \frac{-16pc}{24} = -\frac{2pc}{3}$$

57. Solve $LI^2 + RI + \dfrac{1}{c} = 0$ for I.

$$cLI^2 + cRI + 1 = 0 \quad\textit{Multiply by } c.$$
$$a = cL,\, b = cR,\, c = 1$$

$$I = \frac{-cR \pm \sqrt{(cR)^2 - 4(cL)(1)}}{2(cL)}$$

$$= \frac{-cR \pm \sqrt{c^2R^2 - 4cL}}{2cL}$$

Section 8.5

1. $g(x) = x^2 - 5$ written in the form
$$g(x) = a(x-h)^2 + k \text{ is}$$
$$g(x) = 1(x-0)^2 + (-5).$$

Here, $h = 0$ and $k = -5$, so the vertex (h, k) is $(0, -5)$. The graph is shifted 5 units down from the graph of $f(x) = x^2$. Since $a = 1 > 0$, the graph opens upward. The correct figure is F.

3. $F(x) = (x-1)^2$ written in the form
$$F(x) = a(x-h)^2 + k \text{ is}$$
$$F(x) = 1(x-1)^2 + 0.$$

Here, $h = 1$ and $k = 0$, so the vertex (h, k) is $(1, 0)$. The graph is shifted 1 unit to the right of the graph of $f(x) = x^2$. Since $a = 1 > 0$, the graph opens upward. The correct figure is C.

5. $H(x) = (x-1)^2 + 1$ is written in the form
    $H(x) = a(x-h)^2 + k$.

    Here, $h = 1$ and $k = 1$, so the vertex (h, k) is
    $(1, 1)$. The graph is shifted 1 unit to the right and
    1 unit up from the graph of $f(x) = x^2$. Since
    $a = 1 > 0$, the graph opens upward. The correct
    figure is E.

7. (a) The vertex is the point that contains the
    smallest or largest y-value of the parabola.

    (b) The axis of the parabola is a line through the
    vertex about which the parabola is symmetric.

For Exercises 9–16, we write $f(x)$ in the form
$f(x) = a(x-h)^2 + k$ and then list the vertex
(h, k).

9. $f(x) = -3x^2 = -3(x-0)^2 + 0$
    The vertex (h, k) is $(0, 0)$.

11. $f(x) = x^2 + 4 = 1(x-0)^2 + 4$
    The vertex (h, k) is $(0, 4)$.

13. $f(x) = (x-1)^2 = 1(x-1)^2 + 0$
    The vertex (h, k) is $(1, 0)$.

15. $f(x) = (x+3)^2 - 4 = 1[x-(-3)]^2 - 4$
    The vertex (h, k) is $(-3, -4)$.

17. In Exercise 15, the parabola is shifted 3 units to
    the left and 4 units down. The parabola in
    Exercise 16 is shifted 5 units to the right and 8
    units down.

19. $f(x) = -3x^2 + 1$
    $f(x) = -3(x-0)^2 + 1$

    Since $a = -3 < 0$, the graph opens downward.
    Since $|a| = |-3| = 3 > 1$, the graph is narrower
    than the graph of $f(x) = x^2$.

21. $f(x) = \frac{2}{3}x^2 - 4$
    $f(x) = \frac{2}{3}(x-0)^2 - 4$

    Since $a = \frac{2}{3} > 0$, the graph opens upward. Since
    $|a| = \left|\frac{2}{3}\right| = \frac{2}{3} < 1$, the graph is wider than the
    graph of $f(x) = x^2$.

23. Consider $f(x) = a(x-h)^2 + k$.

    (a) If $h > 0$ and $k > 0$ in $f(x) = a(x-h)^2 + k$,
    the shift is to the right and upward, so the vertex
    is in quadrant I.

    (b) If $h > 0$ and $k < 0$, the shift is to the right
    and downward, so the vertex is in quadrant IV.

    (c) If $h < 0$ and $k > 0$, the shift is to the left and
    upward, so the vertex is in quadrant II.

(d) If $h < 0$ and $k < 0$, the shift is to the left and
downward, so the vertex is in quadrant III.

25. $f(x) = -2x^2$ written in the form
    $f(x) = a(x-h)^2 + k$ is
    $f(x) = -2(x-0)^2 + 0$.

    Here, $h = 0$ and $k = 0$, so the vertex (h, k) is
    $(0, 0)$. Since $a = -2 < 0$, the graph opens
    downward. Since $|a| = |-2| = 2 > 1$, the graph
    is narrower than the graph of $f(x) = x^2$. By
    evaluating the function with $x = 2$ and $x = -2$,
    we see that the points $(2, -8)$ and $(-2, -8)$ are
    on the graph.

    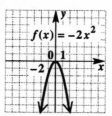

27. $f(x) = x^2 - 1$ written in the form
    $f(x) = a(x-h)^2 + k$ is
    $f(x) = 1(x-0)^2 + (-1)$.

    Here, $h = 0$ and $k = -1$, so the vertex is $(0, -1)$.
    The graph opens upward and has the same shape
    as $f(x) = x^2$ because $a = 1$. Two other points on
    the graph are $(-2, 3)$ and $(2, 3)$.

    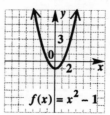

29. $f(x) = -x^2 + 2$ written in the form
    $f(x) = a(x-h)^2 + k$ is
    $f(x) = -1(x-0)^2 + 2$.

    Here, $h = 0$ and $k = 2$, so the vertex (h, k) is
    $(0, 2)$. Since $a = -1 < 0$, the graph opens
    downward. Since $|a| = |-1| = 1$, the graph has
    the same shape as $f(x) = x^2$. The points $(2, -2)$
    and $(-2, -2)$ are on the graph.

    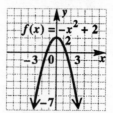

31. $f(x) = .5(x - 4)^2$ written in the form
$f(x) = a(x - h)^2 + k$ is
$f(x) = .5(x - 4)^2 + 0.$

Here, $h = 4$ and $k = 0$, so the vertex (h, k) is $(4, 0)$. The graph opens upward since a is positive and is wider than $f(x) = x^2$ because $|a| = |.5| < 1$. Two other points on the graph are $(2, 2)$ and $(6, 2)$.

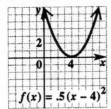

33. $f(x) = (x + 2)^2 - 1$ written in the form
$f(x) = a(x - h)^2 + k$ is
$f(x) = 1[x - (-2)]^2 + (-1).$

Since $h = -2$ and $k = -1$, the vertex (h, k) is $(-2, -1)$. Here, $a = 1$, so the graph opens upward and has the same shape as $f(x) = x^2$. The points $(-1, 0)$ and $(-3, 0)$ are on the graph.

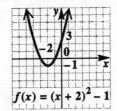

35. $f(x) = 2(x - 2)^2 - 4$ written in the form
$f(x) = a(x - h)^2 + k$ is
$f(x) = 2(x - 2)^2 + (-4).$

Here, $h = 2$ and $k = -4$, so the vertex (h, k) is $(2, -4)$. The graph opens upward and is narrower than $f(x) = x^2$ because $|a| = |2| > 1$. Two other points on the graph are $(0, 4)$ and $(4, 4)$. We can substitute any value for x, so the domain is $(-\infty, \infty)$. The value of y is greater than or equal to -4, so the range is $[-4, \infty)$.

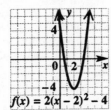

37. $f(x) = -.5(x + 1)^2 + 2$ written in the form
$f(x) = a(x - h)^2 + k$ is
$f(x) = -.5[x - (-1)]^2 + 2.$

Since $h = -1$ and $k = 2$, the vertex (h, k) is $(-1, 2)$. Here, $a = -.5 < 0$, so the graph opens

downward. Also, $|a| = |-.5| = .5 < 1$, so the graph is wider than the graph of $f(x) = x^2$. The points $(1, 0)$ and $(-3, 0)$ are on the graph. We can substitute any value for x, so the domain is $(-\infty, \infty)$. The value of y is less than or equal to 2, so the range is $(-\infty, 2]$.

$f(x) = -.5(x + 1)^2 + 2$

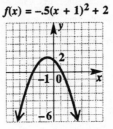

39. $f(x) = 2(x - 2)^2 - 3$ written in the form
$f(x) = a(x - h)^2 + k$ is
$f(x) = 2(x - 2)^2 + (-3).$

Here, $h = 2$ and $k = -3$, so the vertex (h, k) is $(2, -3)$. The graph opens upward and is narrower than $f(x) = x^2$ because $|a| = |2| > 1$. Two other points on the graph are $(3, -1)$ and $(1, -1)$. We can substitute any value for x, so the domain is $(-\infty, \infty)$. The value of y is greater than or equal to -3, so the range is $[-3, \infty)$.

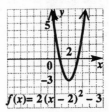

41. The graph of $y = x^2 + 6$ would be shifted 6 units upward from the graph of $y = x^2$.

42. To graph $y = x + 6$, plot the intercepts $(-6, 0)$ and $(0, 6)$, and draw the line through them.

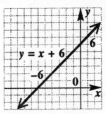

43. When considering the graph of $y = x + 6$, the y-intercept is 6. The graph of $y = x$ has y-intercept 0. Therefore, the graph of $y = x + 6$ is shifted 6 units upward compared to the graph of $y = x$.

44. The graph of $y = (x - 6)^2$ is shifted 6 units to the right compared to the graph of $y = x^2$.

45. To graph $y = x - 6$, plot the intercepts $(6, 0)$ and $(0, -6)$, and draw the line through them.

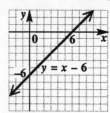

46. When considering the graph of $y = x - 6$, its x-intercept is 6 as compared to the graph of $y = x$ with x-intercept 0. The graph of $y = x - 6$ is shifted 6 units to the right compared to the graph of $y = x$.

47. **(a)** $|x - (-p)| = |x + p|$

(b) The focus should have coordinates $(p, 0)$ because the distance from the focus to the origin should equal the distance from the directrix to the origin.

(c) The distance from (x, y) to $(p, 0)$ is

$$\sqrt{(x - p)^2 + (y - 0)^2} = \sqrt{(x - p)^2 + y^2}.$$

(d) Using the results from parts (a) and (c), these distances should be equal.

$$\sqrt{(x - p)^2 + y^2} = |x + p|$$

Square both sides.

$$(x - p)^2 + y^2 = (x + p)^2$$
$$x^2 - 2px + p^2 + y^2 = x^2 + 2px + p^2$$
$$y^2 = 4px$$

49. **(a)**

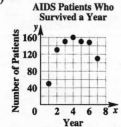

AIDS Patients Who Survived a Year

(b) Since the parabola would be opening downward, the value of a would be *negative*.

(c) Use $ax^2 + bx + c = y$ with $(2, 130)$, $(3, 155)$, and $(7, 115)$.

$$
\begin{array}{ll}
4a + 2b + c = 130 & (1) \\
9a + 3b + c = 155 & (2) \\
49a + 7b + c = 115 & (3)
\end{array}
$$

Eliminate c twice.

$$
\begin{array}{lll}
-4a & - 2b & - c = -130 & -1 \times (1) \\
49a & + 7b & + c = 115 & (3) \\
\hline
45a & + 5b & = -15 & \\
9a & + b & = -3 & (4)
\end{array}
$$

$$
\begin{array}{lll}
-9a & - 3b & - c = -155 & -1 \times (2) \\
49a & + 7b & + c = 115 & (3) \\
\hline
40a & + 4b & = -40 & \\
10a & + b & = -10 & (5)
\end{array}
$$

Now eliminate b from (4) and (5).

$$
\begin{array}{lll}
-9a & - b & = 3 & -1 \times (4) \\
10a & + b & = -10 & (5) \\
\hline
a & & = -7 &
\end{array}
$$

Use (5) to find b.

$$
\begin{aligned}
10(-7) + b &= -10 \\
-70 + b &= -10 \\
b &= 60
\end{aligned}
$$

Use (1) to find c.

$$
\begin{aligned}
4a + 2b + c &= 130 \quad (1) \\
4(-7) + 2(60) + c &= 130 \\
-28 + 120 + c &= 130 \\
c &= 38
\end{aligned}
$$

The quadratic function is

$$y = f(x) = -7x^2 + 60x + 38.$$

(d) Many answers are possible. Since the data give one y-value for each year, $\{1, 2, 3, 4, 5, 6, 7\}$ may be the most appropriate.

51. $x^2 - x - 20 = 0$
From the screens, we see that the x-values of the x-intercepts are -4 and 5, so the solution set is $\{-4, 5\}$.

53. $-2x^2 + 5x + 3 = 0$
From the screens, we see that the x-values of the x-intercepts are $-.5$ and 3, so the solution set is $\{-.5, 3\}$.

55. The only possible choice for the solution set for the equation $f(x) = 0$ would be choice (a) $\{-4, 1\}$. This is because the graph crosses the x-axis on each side of the y-axis, indicating that there is one negative solution and one positive solution. Choice (d) also has a negative and a positive value, but in this case, the negative value must have the greater absolute value of the two numbers.

Section 8.6

1. If there is an x^2-term in the equation, the axis is vertical. If there is a y^2-term, the axis is horizontal.

3. (a) Use the discriminant, $b^2 - 4ac$, of the function. If it is positive, there are two x-intercepts. If it is zero, there is one x-intercept (at the vertex), and if it is negative, there is no x-intercept.

(b) If the vertex is at $(1, -3)$ and the graph opens downward, the vertex is the highest point of the graph. Because the y-coordinate of the vertex is negative, the parabola lies below the x-axis so the graph has no x-intercepts.

5. $$y = 2x^2 + 4x + 5$$

Complete the square on x to find the vertex.

$$\frac{y}{2} = x^2 + 2x + \frac{5}{2} \qquad \text{Divide by 2.}$$

$$\frac{y}{2} - \frac{5}{2} = x^2 + 2x \qquad \text{Get the constant term on the left.}$$

$$\frac{y}{2} - \frac{5}{2} + 1 = x^2 + 2x + 1 \qquad \text{Half of 2 is 1; } (1)^2 = 1. \text{ Add 1 to both sides.}$$

$$\frac{y}{2} - \frac{3}{2} = (x + 1)^2 \qquad \text{Combine terms on the left and factor on the right.}$$

$$\frac{y}{2} = (x + 1)^2 + \frac{3}{2} \qquad \text{Add } \frac{3}{2}.$$

$$y = 2(x + 1)^2 + 3 \qquad \text{Multiply by 2.}$$

The vertex is $(-1, 3)$.
Because $a = 2$, the graph opens upward and is narrower than the graph of $y = x^2$.

For $y = 2x^2 + 4x + 5$, $a = 2$, $b = 4$, and $c = 5$. The discriminant is

$$b^2 - 4ac = 4^2 - 4(2)(5)$$
$$= 16 - 40 = -24.$$

The discriminant is negative, so the parabola has no x-intercepts.

7. $$y = -x^2 + 5x + 3$$

Use the vertex formula.
$a = -1, b = 5, c = 3$

$$\frac{-b}{2a} = \frac{-5}{2(-1)} = \frac{5}{2}$$

$y = f(x)$, so

$$f\left(\frac{-b}{2a}\right) = f\left(\frac{5}{2}\right)$$

$$= -\left(\frac{5}{2}\right)^2 + 5\left(\frac{5}{2}\right) + 3$$

$$= -\frac{25}{4} + \frac{25}{2} + 3$$

$$= \frac{-25 + 50 + 12}{4} = \frac{37}{4}.$$

The vertex is

$$\left(\frac{-b}{2a}, f\left(\frac{-b}{2a}\right)\right) = \left(\frac{5}{2}, \frac{37}{4}\right).$$

Because $a = -1$, the parabola opens downward and has the same shape as the graph of $y = x^2$.

$$b^2 - 4ac = 5^2 - 4(-1)(3)$$
$$= 25 + 12 = 37$$

The discriminant is positive, so the parabola has two x-intercepts.

9. $$x = \frac{1}{3}y^2 + 6y + 24$$

Complete the square on the y-terms to find the vertex.

$$3x = y^2 + 18y + 72$$
$$3x - 72 = y^2 + 18y$$
$$3x - 72 + 81 = y^2 + 18y + 81$$
$$3x + 9 = (y + 9)^2$$
$$3x = (y + 9)^2 - 9$$
$$x = \frac{1}{3}(y + 9)^2 - 3$$

The vertex is at $(-3, -9)$.
The graph is a horizontal parabola. Because of the coefficient, $\frac{1}{3}$, the parabola opens to the right and it is wider than the graph of $y = x^2$.

11. $y = f(x) = x^2 + 8x + 10$
To find the y-intercept, let $x = 0$.
$f(0) = 10$, so the y-intercept is $(0, 10)$.
To find the x-intercepts, let $y = 0$.

$$0 = x^2 + 8x + 10$$

$$x = \frac{-8 \pm \sqrt{64 - 40}}{2} = \frac{-8 \pm \sqrt{24}}{2}$$

$$= \frac{-8 \pm 2\sqrt{6}}{2} = -4 \pm \sqrt{6}$$

Complete the square to find the vertex.

$$y - 10 = x^2 + 8x$$
$$\frac{1}{2}(8) = 4; 4^2 = 16$$
$$y - 10 + 16 = x^2 + 8x + 16$$
$$y + 6 = (x + 4)^2$$
$$y = (x + 4)^2 - 6$$

The vertex is $(-4, -6)$. Since $a = 1$, the graph opens upward and is the same shape as the graph of $y = x^2$. For an additional point on the graph, let $x = -2$ (two units to the right of the axis) to get $f(-2) = -2$. So the point $(-2, -2)$ is on the graph.

continued

By symmetry, the point $(-6, -2)$ (two units to the left of the axis) is on the graph.

The x-intercepts are approximately $(-6.45, 0)$ and $(-1.55, 0)$.

$$f(x) = x^2 + 8x + 10$$

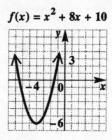

From the graph, we see that the domain is $(-\infty, \infty)$ and the range is $[-6, \infty)$.

13. $y = -2x^2 + 4x - 5$

If $x = 0$, $y = -5$, so the y-intercept is $(0, -5)$.

To find the x-intercepts, let $y = 0$.

$$0 = -2x^2 + 4x - 5$$

$$x = \frac{-4 \pm \sqrt{16 - 40}}{2(-2)}$$

The discriminant is negative, so there are no x-intercepts.

Use the formula to find the x-value of the vertex.

$$\frac{-b}{2a} = \frac{-4}{2(-2)} = 1$$

If $x = 1$, $y = -3$, so the vertex is $(1, -3)$. By symmetry, $(2, -5)$ is also on the graph. Since $a = -2$, the graph opens downward and is narrower than the graph of $y = x^2$.

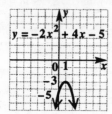

From the graph, we see that the domain is $(-\infty, \infty)$ and the range is $(-\infty, -3]$.

15. $x = -\frac{1}{5}y^2 + 2y - 4$

The roles of x and y are reversed, so this is a horizontal parabola.

To find the x-intercept, let $y = 0$.

If $y = 0$, $x = -4$, so the x-intercept is $(-4, 0)$.

To find the y-intercepts, let $x = 0$.

$$0 = -\frac{1}{5}y^2 + 2y - 4$$

$$0 = y^2 - 10y + 20$$

$$y = \frac{10 \pm \sqrt{100 - 80}}{2} = \frac{10 \pm \sqrt{20}}{2}$$

$$= \frac{10 \pm 2\sqrt{5}}{2} = 5 \pm \sqrt{5}$$

The y-intercepts are approximately $(0, 7.2)$ and $(0, 2.8)$.

Complete the square to find the vertex.

$$x = -\frac{1}{5}y^2 + 2y - 4$$

$$-5x = y^2 - 10y + 20$$

$$-5x - 20 = y^2 - 10y$$

$$-5x - 20 + 25 = y^2 - 10y + 25$$

$$-5x + 5 = (y - 5)^2$$

$$-5x = (y - 5)^2 - 5$$

$$x = -\frac{1}{5}(y - 5)^2 + 1$$

When $y = 5$, $x = 1$, so the vertex is $(1, 5)$. Since $a = -\frac{1}{5}$, the graph opens to the left and is wider than the graph of $y = x^2$. For an additional point on the graph, let $y = 7$ (two units above the axis) to get $x = \frac{1}{5}$. So the point $\left(\frac{1}{5}, 7\right)$ is on the graph. By symmetry, the point $\left(\frac{1}{5}, 3\right)$ (two units below the axis) is on the graph. The graph does not pass the vertical line test, so it does not represent a function.

$$x = -\tfrac{1}{5}y^2 + 2y - 4$$

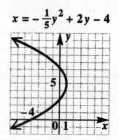

17. $x = 3y^2 + 12y + 5$

If $y = 0$, $x = 5$, so the x-intercept is $(5, 0)$.

To find the y-intercepts, let $x = 0$.

$$0 = 3y^2 + 12y + 5$$

$$y = \frac{-12 \pm \sqrt{144 - 60}}{6} = \frac{-12 \pm \sqrt{84}}{6}$$

$$= \frac{-12 \pm 2\sqrt{21}}{6} = -2 \pm \frac{1}{3}\sqrt{21}$$

The y-intercepts are approximately $(0, -.5)$ and $(0, -3.5)$. Use the formula to find the y-value of the vertex.

$$\frac{-b}{2a} = \frac{-12}{2(3)} = -2$$

If $y = -2$, $x = -7$, so the vertex is $(-7, -2)$. By symmetry, $(5, -4)$ is also on the graph. Since $a = 3$, the graph opens to the right and is narrower than the graph of $y = x^2$. The graph does not pass the vertical line test, so it does not represent a function.

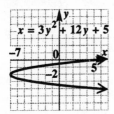

19. The graph of $y = 2x^2 + 4x - 3$ is a vertical parabola opening upward, so choice F is correct. (F)

21. The graph of $y = -\frac{1}{2}x^2 - x + 1$ is a vertical parabola opening downward, so choices A and C are possibilities. The graph in C is wider than the graph in A, so it must correspond to $a = -\frac{1}{2}$ while the graph in A must correspond to $a = -1$. (C)

23. The graph of $x = -y^2 - 2y + 4$ is a horizontal parabola opening to the left, so choice D is correct. (D)

25. Let $x =$ the width of the rectangle, and
$\quad\quad y =$ the length of the rectangle.

The sum of the sides is 100 m, so

$$\begin{aligned} x + x + y + y &= 100 \\ 2x + 2y &= 100 \\ 2y &= 100 - 2x. \\ y &= 50 - x \quad\quad \textit{Divide by 2.} \end{aligned}$$

Since Area = Width • Length,
$\quad\quad A = xy.$
Substitute $50 - x$ for y.
$$\begin{aligned} A &= x(50 - x) \\ &= 50x - x^2 \text{ or } -x^2 + 50x \end{aligned}$$

The width x of the rectangle with maximum area will occur at the vertex of $A = -x^2 + 50x$. Use the vertex formula with $a = -1$ and $b = 50$ to get

$$x = \frac{-b}{2a} = \frac{-50}{2(-1)} = 25.$$

A width of 25 m will produce the maximum area. Note that the length is also 25 m, so the rectangle is a square.

27. $h = 32t - 16t^2$ or $h = -16t^2 + 32t$

Here, $a = -16 < 0$, so the parabola opens downward. The time it takes to reach the maximum height and the maximum height are given by the vertex of the parabola. Use the vertex formula to find that

$$t = \frac{-b}{2a} = \frac{-32}{2(-16)} = \frac{-32}{-32} = 1,$$

and $\quad\quad h = -16(1)^2 + 32(1)$
$$= -16 + 32 = 16.$$

The vertex is $(1, 16)$, so the maximum height is 16 feet which occurs when the time is 1 second. The object hits the ground when $h = 0$.

$$\begin{aligned} 0 &= -16t^2 + 32t \\ 0 &= -16t(t - 2) \end{aligned}$$

$$\begin{array}{lll} -16t = 0 & \text{or} & t - 2 = 0 \\ t = 0 & \text{or} & t = 2 \end{array}$$

It takes 2 seconds for the object to hit the ground.

29. The graph of the height of the projectile,

$$s(t) = -4.9t^2 + 40t,$$

is a parabola that opens downward since $a = -4.9 < 0$. The time is the t-coordinate of the highest point, or vertex; the maximum height is the s-coordinate of the vertex. To find the vertex, use the vertex formula with $a = -4.9$ and $b = 40$.

$$t = \frac{-b}{2a} = \frac{-40}{2(-4.9)} = \frac{40}{9.8} \approx 4.1$$

Therefore, the projectile will reach its maximum height after about 4.1 seconds.
The maximum height is given by

$$\begin{aligned} s(4.1) &= -4.9(4.1)^2 + 40(4.1) \\ &= -4.9(16.81) + 164 \\ &\approx 81.6. \end{aligned}$$

The maximum height is about 81.6 m.

31. $f(x) = -20.57x^2 + 758.9x - 3140$

(a) The coefficient of x^2 is negative because the parabola opens downward.

(b) Use the vertex formula.

$$x = \frac{-b}{2a} = \frac{-758.9}{2(-20.57)} \approx 18.45$$
$$f(18.45) \approx 3860$$

The vertex is approximately $(18.45, 3860)$.

(c) 18 corresponds to 2018, so in 2018 social security assets will reach their maximum value of $3860 billion.

33. The number of people on the plane is $100 - x$ since x is the number of unsold seats. The price per seat is $200 + 4x$.

(a) The total revenue received for the flight is found by multiplying the number of seats by the price per seat. Thus, the revenue is

$$R(x) = (100 - x)(200 + 4x)$$
$$= 20,000 + 200x - 4x^2.$$

(b) Use the formula for the vertex.

$$x = \frac{-b}{2a} = \frac{-200}{2(-4)} = 25$$

$R(25) = 22,500$, so the vertex is $(25, 22,500)$. $R(0) = 20,000$, so the R-intercept is $(0, 20,000)$. From the factored form for R, we see that the positive x-intercept is $(100, 0)$. (The factor $200 + 4x$ leads to a negative x-intercept, meaningless in this problem.)

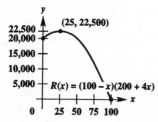

(c) The number of unsold seats x that produce the maximum revenue is 25, the x-value of the vertex.

(d) The maximum revenue is $22,500$, the y-value of the vertex.

35. $f(x) = x^2 - 8x + 18$

$$\frac{-b}{2a} = \frac{-(-8)}{2(1)} = 4$$

$f(4) = 2$, so the vertex is $(4, 2)$, which matches choice B.

37. $f(x) = x^2 - 8x + 14$

$$\frac{-b}{2a} = \frac{-(-8)}{2(1)} = 4$$

$f(4) = -2$, so the vertex is $(4, -2)$, which matches choice A.

Section 8.7

1. $2x^2 + x - 3 = 0$

$(2x + 3)(x - 1) = 0$

$2x + 3 = 0$ or $x - 1 = 0$

$x = -\dfrac{3}{2}$ or $x = 1$

Solution set: $\left\{ -\dfrac{3}{2}, 1 \right\}$

3. $2x^2 + x - 3 \le 0$

Region A: Let $x = -2$.

$2(-2)^2 + (-2) - 3 \le 0$?

$\qquad\qquad 3 \le 0$ *False*

Region B: Let $x = 0$.

$\qquad\qquad -3 \le 0$ *True*

Region C: Let $x = 2$.

$2(2)^2 + (2) - 3 \le 0$?

$\qquad\qquad 7 \le 0$ *False*

The open interval that satisfies the strict inequality corresponds to Region B; that is, the interval $\left(-\dfrac{3}{2}, 1 \right)$.

5. $(x + 1)(x - 5) > 0$

Solve the equation

$(x + 1)(x - 5) = 0$.

$x + 1 = 0$ or $x - 5 = 0$

$x = -1$ or $x = 5$

The numbers -1 and 5 divide the number line into three regions: A, B, and C.

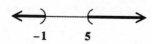

Test a number from each region in the original inequality.

Region A: Let $x = -2$.

$(x + 1)(x - 5) > 0$

$(-2 + 1)(-2 - 5) > 0$?

$-1(-7) > 0$?

$7 > 0$ *True*

Region B: Let $x = 0$.

$(0 + 1)(0 - 5) > 0$?

$-5 > 0$ *False*

Region C: Let $x = 6$.

$(6 + 1)(6 - 5) > 0$?

$7 > 0$ *True*

The solution set includes the numbers in Regions A and C, excluding -1 and 5 because of $>$.

Solution set: $(-\infty, -1) \cup (5, \infty)$

7. $(r + 4)(r - 6) < 0$

Solve the equation

$(r + 4)(r - 6) = 0.$

$r + 4 = 0 \quad$ or $\quad r - 6 = 0$

$r = -4 \quad$ or $\qquad r = 6$

These numbers divide the number line into three regions: A, B, and C.

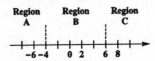

Test a number from each region in the original inequality.

Region A: Let $r = -5$.

$\quad (r + 4)(r - 6) < 0$

$(-5 + 4)(-5 - 6) < 0 \qquad ?$

$\qquad -1(-11) < 0 \qquad ?$

$\qquad\qquad 11 < 0 \qquad\qquad$ *False*

Region B: Let $r = 0$.

$\qquad 4(-6) < 0 \qquad ?$

$\qquad -24 < 0 \qquad\qquad$ *True*

Region C: Let $r = 7$.

$\quad (7 + 4)(7 - 6) < 0 \qquad ?$

$\qquad 11(1) < 0 \qquad ?$

$\qquad\qquad 11 < 0 \qquad\qquad$ *False*

The solution set includes Region B, where the expression is negative.

Solution set: $(-4, 6)$

9. $x^2 - 4x + 3 \geq 0$

Solve the equation

$x^2 - 4x + 3 = 0.$

$(x - 1)(x - 3) = 0$

$x - 1 = 0 \quad$ or $\quad x - 3 = 0$

$x = 1 \quad$ or $\qquad x = 3$

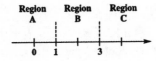

Region A: Let $x = 0$.

$\qquad\qquad 3 \geq 0 \qquad\qquad$ *True*

Region B: Let $x = 2$.

$\quad 2^2 - 4(2) + 3 \geq 0 \qquad ?$

$\qquad\qquad -1 \geq 0 \qquad\qquad$ *False*

Region C: Let $x = 4$.

$\quad 4^2 - 4(4) + 3 \geq 0 \qquad ?$

$\qquad\qquad 3 \geq 0 \qquad\qquad$ *True*

The solution set includes the numbers in Regions A and C, including 1 and 3 because of $\geq$.

Solution set: $(-\infty, 1] \cup [3, \infty)$

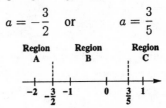

11. $\qquad 10a^2 + 9a \geq 9$

$\qquad 10a^2 + 9a - 9 \geq 0$

Solve the equation

$\qquad 10a^2 + 9a - 9 = 0.$

$(2a + 3)(5a - 3) = 0$

$2a + 3 = 0 \qquad$ or $\quad 5a - 3 = 0$

$\qquad a = -\dfrac{3}{2} \quad$ or $\qquad a = \dfrac{3}{5}$

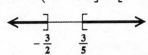

Test a number from each region in the original inequality.

Region A: Let $a = -2$.

$\quad 10(-2)^2 + 9(-2) \geq 9 \qquad ?$

$\qquad\qquad 40 - 18 \geq 9 \qquad ?$

$\qquad\qquad\qquad 22 \geq 9 \qquad\qquad$ *True*

Region B: Let $a = 0$.

$\qquad\qquad 0 \geq 9 \qquad\qquad$ *False*

Region C: Let $a = 1$.

$\quad 10(1)^2 + 9(1) \geq 9 \qquad ?$

$\qquad\qquad 10 + 9 \geq 9 \qquad ?$

$\qquad\qquad\qquad 19 \geq 9 \qquad\qquad$ *True*

The solution set includes the numbers in Region A and C, including $-\dfrac{3}{2}$ and $\dfrac{3}{5}$ because of $\geq$.

Solution set: $\left(-\infty, -\dfrac{3}{2}\right] \cup \left[\dfrac{3}{5}, \infty\right)$

13.
$$9p^2 + 3p < 2$$
Solve the equation
$$9p^2 + 3p - 2 = 0.$$
$$(3p - 1)(3p + 2) = 0$$

$$3p - 1 = 0 \quad \text{or} \quad 3p + 2 = 0$$
$$p = \frac{1}{3} \quad \text{or} \quad p = -\frac{2}{3}$$

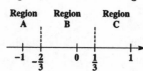

Test a number from each region in the inequality
$$9p^2 + 3p < 2.$$

Region A: Let $p = -1$.
$$9(-1)^2 + 3(-1) < 2 \quad ?$$
$$6 < 2 \qquad \textit{False}$$

Region B: Let $p = 0$.
$$0 < 2 \qquad \textit{True}$$

Region C: Let $p = 1$.
$$9(1)^2 + 3(1) < 2 \quad ?$$
$$12 < 2 \qquad \textit{False}$$

The solution set includes Region B, but not the endpoints.

Solution set: $\left(-\frac{2}{3}, \frac{1}{3}\right)$

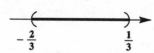

15.
$$6x^2 + x \geq 1$$
$$6x^2 + x - 1 \geq 0$$
Solve the equation
$$6x^2 + x - 1 = 0.$$
$$(2x + 1)(3x - 1) = 0$$

$$2x + 1 = 0 \quad \text{or} \quad 3x - 1 = 0$$
$$x = -\frac{1}{2} \quad \text{or} \quad x = \frac{1}{3}$$

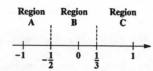

Test a number from each region in the inequality.
$$6x^2 + x \geq 1.$$

Region A: Let $x = -1$.
$$6(-1)^2 + (-1) \geq 1 \quad ?$$
$$5 \geq 1 \qquad \textit{True}$$

Region B: Let $x = 0$.
$$0 \geq 1 \qquad \textit{False}$$

Region C: Let $x = 1$.
$$6(1)^2 + 1 \geq 1 \quad ?$$
$$7 \geq 1 \qquad \textit{True}$$

The solution set includes the numbers in Region A and C, including $-\frac{1}{2}$ and $\frac{1}{3}$ because of $\geq$.

Solution set: $\left(-\infty, -\frac{1}{2}\right] \cup \left[\frac{1}{3}, \infty\right)$

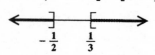

17. $y^2 - 6y + 6 \geq 0$
Solve the equation
$$y^2 - 6y + 6 = 0.$$

Since $y^2 - 6y + 6$ does not factor, let $a = 1$, $b = -6$, and $c = 6$ in the quadratic formula.
$$y = \frac{-(-6) \pm \sqrt{(-6)^2 - 4(1)(6)}}{2(1)}$$
$$= \frac{6 \pm \sqrt{12}}{2} = \frac{6 \pm 2\sqrt{3}}{2}$$
$$= \frac{2\left(3 \pm \sqrt{3}\right)}{2} = 3 \pm \sqrt{3}$$
$$y = 3 + \sqrt{3} \approx 4.7 \text{ or}$$
$$y = 3 - \sqrt{3} \approx 1.3$$

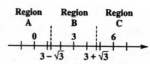

Test a number from each region in the inequality
$$y^2 - 6y + 6 \geq 0.$$

Region A: Let $y = 0$.
$$6 \geq 0 \qquad \textit{True}$$

Region B: Let $y = 3$.
$$3^2 - 6(3) + 6 \geq 0 \quad ?$$
$$-3 \geq 0 \qquad \textit{False}$$

Region C: Let $y = 5$.
$$5^2 - 6(5) + 6 \geq 0 \quad ?$$
$$1 \geq 0 \qquad \textit{True}$$

The solution set includes the numbers in Regions A and C, including $3 - \sqrt{3}$ and $3 + \sqrt{3}$ because of $\geq$.

Solution set: $\left(-\infty, 3 - \sqrt{3}\right] \cup \left[3 + \sqrt{3}, \infty\right)$

$$3 - \sqrt{3} \qquad 3 + \sqrt{3}$$

19. Include the endpoints if the inequality is $\leq$ or $\geq$. Exclude the endpoints if the inequality is $<$ or $>$.

21. $(4 - 3x)^2 \geq -2$

Since $(4 - 3x)^2$ is either 0 or positive, $(4 - 3x)^2$ will always be greater than -2. Therefore, the solution set is $(-\infty, \infty)$.

23. $(3x + 5)^2 \leq -4$

Since $(3x + 5)^2$ is never negative, $(3x + 5)^2$ will never be less than or equal to a negative number. Therefore, the solution set is $\emptyset$.

25. Change the inequality to an equation and solve it. Use the solutions of the equation to determine the regions on the number line where the inequality is true. Use test points to do this. Determine whether or not the endpoints are included and indicate them on the number line. Use interval notation to write the solution set. Examples will vary.

27. $(2r + 1)(3r - 2)(4r + 7) < 0$

The numbers $-\dfrac{1}{2}$, $\dfrac{2}{3}$, and $-\dfrac{7}{4}$ are solutions of the cubic equation

$$(2r + 1)(3r - 2)(4r + 7) = 0.$$

These numbers divide the number line into 4 regions.

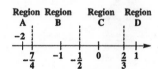

Region A: Let $r = -2$.
$$-3(-8)(-1) < 0 \qquad ?$$
$$-24 < 0 \qquad \textit{True}$$

Region B: Let $r = -1$.
$$-1(-5)(3) < 0 \qquad ?$$
$$15 < 0 \qquad \textit{False}$$

Region C: Let $r = 0$.
$$1(-2)(7) < 0 \qquad ?$$
$$-14 < 0 \qquad \textit{True}$$

Region D: Let $r = 1$.
$$3(1)(11) < 0 \qquad ?$$
$$33 < 0 \qquad \textit{False}$$

The solution set includes numbers in Regions A and C, excluding endpoints.

Solution set: $\left(-\infty, -\dfrac{7}{4}\right) \cup \left(-\dfrac{1}{2}, \dfrac{2}{3}\right)$

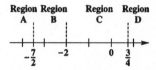

29. $(z + 2)(4z - 3)(2z + 7) \geq 0$

The numbers -2, $\dfrac{3}{4}$, and $-\dfrac{7}{2}$ are solutions of the cubic equation

$$(z + 2)(4z - 3)(2z + 7) = 0.$$

These numbers divide the number line into 4 regions.

Region A: Let $z = -4$.
$$-2(-19)(-1) \geq 0 \qquad ?$$
$$-38 \geq 0 \qquad \textit{False}$$

Region B: Let $z = -3$.
$$-1(-15)(1) \geq 0 \qquad ?$$
$$15 \geq 0 \qquad \textit{True}$$

Region C: Let $z = 0$.
$$2(-3)(7) \geq 0 \qquad ?$$
$$-42 \geq 0 \qquad \textit{False}$$

Region D: Let $z = 1$.
$$3(1)(9) \geq 0 \qquad ?$$
$$27 \geq 0 \qquad \textit{True}$$

The solution set includes numbers in Regions B and D, including the endpoints.

Solution set: $\left[-\dfrac{7}{2}, -2\right] \cup \left[\dfrac{3}{4}, \infty\right)$

$$-\dfrac{7}{2} \qquad -2 \qquad \dfrac{3}{4}$$

31. $\dfrac{x-1}{x-4} > 0$

Solve the equation

$\dfrac{x-1}{x-4} = 0.$

Multiply by the LCD, $x - 4$.

$x - 1 = 0$

$x = 1$

Find the number that makes the denominator 0.

$x - 4 = 0$

$x = 4$

The numbers 1 and 4 divide the number line into three regions.

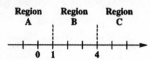

Test a number from each region in the inequality

$\dfrac{x-1}{x-4} > 0.$

Region A: Let $x = 0$.

$\dfrac{0-1}{0-4} > 0$?

$\dfrac{1}{4} > 0$ *True*

Region B: Let $x = 2$.

$\dfrac{2-1}{2-4} > 0$?

$\dfrac{1}{-2} > 0$ *False*

Region C: Let $x = 5$.

$\dfrac{5-1}{5-4} > 0$?

$4 > 0$ *True*

The solution set includes numbers in Regions A and C, excluding endpoints.

Solution set: $(-\infty, 1) \cup (4, \infty)$

33. $\dfrac{2y+3}{y-5} \le 0$

Solve the equation

$\dfrac{2y+3}{y-5} = 0.$

$2y + 3 = 0$

$y = -\dfrac{3}{2}$

Find the number that makes the denominator 0.

$y - 5 = 0$

$y = 5$

The numbers $-\dfrac{3}{2}$ and 5 divide the number line into three regions.

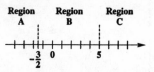

Test a number from each region in the inequality

$\dfrac{2y+3}{y-5} \le 0.$

Region A: Let $y = -2$.

$\dfrac{2(-2)+3}{(-2)-5} \le 0$?

$\dfrac{1}{7} \le 0$ *False*

Region B: Let $y = 0$.

$\dfrac{2(0)+3}{0-5} \le 0$?

$-\dfrac{3}{5} \le 0$ *True*

Region C: Let $y = 6$.

$\dfrac{2(6)+3}{6-5} \le 0$?

$15 \le 0$ *False*

The solution set includes the points in Region B. The endpoint 5 is not included since it makes the left side undefined. The endpoint $-\dfrac{3}{2}$ is included because it makes the left side equal to 0.

Solution set: $\left[-\dfrac{3}{2}, 5\right)$

35. $\dfrac{8}{x-2} \ge 2$

Solve the equation

$\dfrac{8}{x-2} = 2.$

$8 = 2(x - 2)$

$8 = 2x - 4$

$12 = 2x$

$6 = x$

Find the number that makes the denominator 0.

$x - 2 = 0$

$x = 2$

The numbers 2 and 6 divide the number line into three regions.

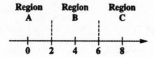

Test a number from each region in the inequality

$$\frac{8}{x-2} \geq 2.$$

Region A: Let $x = 0$.

$$\frac{8}{0-2} \geq 2 \quad ?$$

$$-4 \geq 2 \qquad False$$

Region B: Let $x = 3$.

$$\frac{8}{3-2} \geq 2 \quad ?$$

$$8 \geq 2 \qquad True$$

Region C: Let $x = 7$.

$$\frac{8}{7-2} \geq 2 \quad ?$$

$$\frac{8}{5} \geq 2 \qquad False$$

The solution set includes numbers in Region B, including 6 but excluding 2, which makes the fraction undefined.

Solution set: $(2, 6]$

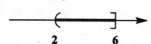

37. $\dfrac{3}{2t-1} < 2$

Solve the equation

$$\frac{3}{2t-1} = 2.$$

$$3 = 2(2t-1)$$

$$3 = 4t - 2$$

$$5 = 4t$$

$$\frac{5}{4} = t$$

Find the number that makes the denominator 0.

$$2t - 1 = 0$$

$$t = \frac{1}{2}$$

The numbers $\dfrac{1}{2}$ and $\dfrac{5}{4}$ divide the number line into three regions.

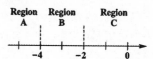

Test a number from each region in the inequality

$$\frac{3}{2t-1} < 2.$$

Region A: Let $t = 0$.

$$\frac{3}{2(0)-1} < 2 \quad ?$$

$$-3 < 2 \qquad True$$

Region B: Let $t = 1$.

$$\frac{3}{2(1)-1} < 2 \quad ?$$

$$3 < 2 \qquad False$$

Region C: Let $t = 2$.

$$\frac{3}{2(2)-1} < 2 \quad ?$$

$$1 < 2 \qquad True$$

The solution set includes numbers in Regions A and C, excluding endpoints.

Solution set: $\left(-\infty, \dfrac{1}{2}\right) \cup \left(\dfrac{5}{4}, \infty\right)$

39. $\dfrac{a}{a+2} \geq 2$

Solve the equation

$$\frac{a}{a+2} = 2.$$

$$a = 2(a+2)$$

$$a = 2a + 4$$

$$-a = 4$$

$$a = -4$$

Find the number that makes the denominator 0.

$$a + 2 = 0$$

$$a = -2$$

The numbers -4 and -2 divide the number line into three regions.

Test a number from each region in the inequality

$$\frac{a}{a+2} \geq 2.$$

Region A: Let $a = -5$.

$$\frac{-5}{-5+2} \geq 2 \quad ?$$

$$\frac{5}{3} \geq 2 \qquad False$$

Region B: Let $a = -3$.

$$\frac{-3}{-3+2} \geq 2 \quad ?$$

$$3 \geq 2 \qquad True$$

continued

Region C: Let $a = 0$.

$$\frac{0}{0+2} \geq 2 \quad ?$$

$$0 \geq 2 \qquad \textit{False}$$

The solution set includes numbers in Region B, including -4 but excluding -2, which makes the fraction undefined.

Solution set: $[-4, -2)$

41. $\dfrac{x}{x-4} < 3$

Solve the equation

$$\frac{x}{x-4} = 3.$$

$$x = 3(x-4)$$

$$x = 3x - 12$$

$$12 = 2x$$

$$6 = x$$

Find the number that makes the denominator 0.

$$x - 4 = 0$$

$$x = 4$$

The numbers 4 and 6 divide the number line into three regions.

Region A	Region B	Region C

$$\begin{array}{ccccc} \vert & \vert & \vert & \vert & \vert \\ 3 & 4 & 5 & 6 & 7 \end{array}$$

Test a number from each region in the inequality

$$\frac{x}{x-4} < 3.$$

Region A: Let $x = 0$.

$$\frac{0}{0-4} < 3 \quad ?$$

$$0 < 3 \qquad \textit{True}$$

Region B: Let $x = 5$.

$$\frac{5}{5-4} < 3 \quad ?$$

$$5 < 3 \qquad \textit{False}$$

Region C: Let $x = 7$.

$$\frac{7}{7-4} < 3 \quad ?$$

$$\frac{7}{3} < 3 \qquad \textit{True}$$

The solution set includes numbers in Regions A and C, excluding endpoints.

Solution set: $(-\infty, 4) \cup (6, \infty)$

$$\begin{array}{cc} \longleftarrow \quad) \qquad (\longrightarrow \\ 4 \qquad\quad 6 \end{array}$$

43. $\dfrac{4k}{2k-1} < k$

Solve the equation

$$\frac{4k}{2k-1} = k.$$

$$4k = k(2k - 1)$$

$$4k = 2k^2 - k$$

$$0 = 2k^2 - 5k$$

$$0 = k(2k - 5)$$

$$k = 0 \quad \text{or} \quad 2k - 5 = 0$$

$$k = \frac{5}{2}$$

Find the number that makes the denominator 0.

$$2k - 1 = 0$$

$$k = \frac{1}{2}$$

The numbers 0, $\dfrac{1}{2}$, and $\dfrac{5}{2}$ divide the number line into four regions.

Region A	Region B	Region C	Region D

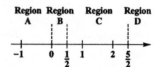

Test a number from each region in the inequality

$$\frac{4k}{2k-1} < k.$$

Region A: Let $k = -1$.

$$\frac{4(-1)}{2(-1)-1} < -1 \quad ?$$

$$\frac{4}{3} < -1 \qquad \textit{False}$$

Region B: Let $k = \frac{1}{4}$.

$$\frac{4\left(\frac{1}{4}\right)}{2\left(\frac{1}{4}\right) - 1} < \frac{1}{4} \quad ?$$

$$-2 < \frac{1}{4} \qquad \textit{True}$$

Region C: Let $k = 1$.

$$\frac{4(1)}{2(1)-1} < 1 \quad ?$$

$$4 < 1 \qquad \textit{False}$$

Region D: Let $k = 3$.

$$\frac{4(3)}{2(3)-1} < 3 \quad ?$$

$$\frac{12}{5} < 3 \qquad \textit{True}$$

The solution set includes numbers in Regions B and D. None of the endpoints are included.

Solution set: $\left(0, \dfrac{1}{2}\right) \cup \left(\dfrac{5}{2}, \infty\right)$

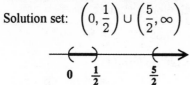

45. $\dfrac{2x - 3}{x^2 + 1} \geq 0$

The denominator is positive for all real numbers x, so it has no effect on the solution set for the inequality.

$$2x - 3 \geq 0.$$
$$2x \geq 3$$
$$x \geq \dfrac{3}{2}$$

Solution set: $\left[\dfrac{3}{2}, \infty\right)$

47. $\dfrac{(3x - 5)^2}{x + 2} > 0$

The numerator is positive for all real numbers x except $x = \dfrac{5}{3}$, which makes it equal to 0. If we solve the inequality $x + 2 > 0$, then we only have to be sure to exclude $\dfrac{5}{3}$ from that solution set to determine the solution set of the original inequality.

$$x + 2 > 0$$
$$x > -2$$

Solution set: $\left(-2, \dfrac{5}{3}\right) \cup \left(\dfrac{5}{3}, \infty\right)$

49. $\dfrac{-1}{p - 3} \geq 1$

$\dfrac{-1}{p - 3} - 1 \geq 0$

50. $\dfrac{-1}{p - 3} - \dfrac{p - 3}{p - 3} \geq 0$

51. $\dfrac{-1 - (p - 3)}{p - 3} \geq 0$

$\dfrac{-1 - p + 3}{p - 3} \geq 0$

$\dfrac{-p + 2}{p - 3} \geq 0$

52. $-p + 2 = 0$

$2 = p$

2 makes the numerator 0.

$p - 3 = 0$

$p = 3$

3 makes the denominator 0.

The solution (see Example 4) gives us the solution set $[2, 3)$.

53. "The height is greater than .140 foot" is described by the inequality

$$y > .140.$$
$$x - 1.727x^2 > .140$$
$$0 > 1.727x^2 - x + .140$$

This is equivalent to

$$1.727x^2 - x + .140 < 0.$$

To solve the corresponding equation,

$$1.727x^2 - x + .140 = 0,$$

use the quadratic formula.

$$x = \dfrac{-(-1) \pm \sqrt{(-1)^2 - 4(1.727)(.140)}}{2(1.727)}$$

$$= \dfrac{1 \pm \sqrt{.03288}}{3.454}$$

$$\approx .342 \quad \text{or} \quad .237$$

For this problem, our common sense tells us that $y > .140$ between approximately .237 and .342 second. We can check a time in this interval, say .3, to verify this.

$$1.727(.3)^2 - (.3) + .140 < 0 \quad ?$$
$$-.00457 < 0 \qquad \textit{True}$$

55. (a) The x-intercepts determine the solutions of the equation $x^2 - 4x + 3 = 0$. From the graph, the solution set is $\{1, 3\}$.

(b) The x-values of the points on the graph that are *above* the x-axis form the solution set of the inequality $x^2 - 4x + 3 > 0$. From the graph, the solution set is $(-\infty, 1) \cup (3, \infty)$.

(c) The x-values of the points on the graph that are *below* the x-axis form the solution set of the inequality $x^2 - 4x + 3 < 0$. From the graph, the solution set is $(1, 3)$.

56. **(a)** The x-intercepts determine the solutions of the equation $3x^2 + 10x - 8 = 0$. From the graph, the solution set is $\left\{-4, \dfrac{2}{3}\right\}$.

(b) The x-values of the points on the graph that are *above* the x-axis form the solution set of the inequality $3x^2 + 10x - 8 > 0$. From the graph, the solution set for $3x^2 + 10x - 8 \geq 0$ is

$(-\infty, -4] \cup \left[\dfrac{2}{3}, \infty\right)$.

(c) The x-values of the points on the graph that are *below* the x-axis form the solution set of the inequality $3x^2 + 10x - 8 < 0$. From the graph, the solution set is $\left(-4, \dfrac{2}{3}\right)$.

57. **(a)** The x-intercepts determine the solutions of the equation $-x^2 + 3x + 10 = 0$. From the graph, the solution set is $\{-2, 5\}$.

(b) The x-values of the points on the graph that are *above* the x-axis form the solution set of the inequality $-x^2 + 3x + 10 > 0$. From the graph, the solution set for $-x^2 + 3x + 10 \geq 0$ is $[-2, 5]$.

(c) The x-values of the points on the graph that are *below* the x-axis form the solution set of the inequality $-x^2 + 3x + 10 < 0$. From the graph, the solution set for $-x^2 + 3x + 10 \leq 0$ is $(-\infty, -2] \cup [5, \infty)$.

58. **(a)** The x-intercepts determine the solutions of the equation $-2x^2 - x + 15 = 0$. From the graph, the solution set is $\left\{-3, \dfrac{5}{2}\right\}$.

(b) The x-values of the points on the graph that are *above* the x-axis form the solution set of the inequality $-2x^2 - x + 15 > 0$. From the graph, the solution set for $-2x^2 - x + 15 \geq 0$ is

$\left[-3, \dfrac{5}{2}\right]$.

(c) The x-values of the points on the graph that are *below* the x-axis form the solution set of the inequality $-2x^2 - x + 15 < 0$. From the graph, the solution set for $-2x^2 - x + 15 \leq 0$ is

$(-\infty, -3] \cup \left[\dfrac{5}{2}, \infty\right)$.

59. The height is less than 10 feet is described by the inequality

$$f(x) < 10.$$
$$x - .0066x^2 < 10$$
$$0 < .0066x^2 - x + 10$$

This is equivalent to

$$.0066x^2 - x + 10 > 0.$$

To solve the corresponding equation,

$$.0066x^2 - x + 10 = 0,$$

use the quadratic formula.

$$x = \dfrac{-(-1) \pm \sqrt{(-1)^2 - 4(.0066)(10)}}{2(.0066)}$$

$$= \dfrac{1 \pm \sqrt{.736}}{.0132}$$

$$\approx 140.8 \quad \text{or} \quad 10.8$$

To determine how long the rocket is in the air, let $f(x) = 0$ and solve for x.

$$0 = x - .0066x^2$$
$$0 = x(1 - .0066x)$$

$$x = 0 \quad \text{or} \quad 1 - .0066x = 0$$

$$x = \dfrac{1}{.0066}$$

$$\approx 151.5$$

Thus, our common sense tells us that $f(x) < 10$ between 0 and 10.8 seconds and between 140.8 and 151.5 seconds.

Chapter 8 Review Exercises

1. $t^2 = 121$

$t = 11 \quad \text{or} \quad t = -11$

Solution set: $\{-11, 11\}$

2. $p^2 = 3$

$p = \sqrt{3} \quad \text{or} \quad p = -\sqrt{3}$

Solution set: $\left\{-\sqrt{3}, \sqrt{3}\right\}$

3. $(2x + 5)^2 = 100$

$2x + 5 = 10 \quad \text{or} \quad 2x + 5 = -10$

$\quad 2x = 5 \qquad\qquad 2x = -15$

$\quad x = \dfrac{5}{2} \quad \text{or} \qquad x = -\dfrac{15}{2}$

Solution set: $\left\{-\dfrac{15}{2}, \dfrac{5}{2}\right\}$

4. $(3k - 2)^2 = -25$

$3k - 2 = \sqrt{-25} \quad \text{or} \quad 3k - 2 = -\sqrt{-25}$

$3k - 2 = 5i \qquad\qquad 3k - 2 = -5i$

$3k = 2 + 5i \qquad\qquad 3k = 2 - 5i$

$k = \dfrac{2 + 5i}{3} \qquad\qquad k = \dfrac{2 - 5i}{3}$

$k = \dfrac{2}{3} + \dfrac{5}{3}i \quad \text{or} \quad k = \dfrac{2}{3} - \dfrac{5}{3}i$

Solution set: $\left\{\dfrac{2}{3} + \dfrac{5}{3}i, \dfrac{2}{3} - \dfrac{5}{3}i\right\}$

5.
$$x^2 + 4x = 15$$
Complete the square.
$$\left(\frac{1}{2} \cdot 4\right)^2 = 2^2 = 4$$
Add 4 to both sides.
$$x^2 + 4x + 4 = 15 + 4$$
$$(x + 2)^2 = 19$$

$$x + 2 = \sqrt{19} \quad \text{or} \quad x + 2 = -\sqrt{19}$$
$$x = -2 + \sqrt{19} \quad \text{or} \quad x = -2 - \sqrt{19}$$
Solution set: $\left\{-2 + \sqrt{19}, -2 - \sqrt{19}\right\}$

6.
$$2m^2 - 3m = -1$$
$$m^2 - \frac{3}{2}m = -\frac{1}{2} \qquad \text{Divide by 2.}$$
Complete the square.
$$\left[\frac{1}{2}\left(-\frac{3}{2}\right)\right]^2 = \left(-\frac{3}{4}\right)^2 = \frac{9}{16}$$
$$m^2 - \frac{3}{2}m + \frac{9}{16} = -\frac{1}{2} + \frac{9}{16} \qquad \text{Add } \frac{9}{16}.$$
$$\left(m - \frac{3}{4}\right)^2 = -\frac{8}{16} + \frac{9}{16}$$
$$\left(m - \frac{3}{4}\right)^2 = \frac{1}{16}$$

$$m - \frac{3}{4} = \sqrt{\frac{1}{16}} \quad \text{or} \quad m - \frac{3}{4} = -\sqrt{\frac{1}{16}}$$
$$m - \frac{3}{4} = \frac{1}{4} \qquad\qquad m - \frac{3}{4} = -\frac{1}{4}$$
$$m = \frac{3}{4} + \frac{1}{4} \qquad\qquad m = \frac{3}{4} - \frac{1}{4}$$
$$m = 1 \quad \text{or} \quad m = \frac{1}{2}$$
Solution set: $\left\{\frac{1}{2}, 1\right\}$

7. $2y^2 + y - 21 = 0$
Here $a = 2$, $b = 1$, and $c = -21$.
$$y = \frac{-b \pm \sqrt{b^2 - 4ac}}{2a}$$
$$y = \frac{-1 \pm \sqrt{1^2 - 4(2)(-21)}}{2(2)}$$
$$= \frac{-1 \pm \sqrt{1 + 168}}{4}$$
$$= \frac{-1 \pm \sqrt{169}}{4} = \frac{-1 \pm 13}{4}$$
$$y = \frac{-1 + 13}{4} = \frac{12}{4} = 3 \text{ or}$$
$$y = \frac{-1 - 13}{4} = -\frac{14}{4} = -\frac{7}{2}$$
Solution set: $\left\{-\frac{7}{2}, 3\right\}$

8.
$$(t + 3)(t - 4) = -2$$
$$t^2 - t - 12 = -2$$
$$t^2 - t - 10 = 0$$
Here $a = 1$, $b = -1$, and $c = -10$.
$$t = \frac{-b \pm \sqrt{b^2 - 4ac}}{2a}$$
$$t = \frac{-(-1) \pm \sqrt{(-1)^2 - 4(1)(-10)}}{2(1)}$$
$$= \frac{1 \pm \sqrt{1 + 40}}{2} = \frac{1 \pm \sqrt{41}}{2}$$
Solution set: $\left\{\frac{1 + \sqrt{41}}{2}, \frac{1 - \sqrt{41}}{2}\right\}$

9.
$$9p^2 = 42p - 49$$
$$9p^2 - 42p + 49 = 0$$
Here $a = 9$, $b = -42$, and $c = 49$.
$$p = \frac{-b \pm \sqrt{b^2 - 4ac}}{2a}$$
$$p = \frac{-(-42) \pm \sqrt{(-42)^2 - 4(9)(49)}}{2(9)}$$
$$= \frac{42 \pm \sqrt{1764 - 1764}}{18}$$
$$= \frac{42 \pm 0}{18} = \frac{7}{3}$$
Solution set: $\left\{\frac{7}{3}\right\}$

10.
$$3p^2 = 2(2p - 1)$$
$$3p^2 = 4p - 2$$
$$3p^2 - 4p + 2 = 0$$
Here $a = 3$, $b = -4$, and $c = 2$.
$$p = \frac{-b \pm \sqrt{b^2 - 4ac}}{2a}$$
$$p = \frac{-(-4) \pm \sqrt{(-4)^2 - 4(3)(2)}}{2(3)}$$
$$= \frac{4 \pm \sqrt{16 - 24}}{6} = \frac{4 \pm \sqrt{-8}}{6}$$
$$= \frac{4 \pm 2i\sqrt{2}}{6} = \frac{2\left(2 \pm i\sqrt{2}\right)}{6}$$
$$= \frac{2 \pm i\sqrt{2}}{3} = \frac{2}{3} \pm \frac{\sqrt{2}}{3}i$$
Solution set: $\left\{\frac{2}{3} + \frac{\sqrt{2}}{3}i, \frac{2}{3} - \frac{\sqrt{2}}{3}i\right\}$

11. The formula $x = -b \pm \dfrac{\sqrt{b^2 - 4ac}}{2a}$ is incorrect. It should be

$$x = \frac{-b \pm \sqrt{b^2 - 4ac}}{2a}.$$

All terms should be divided by $2a$.

12.
$$10 = y$$
$$10 = -.0013x^2 + 1.727x$$
$$.0013x^2 - 1.727x + 10 = 0$$

Here $a = .0013$, $b = -1.727$, and $c = 10$.

$$x = \frac{-(-1.727) \pm \sqrt{(-1.727)^2 - 4(.0013)(10)}}{2(.0013)}$$

$$= \frac{1.727 \pm \sqrt{2.930529}}{.0026}$$

$$\approx 1322.6 \quad \text{or} \quad 5.8 \text{ seconds}$$

13.   Let $x =$ the amount of time for the slower pipe alone and

$x - 1 =$ the amount of time for the faster pipe alone.

Make a chart.

Pipe	Rate	Time Together	Part of Job Done
Slower	$\dfrac{1}{x}$	3	$\dfrac{3}{x}$
Faster	$\dfrac{1}{x-1}$	3	$\dfrac{3}{x-1}$

$$\begin{array}{ccccc} \text{Part done by} & & \text{Part done by} & & \text{1 whole} \\ \text{slower pipe} & + & \text{faster pipe} & = & \text{job.} \\ \dfrac{3}{x} & + & \dfrac{3}{x-1} & = & 1 \end{array}$$

Multiply by the LCD, $x(x - 1)$.

$$x(x-1)\left(\frac{3}{x} + \frac{3}{x-1}\right) = x(x-1) \cdot 1$$
$$3(x-1) + 3x = x^2 - x$$
$$3x - 3 + 3x = x^2 - x$$
$$0 = x^2 - 7x + 3$$

Use the quadratic formula.

$$x = \frac{-b \pm \sqrt{b^2 - 4ac}}{2a}$$

$$x = \frac{-(-7) \pm \sqrt{(-7)^2 - 4(1)(3)}}{2(1)}$$

$$= \frac{7 \pm \sqrt{49 - 12}}{2} = \frac{7 \pm \sqrt{37}}{2}$$

$$x = \frac{7 + \sqrt{37}}{2} \approx 6.5 \text{ or}$$

$$x = \frac{7 - \sqrt{37}}{2} \approx .5$$

Reject .5 as the time for the slower tank, because that would yield a negative time for the faster tank. So, the slower tank takes 6.5 hours and the faster tank takes $x - 1 = 5.5$ hours.

14.   Let $x =$ the amount of time for the old machine alone and

$x - 1 =$ the amount of time for the new machine alone.

Make a chart.

Machine	Rate	Time Together	Part of Job Done
Old	$\dfrac{1}{x}$	2	$\dfrac{2}{x}$
New	$\dfrac{1}{x-1}$	2	$\dfrac{2}{x-1}$

$$\begin{array}{ccccc} \text{Part done by} & & \text{Part done by} & & \text{1 whole} \\ \text{old machine} & + & \text{new machine} & = & \text{job.} \\ \dfrac{2}{x} & + & \dfrac{2}{x-1} & = & 1 \end{array}$$

Multiply by the LCD, $x(x - 1)$.

$$x(x-1)\left(\frac{2}{x} + \frac{2}{x-1}\right) = x(x-1) \cdot 1$$
$$2(x-1) + 2x = x^2 - x$$
$$2x - 2 + 2x = x^2 - x$$
$$0 = x^2 - 5x + 2$$

Use the quadratic formula.

$$x = \frac{-b \pm \sqrt{b^2 - 4ac}}{2a}$$

$$x = \frac{-(-5) \pm \sqrt{(-5)^2 - 4(1)(2)}}{2(1)}$$

$$= \frac{5 \pm \sqrt{25 - 8}}{2} = \frac{5 \pm \sqrt{17}}{2}$$

$$x = \frac{5 + \sqrt{17}}{2} \approx 4.6 \text{ or}$$

$$x = \frac{5 - \sqrt{17}}{2} \approx .4$$

Reject .4 as the time for the old machine, because that would yield a negative time for the new machine. So, the old machine takes 4.6 hours.

15. $a^2 + 5a + 2 = 0$
$$a = 1, b = 5, c = 2$$
$$b^2 - 4ac = 5^2 - 4(1)(2)$$
$$= 25 - 8 = 17$$

Since the discriminant is positive, but not a perfect square, there are two distinct irrational number solutions. The answer is (c).

16.
$$4x^2 = 3 - 4x$$
$$4x^2 + 4x - 3 = 0$$
$$a = 4, b = 4, c = -3$$
$$b^2 - 4ac = 4^2 - 4(4)(-3)$$
$$= 16 + 48$$
$$= 64 \text{ or } 8^2$$

Since the discriminant is positive, and a perfect square, there are two distinct rational number solutions. The answer is (a).

17.
$$4x^2 = 6x - 8$$
$$4x^2 - 6x + 8 = 0$$
$$a = 4, b = -6, c = 8$$
$$b^2 - 4ac = (-6)^2 - 4(4)(8)$$
$$= 36 - 128 = -92$$

Since the discriminant is negative, there are two distinct imaginary number solutions. The answer is (d).

18. $9z^2 + 30z + 25 = 0$
$$a = 9, b = 30, c = 25$$
$$b^2 - 4ac = 30^2 - 4(9)(25)$$
$$= 900 - 900 = 0$$

Since the discriminant is zero, there is exactly one rational number solution. The answer is (b).

19. $24x^2 - 74x + 45$
$$a = 24, b = -74, c = 45$$
$$b^2 - 4ac = (-74)^2 - 4(24)(45)$$
$$= 5476 - 4320$$
$$= 1156 \text{ or } 34^2$$

Since the discriminant is positive and a perfect square, the polynomial can be factored using integer coefficients.
$$24x^2 - 74x + 45 = (6x - 5)(4x - 9)$$

20. $36x^2 + 69x - 34$
$$a = 36, b = 69, c = -34$$
$$b^2 - 4ac = 69^2 - 4(36)(-34)$$
$$= 4761 + 4896$$
$$= 9657$$

Since the discriminant is not a perfect square, the polynomial cannot be factored using integer coefficients.

21.
$$\frac{15}{x} = 2x - 1$$
$$x\left(\frac{15}{x}\right) = x(2x - 1) \qquad \textit{Multiply by the LCD, x.}$$
$$15 = 2x^2 - x$$
$$0 = 2x^2 - x - 15$$
$$0 = (2x + 5)(x - 3)$$
$$2x + 5 = 0 \qquad \text{or} \qquad x - 3 = 0$$
$$x = -\frac{5}{2} \qquad \text{or} \qquad x = 3$$

Check $x = -\dfrac{5}{2}$: $\quad -6 \overset{?}{=} -5 - 1 \quad$ *True*

Check $x = 3$: $\quad 5 \overset{?}{=} 6 - 1 \quad$ *True*

Solution set: $\left\{-\dfrac{5}{2}, 3\right\}$

22.
$$\frac{1}{y} + \frac{2}{y+1} = 2$$
$$y(y+1)\left(\frac{1}{y} + \frac{2}{y+1}\right) \qquad \textit{Multiply by the LCD, y(y+1).}$$
$$= y(y+1) \cdot 2$$
$$(y+1) + 2y = 2y^2 + 2y$$
$$0 = 2y^2 - y - 1$$
$$0 = (2y+1)(y-1)$$
$$2y + 1 = 0 \qquad \text{or} \qquad y - 1 = 0$$
$$y = -\frac{1}{2} \qquad \text{or} \qquad y = 1$$

Check $y = -\dfrac{1}{2}$: $\quad -2 + 4 \overset{?}{=} 2 \quad$ *True*

Check $y = 1$: $\quad 1 + 1 \overset{?}{=} 2 \quad$ *True*

Solution set: $\left\{-\dfrac{1}{2}, 1\right\}$

23. $8(3x + 5)^2 + 2(3x + 5) - 1 = 0$

Let $u = 3x + 5$. The equation becomes
$$8u^2 + 2u - 1 = 0$$
$$(2u + 1)(4u - 1) = 0$$
$$2u + 1 = 0 \qquad \text{or} \qquad 4u - 1 = 0$$
$$u = -\frac{1}{2} \qquad \text{or} \qquad u = \frac{1}{4}$$

To find x, substitute $3x + 5$ for u.
$$3x + 5 = -\frac{1}{2} \qquad \text{or} \qquad 3x + 5 = \frac{1}{4}$$
$$3x = -\frac{11}{2} \qquad\qquad 3x = -\frac{19}{4}$$
$$x = -\frac{11}{6} \qquad \text{or} \qquad x = -\frac{19}{12}$$

Check $x = -\dfrac{11}{6}$: $\quad 2 - 1 - 1 \overset{?}{=} 0 \quad$ *True*

Check $x = -\dfrac{19}{12}$: $\quad .5 + .5 - 1 \overset{?}{=} 0 \quad$ *True*

Solution set: $\left\{-\dfrac{11}{6}, -\dfrac{19}{12}\right\}$

24.
$$-2r = \sqrt{\frac{48 - 20r}{2}}$$

Square both sides.

$$(-2r)^2 = \left(\sqrt{\frac{48 - 20r}{2}}\right)^2$$

$$4r^2 = \frac{48 - 20r}{2}$$

$$4r^2 = 24 - 10r$$

$$4r^2 + 10r - 24 = 0$$

$$2r^2 + 5r - 12 = 0$$

$$(r + 4)(2r - 3) = 0$$

$$r + 4 = 0 \quad \text{or} \quad 2r - 3 = 0$$

$$r = -4 \quad \text{or} \quad r = \frac{3}{2}$$

Check $r = -4$: $8 \overset{?}{=} \sqrt{64}$ *True*

Check $r = \frac{3}{2}$: $-3 \overset{?}{=} \sqrt{9}$ *False*

Solution set: $\{-4\}$

25. $2x^{2/3} - x^{1/3} - 28 = 0$

Let $u = x^{1/3}$, so $u^2 = \left(x^{1/3}\right)^2 = x^{2/3}$.
The equation becomes

$$2u^2 - u - 28 = 0.$$

$$(2u + 7)(u - 4) = 0$$

$$2u + 7 = 0 \quad \text{or} \quad u - 4 = 0$$

$$u = -\frac{7}{2} \quad \text{or} \quad u = 4$$

To find x, substitute $x^{1/3}$ for u.

$$x^{1/3} = -\frac{7}{2} \quad \text{or} \quad x^{1/3} = 4$$

$$\left(x^{1/3}\right)^3 = \left(-\frac{7}{2}\right)^3 \qquad \left(x^{1/3}\right)^3 = 4^3$$

$$x = -\frac{343}{8} \quad \text{or} \quad x = 64$$

Check $x = -\frac{343}{8}$: $24.5 + 3.5 - 28 \overset{?}{=} 0$ *True*

Check $x = 64$: $32 - 4 - 28 \overset{?}{=} 0$ *True*

Solution set: $\left\{-\frac{343}{8}, 64\right\}$

26. $5x^4 - 2x^2 - 3 = 0$
Let $y = x^2$, so $y^2 = x^4$.

$$5y^2 - 2y - 3 = 0$$

$$(5y + 3)(y - 1) = 0$$

$$5y + 3 = 0 \quad \text{or} \quad y - 1 = 0$$

$$y = -\frac{3}{5} \quad \text{or} \quad y = 1$$

To find x, substitute x^2 for y.

$$x^2 = -\frac{3}{5} \qquad \text{or} \quad x^2 = 1$$

$$x = \pm i\sqrt{\frac{3}{5}} \quad \text{or} \quad x = \pm 1$$

$$x = \frac{\pm\sqrt{15}}{5}i$$

Check $x = \pm\frac{\sqrt{15}}{5}i$: $\frac{9}{5} + \frac{6}{5} - 3 \overset{?}{=} 0$ *True*

Check $x = \pm 1$: $5 - 2 - 3 \overset{?}{=} 0$ *True*

Solution set: $\left\{-1, 1, -\frac{\sqrt{15}}{5}i, \frac{\sqrt{15}}{5}i\right\}$

27. Let $x =$ Lisa's speed on the trip to pick up Laurie.
Make a chart. Use $d = rt$, or $t = \dfrac{d}{r}$.

	Distance	Rate	Time
To Laurie	8	x	$\dfrac{8}{x}$
To the mall	11	$x + 15$	$\dfrac{11}{x + 15}$

$$\begin{array}{ccc} \text{Time to pick} & \text{time to} & 24\text{ min} \\ \text{up Laurie} + & \text{mall} = & (\text{or }.4\text{ hr}). \end{array}$$

$$\frac{8}{x} + \frac{11}{x + 15} = .4$$

Multiply by the LCD, $x(x + 15)$.

$$x(x + 15)\left(\frac{8}{x} + \frac{11}{x + 15}\right) = x(x + 15)(.4)$$

$$8(x + 15) + 11x = .4x(x + 15)$$

$$8x + 120 + 11x = .4x^2 + 6x$$

$$0 = .4x^2 - 13x - 120$$

Multiply by 5 to clear the decimal.

$$0 = 2x^2 - 65x - 600$$

$$0 = (x - 40)(2x + 15)$$

$$x - 40 = 0 \quad \text{or} \quad 2x + 15 = 0$$

$$x = 40 \quad \text{or} \qquad x = -\frac{15}{2}$$

Speed cannot be negative, so $-\dfrac{15}{2}$ is not a
solution. Lisa's speed on the trip to pick up Laurie
was 40 mph.

28. Let $x =$ the time for Laketa alone and
$x - 2 =$ the time for Ed alone.

Worker	Rate	Time Together	Part of Job Done
Laketa	$\dfrac{1}{x}$	3	$\dfrac{3}{x}$
Ed	$\dfrac{1}{x - 2}$	3	$\dfrac{3}{x - 2}$

Part by Laketa + part by Ed = 1 whole job.

$$\frac{3}{x} + \frac{3}{x-2} = 1$$

Multiply by the LCD, $x(x-2)$.

$$x(x-2)\left(\frac{3}{x} + \frac{3}{x-2}\right) = x(x-2)\cdot 1$$

$$3x - 6 + 3x = x^2 - 2x$$

$$0 = x^2 - 8x + 6$$

Use the quadratic formula.

$$x = \frac{-b \pm \sqrt{b^2 - 4ac}}{2a}$$

$$x = \frac{8 \pm \sqrt{(-8)^2 - 4(1)(6)}}{2}$$

$$= \frac{8 \pm \sqrt{64 - 24}}{2} = \frac{8 \pm \sqrt{40}}{2}$$

$$= \frac{8 \pm 2\sqrt{10}}{2} = \frac{2\left(4 \pm \sqrt{10}\right)}{2} = 4 \pm \sqrt{10}$$

$$x = 4 + \sqrt{10} \approx 7.2 \text{ or}$$

$$x = 4 - \sqrt{10} \approx .8$$

Reject .8 as Laketa's time, because that would make Ed's time negative. So, Laketa's time alone is about 7.2 hours and Ed's time alone is about $x - 2 = 5.2$ hours.

29. The equation $x = \sqrt{2x + 4}$ can't have a negative solution, because the square root can't be negative.

30. Solve $S = \dfrac{Id^2}{k}$ for d.

Multiply both sides by k, then divide by I.

$$\frac{Sk}{I} = d^2$$

$$d = \pm\sqrt{\frac{Sk}{I}} = \frac{\pm\sqrt{Sk}}{\sqrt{I}}$$

$$= \frac{\pm\sqrt{Sk}}{\sqrt{I}} \cdot \frac{\sqrt{I}}{\sqrt{I}}$$

$$d = \frac{\pm\sqrt{SkI}}{I}$$

31. Solve $k = \dfrac{rF}{wv^2}$ for v.

Multiply both sides by v^2, then divide by k.

$$v^2 = \frac{rF}{kw}$$

$$v = \pm\sqrt{\frac{rF}{kw}} = \frac{\pm\sqrt{rF}}{\sqrt{kw}}$$

$$= \frac{\pm\sqrt{rF}}{\sqrt{kw}} \cdot \frac{\sqrt{kw}}{\sqrt{kw}}$$

$$v = \frac{\pm\sqrt{rFkw}}{kw}$$

32. Solve $2\pi R^2 + 2\pi RH - S = 0$ for R.

$$(2\pi)R^2 + (2\pi H)R - S = 0$$

Use the quadratic formula with

$$a = 2\pi, \, b = 2\pi H, \text{ and } c = -S.$$

$$R = \frac{-b \pm \sqrt{b^2 - 4ac}}{2a}$$

$$R = \frac{-2\pi H \pm \sqrt{(2\pi H)^2 - 4(2\pi)(-S)}}{2(2\pi)}$$

$$= \frac{-2\pi H \pm \sqrt{4\pi^2 H^2 + 8\pi S}}{4\pi}$$

$$= \frac{-2\pi H \pm 2\sqrt{\pi^2 H^2 + 2\pi S}}{4\pi}$$

$$= \frac{2\left(-\pi H \pm \sqrt{\pi^2 H^2 + 2\pi S}\right)}{4\pi}$$

$$R = \frac{-\pi H \pm \sqrt{\pi^2 H^2 + 2\pi S}}{2\pi}$$

33.
$$s = 16t^2 + 15t$$

$$25 = 16t^2 + 15t \qquad \textit{Let } s=25.$$

$$0 = 16t^2 + 15t - 25$$

$$a = 16, \, b = 15, \, c = -25$$

$$t = \frac{-15 \pm \sqrt{15^2 - 4(16)(-25)}}{2(16)}$$

$$= \frac{-15 \pm \sqrt{1825}}{32}$$

$$t = \frac{-15 + \sqrt{1825}}{32} \approx .87 \text{ or}$$

$$t = \frac{-15 - \sqrt{1825}}{32} \approx -1.80$$

Reject the negative solution since time cannot be negative. The object has fallen 25 feet in approximately .87 second.

34. Let $x =$ the length of the longer leg;

$\dfrac{3}{4}x =$ the length of the shorter leg;

$2x - 9 =$ the length of the hypotenuse.

Use the Pythagorean formula.

$$c^2 = a^2 + b^2$$

$$(2x - 9)^2 = x^2 + \left(\frac{3}{4}x\right)^2$$

$$4x^2 - 36x + 81 = x^2 + \frac{9}{16}x^2$$

$$16(4x^2 - 36x + 81) = 16\left(x^2 + \frac{9}{16}x^2\right)$$

$$64x^2 - 576x + 1296 = 16x^2 + 9x^2$$

$$39x^2 - 576x + 1296 = 0$$

continued

Divide by 3.
$$13x^2 - 192x + 432 = 0$$
$$(13x - 36)(x - 12) = 0$$

$$13x - 36 = 0 \quad \text{or} \quad x - 12 = 0$$

$$x = \frac{36}{13} \quad \text{or} \quad x = 12$$

Reject $x = \frac{36}{13}$ since $2\left(\frac{36}{13}\right) - 9$ is negative.

If $x = 12$, then

$$\frac{3}{4}x = \frac{3}{4}(12) = 9$$

and

$$2x - 9 = 2(12) - 9 = 15.$$

The lengths of the three sides are 9 feet, 12 feet, and 15 feet.

35. Let $x =$ the width of the original rectangle and $x + 2 =$ the length of the original rectangle.

Adding 1 m to the width and subtracting 1 m from the length of the rectangle gives $x + 1$ in each case, which is the length of a side of the square.

$$(\text{length of one side})^2 = \text{Area of square}$$
$$(x + 1)^2 = 121$$

$$x + 1 = 11 \quad \text{or} \quad x + 1 = -11$$
$$x = 10 \quad \text{or} \quad x = -12$$

Since width cannot be negative, the original rectangle had a width of 10 m and a length $x + 2 = 12$ m.

36. Let $x =$ the width of the border.

$$\text{Area of mat} = \text{length} \cdot \text{width}$$
$$352 = (2x + 20)(2x + 14)$$
$$352 = 4x^2 + 68x + 280$$
$$0 = 4x^2 + 68x - 72$$
$$0 = x^2 + 17x - 18$$
$$0 = (x + 18)(x - 1)$$
$$x + 18 = 0 \quad \text{or} \quad x - 1 = 0$$
$$x = -18 \quad \text{or} \quad x = 1$$

Reject the negative answer for length. The mat is 1 in wide.

37. Let $x =$ the length of the middle side,
$2x - 110 =$ the length of the longest side (the hypotenuse),
and $50 =$ the length of the shortest side.

Use the Pythagorean formula.
$$a^2 + b^2 = c^2$$
$$x^2 + 50^2 = (2x - 110)^2$$
$$x^2 + 2500 = 4x^2 - 440x + 12,100$$
$$0 = 3x^2 - 440x + 9600$$
$$0 = (x - 120)(3x - 80)$$

$$x - 120 = 0 \quad \text{or} \quad 3x - 80 = 0$$

$$x = 120 \quad \text{or} \quad x = \frac{80}{3}$$

Reject $\frac{80}{3}$ as the length of the middle side, because that is smaller than the length of the shortest side. So, the middle side is 120 m long.

38. $y = 3x^2 - 2$
Write in $y = a(x - h)^2 + k$ form as
$y = 3(x - 0)^2 - 2$. The vertex (h, k) is $(0, -2)$.

39. $y = 6 - 2x^2$
Write in $y = a(x - h)^2 + k$ form as
$y = -2(x - 0)^2 + 6$. The vertex (h, k) is $(0, 6)$.

40. $f(x) = -(x - 1)^2$
Write in $y = a(x - h)^2 + k$ form as
$y = -1(x - 1)^2 + 0$. The vertex (h, k) is $(1, 0)$.

41. $f(x) = (x + 2)^2$
Write in $y = a(x - h)^2 + k$ form as
$y = 1[x - (-2)]^2 + 0$. The vertex (h, k) is $(-2, 0)$.

42. $y = (x - 3)^2 + 7$
The equation is in the form $y = a(x - h)^2 + k$, so the vertex (h, k) is $(3, 7)$.

43.
$$y = -3x^2 + 4x - 2$$
$$\frac{y}{-3} = x^2 - \frac{4}{3}x + \frac{2}{3}$$
$$-\frac{y}{3} - \frac{2}{3} = x^2 - \frac{4}{3}x$$
$$-\frac{y}{3} - \frac{2}{3} + \frac{4}{9} = x^2 - \frac{4}{3}x + \frac{4}{9}$$
$$-\frac{y}{3} - \frac{2}{9} = \left(x - \frac{2}{3}\right)^2$$
$$-\frac{y}{3} = \left(x - \frac{2}{3}\right)^2 + \frac{2}{9}$$
$$y = -3\left(x - \frac{2}{3}\right)^2 - \frac{2}{3}$$

The equation is now in the form

$$y = a(x - h)^2 + k,$$

so the vertex (h, k) is $\left(\dfrac{2}{3}, -\dfrac{2}{3}\right)$.

44. $y = 4x^2 + 4x - 2$

Complete the square to find the vertex.

$$\frac{y}{4} = x^2 + x - \frac{1}{2}$$

$$\frac{y}{4} + \frac{1}{2} = x^2 + x$$

$$\frac{y}{4} + \frac{1}{2} + \frac{1}{4} = x^2 + x + \frac{1}{4}$$

$$\frac{y}{4} + \frac{3}{4} = \left(x + \frac{1}{2}\right)^2$$

$$\frac{y}{4} = \left(x + \frac{1}{2}\right)^2 - \frac{3}{4}$$

$$y = 4\left(x + \frac{1}{2}\right)^2 - 3$$

The equation is now in the form
$y = a(x - h)^2 + k$, so the vertex (h, k) is
$\left(-\dfrac{1}{2}, -3\right)$. Since $a = 4 > 0$, the parabola opens
upward. Also, $|a| = |4| = 4 > 1$, so the graph is
narrower than the graph of $y = x^2$. The points
$(-2, 6)$, $(0, -2)$, and $(1, 6)$ are on the graph.

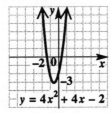

The domain is the set of all real numbers.
The range is $y \geq -3$.

45. $x = 2y^2 + 8y + 3$

Complete the square to find the vertex. Since the
roles of x and y are reversed, this is a horizontal
parabola.

$$\frac{x}{2} = y^2 + 4y + \frac{3}{2}$$

$$\frac{x}{2} - \frac{3}{2} = y^2 + 4y$$

$$\frac{x}{2} - \frac{3}{2} + 4 = y^2 + 4y + 4$$

$$\frac{x}{2} + \frac{5}{2} = (y + 2)^2$$

$$\frac{x}{2} = (y + 2)^2 - \frac{5}{2}$$

$$x = 2(y + 2)^2 - 5$$

The equation is in the form $x = a(y - k)^2 + h$, so
the vertex (h, k) is $(-5, -2)$. Here, $a = 2 > 0$, so
the parabola opens to the right. Also,
$|a| = |2| = 2 > 1$, so the graph is narrower than
the graph of $y = x^2$. The points $(3, 0)$ and $(3, -4)$
are on the graph.

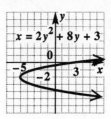

46.   If the discriminant is negative, there are no
x-intercepts.

47. $f(x) = a(x - h)^2 + k$
Since $a < 0$, the parabola opens downward. Since
$h > 0$ and $k < 0$, the x-coordinate is positive and
the y-coordinate is negative. Therefore, the vertex
is in quadrant IV. The correct graph is (a).

48.   The missile reaches its maximum height at the
vertex of the parabola with equation
$f(x) = -.017x^2 + x$. Use the vertex formula.

$$x = \frac{-b}{2a} = \frac{-1}{2(-.017)} \approx 29.41 \text{ seconds}$$

To find the maximum height, evaluate $f(29.41)$,
which is about 14.71 feet.

49.   **(a)** Use $ax^2 + bx + c = y$ with $(0, 12.39)$,
$(4, 15.78)$, and $(7, 22.71)$.

$$c = 12.39 \quad (1)$$
$$16a + 4b + c = 15.78 \quad (2)$$
$$49a + 7b + c = 22.71 \quad (3)$$

(b) Rewrite equations (2) and (3) with $c = 12.39$.

$$16a + 4b + 12.39 = 15.78$$
$$49a + 7b + 12.39 = 22.71$$

$$16a + 4b = 3.39 \quad (4)$$
$$49a + 7b = 10.32 \quad (5)$$

Now eliminate b.

$$
\begin{array}{rl}
-112a - 28b = -23.73 & -7 \times (4) \\
196a + 28b = 41.28 & 4 \times (5) \\
\hline
84a = 17.55 & \\
a \approx .2089 &
\end{array}
$$

Use (4) to approximate b.

$$16\left(\frac{17.55}{84}\right) + 4b = 3.39$$

$$4b \approx .0471$$

$$b \approx .0118$$

Thus, $f(x) = .2089x^2 + .0118x + 12.39$.

50. Let x and $40 - x$ denote the two numbers. The product P is given by

$$P = x(40 - x)$$
$$= -x^2 + 40x.$$

The maximum of P can be found by locating the vertex of the graph of P.

$$x = \frac{-b}{2a} = \frac{-40}{2(-1)} = 20$$

So x is 20 and $40 - x$ is also 20.

51. **(a)** The solution set is $\{-.5, 4\}$ since that is where the graph of

$$y = 2x^2 - 7x - 4$$

crosses the x-axis.

(b) The solution set of $2x^2 - 7x - 4 > 0$ can be found by determining the values of x such that y is greater than zero (above the x-axis). The solution set is $(-\infty, -.5) \cup (4, \infty)$.

(c) The solution set of $2x^2 - 7x - 4 \leq 0$ can be found by determining the values of x such that y is less than or equal to zero (on or below the x-axis). The solution set is $[-.5, 4]$.

52. $(x - 4)(2x + 3) > 0$

Solve the equation
$(x - 4)(2x + 3) = 0.$
$x - 4 = 0$ or $2x + 3 = 0$

$$x = 4 \quad \text{or} \quad x = -\frac{3}{2}$$

The numbers $-\frac{3}{2}$ and 4 divide the number line into three regions.

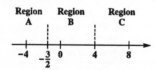

Test a number from each region in the inequality

$$(x - 4)(2x + 3) > 0.$$

Region A: Let $x = -2$.
$$-6(-1) > 0 \qquad ?$$
$$6 > 0 \qquad True$$

Region B: Let $x = 0$.
$$-4(3) > 0 \qquad ?$$
$$-12 > 0 \qquad False$$

Region C: Let $x = 5$.
$$1(13) > 0 \qquad ?$$
$$13 > 0 \qquad True$$

The solution set includes numbers in Regions A and C, excluding endpoints.

Solution set: $\left(-\infty, -\frac{3}{2}\right) \cup (4, \infty)$.

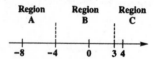

53. $\qquad x^2 + x \leq 12$

Solve the equation
$$x^2 + x = 12.$$
$$x^2 + x - 12 = 0$$
$$(x + 4)(x - 3) = 0$$
$x + 4 = 0$ or $x - 3 = 0$
$\qquad x = -4$ or $x = 3$

The numbers -4 and 3 divide the number line into three regions.

Test a number from each region in the inequality

$$x^2 + x \leq 12.$$

Region A: Let $x = -5$.
$$25 - 5 \leq 12 \qquad ?$$
$$20 \leq 12 \qquad False$$

Region B: Let $x = 0$.
$$0 \leq 12 \qquad True$$

Region C: Let $x = 4$.
$$16 + 4 \leq 12 \qquad ?$$
$$20 \leq 12 \qquad False$$

The numbers in Region B, including -4 and 3, are solutions.

Solution set: $[-4, 3]$

54. $\qquad 2k^2 > 5k + 3$

Solve the equation
$$2k^2 = 5k + 3.$$
$$2k^2 - 5k - 3 = 0$$
$$(2k + 1)(k - 3) = 0$$
$2k + 1 = 0$ or $k - 3 = 0$

$$k = -\frac{1}{2} \quad \text{or} \quad k = 3$$

The numbers $-\frac{1}{2}$ and 3 divide the number line into three regions.

Test a number from each region in the inequality

$$2k^2 > 5k + 3.$$

Region A: Let $k = -1$.
$$2 > -5 + 3 \quad ?$$
$$2 > -2 \qquad \textit{True}$$

Region B: Let $k = 0$.
$$0 > 3 \qquad \textit{False}$$

Region C: Let $k = 4$.
$$32 > 20 + 3 \quad ?$$
$$32 > 23 \qquad \textit{True}$$

The numbers in Regions A and C, not including $-\dfrac{1}{2}$ or 3, are solutions.

Solution set: $\left(-\infty, -\dfrac{1}{2}\right) \cup (3, \infty)$

55. $(4m + 3)^2 \le -4$

The square of a real number is never negative. So, the solution set of this inequality is $\emptyset$.

56. $\dfrac{6}{2z - 1} < 2$

Solve the equation

$$\dfrac{6}{2z - 1} = 2.$$
$$6 = 2(2z - 1)$$
$$6 = 4z - 2$$
$$8 = 4z$$
$$2 = z$$

Find the number that makes the denominator 0.
$$2z - 1 = 0$$
$$z = \dfrac{1}{2}$$

The numbers $\dfrac{1}{2}$ and 2 divide the number line into three regions.

Test a number from each region in the inequality

$$\dfrac{6}{2z - 1} < 2.$$

Region A: Let $z = 0$.
$$-6 < 2 \qquad \textit{True}$$

Region B: Let $z = 1$.
$$6 < 2 \qquad \textit{False}$$

Region C: Let $z = 3$.
$$\dfrac{6}{5} < 2 \qquad \textit{True}$$

The solution set includes numbers in Regions A and C, excluding endpoints.

Solution set: $\left(-\infty, \dfrac{1}{2}\right) \cup (2, \infty)$

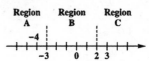

57. $\dfrac{3y + 4}{y - 2} \le 1$

Solve the equation

$$\dfrac{3y + 4}{y - 2} = 1.$$
$$3y + 4 = y - 2$$
$$2y = -6$$
$$y = -3$$

Find the number that makes the denominator 0.
$$y - 2 = 0$$
$$y = 2$$

The numbers -3 and 2 divide the number line into three regions.

Test a number from each region in the inequality

$$\dfrac{3y + 4}{y - 2} \le 1.$$

Region A: Let $y = -4$.
$$\dfrac{-8}{-6} \le 1 \quad ?$$
$$\dfrac{4}{3} \le 1 \qquad \textit{False}$$

Region B: Let $y = 0$.
$$\dfrac{4}{-2} \le 1 \quad ?$$
$$-2 \le 1 \qquad \textit{True}$$

Region C: Let $y = 3$.
$$\dfrac{13}{1} \le 1 \quad ?$$
$$13 \le 1 \qquad \textit{False}$$

The numbers in Region B, including -3 but not 2, are solutions.

Solution set: $[-3, 2)$

continued

(number line showing bracket at −3 with arrow to right, points at −3 and 2)

58. Solve $V = r^2 + R^2 h$ for R.

$$V - r^2 = R^2 h$$

$$R^2 h = V - r^2$$

$$R^2 = \frac{V - r^2}{h}$$

$$R = \pm \sqrt{\frac{V - r^2}{h}} = \frac{\pm \sqrt{V - r^2}}{\sqrt{h}}$$

$$= \frac{\pm \sqrt{V - r^2}}{\sqrt{h}} \cdot \frac{\sqrt{h}}{\sqrt{h}}$$

$$= \frac{\pm \sqrt{Vh - r^2 h}}{h}$$

59.
$$3t^2 - 6t = -4$$

$$3t^2 - 6t + 4 = 0$$

Use the quadratic formula.

$$t = \frac{-b \pm \sqrt{b^2 - 4ac}}{2a}$$

$$t = \frac{-(-6) \pm \sqrt{(-6)^2 - 4(3)(4)}}{2(3)}$$

$$= \frac{6 \pm \sqrt{-12}}{6} = \frac{6 \pm 2i\sqrt{3}}{6}$$

$$= \frac{2\left(3 \pm i\sqrt{3}\right)}{6} = \frac{3 \pm i\sqrt{3}}{3}$$

Solution set: $\left\{ \dfrac{3 + i\sqrt{3}}{3}, \dfrac{3 - i\sqrt{3}}{3} \right\}$

60.
$$x^4 - 1 = 0$$

$$\left(x^2 + 1\right)\left(x^2 - 1\right) = 0$$

$x^2 + 1 = 0$ or $x^2 - 1 = 0$

$x^2 = -1$ $(x+1)(x-1) = 0$

$x = \pm\sqrt{-1}$ $x + 1 = 0$ or $x - 1 = 0$

$x = \pm i$ or $x = -1$ or $x = 1$

Solution set: $\{-i, i, -1, 1\}$

61. $\left(b^2 - 2b\right)^2 = 11\left(b^2 - 2b\right) - 24$

Let $u = b^2 - 2b$. The equation becomes

$$u^2 = 11u - 24.$$

$$u^2 - 11u + 24 = 0$$

$$(u - 8)(u - 3) = 0$$

$u - 8 = 0$ or $u - 3 = 0$

$u = 8$ or $u = 3$

To find b, substitute $b^2 - 2b$ for u.

$b^2 - 2b = 8$ or $b^2 - 2b = 3$

$b^2 - 2b - 8 = 0$ $b^2 - 2b - 3 = 0$

$(b - 4)(b + 2) = 0$ $(b - 3)(b + 1) = 0$

$b - 4 = 0$ or $b + 2 = 0$ $b - 3 = 0$ or $b + 1 = 0$

$b = 4$ or $b = -2$ or $b = 3$ or $b = -1$

The potential solutions all check.

Solution set: $\{-2, -1, 3, 4\}$

62. $(r - 1)(2r + 3)(r + 6) < 0$

Solve the equation

$(r - 1)(2r + 3)(r + 6) = 0.$

$r - 1 = 0$ or $2r + 3 = 0$ or $r + 6 = 0$

$r = 1$ or $r = -\dfrac{3}{2}$ or $r = -6$

The numbers -6, $-\dfrac{3}{2}$, and 1 divide the number line into four regions.

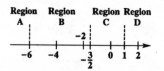

Test a number from each region in the inequality

$$(r - 1)(2r + 3)(r + 6) < 0.$$

Region A: Let $r = -7$.

$-8(-11)(-1) < 0$?

$-88 < 0$ *True*

Region B: Let $r = -2$.

$-3(-1)(4) < 0$?

$12 < 0$ *False*

Region C: Let $r = 0$.

$-1(3)(6) < 0$?

$-18 < 0$ *True*

Region D: Let $r = 2$.

$1(7)(8) < 0$?

$56 < 0$ *False*

The numbers in Regions A and C, not including $-6, -\dfrac{3}{2}$, or 1, are solutions.

Solution set: $(-\infty, -6) \cup \left(-\dfrac{3}{2}, 1\right)$

63.
$$\frac{2}{x-4}+\frac{1}{x}=\frac{11}{5}$$

Multiply by the LCD, $5x(x-4)$.

$$5x(x-4)\left(\frac{2}{x-4}+\frac{1}{x}\right)=5x(x-4)\left(\frac{11}{5}\right)$$
$$10x+5(x-4)=11x(x-4)$$
$$10x+5x-20=11x^2-44x$$
$$0=11x^2-59x+20$$
$$0=(x-5)(11x-4)$$

$$11x-4=0 \quad\text{or}\quad x-5=0$$
$$x=\frac{4}{11} \quad\text{or}\quad x=5$$

Check $x=\frac{4}{11}$: $\quad -\frac{11}{20}+\frac{11}{4}\overset{?}{=}\frac{11}{5}\quad$ *True*

Check $x=5$: $\quad 2+\frac{1}{5}\overset{?}{=}\frac{11}{5}\quad$ *True*

Solution set: $\left\{\frac{4}{11},5\right\}$

64. $(3k+11)^2=7$

$$3k+11=\sqrt{7}\qquad\text{or}\quad 3k+11=-\sqrt{7}$$
$$3k=-11+\sqrt{7}\qquad\qquad 3k=-11-\sqrt{7}$$
$$k=\frac{-11+\sqrt{7}}{3}\quad\text{or}\qquad k=\frac{-11-\sqrt{7}}{3}$$

Solution set: $\left\{\dfrac{-11+\sqrt{7}}{3},\dfrac{-11-\sqrt{7}}{3}\right\}$

65. Solve $p=\sqrt{\dfrac{yz}{6}}$ for y.

$$p^2=\left(\sqrt{\frac{yz}{6}}\right)^2 \qquad \textit{Square}$$
$$p^2=\frac{yz}{6}$$
$$\frac{6p^2}{z}=y \text{ or } y=\frac{6p^2}{z}$$

66. $(8k-7)^2\geq -1$

The square of any real number is always greater than or equal to 0, so any real number satisfies this inequality. The solution set is $(-\infty,\infty)$.

67.
$$-5x^2=-8x+3$$
$$-5x^2+8x-3=0$$
$$5x^2-8x+3=0$$
$$(5x-3)(x-1)=0$$

$$5x-3=0 \quad\text{or}\quad x-1=0$$
$$x=\frac{3}{5} \quad\text{or}\quad x=1$$

Solution set: $\left\{\frac{3}{5},1\right\}$

68.
$$6+\frac{15}{s^2}=-\frac{19}{s}$$

Multiply by the LCD, s^2.

$$s^2\left(6+\frac{15}{s^2}\right)=s^2\left(-\frac{19}{s}\right)$$
$$6s^2+15=-19s$$
$$6s^2+19s+15=0$$
$$(3s+5)(2s+3)=0$$

$$3s+5=0 \quad\text{or}\quad 2s+3=0$$
$$s=-\frac{5}{3} \quad\text{or}\qquad s=-\frac{3}{2}$$

Check $s=-\frac{5}{3}$: $\quad 6+\frac{27}{5}\overset{?}{=}\frac{57}{5}\quad$ *True*

Check $s=-\frac{3}{2}$: $\quad 6+\frac{20}{3}\overset{?}{=}\frac{38}{3}\quad$ *True*

Solution set: $\left\{-\frac{5}{3},-\frac{3}{2}\right\}$

69. $\dfrac{-2}{x+5}\leq -5$

Solve the equation
$$\frac{-2}{x+5}=-5.$$
$$-2=-5(x+5)$$
$$-2=-5x-25$$
$$5x=-23$$
$$x=-\frac{23}{5}$$

Find the number that makes the denominator 0.
$$x+5=0$$
$$x=-5$$

The numbers -5 and $-\dfrac{23}{5}$ divide the number line into three regions.

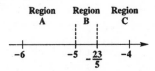

Test a number from each region in the inequality
$$\frac{-2}{x+5}\leq -5.$$

Region A: Let $x=-6$.
$$\frac{-2}{-1}\leq -5 \qquad ?$$
$$2\leq -5 \qquad \textit{False}$$

Region B: Let $x=-\dfrac{24}{5}$.
$$-\frac{2}{\frac{1}{5}}\leq -5 \qquad ?$$
$$-10\leq -5 \qquad \textit{True}$$

continued

Region C: Let $x = 0$.

$$\frac{-2}{5} \leq -5 \quad \textit{False}$$

The numbers in Region B, including $-\dfrac{23}{5}$ but not -5, are solutions.

Solution set: $\left(-5, -\dfrac{23}{5}\right]$

70. $y = \dfrac{4}{3}(x - 2)^2 + 1$

The equation is in the form $y = a(x - h)^2 + k$, so the vertex (h, k) is $(2, 1)$. $a = \dfrac{4}{3}$, so the parabola opens upward and is narrower than the graph of $y = x^2$. Two other points on the graph are $\left(0, \dfrac{19}{3}\right)$ and $\left(4, \dfrac{19}{3}\right)$.

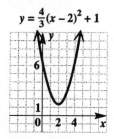

$y = \dfrac{4}{3}(x - 2)^2 + 1$

71. $x = 2(y + 3)^2 - 4$

$x = 2[y - (-3)]^2 + (-4)$

The equation is in the form

$$x = a(y - k)^2 + h,$$

so the vertex (h, k) is $(-4, -3)$. $a = 2$, so the parabola opens to the right and is narrower than the graph of $y = x^2$. Two other points on the graph are $(4, -1)$ and $(4, -5)$.

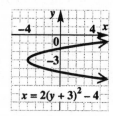

$x = 2(y + 3)^2 - 4$

72. $f(x) = -2x^2 + 8x - 5$

Replace $f(x)$ with y, and then complete the square to find the vertex.

$$y = -2x^2 + 8x - 5$$
$$\frac{y}{-2} = x^2 - 4x + \frac{5}{2}$$
$$-\frac{y}{2} - \frac{5}{2} = x^2 - 4x$$
$$-\frac{y}{2} - \frac{5}{2} + 4 = x^2 - 4x + 4$$
$$-\frac{y}{2} + \frac{3}{2} = (x - 2)^2$$
$$-\frac{y}{2} = (x - 2)^2 - \frac{3}{2}$$
$$y = -2(x - 2)^2 + 3$$

The equation is in the form $y = a(x - h)^2 + k$, so the vertex (h, k) is $(2, 3)$. Here, $a = -2 < 0$, so the parabola opens downward.

Also, $|a| = |-2| = 2 > 1$, so the graph is narrower than the graph of $y = x^2$. The points $(0, -5)$, $(1, 1)$, and $(3, 1)$ are on the graph.

$f(x) = -2x^2 + 8x - 5$

73. $x = -\dfrac{1}{2}y^2 + 6y - 14$

Complete the square to find the vertex. Since the roles of x and y are reversed, this is a horizontal parabola.

$$x = -\frac{1}{2}y^2 + 6y - 14$$
$$-2x = y^2 - 12y + 28$$
$$-2x - 28 = y^2 - 12y$$
$$-2x - 28 + 36 = y^2 - 12y + 36$$
$$-2x + 8 = (y - 6)^2$$
$$-2x = (y - 6)^2 - 8$$
$$x = -\frac{1}{2}(y - 6)^2 + 4$$

The equation is in the form $x = a(y - k)^2 + h$, so the vertex (h, k) is $(4, 6)$. Here, $a = -\dfrac{1}{2} < 0$, so the parabola opens to the left.

Also, $|a| = \left|-\dfrac{1}{2}\right| = \dfrac{1}{2} < 1$, so the graph is wider than the graph of $y = x^2$. The points $(-14, 0)$, $(2, 4)$, and $(2, 8)$ are on the graph.

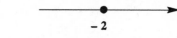

$$x = -\frac{1}{2}y^2 + 6y - 14$$

74. The student improperly used the square root property. If $b^2 = 12$, then $b = \pm\sqrt{12}$ $= \pm 2\sqrt{3}$. The student forgot the "$\pm$" sign.

75. To write the equation, work backwards. First we know that

$$x = -5 \quad \text{or} \quad x = 6.$$

Then,

$$x + 5 = 0 \quad \text{or} \quad x - 6 = 0,$$

and

$$(x + 5)(x - 6) = 0.$$

Multiply the factors. In standard form the equation is,

$$x^2 - x - 30 = 0.$$

76. $s(t) = -16t^2 + 160t$
The equation represents a parabola. Since $a = -16 < 0$, the parabola opens downward. The time and maximum height occur at the vertex (h, k) of the parabola, given by

$$\left(\frac{-b}{2a}, s\left(\frac{-b}{2a}\right)\right).$$

Using the standard form of the equation, $a = -16$ and $b = 160$, so

$$h = \frac{-b}{2a} = \frac{-160}{2(-16)} = 5,$$

and $k = s(h) = -16(5)^2 + 160(5)$
$= -400 + 800 = 400.$

The vertex is $(5, 400)$.

(a) The time at which the maximum height is reached is 5 seconds.

(b) The maximum height is 400 feet.

77. Let $L =$ the length of the rectangle and $W =$ the width.

The perimeter of the rectangle is 600 m, so

$$2L + 2W = 600$$
$$2W = 600 - 2L$$
$$W = 300 - L.$$

Since the area is length times width, substitute $300 - L$ for W.

$$A = LW$$
$$= L(300 - L)$$
$$= 300L - L^2 \text{ or } -L^2 + 300L$$

Use the vertex formula.

$$L = \frac{-b}{2a} = \frac{-300}{2(-1)} = 150$$

So $L = 150$ meters and
$W = 300 - L = 300 - 150 = 150$ meters.

78. The quadratic formula could be used to solve for x^2 (not x) in the equation $x^4 - 5x^2 + 6 = 0$. After you solve for x^2, you would have to remember to use the square root property to solve for x.

79. (a) $3x - (4x + 2) = 0$
$$3x - 4x - 2 = 0$$
$$-x - 2 = 0$$
$$x = -2$$
Solution set: $\{-2\}$

(b) $3x - (4x + 2) > 0$
$$3x - 4x - 2 > 0$$
$$-x - 2 > 0$$
$$-x > 2$$
$$x < -2$$
Solution set: $(-\infty, -2)$

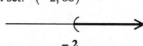

(c) $3x - (4x + 2) < 0$
$$3x - 4x - 2 < 0$$
$$-x - 2 < 0$$
$$-x < 2$$
$$x > -2$$
Solution set: $(-2, \infty)$

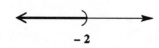

80. (a) $x^2 - 6x + 5 = 0$
$$(x - 1)(x - 5) = 0$$

$$x - 1 = 0 \quad \text{or} \quad x - 5 = 0$$
$$x = 1 \quad \text{or} \quad x = 5$$

Solution set: $\{1, 5\}$

(b) $x^2 - 6x + 5 > 0$

From the equation

$$x^2 - 6x + 5 = 0$$

in part (a), the solutions are $x = 1$ or $x = 5$. These numbers divide the number line into three regions.

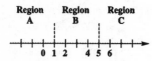

Test a point from each region in the inequality

$$x^2 - 6x + 5 > 0.$$

Region A: Let $x = 0$.
                $5 > 0$ *True*

Region B: Let $x = 2$.
          $4 - 12 + 5 > 0$?
             $-3 > 0$ *False*

Region C: Let $x = 6$.
         $36 - 36 + 5 > 0$?
             $5 > 0$ *True*

The numbers in Regions A and C, not including the endpoints, are solutions.
Solution set: $(-\infty, 1) \cup (5, \infty)$

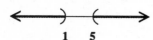

(c) $x^2 - 6x + 5 < 0$
The regions are the same as in part (b). Solutions to this inequality would be the numbers in Region B, again not including the endpoints.
Solution set: $(1, 5)$

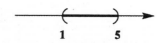

81. (a) $\dfrac{-5x + 20}{x - 2} = 0$

$$(x - 2)\left(\dfrac{-5x + 20}{x - 2}\right) = (x - 2) \cdot 0$$

$$-5x + 20 = 0$$
$$-5x = -20$$
$$x = 4$$

Solution set: $\{4\}$

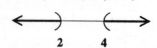

(b) $\dfrac{-5x + 20}{x - 2} > 0$

From part (a), the solution of the corresponding equation is $x = 4$.

The number that makes the denominator zero is $x = 2$.
These numbers divide the number line into three regions.

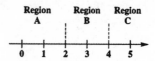

Test a number from each region in the inequality

$$\dfrac{-5x + 20}{x - 2} > 0.$$

continued

Region A: Let $x = 0$.
           $\dfrac{20}{-2} > 0$?
         $-10 > 0$ *False*

Region B: Let $x = 3$.
           $\dfrac{5}{1} > 0$?
            $5 > 0$ *True*

Region C: Let $x = 5$.
           $\dfrac{-5}{3} > 0$?
          $-\dfrac{5}{3} > 0$ *False*

The numbers in Region B, not including the endpoints, are solutions.
Solution set: $(2, 4)$

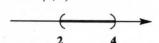

(c) $\dfrac{-5x + 20}{x - 2} < 0$

The regions are the same as in part (b). Solutions to this inequality would be the numbers in Regions A and C, again not including the endpoints.
Solution set: $(-\infty, 2) \cup (4, \infty)$

82. $\{-2\} \cup (-\infty, -2) \cup (-2, \infty) = (-\infty, \infty)$

83. $\{1, 5\} \cup (-\infty, 1) \cup (5, \infty) \cup (1, 5) = (-\infty, \infty)$

84.   The number 2 cannot possibly be part of any of the solution sets since it makes the denominator zero.
$$\{4\} \cup (2, 4) \cup (-\infty, 2) \cup (4, \infty)$$
$$= (-\infty, 2) \cup (2, \infty)$$

85. If we solve a linear, quadratic, or rational equation and the two inequalities associated with it, the union of the three sets will be $(-\infty, \infty)$; the only exception will be in the case of the rational equation and inequalities, where the number or numbers that cause the _denominator_ to be zero will be deleted.

86. If set S is the solution of the other inequality, then

$$S \cup \{-5, 3\} \cup (-\infty, -5) \cup (3, \infty) = (-\infty, \infty).$$

Therefore, $S = (-5, 3)$.

Chapter 8 Test

1. $(7x + 3)^2 = 25$

$$7x + 3 = 5 \quad \text{or} \quad 7x + 3 = -5$$
$$7x = 2 \qquad\qquad 7x = -8$$
$$x = \frac{2}{7} \quad \text{or} \quad x = -\frac{8}{7}$$

Solution set: $\left\{-\frac{8}{7}, \frac{2}{7}\right\}$

2. $2x^2 + 4x = 8$

$$x^2 + 2x = 4 \qquad \text{Divide by 2.}$$
$$x^2 + 2x + 1 = 4 + 1$$
$$(x + 1)^2 = 5$$
$$x + 1 = \sqrt{5} \quad \text{or} \quad x + 1 = -\sqrt{5}$$
$$x = -1 + \sqrt{5} \quad \text{or} \quad x = -1 - \sqrt{5}$$

Solution set: $\left\{-1 + \sqrt{5}, -1 - \sqrt{5}\right\}$

3. $2x^2 - 3x - 1 = 0$

Here $a = 2$, $b = -3$, and $c = -1$.

$$x = \frac{-b \pm \sqrt{b^2 - 4ac}}{2a}$$
$$x = \frac{-(-3) \pm \sqrt{(-3)^2 - 4(2)(-1)}}{2(2)}$$
$$= \frac{3 \pm \sqrt{17}}{4}$$

Solution set: $\left\{\frac{3 + \sqrt{17}}{4}, \frac{3 - \sqrt{17}}{4}\right\}$

4. $3t^2 - 4t = -5$

$$3t^2 - 4t + 5 = 0$$

Here $a = 3$, $b = -4$, and $c = 5$.

$$t = \frac{-b \pm \sqrt{b^2 - 4ac}}{2a}$$
$$t = \frac{-(-4) \pm \sqrt{(-4)^2 - 4(3)(5)}}{2(3)}$$

$$= \frac{4 \pm \sqrt{-44}}{6} = \frac{4 \pm 2i\sqrt{11}}{6}$$
$$= \frac{2\left(2 \pm i\sqrt{11}\right)}{6} = \frac{2 \pm i\sqrt{11}}{3}$$

Solution set: $\left\{\frac{2 + i\sqrt{11}}{3}, \frac{2 - i\sqrt{11}}{3}\right\}$

5. $$3x = \sqrt{\frac{9x + 2}{2}}$$

Square both sides.

$$9x^2 = \frac{9x + 2}{2}$$
$$18x^2 = 9x + 2$$
$$18x^2 - 9x - 2 = 0$$

As directed, use the quadratic formula with $a = 18$, $b = -9$, and $c = -2$.

$$x = \frac{-b \pm \sqrt{b^2 - 4ac}}{2a}$$
$$x = \frac{-(-9) \pm \sqrt{(-9)^2 - 4(18)(-2)}}{2(18)}$$
$$= \frac{9 \pm \sqrt{225}}{36} = \frac{9 \pm 15}{36}$$
$$x = \frac{9 + 15}{36} = \frac{24}{36} = \frac{2}{3} \quad \text{or}$$
$$x = \frac{9 - 15}{36} = \frac{-6}{36} = -\frac{1}{6}$$

Check $x = \frac{2}{3}$: $2 \overset{?}{=} \sqrt{4}$ _True_

Check $x = -\frac{1}{6}$: $-\frac{1}{2} \overset{?}{=} \sqrt{\frac{1}{4}}$ _False_

Solution set: $\left\{\frac{2}{3}\right\}$

6. **(a)** The radius of the circular orbit is the sum of the altitude and the radius of Earth.

$$1.60 \times 10^5 + 6.37 \times 10^6$$
$$= .16 \times 10^6 + 6.37 \times 10^6$$
$$= 6.53 \times 10^6$$

The radius is about 6.53×10^6 meters.

(b) $v^2 = \frac{GM}{r}$

$$v^2 = \frac{3.99 \times 10^{11}}{6.53 \times 10^6}$$
$$\approx 61,102.60$$
$$v \approx \sqrt{61,102.60} \approx 247$$

The velocity of the spacecraft is approximately 247 meters per second.

7. If k is a negative number, then $4k$ is also negative, so the equation $x^2 = 4k$ will have two imaginary solutions. The answer is (a).

8. $2x^2 - 8x - 3 = 0$

$$b^2 - 4ac = (-8)^2 - 4(2)(-3)$$
$$= 64 + 24 = 88$$

The discriminant is positive but not a perfect square, so there will be two distinct irrational number solutions.

9. $$3 - \frac{16}{x} - \frac{12}{x^2} = 0$$

Multiply by the LCD, x^2.

$$x^2\left(3 - \frac{16}{x} - \frac{12}{x^2}\right) = x^2 \cdot 0$$
$$3x^2 - 16x - 12 = 0$$
$$(3x + 2)(x - 6) = 0$$
$$3x + 2 = 0 \quad \text{or} \quad x - 6 = 0$$
$$x = -\frac{2}{3} \quad \text{or} \quad x = 6$$

Check $x = -\frac{2}{3}$: $3 + 24 - 27 \overset{?}{=} 0$ *True*

Check $x = 6$: $3 - \frac{8}{3} - \frac{1}{3} \overset{?}{=} 0$ *True*

Solution set: $\left\{-\frac{2}{3}, 6\right\}$

10. $$9x^4 + 4 = 37x^2$$
$$9x^4 - 37x^2 + 4 = 0$$

Let $u = x^2$, so $u^2 = \left(x^2\right)^2 = x^4$.
The equation becomes
$$9u^2 - 37u + 4 = 0.$$
$$(9u - 1)(u - 4) = 0$$
$$9u - 1 = 0 \quad \text{or} \quad u - 4 = 0$$
$$u = \frac{1}{9} \quad \text{or} \quad u = 4$$

To find x, substitute x^2 for u.

$$x^2 = \frac{1}{9} \qquad \text{or} \qquad x^2 = 4$$
$$x = \frac{1}{3} \text{ or } x = -\frac{1}{3} \quad \text{or} \quad x = 2 \text{ or } x = -2$$

Check $x = \pm\frac{1}{3}$: $\frac{1}{9} + 4 \overset{?}{=} \frac{37}{9}$ *True*

Check $x = \pm 2$: $144 + 4 \overset{?}{=} 37(4)$ *True*

Solution set: $\left\{-2, -\frac{1}{3}, \frac{1}{3}, 2\right\}$

11. $12 = (2d + 1)^2 + (2d + 1)$
Let $u = 2d + 1$. The equation becomes
$$12 = u^2 + u.$$
$$0 = u^2 + u - 12$$
$$0 = (u + 4)(u - 3)$$

$$u + 4 = 0 \quad \text{or} \quad u - 3 = 0$$
$$u = -4 \quad \text{or} \quad u = 3$$

To find d, substitute $2d + 1$ for u.

$$2d + 1 = -4 \quad \text{or} \quad 2d + 1 = 3$$
$$2d = -5 \qquad\qquad 2d = 2$$
$$d = -\frac{5}{2} \quad \text{or} \qquad d = 1$$

Check $d = -\frac{5}{2}$: $12 \overset{?}{=} 16 - 4$ *True*

Check $d = 1$: $12 \overset{?}{=} 9 + 3$ *True*

Solution set: $\left\{-\frac{5}{2}, 1\right\}$

12. Solve $S = 4\pi r^2$ for r.
$$\frac{S}{4\pi} = r^2$$

$$r = \pm\sqrt{\frac{S}{4\pi}} = \frac{\pm\sqrt{S}}{2\sqrt{\pi}}$$
$$= \frac{\pm\sqrt{S}}{2\sqrt{\pi}} \cdot \frac{\sqrt{\pi}}{\sqrt{\pi}}$$
$$r = \frac{\pm\sqrt{\pi S}}{2\pi}$$

13. Let x = the width of the walk.
The area of the walk is equal to the area of the outer figure minus the area of the pool.

$$152 = (10 + 2x)(24 + 2x) - (24)(10)$$
$$152 = 240 + 68x + 4x^2 - 240$$
$$0 = 4x^2 + 68x - 152$$
$$0 = x^2 + 17x - 38$$
$$0 = (x + 19)(x - 2)$$

$$x + 19 = 0 \quad \text{or} \quad x - 2 = 0$$
$$x = -19 \quad \text{or} \quad x = 2$$

Reject -19 since width can't be negative. The walk is 2 feet wide.

14. Let x = the height of the tower, and $2x + 2$ = the distance from the point to the top.

The distance from the base to the point is 30 m. These three segments form a right triangle, so the Pythagorean formula applies.

$$a^2 + b^2 = c^2$$
$$x^2 + 30^2 = (2x + 2)^2$$
$$x^2 + 900 = 4x^2 + 8x + 4$$
$$0 = 3x^2 + 8x - 896$$
$$0 = (x - 16)(3x + 56)$$

$$x - 16 = 0 \quad \text{or} \quad 3x + 56 = 0$$
$$x = 16 \quad \text{or} \quad x = -\frac{56}{3}$$

Reject $-\dfrac{56}{3}$ since height can't be negative. The tower is 16 m high.

15. $f(x) = \dfrac{1}{2}x^2 - 2$

$f(x) = \dfrac{1}{2}(x - 0)^2 - 2$

The graph is a parabola in $f(x) = a(x - h)^2 + k$ form with vertex (h, k) at $(0, -2)$. Since

$a = \dfrac{1}{2} > 0$, the parabola opens upward. Also,

$|a| = \left|\dfrac{1}{2}\right| = \dfrac{1}{2} < 1$, so the graph of the parabola

is wider than the graph of $f(x) = x^2$. The points $(2, 0)$ and $(-2, 0)$ are on the graph.

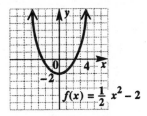

From the graph, we see that the x-values can be any real number, so the domain is $(-\infty, \infty)$. The y-values are greater than or equal to -2, so the range is $[-2, \infty)$.

16. $f(x) = -x^2 + 4x - 1$

Replace $f(x)$ with y and complete the square to find the vertex.

$$y = -x^2 + 4x - 1$$
$$-y = x^2 - 4x + 1$$
$$-y - 1 = x^2 - 4x$$
$$-y - 1 + 4 = x^2 - 4x + 4$$
$$-y + 3 = (x - 2)^2$$
$$-y = (x - 2)^2 - 3$$
$$y = -(x - 2)^2 + 3$$

The graph is a parabola with vertex (h, k) at $(2, 3)$. Since $a = -1 < 0$, the parabola opens downward.

Also, $|a| = |-1| = 1$, so the graph has the same shape as the graph of $f(x) = x^2$. The points $(0, -1)$ and $(4, -1)$ are on the graph.

$f(x) = -x^2 + 4x - 1$

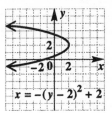

17. $x = -(y - 2)^2 + 2$

The equation is in $x = a(y - k)^2 + h$ form. The graph is a horizontal parabola with vertex (h, k) at $(2, 2)$. Since $a = -1 < 0$, the graph opens to the left. Also, $|a| = |-1| = 1$, so the graph has the same shape as the graph of $y = x^2$. The points $(-2, 0)$ and $(-2, 4)$ are on the same graph.

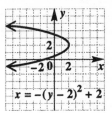

$x = -(y - 2)^2 + 2$

18. $f(x) = 22.56x^2 - 129.8x + 611.8$

(a) 1992 corresponds to $x = 2$.

$$f(2) = 442.44 \approx 442$$

Based on the model, about 442 million salmon were caught in 1992.

(b) Find the vertex.

$$x = \dfrac{-b}{2a} = \dfrac{-(-129.8)}{2(22.56)} \approx 2.877$$
$$f(2.88) \approx 425$$

Rounding 2.88 to 3 gives us a minimum of 425 million salmon in 1993.

19.
$$2x^2 + 7x > 15$$
$$2x^2 + 7x - 15 > 0$$
Solve the equation
$$2x^2 + 7x - 15 = 0.$$
$$(2x - 3)(x + 5) = 0$$

$$2x - 3 = 0 \quad \text{or} \quad x + 5 = 0$$
$$x = \dfrac{3}{2} \quad \text{or} \quad x = -5$$

The numbers -5 and $\dfrac{3}{2}$ divide the number line into three regions.

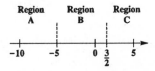

Test a number from each region in the inequality

$$2x^2 + 7x > 15.$$

continued

Region A: Let $x = -6$.

$$72 - 42 > 15 \quad ?$$

$$30 > 15 \qquad \textit{True}$$

Region B: Let $x = 0$.

$$0 > 15 \qquad \textit{False}$$

Region C: Let $x = 2$.

$$8 + 14 > 15 \quad ?$$

$$22 > 15 \qquad \textit{True}$$

The numbers in Regions A and C, not including -5 and $\dfrac{3}{2}$, are solutions.

Solution set: $(-\infty, -5) \cup \left(\dfrac{3}{2}, \infty\right)$

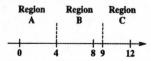

20. $\dfrac{5}{t-4} \leq 1$

Solve the equation

$$\dfrac{5}{t-4} = 1.$$

$$5 = t - 4$$

$$t = 9$$

Find the number that makes the denominator 0.

$$t - 4 = 0$$

$$t = 4$$

The numbers 4 and 9 divide the number line into three regions.

Region A	Region B	Region C

```
  +----+------++--+-->
  0    4     8 9  12
```

Test a number from each region in the inequality

$$\dfrac{5}{t-4} \leq 1.$$

Region A: Let $t = 0$.

$$\dfrac{5}{-4} \leq 1 \qquad \textit{True}$$

Region B: Let $t = 7$.

$$\dfrac{5}{3} \leq 1 \qquad \textit{False}$$

Region C: Let $t = 10$.

$$\dfrac{5}{6} \leq 1 \qquad \textit{True}$$

The numbers in Regions A and C, including 9 but not 4, are solutions.

Solution set: $(-\infty, 4) \cup [9, \infty)$

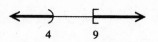

Cumulative Review Exercises Chapters 1–8

1. $S = \left\{-\dfrac{7}{3}, -2, -\sqrt{3}, 0, .7, \sqrt{12}, \sqrt{-8}, 7, \dfrac{32}{3}\right\}$

(a) The elements of S that are integers are -2, 0, and 7.

(b) The elements of S that are rational numbers are $-\dfrac{7}{3}, -2, 0, .7, 7,$ and $\dfrac{32}{3}$.

2. $2(-3)^2 + (-8)(-5) + (-17)$

$$= 2(9) + 40 - 17$$

$$= 18 + 23 = 41$$

3. $7 - (4 + 3t) + 2t = -6(t - 2) - 5$

$$7 - 4 - 3t + 2t = -6t + 12 - 5$$

$$3 - t = -6t + 7$$

$$5t = 4$$

$$t = \dfrac{4}{5}$$

Check $t = \dfrac{4}{5}$: $\dfrac{11}{5} = \dfrac{11}{5}$

Solution set: $\left\{\dfrac{4}{5}\right\}$

4. $|6x - 9| = |-4x + 2|$

$6x - 9 = -4x + 2$ or $6x - 9 = -(-4x + 2)$

$10x = 11$ $\qquad\qquad\quad 6x - 9 = 4x - 2$

$\qquad\qquad\qquad\qquad\qquad 2x = 7$

$x = \dfrac{11}{10}$ or $x = \dfrac{7}{2}$

Check $x = \dfrac{11}{10}$: $\left|-\dfrac{24}{10}\right| \overset{?}{=} \left|-\dfrac{24}{10}\right|$ *True*

Check $x = \dfrac{7}{2}$: $|12| \overset{?}{=} |-12|$ *True*

Solution set: $\left\{\dfrac{11}{10}, \dfrac{7}{2}\right\}$

5. $2x = \sqrt{\dfrac{5x + 2}{3}}$

Square both sides.

$$(2x)^2 = \left(\sqrt{\dfrac{5x + 2}{3}}\right)^2$$

$$4x^2 = \dfrac{5x + 2}{3}$$

$$12x^2 = 5x + 2$$

$$12x^2 - 5x - 2 = 0$$

$$(3x - 2)(4x + 1) = 0$$

$$3x - 2 = 0 \quad \text{or} \quad 4x + 1 = 0$$

$$x = \frac{2}{3} \quad \text{or} \quad x = -\frac{1}{4}$$

Check $x = \frac{2}{3}$: $\frac{4}{3} \stackrel{?}{=} \sqrt{\frac{16}{9}}$ *True*

Check $x = -\frac{1}{4}$: $-\frac{1}{2} \stackrel{?}{=} \sqrt{\frac{1}{4}}$ *False*

Solution set: $\left\{ \dfrac{2}{3} \right\}$

6.

$$\frac{3}{x - 3} - \frac{2}{x - 2} = \frac{3}{x^2 - 5x + 6}$$

$$\frac{3}{x - 3} - \frac{2}{x - 2} = \frac{3}{(x - 3)(x - 2)}$$

Multiply by the LCD, $(x - 3)(x - 2)$.

$$(x - 3)(x - 2)\left(\frac{3}{x - 3} - \frac{2}{x - 2} \right)$$

$$= (x - 3)(x - 2)\left[\frac{3}{(x - 3)(x - 2)} \right]$$

$$3(x - 2) - 2(x - 3) = 3$$

$$3x - 6 - 2x + 6 = 3$$

$$x = 3$$

The number 3 is not allowed as a solution since it makes the denominator 0. The solution set is $\emptyset$.

7. $(r - 5)(2r + 3) = 1$

$$2r^2 - 7r - 15 = 1$$

$$2r^2 - 7r - 16 = 0$$

Use the quadratic formula.

$$t = \frac{-b \pm \sqrt{b^2 - 4ac}}{2a}$$

$$t = \frac{-(-7) \pm \sqrt{(-7)^2 - 4(2)(-16)}}{2(2)}$$

$$= \frac{7 \pm \sqrt{49 + 128}}{4} = \frac{7 \pm \sqrt{177}}{4}$$

Solution set: $\left\{ \dfrac{7 + \sqrt{177}}{4}, \dfrac{7 - \sqrt{177}}{4} \right\}$

8. $b^4 - 5b^2 + 4 = 0$

Let $u = b^2$, so $u^2 = \left(b^2 \right)^2 = b^4$.

The equation becomes

$$u^2 - 5u + 4 = 0.$$

$$(u - 4)(u - 1) = 0$$

$$u - 4 = 0 \quad \text{or} \quad u - 1 = 0$$

$$u = 4 \quad \text{or} \quad u = 1$$

To find b, substitute b^2 for u.

$$b^2 = 4 \qquad \text{or} \qquad b^2 = 1$$

$$b = 2 \text{ or } b = -2 \quad \text{or} \quad b = 1 \text{ or } b = -1$$

The potential solutions check.

Solution set: $\{ -2, -1, 1, 2 \}$

9. $-2x + 4 \le -x + 3$

$$-x \le -1$$

Multiply by -1, and reverse the direction of the inequality.

$$x \ge 1$$

Solution set: $[1, \infty)$

10. $|3y - 7| \le 1$

$$-1 \le 3y - 7 \le 1$$

$$6 \le 3y \le 8$$

$$2 \le y \le \frac{8}{3}$$

Solution set: $\left[2, \dfrac{8}{3} \right]$

11. $x^2 - 4x + 3 < 0$

Solve the equation

$$x^2 - 4x + 3 = 0.$$

$$(x - 3)(x - 1) = 0$$

$$x - 3 = 0 \quad \text{or} \quad x - 1 = 0$$

$$x = 3 \quad \text{or} \quad x = 1$$

The numbers 1 and 3 divide the number line into three regions.

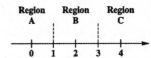

Test a number from each region in the inequality

$$x^2 - 4x + 3 < 0.$$

Region A: Let $x = 0$.

$$3 < 0 \qquad \textit{False}$$

Region B: Let $x = 2$.

$$4 - 8 + 3 < 0 \quad ?$$

$$-1 < 0 \qquad \textit{True}$$

Region C: Let $x = 4$.

$$16 - 16 + 3 < 0 \quad ?$$

$$3 < 0 \qquad \textit{False}$$

The numbers from Region B, not including 1 or 3, are solutions.

Solution set: $(1, 3)$

12. $\dfrac{3}{y+2} > 1$

Solve the equation

$\dfrac{3}{y+2} = 1.$

$3 = y+2$

$y = 1$

Find the number that makes the denominator 0.

$y+2 = 0$

$y = -2$

The numbers -2 and 1 divide the number line into three regions.

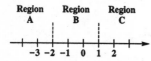

Test a number from each region in the inequality

$\dfrac{3}{y+2} > 1.$

Region A: Let $y = -4.$

$\dfrac{3}{-2} > 1$ *False*

Region B: Let $y = 0.$

$\dfrac{3}{2} > 1$ *True*

Region C: Let $y = 2.$

$\dfrac{3}{4} > 1$ *False*

The numbers from Region B, not including -2 or 1, are solutions.

Solution set: $(-2, 1)$

13. $4x - 5y = 15$

Draw the line through its intercepts, $\left(\dfrac{15}{4}, 0\right)$ and

$(0, -3)$. The graph passes the vertical line test, so the relation is a function. As with any line that is not horizontal or vertical, the domain and range are both $(-\infty, \infty)$.

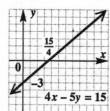

14. $4x - 5y < 15$

Draw a dashed line through the points $\left(\dfrac{15}{4}, 0\right)$

and $(0, -3)$. Check the origin:

$4(0) - 5(0) < 15$?

$0 < 15$ *True*

Shade the region that contains the origin.

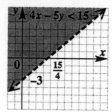

The graph does not pass the vertical line test, so the relation is not a function.

15. $f(x) = -2(x-1)^2 + 3$

The equation is in $f(x) = a(x-h)^2 + k$ form, so the graph is a parabola with vertex (h, k) at $(1, 3)$. Since $a = -2 < 0$, the parabola opens downward. Also $|a| = |-2| = 2 > 1$, so the graph is narrower than the graph of $f(x) = x^2$. The points $(0, 1)$ and $(2, 1)$ are on the graph.

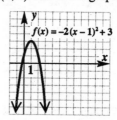

The graph passes the vertical line test, so the relation is a function. The domain is $(-\infty, \infty)$ and the range is $(-\infty, 3]$.

$x + 7y = 16$

Solve the equation for y.

$7y = 2x + 16$

$y = \dfrac{2}{7}x + \dfrac{16}{7}$

So the slope is $\dfrac{2}{7}$ and the y-intercept is $\left(0, \dfrac{16}{7}\right)$.

Let $y = 0$ in $-2x + 7y = 16$ to find the x-intercept.

$-2x + 7(0) = 16$

$-2x = 16$

$x = -8$

The x-intercept is $(-8, 0)$.

17.   Solve $5x + 2y = 6$ for y.

$2y = -5x + 6$

$y = -\dfrac{5}{2}x + 3$

So the slope of the given line is $-\dfrac{5}{2}$ and the required form is

$y = -\dfrac{5}{2}x + b.$

$-3 = -\dfrac{5}{2}(2) + b$ *Let x=2, y=−3.*

$-3 = -5 + b$

$2 = b$

The equation is $y = -\dfrac{5}{2}x + 2$.

18. The negative reciprocal of the slope in Exercise 17 is

$$-\frac{1}{-\dfrac{5}{2}} = \frac{2}{5}.$$

So $y = \dfrac{2}{5}x + b$.

$1 = \dfrac{2}{5}(-4) + b$ *Let x= –4, y=1.*

$1 = -\dfrac{8}{5} + b$

$\dfrac{13}{5} = b$

The equation is $y = \dfrac{2}{5}x + \dfrac{13}{5}$.

19. **(a)** The points are $(0, 116.26)$, $(10, 132.02)$, $(20, 141.85)$, $(30, 155.38)$, and $(33, 156.36)$.

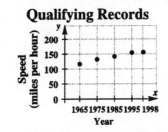

Qualifying Records

(b) The set of ordered pairs is a function since each year corresponds to a unique speed.

(c) The slope m of the line passing through $(0, 116.26)$ and $(20, 141.85)$ is

$$m = \frac{141.85 - 116.26}{20 - 0} = \frac{25.59}{20} = 1.2795.$$

Note that $x = 0$ represents 1965, so the point $(0, 116.26)$ can be considered to be the y-intercept.

The slope-intercept form of the linear equation that models these data is

$$y = 1.2795x + 116.26.$$

(d) 1998 corresponds to $x = 33$.

$$y = 1.2795(33) + 116.26$$
$$= 158.4835,$$

which is an overestimate of the actual 156.36.

20. The relation $x = 5$ does not define a function because its graph is a vertical line, which is not the graph of a function by the vertical line test.

21. $f(x) = 2(x - 1)^2 - 5$

(a) $f(-2) = 2(-2 - 1)^2 - 5$
$\qquad\quad = 2(-3)^2 - 5$
$\qquad\quad = 2(9) - 5 = 13$

(b) Any value can be substituted for x, so the domain is $(-\infty, \infty)$. The graph of f is a parabola that opens upward with vertex $(1, -5)$. The vertex is a minimum point so the y-values are all greater than or equal to -5. Thus, the range is $[-5, \infty)$.

22. $\begin{array}{rcll} 2x & - & 4y & = & 10 \quad (1) \\ 9x & + & 3y & = & 3 \quad (2) \end{array}$

Simplify the equations.

$$x - 2y = 5 \quad (3) \; \tfrac{1}{2} \times (1)$$
$$3x + y = 1 \quad (4) \; \tfrac{1}{3} \times (2)$$

To eliminate y, multiply (4) by 2 and add the result to (3).

$$\begin{array}{rl} x - 2y = 5 & (3) \\ 6x + 2y = 2 & 2 \times (4) \\ \hline 7x \quad\quad = 7 & \\ x = 1 & \end{array}$$

Substitute $x = 1$ into (4).

$$3x + y = 1 \qquad (4)$$
$$3(1) + y = 1$$
$$y = -2$$

Solution set: $\{(1, -2)\}$

23. $\begin{array}{rcl} x + y + 2z & = & 3 \quad (1) \\ -x + y + z & = & -5 \quad (2) \\ 2x + 3y - z & = & -8 \quad (3) \end{array}$

Eliminate z by adding (2) and (3).

$$\begin{array}{rl} -x + y + z = -5 & (2) \\ 2x + 3y - z = -8 & (3) \\ \hline x + 4y \quad\quad = -13 & (4) \end{array}$$

To get another equation without z, multiply equation (3) by 2 and add the result to equation (1).

$$\begin{array}{rl} x + y + 2z = 3 & (1) \\ 4x + 6y - 2z = -16 & 2 \times (3) \\ \hline 5x + 7y \quad\quad = -13 & (5) \end{array}$$

continued

To eliminate x, multiply (4) by -5 and add the result to (5).

$$
\begin{array}{ll}
-5x - 20y = 65 & -5 \times (4) \\
\underline{5x + 7y = -13} & (5) \\
{-}13y = 52 & \\
y = -4 &
\end{array}
$$

Use (4) to find x.

$$
\begin{aligned}
x + 4y &= -13 \quad (4) \\
x + 4(-4) &= -13 \\
x - 16 &= -13 \\
x &= 3
\end{aligned}
$$

Use (2) to find z.

$$
\begin{aligned}
-x + y + z &= -5 \quad (2) \\
-3 - 4 + z &= -5 \\
-7 + z &= -5 \\
z &= 2
\end{aligned}
$$

Solution set: $\{(3, -4, 2)\}$

24. Let $x =$ the speed of the boat in still water and $y =$ the speed of the current.

Use the facts that the rate upriver is $x - y$ and that the rate downriver is $x + y$ along with the given information to make a chart.

	d	r	t
Upriver	20	$x - y$	1
Downriver	20	$x + y$	.5

Use $d = rt$ to write a system of equations.

$$
\begin{aligned}
20 &= (x - y)(1) \quad (1) \\
20 &= (x + y)(.5) \quad (2)
\end{aligned}
$$

Multiply (2) by 2 and add the result to (1)

$$
\begin{array}{ll}
20 = x - y & (1) \\
\underline{40 = x + y} & 2 \times (2) \\
60 = 2x & \\
30 = x &
\end{array}
$$

From (1), $20 = 30 - y$, so $y = 10$.
The speed of the boat is 30 mph and the speed of the current is 10 mph.

25. $(7x + 4)(2x - 3)$

$$
\begin{aligned}
&= 14x^2 - 21x + 8x - 12 \\
&= 14x^2 - 13x - 12
\end{aligned}
$$

26. $\left(\dfrac{2}{3}t + 9\right)^2 = \left(\dfrac{2}{3}t\right)^2 + 2\left(\dfrac{2}{3}t\right)(9) + 9^2$

$$
= \dfrac{4}{9}t^2 + 12t + 81
$$

27. $\left(3t^3 + 5t^2 - 8t + 7\right) - \left(6t^3 + 4t - 8\right)$

$$
\begin{aligned}
&= 3t^3 + 5t^2 - 8t + 7 - 6t^3 - 4t + 8 \\
&= -3t^3 + 5t^2 - 12t + 15
\end{aligned}
$$

28. Divide $4x^3 + 2x^2 - x + 26$ by $x + 2$. Use synthetic division.

$$
\begin{array}{r|rrrr}
-2 & 4 & 2 & -1 & 26 \\
   &   & -8 & 12 & -22 \\
\hline
   & 4 & -6 & 11 & 4
\end{array}
$$

The answer is

$$
4x^2 - 6x + 11 + \dfrac{4}{x + 2}.
$$

29. $16x - x^3 = x\left(16 - x^2\right)$

$$
= x(4 + x)(4 - x)
$$

30. $(3x + 2)^2 - 4(3x + 2) - 5$

Let $m = 3x + 2$.

$$
\begin{aligned}
&= m^2 - 4m - 5 \\
&= (m - 5)(m + 1)
\end{aligned}
$$

Substitute $3x + 2$ for m.

$$
\begin{aligned}
&= [(3x + 2) - 5][(3x + 2) + 1] \\
&= (3x - 3)(3x + 3) \\
&= 3(x - 1) \cdot 3(x + 1) \\
&= 9(x - 1)(x + 1)
\end{aligned}
$$

31. $8x^3 + 27y^3$

Use the sum of two cubes formula,

$$
a^3 + b^3 = (a + b)\left(a^2 - ab + b^2\right),
$$

with $a = 2x$ and $b = 3y$.

$$
\begin{aligned}
&8x^3 + 27y^3 \\
&= (2x + 3y)[(2x)^2 - (2x)(3y) + (3y)^2] \\
&= (2x + 3y)\left(4x^2 - 6xy + 9y^2\right)
\end{aligned}
$$

32. $9x^2 - 30xy + 25y^2$

Use the perfect square formula,

$$
a^2 - 2ab + b^2 = (a - b)^2,
$$

with $a = 3x$ and $b = 5y$.

$$
\begin{aligned}
&9x^2 - 30xy + 25y^2 \\
&= [(3x)^2 - 2(3x)(5y) + (5y)^2] \\
&= (3x - 5y)^2
\end{aligned}
$$

33. $\dfrac{x^2 - 3x - 10}{x^2 + 3x + 2} \cdot \dfrac{x^2 - 2x - 3}{x^2 + 2x - 15}$

$$
\begin{aligned}
&= \dfrac{(x - 5)(x + 2)}{(x + 2)(x + 1)} \cdot \dfrac{(x - 3)(x + 1)}{(x + 5)(x - 3)} \\
&= \dfrac{x - 5}{x + 5}
\end{aligned}
$$

34.

$$\frac{3}{2-k} - \frac{5}{k} + \frac{6}{k^2 - 2k}$$

$$= \frac{3}{2-k} - \frac{5}{k} + \frac{6}{k(k-2)}$$

$$= \frac{-3}{k-2} - \frac{5}{k} + \frac{6}{k(k-2)}$$

The LCD is $k(k-2)$.

$$= \frac{-3k}{(k-2)k} - \frac{5(k-2)}{k(k-2)} + \frac{6}{k(k-2)}$$

$$= \frac{-3k - 5(k-2) + 6}{k(k-2)}$$

$$= \frac{-3k - 5k + 10 + 6}{k(k-2)}$$

$$= \frac{-8k + 16}{k(k-2)}$$

$$= \frac{-8(k-2)}{k(k-2)} = -\frac{8}{k}$$

35.

$$\frac{\dfrac{r}{s} - \dfrac{s}{r}}{\dfrac{r}{s} + 1}$$

Multiply the numerator and denominator by the LCD of all the fractions, rs.

$$= \frac{\left(\dfrac{r}{s} - \dfrac{s}{r}\right)rs}{\left(\dfrac{r}{s} + 1\right)rs} = \frac{r^2 - s^2}{r^2 + rs}$$

$$= \frac{(r-s)(r+s)}{r(r+s)} = \frac{r-s}{r}$$

36. $\sqrt[3]{\dfrac{27}{16}} = \dfrac{\sqrt[3]{27}}{\sqrt[3]{16}} = \dfrac{\sqrt[3]{3^3}}{\sqrt[3]{8 \cdot 2}} = \dfrac{3}{2\sqrt[3]{2}}$

$$= \frac{3 \cdot \sqrt[3]{4}}{2\sqrt[3]{2} \cdot \sqrt[3]{4}} = \frac{3\sqrt[3]{4}}{2\sqrt[3]{8}}$$

$$= \frac{3\sqrt[3]{4}}{2 \cdot 2} = \frac{3\sqrt[3]{4}}{4}$$

37. $\dfrac{2}{\sqrt{7} - \sqrt{5}} = \dfrac{2\left(\sqrt{7} + \sqrt{5}\right)}{\left(\sqrt{7} - \sqrt{5}\right)\left(\sqrt{7} + \sqrt{5}\right)}$

$$= \frac{2\left(\sqrt{7} + \sqrt{5}\right)}{7 - 5}$$

$$= \frac{2\left(\sqrt{7} + \sqrt{5}\right)}{2} = \sqrt{7} + \sqrt{5}$$

38. Let x = the width of the rectangle.
The perimeter of the rectangle is 20 inches. Since the formula for perimeter is $P = 2L + 2W$,

$$20 = 2L + 2x.$$

Solve the equation for L.

$$20 - 2x = 2L$$
$$10 - x = L$$

The area of the rectangle is 21 in^2.
Since the formula for area is LW,

$$21 = (10 - x)x$$
$$21 = 10x - x^2$$
$$x^2 - 10x + 21 = 0$$
$$(x - 7)(x - 3) = 0$$
$$x - 7 = 0 \quad \text{or} \quad x - 3 = 0$$
$$x = 7 \quad \text{or} \qquad x = 3.$$

Reject 7 for the width, since width is not longer than length. Thus, the width is 3 inches, and the length is $10 - x = 7$ inches.

39. Let x = Tri's rate on the bicycle and
$x - 10$ = Tri's rate while walking.

Make a chart. Use $d = rt$, or $t = \dfrac{d}{r}$.

	Distance	Rate	Time
Bicycle	12	x	$\dfrac{12}{x}$
Walking	8	$x - 10$	$\dfrac{8}{x - 10}$

$$\begin{array}{ccc} \text{Tri's time on} & \text{Tri's time} & 5 \\ \text{the bicycle} + \text{walking} = \text{hours.} \end{array}$$

$$\frac{12}{x} + \frac{8}{x - 10} = 5$$

Multiply by the LCD, $x(x - 10)$.

$$x(x - 10)\left(\frac{12}{x} + \frac{8}{x - 10}\right) = x(x - 10) \cdot 5$$
$$12(x - 10) + 8x = 5x(x - 10)$$
$$12x - 120 + 8x = 5x^2 - 50x$$
$$0 = 5x^2 - 70x + 120$$
$$0 = x^2 - 14x + 24$$
$$0 = (x - 12)(x - 2)$$

$$x - 12 = 0 \quad \text{or} \quad x - 2 = 0$$
$$x = 12 \quad \text{or} \qquad x = 2$$

Reject 2 for Tri's bicycle speed, since it would yield a negative walking speed. Thus, his bicycle speed was 12 mph, and his walking speed was $x - 10 = 2$ mph.

40. Let $x =$ the distance traveled by the southbound car and

$2x - 38 =$ the distance traveled by the eastbound car.

Since the cars are traveling at right angles with one another, the Pythagorean formula can be applied.

$$a^2 + b^2 = c^2$$
$$x^2 + (2x - 38)^2 = 95^2$$
$$x^2 + 4x^2 - 152x + 1444 = 9025$$
$$5x^2 - 152x - 7581 = 0$$
$$x = \frac{-(-152) \pm \sqrt{(-152)^2 - 4(5)(-7581)}}{2(5)}$$
$$= \frac{152 \pm \sqrt{174,724}}{10} = \frac{152 \pm 418}{10}$$

So $x = \dfrac{152 \pm 418}{10} = 57$ (the other value is negative). The southbound car traveled 57 miles, and the eastbound car traveled $2x - 38 = 2(57) - 38 = 76$ miles.

CHAPTER 9   INVERSE, EXPONENTIAL, AND LOGARITHMIC FUNCTIONS

Section 9.1

1. This function is not one-to-one because both Illinois and Wisconsin are paired with the same range element, 40.

3. The function in the table that pairs a city with a distance is a one-to-one function because for each city there is one distance and each distance has only one city to which it is paired.

    If the distance from Indianapolis to Denver had 1 mile added to it, it would be $1058 + 1 = 1059$ mi, the same as the distance from Los Angeles to Denver. In this case, one distance would have two cities to which it is paired, and the function would not be one-to-one.

5. If a function is made up of ordered pairs in such a way that the same y-value appears in a correspondence with two different x-values, then the function is not one-to-one. **(b)**

7. All of the graphs pass the vertical line test, so they all represent functions. The graph in choice (a) is the only one that passes the horizontal line test, so it is the one-to-one function.

9. $\{(3, 6), (2, 10), (5, 12)\}$ is a one-to-one function, since each x-value corresponds to only one y-value and each y-value corresponds to only one x-value. To find the inverse, interchange x and y in each ordered pair. The inverse is

    $$\{(6, 3), (10, 2), (12, 5)\}.$$

11. $\{(-1, 3), (2, 7), (4, 3), (5, 8)\}$ is not a one-to-one function. The ordered pairs $(-1, 3)$ and $(4, 3)$ have the same y-values for two different x-values.

13. The graph of $f(x) = 2x + 4$ is a nonvertical, nonhorizontal line. By the horizontal line test, $f(x)$ is a one-to-one function. To find the inverse, replace $f(x)$ with y.

    $$y = 2x + 4$$

    Interchange x and y.

    $$x = 2y + 4$$

    Solve for y.

    $$2y = x - 4$$
    $$y = \frac{x - 4}{2}$$

    Replace y with $f^{-1}(x)$.

    $$f^{-1}(x) = \frac{x - 4}{2}$$

15. Write $g(x) = \sqrt{x - 3}$ as $y = \sqrt{x - 3}$. Since $x \geq 3$, $y \geq 0$. The graph of g is half of a horizontal parabola that opens to the right. The graph passes the horizontal line test, so g is one-to-one. To find the inverse, interchange x and y to get

    $$x = \sqrt{y - 3}.$$

    Note that now $y \geq 3$, so $x \geq 0$.
    Solve for y by squaring both sides.

    $$x^2 = y - 3$$
    $$x^2 + 3 = y$$

    Replace y with $g^{-1}(x)$.

    $$g^{-1}(x) = x^2 + 3, \; x \geq 0$$

17. $f(x) = 3x^2 + 2$ is not a one-to-one function because two x-values, such as 1 and -1, both have the same y-value, in this case 5. The graph of this function is a vertical parabola which does not pass the horizontal line test.

19. The graph of $f(x) = x^3 - 4$ is the graph of $g(x) = x^3$ shifted down 4 units. (Recall that $g(x) = x^3$ is the elongated S-shaped curve.) The graph of f passes the horizontal line test, so f is one-to-one.

    Replace $f(x)$ with y.

    $$y = x^3 - 4$$

    Interchange x and y.

    $$x = y^3 - 4$$

    Solve for y.

    $$x + 4 = y^3$$

    Take the cube root of each side.

    $$\sqrt[3]{x + 4} = y$$

    Replace y with $f^{-1}(x)$.

    $$f^{-1}(x) = \sqrt[3]{x + 4}$$

21. **(a)** $f(x) = 2^x$
    To find $f(3)$, substitute 3 for x.
    $$f(3) = 2^3 = 8$$

    (b) Since f is one-to-one and $f(3) = 8$, it follows that $f^{-1}(8) = 3$.

23. **(a)** $f(x) = 2^x$
    To find $f(0)$, substitute 0 for x.
    $$f(0) = 2^0 = 1$$

    (b) Since f is one-to-one and $f(0) = 1$, it follows that $f^{-1}(1) = 0$.

25. **(a)** The function is one-to-one since any horizontal line intersects the graph at most once.

(b) In the graph, the two points marked on the line are $(-1, 5)$ and $(2, -1)$. Interchange x and y in each ordered pair to get $(5, -1)$ and $(-1, 2)$. Plot these points, then draw a dashed line through them to obtain the graph of the inverse function.

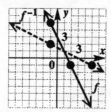

27. **(a)** The function is not one-to-one since there are horizontal lines that intersect the graph more than once. For example, the line $y = 1$ intersects the graph twice.

29. **(a)** The function is one-to-one since any horizontal line intersects the graph at most once.

(b) In the graph, the four points marked on the curve are $(-4, 2)$, $(-1, 1)$, $(1, -1)$, and $(4, -2)$. Interchange x and y in each ordered pair to get $(2, -4)$, $(1, -1)$, $(-1, 1)$, and $(-2, 4)$. Plot these points, then draw a dashed curve (symmetric to the original graph about the line $y = x$) through them to obtain the graph of the inverse.

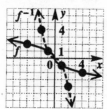

31. $f(x) = 2x - 1$ or $y = 2x - 1$

The graph is a line through $(-2, -5)$, $(0, -1)$, and $(3, 5)$. Plot these points and draw the solid line through them. Then the inverse will be a line through $(-5, -2)$, $(-1, 0)$, and $(5, 3)$. Plot these points and draw the dashed line through them.

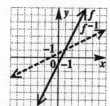

33. $g(x) = -4x$ or $y = -4x$

The graph is a line through $(0, 0)$ and $(1, -4)$. For the inverse, interchange x and y in each ordered pair to get the points $(0, 0)$ and $(-4, 1)$. Draw a dashed line through these points to obtain the graph of the inverse function.

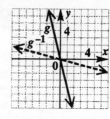

35. $f(x) = \sqrt{x}$, $x \geq 0$

Complete the table of values.

x	$f(x)$
0	0
1	1
4	2

Plot these points and connect them with a solid smooth curve.
Since $f(x)$ is one-to-one, make a table of values for $f^{-1}(x)$ by interchanging x and y.

x	$f^{-1}(x)$
0	0
1	1
2	4

Plot these points and connect them with a dashed smooth curve.

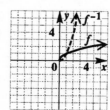

37. $f(x) = x^3 - 2$

Complete the table of values.

x	$f(x)$
-1	-3
0	-2
1	-1
2	6

Plot these points and connect them with a solid smooth curve.
Make a table of values for f^{-1}.

x	$f^{-1}(x)$
-3	-1
-2	0
-1	1
6	2

Plot these points and connect them with a dashed smooth curve.

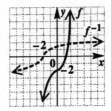

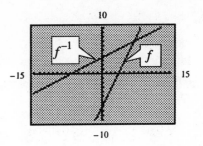

39.
$$f(x) = 4x - 5$$

Replace $f(x)$ with y.
$$y = 4x - 5$$
Interchange x and y.
$$x = 4y - 5$$
Solve for y.
$$x + 5 = 4y$$
$$\frac{x + 5}{4} = y$$
Replace y with $f^{-1}(x)$.
$$\frac{x + 5}{4} = f^{-1}(x)$$

40. Insert each number in the inverse function found in Exercise 39,
$$f^{-1}(x) = \frac{x + 5}{4}.$$
$$f^{-1}(47) = \frac{47 + 5}{4} = \frac{52}{4} = 13 = \text{M},$$
$$f^{-1}(95) = \frac{95 + 5}{4} = \frac{100}{4} = 25 = \text{Y},$$
and so on.

The decoded message is as follows:
My graphing calculator is the greatest thing since sliced bread.

41. A one-to-one code is essential to this process because if the code is not one-to-one, an encoded number would refer to two different letters.

42. Answers will vary according to the student's name. For example, Jane Doe is encoded as follows:

1004 5 2748 129 68 3379 129.

43. $Y_1 = f(x) = 2x - 7$
Replace $f(x)$ with y.
$$y = 2x - 7$$
Interchange x and y.
$$x = 2y - 7$$
Solve for y.
$$x + 7 = 2y$$
$$\frac{x + 7}{2} = y$$
Replace y with $f^{-1}(x)$.
$$\frac{x + 7}{2} = f^{-1}(x) = Y_2$$

Now graph Y_1 and Y_2.

45. $Y_1 = f(x) = x^3 + 5$
Replace $f(x)$ with y.
$$y = x^3 + 5$$
Interchange x and y.
$$x = y^3 + 5$$
Solve for y.
$$x - 5 = y^3$$
Take the cube root of each side.
$$\sqrt[3]{x - 5} = y$$
Replace y with $f^{-1}(x)$.
$$\sqrt[3]{x - 5} = f^{-1}(x) = Y_2$$

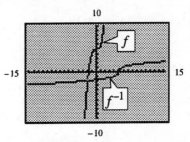

47. $Y_1 = x^2 + 3x + 4$

Graph Y_1 and its inverse in the same square window on a graphing calculator. On a TI-83, graph Y_1 and then enter

DrawInv Y_1

on the home screen. DrawInv is choice 8 under the DRAW menu. Y_1 is choice 1 under VARS, Y-VARS, Function.

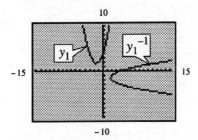

49. The "inverse" of the function in Exercise 47 does not actually satisfy the definition of inverse because the original function is required to be one-to-one. The original function $y = x^2 + 3x + 4$ is not one-to-one as shown by the fact that $(0, 4)$ and $(-3, 4)$ lie on the graph.

Section 9.2

1. Since the graph of $F(x) = a^x$ always contains the point $(0, 1)$, the correct response is (c).

3. Since the graph of $F(x) = a^x$ always approaches the x-axis, the correct response is (a).

5. $f(x) = 3^x$
   Make a table of values.
   $$f(-2) = 3^{-2} = \frac{1}{3^2} = \frac{1}{9},$$
   $$f(-1) = 3^{-1} = \frac{1}{3^1} = \frac{1}{3}, \text{ and so on.}$$

x	-2	-1	0	1	2
$f(x)$	$\frac{1}{9}$	$\frac{1}{3}$	1	3	9

   Plot the points from the table and draw a smooth curve through them.

   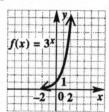

7. $g(x) = \left(\frac{1}{3}\right)^x$
   Make a table of values.
   $$g(-2) = \left(\frac{1}{3}\right)^{-2} = \left(\frac{3}{1}\right)^2 = 9,$$
   $$g(-1) = \left(\frac{1}{3}\right)^{-1} = \left(\frac{3}{1}\right)^1 = 3, \text{ and so on.}$$

x	-2	-1	0	1	2
$g(x)$	9	3	1	$\frac{1}{3}$	$\frac{1}{9}$

   Plot the points from the table and draw a smooth curve through them.

   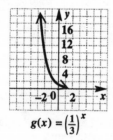

9. $y = 4^{-x}$
   This equation can be rewritten as
   $$y = \left(4^{-1}\right)^x = \left(\frac{1}{4}\right)^x,$$

which shows that it is *falling* from left to right. Make a table of values.

x	-2	-1	0	1	2
y	16	4	1	$\frac{1}{4}$	$\frac{1}{16}$

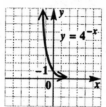

11. $y = 2^{2x-2}$
    Make a table of values. It will help to find values for $2x - 2$ before you find y.

x	-2	-1	0	1	2	3
$2x - 2$	-6	-4	-2	0	2	4
y	$\frac{1}{64}$	$\frac{1}{16}$	$\frac{1}{4}$	1	4	16

    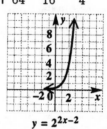
    $y = 2^{2x-2}$

13. (a) For an exponential function defined by $f(x) = a^x$, if $a > 1$, the graph *rises* from left to right. (See Example 1, $f(x) = 2^x$, in your text.) If $0 < a < 1$, the graph *falls* from left to right. (See Example 2, $g(x) = \left(\frac{1}{2}\right)^x = 2^{-x}$, in your text.)

    (b) An exponential function defined by $f(x) = a^x$ is one-to-one and has an inverse, since each value of $f(x)$ corresponds to one and only one value of x.

15. $$100^x = 1000$$
    Write each side as a power of 10.
    $$\left(10^2\right)^x = 10^3$$
    $$10^{2x} = 10^3$$
    For $a > 0$ and $a \neq 1$, if $a^x = a^y$, then $x = y$. Set the exponents equal to each other.
    $$2x = 3$$
    $$x = \frac{3}{2}$$

    Check $x = \frac{3}{2}$: $100^{3/2} \stackrel{?}{=} 1000$ *True*

    Solution set: $\left\{\frac{3}{2}\right\}$

17. $16^{2x+1} = 64^{x+3}$

Write each side as a power of 4.

$$\left(4^2\right)^{2x+1} = \left(4^3\right)^{x+3}$$
$$4^{4x+2} = 4^{3x+9}$$

Set the exponents equal.

$$4x + 2 = 3x + 9$$
$$x = 7$$

Check $x = 7$: $16^{15} \overset{?}{=} 64^{10}$ *True*

Solution set: $\{7\}$

19. $5^x = \dfrac{1}{125}$

$$5^x = \left(\dfrac{1}{5}\right)^3$$

Write each side as a power of 5.

$$5^x = 5^{-3}$$

Set the exponents equal.

$$x = -3$$

Check $x = -3$: $5^{-3} \overset{?}{=} \dfrac{1}{125}$ *True*

Solution set: $\{-3\}$

21. $5^x = .2$

$$5^x = \dfrac{2}{10} = \dfrac{1}{5}$$

Write each side as a power of 5.

$$5^x = 5^{-1}$$

Set the exponents equal.

$$x = -1$$

Check $x = -1$: $5^{-1} \overset{?}{=} .2$ *True*

Solution set: $\{-1\}$

23. $\left(\dfrac{3}{2}\right)^x = \dfrac{8}{27}$

$$\left(\dfrac{3}{2}\right)^x = \left(\dfrac{2}{3}\right)^3$$

Write each side as a power of $\dfrac{3}{2}$.

$$\left(\dfrac{3}{2}\right)^x = \left(\dfrac{3}{2}\right)^{-3}$$

Set the exponents equal.

$$x = -3$$

Check $x = -3$: $\left(\dfrac{3}{2}\right)^{-3} \overset{?}{=} \dfrac{8}{27}$ *True*

Solution set: $\{-3\}$

25. $12^{2.6} \approx 639.545$

27. $.5^{3.921} \approx .066$

29. $2.718^{2.5} \approx 12.179$

31. The reason many scientific calculators cannot calculate something like $(-2)^4$ is that in an exponential function, the base must be positive. This is necessary so the function can be defined for all real numbers. For example, $f(x) = (-4)^x$ would not be defined for $x = \dfrac{1}{2}$ since it is impossible to take the square root of a negative number.

33. **(a)** The increase for the exponential-type curve in the year 2000 is about .5°C.

(b) The increase for the linear graph in the year 2000 is about .35°C.

35. **(a)** The increase for the exponential-type curve in the year 2020 is about 1.6°C.

(b) The increase for the linear graph in the year 2020 is about .5°C.

37. $f(x) = 7147(1.0366)^x$

(a) 1950 corresponds to $x = 0$.

$$f(0) = 7147(1.0366)^0$$
$$= 7147(1) = 7147$$

The answer has units in millions of short tons. In case you were wondering, a short ton is 2000 pounds and a long ton is 2240 pounds.

(b) 1985 corresponds to $x = 35$.

$$f(35) = 7147(1.0366)^{35} \approx 25,149$$

(c) 1990 corresponds to $x = 40$.

$$f(40) = 7147(1.0366)^{40} \approx 30,100$$

The actual amount, 25,010 million short tons, is less than the 30,100 million short tons that the model provides.

39. $A(t) = 100(3.2)^{-.5t}$

(a) The initial measurement is when $t = 0$.

$$A(0) = 100(3.2)^{-.5(0)}$$
$$= 100(3.2)^0 = 100(1) = 100$$

The initial measurement was 100 g.

(b) The measurement after 2 months is when $t = 2$.

$$A(2) = 100(3.2)^{-.5(2)}$$
$$= 100(3.2)^{-1} = 31.25$$

The measurement after 2 months was 31.25 g.

(c) The measurement after 10 months is when $t = 10$.

$$A(10) = 100(3.2)^{-.5(10)}$$
$$= 100(3.2)^{-5} \approx .30$$

The measurement after 10 months was about .30 g.

(d) Use the results of parts (a) – (c) to make a table of values.

t	0	2	10
$A(t)$	100	31.25	.3

Plot the points from the table and draw a smooth curve through them.

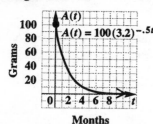

41. $V(t) = 5000(2)^{-.15t}$

$$2500 = 5000(2)^{-.15t} \qquad \text{Let } V(t) = 2500.$$

$$\frac{1}{2} = (2)^{-.15t} \qquad \text{Divide by 5000.}$$

$$2^{-1} = 2^{-.15t}$$

$$-1 = -.15t \qquad \text{Equate exponents}$$

$$t = \frac{-1}{-.15} \approx 6.67$$

The value of the machine will be $2500 in approximately 6.67 years after it was purchased.

43. $S(x) = 74,741(1.17)^x$

Since the year 1976 corresponds to $x = 0$, the year 1986 corresponds to $x = 10$ $(1986 - 1976 = 10)$.

$$S(10) = 74,741(1.17)^{10}$$
$$\approx 359,267$$

The average salary in 1986 was about $360,000.

45. $16^{3/4} = \left(\sqrt[4]{16}\right)^3 = (2)^3 = 8$

46. $16^{3/4} = \sqrt[4]{16^3} = \sqrt[4]{4096} = 8$

47. $\sqrt{16^3} = 64$ and $\sqrt{64} = 8.$

48. In Exercise 47, we are finding $\sqrt{\sqrt{16^3}}$, that is, taking the square root twice.

$$\sqrt{\sqrt{16^3}} = \sqrt{(16^3)^{1/2}} = \sqrt{16^{3/2}}$$
$$= \left(16^{3/2}\right)^{1/2} = 16^{3/4}$$

49. Since $16^{.5} = \sqrt{16} = 4$ and $16^1 = 16$, a reasonable prediction for $16^{.75}$ must be between 4 and 16. From a calculator,

$$16^{.75} = 8.$$

50. $\sqrt[100]{16^{75}} = 16^{75/100} = 16^{3/4}$
$$= \left(\sqrt[4]{16}\right)^3 = (2)^3 = 8$$

51. The display indicates that for the year 1965, the model gives a value of 102.287 million tons, which is slightly less than the actual value of 103.4 million tons.

Section 9.3

1. **(a)** $\log_4 16$ is equal to 2, because 2 is the exponent to which 4 must be raised in order to obtain 16. **(C)**

(b) $\log_3 81$ is equal to 4, because 4 is the exponent to which 3 must be raised in order to obtain 81. **(F)**

(c) $\log_3 \left(\frac{1}{3}\right)$ is equal to -1, because -1 is the exponent to which 3 must be raised in order to obtain $\frac{1}{3}$. **(B)**

(d) $\log_{10} .01$ is equal to -2, because -2 is the exponent to which 10 must be raised in order to obtain .01. **(A)**

(e) $\log_5 \sqrt{5}$ is equal to $\frac{1}{2}$, because $\frac{1}{2}$ is the exponent to which 5 must be raised in order to obtain $\sqrt{5}$. **(E)**

(f) $\log_{13} 1$ is equal to 0, because 0 is the exponent to which 13 must be raised in order to obtain 1. **(D)**

3. The base is 4, the exponent (logarithm) is 5, and the number is 1024, so $4^5 = 1024$ becomes $\log_4 1024 = 5$ in logarithmic form.

5. $\frac{1}{2}$ is the base and -3 is the exponent, so

$$\left(\frac{1}{2}\right)^{-3} = 8 \text{ becomes } \log_{1/2} 8 = -3 \text{ in}$$

logarithmic form.

7. The base is 10, the exponent (logarithm) is -3, and the number is .001, so $10^{-3} = .001$ becomes $\log_{10} .001 = -3$ in logarithmic form.

9. In $\log_4 64 = 3$, 4 is the base and 3 is the logarithm (exponent), so $\log_4 64 = 3$ becomes $4^3 = 64$ in exponential form.

11. The base is 10, logarithm (exponent) is -4, and the number is $\dfrac{1}{10,000}$, so $\log_{10}\dfrac{1}{10,000} = -4$ becomes $10^{-4} = \dfrac{1}{10,000}$ in exponential form.

13. In $\log_6 1 = 0$, 6 is the base and 0 is the logarithm (exponent), so $\log_6 1 = 0$ becomes $6^0 = 1$ in exponential form.

15. To evaluate $\log_9 3$, one has to ask "9 raised to what power gives you a result of 3?" We know that the square root (a radical) of 9 is 3. Therefore, the teacher's hint was to see what root of 9 equals 3. The answer is the reciprocal of the root index.

17. $x = \log_{27} 3$
Write in exponential form.
$$27^x = 3$$
Write each side as a power of 3.
$$\left(3^3\right)^x = 3$$
$$3^{3x} = 3^1$$
Set the exponents equal.
$$3x = 1$$
$$x = \frac{1}{3}$$
Solution set: $\left\{\dfrac{1}{3}\right\}$

19. $\log_x 9 = \dfrac{1}{2}$
Change to exponential form.
$$x^{1/2} = 9$$
$$\left(x^{1/2}\right)^2 = 9^2 \qquad \textit{Square}$$
$$x^1 = 81$$
$$x = 81$$
Solution set: $\{81\}$

21. $\log_x 125 = -3$
Write in exponential form.
$$x^{-3} = 125$$
$$\frac{1}{x^3} = 125$$
$$1 = 125\left(x^3\right)$$
$$\frac{1}{125} = x^3$$
Take the cube root of both sides.
$$\sqrt[3]{\frac{1}{125}} = \sqrt[3]{x^3}$$
$$x = \sqrt[3]{\frac{1}{5^3}} = \frac{1}{5}$$
Solution set: $\left\{\dfrac{1}{5}\right\}$

23. $\log_{12} x = 0$
Write in exponential form.
$$12^0 = x$$
$$1 = x$$
Solution set: $\{1\}$

25. $\log_x x = 1$
Write in exponential form.
$$x^1 = x$$
This equation is true for all the numbers x that are allowed as the base of a logarithm; that is, all positive numbers x, $x \neq 1$.
Solution set: $\{x \mid x > 0,\ x \neq 1\}$

27. $\log_x \dfrac{1}{25} = -2$
Write in exponential form.
$$x^{-2} = \frac{1}{25}$$
$$\frac{1}{x^2} = \frac{1}{25}$$
$$x^2 = 25 \qquad \begin{array}{l}\textit{Denominators must}\\ \textit{be equal.}\end{array}$$
$$x = \pm 5$$
Reject $x = -5$ since the base of a logarithm must be positive.
Solution set: $\{5\}$

29. $\log_8 32 = x$
$$8^x = 32 \qquad \textit{Exponential form}$$
Write each side as a power of 2.
$$\left(2^3\right)^x = 2^5$$
$$2^{3x} = 2^5$$
$$3x = 5 \qquad \textit{Equate exponents}$$
$$x = \frac{5}{3}$$
Solution set: $\left\{\dfrac{5}{3}\right\}$

31. $\log_\pi \pi^4 = x$
$$\pi^x = \pi^4 \quad \textit{Exponential form}$$
$$x = 4 \quad \textit{Equate exponents}$$
Solution set: $\{4\}$

33. $\log_6 \sqrt{216} = x$
$$\log_6 216^{1/2} = x \qquad \textit{Equivalent form}$$
$$6^x = 216^{1/2} \qquad \textit{Exponential form}$$
$$6^x = \left(6^3\right)^{1/2} \qquad \textit{Same base}$$
$$6^x = 6^{3/2}$$
$$x = \frac{3}{2} \qquad \textit{Equate exponents}$$
Solution set: $\left\{\dfrac{3}{2}\right\}$

35.
$$y = \log_3 x$$

Change to exponential form.
$$3^y = x$$

Refer to Section 9.2, Exercise 5, for the graph of $f(x) = 3^x$. Since $y = \log_3 x$ (or $3^y = x$) is the inverse of $f(x) = y = 3^x$, its graph is symmetric about the line $y = x$ to the graph of $f(x) = 3^x$. The graph can be plotted by reversing the ordered pairs in the table of values belonging to $f(x) = 3^x$.

x	$\frac{1}{9}$	$\frac{1}{3}$	1	3	9
y	-2	-1	0	1	2

Plot the points, and draw a smooth curve through them.

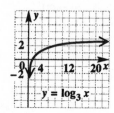

37. $y = \log_{1/3} x$

Change to exponential form.

$$\left(\frac{1}{3}\right)^y = x$$

Refer to Section 9.2, Exercise 7, for the graph of $g(x) = \left(\frac{1}{3}\right)^x$. Since $y = \log_{1/3} x$

$\left(\text{or } \left(\frac{1}{3}\right)^y = x\right)$ is the inverse of $y = \left(\frac{1}{3}\right)^x$,

its graph is symmetric about the line $y = x$ to the

graph of $y = \left(\frac{1}{3}\right)^x$. The graph can be plotted by

reversing the ordered pairs in the table of values

belonging to $g(x) = \left(\frac{1}{3}\right)^x$.

x	9	3	1	$\frac{1}{3}$	$\frac{1}{9}$
y	-2	-1	0	1	2

Plot the points, and draw a smooth curve through them.

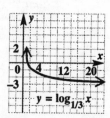

39. The number 1 is not used as a base for a logarithmic function since the function would look like $x = 1^y$ in exponential form. Then, for any real value of y, the statement $1 = 1$ would always be the result since every power of 1 is equal to 1.

41. The range of $F(x) = a^x$ is the domain of $G(x) = \log_a x$, that is, $(0, \infty)$.
The domain of $F(x) = a^x$ is the range of $G(x) = \log_a x$, that is, $(-\infty, \infty)$.

43. The values of t are on the horizontal axis, and the values of $f(t)$ are on the vertical axis. Read the value of $f(t)$ from the graph for the given value of t. At $t = 0$, $f(0) = 8$.

45. To find $f(60)$, find 60 on the t-axis, then go up to the graph and across to the $f(t)$ axis to read the value of $f(60)$. At $t = 60$, $f(60) = 24$.

47. $f(x) = 11.34 + 317.01 \log_2 x$

(a) 1984 corresponds to $x = 4$.

$$f(4) = 11.34 + 317.01 \log_2 4$$
$$= 11.34 + 317.01(2)$$
$$= 11.34 + 634.02 = 645.36$$

The model gives 645 sites for 1984.

(b) 1988 corresponds to $x = 8$.

$$f(8) = 11.34 + 317.01 \log_2 8$$
$$= 11.34 + 317.01(3)$$
$$= 11.34 + 951.03 = 962.37$$

The model gives 962 sites for 1988.

49. $M(t) = 6 \log_4 (2t + 4)$

(a) $t = 0$ corresponds to January 1998.

$$M(0) = 6 \log_4 (2 \cdot 0 + 4)$$
$$= 6 \log_4 4$$
$$= 6(1) = 6$$

There were 6 mice in the house in January 1998.

(b) $t = 1$ corresponds to February 1998, so $t = 6$ corresponds to July 1998.

$$M(6) = 6 \log_4 (2 \cdot 6 + 4)$$
$$= 6 \log_4 16$$
$$= 6(2) = 12$$

There were 12 mice in the house in July 1998.

(c) $t = 30$ corresponds to July 2000.

$$M(30) = 6 \log_4 (2 \cdot 30 + 4)$$
$$= 6 \log_4 64$$
$$= 6(3) = 18$$

There were 18 mice in the house in July 2000.

(d) Make a table for M using parts (a) – (c).

t	0	6	30
$M(t)$	6	12	18

Plot the points, and draw a smooth logarithmic curve through them.

$$M(t) = 6 \log_4 (2t + 4)$$

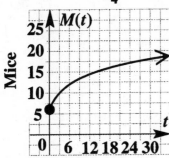

Months since January 1998

51. $g(x) = \log_3 x$

On a TI-83, assign 3^x to Y_1. Then enter

DrawInv Y_1

on the home screen to obtain the figure that follows. (See Exercise 47 in Section 9.1 for TI-83 specifics.)

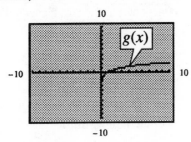

53. $g(x) = \log_{1/3} x$

Assign $(1/3)$^x to Y_1 and enter **DrawInv Y_1**.

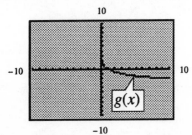

Section 9.4

1. By the product rule,

$$\log_{10} (3 \cdot 4) = \log_{10} 3 + \log_{10} 4.$$

3. By the power rule,

$$\log_{10} 3^4 = 4 \log_{10} 3.$$

5. By a special property (see page 613 in the text),

$$\log_3 3^4 = 4.$$

7. Use the quotient rule for logarithms.

$$\log_7 \frac{4}{5} = \log_7 4 - \log_7 5$$

9. $\log_2 8^{1/4}$

$$= \frac{1}{4} \log_2 8 \qquad \textit{Power rule}$$
$$= \frac{1}{4} \log_2 2^3$$
$$= \frac{1}{4}(3) = \frac{3}{4}$$

11. $\log_4 \dfrac{3\sqrt{x}}{y}$

$$= \log_4 \frac{3 \cdot x^{1/2}}{y}$$

Use the quotient rule for logarithms.

$$= \log_4 \left(3 \cdot x^{1/2}\right) - \log_4 y$$

Use the product and power rules for logarithms.

$$= \log_4 3 + \frac{1}{2} \log_4 x - \log_4 y$$

13. $\log_3 \dfrac{\sqrt[3]{4}}{x^2 y} = \log_3 \dfrac{4^{1/3}}{x^2 y}$

Use the quotient rule for logarithms.

$$= \log_3 4^{1/3} - \log_3 \left(x^2 y\right)$$

Use the product rule for logarithms.

$$= \log_3 4^{1/3} - \left(\log_3 x^2 + \log_3 y \right)$$
$$= \log_3 4^{1/3} - \log_3 x^2 - \log_3 y$$

Use the power rule for logarithms.

$$= \frac{1}{3} \log_3 4 - 2 \log_3 x - \log_3 y$$

15. $\log_3 \sqrt{\dfrac{xy}{5}}$

$= \log_3 \left(\dfrac{xy}{5}\right)^{1/2}$

$= \dfrac{1}{2} \log_3 \left(\dfrac{xy}{5}\right)$ *Power rule*

$= \dfrac{1}{2} \left[\log_3 (xy) - \log_3 5 \right]$ *Quotient rule*

$= \dfrac{1}{2} \left(\log_3 x + \log_3 y - \log_3 5 \right)$ *Product rule*

$= \dfrac{1}{2} \log_3 x + \dfrac{1}{2} \log_3 y - \dfrac{1}{2} \log_3 5$

17. $\log_2 \dfrac{\sqrt[3]{x} \cdot \sqrt[5]{y}}{r^2}$

$= \log_2 \dfrac{x^{1/3} y^{1/5}}{r^2}$

$= \log_2 \left(x^{1/3} y^{1/5} \right) - \log_2 r^2$ *Quotient rule*

$= \log_2 x^{1/3} + \log_2 y^{1/5} - \log_2 r^2$ *Product rule*

$= \dfrac{1}{3} \log_2 x + \dfrac{1}{5} \log_2 y - 2 \log_2 r$ *Power rule*

19. The distributive property tells us that the *product* $a(x + y)$ equals the sum $ax + ay$. In the notation $\log_a (x + y)$, the parentheses do not indicate multiplication. They indicate that $x + y$ is the result of raising a to some power.

21. By the product rule for logarithms,

$$\log_b x + \log_b y = \log_b xy.$$

23. $3 \log_a m - \log_a n$

$= \log_a m^3 - \log_a n$ *Power rule*

$= \log_a \dfrac{m^3}{n}$ *Quotient rule*

25. $\left(\log_a r - \log_a s \right) + 3 \log_a t$

Use the quotient and power rules for logarithms.

$= \log_a \dfrac{r}{s} + \log_a t^3$

$= \log_a \dfrac{rt^3}{s}$ *Product rule*

27. $3 \log_a 5 - 4 \log_a 3$

$= \log_a 5^3 - \log_a 3^4$ *Power rule*

$= \log_a \dfrac{5^3}{3^4}$ *Quotient rule*

$= \log_a \dfrac{125}{81}$

29. $\log_{10} (x + 3) + \log_{10} (x - 3)$

$= \log_{10} (x + 3)(x - 3)$ *Product rule*

$= \log_{10} (x^2 - 9)$

31. By the power rule for logarithms,

$3 \log_p x + \dfrac{1}{2} \log_p y - \dfrac{3}{2} \log_p z - 3 \log_p a$

$= \log_p x^3 + \log_p y^{1/2} - \log_p z^{3/2} - \log_p a^3$

Group the terms into sums.

$= \left(\log_p x^3 + \log_p y^{1/2} \right)$

$\quad - \left(\log_p z^{3/2} + \log_p a^3 \right)$

$= \log_p x^3 y^{1/2} - \log_p z^{3/2} a^3$ *Product rule*

$= \log_p \dfrac{x^3 y^{1/2}}{z^{3/2} a^3}$ *Quotient rule*

In Exercises 33–40, $\log_{10} 2 \approx .3010$ and $\log_{10} 9 \approx .9542$.

33. By the product rule for logarithms,

$\log_{10} 18 = \log_{10} (2 \cdot 9)$

$\quad = \log_{10} 2 + \log_{10} 9$

$\quad \approx .3010 + .9542$

$\quad = 1.2552.$

35. By the quotient rule for logarithms,

$\log_{10} \dfrac{2}{9} = \log_{10} 2 - \log_{10} 9$

$\quad \approx .3010 - .9542$

$\quad = -.6532$

37. By the product and power rules for logarithms,

$\log_{10} 36 = \log_{10} 2^2 \cdot 9$

$\quad = 2 \log_{10} 2 + \log_{10} 9$

$\quad \approx 2(.3010) + .9542$

$\quad = 1.5562.$

39. By the power rule for logarithms,

$\log_{10} 3 = \log_{10} 9^{1/2}$

$\quad = \dfrac{1}{2} \log_{10} 9$

$\quad \approx \dfrac{1}{2} (.9542)$

$\quad = .4771.$

41. $\log_6 60 - \log_6 10 = \log_6 \dfrac{60}{10}$

$\quad = \log_6 6 = 1$

The statement is true.

43. $\dfrac{\log_{10} 7}{\log_{10} 14} \overset{?}{=} \dfrac{1}{2}$

$2 \log_{10} 7 \overset{?}{=} 1 \log_{10} (7 \cdot 2)$

Cross products are equal

$2 \log_{10} 7 \overset{?}{=} \log_{10} 7 + \log_{10} 2$

$\log_{10} 7 \overset{?}{=} \log_{10} 2$ *Subtract* $\log_{10} 7$

The statement is false.

45. The exponent of a quotient is the difference between the exponent of the numerator and the exponent of the denominator.

47. $\log_2 8 - \log_2 4 = \log_2 \dfrac{8}{4}$

$= \log_2 2 = 1$

49. $\log_3 81 = \log_3 3^4 = 4$

50. $\log_3 81$ is the exponent to which 3 must be raised in order to obtain 81.

51. Using the result from Exercise 49,

$$3^{\log_3 81} = 3^4 = 81.$$

52. $\log_2 19$ is the exponent to which 2 must be raised in order to obtain 19.

53. Keeping in mind the result from Exercise 51,

$$2^{\log_2 19} = 19.$$

54. To find $k^{\log_k m}$, first assume $\log_k m = y$. This means, changing to an exponential equation, $k^y = m$. Therefore,

$$k^{\log_k m} = k^y = m.$$

Section 9.5

1. Since $\log x = \log_{10} x$, the base is 10. The correct response is (c).

3. $10^0 = 1$ and $10^1 = 10$, so $\log 1 = 0$ and $\log 10 = 1$. Thus, the value of $\log 5.6$ must lie between 0 and 1. The correct response is (c).

5. $\log 10^{19.2} = \log_{10} 10^{19.2} = 19.2$ by the special property, $\log_b b^x = x$.

7. To four decimal places,

$$\log 43 \approx 1.6335.$$

9. $\log 328.4 \approx 2.5164$

11. $\log .0326 \approx -1.4868$

13. $\log (4.76 \times 10^9) \approx 9.6776$
On a TI-83, enter

$$\boxed{\text{LOG}}\ 4.76\ \boxed{\text{2nd}}\ \boxed{\text{EE}}\ 9\ ).$$

15. $\ln 7.84 \approx 2.0592$

17. $\ln .0556 \approx -2.8896$

19. $\ln 388.1 \approx 5.9613$

21. $\ln (8.59 \times e^2) \approx 4.1506$
On a TI-83, enter

$$\boxed{\text{LN}}\ 8.59\ \boxed{\text{X}}\ \boxed{\text{2nd}}\ \boxed{e^x}\ 2\ )\ ).$$

23. $\ln 10 \approx 2.3026$

25. (a) $\log 356.8 \approx 2.552424846$

(b) $\log 35.68 \approx 1.552424846$

(c) $\log 3.568 \approx 0.552424846$

(d) The whole number part of the answers (2, 1, or 0) varies, whereas the decimal part (.552424846) remains the same, indicating that the whole number part corresponds to the placement of the decimal point and the decimal part corresponds to the digits 3, 5, 6, and 8.

27. When you try to find $\log (-1)$ on a calculator, an error message is displayed. This is because the domain of the logarithmic function is $(0, \infty)$; -1 is not in the domain.

29. $\mathrm{pH} = -\log [H_3O^+]$

$= -\log (2.5 \times 10^{-2})$

≈ 1.6

Since the pH is less than 3.0, the wetland is classified as a *bog*.

31. Ammonia has a hydronium ion concentration of 2.5×10^{-12}.

$\mathrm{pH} = -\log [H_3O^+]$

$\mathrm{pH} = -\log (2.5 \times 10^{-12}) \approx 11.6$

33. Grapes have a hydronium ion concentration of 5.0×10^{-5}.

$\mathrm{pH} = -\log [H_3O^+]$

$\mathrm{pH} = -\log (5.0 \times 10^{-5}) \approx 4.3$

35. Human blood plasma has a pH of 7.4.

$\mathrm{pH} = -\log [H_3O^+]$

$7.4 = -\log [H_3O^+]$

$\log_{10} [H_3O^+] = -7.4$

$[H_3O^+] = 10^{-7.4} \approx 4.0 \times 10^{-8}$

37. Spinach has a pH value of 5.4.

$$\text{pH} = -\log [\text{H}_3\text{O}^+]$$

$$5.4 = -\log [\text{H}_3\text{O}^+]$$

$$\log_{10} [\text{H}_3\text{O}^+] = -5.4$$

$$[\text{H}_3\text{O}^+] = 10^{-5.4} \approx 4.0 \times 10^{-6}$$

39. $P(x) = 70,967e^{.0526x}$

(a) 1987 corresponds to $x = 2$.

$$P(2) = 70,967e^{.0526(2)}$$

$$\approx 78,839.6$$

The approximate expenditures for 1987 is $78,840$ million dollars.

(b) 1990 corresponds to $x = 5$.

$$P(5) \approx 92,316$$

(c) 1993 corresponds to $x = 8$.

$$P(8) \approx 108,095$$

(d) 1985 corresponds to $x = 0$.

$$P(0) = 70,967 \quad e^0 = 1$$

41. $B(x) = 8768e^{.072x}$
1998 corresponds to $x = 18$.
$B(18) = 8768e^{.072(18)}$

$$\approx 32,044$$

The approximate consumer expenditures for 1998 is $32,044$ million dollars.

43. $A(t) = 2.00e^{-.053t}$

(a) Let $t = 4$.

$$A(4) = 2.00e^{-.053(4)}$$

$$= 2.00e^{-.212}$$

$$\approx 1.62$$

About 1.62 grams would be present.

(b) $A(10) = 2.00e^{-.053(10)}$

$$\approx 1.18$$

About 1.18 grams would be present.

(c) $A(20) = 2.00e^{-.053(20)}$

$$\approx .69$$

About .69 grams would be present.

(d) The initial amount is the amount $A(t)$ present at time $t = 0$.

$$A(0) = 2.00e^{-.053(0)}$$

$$= 2.00e^0 = 2.00(1) = 2.00$$

Initially, 2.00 grams were present.

45. $N(r) = -5000 \ln r$
$N(.9) = -5000 \ln .9$

$$\approx -5000(-.1054) = 527$$

About 527 years have elapsed.

47. $N(r) = -5000 \ln r$
$N(.5) = -5000 \ln .5$

$$\approx -5000(-.6931) \approx 3466$$

About 3466 years have elapsed.

49. The change-of-base rule is

$$\log_a x = \frac{\log_b x}{\log_b a}.$$

Use common logarithms ($b = 10$).

$$\log_6 12 = \frac{\log_{10} 12}{\log_{10} 6} = \frac{\log 12}{\log 6} \approx 1.3869$$

51. Use natural logarithms ($b = e$).

$$\log_{12} 6 = \frac{\log_e 6}{\log_e 12} = \frac{\ln 6}{\ln 12} \approx .7211$$

53. $\log_{\sqrt{2}} \pi = \dfrac{\ln \pi}{\ln \sqrt{2}} \approx 3.3030$

55. Let $m =$ the number of letters in your first name and $n =$ the number of letters in your last name.

Answers will vary, but suppose the name is Paul Bunyan, with $m = 4$ and $n = 6$.

(a) $\log_m n = \log_4 6$ is the exponent to which 4 must be raised in order to obtain 6.

(b) Use the change-of-base rule.

$$\log_4 6 = \frac{\log 6}{\log 4}$$

$$\approx 1.29248125$$

(c) Here, $m = 4$. Use the power key (y^x, x^y, ^) on your calculator.

$$4^{1.29248125} \approx 6$$

The result is 6, the value of n.

57. $x = 3$ corresponds to 1988. So in 1988, total expenditures for pollution abatement and control were approximately $83,097.53$ million dollars.

59. **(a)** The expression $\dfrac{1}{x}$ in the base cannot be evaluated, since division by 0 is not defined.

(b) When $x = 100,000$, $Y_1 \approx 2.7183$. The decimal approximation appears to be close to the decimal approximation for e, $2.71828\ldots$.

(c) $\left(1 + \dfrac{1}{1,000,000}\right)^{1,000,000} \approx 2.718280469$

$$e = e^1 \approx 2.718281828$$

The two values differ in the sixth decimal place.

(d) As the values of x approach infinity, the value of $\left(1 + \dfrac{1}{x}\right)^x$ approaches $\underline{e}$.

61. To graph $g(x) = \log_3 x$, assign either $\dfrac{\log x}{\log 3}$ or $\dfrac{\ln x}{\ln 3}$ to Y_1.

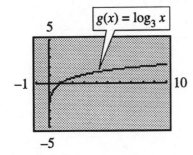

63. To graph $g(x) = \log_{1/3} x$, assign either $\dfrac{\log x}{\log \frac{1}{3}}$ or $\dfrac{\ln x}{\ln \frac{1}{3}}$ to Y_1.

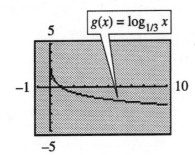

Section 9.6

1. $5^x = 125$

$\log 5^x = \log 125$

2. $x \log 5 = \log 125$

3. $\dfrac{x \log 5}{\log 5} = \dfrac{\log 125}{\log 5}$

$x = \dfrac{\log 125}{\log 5}$

4. $\dfrac{\log 125}{\log 5} = 3$ (calculator)

Solution set: $\{3\}$

5. $7^x = 5$

Take the logarithm of each side.

$\log 7^x = \log 5$

Use the power rule for logarithms.

$x \log 7 = \log 5$

$x = \dfrac{\log 5}{\log 7} \approx .827$

Solution set: $\{.827\}$

7. $9^{-x+2} = 13$

$\log 9^{-x+2} = \log 13$

$(-x + 2) \log 9 = \log 13$ (∗)

$-x \log 9 + 2 \log 9 = \log 13$

$-x \log 9 = \log 13 - 2 \log 9$

$x \log 9 = 2 \log 9 - \log 13$

$x = \dfrac{2 \log 9 - \log 13}{\log 9}$

$\approx .833$

(∗) Alternative solutions steps:

$(-x + 2) \log 9 = \log 13$

$-x + 2 = \dfrac{\log 13}{\log 9}$

$2 - \dfrac{\log 13}{\log 9} = x$

Solution set: $\{.833\}$

9. $2^{y+3} = 5^y$

$\log 2^{y+3} = \log 5^y$

$(y + 3) \log 2 = y \log 5$

$y \log 2 + 3 \log 2 = y \log 5$

Get y-terms on one side.

$y \log 2 - y \log 5 = -3 \log 2$

$y (\log 2 - \log 5) = -3 \log 2$ *Factor out y.*

$y = \dfrac{-3 \log 2}{\log 2 - \log 5}$

≈ 2.269

Solution set: $\{2.269\}$

11. $e^{.006x} = 30$

Take base e logarithms on both sides.

$\ln e^{.006x} = \ln 30$

$.006x \ln e = \ln 30$

$.006x = \ln 30 \quad \ln e = 1$

$x = \dfrac{\ln 30}{.006} \approx 566.866$

Solution set: $\{566.866\}$

13.
$$e^{-.103x} = 7$$
$$\ln e^{-.103x} = \ln 7$$
$$-.103x \ln e = \ln 7$$
$$-.103x = \ln 7 \quad \ln e = 1$$
$$x = \frac{\ln 7}{-.103} \approx -18.892$$
Solution set: $\{-18.892\}$

15.
$$100e^{.045x} = 300$$
$$e^{.045x} = \frac{300}{100} = 3$$
$$\ln e^{.045x} = \ln 3$$
$$.045x\,(\ln e) = \ln 3$$
$$.045x = \ln 3$$
$$x = \frac{\ln 3}{.045} \approx 24.414$$
Solution set: $\{24.414\}$

17. Let's try Exercise 11.
$$e^{.006x} = 30$$
$$\log e^{.006x} = \log 30$$
$$.006x\,(\log e) = \log 30$$
$$x = \frac{\log 30}{.006 \log e} \approx 566.866$$

The natural logarithm is easier because $\ln e = 1$, whereas $\log e$ needs to be calculated.

19. $\log_3 (6x + 5) = 2$
$$6x + 5 = 3^2 \quad \textit{Exponential form}$$
$$6x + 5 = 9$$
$$6x = 4$$
$$x = \frac{4}{6} = \frac{2}{3}$$

Solution set: $\left\{\dfrac{2}{3}\right\}$

21. $\log_7 (x + 1)^3 = 2$
$$(x + 1)^3 = 7^2 \quad \textit{Exponential form}$$
$$x + 1 = \sqrt[3]{49} \quad \textit{Cube root}$$
$$x = -1 + \sqrt[3]{49}$$

Solution set: $\left\{-1 + \sqrt[3]{49}\right\}$

23. The apparent solution, 2, causes the expression $\log_4 (x - 3)$ to be $\log_4 (-1)$, and we cannot take the logarithm of a negative number.

25. $\log (6x + 1) = \log 3$
$$6x + 1 = 3 \quad \textit{Property 4}$$
$$6x = 2$$
$$x = \frac{2}{6} = \frac{1}{3}$$
Solution set: $\left\{\dfrac{1}{3}\right\}$

27. $\log_5 (3t + 2) - \log_5 t = \log_5 4$
$$\log_5 \frac{3t + 2}{t} = \log_5 4$$
$$\frac{3t + 2}{t} = 4$$
$$3t + 2 = 4t$$
$$2 = t$$
Solution set: $\{2\}$

29. $\log 4x - \log (x - 3) = \log 2$
$$\log \frac{4x}{x - 3} = \log 2$$
$$\frac{4x}{x - 3} = 2$$
$$4x = 2(x - 3)$$
$$4x = 2x - 6$$
$$2x = -6$$
$$x = -3$$
Reject $x = -3$, because $4x = -12$, which yields an equation in which the logarithm of a negative number must be found.
Solution set: $\emptyset$

31. $\log_2 x + \log_2 (x - 7) = 3$
$$\log_2 [x(x - 7)] = 3$$
$$x(x - 7) = 2^3 \quad \textit{Exponential form}$$
$$x^2 - 7x = 8$$
$$x^2 - 7x - 8 = 0$$
$$(x - 8)(x + 1) = 0$$

$$x - 8 = 0 \quad \text{or} \quad x + 1 = 0$$
$$x = 8 \quad \text{or} \quad x = -1$$
Reject $x = -1$, because it yields an equation in which the logarithm of a negative number must be found.
Solution set: $\{8\}$

33. $\log 5x - \log (2x - 1) = \log 4$

$$\log \frac{5x}{2x - 1} = \log 4$$

$$\frac{5x}{2x - 1} = 4$$

$$5x = 8x - 4$$

$$4 = 3x$$

$$\frac{4}{3} = x$$

Solution set: $\left\{\dfrac{4}{3}\right\}$

35. $\log_2 x + \log_2 (x - 6) = 4$

$$\log_2 [x(x - 6)] = 4$$

$$x(x - 6) = 2^4 \quad \textit{Exponential form}$$

$$x^2 - 6x = 16$$

$$x^2 - 6x - 16 = 0$$

$$(x - 8)(x + 2) = 0$$

$$x - 8 = 0 \quad \text{or} \quad x + 2 = 0$$

$$x = 8 \quad \text{or} \quad x = -2$$

Reject $x = -2$, because it yields an equation in which the logarithm of a negative number must be found.

Solution set: $\{8\}$

37. Use the formula $A = P\left(1 + \dfrac{r}{n}\right)^{nt}$ with $P = 2000$, $r = .04$, $n = 4$, and $t = 6$.

$$A = 2000\left(1 + \frac{.04}{4}\right)^{4 \cdot 6}$$

$$= 2000(1.01)^{24} \approx 2539.47$$

The account will contain \$2539.47.

39. Find $A(t) = 400e^{-.032t}$ when $t = 25$.

$$A(25) = 400e^{-.032(25)}$$

$$= 400e^{-.8} \approx 179.73$$

About 180 grams of lead will be left.

41. Use $A = P\left(1 + \dfrac{r}{n}\right)^{nt}$ with $P = 5000$, $r = .07$, and $t = 12$.

(a) If the interest is compounded annually, $n = 1$.

$$A = 5000\left(1 + \frac{.07}{1}\right)^{1 \cdot 12}$$

$$= 5000(1.07)^{12} \approx 11,260.96$$

There will be \$11,260.96 in the account.

(b) If the interest is compounded semiannually, $n = 2$.

$$A = 5000\left(1 + \frac{.07}{2}\right)^{2 \cdot 12}$$

$$= 5000(1.035)^{24} \approx 11,416.64$$

There will be \$11,416.64 in the account.

(c) If the interest is compounded quarterly, $n = 4$.

$$A = 5000\left(1 + \frac{.07}{4}\right)^{4 \cdot 12}$$

$$= 5000(1.0175)^{48} \approx 11,497.99$$

There will be \$11,497.99 in the account.

(d) If the interest is compounded daily, $n = 365$.

$$A = 5000\left(1 + \frac{.07}{365}\right)^{365 \cdot 12}$$

$$\approx 11,580.90$$

There will be \$11,580.90 in the account.

(e) Use the continuous compound interest formula.

$$A = Pe^{rt}$$

$$A = 5000e^{.07(12)}$$

$$= 5000e^{.84} \approx 11,581.83$$

There will be \$11,581.83 in the account.

43. In the continuous compound interest formula, let $A = 1850$, $r = .065$, and $t = 40$.

$$A = Pe^{rt}$$

$$1850 = Pe^{.065(40)}$$

$$1850 = Pe^{2.6}$$

$$P = \frac{1850}{e^{2.6}} \approx 137.41$$

Deposit \$137.41 today.

45.

$$P(x) = 70,967e^{.0526x}$$

$$133,500 = 70,967e^{.0526x}$$

$$e^{.0526x} = \frac{133,500}{70,967}$$

$$\ln e^{.0526x} = \ln \frac{133,500}{70,967}$$

$$.0526x\,(\ln e) = \ln \frac{133,500}{70,967}$$

$$.0526x = \ln \frac{133,500}{70,967}$$

$$x = \frac{\ln \dfrac{133,500}{70,967}}{.0526} \approx 12.0$$

Since $x = 0$ corresponds to 1985, $x = 12$ corresponds to 1997.

47. $y = 2e^{-.125t}$

When $t = 0$, $y = 2e^0 = 2(1) = 2$. Thus, the original value is 2. Half the original value is 1.

$$1 = 2e^{-.125t}$$
$$.5 = e^{-.125t}$$
$$\ln .5 = \ln e^{-.125t}$$
$$\ln .5 = -.125t (\ln e)$$
$$-.125t = \ln .5$$
$$t = \frac{\ln .5}{-.125} \approx 5.55$$

It will take about 5.55 hours.

49. $y = y_0 e^{-.0239t}$

(a) Let $y_0 = 5$ and $t = 20$.

$$y = 5e^{-.0239(20)}$$
$$= 5e^{-.478} \approx 3.10$$

About 3.10 grams will be present after 20 years.

(b) Let $y_0 = 5$ and $t = 60$.

$$y = 5e^{-.0239(60)}$$
$$= 5e^{-1.434} \approx 1.19$$

About 1.19 grams will be present after 60 years.

(c) To find the half-life of radioactive strontium, determine how long it takes until 2.5 grams of the original 5 grams remain. Let $y = 2.5$ and $y_0 = 5$.

$$2.5 = 5e^{-.0239t}$$
$$.5 = e^{-.0239t}$$
$$\ln .5 = \ln e^{-.0239t}$$
$$\ln .5 = -.0239t (\ln e)$$
$$-.0239t = \ln .5$$
$$t = \frac{\ln .5}{-.0239} \approx 29.002$$

The half-life is about 29 years.

51. $7^x = 5$

$7^x - 5 = 0$

Graph $Y_1 = 7^x - 5$ and find the x-intercept.

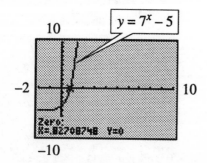

Solution set: $\{.827\}$

53. $\log (6x + 1) = \log 3$

$\log (6x + 1) - \log 3 = 0$

Graph $Y_1 = \log (6x + 1) - \log 3$ and find the x-intercept.

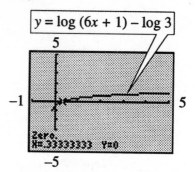

Solution set: $\{.333\}$

Chapter 9 Review Exercises

1. Since a horizontal line intersects the graph in two points, the function is not one-to-one.

2. Since every horizontal line intersects the graph in no more than one point, the function is one-to-one.

3. Each element of the domain corresponds to only one element of the range, and vice versa. Thus, the function is one-to-one.

4. The function $f(x) = -3x + 7$ is a linear function. By the horizontal line test, it is a one-to-one function. To find the inverse, replace $f(x)$ with y.

$$y = -3x + 7$$

Interchange x and y.

$$x = -3y + 7$$

Solve for y.

$$x - 7 = -3y$$
$$\frac{x - 7}{-3} = y$$
$$\text{or } \frac{7 - x}{3} = y$$

Replace y with $f^{-1}(x)$.

$$f^{-1}(x) = \frac{x - 7}{-3} \text{ or } \frac{7 - x}{3}$$

5. $f(x)=\sqrt[3]{6x-4}$

The cube root causes each value of x to be matched with only one value of $f(x)$. The function is one-to-one.

To find the inverse, replace $f(x)$ with y.

$$y = \sqrt[3]{6x-4}$$

Cube both sides.

$$y^3 = 6x - 4$$

Interchange x and y.

$$x^3 = 6y - 4$$

Solve for y.

$$x^3 + 4 = 6y$$

$$\frac{x^3 + 4}{6} = y$$

Replace y with $f^{-1}(x)$

$$\frac{x^3 + 4}{6} = f^{-1}(x)$$

6. $f(x) = -x^2 + 3$

This is an equation of a vertical parabola which opens downward.

Since a horizontal line will intersect the graph in two points, the function is not one-to-one.

7.   The graph is a linear function through $(0, 1)$ and $(3, 0)$. The graph of $f^{-1}(x)$ will include the points $(1, 0)$ and $(0, 3)$, found by interchanging x and y. Plot these points, and draw a straight line through them.

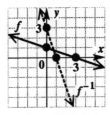

8.   The graph is a curve through $(1, 2)$, $(0, 1)$, and $\left(-1, \dfrac{1}{2}\right)$. Interchange x and y to get $(2, 1)$, $(1, 0)$, and $\left(\dfrac{1}{2}, -1\right)$, which are on the graph of $f^{-1}(x)$. Plot these points, and draw a smooth curve through them.

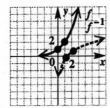

9. $f(x) = 3^x$

Make a table of values.

x	-2	-1	0	1	2
$f(x)$	$\dfrac{1}{9}$	$\dfrac{1}{3}$	1	3	9

Plot the points from the table and draw a smooth curve through them.

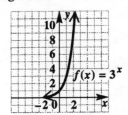

10. $f(x) = \left(\dfrac{1}{3}\right)^x$

Make a table of values.

x	-2	-1	0	1	2
$f(x)$	9	3	1	$\dfrac{1}{3}$	$\dfrac{1}{9}$

Plot the points from the table and draw a smooth curve through them.

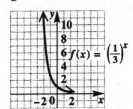

11. $y = 3^{x+1}$

Make a table of values.

x	-3	-2	-1	0	1
y	$\dfrac{1}{9}$	$\dfrac{1}{3}$	1	3	9

Plot the points from the table and draw a smooth curve through them.

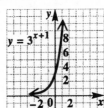

12. $y = 2^{2x+3}$

Make a table of values.

x	-2	$-\dfrac{3}{2}$	-1	0	$\dfrac{1}{2}$
y	$\dfrac{1}{2}$	1	2	8	16

Plot the points from the table and draw a smooth curve through them.

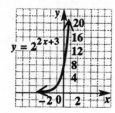

13. $4^{3x} = 8^{x+4}$

Write each side as a power of 2.

$\left(2^2\right)^{3x} = \left(2^3\right)^{(x+4)}$

$2^{6x} = 2^{(3x+12)}$

$6x = 3x + 12$ *Equate exponents*

$3x = 12$

$x = 4$

Solution set: $\{4\}$

14. $\left(\dfrac{1}{27}\right)^{x-1} = 9^{2x}$

$\left[\left(\dfrac{1}{3}\right)^3\right]^{x-1} = \left(3^2\right)^{2x}$

Write each side as a power of 3.

$\left(3^{-3}\right)^{x-1} = \left(3^2\right)^{2x}$

$3^{(-3x+3)} = 3^{4x}$

$-3x + 3 = 4x$ *Equate exponents*

$3 = 7x$

$\dfrac{3}{7} = x$

Solution set: $\left\{\dfrac{3}{7}\right\}$

15. $W(x) = .67(1.123)^x$

(a) 1965 corresponds to $x = 5$.

$W(5) = .67(1.123)^5 \approx 1.2,$

which is less than the actual value of 1.4 million tons.

(b) 1975 corresponds to $x = 15$.

$W(15) = .67(1.123)^{15} \approx 3.8,$

which is less than the actual value of 4.5 million tons.

(c) 1990 corresponds to $x = 30$.

$W(30) = .67(1.123)^{30} \approx 21.8,$

which is more than the actual value of 16.2 million tons.

16. $g(x) = \log_3 x$

Replace $g(x)$ with y, and write in exponential form.

$y = \log_3 x$

$3^y = x$

Make a table of values. Since $x = 3^y$ is the inverse of $f(x) = y = 3^x$ in Exercise 9, simply reverse the ordered pairs in the table of values belonging to $f(x) = 3^x$.

x	$\dfrac{1}{9}$	$\dfrac{1}{3}$	1	3	9
y	-2	-1	0	1	2

Plot the points from the table and draw a smooth curve through them.

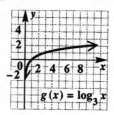

17. $g(x) = \log_{1/3} x$

Replace $g(x)$ with y, and write in exponential form.

$y = \log_{1/3} x$

$\left(\dfrac{1}{3}\right)^y = x$

Make a table of values. Since $x = \left(\dfrac{1}{3}\right)^y$ is the inverse of $f(x) = y = \left(\dfrac{1}{3}\right)^x$ in Exercise 10, simply reverse the ordered pairs in the table of values belonging to $f(x) = \left(\dfrac{1}{3}\right)^x$.

x	9	3	1	$\dfrac{1}{3}$	$\dfrac{1}{9}$
y	-2	-1	0	1	2

Plot the points from the table and draw a smooth curve through them.

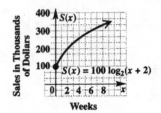

$g(x) = \log_{1/3} x$

18. $\log_8 64 = x$

$\quad 8^x = 64 \quad$ *Exponential form*

Write each side as a power of 8.

$\quad 8^x = 8^2$

$\quad x = 2 \quad$ *Equate exponents*

Solution set: $\{2\}$

19. $\log_2 \sqrt{8} = x$

$\quad 2^x = \sqrt{8} \quad$ *Exponential form*

$\quad 2^x = 8^{1/2}$

Write each side as a power of 2.

$\quad 2^x = \left(2^3\right)^{1/2}$

$\quad 2^x = 2^{3/2}$

$\quad x = \dfrac{3}{2} \quad$ *Equate exponents*

Solution set: $\left\{\dfrac{3}{2}\right\}$

20. $\log_7 \dfrac{1}{49} = x$

$\quad 7^x = \dfrac{1}{49} \quad$ *Exponential form*

Write each side as a power of 7.

$\quad 7^x = 7^{-2}$

$\quad x = -2 \quad$ *Equate exponents*

Solution set: $\{-2\}$

21. $\log_4 x = \dfrac{3}{2}$

$\quad x = 4^{3/2} \quad$ *Exponential form*

$\quad x = \left(\sqrt{4}\right)^3 = 2^3 = 8$

Solution set: $\{8\}$

22. $\log_k 4 = 1$

$\quad k^1 = 4 \quad$ *Exponential form*

$\quad k = 4$

Solution set: $\{4\}$

23. $\log_b b^2 = 2$

$\quad b^2 = b^2 \quad$ *Exponential form*

This is an identity. Thus, b can be any real number, $b > 0$ and $b \neq 1$.

Solution set: $\{b \mid b > 0, b \neq 1\}$

24. $\log_b a$ is the exponent to which b must be raised in order to obtain a.

25. From Exercise 24,

$$b^{\log_b a} = a.$$

26. $S(x) = 100 \log_2 (x + 2)$

When $x = 6$,

$$S(x) = 100 \log_2 (x + 2)$$
$$= 100(3) = 300.$$

After 6 weeks the sales were 300 thousand dollars or $300,000.

To graph the function, make a table of values that includes the ordered pair from above.

x	0	2	6
$S(x)$	100	200	300

Plot the ordered pairs and draw the graph through them.

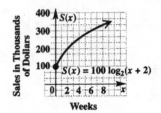

27. $\log_2 3xy^2$

$= \log_2 3 + \log_2 x + \log_2 y^2 \quad$ *Product rule*

$= \log_2 3 + \log_2 x + 2\log_2 y \quad$ *Power rule*

28. $\log_4 \dfrac{\sqrt{x} \cdot w^2}{z}$

$= \log_4 \left(\sqrt{x} \cdot w^2\right) - \log_4 z \quad$ *Quotient rule*

$= \log_4 x^{1/2} + \log_4 w^2 - \log_4 z \quad$ *Product rule*

$= \dfrac{1}{2}\log_4 x + 2\log_4 w - \log_4 z \quad$ *Power rule*

29. $\log_b 3 + \log_b x - 2\log_b y$

Use the product and power rules for logarithms.

$= \log_b (3 \cdot x) - \log_b y^2$

$= \log_b \dfrac{3x}{y^2} \quad\quad$ *Quotient rule*

30. $\log_3 (x + 7) - \log_3 (4x + 6)$

$= \log_3 \dfrac{x + 7}{4x + 6} \quad$ *Quotient rule*

31. $\log 28.9 \approx 1.4609$

32. $\log .257 \approx -.5901$

33. $\ln 28.9 \approx 3.3638$

34. $\ln .257 \approx -1.3587$

35. $\log_{16} 13 = \dfrac{\log 13}{\log 16} \approx .9251$

36. $\log_4 12 = \dfrac{\log 12}{\log 4} \approx 1.7925$

37.   Milk has a hydronium ion concentration of 4.0×10^{-7}.

$$pH = -\log [H_3O^+]$$
$$pH = -\log (4.0 \times 10^{-7}) \approx 6.4$$

38.   Crackers have a hydronium ion concentration of 3.8×10^{-9}.

$$pH = -\log [H_3O^+]$$
$$pH = -\log (3.8 \times 10^{-9}) \approx 8.4$$

39.   Orange juice has a pH of 4.6.

$$pH = -\log [H_3O^+]$$
$$4.6 = -\log [H_3O^+]$$
$$\log_{10} [H_3O^+] = -4.6$$
$$[H_3O^+] = 10^{-4.6} \approx 2.5 \times 10^{-5}$$

40. $Q(t) = 500e^{-.05t}$

(a) Let $t = 0$.

$$Q(0) = 500e^{-.05(0)}$$
$$= 500e^0 = 500(1) = 500$$

There are 500 grams.

(b) Let $t = 4$.

$$Q(4) = 500e^{-.05(4)}$$
$$= 500e^{-.2} \approx 409.4$$

There will be about 409 grams in 4 days.

41. $3^x = 9.42$

$$\log 3^x = \log 9.42$$
$$x \log 3 = \log 9.42$$
$$x = \dfrac{\log 9.42}{\log 3} \approx 2.042$$

Solution set: $\{2.042\}$

42. $e^{.06x} = 3$

Take base e logarithms on both sides.

$$\ln e^{.06x} = \ln 3$$
$$.06x \ln e = \ln 3$$
$$.06x = \ln 3 \qquad \ln e = 1$$
$$x = \dfrac{\ln 3}{.06} \approx 18.310$$

Solution set: $\{18.310\}$

43.   **(a)** Solve $7^x = 23$ by using property 3 with common logarithms.

$$7^x = 23$$
$$\log 7^x = \log 23$$
$$x \log 7 = \log 23$$
$$x = \dfrac{\log 23}{\log 7}$$

(b) Solve $7^x = 23$ by using property 3 with natural logarithms.

$$7^x = 23$$
$$\ln 7^x = \ln 23$$
$$x \ln 7 = \ln 23$$
$$x = \dfrac{\ln 23}{\ln 7}$$

(c) Use the change-of-base rule with the solution in part (a).

$$x = \dfrac{\log 23}{\log 7} = \log_7 23$$

(d) $x = \dfrac{\log 23}{\log 7} \neq \log_{23} 7$

The answer is (d).

44. $\log_3 (9x + 8) = 2$

$$9x + 8 = 3^2 \quad \textit{Exponential form}$$
$$9x + 8 = 9$$
$$9x = 1$$
$$x = \dfrac{1}{9}$$

Solution set: $\left\{ \dfrac{1}{9} \right\}$

45. $\log_3 (p + 2) - \log_3 p = \log_3 2$

$$\log_3 \dfrac{p+2}{p} = \log_3 2 \quad \textit{Quotient rule}$$
$$\dfrac{p+2}{p} = 2 \qquad \textit{Property 4}$$
$$p + 2 = 2p$$
$$2 = p$$

Solution set: $\{2\}$

46.
$$\log(2x + 3) = \log 3x + 2$$
$$\log(2x + 3) - \log 3x = 2$$
$$\log_{10}\frac{2x + 3}{3x} = 2 \quad \textit{Quotient rule}$$
$$\frac{2x + 3}{3x} = 10^2 \quad \textit{Exponential form}$$
$$\frac{2x + 3}{3x} = 100$$
$$2x + 3 = 300x$$
$$3 = 298x$$
$$x = \frac{3}{298}$$

Solution set: $\left\{\dfrac{3}{298}\right\}$

47.
$$\log_4 x + \log_4(8 - x) = 2$$
$$\log_4[x(8 - x)] = 2 \quad \textit{Product rule}$$
$$x(8 - x) = 4^2 \quad \textit{Exponential form}$$
$$8x - x^2 = 16$$
$$x^2 - 8x + 16 = 0$$
$$(x - 4)(x - 4) = 0$$
$$x - 4 = 0$$
$$x = 4$$

Solution set: $\{4\}$

48.
$$\log_2 x + \log_2(x + 15) = 4$$
$$\log_2[x(x + 15)] = 4 \quad \textit{Product rule}$$
$$x(x + 15) = 2^4 \quad \textit{Exponential form}$$
$$x^2 + 15x = 16$$
$$x^2 + 15x - 16 = 0$$
$$(x + 16)(x - 1) = 0$$
$$x + 16 = 0 \quad \text{or} \quad x - 1 = 0$$
$$x = -16 \quad \text{or} \quad x = 1$$

Reject $x = -16$, because it yields an equation in which the logarithm of a negative number must be found.
Solution set: $\{1\}$

49. (a)
$$\log(2x + 3) = \log x + 1$$
$$\log(2x + 3) - \log x = 1$$
$$\log_{10}\frac{2x + 3}{x} = 1 \quad \textit{Quotient rule}$$
$$\frac{2x + 3}{x} = 10^1 \quad \textit{Exponential form}$$
$$2x + 3 = 10x$$
$$3 = 8x$$
$$\frac{3}{8} = x$$

Solution set: $\left\{\dfrac{3}{8}\right\}$

(b) From the graph, the x-value of the x-intercept is .375, the decimal equivalent of $\dfrac{3}{8}$. Note that the solutions of $Y_1 - Y_2 = 0$ are the same as the solutions of $Y_1 = Y_2$.

50. When the power rule was applied in the second step, the domain was changed from $\{x \mid x \neq 0\}$ to $\{x \mid x > 0\}$. Instead of using the power rule for logarithms, we can change the original equation to the exponential form $x^2 = 10^2$ and get $x = \pm 10$. As you can see in the erroneous solution, the valid solution -10 was "lost." The solution set is $\{\pm 10\}$.

51. $A = P\left(1 + \dfrac{r}{n}\right)^{nt}$

Let $P = 20,000$, $r = .07$, and $t = 5$. For $n = 4$ (quarterly compounding),

$$A = 20,000\left(1 + \frac{.07}{4}\right)^{4 \cdot 5} \approx 28,295.56.$$

There will be $28,295.56 in the account after 5 years.

52. In the continuous compounding formula, let $P = 10,000$, $r = .06$, and $t = 3$.

$$A = Pe^{rt}$$
$$A = 10,000e^{.06(3)} \approx 11,972.17$$

There will be $11,972.17 in the account after 3 years.

53. Use $A = P\left(1 + \dfrac{r}{n}\right)^{nt}$.

Plan A: Let $P = 1000$, $r = .04$, $n = 4$, and $t = 3$.

$$A = 1000\left(1 + \frac{.04}{4}\right)^{4 \cdot 3} \approx 1126.83$$

Plan B: Let $P = 1000$, $r = .039$, $n = 12$, and $t = 3$.

$$A = 1000\left(1 + \frac{.039}{12}\right)^{12 \cdot 3} \approx 1123.91$$

Plan A is the better plan by $2.92.

54. Let $Q(t) = \frac{1}{2}(500) = 250$ to find the half-life of the radioactive substance.

$$Q(t) = 500e^{-.05t}$$
$$250 = 500e^{-.05t}$$
$$.5 = e^{-.05t}$$
$$\ln .5 = \ln e^{-.05t}$$
$$\ln .5 = -.05t \,(\ln e)$$
$$-.05t = \ln .5$$
$$t = \frac{\ln .5}{-.05} \approx 13.9$$

The half-life is about 13.9 days.

55. $R(x) = .0597e^{.0553x}$

1990 corresponds to $x = 20$.

$$R(20) = .0597e^{.0553(20)}$$
$$\approx .1804$$

In 1990, about 18.04% of municipal solid waste was recovered.

56. $N(r) = -5000 \ln r$

Replace $N(r)$ by 2000, and solve for r.

$$2000 = -5000 \ln r$$
$$\ln r = -\frac{2000}{5000}$$
$$\log_e r = -.4$$

Change to exponential form and approximate.

$$r = e^{-.4} \approx .67$$

About 67% of the words are common to both of the evolving languages.

57. $S = C(1 - r)^n$

Let $C = 30,000$, $r = .15$, and $n = 12$.

$$S = 30,000(1 - .15)^{12}$$
$$= 30,000(.85)^{12} \approx 4267$$

The scrap value is about $4267.

58. Let $S = \frac{1}{2}C$ and $n = 6$.

$$S = C(1 - r)^n$$
$$\frac{1}{2}C = C(1 - r)^6$$
$$.5 = (1 - r)^6 \,(*)$$
$$\ln .5 = \ln (1 - r)^6$$
$$\ln .5 = 6 \ln (1 - r)$$
$$\ln (1 - r) = \frac{\ln .5}{6}$$
$$\ln (1 - r) \approx -.1155$$
$$1 - r = e^{-.1155}$$
$$1 - r \approx .89$$
$$r = .11$$

The rate is approximately 11%.

$(*)$ Alternative solution steps without logarithms:

$$.5 = (1 - r)^6$$
$$\sqrt[6]{.5} = 1 - r$$
$$r = 1 - \sqrt[6]{.5} \approx .11$$

59. $f(x) = 2^x$

x	$f(x)$
-2	$2^{-2} = \dfrac{1}{4}$
-1	$2^{-1} = \dfrac{1}{2}$
0	$2^0 = 1$
1	$2^1 = 2$
2	$2^2 = 4$
3	$2^3 = 8$

Plot the ordered pairs from the table, and draw a smooth curve through them.

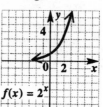

60. $g(x) = \log_2 x$

x	$g(x)$
$\dfrac{1}{4}$	$\log_2 \dfrac{1}{4} = \log_2 2^{-2} = -2$
$\dfrac{1}{2}$	$\log_2 \dfrac{1}{2} = \log_2 2^{-1} = -1$
1	$\log_2 1 = \log_2 2^0 = 0$
2	$\log_2 2 = \log_2 2^1 = 1$
4	$\log_2 4 = \log_2 2^2 = 2$
8	$\log_2 8 = \log_2 2^3 = 3$

Plot the ordered pairs from the table, and draw a smooth curve through them.

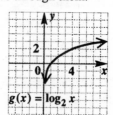

61. The ordered pairs in Exercises 59 and 60 are reverses of each other. In other words, if (x, y) is an ordered pair in Exercise 59, (y, x) is an ordered pair in Exercise 60. Functions f and g are inverses of each other.

62. The graph of f in Exercise 59 has a *horizontal* asymptote, $y = 0$, while the graph of g in Exercise 60 has a *vertical* asymptote, $x = 0$.

63. $2^2 \cdot 2^3 = 2^5$ because _2_ + _3_ = _5_.

64. It is a fact that $32 = 4 \cdot 8$. Therefore, using properties of logarithms,
$\log_2 32 = \log_2 \underline{4} + \log_2 \underline{8}$.

65. $\log_2 13 = \dfrac{\log 13}{\log 2} = 3.700439718$

66. Based on the result of Exercise 65, $\log_2 13$ means that $2^{3.700439718} = 13$. $\log_2 13$ is the exponent to which 2 must be raised in order to obtain 13.

67. $2^{\log_2 13} = 13$ since (using the results of Exercises 65 and 66)
$$2^{\log_2 13} = 2^{3.700439718} = 13.$$

68. Using a calculator,
$$2^{3.700439718} = 13.$$

The number in Exercise 65 is the exponent to which 2 must be raised in order to obtain 13.

69. Based on the result in Exercise 65, the point $(13,\ \underline{3.700439718}\ )$ lies on the graph of $g(x) = \log_2 x$.

70. $2^{x+1} = 8^{2x+3}$
$2^{x+1} = \left(2^3\right)^{2x+3}$
$2^{x+1} = 2^{6x+9}$
$x + 1 = 6x + 9$
$-8 = 5x$
$-\dfrac{8}{5} = x$
Solution set: $\left\{-\dfrac{8}{5}\right\}$

71. $\log_3 (x + 9) = 4$
$x + 9 = 3^4$ *Exponential form*
$x + 9 = 81$
$x = 72$
Solution set: $\{72\}$

72. $\log_2 32 = x$
$2^x = 32$ *Exponential form*
Write each side as a power of 2.
$2^x = 2^5$
$x = 5$ *Equate exponents*
Solution set: $\{5\}$

73. $\log_x \dfrac{1}{81} = 2$
$x^2 = \dfrac{1}{81}$ *Exponential form*
$x^2 = \left(\dfrac{1}{9}\right)^2$
$x = \pm \dfrac{1}{9}$ *Square root property*
The base x cannot be negative.
Solution set: $\left\{\dfrac{1}{9}\right\}$

74. $27^x = 81$
Write each side as a power of 3.
$\left(3^3\right)^x = 3^4$
$3^{3x} = 3^4$
$3x = 4$ *Equate exponents*
$x = \dfrac{4}{3}$
Solution set: $\left\{\dfrac{4}{3}\right\}$

75. $2^{2x-3} = 8$
Write each side as a power of 2.
$2^{2x-3} = 2^3$
$2x - 3 = 3$ *Equate exponents*
$2x = 6$
$x = 3$
Solution set: $\{3\}$

76. $\log_3 (x + 1) - \log_3 x = 2$
$\log_3 \dfrac{x+1}{x} = 2$ *Quotient rule*
$\dfrac{x+1}{x} = 3^2$ *Exponential form*
$9x = x + 1$
$8x = 1$
$x = \dfrac{1}{8}$
Solution set: $\left\{\dfrac{1}{8}\right\}$

77. $\log (3x - 1) = \log 10$
$3x - 1 = 10$
$3x = 11$
$x = \dfrac{11}{3}$
Solution set: $\left\{\dfrac{11}{3}\right\}$

78. $f(t) = 5000(2)^{-.15t}$

(a) The original value is found when $t = 0$.

$$f(0) = 5000(2)^{-.15(0)}$$
$$= 5000(2)^0 = 5000(1) = 5000$$

The original value is $5000.

(b) The value after 5 years is found when $t = 5$.

$$f(5) = 5000(2)^{-.15(5)} \approx 2973.018$$

The value after 5 years is about $2973.

(c) The value after 10 years is found when $t = 10$.

$$f(10) = 5000(2)^{-.15(10)} \approx 1767.767$$

The value after 10 years is about $1768.

Chapter 9 Test

1. **(a)** $f(x) = x^2 + 9$

This function is not one-to-one. The graph of $f(x)$ is a vertical parabola. A horizontal line will intersect the graph more than once.

(b) This function is one-to-one. A horizontal line will not intersect the graph in more than one point.

2. $f(x) = \sqrt[3]{x + 7}$
Replace $f(x)$ with y.
$$y = \sqrt[3]{x + 7}$$
Interchange x and y.
$$x = \sqrt[3]{y + 7}$$
Solve for y.
$$x^3 = y + 7$$
$$x^3 - 7 = y$$
Replace y with $f^{-1}(x)$.
$$f^{-1}(x) = x^3 - 7$$

3. By the horizontal line test, $f(x)$ is a one-to-one function and has an inverse. Choose some points on the graph of $f(x)$, such as $(4, 0)$, $(3, -1)$, and $(0, -2)$. To graph the inverse, interchange the x- and y-values to get $(0, 4)$, $(-1, 3)$, and $(-2, 0)$. Plot these points and draw a smooth curve through them.

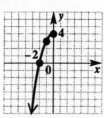

4. $f(x) = 6^x$
Make a table of values.

x	-2	-1	0	1
$f(x)$	$\dfrac{1}{36}$	$\dfrac{1}{6}$	1	6

Plot these points and draw a smooth exponential curve through them.

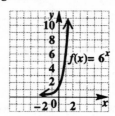

5. $g(x) = \log_6 x$
Make a table of values.

Powers of 6	6^{-2}	6^{-1}	6^0	6^1
x	$\dfrac{1}{36}$	$\dfrac{1}{6}$	1	6
$g(x)$	-2	-1	0	1

Plot these points and draw a smooth logarithmic curve through them.

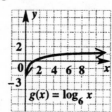

6. $y = 6^x$ and $y = \log_6 x$ are inverse functions. To use the graph from Exercise 4 to obtain the graph of the function in Exercise 5, interchange the x- and y-coordinates of the ordered pairs

$$\left(-2, \frac{1}{36}\right), \left(-1, \frac{1}{6}\right), (0, 1), \text{ and } (1, 6) \text{ to get}$$

$$\left(\frac{1}{36}, -2\right), \left(\frac{1}{6}, -1\right), (1, 0), \text{ and } (6, 1). \text{ Plot}$$

these points and draw a smooth logarithmic curve through them.

7. $5^x = \dfrac{1}{625}$

$$5^x = \left(\frac{1}{5}\right)^4$$

Write each side as a power of 5.

$$5^x = 5^{-4}$$

$x = -4$ *Equate exponents*
Solution set: $\{-4\}$

8. $2^{3x-7} = 8^{2x+2}$

Write each side as a power of 2.

$2^{3x-7} = 2^{3(2x+2)}$

$3x - 7 = 3(2x + 2)$ *Equate exponents*

$3x - 7 = 6x + 6$

$-13 = 3x$

$-\dfrac{13}{3} = x$

Solution set: $\left\{-\dfrac{13}{3}\right\}$

9. $R(x) = 2821(.9195)^x$

(a) 1990 corresponds to $x = 0$.

$R(0) = 2821(.9195)^0$
$= 2821(1) = 2821$

In 1990, the toxic release inventory was 2821 million pounds.

(b) 1992 corresponds to $x = 2$.

$R(2) = 2821(.9195)^2$
≈ 2385 million pounds

(c) 1993 corresponds to $x = 3$.

$R(3) = 2821(.9195)^3$
≈ 2193 million pounds

The actual amount of 2157.4 million pounds is less than the amount provided by the model.

10. The base is 4, the exponent (logarithm) is -2, and the number is .0625, so $4^{-2} = .0625$ becomes $\log_4 .0625 = -2$ in logarithmic form.

11. The base is 7, the logarithm (exponent) is 2, and the number is 49, so $\log_7 49 = 2$ becomes $7^2 = 49$ in exponential form.

12. $\log_{1/2} x = -5$

$x = \left(\dfrac{1}{2}\right)^{-5}$ *Exponential form*

$x = \left(\dfrac{2}{1}\right)^5 = 32$

Solution set: $\{32\}$

13. $x = \log_9 3$

$9^x = 3$ *Exponential form*

Write each side as a power of 3.

$(3^2)^x = 3$

$3^{2x} = 3^1$

$2x = 1$ *Equate exponents*

$x = \dfrac{1}{2}$

Solution set: $\left\{\dfrac{1}{2}\right\}$

14. $\log_x 16 = 4$

$x^4 = 16$ *Exponential form*

$x^2 = \pm 4$ *Square root property*

Reject -4 since $x^2 \geq 0$.

$x^2 = 4$

$x = \pm 2$ *Square root property*

Reject -2 since the base cannot be negative.

Solution set: $\{2\}$

15. The value of $\log_2 32$ is _5_. This means that if we raise _2_ to the *fifth* power, we result is _32_.

16. $\log_3 x^2 y$

$= \log_3 x^2 + \log_3 y$ *Product rule*

$= 2\log_3 x + \log_3 y$ *Power rule*

17. $\log_5 \left(\dfrac{\sqrt{x}}{yz}\right)$

$= \log_5 \sqrt{x} - \log_5 yz$ *Quotient rule*

$= \log_5 x^{1/2} - \left(\log_5 y + \log_5 z\right)$ *Product rule*

$= \dfrac{1}{2}\log_5 x - \log_5 y - \log_5 z$ *Power rule*

18. $3\log_b s - \log_b t$

$= \log_b s^3 - \log_b t$ *Power rule*

$= \log_b \dfrac{s^3}{t}$ *Quotient rule*

19. $\dfrac{1}{4}\log_b r + 2\log_b s - \dfrac{2}{3}\log_b t$

Use the power rule for logarithms.

$= \log_b r^{1/4} + \log_b s^2 - \log_b t^{2/3}$

Use the product and quotient rules for logarithms.

$= \log_b \dfrac{r^{1/4} s^2}{t^{2/3}}$

20. **(a)** $\log 23.1 \approx 1.3636$

(b) $\ln .82 \approx -.1985$

21. **(a)** $\log_3 19 = \dfrac{\log_{10} 19}{\log_{10} 3} = \dfrac{\log 19}{\log 3}$

(b) $\log_3 19 = \dfrac{\log_e 19}{\log_e 3} = \dfrac{\ln 19}{\ln 3}$

(c) The four-decimal-place approximation of either fraction is 2.6801.

22. $3^x = 78$

$\ln 3^x = \ln 78$

$x \ln 3 = \ln 78$ *Power rule*

$x = \dfrac{\ln 78}{\ln 3} \approx 3.9656$

Solution set: $\{3.9656\}$

23. $\log_8 (x + 5) + \log_8 (x - 2) = 1$

Use the product rule for logarithms.

$\log_8 [(x + 5)(x - 2)] = 1$

$(x + 5)(x - 2) = 8^1$

$x^2 + 3x - 10 = 8$

$x^2 + 3x - 18 = 0$

$(x + 6)(x - 3) = 0$

$x + 6 = 0$ or $x - 3 = 0$

$x = -6$ or $x = 3$

Reject $x = -6$, because $x + 5 = -1$, which yields an equation in which the logarithm of a negative number must be found.

Solution set: $\{3\}$

24. $A = P\left(1 + \dfrac{r}{n}\right)^{nt}$

$A = 10,000\left(1 + \dfrac{.045}{4}\right)^{4 \cdot 5} \approx 12,507.51$

$10,000$ invested at 4.5% annual interest, compounded quarterly, will increase to $12,507.51 in 5 years.

25. $A = Pe^{rt}$

(a) $A = 15,000e^{.05(5)} \approx 19,260.38$

There will be $19,260.38 in the account.

(b) Let $A = 2(15,000) = 30,000$ and solve for t.

$30,000 = 15,000e^{.05t}$

$2 = e^{.05t}$

$\ln 2 = \ln e^{.05t}$

$\ln 2 = .05t (\ln e)$

$.05t = \ln 2$

$t = \dfrac{\ln 2}{.05} \approx 13.9$

The principal will double in about 13.9 years.

Cumulative Review Exercises Chapters 1–9

For Exercises 1–4,

$$S = \left\{-\dfrac{9}{4}, -2, -\sqrt{2}, 0, .6, \sqrt{11}, \sqrt{-8}, 6, \dfrac{30}{3}\right\}.$$

1. The integers are $-2, 0, 6,$ and $\dfrac{30}{3}$ (or 10).

2. The rational numbers are $-\dfrac{9}{4}, -2, 0, .6, 6,$ and $\dfrac{30}{3}$ (or 10).

Each can be expressed as a quotient of two integers.

3. The irrational numbers are $-\sqrt{2}$ and $\sqrt{11}$.

4. All are real numbers except $\sqrt{-8}$.

5. $|-8| + 6 - |-2| - (-6 + 2)$

$= 8 + 6 - 2 - (-4)$

$= 14 - 2 + 4 = 16$

6. $-12 - |-3| - 7 - |-5|$

$= -12 - 3 - 7 - 5 = -27$

7. $2(-5) + (-8)(4) - (-3)$

$= -10 - 32 + 3 = -39$

8. $7 - (3 + 4a) + 2a = -5(a - 1) - 3$

$7 - 3 - 4a + 2a = -5a + 5 - 3$

$4 - 2a = -5a + 2$

$3a = -2$

$a = -\dfrac{2}{3}$

Solution set: $\left\{-\dfrac{2}{3}\right\}$

9. $2m + 2 \le 5m - 1$

$-3m \le -3$

Divide by -3; reverse the inequality.

$m \ge 1$

Solution set: $[1, \infty)$

10. $|2x - 5| = 9$

$2x - 5 = 9$ or $2x - 5 = -9$

$2x = 14$ $2x = -4$

$x = 7$ or $x = -2$

Solution set: $\{-2, 7\}$

11. $|3p| - 4 = 12$

$|3p| = 16$

$3p = 16$ or $3p = -16$

$p = \dfrac{16}{3}$ or $p = -\dfrac{16}{3}$

Solution set: $\left\{\pm \dfrac{16}{3}\right\}$

12. $|3k - 8| \le 1$

$-1 \le 3k - 8 \le 1$

$7 \le 3k \le 9$

$\dfrac{7}{3} \le k \le 3$

Solution set: $\left[\dfrac{7}{3}, 3\right]$

13. $|4m + 2| > 10$

$$4m + 2 > 10 \quad \text{or} \quad 4m + 2 < -10$$
$$4m > 8 \qquad\qquad 4m < -12$$
$$m > 2 \quad \text{or} \qquad m < -3$$

Solution set: $(-\infty, -3) \cup (2, \infty)$

14. $5x + 2y = 10$

Find the x- and y-intercepts. To find the x-intercept, let $y = 0$.

$$5x + 2(0) = 10$$
$$5x = 10$$
$$x = 2$$

The x-intercept is $(2, 0)$.
To find the y-intercept, let $x = 0$.

$$5(0) + 2y = 10$$
$$2y = 10$$
$$y = 5$$

The y-intercept is $(0, 5)$.
Plot the intercepts and draw the line through them.

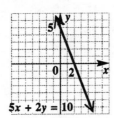

15. $-4x + y \leq 5$

Graph the line $-4x + y = 5$, which has intercepts $(0, 5)$ and $\left(-\dfrac{5}{4}, 0\right)$, as a solid line because the inequality involves $\leq$. Test $(0, 0)$, which yields $0 \leq 5$, a true statement. Shade the region that includes $(0, 0)$.

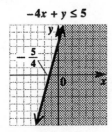

16.   The points are $(1986, 70{,}000{,}000)$ and $(1994, 25{,}300{,}000)$.

$$m = \frac{70{,}000{,}000 - 25{,}300{,}000}{1986 - 1994}$$
$$= \frac{44{,}700{,}000}{-8}$$
$$= -5{,}587{,}500$$

The slope is $-5{,}587{,}500$ (in \$/year).

17.   Through $(5, -1)$; parallel to $3x - 4y = 12$
Find the slope of

$$3x - 4y = 12$$
$$-4y = -3x + 12$$
$$y = \frac{3}{4}x - 3.$$

The slope is $\dfrac{3}{4}$, so a line parallel to it also has slope $\dfrac{3}{4}$.

Let $m = \dfrac{3}{4}$ and $(x_1, y_1) = (5, -1)$ in the point-slope form.

$$y - y_1 = m(x - x_1)$$
$$y - (-1) = \frac{3}{4}(x - 5)$$
$$y + 1 = \frac{3}{4}(x - 5)$$

Multiply by 4 to clear the fraction.

$$4(y + 1) = 3(x - 5)$$
$$4y + 4 = 3x - 15$$
$$19 = 3x - 4y$$

Write in standard form.

$$3x - 4y = 19$$

18. $5x - 3y = 14$ (1)
      $2x + 5y = 18$ (2)

Multiply equation (1) by 5 and equation (2) by 3. Then add the results.

$$
\begin{array}{rcl r}
25x \;-\; 15y & = & 70 & 5 \times (1) \\
6x \;+\; 15y & = & 54 & 3 \times (2) \\
\hline
31x \qquad\quad & = & 124 & Add \\
x & = & 4 &
\end{array}
$$

Substitute 4 for x in equation (1) to find y.

$$5x - 3y = 14 \quad (1)$$
$$5(4) - 3y = 14$$
$$20 - 3y = 14$$
$$-3y = -6$$
$$y = 2$$

Solution set: $\{(4, 2)\}$

19.
$$x + 2y + 3z = 11 \quad (1)$$
$$3x - y + z = 8 \quad (2)$$
$$2x + 2y - 3z = -12 \quad (3)$$

To eliminate z, add equations (1) and (3).

$$
\begin{array}{rl}
x + 2y + 3z = & 11 \quad (1) \\
2x + 2y - 3z = & -12 \quad (3) \\
\hline
3x + 4y = & -1 \quad (4)
\end{array}
$$

To eliminate z again, multiply equation (2) by 3 and add the result to equation (3).

$$
\begin{array}{rll}
9x - 3y + 3z = & 24 & 3 \times (2) \\
2x + 2y - 3z = & -12 & (3) \\
\hline
11x - y = & 12 & (5)
\end{array}
$$

Multiply equation (5) by 4 and add the result to equation (4).

$$
\begin{array}{rll}
44x - 4y = & 48 & 4 \times (5) \\
3x + 4y = & -1 & (4) \\
\hline
47x = & 47 & \\
x = & 1 &
\end{array}
$$

Substitute 1 for x in equation (5) to find y.

$$11x - y = 12 \quad (5)$$
$$11(1) - y = 12$$
$$11 - y = 12$$
$$-y = 1$$
$$y = -1$$

Substitute 1 for x and -1 for y in equation (2) to find z.

$$3x - y + z = 8 \quad (2)$$
$$3(1) - (-1) + z = 8$$
$$3 + 1 + z = 8$$
$$4 + z = 8$$
$$z = 4$$

Solution set: $\{(1, -1, 4)\}$

20.
$$\begin{vmatrix} -2 & -1 \\ 5 & 3 \end{vmatrix} = (-2)(3) - (-1)(5)$$
$$= -6 + 5 = -1$$

21. Let $x =$ the amount of candy at \$1.00 per pound.

	Amount of Candy	Price per Pound	Total Price
First Candy	x	1.00	$1.00x$
Second Candy	10	1.96	1.96(10)
Mixture	$x + 10$	1.60	$1.60(x + 10)$

Solve the equation:

$$1.00x + 1.96(10) = 1.60(x + 10)$$

Multiply by 10 to clear the decimals.

$$10x + 196 = 16(x + 10)$$
$$10x + 196 = 16x + 160$$
$$36 = 6x$$
$$6 = x$$

Use 6 pounds of the \$1.00 candy.

22.
$$(2p + 3)(3p - 1) = 6p^2 - 2p + 9p - 3$$
$$= 6p^2 + 7p - 3$$

23.
$$(4k - 3)^2 = (4k)^2 - 2(4k)(3) + 3^2$$
$$= 16k^2 - 24k + 9$$

24.
$$(3m^3 + 2m^2 - 5m) - (8m^3 + 2m - 4)$$
$$= 3m^3 + 2m^2 - 5m - 8m^3 - 2m + 4$$
$$= 3m^3 - 8m^3 + 2m^2 - 5m - 2m + 4$$
$$= -5m^3 + 2m^2 - 7m + 4$$

25.

$$
\begin{array}{r}
2t^3 + 5t^2 - 3t + 4 \\
3t + 1 \overline{\smash{)}\, 6t^4 + 17t^3 - 4t^2 + 9t + 4} \\
\underline{6t^4 + 2t^3} \\
15t^3 - 4t^2 \\
\underline{15t^3 + 5t^2} \\
-9t^2 + 9t \\
\underline{-9t^2 - 3t} \\
12t + 4 \\
\underline{12t + 4} \\
0
\end{array}
$$

The quotient is

$$2t^3 + 5t^2 - 3t + 4.$$

26. $8x + x^3 = x(8 + x^2)$

27. $24y^2 - 7y - 6 = (8y + 3)(3y - 2)$

28.
$$5z^3 - 19z^2 - 4z = z(5z^2 - 19z - 4)$$
$$= z(5z + 1)(z - 4)$$

29. $16a^2 - 25b^4$

Use the difference of two squares formula,

$$x^2 - y^2 = (x + y)(x - y),$$

where $x = 4a$ and $y = 5b^2$.

$$16a^2 - 25b^4 = (4a + 5b^2)(4a - 5b^2)$$

30. $8c^3 + d^3$

Use the sum of two cubes formula,

$$x^3 + y^3 = (x + y)(x^2 - xy + y^2),$$

where $x = 2c$ and $y = d$.

$$8c^3 + d^3 = (2c + d)(4c^2 - 2cd + d^2)$$

31. $16r^2 + 56rq + 49q^2$
$$= (4r)^2 + 2(4r)(7q) + (7q)^2$$

Use the perfect square formula,

$$x^2 + 2xy + y^2 = (x + y)^2,$$

where $x = 4r$ and $y = 7q$.

$$16r^2 + 56rq + 49q^2 = (4r + 7q)^2$$

32. $\dfrac{(5p^3)^4(-3p^7)}{2p^2(4p^4)} = \dfrac{(5^4 p^{12})(-3p^7)}{8p^6}$

$$= \dfrac{(625)(-3)p^{19}}{8p^6}$$

$$= -\dfrac{1875p^{13}}{8}$$

33. $\dfrac{x^2 - 9}{x^2 + 7x + 12} \div \dfrac{x - 3}{x + 5}$

Multiply by the reciprocal.

$$= \dfrac{x^2 - 9}{x^2 + 7x + 12} \cdot \dfrac{x + 5}{x - 3}$$

$$= \dfrac{(x + 3)(x - 3)}{(x + 3)(x + 4)} \cdot \dfrac{(x + 5)}{(x - 3)} \quad Factor$$

$$= \dfrac{x + 5}{x + 4}$$

34. $\dfrac{2}{k + 3} - \dfrac{5}{k - 2}$

The LCD is $(k + 3)(k - 2)$.

$$= \dfrac{2(k - 2)}{(k + 3)(k - 2)} - \dfrac{5(k + 3)}{(k - 2)(k + 3)}$$

$$= \dfrac{2k - 4 - 5k - 15}{(k + 3)(k - 2)}$$

$$= \dfrac{-3k - 19}{(k + 3)(k - 2)}$$

35. $\dfrac{3}{p^2 - 4p} - \dfrac{4}{p^2 + 2p}$

$$= \dfrac{3}{p(p - 4)} - \dfrac{4}{p(p + 2)}$$

The LCD is $p(p - 4)(p + 2)$.

$$= \dfrac{3(p + 2)}{p(p - 4)(p + 2)} - \dfrac{4(p - 4)}{p(p + 2)(p - 4)}$$

$$= \dfrac{3p + 6 - 4p + 16}{p(p - 4)(p + 2)}$$

$$= \dfrac{22 - p}{p(p - 4)(p + 2)}$$

36. $\sqrt{288} = \sqrt{144 \cdot 2} = \sqrt{144}\sqrt{2} = 12\sqrt{2}$

37. $2\sqrt{32} - 5\sqrt{98} = 2\sqrt{16 \cdot 2} - 5\sqrt{49 \cdot 2}$

$$= 2 \cdot 4\sqrt{2} - 5 \cdot 7\sqrt{2}$$

$$= 8\sqrt{2} - 35\sqrt{2}$$

$$= -27\sqrt{2}$$

38. $\sqrt{2x + 1} - \sqrt{x} = 1$

$$\sqrt{2x + 1} = 1 + \sqrt{x}$$

$$\left(\sqrt{2x + 1}\right)^2 = \left(1 + \sqrt{x}\right)^2$$

$$2x + 1 = 1 + 2\sqrt{x} + x$$

$$x = 2\sqrt{x}$$

$$(x)^2 = \left(2\sqrt{x}\right)^2$$

$$x^2 = 4x$$

$$x^2 - 4x = 0$$

$$x(x - 4) = 0$$

$x = 0$ or $x = 4$

Check $x = 0$: $\sqrt{1} - \sqrt{0} \overset{?}{=} 1$ *True*

Check $x = 4$: $\sqrt{9} - \sqrt{4} \overset{?}{=} 1$ *True*

Solution set: $\{0, 4\}$

39. $(5 + 4i)(5 - 4i) = 5^2 - (4i)^2$

$$= 25 - 16i^2$$

$$= 25 - 16(-1)$$

$$= 25 + 16 = 41$$

40. $3x^2 - x - 1 = 0$

Here, $a = 3$, $b = -1$, $c = -1$.

Use the quadratic formula.

$$x = \dfrac{-b \pm \sqrt{b^2 - 4ac}}{2a}$$

$$x = \dfrac{-(-1) \pm \sqrt{(-1)^2 - 4(3)(-1)}}{2(3)}$$

$$= \dfrac{1 \pm \sqrt{1 + 12}}{6} = \dfrac{1 \pm \sqrt{13}}{6}$$

Solution set: $\left\{ \dfrac{1 + \sqrt{13}}{6}, \dfrac{1 - \sqrt{13}}{6} \right\}$

41. $k^2 + 2k - 8 > 0$

Solve the equation

$$k^2 + 2k - 8 = 0.$$
$$(k + 4)(k - 2) = 0$$

$$k + 4 = 0 \quad \text{or} \quad k - 2 = 0$$
$$k = -4 \quad \text{or} \quad k = 2$$

The numbers -4 and 2 divide the number line into three regions.

Region A	Region B	Region C

$$\xleftarrow{\hspace{0.3cm}} \underset{-6}{+} \quad \underset{-4}{+} \quad \underset{-2}{+} \quad \underset{0}{+} \quad \underset{2}{+} \quad \underset{4}{+} \xrightarrow{\hspace{0.3cm}}$$

Test a number from each region in the inequality

$$k^2 + 2k - 8 > 0.$$

continued

Region A: Let $k = -5$.

$$25 - 10 - 8 > 0 \qquad ?$$
$$7 > 0 \qquad \qquad \textit{True}$$

Region B: Let $k = 0$.

$$-8 > 0 \qquad \qquad \textit{False}$$

Region C: Let $k = 3$.

$$9 + 6 - 8 > 0 \qquad ?$$
$$7 > 0 \qquad \qquad \textit{True}$$

The numbers in Regions A and C, not including -4 or 2 because of $>$, are solutions.

Solution set: $(-\infty, -4) \cup (2, \infty)$

42. $x^4 - 5x^2 + 4 = 0$

Let $u = x^2$, so $u^2 = \left(x^2\right)^2 = x^4$.

$$u^2 - 5u + 4 = 0$$
$$(u - 1)(u - 4) = 0$$

$$u - 1 = 0 \quad \text{or} \quad u - 4 = 0$$
$$u = 1 \quad \text{or} \quad u = 4$$

To find x, substitute x^2 for u.

$$x^2 = 1 \quad \text{or} \quad x^2 = 4$$
$$x = \pm 1 \quad \text{or} \quad x = \pm 2$$

Solution set: $\{\pm 1, \pm 2\}$

43.   Let $x =$ one of the numbers;

$300 - x =$ the other number.

The product of the two numbers is given by

$$P = x(300 - x).$$

Writing this equation in standard form gives us

$$P = -x^2 + 300x.$$

Finding the maximum of the product is the same as finding the vertex of the graph of P. The x-value of the vertex is

$$x = -\frac{b}{2a} = -\frac{300}{2(-1)} = 150.$$

If x is 150, then $300 - x$ must also be 150. The two numbers are 150 and 150 and the product is $150 \cdot 150 = 22{,}500$.

44. $f(x) = \frac{1}{3}(x - 1)^2 + 2$ is in

$f(x) = a(x - h)^2 + k$ form. The graph is a vertical parabola with vertex (h, k) at $(1, 2)$. Since $a = \frac{1}{3} > 0$, the graph opens upward.

Also, $|a| = \left|\frac{1}{3}\right| = \frac{1}{3} < 1$, so the graph is wider than the graph of $f(x) = x^2$. The points $\left(0, 2\frac{1}{3}\right)$, $(-2, 5)$, and $(4, 5)$ are also on the graph.

$$f(x) = \frac{1}{3}(x - 1)^2 + 2$$

45. $f(x) = 2^x$

Make a table of values.

x	-2	-1	0	1	2
$f(x)$	$\dfrac{1}{4}$	$\dfrac{1}{2}$	1	2	4

Plot the ordered pairs from the table, and draw a smooth exponential curve through the points.

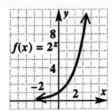

46. $5^{x+3} = \left(\dfrac{1}{25}\right)^{3x+2}$

$5^{x+3} = \left[\left(\dfrac{1}{5}\right)^2\right]^{3x+2}$

Write each side to the power of 5.

$5^{x+3} = \left(5^{-2}\right)^{(3x+2)}$

$5^{x+3} = 5^{-2(3x+2)}$

$x + 3 = -2(3x + 2)$

$x + 3 = -6x - 4$

$7x = -7$

$x = -1$

Solution set: $\{-1\}$

47. $f(x) = \log_3 x$

Make a table of values.

Powers of 3	3^{-2}	3^{-1}	3^0	3^1	3^2
x	$\dfrac{1}{9}$	$\dfrac{1}{3}$	1	3	9
y	-2	-1	0	1	2

Plot the ordered pairs and draw a smooth logarithmic curve through the points.

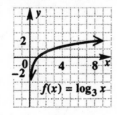

48. $\log_2 81 = \log_2 9^2 = 2\log_2 9$

$\approx 2(3.1699) = 6.3398$

49. $\log \dfrac{x^3\sqrt{y}}{z}$

$= \log \dfrac{x^3 y^{1/2}}{z}$

$= \log\left(x^3 y^{1/2}\right) - \log z$ *Quotient rule*

$= \log x^3 + \log y^{1/2} - \log z$ *Product rule*

$= 3\log x + \dfrac{1}{2}\log y - \log z$ *Power rule*

50. $B(t) = 25,000e^{.2t}$

(a) At noon, $t = 0$.

$B(0) = 25,000e^{.2(0)}$

$= 25,000e^0 = 25,000(1) = 25,000$

25,000 bacteria are present at noon.

(b) At 1 P.M., $t = 1$.

$B(1) = 25,000e^{.2(1)} \approx 30,535$

About 30,500 bacteria are present at 1 P.M.

(c) At 2 P.M., $t = 2$.

$B(2) = 25,000e^{.2(2)} \approx 37,296$

About 37,300 bacteria are present at 2 P.M.

(d) At 5 P.M., $t = 5$.

$B(5) = 25,000e^{.2(5)} \approx 67,957$

About 68,000 bacteria are present at 5 P.M.

CHAPTER 10 CONIC SECTIONS

Section 10.1

1. $x^2 + y^2 = 100$ can be written as

$$(x-0)^2 + (y-0)^2 = 10^2.$$

The center of the circle is $(0,0)$ and the radius of the circle is 10.

3. $$\frac{(y+2)^2}{25} + \frac{(x+3)^2}{49} = 1$$

can be written as

$$\frac{[x-(-3)]^2}{49} + \frac{[y-(-2)]^2}{25} = 1.$$

The center is $(-3, -2)$.

5. Center $(0,0)$; $r = 6$

Use the equation of a circle with radius r and center (h, k). Here, $h = 0$, $k = 0$, and $r = 6$.

$$(x-h)^2 + (y-k)^2 = r^2$$
$$(x-0)^2 + (y-0)^2 = 6^2$$
$$x^2 + y^2 = 36$$

7. Center $(-1, 3)$; $r = 4$

Substitute $h = -1$, $k = 3$, and $r = 4$ in the center-radius form of the equation of a circle.

$$(x-h)^2 + (y-k)^2 = r^2$$
$$[x-(-1)]^2 + (y-3)^2 = 4^2$$
$$(x+1)^2 + (y-3)^2 = 16$$

9. Center $(0, 4)$; $r = \sqrt{3}$

Substitute $h = 0$, $k = 4$, and $r = \sqrt{3}$ in the center-radius form of the equation of the circle.

$$(x-h)^2 + (y-k)^2 = r^2$$
$$(x-0)^2 + (y-4)^2 = \left(\sqrt{3}\right)^2$$
$$x^2 + (y-4)^2 = 3$$

11. By the vertical line test the set is not a function, because a vertical line may intersect the graph of an ellipse in two points.

13. $$x^2 + y^2 + 4x + 6y + 9 = 0$$

Rewrite the equation keeping only the variable terms on the left and grouping the x-terms and y-terms.

$$x^2 + 4x + y^2 + 6y = -9$$

Complete both squares on the left, and add the same constants to the right.

$$(x^2 + 4x + \underline{4}) + (y^2 + 6y + \underline{9}) = -9 + \underline{4} + \underline{9}$$
$$(x+2)^2 + (y+3)^2 = 4$$

From the form $(x-h)^2 + (y-k)^2 = r^2$, we have $h = -2$, $k = -3$, and $r = 2$. The center is $(-2, -3)$, and the radius r is 2.

15. $x^2 + y^2 + 10x - 14y - 7 = 0$

$$(x^2 + 10x \quad) + (y^2 - 14y \quad) = 7$$
$$(x^2 + 10x + \underline{25}) + (y^2 - 14y + \underline{49})$$
$$= 7 + \underline{25} + \underline{49}$$
$$(x+5)^2 + (y-7)^2 = 81$$

The center is $(-5, 7)$, and the radius is $\sqrt{81} = 9$.

17. $$3x^2 + 3y^2 - 12x - 24y + 12 = 0$$
$$3(x^2 - 4x \quad) + 3(y^2 - 8y \quad) = -12$$

Divide by 3.

$$(x^2 - 4x \quad) + (y^2 - 8y \quad) = -4$$
$$(x^2 - 4x + \underline{4}) + (y^2 - 8y + \underline{16})$$
$$= -4 + \underline{4} + \underline{16}$$
$$(x-2)^2 + (y-4)^2 = 16$$

The center is $(2, 4)$, and the radius is $\sqrt{16} = 4$.

19. $$x^2 + y^2 = 9$$
$$(x-0)^2 + (y-0)^2 = 3^2$$

Here, $h = 0$, $k = 0$, and $r = 3$, so the graph is a circle with center $(0, 0)$ and radius 3.

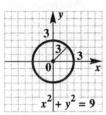

21. $$2y^2 = 10 - 2x^2$$
$$2x^2 + 2y^2 = 10$$
$$x^2 + y^2 = 5 \qquad \textit{Divide by 2.}$$
$$(x-0)^2 + (y-0)^2 = \left(\sqrt{5}\right)^2$$

Here, $h = 0$, $k = 0$, and $r = \sqrt{5} \approx 2.2$, so the graph is a circle with center $(0, 0)$ and radius $\sqrt{5}$.

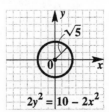

23. $(x+3)^2 + (y-2)^2 = 9$

Here, $h = -3$, $k = 2$, and $r = \sqrt{9} = 3$. The graph is a circle with center $(-3, 2)$ and radius 3.

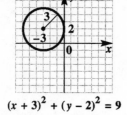

$$(x + 3)^2 + (y - 2)^2 = 9$$

25. $x^2 + y^2 - 4x - 6y + 9 = 0$

$(x^2 - 4x \quad) + (y^2 - 6y \quad) = -9$

Complete the square for each variable.

$\left(x^2 - 4x + \underline{4}\right) + \left(y^2 - 6y + \underline{9}\right)$

$$= -9 + \underline{4} + \underline{9}$$

$(x - 2)^2 + (y - 3)^2 = 4$

Here, $h = 2$, $k = 3$, and $r = \sqrt{4} = 2$. The graph is a circle with center $(2, 3)$ and radius 2.

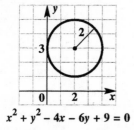

$$x^2 + y^2 - 4x - 6y + 9 = 0$$

27.   This method works because the pencil is always the same distance from the fastened end. The fastened end works as the center, and the length of the string from the fastened end to the pencil is the radius.

29.   A circular racetrack is most appropriate because the crawfish can move in any direction. Distance from the center determines the winner.

31.   The equation $\dfrac{x^2}{9} + \dfrac{y^2}{25} = 1$ is of the form

$\dfrac{x^2}{a^2} + \dfrac{y^2}{b^2} = 1$. The graph is an ellipse with

$a^2 = 9$ and $b^2 = 25$, so $a = 3$ and $b = 5$. The x-intercepts are $(3, 0)$ and $(-3, 0)$. The y-intercepts are $(0, 5)$ and $(0, -5)$. Plot the intercepts, and draw the ellipse through them.

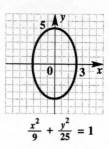

$$\frac{x^2}{9} + \frac{y^2}{25} = 1$$

33. $\dfrac{x^2}{36} = 1 - \dfrac{y^2}{16}$

$\dfrac{x^2}{36} + \dfrac{y^2}{16} = 1$ is in the form $\dfrac{x^2}{a^2} + \dfrac{y^2}{b^2} = 1$.

The graph is an ellipse with $a^2 = 36$ and $b^2 = 16$, so $a = 6$ and $b = 4$. The x-intercepts are $(6, 0)$ and $(-6, 0)$. The y-intercepts are $(0, 4)$ and $(0, -4)$. Plot the intercepts, and draw the ellipse through them.

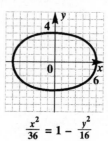

$$\frac{x^2}{36} = 1 - \frac{y^2}{16}$$

35. $\dfrac{y^2}{25} = 1 - \dfrac{x^2}{49}$

$\dfrac{x^2}{49} + \dfrac{y^2}{25} = 1$ is in the form $\dfrac{x^2}{a^2} + \dfrac{y^2}{b^2} = 1$.

The graph is an ellipse with $a^2 = 49$ and $b^2 = 25$, so $a = 7$ and $b = 5$. The x-intercepts are $(7, 0)$ and $(-7, 0)$. The y-intercepts are $(0, 5)$ and $(0, -5)$. Plot the intercepts, and draw the ellipse through them.

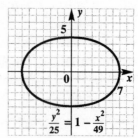

$$\frac{y^2}{25} = 1 - \frac{x^2}{49}$$

37. $\dfrac{(x+1)^2}{64} + \dfrac{(y-2)^2}{49} = 1$

This equation is of the form

$$\frac{(x-h)^2}{a^2} + \frac{(y-k)^2}{b^2} = 1,$$

so the center of the ellipse is at $(-1, 2)$. Since $a^2 = 64$, $a = 8$. Since $b^2 = 49$, $b = 7$. Add ± 8 to -1, and add ± 7 to 2 to find the points $(7, 2)$, $(-9, 2)$, $(-1, 9)$, and $(-1, -5)$. Plot the points, and draw the ellipse through them.

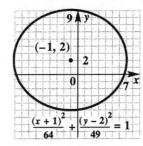

39. $\dfrac{(x-2)^2}{16} + \dfrac{(y-1)^2}{9} = 1$

The center of the ellipse is at $(2, 1)$. Since $a^2 = 16$, $a = 4$. Since $b^2 = 9$, $b = 3$. Add ± 4 to 2, and add ± 3 to 1 to find the points $(6, 1)$, $(-2, 1)$, $(2, 4)$, and $(2, -2)$. Plot the points, and draw the ellipse through them.

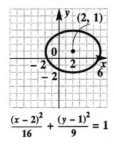

41. $(x+2)^2 + (y-4)^2 = 16$

$\qquad\quad (y-4)^2 = 16 - (x+2)^2$

Take the square root of each side.

$$y - 4 = \pm\sqrt{16 - (x+2)^2}$$
$$y = 4 \pm \sqrt{16 - (x+2)^2}$$

Therefore, the two functions used to obtain the graph were

$$y_1 = 4 + \sqrt{16 - (x+2)^2} \quad \text{and}$$
$$y_2 = 4 - \sqrt{16 - (x+2)^2}.$$

43. $\qquad x^2 + y^2 = 36$

$\qquad\qquad y^2 = 36 - x^2$

Take the square root of both sides.

$$y = \pm\sqrt{36 - x^2}$$

Therefore,

$$y_1 = \sqrt{36 - x^2} \quad \text{and} \quad y_2 = -\sqrt{36 - x^2}.$$

Use these two functions to obtain the graph.

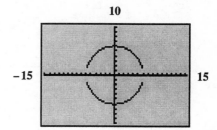

45. $\dfrac{x^2}{16} + \dfrac{y^2}{4} = 1$

$$\frac{y^2}{4} = 1 - \frac{x^2}{16}$$

$$y^2 = 4\left(1 - \frac{x^2}{16}\right)$$

Take the square root of both sides.

$$y = \pm 2\sqrt{1 - \frac{x^2}{16}}$$

Therefore,

$$y_1 = 2\sqrt{1 - \frac{x^2}{16}} \quad \text{and} \quad y_2 = -2\sqrt{1 - \frac{x^2}{16}}.$$

Use these two functions to obtain the graph.

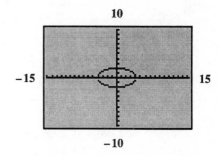

47. $\dfrac{x^2}{141.7^2} + \dfrac{y^2}{141.1^2} = 1$

(a) $c^2 = a^2 - b^2$, so

$$c = \sqrt{a^2 - b^2} = \sqrt{141.7^2 - 141.1^2}$$
$$= \sqrt{169.68} \approx 13.0$$

From the figure, the apogee is
$a + c = 141.7 + 13.0 = 154.7$ million miles.

(b) The perigee is $a - c = 141.7 - 13.0 = 128.7$ million miles.

49. $\dfrac{x^2}{36} + \dfrac{y^2}{9} = 1$

$c^2 = a^2 - b^2 = 36 - 9 = 27$, so

$c = \sqrt{27} = 3\sqrt{3}$.

The kidney stone and the source of the beam must be placed $3\sqrt{3}$ units from the center of the ellipse.

51. **(a)** The foci are $(c, 0)$ and $(-c, 0)$. The sum of the distances is $2a$. By the definition of an ellipse,

$$\sqrt{(x-c)^2 + (y-0)^2}$$
$$+ \sqrt{(x+c)^2 + (y-0)^2} = 2a$$
$$\sqrt{(x-c)^2 + y^2} + \sqrt{(x+c)^2 + y^2} = 2a$$
$$\sqrt{(x-c)^2 + y^2} = 2a - \sqrt{(x+c)^2 + y^2}.$$

Square both sides.

$$(x-c)^2 + y^2 = 4a^2 - 4a\sqrt{(x+c)^2 + y^2}$$
$$+ (x+c)^2 + y^2$$
$$x^2 - 2xc + c^2 + y^2$$
$$= 4a^2 - 4a\sqrt{(x+c)^2 + y^2}$$
$$+ x^2 + 2xc + c^2 + y^2$$
$$-4xc = 4a^2 - 4a\sqrt{(x+c)^2 + y^2}$$
$$-4xc - 4a^2 = -4a\sqrt{(x+c)^2 + y^2}$$

Divide by -4.

$$xc + a^2 = a\sqrt{(x+c)^2 + y^2}$$

Square both sides.

$$x^2c^2 + 2xca^2 + a^4 = a^2\left[(x+c)^2 + y^2\right]$$
$$x^2c^2 + 2xca^2 + a^4 = a^2\left(x^2 + 2xc + c^2 + y^2\right)$$
$$x^2c^2 + 2xca^2 + a^4 = a^2x^2 + 2a^2xc + a^2c^2 + a^2y^2$$
$$x^2c^2 + a^4 = a^2x^2 + a^2c^2 + a^2y^2$$

Isolate the x^2- and y^2-terms on one side of the equals sign.

$$a^4 - a^2c^2 = a^2x^2 - x^2c^2 + a^2y^2$$
$$a^2\left(a^2 - c^2\right) = x^2\left(a^2 - c^2\right) + a^2y^2$$
$$1 = \frac{x^2(a^2 - c^2)}{a^2(a^2 - c^2)}$$
$$+ \frac{a^2y^2}{a^2(a^2 - c^2)}$$
$$1 = \frac{x^2}{a^2} + \frac{y^2}{a^2 - c^2}$$

(b) $\dfrac{x^2}{a^2} + \dfrac{y^2}{a^2 - c^2} = 1$

To find the x-intercept, let $y = 0$.

$$\frac{x^2}{a^2} + \frac{0^2}{a^2 - c^2} = 1$$
$$\frac{x^2}{a^2} = 1$$
$$x^2 = a^2$$
$$x = \pm a$$

The x-intercepts are $(a, 0)$ and $(-a, 0)$.

(c) Let $b^2 = a^2 - c^2$.

Therefore, the equation is

$$\frac{x^2}{a^2} + \frac{y^2}{b^2} = 1.$$

To find the y-intercepts, let $x = 0$.

$$\frac{0^2}{a^2} + \frac{y^2}{b^2} = 1$$
$$\frac{y^2}{b^2} = 1$$
$$y^2 = b^2$$
$$y = \pm b$$

The y-intercepts are $(0, b)$ and $(0, -b)$.

Section 10.2

1. $\dfrac{x^2}{25} + \dfrac{y^2}{9} = 1$

This is the standard form for the equation of an ellipse with x-intercepts $(5, 0)$ and $(-5, 0)$ and y-intercepts $(0, 3)$ and $(0, -3)$. This is graph C.

3. $\dfrac{x^2}{9} - \dfrac{y^2}{25} = 1$

This is the standard form for the equation of a hyperbola that opens left and right. Its x-intercepts are $(3, 0)$ and $(-3, 0)$. This is graph D.

5. If the equation of a hyperbola is in standard form (that is, equal to one), the hyperbola would open to the left and right if the x^2-term was positive. It would open up and down if the y^2-term was positive.

7. The equation $\dfrac{x^2}{16} - \dfrac{y^2}{9} = 1$ is in the form $\dfrac{x^2}{a^2} - \dfrac{y^2}{b^2} = 1$. The graph is a hyperbola with $a = 4$ and $b = 3$. The x-intercepts are $(4, 0)$ and $(-4, 0)$. There are no y-intercepts. The corners of the fundamental rectangle are $(4, 3)$, $(4, -3)$, $(-4, -3)$, and $(-4, 3)$. Extend the diagonals of the rectangle through these points to get the asymptotes. Graph a branch of the hyperbola through each intercept and approaching the asymptotes.

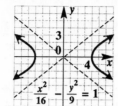

$$\frac{x^2}{16} - \frac{y^2}{9} = 1$$

9. $\frac{y^2}{9} - \frac{x^2}{9} = 1$ is a hyperbola with $a = 3$ and $b = 3$. The y-intercepts are $(0, 3)$ and $(0, -3)$. There are no x-intercepts. One asymptote passes through $(3, 3)$ and $(-3, -3)$. The other asymptote passes through $(-3, 3)$ and $(3, -3)$. Draw the asymptotes and sketch the hyperbola through the intercepts and approaching the asymptotes.

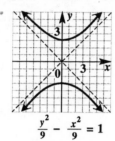

$$\frac{y^2}{9} - \frac{x^2}{9} = 1$$

11. $\frac{x^2}{25} - \frac{y^2}{36} = 1$ is a hyperbola with $a = 5$ and $b = 6$. The x-intercepts are $(5, 0)$ and $(-5, 0)$. There are no y-intercepts. To sketch the graph, draw the diagonals of the fundamental rectangle with corners $(5, 6)$, $(5, -6)$, $(-5, -6)$ and $(-5, 6)$. Graph a branch of the hyperbola through each intercept and approaching the asymptotes.

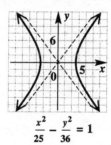

$$\frac{x^2}{25} - \frac{y^2}{36} = 1$$

13. $x^2 - y^2 = 16$

$$\frac{x^2}{16} - \frac{y^2}{16} = 1 \qquad \textit{Divide by 16.}$$

This equation is in the form $\frac{x^2}{a^2} - \frac{y^2}{b^2} = 1$ with $a = 4$ and $b = 4$. The graph is a hyperbola with x-intercepts $(4, 0)$ and $(-4, 0)$ and no y-intercepts. One asymptote passes through $(4, 4)$ and $(-4, -4)$. The other asymptote passes through $(-4, 4)$ and $(4, -4)$. Sketch the graph through the intercepts and approaching the asymptotes.

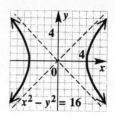

$$x^2 - y^2 = 16$$

15. $4x^2 + y^2 = 16$

$$\frac{x^2}{4} + \frac{y^2}{16} = 1 \qquad \textit{Divide by 16.}$$

This equation is in the form $\frac{x^2}{a^2} + \frac{y^2}{b^2} = 1$ with $a = 2$ and $b = 4$. The graph is an ellipse. The x-intercepts $(2, 0)$ and $(-2, 0)$. The y-intercepts are $(0, 4)$ and $(0, -4)$. Plot the intercepts and draw the ellipse through them.

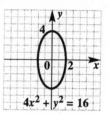

$$4x^2 + y^2 = 16$$

17. $$y^2 = 36 - x^2$$
$$x^2 + y^2 = 36$$
$$(x - 0)^2 + (y - 0)^2 = 36$$

The graph is a circle with center at $(0, 0)$ and radius $\sqrt{36} = 6$.

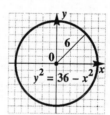

$$y^2 = 36 - x^2$$

19. $$9x^2 = 144 + 16y^2$$
$$9x^2 - 16y^2 = 144$$
$$\frac{x^2}{16} - \frac{y^2}{9} = 1 \qquad \textit{Divide by 144.}$$

The equation is a hyperbola in the form $\frac{x^2}{a^2} - \frac{y^2}{b^2} = 1$ with $a = 4$ and $b = 3$. The x-intercepts are $(4, 0)$ and $(-4, 0)$. There are no y-intercepts. To sketch the graph, draw the diagonals of the fundamental rectangle with corners $(4, 3)$, $(4, -3)$, $(-4, -3)$ and $(-4, 3)$. These are the asymptotes. Graph a branch of the hyperbola through each intercept approaching the asymptotes.

continued

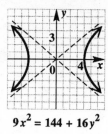

$$9x^2 = 144 + 16y^2$$

21.
$$y^2 = 4 + x^2$$
$$y^2 - x^2 = 4$$
$$\frac{y^2}{4} - \frac{x^2}{4} = 1 \qquad \textit{Divide by 4.}$$

The graph is a hyperbola with y-intercepts $(0, 2)$ and $(0, -2)$. One asymptote passes through $(2, 2)$ and $(-2, -2)$. The other asymptote passes through $(-2, 2)$ and $(2, -2)$.

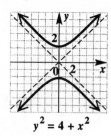

$$y^2 = 4 + x^2$$

23.
$$f(x) = -\sqrt{36 - x^2}$$

Replace $f(x)$ with y, and square both sides of the equation.

$$y = -\sqrt{36 - x^2}$$
$$y^2 = 36 - x^2$$
$$x^2 + y^2 = 36$$

This is a circle centered at the origin with radius $\sqrt{36} = 6$. Since $f(x)$, or y, represents a nonpositive square root in the original equation, $f(x)$ must be nonpositive. This restricts the graph to the bottom half of the circle.

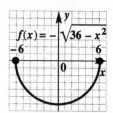

25.
$$f(x) = \sqrt{\frac{x + 4}{2}}$$

Replace $f(x)$ with y.

$$y = \sqrt{\frac{x + 4}{2}}$$

Square both sides.

$$y^2 = \frac{x + 4}{2}$$
$$2y^2 = x + 4$$
$$2y^2 - 4 = x$$
$$2(y - 0)^2 - 4 = x$$

This is a parabola that opens to the right with vertex $(-4, 0)$. However, $f(x)$, or y, is nonnegative in the original equation, so only the top half of the parabola is included in the graph.

x	-2	0	4
y	1	$\sqrt{2}$	2

$$f(x) = \sqrt{\frac{x + 4}{2}}$$

27. $\dfrac{(x - 2)^2}{4} - \dfrac{(y + 1)^2}{9} = 1$

is a hyperbola centered at $(2, -1)$, with $a = 2$ and $b = 3$. The x-intercepts are $(2 \pm 2, -1)$ or $(4, -1)$ and $(0, -1)$. The asymptotes are the extended diagonals of the rectangle with corners $(2, 3)$, $(2, -3)$, $(-2, -3)$ and $(-2, 3)$ shifted 2 units right and 1 unit down, or $(4, 2)$, $(4, -4)$, $(0, -4)$ and $(0, 2)$. Draw the hyperbola.

$$\frac{(x - 2)^2}{4} - \frac{(y + 1)^2}{9} = 1$$

29. $\dfrac{y^2}{36} - \dfrac{(x - 2)^2}{49} = 1$

is a hyperbola centered at $(2, 0)$ with $a = 7$, and $b = 6$. The y-intercepts are $(0, 0 \pm 6)$ or $(0, 6)$ and $(0, -6)$. The asymptotes are the extended diagonals of the rectangle with corners $(7, 6)$, $(7, -6)$, $(-7, -6)$, and $(-7, 6)$ shifted right 2 units, or $(9, 6)$, $(9, -6)$, $(-5, -6)$, and $(-5, 6)$. Draw the hyperbola.

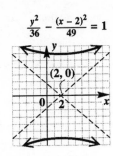

$$\frac{y^2}{36} - \frac{(x-2)^2}{49} = 1$$

31. (a) $100x^2 + 324y^2 = 32{,}400$

Divide by $32{,}400$.

$$\frac{x^2}{324} + \frac{y^2}{100} = 1$$

$$\frac{x^2}{18^2} + \frac{y^2}{10^2} = 1$$

The height in the center is the y-coordinate of the positive y-intercept. The height is 10 meters.

(b) The width of the ellipse is the distance between the x-intercepts, $(-18, 0)$ and $(18, 0)$. The width across the bottom of the arch is $18 + 18 = 36$ meters.

33. $\dfrac{x^2}{g^2} - \dfrac{y^2}{g^2} = 1$

Think of the crossbar as the x-axis and the goal posts at $(g, 0)$ and $(-g, 0)$ as the x-intercepts. The asymptotes go through the center of the goal posts, $(0, 0)$, and on the line from $(-g, -g)$ to (g, g) and on the line from $(g, -g)$ to $(-g, g)$. Since $a = b = g$, the asymptotes are

$y = \pm \dfrac{b}{a}x = \pm \dfrac{g}{g}x = \pm x$. These form a 45°

angle with the line through the goal posts. Most people can estimate a 45° angle fairly easily.

35. $y = .607\sqrt{383.9 + x^2}$

(a) According to the graph, about 55% of women worked in 1985.

(b) 1985 corresponds to $x = 85$.

$y = .607\sqrt{383.9 + 85^2} \approx 52.95$

According to the equation, about 53% of women worked in 1985.

37. $\dfrac{x^2}{9} - y^2 = 1$

$$-y^2 = 1 - \frac{x^2}{9}$$

$$y^2 = \frac{x^2}{9} - 1 \qquad \textit{Multiply by } -1.$$

Take the square root of both sides.

$$y = \pm\sqrt{\frac{x^2}{9} - 1}$$

The two functions used to obtain the graph were

$$y_1 = \sqrt{\frac{x^2}{9} - 1} \quad \text{and} \quad y_2 = -\sqrt{\frac{x^2}{9} - 1}.$$

39. $\dfrac{x^2}{25} - \dfrac{y^2}{49} = 1$

$$-\frac{y^2}{49} = 1 - \frac{x^2}{25}$$

$$\frac{y^2}{49} = \frac{x^2}{25} - 1 \qquad \textit{Multiply by } -1.$$

$$y^2 = 49\left(\frac{x^2}{25} - 1\right)$$

Take the square root of both sides.

$$y = \pm 7\sqrt{\frac{x^2}{25} - 1}$$

To obtain the graph, use the two functions

$$y_1 = 7\sqrt{\frac{x^2}{25} - 1} \quad \text{and} \quad y_2 = -7\sqrt{\frac{x^2}{25} - 1}.$$

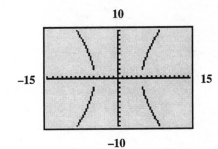

41. $\dfrac{y^2}{9} - x^2 = 1$

$$y^2 - 9x^2 = 9 \qquad \textit{Multiply by } 9.$$

$$y^2 = 9 + 9x^2$$

$$y^2 = 9\left(1 + x^2\right)$$

Take the square root of both sides.

$$y = \pm 3\sqrt{1 + x^2}$$

To obtain the graphs, use the two functions

$$y_1 = 3\sqrt{1 + x^2} \quad \text{and} \quad y_2 = -3\sqrt{1 + x^2}.$$

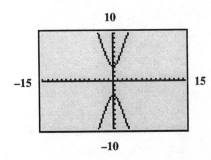

43. $\dfrac{x^2}{4} - y^2 = 1$

$$-y^2 = 1 - \dfrac{x^2}{4}$$

$$y^2 = \dfrac{x^2}{4} - 1 \qquad \textit{Multiply by } -1.$$

Take the square root of both sides.

$$y = \pm \sqrt{\dfrac{x^2}{4} - 1}$$

The positive square root is

$$y = \sqrt{\dfrac{x^2}{4} - 1}.$$

44. $\dfrac{x^2}{4} - y^2 = 1$

Here, $a^2 = 4$ and $b^2 = 1$, so $a = 2$ and $b = 1$.
The equation of the asymptote with positive slope
is $y = \dfrac{b}{a}x$, or $y = \dfrac{1}{2}x$.

45. $y = \sqrt{\dfrac{x^2}{4} - 1}$

$$y = \sqrt{\dfrac{50^2}{4} - 1} \qquad \textit{Let } x = 50.$$

$$= \sqrt{625 - 1} = \sqrt{624} \approx 24.98$$

46. $y = \dfrac{1}{2}x$

$$y = \dfrac{1}{2}(50) = 25 \quad \textit{Let } x = 50.$$

47. Because $24.98 < 25$, the graph of $y = \sqrt{\dfrac{x^2}{4} - 1}$

lies below the graph of $y = \dfrac{1}{2}x$ when $x = 50$.

48. If x-values larger than 50 are chosen, the y-values
of the hyperbola will get closer to the y-values of
the asymptote. The y-values on the hyperbola will
always be less than the y-values on the line.

49. $\sqrt{(x - c)^2 + (y - 0)^2}$

$$- \sqrt{[x - (-c)]^2 + (y - 0)^2} = 2a$$

$$\sqrt{(x - c)^2 + y^2} = 2a + \sqrt{(x + c)^2 + y^2}$$

Square both sides.

$$(x - c)^2 + y^2 = 4a^2 + 4a\sqrt{(x + c)^2 + y^2}$$
$$+ (x + c)^2 + y^2$$
$$x^2 - 2xc + c^2 + y^2 = 4a^2 + 4a\sqrt{(x + c)^2 + y^2}$$
$$+ x^2 + 2xc + c^2 + y^2$$
$$-4xc = 4a^2 + 4a\sqrt{(x + c)^2 + y^2}$$

Divide by 4.

$$-xc = a^2 + a\sqrt{(x + c)^2 + y^2}$$
$$-xc - a^2 = a\sqrt{(x + c)^2 + y^2}$$

Square both sides again.

$$x^2c^2 + 2xca^2 + a^4 = a^2\left[(x + c)^2 + y^2\right]$$
$$x^2c^2 + 2xca^2 + a^4 = a^2\left(x^2 + 2xc + c^2 + y^2\right)$$
$$x^2c^2 + 2xca^2 + a^4 = a^2x^2 + 2a^2xc + a^2c^2 + a^2y^2$$
$$x^2c^2 + a^4 = a^2x^2 + a^2c^2 + a^2y^2$$

Isolate the x^2- and y^2-terms on one side of the
equals sign.

$$x^2c^2 - a^2x^2 - a^2y^2 = a^2c^2 - a^4$$
$$x^2\left(c^2 - a^2\right) - a^2y^2 = a^2\left(c^2 - a^2\right)$$

Since $b^2 = c^2 - a^2$,

$$x^2b^2 - a^2y^2 = a^2b^2$$
$$\dfrac{x^2b^2}{a^2b^2} - \dfrac{a^2y^2}{a^2b^2} = \dfrac{a^2b^2}{a^2b^2} \qquad \textit{Divide by } a^2b^2.$$
$$\dfrac{x^2}{a^2} - \dfrac{y^2}{b^2} = 1.$$

Section 10.3

1. Substitute $x - 1$ for y in the first equation. Then
solve for x. Find the corresponding y-values by
substituting back into $y = x - 1$. In the first
equation, both variables are squared and in the
second, both variables are to the first power, so the
elimination method is not appropriate.

3. The line intersects the ellipse in exactly one point,
so there is *one* point in the solution set of the
system.

5. The line does not intersect the hyperbola, so the
solution set is the empty set.

7. A line and a circle; no points

Draw any circle, and then draw a line that does not
cross the circle.

9. A line and an ellipse; no points

The line does not intersect the ellipse.

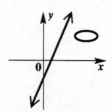

11. A line and a hyperbola; no points

Draw any hyperbola, and then draw a line between the two branches of the hyperbola that does not cross either branch.

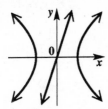

13. A line and a hyperbola; two points

Draw any hyperbola, and then draw a line that crosses both branches.

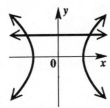

15. A circle and an ellipse; four points

Draw any ellipse, and then draw a circle with the same center whose radius is just large enough so that there are four points of intersection. (If the radius of the circle is too large or too small, there may be fewer points of intersection.)

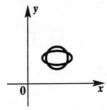

17. A parabola and an ellipse; four points

Draw any parabola, and then draw an ellipse large enough so that there are four points of intersection. (If the ellipse is too large or too small, there may be fewer points of intersection.)

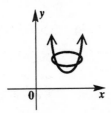

19. $y = 4x^2 - x$ (1)
$y = x$ (2)

Substitute x for y in equation (1).

$$y = 4x^2 - x \quad (1)$$
$$x = 4x^2 - x$$
$$0 = 4x^2 - 2x$$
$$0 = 2x(2x - 1)$$

$$2x = 0 \quad \text{or} \quad 2x - 1 = 0$$
$$x = 0 \quad \text{or} \quad x = \frac{1}{2}$$

Use equation (2) to find y for each x-value.

If $x = 0$, then $y = 0$.

If $x = \frac{1}{2}$, then $y = \frac{1}{2}$.

Solution set: $\left\{ (0,0), \left(\frac{1}{2}, \frac{1}{2} \right) \right\}$

21. $y = x^2 + 6x + 9$ (1)
$x + y = 3$ (2)

Substitute $x^2 + 6x + 9$ for y in equation (2).

$$x + y = 3 \quad (2)$$
$$x + \left(x^2 + 6x + 9 \right) = 3$$
$$x^2 + 7x + 9 = 3$$
$$x^2 + 7x + 6 = 0$$
$$(x + 6)(x + 1) = 0$$

$$x + 6 = 0 \quad \text{or} \quad x + 1 = 0$$
$$x = -6 \quad \text{or} \quad x = -1$$

Substitute these values for x in equation (2) and solve for y.

If $x = -6$, then

$$x + y = 3 \quad (2)$$
$$-6 + y = 3$$
$$y = 9.$$

If $x = -1$, then

$$x + y = 3 \quad (2)$$
$$-1 + y = 3$$
$$y = 4.$$

Solution set: $\{(-6, 9), (-1, 4)\}$

23. $x^2 + y^2 = 2$ (1)
$\quad 2x + y = 1$ (2)

Solve equation (2) for y.

$$y = 1 - 2x \quad (3)$$

Substitute $1 - 2x$ for y in equation (1).

$$x^2 + y^2 = 2 \quad (1)$$
$$x^2 + (1 - 2x)^2 = 2$$
$$x^2 + 1 - 4x + 4x^2 = 2$$
$$5x^2 - 4x - 1 = 0$$
$$(5x + 1)(x - 1) = 0$$

$$5x + 1 = 0 \quad \text{or} \quad x - 1 = 0$$
$$x = -\frac{1}{5} \quad \text{or} \quad x = 1$$

Use equation (3) to find y for each x-value.

If $x = -\frac{1}{5}$, then

$$y = 1 - 2\left(-\frac{1}{5}\right) = 1 + \frac{2}{5} = \frac{7}{5}.$$

If $x = 1$, then

$$y = 1 - 2(1) = -1.$$

Solution set: $\left\{\left(-\frac{1}{5}, \frac{7}{5}\right), (1, -1)\right\}$

25. $\quad xy = 4$ (1)
$\quad 3x + 2y = -10$ (2)

Solve equation (1) for y to get $y = \frac{4}{x}$.

Substitute $\frac{4}{x}$ for y in equation (2) to find x.

$$3x + 2y = -10 \quad (2)$$
$$3x + 2\left(\frac{4}{x}\right) = -10$$

Multiply by the LCD, x.

$$3x^2 + 8 = -10x$$
$$3x^2 + 10x + 8 = 0$$
$$(3x + 4)(x + 2) = 0$$

$$3x + 4 = 0 \quad \text{or} \quad x + 2 = 0$$
$$x = -\frac{4}{3} \quad \text{or} \quad x = -2$$

Since $y = \frac{4}{x}$, if $x = -\frac{4}{3}$, then $y = \frac{4}{-\frac{4}{3}} = -3$.

If $x = -2$, then

$$y = \frac{4}{-2} = -2.$$

Solution set: $\left\{(-2, -2), \left(-\frac{4}{3}, -3\right)\right\}$

27. $\quad xy = -3$ (1)
$\quad x + y = -2$ (2)

Solve equation (2) for y.

$$y = -x - 2. \quad (3)$$

Substitute $-x - 2$ for y in equation (1).

$$xy = -3 \quad (1)$$
$$x(-x - 2) = -3$$
$$-x^2 - 2x = -3$$
$$-x^2 - 2x + 3 = 0$$
$$x^2 + 2x - 3 = 0 \qquad \text{Multiply by } -1.$$
$$(x + 3)(x - 1) = 0$$

$$x + 3 = 0 \quad \text{or} \quad x - 1 = 0$$
$$x = -3 \quad \text{or} \quad x = 1$$

Use equation (3) to find y for each x-value.

If $x = -3$, then
$$y = -(-3) - 2 = 1.$$

If $x = 1$, then
$$y = -(1) - 2 = -3.$$

Solution set: $\{(-3, 1), (1, -3)\}$

29. $y = 3x^2 + 6x$ (1)
$\quad y = x^2 - x - 6$ (2)

Substitute $x^2 - x - 6$ for y in equation (1) to find x.

$$y = 3x^2 + 6x \qquad (1)$$
$$x^2 - x - 6 = 3x^2 + 6x$$
$$0 = 2x^2 + 7x + 6$$
$$0 = (2x + 3)(x + 2)$$

$$2x + 3 = 0 \quad \text{or} \quad x + 2 = 0$$
$$x = -\frac{3}{2} \quad \text{or} \quad x = -2$$

Substitute $-\frac{3}{2}$ for x in equation (1) to find y.

$$y = 3x^2 + 6x \qquad (1)$$

$$y = 3\left(-\frac{3}{2}\right)^2 + 6\left(-\frac{3}{2}\right)$$

$$= 3\left(\frac{9}{4}\right) + 6\left(-\frac{6}{4}\right)$$

$$= \frac{27}{4} - \frac{36}{4} = -\frac{9}{4}$$

Substitute -2 for x in equation (1) to find y.

$$y = 3x^2 + 6x \qquad (1)$$

$$y = 3(-2)^2 + 6(-2)$$

$$= 12 - 12 = 0$$

Solution set: $\left\{\left(-\frac{3}{2}, -\frac{9}{4}\right), (-2, 0)\right\}$

31. $\quad 2x^2 - y^2 = 6 \qquad (1)$

$\qquad y = x^2 - 3 \qquad (2)$

Substitute $x^2 - 3$ for y in equation (1).

$$2x^2 - y^2 = 6 \qquad (1)$$

$$2x^2 - \left(x^2 - 3\right)^2 = 6$$

$$2x^2 - \left(x^4 - 6x^2 + 9\right) = 6$$

$$-x^4 + 8x^2 - 9 = 6$$

$$-x^4 + 8x^2 - 15 = 0$$

$$x^4 - 8x^2 + 15 = 0 \qquad \text{\textit{Multiply by } -1.}$$

Let $z = x^2$, so $z^2 = x^4$.

$$z^2 - 8z + 15 = 0$$

$$(z - 3)(z - 5) = 0$$

$$z - 3 = 0 \quad \text{or} \quad z - 5 = 0$$

$$x = 3 \quad \text{or} \qquad z = 5$$

Since $z = x^2$,

$$x^2 = 3 \quad \text{or} \quad x^2 = 5.$$

Use the square root property.

$$x = \pm\sqrt{3} \quad \text{or} \quad x = \pm\sqrt{5}$$

Use equation (2) to find y for each x-value.

If $x = \sqrt{3}$ or $-\sqrt{3}$, then

$$y = \left(\pm\sqrt{3}\right)^2 - 3 = 0.$$

(Note: We could substitute 3 for x^2 in (2) to obtain the same result.)

If $x = \sqrt{5}$ or $-\sqrt{5}$, then

$$y = \left(\pm\sqrt{5}\right)^2 - 3 = 2.$$

Solution set:

$$\left\{\left(-\sqrt{3}, 0\right), \left(\sqrt{3}, 0\right), \left(-\sqrt{5}, 2\right), \left(\sqrt{5}, 2\right)\right\}$$

33. $\quad 3x^2 + 2y^2 = 12 \qquad (1)$

$\qquad x^2 + 2y^2 = 4 \qquad (2)$

Multiply equation (2) by -1 and add the result to equation (1).

$$
\begin{array}{rcll}
3x^2 + 2y^2 &=& 12 & (1) \\
-x^2 - 2y^2 &=& -4 & -1 \times (2) \\
\hline
2x^2 &=& 8 & \\
x^2 &=& 4 & \\
x &=& \pm 2 &
\end{array}
$$

Substitute ± 2 for x in equation (2) to find y.

$$x^2 + 2y^2 = 4 \qquad (2)$$

$$(\pm 2)^2 + 2y^2 = 4$$

$$4 + 2y^2 = 4$$

$$2y^2 = 0$$

$$y^2 = 0$$

$$y = 0$$

Solution set: $\{(-2, 0), (2, 0)\}$

35. $\quad 2x^2 + 3y^2 = 6 \qquad (1)$

$\qquad x^2 + 3y^2 = 3 \qquad (2)$

Multiply equation (2) by -1 and add the result to equation (1).

$$
\begin{array}{rcll}
2x^2 + 3y^2 &=& 6 & (1) \\
-x^2 - 3y^2 &=& -3 & -1 \times (2) \\
\hline
x^2 &=& 3 & \\
x &=& \pm\sqrt{3} &
\end{array}
$$

Substitute $\pm\sqrt{3}$ for x in equation (2).

$$x^2 + 3y^2 = 3 \qquad (2)$$

$$\left(\pm\sqrt{3}\right)^2 + 3y^2 = 3$$

$$3 + 3y^2 = 3$$

$$y^2 = 0$$

$$y = 0$$

Solution set: $\left\{\left(\sqrt{3}, 0\right), \left(-\sqrt{3}, 0\right)\right\}$

37. $2x^2 = 8 - 2y^2$ (1)
$\quad\ \ 3x^2 = 24 - 4y^2$ (2)

Multiply equation (1) by -2 and add the result to equation (2).

$$\begin{array}{rl}
-4x^2 = & -16 + 4y^2 \qquad -2 \times (1) \\
\underline{3x^2 = \quad 24 - 4y^2 \ \ (2)} \\
-x^2 = \quad 8 \\
x^2 = \quad -8 \\
x = \pm\sqrt{-8} = \pm 2i\sqrt{2}
\end{array}$$

Substitute $\pm 2i\sqrt{2}$ for x in equation (1).

$$2x^2 = 8 - 2y^2 \qquad\qquad (1)$$
$$2\left(\pm 2i\sqrt{2}\right)^2 = 8 - 2y^2$$
$$2(-8) = 8 - 2y^2$$
$$-16 = 8 - 2y^2$$
$$2y^2 = 24$$
$$y^2 = 12$$
$$y = \pm\sqrt{12} = \pm 2\sqrt{3}$$

Since $2i\sqrt{2}$ can be paired with either $2\sqrt{3}$ or $-2\sqrt{3}$ and $-2i\sqrt{2}$ can be paired with either $2\sqrt{3}$ or $-2\sqrt{3}$, there are four possible solutions.

Solution set:

$$\left\{\left(-2i\sqrt{2}, -2\sqrt{3}\right), \left(-2i\sqrt{2}, 2\sqrt{3}\right),\right.$$
$$\left.\left(2i\sqrt{2}, -2\sqrt{3}\right), \left(2i\sqrt{2}, 2\sqrt{3}\right)\right\}$$

39. $x^2 + xy + y^2 = 15$ (1)
$\quad\quad\ \ x^2 + y^2 = 10$ (2)

Multiply equation (2) by -1 and add the result to equation (1).

$$\begin{array}{rl}
x^2 + xy + y^2 = & 15 \ \ (1) \\
\underline{-x^2 \qquad - y^2 = -10 \qquad -1 \times (2)} \\
xy \qquad\quad = \quad 5 \\
y = \dfrac{5}{x}
\end{array}$$

Substitute $\dfrac{5}{x}$ for y in equation (2).

$$x^2 + y^2 = 10 \qquad (2)$$
$$x^2 + \left(\frac{5}{x}\right)^2 = 10$$
$$x^2 + \frac{25}{x^2} = 10$$
$$x^4 + 25 = 10x^2 \quad \textit{Multiply by } x^2.$$
$$x^4 - 10x^2 + 25 = 0$$

Let $z = x^2$, so $z^2 = x^4$.

$$z^2 - 10z + 25 = 0$$
$$(z - 5)^2 = 0$$
$$z - 5 = 0$$
$$z = 5$$

Since $z = x^2$,

$$x^2 = 5, \text{ and } x = \pm\sqrt{5}.$$

Using the equation $y = \dfrac{5}{x}$, we get the following.

If $x = -\sqrt{5}$, then

$$y = \frac{5}{-\sqrt{5}} = \frac{5 \cdot \sqrt{5}}{-\sqrt{5} \cdot \sqrt{5}} = \frac{5\sqrt{5}}{-5} = -\sqrt{5}.$$

Similarly, if $x = \sqrt{5}$, then $y = \sqrt{5}$.

Solution set: $\left\{\left(-\sqrt{5}, -\sqrt{5}\right), \left(\sqrt{5}, \sqrt{5}\right)\right\}$

41. $3x^2 + 2xy - 3y^2 = 5$ (1)
$\quad\quad -x^2 - 3xy + y^2 = 3$ (2)

Multiply equation (2) by 3 and add the result to equation (1).

$$\begin{array}{rl}
3x^2 + 2xy - 3y^2 = & 5 \ (1) \\
\underline{-3x^2 - 9xy + 3y^2 = \quad 9 \qquad 3 \times (2)} \\
-7xy \qquad\quad = 14 \\
x = \dfrac{14}{-7y} = -\dfrac{2}{y}
\end{array}$$

Substitute $-\dfrac{2}{y}$ for x in equation (2).

$$-x^2 - 3xy + y^2 = 3 \qquad (2)$$
$$-\left(-\frac{2}{y}\right)^2 - 3\left(-\frac{2}{y}\right)y + y^2 = 3$$
$$-\left(\frac{4}{y^2}\right) + 6 + y^2 = 3$$
$$y^2 + 3 - \frac{4}{y^2} = 0$$
$$y^4 + 3y^2 - 4 = 0 \quad \textit{Multiply by } y^2.$$
$$(y^2 + 4)(y^2 - 1) = 0$$

$$\begin{array}{rcl}
y^2 + 4 = 0 & \text{or} & y^2 - 1 = 0 \\
y^2 = -4 & & y^2 = 1 \\
y = \pm 2i & \text{or} & y = \pm 1
\end{array}$$

Since $x = -\dfrac{2}{y}$, substitute these values for y to find the values of x.

If $y = 2i$, then

$$x = -\frac{2}{2i} = -\frac{1}{i} = -\frac{1}{i} \cdot \frac{i}{i} = \frac{-i}{-1} = i.$$

If $y = -2i$, then

$$x = -\frac{2}{-2i} = \frac{1}{i} = \frac{1}{i} \cdot \frac{i}{i} = \frac{i}{-1} = -i.$$

If $y = 1$, then

$$x = -\frac{2}{1} = -2.$$

If $y = -1$, then

$$x = -\frac{2}{-1} = 2.$$

Solution set: $\{(i, 2i), (-i, -2i), (2, -1), (-2, 1)\}$

43. $y = x^2 + 1$ (1)
$x + y = 1$ (2)

Substitute $x^2 + 1$ for y in equation (2) to get

$$x + (x^2 + 1) = 1$$
$$x^2 + x = 0$$
$$x(x + 1) = 0$$
$$x = 0 \quad \text{or} \quad x + 1 = 0$$
$$x = -1$$

From the graphing calculator screen, we already know one solution is $(0, 1)$.
Use equation (1) to find y for each $x = -1$.

$$y = (-1)^2 + 1 = 1 + 1 = 2$$

The other solution is $(-1, 2)$.

45. $y = \frac{1}{2}x^2$ (1)
$x + y = 4$ (2)

Replace y by $\frac{1}{2}x^2$ in equation (2).

$$x + \frac{1}{2}x^2 = 4$$
$$2x + x^2 = 8 \quad \textit{Multiply by 2.}$$
$$x^2 + 2x - 8 = 0$$
$$(x + 4)(x - 2) = 0$$
$$x + 4 = 0 \quad \text{or} \quad x - 2 = 0$$
$$x = -4 \quad \text{or} \quad x = 2$$

The graphing calculator screen shows the solution $(2, 2)$. If $x = -4$ in equation (1), then

$$y = \frac{1}{2}(-4)^2 = \frac{1}{2}(16) = 8,$$

and the other solution is $(-4, 8)$.

47. $xy = -6$ (1)
$x + y = -1$ (2)

Solve both equations for y.

$$y = -\frac{6}{x}$$
$$y = -x - 1$$

Graph

$$Y_1 = -\frac{6}{x} \quad \text{and} \quad Y_2 = -x - 1$$

on a graphing calculator to obtain the solution set
$\{(2, -3), (-3, 2)\}$.

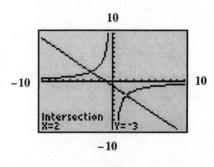

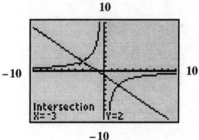

Now solve the system algebraically using substitution.

$$xy = -6 \quad (1)$$
$$x + y = -1 \quad (2)$$

Solve equation (2) for y.

$$y = -1 - x \quad (3)$$

Substitute $-1 - x$ for y in equation (1).

$$x(-1 - x) = -6$$
$$-x - x^2 = -6$$
$$0 = x^2 + x - 6$$
$$0 = (x + 3)(x - 2)$$
$$x + 3 = 0 \quad \text{or} \quad x - 2 = 0$$
$$x = -3 \quad \text{or} \quad x = 2$$

Substitute these values for x in (3) to find y.

If $x = -3$, then $y = -1 - (-3) = 2$.

If $x = 2$, then $y = -1 - 2 = -3$.

The solution set, $\{(2, -3), (-3, 2)\}$, is the same as that obtained using a graphing calculator.

49.
$$x^2 = 3x + 10$$
$$x^2 - 3x - 10 = 0$$
$$(x - 5)(x + 2) = 0$$

$$x - 5 = 0 \quad \text{or} \quad x + 2 = 0$$
$$x = 5 \quad \text{or} \quad x = -2$$

Solution set: $\{-2, 5\}$

50. The graph of $y = x^2$ is a parabola with vertex $(0, 0)$ that opens upward. The points $(1, 1)$, $(2, 4)$, and $(-2, 4)$ are also on the graph. The graph of $y = 3x + 10$ is a line through intercepts $\left(-\dfrac{10}{3}, 0\right)$ and $(0, 10)$.

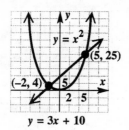

$$y = 3x + 10$$

51. $y = x^2 \qquad (1)$
$y = 3x + 10 \quad (2)$

Substitute x^2 for y in equation (2).

$$y = 3x + 10 \quad (2)$$
$$x^2 = 3x + 10$$
$$x^2 - 3x - 10 = 0$$
$$(x - 5)(x + 2) = 0$$

$$x - 5 = 0 \quad \text{or} \quad x + 2 = 0$$
$$x = 5 \quad \text{or} \quad x = -2$$

If $x = 5$, then $y = 5^2 = 25$.

If $x = -2$, then $y = (-2)^2 = 4$.

Solution set: $\{(-2, 4), (5, 25)\}$

52. They are exactly the same.

53. $y = x^2 - 3x - 10$

Complete the square.

$$\left[\frac{1}{2}(-3)\right]^2 = \left(-\frac{3}{2}\right)^2 = \frac{9}{4}$$

$$y = \left(x^2 - 3x + \frac{9}{4}\right) - \frac{9}{4} - 10$$

$$y = \left(x - \frac{3}{2}\right)^2 - \frac{49}{4}$$

This is the graph of a vertical parabola with vertex $\left(\dfrac{3}{2}, -\dfrac{49}{4}\right)$ that opens upward. The y-intercept is

$(0, -10)$, and the x-intercepts are $(-2, 0)$ and $(5, 0)$.

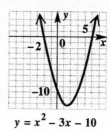

$$y = x^2 - 3x - 10$$

54. They are exactly the same.

55. Let $W =$ the width, and
$L =$ the length.

The formula for the area of a rectangle is $LW = A$, so

$$LW = 84. \quad (1)$$

The perimeter of a rectangle is given by $2L + 2W = P$, so

$$2L + 2W = 38. \quad (2)$$

Solve equation (2) for L to get

$$L = 19 - W. \quad (3)$$

Substitute $19 - W$ for L in equation (1).

$$LW = 84 \quad (1)$$
$$(19 - W)W = 84$$
$$19W - W^2 = 84$$
$$-W^2 + 19W - 84 = 0$$
$$W^2 - 19W + 84 = 0 \qquad \textit{Multiply by } -1.$$
$$(W - 7)(W - 12) = 0$$

$$W - 7 = 0 \quad \text{or} \quad W - 12 = 0$$
$$W = 7 \quad \text{or} \quad W = 12$$

Using equation (3), with $W = 7$,

$$L = 19 - 7 = 12.$$

If $W = 12$, then $L = 7$, which are the same two numbers. Length must be greater than width, so the length is 12 feet and the width is 7 feet.

57. Men: $\quad y = .138x^2 + .064x + 451 \quad (1)$
Women: $y = 12.1x + 334 \qquad\qquad (2)$

Substitute $12.1x + 334$ for y in (1).

$$12.1x + 334 = .138x^2 + .064x + 451$$
$$0 = .138x^2 - 12.036x + 117$$

Use the quadratic formula.

$$x = \frac{-(-12.036) \pm \sqrt{(-12.036)^2 - 4(.138)(117)}}{2(.138)}$$

$$= \frac{12.036 \pm \sqrt{80.28196}}{.276}$$

These values are approximately 76 and 11. $x = 11$ corresponds to 1981 and $x = 76$ is out of the range for these models. Substituting $x = 11$ into (2) gives us approximately 467 thousand bachelor's degrees awarded to women in 1981.

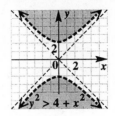

Section 10.4

1. $x^2 + y^2 < 25$
   $\quad y > -2$

   The boundary, $x^2 + y^2 = 25$, is a dashed circle with center $(0,0)$ and radius 5. When $(0,0)$ is tested, a true statement, $0 < 25$, results, so the inside of the circle is shaded. The graph of $y = -2$ is a dashed horizontal line through $(0,-2)$ with shading above the line, since $y > -2$. The correct answer is (c).

3. To graph the solution set of a nonlinear inequality, first graph the corresponding equality. This graph should be a dashed curve for $<$ or $>$ inequalities or a solid curve for $\leq$ or $\geq$ inequalities. Next, decide which region to shade by substituting any point not on the boundary (usually $(0,0)$ is the easiest) into the inequality. If the statement is true, then shade that area. If the statement is false, then shade the other area.

5. $y \geq x^2 + 4$

   This is an inequality whose boundary is a solid parabola, opening upward, with vertex $(0,4)$. The inside of the parabola is shaded since $(0,0)$ makes the inequality false. This is graph B.

7. $y < x^2 + 4$

   This is an inequality whose boundary is a dashed parabola, opening upward, with vertex $(0,4)$. The outside of the parabola is shaded since $(0,0)$ makes the inequality true. This is graph A.

9. $\quad y^2 > 4 + x^2$
   $\quad y^2 - x^2 > 4$
   $\quad \dfrac{y^2}{4} - \dfrac{x^2}{4} > 1$

   The boundary, $\dfrac{y^2}{4} - \dfrac{x^2}{4} = 1$, is a hyperbola with y-intercepts $(0,2)$ and $(0,-2)$ and asymptotes formed by the diagonals of the fundamental rectangle with corners at $(2,2)$, $(2,-2)$, $(-2,-2)$, and $(-2,2)$. The hyperbola has dashed branches because of $>$. Test $(0,0)$.

   $$0^2 > 4 + 0^2 \quad ?$$
   $$0 > 4 \qquad \textit{False}$$

   Shade the sides of the hyperbola that do not contain $(0,0)$. These are the regions inside the branches of the hyperbola.

11. $y + 2 \geq x^2$
    $\quad y \geq x^2 - 2$
    $\quad y \geq (x - 0)^2 - 2$

    Graph the solid vertical parabola $y = x^2 - 2$ with vertex $(0, -2)$. Two other points on the parabola are $(2, 2)$ and $(-2, 2)$. Test a point not on the parabola, say $(0,0)$, in $y \geq x^2 - 2$ to get $0 \geq -2$, a true statement. Shade that portion of the graph that contains the point $(0,0)$. This is the region inside the parabola.

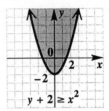

13. $\quad\quad 2y^2 \geq 8 - x^2$
    $\quad x^2 + 2y^2 \geq 8$
    $\quad \dfrac{x^2}{8} + \dfrac{y^2}{4} \geq 1$

    The boundary, $\dfrac{x^2}{8} + \dfrac{y^2}{4} = 1$, is the ellipse with intercepts $\left(2\sqrt{2}, 0\right)$, $\left(-2\sqrt{2}, 0\right)$, $(0, 2)$, and $(0, -2)$, drawn as a solid curve because of $\geq$. Test $(0,0)$.

    $$2(0)^2 \geq 8 - 0^2 \quad ?$$
    $$0 \geq 8 \qquad \textit{False}$$

    Shade the region of the ellipse that does not contain $(0,0)$. This is the region outside the ellipse.

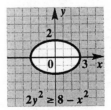

15. $y \le x^2 + 4x + 2$

Graph the solid vertical parabola

$y = x^2 + 4x + 2$. Use the vertex formula

$x = \dfrac{-b}{2a}$ to obtain the vertex $(-2, -2)$. Two

other points on the parabola are $(0, 2)$ and $(1, 7)$.
Test a point not on the parabola, say $(0, 0)$, in

$y \le x^2 + 4x + 2$ to get $0 \le 2$, a true statement.
Shade outside the parabola, since this region
contains $(0, 0)$.

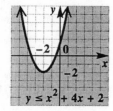

17. $9x^2 > 16y^2 + 144$
$9x^2 - 16y^2 > 144$
$\dfrac{x^2}{16} - \dfrac{y^2}{9} > 1$

The boundary, $\dfrac{x^2}{16} - \dfrac{y^2}{9} = 1$, is a hyperbola with

x-intercepts $(4, 0)$ and $(-4, 0)$ and asymptotes
formed by the diagonals of the fundamental
rectangle with corners at $(4, 3)$, $(4, -3)$, $(-4, -3)$,
and $(-4, 3)$. The hyperbola has dashed branches
because of $>$. Test $(0, 0)$.

$9(0)^2 > 16(0)^2 + 144$?
$0 > 144$ *False*

Shade the sides of the hyperbola that do not
contain $(0, 0)$. These are the regions inside the
branches of the hyperbola.

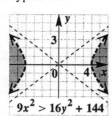

19. $x^2 - 4 \ge -4y^2$
$x^2 + 4y^2 \ge 4$
$\dfrac{x^2}{4} + \dfrac{y^2}{1} \ge 1$

Graph the solid ellipse $\dfrac{x^2}{4} + \dfrac{y^2}{1} = 1$ through the

x-intercepts $(2, 0)$ and $(-2, 0)$ and y-intercepts
$(0, 1)$ and $(0, -1)$. Test a point not on the
ellipse, say $(0, 0)$, in $x^2 - 4 \ge -4y^2$ to get
$-4 \ge 0$, a false statement. Shade outside the
ellipse, since this region does *not* include $(0, 0)$.

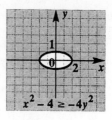

21. $x \le -y^2 + 6y - 7$

Complete the square to find the vertex.

$$x = -y^2 + 6y - 7$$
$$\frac{x}{-1} = y^2 - 6y + 7$$
$$\frac{x}{-1} - 7 = y^2 - 6y$$
$$\frac{x}{-1} - 7 + 9 = y^2 - 6y + 9$$
$$\frac{x}{-1} + 2 = (y - 3)^2$$
$$\frac{x}{-1} = (y - 3)^2 - 2$$
$$x = -(y - 3)^2 + 2$$

The boundary, $x = -(y - 3)^2 + 2$, is a solid
horizontal parabola with vertex at $(2, 3)$ that opens
to the left. Test $(0, 0)$.

$0 \le -0^2 + 6(0) - 7$?
$0 \le -7$ *False*

Shade the region of the parabola that does not
contain $(0, 0)$. This is the region inside the
parabola.

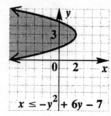

23. $2x + 5y < 10$
  $x - 2y < 4$

Graph $2x + 5y = 10$ as a dashed line through
$(5, 0)$ and $(0, 2)$. Test $(0, 0)$.

$2x + 5y < 10$
$2(0) + 5(0) < 10$?
$0 < 10$ *True*

Shade the region containing $(0, 0)$. Graph
$x - 2y = 4$ as a dashed line through $(4, 0)$ and
$(0, -2)$. Test $(0, 0)$.

$x - 2y < 4$
$0 - 2(0) < 4$?
$0 < 4$ *True*

Shade the region containing $(0, 0)$. The graph of the system is the intersection of the two shaded regions.

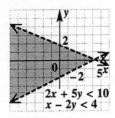

25. $5x - 3y \leq 15$
$4x + y \geq 4$

The boundary, $5x - 3y = 15$, is a solid line with intercepts $(3, 0)$ and $(0, -5)$. Test $(0, 0)$.

$$5(0) - 3(0) \leq 15 \quad ?$$
$$0 \leq 15 \quad True$$

Shade the side of the line that contains $(0, 0)$.
The boundary, $4x + y = 4$, is a solid line with intercepts $(1, 0)$ and $(0, 4)$. Test $(0, 0)$.

$$4(0) + 0 \geq 4 \quad ?$$
$$0 \geq 4 \quad False$$

Shade the side of the line that does not contain $(0, 0)$.
The graph of the system is the intersection of the two shaded regions.

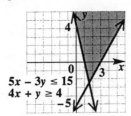

27. $x \leq 5$
$y \leq 4$

Graph $x = 5$ as a solid vertical line through $(5, 0)$. Since $x \leq 5$, shade the left side of the line.
Graph $y = 4$ as a solid horizontal line through $(0, 4)$. Since $y \leq 4$, shade below the line.
The graph of the system is the intersection of the two shaded regions.

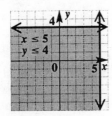

29. $y > x^2 - 4$
$y < -x^2 + 3$

The boundary, $y = x^2 - 4$, is a dashed parabola with vertex $(0, -4)$ that opens upward. Test $(0, 0)$.

$$0 > 0^2 - 4 \quad ?$$
$$0 > -4 \quad True$$

Shade the side of the parabola that contains $(0, 0)$. This is the region inside the parabola.
The boundary, $y = -x^2 + 3$, is a dashed parabola with vertex $(0, 3)$ that opens downward. Test $(0, 0)$.

$$0 < -0^2 + 3 \quad ?$$
$$0 < 3 \quad True$$

Shade the side of the parabola that contains $(0, 0)$. This is the region inside the parabola.
The graph of the system is the intersection of the two shaded regions.

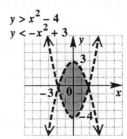

31. $y^2 - x^2 \geq 4$
$-5 \leq y \leq 5$

Rewrite $y^2 - x^2 \geq 4$ as $\dfrac{y^2}{4} - \dfrac{x^2}{4} \geq 1$. Graph the solid hyperbola through the y-intercepts $(0, 2)$ and $(0, -2)$. The asymptotes go through $(2, 2)$ and $(-2, -2)$ and through $(-2, 2)$ and $(2, -2)$. Test $(0, 0)$ in $y^2 - x^2 \geq 4$ to get $0 \geq 4$, a false statement. Shade above and below the two branches of the hyperbola.
Graph solid horizontal lines through $(0, 5)$ and $(0, -5)$. Since $-5 \leq y \leq 5$, shade the region between the two horizontal lines.
The graph of the system is the intersection of the shaded regions.

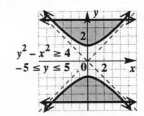

33. $y \leq -x^2$
$y \geq x - 3$
$y \leq -1$
$x < 1$

The boundary, $y = -x^2$, is a solid parabola with vertex at $(0, 0)$ that opens downward. Test $(0, -1)$.

$$-1 \leq -(0)^2 \quad ?$$
$$-1 \leq 0 \qquad \textit{True}$$

Shade the side of the parabola that contains $(0, -1)$. This is the region inside the parabola. The boundary, $y = x - 3$, is a solid line with intercepts $(3, 0)$ and $(0, -3)$. Test $(0, 0)$.

$$0 \geq 0 - 3 \quad ?$$
$$0 \geq -3 \qquad \textit{True}$$

Shade the side of the line that contains $(0, 0)$. For $y \leq -1$, shade below the solid horizontal line $y = -1$.
For $x < 1$, shade to the left of the dashed vertical line $x = 1$.
The intersection of the four shaded regions is the graph of the system.

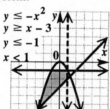

35. $x^2 + y^2 > 36$, $x \geq 0$

This is a circle of radius 6 centered at the origin. The graph is a dashed curve. Since $(0, 0)$ does not satisfy the inequality, the shading will be outside the circle. By including the restriction $x \geq 0$, we consider only the shading to the right of, and including, the y-axis.

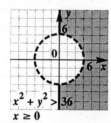

37. $x < y^2 - 3$, $x < 0$

Consider the equation.

$$x = y^2 - 3, \text{ or}$$
$$x = (y - 0)^2 - 3.$$

This is a parabola with vertex $(-3, 0)$, opening to the right having the same shape as $x = y^2$. The graph is a dashed curve. The shading will be outside the parabola, since $(0, 0)$ does not satisfy the inequality. By including the restriction $x < 0$, we consider only the shading to the left of, but not including, the y-axis. The y-axis is a dashed line.

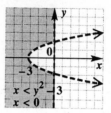

39. $4x^2 - y^2 > 16$, $x < 0$

Consider the equation

$$4x^2 - y^2 = 16, \text{ or}$$
$$\frac{x^2}{4} - \frac{y^2}{16} = 1.$$

This is a hyperbola with x-intercepts $(2, 0)$ and $(-2, 0)$. The graph is a dashed curve. The shading will be to the left and to the right of the two branches of the hyperbola, since $(0, 0)$ does not satisfy the inequality. By including the restriction $x < 0$, we consider only the shading to the left of the y-axis.

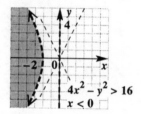

41. $x^2 + 4y^2 \geq 1$, $x \geq 0$, $y \geq 0$

Consider the equation

$$x^2 + 4y^2 = 1, \text{ or}$$
$$\frac{x^2}{1} + \frac{y^2}{\frac{1}{4}} = 1.$$

The graph is a solid ellipse with x-intercepts $(1, 0)$ and $(-1, 0)$ and y-intercepts

$\left(0, \dfrac{1}{2}\right)$ and $\left(0, -\dfrac{1}{2}\right)$. The shading will be

outside the ellipse, since $(0, 0)$ does not satisfy the inequality. By including the restrictions $x \geq 0$ and $y \geq 0$, we consider only the shading in quadrant I and portions of the x- and y-axis.

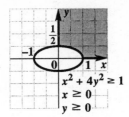

$$x^2 + 4y^2 \geq 1$$
$$x \geq 0$$
$$y \geq 0$$

43. $y > x^2 + 2$

$y = x^2 + 2$ is the graph of a parabola that opens upward and has been shifted up 2 units. The graph of $y > x^2 + 2$ consists of all points that are above the graph of $y = x^2 + 2$. The graph is shown in A.

45. $y > x^2 + 2$
$y < 5$

The graph of $y > x^2 + 2$ will be the region inside a parabola that opens upward and has been shifted two units up. The graph of $y < 5$ will be the region below the line $y = 5$. The graph of the system is choice B.

47. $y \geq x - 3$
$y \leq -x + 4$

The graphs of both inequalities include the points on the lines as part of the solution because of $\geq$ and $\leq$.
To produce the graph of the system on the TI-83, make the following Y-assignments:

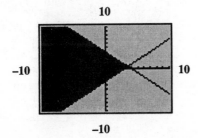

To get the upper and lower darkened triangles to the left of Y_1 and Y_2, simply place the cursor in that spot and press $\boxed{\text{ENTER}}$ to cycle through the graphing choices.

49. $y < x^2 + 4x + 4$
$y > -3$

The graphs do *not* include the points on the parabola or the line as part of the solution because of $<$ and $>$.

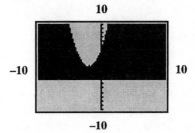

Chapter 10 Review Exercises

1. Center $(-2, 4)$, radius 3
Here $h = -2$, $k = 4$, and $r = 3$, so the equation of the circle is

$$(x - h)^2 + (y - k)^2 = r^2$$
$$[x - (-2)]^2 + (y - 4)^2 = 3^2$$
$$(x + 2)^2 + (y - 4)^2 = 9.$$

2. The graphed circle has center $(2, 1)$ and radius 1, so $h = 2$, $k = 1$, and $r = 1$. Then

$$(x - h)^2 + (y - k)^2 = r^2$$
$$(x - 2)^2 + (y - 1)^2 = 1^2$$
$$(x - 2)^2 + (y - 1)^2 = 1.$$

3. $x^2 + y^2 + 6x - 4y - 3 = 0$

Write the equation in
$$(x - h)^2 + (y - k)^2 = r^2$$
form by completing the squares on x and y.
$$\left(x^2 + 6x \quad \right) + \left(y^2 - 4y \quad \right) = 3$$
$$\left(x^2 + 6x + \underline{9}\right) + \left(y^2 - 4y + \underline{4}\right)$$
$$= 3 + \underline{9} + \underline{4}$$
$$(x + 3)^2 + (y - 2)^2 = 16$$
$$[x - (-3)]^2 + (y - 2)^2 = 16$$

The circle has center (h, k) at $(-3, 2)$ and radius $\sqrt{16} = 4$.

4. $x^2 + y^2 - 8x - 2y + 13 = 0$

Write the equation in

$$(x - h)^2 + (y - k)^2 = r^2$$

form by completing the squares on x and y.

$$\left(x^2 - 8x \quad\right) + \left(y^2 - 2y \quad\right) = -13$$
$$\left(x^2 - 8x + \underline{16}\right) + \left(y^2 - 2y + \underline{1}\right)$$
$$= -13 + \underline{16} + \underline{1}$$
$$(x - 4)^2 + (y - 1)^2 = 4$$

The circle has center (h, k) at $(4, 1)$ and radius $\sqrt{4} = 2$.

5. $2x^2 + 2y^2 + 4x + 20y = -34$
$$x^2 + y^2 + 2x + 10y = -17$$

Write the equation in

$$(x - h)^2 + (y - k)^2 = r^2$$

form by completing the squares on x and y.

$$\left(x^2 + 2x \quad\right) + \left(y^2 + 10y \quad\right) = -17$$
$$\left(x^2 + 2x + \underline{1}\right) + \left(y^2 + 10y + \underline{25}\right)$$
$$= -17 + \underline{1} + \underline{25}$$
$$(x + 1)^2 + (y + 5)^2 = 9$$
$$[x - (-1)]^2 + [y - (-5)]^2 = 9$$

The circle has center (h, k) at $(-1, -5)$ and radius $\sqrt{9} = 3$.

6. $4x^2 + 4y^2 - 24x + 16y = 48$
$$x^2 + y^2 - 6x + 4y = 12$$

Write the equation in

$$(x - h)^2 + (y - k)^2 = r^2$$

form by completing the squares on x and y.

$$\left(x^2 - 6x \quad\right) + \left(y^2 + 4y \quad\right) = 12$$
$$\left(x^2 - 6x + \underline{9}\right) + \left(y^2 + 4y + \underline{4}\right)$$
$$= 12 + \underline{9} + \underline{4}$$
$$(x - 3)^2 + (y + 2)^2 = 25$$
$$(x - 3)^2 + [y - (-2)]^2 = 25$$

The circle has center (h, k) at $(3, -2)$ and radius $\sqrt{25} = 5$.

7. $\dfrac{x^2}{16} + \dfrac{y^2}{9} = 1$ is in $\dfrac{x^2}{a^2} + \dfrac{y^2}{b^2} = 1$ form with $a = 4$ and $b = 3$. The graph is an ellipse with x-intercepts $(4, 0)$ and $(-4, 0)$ and y-intercepts $(0, 3)$ and $(0, -3)$. Plot the intercepts, and draw the ellipse through them.

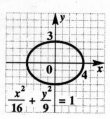

8. $\dfrac{x^2}{49} + \dfrac{y^2}{25} = 1$ is in $\dfrac{x^2}{a^2} + \dfrac{y^2}{b^2} = 1$ form with $a = 7$ and $b = 5$. The graph is an ellipse with x-intercepts $(7, 0)$ and $(-7, 0)$ and y-intercepts $(0, 5)$ and $(0, -5)$. Plot the intercepts, and draw the ellipse through them.

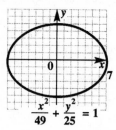

9. $\dfrac{x^2}{16} - \dfrac{y^2}{25} = 1$ is in $\dfrac{x^2}{a^2} - \dfrac{y^2}{b^2} = 1$ form with $a = 4$ and $b = 5$. The graph is a hyperbola with x-intercepts $(4, 0)$ and $(-4, 0)$ and asymptotes that are the extended diagonals of the rectangle with corners $(4, 5)$, $(4, -5)$, $(-4, -5)$, and $(-4, 5)$. Graph a branch of the hyperbola through each intercept approaching the asymptotes.

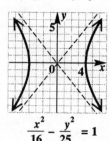

10. $\dfrac{y^2}{25} - \dfrac{x^2}{4} = 1$ is in $\dfrac{y^2}{b^2} - \dfrac{x^2}{a^2} = 1$ form with $a = 2$ and $b = 5$. The graph is a hyperbola with y-intercepts $(0, 5)$ and $(0, -5)$ and asymptotes that are the extended diagonals of the rectangle with corners $(2, 5)$, $(2, -5)$, $(-2, -5)$, and $(-2, 5)$. Graph a branch of the hyperbola through each intercept approaching the asymptotes.

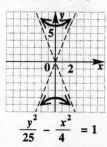

11. $x^2 + 9y^2 = 9$ in $\dfrac{x^2}{a^2} + \dfrac{y^2}{b^2} = 1$ form is

$\dfrac{x^2}{9} + \dfrac{y^2}{1} = 1$ with $a = 3$ and $b = 1$. The graph is an ellipse with x-intercepts $(3, 0)$ and $(-3, 0)$ and y-intercepts $(0, 1)$ and $(0, -1)$. Plot the intercepts, and draw the ellipse through them.

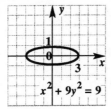

12. $$f(x) = \sqrt{4 + x^2}$$
Replace $f(x)$ with y.
$$y = \sqrt{4 + x^2}$$
Square both sides.
$$y^2 = 4 + x^2$$
$$y^2 - x^2 = 4$$
$$\dfrac{y^2}{4} - \dfrac{x^2}{4} = 1$$

The graph is a hyperbola that opens vertically with vertices $(0, 2)$ and $(0, -2)$. The original equation $f(x) = y = \sqrt{4 + x^2}$ indicates that y must be positive. Therefore, only the upper half of the hyperbola is part of the graph.

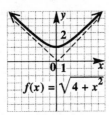

13. $$(x - 2)^2 + (y + 3)^2 = 16$$
$$(x - 2)^2 + [y - (-3)]^2 = 4^2$$

The graph is a circle with center $(2, -3)$ and radius 4.

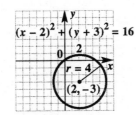

14. $y = 2x^2 - 3$
$y = 2(x - 0)^2 - 3$

The last equation is in $y = a(x - h)^2 + k$ form. The graph is a parabola.

15. $$y^2 = 2x^2 - 8$$
$$2x^2 - y^2 = 8$$
$$\dfrac{x^2}{4} - \dfrac{y^2}{8} = 1$$

The last equation is in $\dfrac{x^2}{a^2} - \dfrac{y^2}{b^2} = 1$ form, so the graph is a hyperbola.

16. $$y^2 = 8 - 2x^2$$
$$2x^2 + y^2 = 8$$
$$\dfrac{x^2}{4} + \dfrac{y^2}{8} = 1$$

The last equation is in $\dfrac{x^2}{a^2} + \dfrac{y^2}{b^2} = 1$ form, so the graph is an ellipse.

17. $x = y^2 + 4$
$x = (y - 0)^2 + 4$

The last equation is in $x = a(y - k)^2 + h$ form, so the graph is a parabola.

18. $$x^2 + y^2 = 64$$
$$(x - 0)^2 + (y - 0)^2 = 8^2$$

The last equation is in $(x - h)^2 + (y - k)^2 = r^2$ form. The graph is a circle.

19. The total distance on the horizontal axis is $160 + 16{,}000 = 16{,}160$ km. This represents $2a$, so $a = \dfrac{1}{2}(16{,}160) = 8080$. The distance from Earth to the center of the ellipse is

$$8080 - 160 = 7920,$$

which is the value of c. From Exercise 47 in Section 10.1, we know that $c^2 = a^2 - b^2$, so

$$b^2 = a^2 - c^2.$$
$$b^2 = 8080^2 - 7920^2$$
$$= 2{,}560{,}000$$

Thus, $b = \sqrt{2{,}560{,}000} = 1600$ and the equation is

$$\dfrac{x^2}{8080^2} + \dfrac{y^2}{1600^2} = 1$$

or $\dfrac{x^2}{65{,}286{,}400} + \dfrac{y^2}{2{,}560{,}000} = 1.$

20. This hyperbola has center at the origin with foci $(-50, 0)$ and $(50, 0)$. The difference between the distances from any point on the hyperbola to the two foci is equal to $80 - 30 = 50$. From Exercise 49 in Section 10.2, the foci are $(-c, 0)$ and $(c, 0)$. Hence, $c = 50$. Since the difference between the distances from any points to the two foci is $2a$, $2a = 50$, or $a = 25$. Since

$$b^2 = c^2 - a^2,$$
$$b^2 = 50^2 - 25^2$$
$$= 2500 - 625 = 1875.$$

Since the equation of a hyperbola is

$$\frac{x^2}{a^2} - \frac{y^2}{b^2} = 1,$$

the equation for this hyperbola is

$$\frac{x^2}{25^2} - \frac{y^2}{1875} = 1 \quad \text{or} \quad \frac{x^2}{625} - \frac{y^2}{1875} = 1.$$

21. $\quad 2y = 3x - x^2 \quad (1)$
$\quad x + 2y = -12 \quad (2)$

Substitute $3x - x^2$ for $2y$ in equation (2).

$$x + 2y = -12 \quad (2)$$
$$x + (3x - x^2) = -12$$
$$-x^2 + 4x + 12 = 0$$
$$x^2 - 4x - 12 = 0$$
$$(x - 6)(x + 2) = 0$$

$$x - 6 = 0 \quad \text{or} \quad x + 2 = 0$$
$$x = 6 \quad \text{or} \quad x = -2$$

Substitute these values for x in equation (2) to find y.

If $x = 6$, then

$$x + 2y = -12 \quad (2)$$
$$6 + 2y = -12$$
$$2y = -18$$
$$y = -9.$$

If $x = -2$ then

$$x + 2y = -12 \quad (2)$$
$$-2 + 2y = -12$$
$$2y = -10$$
$$y = -5$$

Solution set: $\{(6, -9), (-2, -5)\}$

22. $\quad y + 1 = x^2 + 2x \quad (1)$
$\quad y + 2x = 4 \quad\quad (2)$

Solve equation (2) for y.

$$y = 4 - 2x \quad (3)$$

Substitute $4 - 2x$ for y in (1).

$$(4 - 2x) + 1 = x^2 + 2x$$
$$0 = x^2 + 4x - 5$$
$$0 = (x + 5)(x - 1)$$

$$x + 5 = 0 \quad \text{or} \quad x - 1 = 0$$
$$x = -5 \quad \text{or} \quad x = 1$$

Substitute these values for x in equation (3) to find y.

If $x = -5$, then $y = 4 - 2(-5) = 14$.
If $x = 1$, then $y = 4 - 2(1) = 2$.

Solution set: $\{(1, 2), (-5, 14)\}$

23. $\quad x^2 + 3y^2 = 28 \quad (1)$
$\quad y - x = -2 \quad\quad (2)$

Solve equation (2) for y.

$$y = x - 2$$

Substitute $x - 2$ for y in equation (1).

$$x^2 + 3y^2 = 28 \quad (1)$$
$$x^2 + 3(x - 2)^2 = 28$$
$$x^2 + 3(x^2 - 4x + 4) - 28 = 0$$
$$x^2 + 3x^2 - 12x + 12 - 28 = 0$$
$$4x^2 - 12x - 16 = 0$$
$$4(x^2 - 3x - 4) = 0$$
$$4(x - 4)(x + 1) = 0$$

$$x - 4 = 0 \quad \text{or} \quad x + 1 = 0$$
$$x = 4 \quad \text{or} \quad x = -1$$

Since $y = x - 2$, if $x = 4$, then $y = 4 - 2 = 2$.
If $x = -1$, then $y = -1 - 2 = -3$.

Solution set: $\{(4, 2), (-1, -3)\}$

24. $\quad xy = 8 \quad\quad (1)$
$\quad x - 2y = 6 \quad (2)$

Solve equation (2) for x.

$$x = 2y + 6 \quad (3)$$

Substitute $2y + 6$ for x in equation (1) to find y.

$$xy = 8 \quad (1)$$
$$(2y + 6)y = 8$$
$$2y^2 + 6y - 8 = 0$$
$$2(y^2 + 3y - 4) = 0$$
$$2(y + 4)(y - 1) = 0$$

$$y + 4 = 0 \quad \text{or} \quad y - 1 = 0$$
$$y = -4 \quad \text{or} \quad y = 1$$

Substitute these values for y in equation (3) to find x.

If $y = -4$, then $x = 2(-4) + 6 = -2$.
If $y = 1$, then $x = 2(1) + 6 = 8$.

Solution set: $\{(-2, -4), (8, 1)\}$

25. $\begin{array}{rcll} x^2 + y^2 &=& 6 & (1) \\ x^2 - 2y^2 &=& -6 & (2) \end{array}$

Multiply equation (2) by -1 and add the result to equation (1).

$$\begin{array}{rcll} x^2 + y^2 &=& 6 & (1) \\ -x^2 + 2y^2 &=& 6 & -1 \times (2) \\ \hline 3y^2 &=& 12 & \\ y^2 &=& 4 & \end{array}$$

$$y = 2 \quad \text{or} \quad y = -2$$

Substitute these values for y in equation (1) to find x.

If $y = \pm 2$, then

$$x^2 + y^2 = 6 \quad (1)$$
$$x^2 + (\pm 2)^2 = 6$$
$$x^2 + 4 = 6$$
$$x^2 = 2.$$

$$x = \sqrt{2} \quad \text{or} \quad x = -\sqrt{2}$$

Since each value of x can be paired with each value of y, there are four points and the solution set is

$$\left\{ \left(\sqrt{2}, 2 \right), \left(-\sqrt{2}, 2 \right), \left(\sqrt{2}, -2 \right), \left(-\sqrt{2}, -2 \right) \right\}.$$

26. $\begin{array}{rcll} 3x^2 - 2y^2 &=& 12 & (1) \\ x^2 + 4y^2 &=& 18 & (2) \end{array}$

Multiply equation (1) by 2 and add the result to equation (2).

$$\begin{array}{rcll} 6x^2 - 4y^2 &=& 24 & 2 \times (1) \\ x^2 + 4y^2 &=& 18 & (2) \\ \hline 7x^2 &=& 42 & \\ x^2 &=& 6 & \end{array}$$

$$x = \sqrt{6} \quad \text{or} \quad x = -\sqrt{6}$$

Substitute these values for x in equation (2) to find y.

If $x = \pm\sqrt{6}$, then

$$x^2 + 4y^2 = 18. \quad (2)$$
$$\left(\pm\sqrt{6} \right)^2 + 4y^2 = 18$$
$$6 + 4y^2 = 18$$
$$4y^2 = 12$$
$$y^2 = 3$$

$$y = \sqrt{3} \quad \text{or} \quad y = -\sqrt{3}$$

The solution set is

$$\left\{ \left(\sqrt{6}, \sqrt{3} \right), \left(\sqrt{6}, -\sqrt{3} \right), \right.$$
$$\left. \left(-\sqrt{6}, \sqrt{3} \right), \left(-\sqrt{6}, -\sqrt{3} \right) \right\}.$$

27. A circle and a line can intersect in zero, one, or two points, so zero, one, or two solutions are possible.

28. A parabola and a hyperbola can intersect in zero, one, two, three, or four points, so zero, one, two, three, or four solutions are possible.

29. The graph shows that in about 1950 the altitude was reduced to approximately 4005 feet above sea level, causing the salinity to increase to about 6000 milligrams per liter.

30. $$9x^2 \geq 16y^2 + 144$$
$$9x^2 - 16y^2 \geq 144$$
$$\frac{x^2}{16} - \frac{y^2}{9} \geq 1$$

The boundary, $\frac{x^2}{16} - \frac{y^2}{9} = 1$, is a solid hyperbola with x-intercepts $(4, 0)$ and $(-4, 0)$. The asymptotes are the extended diagonals of the rectangle with corners $(4, 3)$, $(4, -3)$, $(-4, -3)$, and $(-4, 3)$. Test $(0, 0)$.

$$\begin{array}{rl} 9(0)^2 \geq 16(0)^2 + 144 & ? \\ 0 \geq 144 & \textit{False} \end{array}$$

Shade the sides of the hyperbola that do not contain $(0, 0)$. These are the regions inside the branches of the hyperbola.

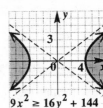

31. $$4x^2 + y^2 \geq 16$$
$$\frac{x^2}{4} + \frac{y^2}{16} \geq 1$$

The boundary, $\frac{x^2}{4} + \frac{y^2}{16} = 1$, is a solid ellipse with intercepts $(2, 0)$, $(-2, 0)$, $(0, 4)$, and $(0, -4)$. Test $(0, 0)$.

$$\begin{array}{rl} 4(0)^2 + 0^2 \geq 16 & ? \\ 0 \geq 16 & \textit{False} \end{array}$$

continued

Shade the side of ellipse that does not contain $(0,0)$. This is the region outside the ellipse.

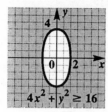

32. $y < -(x+2)^2 + 1$

The boundary, $y = -(x+2)^2 + 1$, is a dashed vertical parabola with vertex $(-2, 1)$. Since $a = -1 < 0$, the parabola opens downward. Also, $|a| = |-1| = 1$, so the graph has the same shape as the graph of $y = x^2$. Test $(0,0)$.

$$0 < -(0+2)^2 + 1 \quad ?$$
$$0 < -(4) + 1 \quad ?$$
$$0 < -3 \quad \textit{False}$$

Shade the side of the parabola that does not contain $(0,0)$. This is the region inside the parabola.

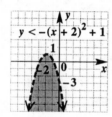

33. $2x + 5y \leq 10$
$3x - y \leq 6$

The boundary, $2x + 5y = 10$ is a solid line with intercepts $(5, 0)$ and $(0, 2)$. Test $(0,0)$.

$$2(0) + 5(0) \leq 10 \quad ?$$
$$0 \leq 10 \quad \textit{True}$$

Shade the side of the line that contains $(0,0)$. The boundary, $3x - y = 6$, is a solid line with intercepts $(2, 0)$ and $(0, -6)$. Test $(0,0)$.

$$3(0) - 0 \leq 6 \quad ?$$
$$0 \leq 6 \quad \textit{True}$$

Shade the side of the line that contains $(0,0)$. The graph of the system is the intersection of the two shaded regions.

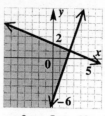

$$2x + 5y \leq 10$$
$$3x - y \leq 6$$

34. $\qquad |x| \leq 2$
$\qquad |y| > 1$
$4x^2 + 9y^2 \leq 36$

The equation of the boundary, $|x| = 2$, can be written as

$$x = -2 \quad \text{or} \quad x = 2.$$

The graph is these two solid vertical lines. Since $0 \leq 2$ is true, the region between the lines, containing $(0,0)$, is shaded.
The boundary, $|y| = 1$, consists of the two dashed horizontal lines $y = 1$ and $y = -1$. Since $0 > 1$ is false, the regions above and below the lines, not containing $(0,0)$, are shaded.
The boundary given by

$$4x^2 + 9y^2 = 36$$
$$\text{or} \quad \frac{x^2}{9} + \frac{y^2}{4} = 1$$

is graphed as a solid ellipse with intercepts $(3, 0)$, $(-3, 0)$, $(0, 2)$, and $(0, -2)$. Test $(0,0)$.

$$4(0)^2 + 9(0)^2 \leq 36 \quad ?$$
$$0 \leq 36 \quad \textit{True}$$

The region inside the ellipse, containing $(0,0)$, is shaded.
The graph of the system consists of the regions that include the common points of the three shaded regions.

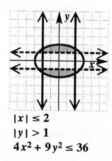

$$|x| \leq 2$$
$$|y| > 1$$
$$4x^2 + 9y^2 \leq 36$$

35. $\qquad 9x^2 \leq 4y^2 + 36$
$\qquad x^2 + y^2 \leq 16$

The equation of the first boundary is

$$9x^2 = 4y^2 + 36$$
$$9x^2 - 4y^2 = 36$$
$$\frac{x^2}{4} - \frac{y^2}{9} = 1.$$

The graph is a solid hyperbola with x-intercepts $(2,0)$ and $(-2,0)$. The asymptotes are the extended diagonals of the rectangle with corners $(2,3)$, $(2,-3)$, $(-2,-3)$, and $(-2,3)$. Test $(0,0)$.

$$9(0)^2 \leq 4(0)^2 + 36 \quad ?$$
$$0 \leq 36 \qquad \textit{True}$$

Shade the region between the branches of the hyperbola that contains $(0,0)$.
The equation of the second boundary is
$x^2 + y^2 = 16$. This is a solid circle with center $(0,0)$ and radius 4. Test $(0,0)$.

$$0^2 + 0^2 \leq 16 \quad ?$$
$$0 \leq 16 \qquad \textit{True}$$

Shade the region inside the circle.
The graph of the system is the intersection of the shaded regions which is between the two branches of the hyperbola and inside the circle.

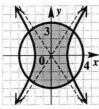

$$9x^2 \leq 4y^2 + 36$$
$$x^2 + y^2 \leq 16$$

36. $\frac{x^2}{64} + \frac{y^2}{25} = 1$ is in $\frac{x^2}{a^2} + \frac{y^2}{b^2} = 1$ form with $a = 8$ and $b = 5$. The graph is an ellipse with intercepts $(8,0)$, $(-8,0)$, $(0,5)$, and $(0,-5)$. Plot the intercepts, and draw the ellipse through them.

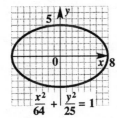

$$\frac{x^2}{64} + \frac{y^2}{25} = 1$$

37. $\frac{y^2}{4} - 1 = \frac{x^2}{9}$

$$\frac{y^2}{4} - \frac{x^2}{9} = 1$$

The equation is in $\frac{y^2}{b^2} - \frac{x^2}{a^2} = 1$ form with $a = 3$ and $b = 2$. The graph is a hyperbola with y-intercepts $(0,2)$ and $(0,-2)$ and asymptotes that are the extended diagonals of the rectangle with

corners $(3,2)$, $(3,-2)$, $(-3,-2)$, and $(-3,2)$. Draw a branch of the hyperbola through each intercept and approaching the asymptotes.

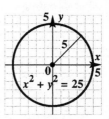

$$\frac{y^2}{4} - 1 = \frac{x^2}{9}$$

38. $x^2 + y^2 = 25$ is in $(x-h)^2 + (y-k)^2 = r^2$ form. The graph is a circle with center at $(0,0)$ and radius 5.

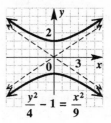

$$x^2 + y^2 = 25$$

39. $y = 2(x-2)^2 - 3$ is in $y = a(x-h)^2 + k$ form. The graph is a parabola with vertex $(2,-3)$. Since $a = 2 > 0$, the parabola opens upward. Also, $|a| = |2| = 2 > 1$, so the graph is narrower than the graph of $y = x^2$. The points $(0,5)$ and $(4,5)$ are on the graph.

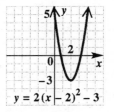

$$y = 2(x-2)^2 - 3$$

40. $$f(x) = -\sqrt{16 - x^2}$$
Replace $f(x)$ with y.
$$y = -\sqrt{16 - x^2}$$
Square both sides.
$$y^2 = 16 - x^2$$
$$x^2 + y^2 = 16$$

This equation is the graph of a circle with center $(0,0)$ and radius 4. Since $f(x)$ represents a nonpositive square root, $f(x)$ is nonpositive and its graph is the lower half of the circle.

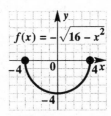

$$f(x) = -\sqrt{16 - x^2}$$

41.
$$f(x) = \sqrt{4 - x}$$
Replace $f(x)$ with y.
$$y = \sqrt{4 - x}$$
Square both sides.
$$y^2 = 4 - x$$
$$x = -y^2 + 4$$
$$x = -1(y - 0)^2 + 4$$

This equation is the graph of a horizontal parabola with vertex $(4, 0)$. Since $a = -1 < 0$, the graph opens to the left. Also, $|a| = |-1| = 1$, so the graph has the same shape as the graph of $y = x^2$. The points $(0, 2)$ and $(3, 1)$ are on the graph. Since $f(x)$ represents a square root, $f(x)$ is nonnegative and its graph is the upper half of the parabola.

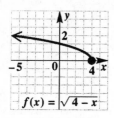

42. $3x + 2y \geq 0$
$$y \leq 4$$
$$x \leq 4$$

The boundary $3x + 2y = 0$ is a solid line through $(0, 0)$ and $(2, -3)$. Test $(0, 1)$.

$$3(0) + 2(1) \geq 0 \quad ?$$
$$2 \geq 0 \quad \text{True}$$

Shade the side of the line that contains $(0, 1)$. The boundary $y = 4$ is a solid horizontal line through $(0, 4)$. Since $y \leq 4$, shade below the line.
The boundary $x = 4$ is a solid vertical line through $(4, 0)$. Since $x \leq 4$, shade the region to the left of the line.
The graph of the system is the intersection of the three shaded regions.

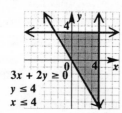

43. $4y > 3x - 12$
$$x^2 < 16 - y^2$$

The boundary $4y = 3x - 12$ is a dashed line with intercepts $(0, -3)$ and $(4, 0)$. Test $(0, 0)$.

$$4(0) > 3(0) - 12 \quad ?$$
$$0 > -12 \quad \text{True}$$

Shade the side of the line that contains $(0, 0)$.
The boundary $x^2 = 16 - y^2$, or $x^2 + y^2 = 16$, is a dashed circle with center at $(0, 0)$ and radius 4.
Test $(0, 0)$.

$$0^2 < 16 - 0^2 \quad ?$$
$$0 < 16 \quad \text{True}$$

Shade the region inside the circle.
The graph of the system is the intersection of the two shaded regions.

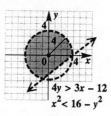

44. $x^2 + y^2 + ax + by + c = 0$

Let $x = 2$ and $y = 4$.

$$2^2 + 4^2 + a(2) + b(4) + c = 0$$
$$4 + 16 + 2a + 4b + c = 0$$
$$2a + 4b + c = -20$$

45. $x^2 + y^2 + ax + by + c = 0$

Let $x = 5$ and $y = 1$.

$$5^2 + 1^2 + a(5) + b(1) + c = 0$$
$$25 + 1 + 5a + b + c = 0$$
$$5a + b + c = -26$$

46. $x^2 + y^2 + ax + by + c = 0$

Let $x = -1$ and $y = 1$.

$$(-1)^2 + 1^2 + a(-1) + b(1) + c = 0$$
$$1 + 1 - a + b + c = 0$$
$$-a + b + c = -2$$

47.
$$2a + 4b + c = -20 \quad (1)$$
$$5a + b + c = -26 \quad (2)$$
$$-a + b + c = -2 \quad (3)$$

Multiply equation (3) by 2 and add the result to equation (1).

$$2a + 4b + c = -20 \quad (1)$$
$$\underline{-2a + 2b + 2c = -4} \quad 2 \times (3)$$
$$6b + 3c = -24 \quad (4)$$

Multiply equation (3) by 5 and add the result to equation (2).

$$5a + b + c = -26 \quad (2)$$
$$\underline{-5a + 5b + 5c = -10} \quad 5 \times (3)$$
$$6b + 6c = -36 \quad (5)$$

Multiply equation (4) by -1 and add the result to equation (5).

$$-6b - 3c = 24 \qquad -1 \times (4)$$
$$\underline{6b + 6c = -36} \quad (5)$$
$$3c = -12$$
$$c = -4$$

Substitute $c = -4$ into equation (4), and solve for b.

$$6b + 3c = -24 \quad (4)$$
$$6b + 3(-4) = -24$$
$$6b - 12 = -24$$
$$6b = -12$$
$$b = -2$$

Substitute $c = -4$ and $b = -2$ into equation (3), and solve for a.

$$-a + b + c = -2 \quad (3)$$
$$-a + (-2) + (-4) = -2$$
$$-a - 6 = -2$$
$$-a = 4$$
$$a = -4$$

The solution set is $\{(-4, -2, -4)\}$.

Therefore, the equation of the circle is

$$x^2 + y^2 - 4x - 2y - 4 = 0.$$

48. $x^2 + y^2 - 4x - 2y - 4 = 0$
$$\left(x^2 - 4x \qquad\right) + \left(y^2 - 2y \qquad\right) = 4$$

Complete the square for x and y.

$$\left(x^2 - 4x + \underline{4}\right) + \left(y^2 - 2y + \underline{1}\right) = 4 + \underline{4} + \underline{1}$$
$$(x - 2)^2 + (y - 1)^2 = 9$$

The center is $(2, 1)$ and the radius is $\sqrt{9} = 3$.

49.   From Exercise 48, the equation

$$x^2 + y^2 - 4x - 2y - 4 = 0$$

is a circle with center $(2, 1)$ and radius 3.

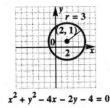

$$x^2 + y^2 - 4x - 2y - 4 = 0$$

Chapter 10 Test

1.   The circle has center (h, k) at $(-4, 4)$ and radius 4. Here, $h = -4$, $k = 4$, and $r = 4$, so an equation is

$$(x - h)^2 + (y - k)^2 = r^2$$
$$[x - (-4)]^2 + (y - 4)^2 = 4^2$$
$$(x + 4)^2 + (y - 4)^2 = 16.$$

2. $x^2 + y^2 + 8x - 2y = 8$

To find the center and radius, complete the squares on x and y.

$$\left(x^2 + 8x \qquad\right) + \left(y^2 - 2y \qquad\right) = 8$$
$$\left(x^2 + 8x + \underline{16}\right) + \left(y^2 - 2y + \underline{1}\right) = 8 + \underline{16} + \underline{1}$$
$$(x + 4)^2 + (y - 1)^2 = 25$$

The graph is a circle with center $(-4, 1)$ and radius $\sqrt{25} = 5$.

3. $3x^2 + 3y^2 = 27$
$$x^2 + y^2 = 9 \quad \textit{Divide by 3.}$$

This is a *circle* centered at the origin with radius $\sqrt{9} = 3$.

4. $9x^2 + 4y^2 = 36$
$$\frac{x^2}{4} + \frac{y^2}{9} = 1 \quad \textit{Divide by 36.}$$

This is an *ellipse* with intercepts $(2, 0)$, $(-2, 0)$, $(0, 3)$, and $(0, -3)$.

5. $9x^2 = 36 + 4y^2$
$$9x^2 - 4y^2 = 36$$
$$\frac{x^2}{4} - \frac{y^2}{9} = 1 \qquad \textit{Divide by 36.}$$

This is a *hyperbola* that opens right and left with vertices $(2, 0)$ and $(-2, 0)$.

6. $x = 36 - 4y^2$ or $x = -4y^2 + 36$

This is a *parabola* that opens to the left. Its vertex is $(36, 0)$.

7. $x^2 + y^2 = 64$

This is a circle centered at the origin with radius $\sqrt{64} = 8$.

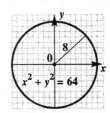

8. $4x^2 + 9y^2 = 36$

$$\frac{x^2}{9} + \frac{y^2}{4} = 1$$

The equation is in $\frac{x^2}{a^2} + \frac{y^2}{b^2} = 1$ form with $a = 3$ and $b = 2$. The graph is an ellipse with intercepts $(3, 0)$, $(-3, 0)$, $(0, 2)$, and $(0, -2)$. Plot these intercepts, and draw the ellipse through them.

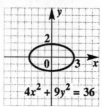

9. $16y^2 - 4x^2 = 64$

$$\frac{y^2}{4} - \frac{x^2}{16} = 1$$

The equation is in $\frac{y^2}{b^2} - \frac{x^2}{a^2} = 1$ form with $a = 4$ and $b = 2$. The graph is a hyperbola with y-intercepts $(0, 2)$ and $(0, -2)$ and asymptotes that are the extended diagonals of the rectangle with corners $(4, 2)$, $(4, -2)$, $(-4, -2)$, and $(-4, 2)$. Draw a branch of the hyperbola through each intercept and approaching the asymptotes.

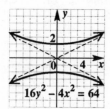

10. $$f(x) = \sqrt{16 - x^2}$$

Replace $f(x)$ with y.

$$y = \sqrt{16 - x^2}$$

Square both sides.

$$y^2 = 16 - x^2$$
$$x^2 + y^2 = 16$$

The graph of $x^2 + y^2 = 16$ is a circle of radius $\sqrt{16} = 4$ centered at the origin. Since $f(x)$ is nonnegative, only the top half of the circle is graphed.

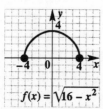

11. If a parabola and ellipse are graphed in the same plane, they can intersect at zero, one, two, three, or four points.

12. Draw any parabola, and then draw an ellipse that intersects the parabola in two points.

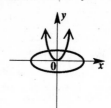

13. $2x - y = 9$ (1)

     $xy = 5$ (2)

Solve equation (1) for y.

$$y = 2x - 9 \quad (3)$$

Substitute $2x - 9$ for y in equation (2).

$$xy = 5 \quad (2)$$
$$x(2x - 9) = 5$$
$$2x^2 - 9x = 5$$
$$2x^2 - 9x - 5 = 0$$
$$(2x + 1)(x - 5) = 0$$

$$2x + 1 = 0 \quad \text{or} \quad x - 5 = 0$$
$$x = -\frac{1}{2} \quad \text{or} \quad x = 5$$

Substitute these values for x in equation (3) to find y.

If $x = -\frac{1}{2}$, then $y = 2\left(-\frac{1}{2}\right) - 9 = -10$.

If $x = 5$, then $y = 2(5) - 9 = 1$.

Solution set: $\left\{\left(-\frac{1}{2}, -10\right), (5, 1)\right\}$

14. $x - 4 = 3y$ (1)

   $x^2 + y^2 = 8$ (2)

Solve equation (1) for x.

$$x = 3y + 4$$

Substitute $3y + 4$ for x in equation (2).

$$x^2 + y^2 = 8 \quad (2)$$
$$(3y + 4)^2 + y^2 = 8$$
$$9y^2 + 24y + 16 + y^2 = 8$$
$$10y^2 + 24y + 8 = 0$$
$$2\left(5y^2 + 12y + 4\right) = 0$$
$$2(5y + 2)(y + 2) = 0$$

$$5y + 2 = 0 \quad \text{or} \quad y + 2 = 0$$
$$y = -\frac{2}{5} \quad \text{or} \quad y = -2$$

Since $x = 3y + 4$, substitute these values for y to find x.

If $y = -\frac{2}{5}$, then

$$x = 3\left(-\frac{2}{5}\right) + 4 = -\frac{6}{5} + 4 = \frac{14}{5}.$$

If $y = -2$, then $x = 3(-2) + 4 = -2.$

Solution set: $\left\{(-2, -2), \left(\frac{14}{5}, -\frac{2}{5}\right)\right\}$

15. $\quad x^2 + y^2 = 25 \quad (1)$
$\quad\quad\; x^2 - 2y^2 = 16 \quad (2)$

Multiply equation (1) by 2 and add the result to equation (2).

$$
\begin{array}{rll}
2x^2 + 2y^2 &= 50 & 2 \times (1) \\
x^2 - 2y^2 &= 16 & (2) \\
\hline
3x^2 &= 66 & \\
x^2 &= 22 &
\end{array}
$$

$$x = \sqrt{22} \quad \text{or} \quad x = -\sqrt{22}$$

Substitute these values for x in equation (1).
If $x = \pm\sqrt{22}$, then

$$x^2 + y^2 = 25 \quad (1)$$
$$\left(\pm\sqrt{22}\right)^2 + y^2 = 25$$
$$22 + y^2 = 25$$
$$y^2 = 3$$
$$y = \sqrt{3} \quad \text{or} \quad y = -\sqrt{3},$$

Solution set: $\left\{\left(\sqrt{22}, \sqrt{3}\right), \left(\sqrt{22}, -\sqrt{3}\right),\right.$
$\left.\left(-\sqrt{22}, \sqrt{3}\right), \left(-\sqrt{22}, -\sqrt{3}\right)\right\}$

16. $\quad y \le x^2 - 2$

The boundary, $y = x^2 - 2$, is a solid parabola in $y = a(x - h)^2 + k$ form with vertex (h, k) at $(0, -2)$. Since $a = 1 > 0$, the parabola opens upward. It also has the same shape as $y = x^2$. The points $(2, 2)$ and $(-2, 2)$ are on the graph. Test $(0, 0)$.

$$0 \le (0)^2 - 2 \quad ?$$
$$0 \le -2 \quad\quad \textit{False}$$

Shade the side of the parabola that does not contain $(0, 0)$. This is the region outside the parabola.

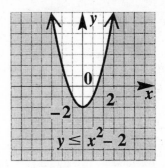

17. $\quad 2x - 5y \ge 12$
$\quad\quad\; 3x + 4y \le 12$

To graph $2x - 5y \ge 12$, draw $2x - 5y = 12$ as a solid line through $(6, 0)$ and $\left(0, -\frac{12}{5}\right)$. Test $(0, 0)$ to get $0 \ge 12$, a false statement. Shade the side of the line not containing the origin. To graph $3x + 4y \le 12$, draw $3x + 4y = 12$ as a solid line through $(4, 0)$ and $(0, 3)$. Test $(0, 0)$ to get $0 \le 12$, a true statement. Shade the side of the line containing the origin.
The intersection of the shaded regions is the graph of the system.

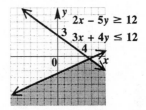

18. $\quad x^2 + 25y^2 \le 25$
$\quad\quad\; x^2 + y^2 \le 9$

The first boundary, $\dfrac{x^2}{25} + \dfrac{y^2}{1} = 1$, is a solid ellipse with intercepts $(5, 0)$, $(-5, 0)$, $(0, 1)$, and $(0, -1)$. Test $(0, 0)$.

$$0^2 + 25 \cdot 0^2 \le 25 \quad ?$$
$$0 \le 25 \quad\quad \textit{True}$$

Shade the region inside the ellipse.
The second boundary, $x^2 + y^2 = 9$, is a solid circle with center $(0, 0)$ and radius 3. Test $(0, 0)$.

$$0^2 + 0^2 \le 9 \quad ?$$
$$0 \le 9 \quad\quad \textit{True}$$

Shade the region inside the circle. The solution of the system is the intersection of the two shaded regions.

continued

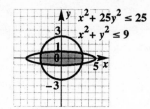

19. $\dfrac{x^2}{3352} + \dfrac{y^2}{3211} = 1$

$c^2 = a^2 - b^2 = 3352 - 3211,$

so $c^2 = 141$ and $c = \sqrt{141}$.

From Exercise 47 in Section 10.1, the apogee is $a + c = \sqrt{3352} + \sqrt{141} \approx 69.8$ million kilometers.

20. The perigee is $a - c = \sqrt{3352} - \sqrt{141} \approx 46.0$ million kilometers.

Cumulative Review Exercises Chapters 1–10

1. $-10 + |-5| - |3| + 4$
$= -10 + 5 - 3 + 4$
$= -5 + 1 = -4$

2. $4 - (2x + 3) + x = 5x - 3$
$4 - 2x - 3 + x = 5x - 3$
$-x + 1 = 5x - 3$
$-6x = -4$
$x = \dfrac{2}{3}$

Solution set: $\left\{ \dfrac{2}{3} \right\}$

3. $-4k + 7 \geq 6k + 1$
$-10k \geq -6$

Divide by -10; reverse the direction of the inequality.

$k \leq \dfrac{-6}{-10}$

$k \leq \dfrac{3}{5}$

Solution set: $\left(-\infty, \dfrac{3}{5} \right]$

4. $|5m| - 6 = 14$
$|5m| = 20$

$5m = 20$ or $5m = -20$
$m = 4$ or $m = -4$

Solution set: $\{-4, 4\}$

5. $|2p - 5| > 15$

$2p - 5 > 15$ or $2p - 5 < -15$
$2p > 20$ $\qquad\qquad$ $2p < -10$
$p > 10$ or $\qquad\quad$ $p < -5$

Solution set: $(-\infty, -5) \cup (10, \infty)$

6. Let $(x_1, y_1) = (2, 5)$ and $(x_2, y_2) = (-4, 1)$.

$$m = \frac{y_2 - y_1}{x_2 - x_1} = \frac{1 - 5}{-4 - 2} = \frac{-4}{-6} = \frac{2}{3}$$

7. Through $(-3, -2)$; perpendicular to $2x - 3y = 7$

Write $2x - 3y = 7$ in slope-intercept form.

$$-3y = -2x + 7$$
$$y = \frac{2}{3}x - \frac{7}{3}$$

The slope is $\dfrac{2}{3}$. Perpendicular lines have slopes that are negative reciprocals of each other, so a line perpendicular to the given line will have slope $-\dfrac{3}{2}$. Let $m = -\dfrac{3}{2}$ and $(x_1, y_1) = (-3, -2)$ in the point-slope form.

$$y - y_1 = m(x - x_1)$$
$$y - (-2) = -\frac{3}{2}[x - (-3)]$$
$$y + 2 = -\frac{3}{2}(x + 3)$$

Multiply by 2 to clear the fraction.

$$2y + 4 = -3(x + 3)$$
$$2y + 4 = -3x - 9$$
$$3x + 2y = -13$$

8. $3x - y = 12$ $\quad (1)$
$2x + 3y = -3$ $\quad (2)$

Multiply equation (1) by 3 and add the result to equation (2).

$$
\begin{array}{rll}
9x - 3y &= 36 & 3 \times (1) \\
2x + 3y &= -3 & (2) \\
\hline
11x &= 33 & \\
x &= 3 &
\end{array}
$$

Substitute 3 for x in equation (1) to find y.

$$3x - y = 12 \quad (1)$$
$$3(3) - y = 12$$
$$9 - y = 12$$
$$-y = 3$$
$$y = -3$$

Solution set: $\{(3, -3)\}$

9.
$$x + y - 2z = 9 \quad (1)$$
$$2x + y + z = 7 \quad (2)$$
$$3x - y - z = 13 \quad (3)$$

Add equation (2) and equation (3).

$$2x + y + z = 7 \quad (2)$$
$$\underline{3x - y - z = 13 \quad (3)}$$
$$5x \qquad\quad = 20$$
$$x = 4$$

Multiply equation (1) by -1 and add the result to equation (2).

$$-x - y + 2z = -9$$
$$\underline{2x + y + z = 7 \quad (2)}$$
$$x \qquad + 3z = -2 \quad (4)$$

Substitute 4 for x in equation (4) to find z.

$$x + 3z = -2 \quad (4)$$
$$4 + 3z = -2$$
$$3z = -6$$
$$z = -2$$

Substitute 4 for x and -2 for z in equation (2) to find y.

$$2x + y + z = 7 \quad (2)$$
$$2(4) + y - 2 = 7$$
$$y + 6 = 7$$
$$y = 1$$

Solution set: $\{(4, 1, -2)\}$

10.
$$xy = -5 \quad (1)$$
$$2x + y = 3 \quad (2)$$

Solve equation (2) for y.

$$y = -2x + 3 \quad (3)$$

Substitute $-2x + 3$ for y in equation (1).

$$xy = -5 \quad (1)$$
$$x(-2x + 3) = -5$$
$$-2x^2 + 3x = -5$$
$$-2x^2 + 3x + 5 = 0$$
$$2x^2 - 3x - 5 = 0$$
$$(2x - 5)(x + 1) = 0$$

$$2x - 5 = 0 \quad \text{or} \quad x + 1 = 0$$
$$x = \frac{5}{2} \quad \text{or} \qquad x = -1$$

Substitute these values for x in equation (3) to find y.

If $x = \dfrac{5}{2}$, then $y = -2\left(\dfrac{5}{2}\right) + 3 = -2$.

If $x = -1$, then $y = -2(-1) + 3 = 5$.

Solution set: $\left\{(-1, 5), \left(\dfrac{5}{2}, -2\right)\right\}$

11. $(5y - 3)^2 = (5y)^2 - 2(5y)3 + 3^2$
$$= 25y^2 - 30y + 9$$

12. $(2r + 7)(6r - 1)$
$$= 12r^2 - 2r + 42r - 7$$
$$= 12r^2 + 40r - 7$$

13. $\left(8x^4 - 4x^3 + 2x^2 + 13x + 8\right) \div (2x + 1)$

$$
\begin{array}{r}
4x^3 - 4x^2 + 3x + 5 \\
2x + 1\overline{)8x^4 - 4x^3 + 2x^2 + 13x + 8} \\
\underline{8x^4 + 4x^3} \\
-8x^3 + 2x^2 \\
\underline{-8x^3 - 4x^2} \\
6x^2 + 13x \\
\underline{6x^2 + 3x} \\
10x + 8 \\
\underline{10x + 5} \\
3
\end{array}
$$

The answer is

$$4x^3 - 4x^2 + 3x + 5 + \frac{3}{2x + 1}.$$

14. $12x^2 - 7x - 10 = (4x - 5)(3x + 2)$

15. $2y^4 + 5y^2 - 3$

Let $p = y^2$, so $p^2 = y^4$.

$$2y^4 + 5y^2 - 3 = 2p^2 + 5p - 3$$
$$= (2p - 1)(p + 3)$$

Now substitute y^2 for p.

$$= \left(2y^2 - 1\right)\left(y^2 + 3\right)$$

16. $z^4 - 1 = \left(z^2 + 1\right)\left(z^2 - 1\right)$
$$= \left(z^2 + 1\right)(z + 1)(z - 1)$$

17. $a^3 - 27b^3 = a^3 - (3b)^3$
$$= (a - 3b)\left(a^2 + 3ab + 9b^2\right)$$

18. $\dfrac{5x - 15}{24} \cdot \dfrac{64}{3x - 9} = \dfrac{5(x - 3)}{3 \cdot 8} \cdot \dfrac{8 \cdot 8}{3(x - 3)}$
$$= \frac{5 \cdot 8}{3 \cdot 3} = \frac{40}{9}$$

19. $\dfrac{y^2-4}{y^2-y-6} \div \dfrac{y^2-2y}{y-1}$

Multiply by the reciprocal.

$= \dfrac{y^2-4}{y^2-y-6} \cdot \dfrac{y-1}{y^2-2y}$

Factor and simplify.

$= \dfrac{(y+2)(y-2)}{(y-3)(y+2)} \cdot \dfrac{(y-1)}{y(y-2)}$

$= \dfrac{y-1}{y(y-3)}$

20. $\dfrac{5}{c+5} - \dfrac{2}{c+3}$

The LCD is $(c+5)(c+3)$.

$= \dfrac{5(c+3)}{(c+5)(c+3)} - \dfrac{2(c+5)}{(c+3)(c+5)}$

$= \dfrac{5c+15-2c-10}{(c+5)(c+3)}$

$= \dfrac{3c+5}{(c+5)(c+3)}$

21. $\dfrac{p}{p^2+p} + \dfrac{1}{p^2+p} = \dfrac{p+1}{p^2+p}$

$= \dfrac{p+1}{p(p+1)} = \dfrac{1}{p}$

22. Let x = the time to do the job working together.

Make a chart.

Worker	Rate	Time Together	Part of Job Done
Kareem	$\dfrac{1}{3}$	x	$\dfrac{x}{3}$
Jamal	$\dfrac{1}{2}$	x	$\dfrac{x}{2}$

Part done by Kareem	plus	part done by Jamaal	equals	1 whole job.
$\dfrac{x}{3}$	$+$	$\dfrac{x}{2}$	$=$	1

Multiply by the LCD, 6.

$6\left(\dfrac{x}{3} + \dfrac{x}{2}\right) = 6 \cdot 1$

$2x + 3x = 6$

$5x = 6$

$x = \dfrac{6}{5} = 1\dfrac{1}{5}$

It takes $\dfrac{6}{5}$ or $1\dfrac{1}{5}$ hours to do the job together.

23. $\left(\dfrac{4}{3}\right)^{-1} = \left(\dfrac{3}{4}\right)^{1} = \dfrac{3}{4}$

24. $\dfrac{(2a)^{-2}a^4}{a^{-3}} = \dfrac{2^{-2}a^{-2}a^4}{a^{-3}} = \dfrac{2^{-2}a^2}{a^{-3}}$

$= \dfrac{a^2 a^3}{2^2} = \dfrac{a^5}{4}$

25. $4\sqrt[3]{16} - 2\sqrt[3]{54} = 4\sqrt[3]{8 \cdot 2} - 2\sqrt[3]{27 \cdot 2}$

$= 4 \cdot 2\sqrt[3]{2} - 2 \cdot 3\sqrt[3]{2}$

$= 8\sqrt[3]{2} - 6\sqrt[3]{2} = 2\sqrt[3]{2}$

26. $\dfrac{3\sqrt{5x}}{\sqrt{2x}} = \dfrac{3\sqrt{5x} \cdot \sqrt{2x}}{\sqrt{2x} \cdot \sqrt{2x}} = \dfrac{3\sqrt{10x^2}}{2x}$

$= \dfrac{3x\sqrt{10}}{2x} = \dfrac{3\sqrt{10}}{2}$

27. $\dfrac{5+3i}{2-i}$

Multiply by the conjugate of the denominator.

$= \dfrac{(5+3i)(2+i)}{(2-i)(2+i)}$

$= \dfrac{10+5i+6i+3i^2}{4-i^2}$

$= \dfrac{10+11i+3(-1)}{4-(-1)}$

$= \dfrac{7+11i}{5} = \dfrac{7}{5} + \dfrac{11}{5}i$

28. $2\sqrt{k} = \sqrt{5k+3}$

Square both sides.

$4k = 5k+3$

$-k = 3$

$k = -3$

Since k must be nonnegative so that $\sqrt{k}$ is a real number, -3 cannot be a solution. The solution set is $\emptyset$.

29.

$$10q^2 + 13q = 3$$
$$10q^2 + 13q - 3 = 0$$
$$(5q-1)(2q+3) = 0$$

$5q - 1 = 0$ or $2q + 3 = 0$

$q = \dfrac{1}{5}$ or $q = -\dfrac{3}{2}$

Solution set: $\left\{\dfrac{1}{5}, -\dfrac{3}{2}\right\}$

30. $(4x-1)^2 = 8$

$4x - 1 = \pm\sqrt{8}$

$4x = 1 \pm 2\sqrt{2}$

$x = \dfrac{1 \pm 2\sqrt{2}}{4}$

Solution set: $\left\{\dfrac{1+2\sqrt{2}}{4}, \dfrac{1-2\sqrt{2}}{4}\right\}$

31. $\log(2x) - \log(x - 1) = \log 3$

$$\log \frac{2x}{x - 1} = \log 3$$

$$\frac{2x}{x - 1} = 3$$

$$2x = 3(x - 1)$$

$$2x = 3x - 3$$

$$3 = x$$

Check $x = 3$: $\log 6 - \log 2 \overset{?}{=} \log 3$ *True*

Solution set: $\{3\}$

32. $2(x^2 - 3)^2 - 5(x^2 - 3) = 12$

Let $u = (x^2 - 3)$.

$$2u^2 - 5u = 12$$

$$2u^2 - 5u - 12 = 0$$

$$(2u + 3)(u - 4) = 0$$

$$2u + 3 = 0 \quad \text{or} \quad u - 4 = 0$$

$$u = -\frac{3}{2} \quad \text{or} \quad u = 4$$

Substitute $x^2 - 3$ for u to find x.

If $u = -\frac{3}{2}$, then

$$x^2 - 3 = -\frac{3}{2}$$

$$x^2 = \frac{3}{2}$$

$$x = \pm\sqrt{\frac{3}{2}} = \pm\frac{\sqrt{3}}{\sqrt{2}} \cdot \frac{\sqrt{2}}{\sqrt{2}} = \pm\frac{\sqrt{6}}{2}.$$

If $u = 4$, then

$$x^2 - 3 = 4$$

$$x^2 = 7$$

$$x = \pm\sqrt{7}.$$

Solution set: $\left\{ -\dfrac{\sqrt{6}}{2}, \dfrac{\sqrt{6}}{2}, -\sqrt{7}, \sqrt{7} \right\}$

33. Solve $F = \dfrac{kwv^2}{r}$ for v.

$$Fr = kwv^2$$

$$v^2 = \frac{Fr}{kw}$$

Take the square root of each side.

$$v = \pm\sqrt{\frac{Fr}{kw}} = \frac{\pm\sqrt{Fr}}{\sqrt{kw}} \cdot \frac{\sqrt{kw}}{\sqrt{kw}} = \frac{\pm\sqrt{Frkw}}{kw}$$

34. $3x + y = 5$

Write the equation in slope-intercept form, $y = mx + b$.

$$y = -3x + 5$$

The y-intercept is $(0, 5)$ and $m = -3$ or $\dfrac{-3}{1}$.

Plot $(0, 5)$. From $(0, 5)$, move down 3 units and right 1 unit. Draw the line through these two points.

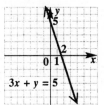

35. $f(x) = x^2 - 4x + 5$

The graph of f is a parabola. The x-value of the vertex is

$$x = \frac{-b}{2a} = \frac{-(-4)}{2(1)} = 2.$$

The y-value of the vertex is

$$f(2) = 2^2 - 4(2) + 5 = 1.$$

The vertex is at $(2, 1)$ and the parabola opens upward since $a = 1 > 0$. The y-intercept is $(0, 5)$ and by symmetry, the point $(4, 5)$ is also on the graph.

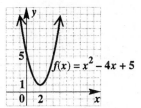

36. $f(x) = 3^{x-1}$

The graph of f is an increasing exponential.

x	-1	0	1	2	3
$f(x)$	$\frac{1}{9}$	$\frac{1}{3}$	1	3	9

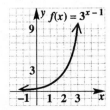

37. $\dfrac{x^2}{4} - \dfrac{y^2}{16} = 1$ is in $\dfrac{x^2}{a^2} - \dfrac{y^2}{b^2} = 1$ form.

The graph is a hyperbola with x-intercepts $(2, 0)$ and $(-2, 0)$ and asymptotes that are the extended diagonals of the rectangle with corners $(2, 4)$, $(2, -4)$, $(-2, -4)$, and $(-2, 4)$. Draw a branch of the hyperbola through each intercept approaching the asymptotes.

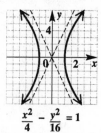

38. $\dfrac{x^2}{25} + \dfrac{y^2}{16} \leq 1$

The boundary, $\dfrac{x^2}{25} + \dfrac{y^2}{16} = 1$, is a solid ellipse in

$\dfrac{x^2}{a^2} + \dfrac{y^2}{b^2} = 1$ form with intercepts $(5, 0)$, $(-5, 0)$, $(0, 4)$, and $(0, -4)$. Test $(0, 0)$.

$$\dfrac{0^2}{25} + \dfrac{0^2}{16} \leq 1 \quad ?$$
$$0 \leq 1 \quad \textit{True}$$

Shade the region inside the ellipse.

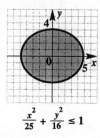

39. (a) $\quad y = ax^2 + bx$
$61.7 = a(25)^2 + b(25)$
$61.7 = 625a + 25b \qquad (1)$
$106 = a(35)^2 + b(35)$
$106 = 1225a + 35b \qquad (2)$

(b) Multiply (1) by -7 and (2) by 5 and add the results.

$$
\begin{array}{rcll}
-431.9 & = & -4375a - 175b & -7 \times (1) \\
530 & = & 6125a + 175b & 5 \times (2) \\
\hline
98.1 & = & 1750a &
\end{array}
$$

$$a = \dfrac{98.1}{1750} \approx .056$$

To find b, substitute $\dfrac{98.1}{1750}$ for a in (1).

$$61.7 = 625\left(\dfrac{98.1}{1750}\right) + 25b$$

$$25b = 61.7 - 625\left(\dfrac{98.1}{1750}\right)$$

$$b = \dfrac{61.7 - 625\left(\dfrac{98.1}{1750}\right)}{25}$$

$$\approx 1.067$$

(c) $\quad y = ax^2 + bx$
$y = .056x^2 + 1.067x$

(d) Let $x = 55$.

$$y = .056(55)^2 + 1.067(55)$$
$$= 228.085 \approx 228.1 \text{ feet}$$

40. (a) $\quad y = 1.38(1.65)^x$
$y = 1.38(1.65)^5 \quad$ *Let x=5.*
≈ 16.9 billion dollars

(b) $\quad 2003 - 1995 = 8$

$y = 1.38(1.65)^8 \quad$ *Let x=8.*
≈ 75.8 billion dollars

CHAPTER 11   SEQUENCES AND SERIES

Section 11.1

1. The similarities are that both are defined by the same linear expression, and that points satisfying both lie in a straight line. The difference is that the domain of f consists of all real numbers, but the domain of the sequence is $\{1, 2, 3, \dots\}$. Some examples are $f(1) = 6$ and $a_1 = 6$ is a similarity, while $f\left(\dfrac{3}{2}\right) = 7$, but $a_{3/2}$ is not allowed.

3. $a_n = \dfrac{n+3}{n}$

To get a_1, the first term, replace n with 1.
$$a_1 = \frac{1+3}{1} = \frac{4}{1} = 4$$
To get a_2, the second term, replace n with 2.
$$a_2 = \frac{2+3}{2} = \frac{5}{2}$$
To get a_3, the third term, replace n with 3.
$$a_3 = \frac{3+3}{3} = \frac{6}{3} = 2$$
To get a_4, the fourth term, replace n with 4.
$$a_4 = \frac{4+3}{4} = \frac{7}{4}$$
To get a_5, the fifth term, replace n with 5.
$$a_5 = \frac{5+3}{5} = \frac{8}{5}$$

Answer: $4, \dfrac{5}{2}, 2, \dfrac{7}{4}, \dfrac{8}{5}$

5. $a_n = 3^n$
$a_1 = 3^1 = 3$
$a_2 = 3^2 = 9$
$a_3 = 3^3 = 27$
$a_4 = 3^4 = 81$
$a_5 = 3^5 = 243$

Answer: 3, 9, 27, 81, 243

7. $a_n = \dfrac{1}{n^2}$
$a_1 = \dfrac{1}{1^2} = 1$
$a_2 = \dfrac{1}{2^2} = \dfrac{1}{4}$
$a_3 = \dfrac{1}{3^2} = \dfrac{1}{9}$
$a_4 = \dfrac{1}{4^2} = \dfrac{1}{16}$
$a_5 = \dfrac{1}{5^2} = \dfrac{1}{25}$

Answer: $1, \dfrac{1}{4}, \dfrac{1}{9}, \dfrac{1}{16}, \dfrac{1}{25}$

9. $a_n = (-1)^n$
$a_1 = (-1)^1 = -1$
$a_2 = (-1)^2 = 1$
$a_3 = (-1)^3 = -1$
$a_4 = (-1)^4 = 1$
$a_5 = (-1)^5 = -1$

Answer: $-1, 1, -1, 1, -1$

11. $a_n = -9n + 2$
$a_8 = -9(8) + 2 = -70$

13. $a_n = \dfrac{3n+7}{2n-5}$
$a_{14} = \dfrac{3(14)+7}{2(14)-5} = \dfrac{49}{23}$

15. $a_n = (n+1)(2n+3)$
$a_8 = (8+1)[2(8)+3]$
$= 9(19) = 171$

17. $4, 8, 12, 16, \dots$ can be written as $4 \cdot 1, 4 \cdot 2, 4 \cdot 3, 4 \cdot 4, \dots$, so $a_n = 4n$.

19. $\dfrac{1}{3}, \dfrac{1}{9}, \dfrac{1}{27}, \dfrac{1}{81}, \dots$ can be written as $\dfrac{1}{3^1}, \dfrac{1}{3^2}, \dfrac{1}{3^3}, \dfrac{1}{3^4}, \dots$, so $a_n = \dfrac{1}{3^n}$.

21. Make a table as follows:

Month	Interest	Payment	Unpaid balance
0			1000
1	$1000(.01) = 10$	$100 + 10 = 110$	$1000 - 100 = 900$
2	$900(.01) = 9$	$100 + 9 = 109$	$900 - 100 = 800$
3	$800(.01) = 8$	$100 + 8 = 108$	$800 - 100 = 700$
4	$700(.01) = 7$	$100 + 7 = 107$	$700 - 100 = 600$
5	$600(.01) = 6$	$100 + 6 = 106$	$600 - 100 = 500$
6	$500(.01) = 5$	$100 + 5 = 105$	$500 - 100 = 400$

The payments are $110, $109, $108, $107, $106, and $105; the unpaid balance is $400.

23. When new, the car is worth $20,000. Let $a_n =$ the value of the car after the nth year. The car retains $\dfrac{4}{5}$ of its value each year.
$a_1 = \dfrac{4}{5}(20,000) = \$16,000$ (value after the first year)
$a_2 = \dfrac{4}{5}(16,000) = \$12,800$
$a_3 = \dfrac{4}{5}(12,800) = \$10,240$
$a_4 = \dfrac{4}{5}(10,240) = \$8192$
$a_5 = \dfrac{4}{5}(8192) = \$6553.60$

The value of the car after 5 years is about $6554.

25. $\displaystyle\sum_{i=1}^{3} \left(i^2 + 2\right)$

$= \left(1^2 + 2\right) + \left(2^2 + 2\right) + \left(3^2 + 2\right)$

$= 3 + 6 + 11 = 20$

27. $\displaystyle\sum_{i=2}^{5} \frac{1}{i} = \frac{1}{2} + \frac{1}{3} + \frac{1}{4} + \frac{1}{5}$

$= \frac{30}{60} + \frac{20}{60} + \frac{15}{60} + \frac{12}{60} = \frac{77}{60}$

29. $\displaystyle\sum_{i=1}^{6} (-1)^i$

$= (-1)^1 + (-1)^2 + (-1)^3 + (-1)^4$
$+ (-1)^5 + (-1)^6$

$= -1 + 1 - 1 + 1 - 1 + 1 = 0$

31. $\displaystyle\sum_{i=3}^{7} (i - 3)(i + 2)$

$= (3 - 3)(3 + 2) + (4 - 3)(4 + 2)$
$+ (5 - 3)(5 + 2) + (6 - 3)(6 + 2)$
$+ (7 - 3)(7 + 2)$

$= 0(5) + (1)(6) + (2)(7) + (3)(8)$
$+ (4)(9)$

$= 0 + 6 + 14 + 24 + 36 = 80$

33. $\displaystyle\sum_{i=1}^{5} 2x \cdot i$

$= 2x \cdot 1 + 2x \cdot 2 + 2x \cdot 3 + 2x \cdot 4$
$+ 2x \cdot 5$

$= 2x + 4x + 6x + 8x + 10x$

35. $\displaystyle\sum_{i=1}^{5} i \cdot x^i = 1\left(x^1\right) + 2\left(x^2\right) + 3\left(x^3\right) + 4\left(x^4\right)$

$+ 5\left(x^5\right)$

$= x + 2x^2 + 3x^3 + 4x^4 + 5x^5$

37. $3 + 4 + 5 + 6 + 7$

$= (1 + 2) + (2 + 2) + (3 + 2)$
$+ (4 + 2) + (5 + 2)$

$= \displaystyle\sum_{i=1}^{5} (i + 2)$

39. $\dfrac{1}{2} + \dfrac{1}{3} + \dfrac{1}{4} + \dfrac{1}{5} + \dfrac{1}{6}$

$= \dfrac{1}{1 + 1} + \dfrac{1}{2 + 1} + \dfrac{1}{3 + 1} + \dfrac{1}{4 + 1}$

$+ \dfrac{1}{5 + 1}$

$= \displaystyle\sum_{i=1}^{5} \dfrac{1}{i + 1}$

41. A sequence is a list of terms in a specific order, while a series is the indicated sum of the terms of a sequence.

43. $\bar{x} = \dfrac{\displaystyle\sum_{i=1}^{n} x_i}{n} = \dfrac{\displaystyle\sum_{i=1}^{7} x_i}{7}$

$= \dfrac{8 + 11 + 14 + 9 + 3 + 6 + 8}{7} = \dfrac{59}{7}$

45. $\bar{x} = \dfrac{5 + 9 + 8 + 2 + 4 + 7 + 3 + 2}{8} = \dfrac{40}{8} = 5$

47. $\bar{x} = \dfrac{3427 + 3850 + 4558 + 5357 + 5761}{5}$

$= \dfrac{22{,}953}{5} = 4590.6$

The average number of funds available for this five-year period was about 4591.

49. $\displaystyle\sum_{i=1}^{6} \left(i^2 + 3i + 5\right)$

$= \displaystyle\sum_{i=1}^{6} i^2 + \sum_{i=1}^{6} 3i + \sum_{i=1}^{6} 5$

50. From Exercise 49, the second summation is

$\displaystyle\sum_{i=1}^{6} 3i = 3\sum_{i=1}^{6} i.$

51. From Exercise 49, the third summation is

$\displaystyle\sum_{i=1}^{6} 5 = 6 \cdot 5 = 30.$

52. $1 + 2 + 3 + 4 + \cdots + n = \dfrac{n(n + 1)}{2}$

written in summation notation is

$\displaystyle\sum_{i=1}^{n} i = \dfrac{n(n + 1)}{2}.$

53. $1^2 + 2^2 + 3^2 + 4^2 + \cdots + n^2$

$= \dfrac{n(n + 1)(2n + 1)}{6}$

written in summation notation is

$\displaystyle\sum_{i=1}^{n} i^2 = \dfrac{n(n + 1)(2n + 1)}{6}.$

54. From Exercises 49–51,

$\displaystyle\sum_{i=1}^{6} \left(i^2 + 3i + 5\right)$

$= \displaystyle\sum_{i=1}^{6} i^2 + 3\sum_{i=1}^{6} i + 30.$

Now apply the results of Exercises 52 & 53.

$= \dfrac{6(6 + 1)(2 \cdot 6 + 1)}{6} + 3 \cdot \dfrac{6(6 + 1)}{2} + 30$

$= \dfrac{6(7)(13)}{6} + 3 \cdot \dfrac{6(7)}{2} + 30$

$= 91 + 63 + 30 = 184$

55. $\displaystyle\sum_{i=1}^{12} \left(i^2 - i\right)$

$$= \sum_{i=1}^{12} i^2 - \sum_{i=1}^{12} i$$

$$= \frac{12(12+1)(2 \cdot 12 + 1)}{6} - \frac{12(12+1)}{2}$$

$$= \frac{12(13)(25)}{6} - \frac{12(13)}{2}$$

$$= 650 - 78 = 572$$

56. $\displaystyle\sum_{i=1}^{20} \left(2 + i - i^2\right)$

$$\sum_{i=1}^{20} 2 + \sum_{i=1}^{20} i - \sum_{i=1}^{20} i^2$$

$$= 20 \cdot 2 + \frac{20(20+1)}{2} - \frac{20(20+1)(2 \cdot 20 + 1)}{6}$$

$$= 40 + 210 - 2870 = -2620$$

Section 11.2

1. An arithmetic sequence is a sequence (list) of numbers in a specific order such that there is a common difference between any two successive terms. For example, the sequence $1, 5, 9, 13, \ldots$ is arithmetic with difference $d = 5 - 1 = 9 - 5 = 13 - 9 = 4$. As another example, $2, -1, -4, -7, \ldots$ is an arithmetic sequence with $d = -3$.

3. $1, 2, 3, 4, 5, \ldots$
d is the difference between any two adjacent terms. Choose the terms 3 and 2.

$$d = 3 - 2 = 1$$

The terms 2 and 1 would give

$$d = 2 - 1 = 1,$$

the same result. Therefore, the common difference is $d = 1$.

Note: You should find the difference for all pairs of adjacent terms to determine if the sequence is arithmetic.

5. $2, -4, 6, -8, 10, -12, \ldots$
The difference between the first two terms is $-4 - 2 = -6$, but the difference between the second and third terms is $6 - (-4) = 10$. The differences are not the same so the sequence is *not arithmetic*.

7. $-10, -5, 0, 5, 10, \ldots$
Choose the terms 10 and 5, and find the difference.

$$d = 10 - 5 = 5$$

The terms -5 and -10 would give

$$d = -5 - (-10) = -5 + 10 = 5,$$

the same result. Therefore, the common difference is $d = 5$.

9. $3.42, 5.57, 7.72, 9.87, \ldots$
Choose the terms 5.57 and 3.42.

$$d = 5.57 - 3.42 = 2.15$$

The terms 7.72 and 5.57 give

$$d = 7.72 - 5.57 = 2.15,$$

the same result. Therefore, the common difference is $d = 2.15$.

11. $-\dfrac{5}{3}, -1, -\dfrac{1}{3}, \dfrac{1}{3}, \ldots$
Choose the terms $\dfrac{1}{3}$ and $-\dfrac{1}{3}$.

$$d = \frac{1}{3} - \left(-\frac{1}{3}\right) = \frac{2}{3}$$

The terms $-\dfrac{1}{3}$ and -1 give

$$d = -\frac{1}{3} - (-1) = -\frac{1}{3} + 1 = \frac{2}{3},$$

the same result. Therefore, the common difference is $d = \dfrac{2}{3}$.

13. $a_1 = 2, d = 5$
$\begin{aligned} a_n &= a_1 + (n-1)d \\ &= 2 + (n-1)5 \\ &= 2 + 5n - 5 \\ &= 5n - 3 \end{aligned}$

15. $3, \dfrac{15}{4}, \dfrac{9}{2}, \dfrac{21}{4}, \ldots$
To find d, subtract any two adjacent terms.

$$d = \frac{15}{4} - 3 = \frac{15}{4} - \frac{12}{4} = \frac{3}{4}$$

The first term is $a_1 = 3$. Now find a_n.
$\begin{aligned} a_n &= a_1 + (n-1)d \\ &= 3 + (n-1)\left(\frac{3}{4}\right) \\ &= 3 + \frac{3}{4}n - \frac{3}{4} \\ &= \frac{3}{4}n + \frac{9}{4} \end{aligned}$

17. $-3, 0, 3, \ldots$
To find d, subtract any two adjacent terms.

$$d = 0 - (-3) = 3$$

The first term is $a_1 = -3$. Now find a_n.
$$\begin{aligned} a_n &= a_1 + (n-1)d \\ &= -3 + (n-1)3 \\ &= -3 + 3n - 3 \\ &= 3n - 6 \end{aligned}$$

19.   Given $a_1 = 4$ and $d = 3$; find a_{25}.
$$\begin{aligned} a_n &= a_1 + (n-1)d \\ a_{25} &= 4 + (25-1)3 \\ &= 4 + 72 = 76 \end{aligned}$$

21.   Given $2, 4, 6, \ldots$; find a_{24}.
Here, $a_1 = 2$ and $d = 4 - 2 = 2$.
Now find a_{24}.
$$\begin{aligned} a_n &= a_1 + (n-1)d \\ a_{24} &= 2 + (24-1)2 \\ &= 2 + 46 = 48 \end{aligned}$$

23.   Given $a_{12} = -45$ and $a_{10} = -37$; find a_1. Use $a_n = a_1 + (n-1)d$ to write a system of equations.
$$\begin{aligned} a_{12} &= a_1 + (12-1)d \\ -45 &= a_1 + 11d \qquad (1) \\ a_{10} &= a_1 + (10-1)d \\ -37 &= a_1 + 9d \qquad (2) \end{aligned}$$
To eliminate d, multiply equation (1) by -9 and equation (2) by 11. Then add the results.
$$\begin{array}{rll} 405 = & -9a_1 - 99d & -9 \times (1) \\ -407 = & 11a_1 + 99d & 11 \times (2) \\ \hline -2 = & 2a_1 & \\ -1 = & a_1 & \end{array}$$

25. $3, 5, 7, \ldots, 33$
Let n represent the number of terms in the sequence. So, $a_n = 33$, $a_1 = 3$, and $d = 5 - 3 = 2$.
$$\begin{aligned} a_n &= a_1 + (n-1)d \\ 33 &= 3 + (n-1)2 \\ 33 &= 3 + 2n - 2 \\ 33 &= 2n + 1 \\ 32 &= 2n \\ n &= 16 \end{aligned}$$
The sequence has 16 terms.

27. $\dfrac{3}{4}, 3, \dfrac{21}{4}, \ldots, 12$
Let n represent the number of terms in the sequence. So, $a_n = 12$,
$a_1 = \dfrac{3}{4}$, and $d = 3 - \dfrac{3}{4} = \dfrac{9}{4}$.

$$\begin{aligned} a_n &= a_1 + (n-1)d \\ 12 &= \frac{3}{4} + (n-1)\left(\frac{9}{4}\right) \\ \frac{45}{4} &= (n-1)\left(\frac{9}{4}\right) \\ 5 &= n - 1 \qquad \textit{Multiply by } \frac{4}{9}. \\ 6 &= n \end{aligned}$$
The sequence has 6 terms.

29. n represents the number of terms.

31.   The sum of the left sides is $S + S = 2S$. For the right sides, $1 + 100 = 101$, $2 + 99 = 101$, $3 + 98 = 101$, and so on. The sum 101 appears 100 times. By multiplying, the sum of the right sides is $101 \cdot 100 = 10,100$.

32. $2S = 10,100$

33.   Divide by 2 to obtain $S = 5050$.

34.   The right sides now have a sum of 201 appearing 200 times. Thus, the sum is $201 \cdot 200 = 40,200$.
$$\begin{aligned} 2S &= 40,200 \\ S &= 20,100 \end{aligned}$$
Note that the pattern is $S = \dfrac{(n+1)(n)}{2}$.

35.   Find S_6 given $a_1 = 6$, $d = 3$, and $n = 6$.
$$\begin{aligned} S_n &= \frac{n}{2}\big[2a_1 + (n-1)d\big] \\ S_6 &= \frac{6}{2}\big[2 \cdot 6 + (6-1)3\big] \\ &= 3(12 + 5 \cdot 3) \\ &= 3(27) = 81 \end{aligned}$$

37.   Find S_6 given $a_1 = 7$, $d = -3$, and $n = 6$.
$$\begin{aligned} S_n &= \frac{n}{2}\big[2a_1 + (n-1)d\big] \\ S_6 &= \frac{6}{2}\big[2 \cdot 7 + (6-1)(-3)\big] \\ &= 3\big[14 + 5(-3)\big] \\ &= 3(14 - 15) = 3(-1) = -3 \end{aligned}$$

39.   Find S_6 given $a_n = 4 + 3n$.
$$\begin{aligned} a_1 &= 4 + 3(1) = 7 \\ a_6 &= 4 + 3(6) = 22 \\ S_n &= \frac{n}{2}(a_1 + a_n) \\ S_6 &= \frac{6}{2}(7 + 22) \\ &= 3(29) = 87 \end{aligned}$$

41. $\displaystyle\sum_{i=1}^{10} (8i - 5)$
$$\begin{aligned} a_n &= 8n - 5 \\ a_1 &= 8(1) - 5 = 3 \\ a_{10} &= 8(10) - 5 = 75 \end{aligned}$$

Use $S_n = \dfrac{n}{2}(a_1 + a_n)$ with $n = 10$,
$a_1 = 3$, and $a_{10} = 75$.
$$S_{10} = \frac{10}{2}(3 + 75)$$
$$= 5(78) = 390$$

43. $\displaystyle\sum_{i=1}^{20}(2i - 5)$
$a_n = 2n - 5$
$a_1 = 2(1) - 5 = -3$
$a_{20} = 2(20) - 5 = 35$
Use $S_n = \dfrac{n}{2}(a_1 + a_n)$ with $n = 20$,
$a_1 = -3$, and $a_{20} = 35$.
$$S_{20} = \frac{20}{2}(-3 + 35)$$
$$= 10(32) = 320$$

45. $\displaystyle\sum_{i=1}^{250} i$
$a_n = n$, $a_1 = 1$, $a_{250} = 250$
Use $S_n = \dfrac{n}{2}(a_1 + a_n)$ with $n = 250$,
$a_1 = 1$, and $a_{250} = 250$.
$$S_{250} = \frac{250}{2}(1 + 250)$$
$$= 125(251) = 31,375$$

47. The sequence is $1, 2, 3, \ldots, 30$.
$$S_n = \frac{n}{2}(a_1 + a_n)$$
$$= \frac{30}{2}(1 + 30)$$
$$= 15(31) = 465$$
The account will have \$465 deposited in it over the entire month.

49. Your salaries at six-month intervals form an arithmetic sequence with $a_1 = 1600$ and $d = 50$. Since your salary is increased every 6 months, or $2(5) = 10$ times, after 5 years your salary will equal the term a_{11}.
$$a_n = a_1 + (n - 1)d$$
$$a_{11} = 1600 + (11 - 1)50$$
$$= 1600 + 500 = 2100$$
Your salary will be \$2100/month.

51. Given the sequence $20, 22, 24, \ldots$, for 25 terms, find a_{25}. Here,
$a_1 = 20$, $d = 22 - 20 = 2$, and $n = 25$.
$$a_n = a_1 + (n - 1)d$$
$$a_{25} = 20 + (25 - 1)2$$
$$= 20 + 48 = 68$$
There are 68 seats in the last row.
Now find S_{25}.

$$S_n = \frac{n}{2}(a_1 + a_n)$$
$$S_{25} = \frac{25}{2}(20 + 68)$$
$$= \frac{25}{2}(88) = 25(44) = 1100$$
There are 1100 seats in the section.

53. Given the sequence $35, 31, 27, \ldots$, can the sequence end in 1? If not, find the last positive value. If the sequence ends in 1, we can find n, a whole number.
$$d = 31 - 35 = -4$$
$$a_n = a_1 + (n - 1)d$$
$$1 = 35 + (n - 1)(-4)$$
$$1 = 35 - 4n + 4$$
$$-38 = -4n$$
$$9.5 = n$$
Since n is not a whole number, the sequence cannot end in 1. The largest n possible is $n = 9$.
$$a_n = a_1 + (n - 1)d$$
$$a_9 = 35 + (9 - 1)(-4)$$
$$= 35 - 32 = 3$$
She can build 9 rows. There are 3 blocks in the last row.

55. $f(x) = mx + b$
$f(1) = m(1) + b = m + b$
$f(2) = m(2) + b = 2m + b$
$f(3) = m(3) + b = 3m + b$

56. The sequence $f(1)$, $f(2)$, $f(3)$ is an arithmetic sequence since the difference between any two adjacent terms is m.

57. The common difference is the difference between any two adjacent terms.
$$d = (3m + b) - (2m + b) = m$$

58. From Exercise 55, we know that $a_1 = m + b$. From Exercise 57, we know that $d = m$. Therefore,
$$a_n = a_1 + (n - 1)d$$
$$= (m + b) + (n - 1)m$$
$$= m + b + nm - m$$
$$a_n = mn + b.$$

Section 11.3

1. A geometric sequence is an ordered list of numbers such that each term after the first is obtained by multiplying the previous term by a constant, r, called the common ratio. For example, if the first term is 3 and $r = 4$, the sequence is $3, 12, 48, 192, \ldots$. If the first term is 2 and $r = -1$, the sequence is $2, -2, 2, -2, \ldots$.

3. $4, 8, 16, 32, \ldots$

To find r, choose any two adjacent terms and divide the second one by the first one.

$$r = \frac{8}{4} = 2$$

Notice that any two other adjacent terms could have been used with the same result. The common ratio is $r = 2$.

5. $\frac{1}{3}, \frac{2}{3}, \frac{3}{3}, \frac{4}{3}, \frac{5}{3}, \ldots$

Choose any two adjacent terms and divide the second by the first.

$$r = \frac{\frac{2}{3}}{\frac{1}{3}} = \frac{2}{3} \cdot 3 = 2$$

Confirm this result with any two other adjacent terms.

$$r = \frac{\frac{3}{3}}{\frac{2}{3}} = 1 \cdot \frac{3}{2} = \frac{3}{2}$$

Since $2 \neq \frac{3}{2}$, the ratios are not the same. The sequence is *not geometric*.

7. $1, -3, 9, -27, 81, \ldots$

$$r = \frac{-3}{1} = -3$$
$$r = \frac{9}{-3} = -3$$

The common ratio is $r = -3$.

9. $1, -\frac{1}{2}, \frac{1}{4}, -\frac{1}{8}, \frac{1}{16}, \ldots$

$$r = \frac{-\frac{1}{2}}{1} = -\frac{1}{2}$$

$$r = \frac{\frac{1}{4}}{-\frac{1}{2}} = \frac{1}{4} \cdot (-2) = -\frac{1}{2}$$

The common ratio is $r = -\frac{1}{2}$.

11.   Find a general term for $5, 10, \ldots$.
First, find r.

$$r = \frac{10}{5} = 2$$

Use $a_1 = 5$ and $r = 2$ to find a_n.

$$a_n = a_1 r^{n-1}$$
$$a_n = 5(2)^{n-1}$$

13.   Find a general term for $\frac{1}{9}, \frac{1}{3}, \ldots$.

Here, $a_1 = \frac{1}{9}$. Find r.

$$r = \frac{\frac{1}{3}}{\frac{1}{9}} = \frac{1}{3} \cdot \frac{9}{1} = 3$$

Now find a_n.

$$a_n = a_1 r^{n-1}$$
$$a_n = \frac{1}{9}(3)^{n-1}$$
$$\text{or} \quad a_n = \frac{3^{n-1}}{9}$$

15.   Find a general term for $10, -2, \ldots$.
Here, $a_1 = 10$. Find r.

$$r = \frac{-2}{10} = -\frac{1}{5}$$

Now find a_n.

$$a_n = a_1 r^{n-1}$$
$$a_n = 10\left(-\frac{1}{5}\right)^{n-1}$$

17.   Given $2, 10, 50, \ldots$; find a_{10}.
First find the common ratio.

$$r = \frac{10}{2} = 5$$

Substitute $a_1 = 2$, $r = 5$, and $n = 10$ in the nth-term formula.

$$a_n = a_1 r^{n-1}$$
$$a_{10} = a_1 (r)^{10-1}$$
$$= 2(5)^9$$

19.   Given $\frac{1}{2}, \frac{1}{6}, \frac{1}{18}, \ldots$; find a_{12}.
First find the common ratio.

$$r = \frac{\frac{1}{6}}{\frac{1}{2}} = \frac{1}{6} \cdot 2 = \frac{1}{3}$$

Substitute $a_1 = \frac{1}{2}$, $r = \frac{1}{3}$, and $n = 12$ in the formula.

$$a_n = a_1 r^{n-1}$$
$$a_{12} = a_1 (r)^{12-1}$$
$$= \left(\frac{1}{2}\right)\left(\frac{1}{3}\right)^{11}$$

21. Given $a_3 = \dfrac{1}{2}, a_7 = \dfrac{1}{32}$; find a_{25}.

Find a_1 and r using the general term $a_n = a_1 r^{n-1}$.

$$a_3 = a_1 r^{3-1}$$
$$\frac{1}{2} = a_1 r^2 \quad (1)$$
$$a_7 = a_1 r^{7-1}$$
$$\frac{1}{32} = a_1 r^6 \quad (2)$$

Solve (1) for a_1.

$$a_1 = \frac{1}{2r^2}$$

Substitute for a_1 in (2).

$$\frac{1}{32} = \frac{1}{2r^2} r^6$$
$$\frac{1}{16} = r^4$$
$$r^2 = \pm\frac{1}{4}$$

Since r^2 is positive,

$$r^2 = \frac{1}{4}.$$

Substitute $\dfrac{1}{4}$ for r^2 in (1).

$$\frac{1}{2} = a_1\left(\frac{1}{4}\right)$$
$$2 = a_1$$

Use $a_1 = 2$ and $r = \dfrac{1}{2}$ $\left(\text{or } -\dfrac{1}{2}\right)$ to find a_{25}.

$$a_{25} = a_1(r)^{25-1}$$
$$= 2\left(\frac{1}{2}\right)^{24}$$
$$= \frac{1}{2^{23}}$$

23. $\dfrac{1}{3} = .33333\ldots$

24. $\dfrac{2}{3} = .66666\ldots$

25. $\quad .33333\ldots$
$\dfrac{+ .66666\ldots}{.99999\ldots}$

26. $S = \dfrac{a_1}{1-r}$
$$= \frac{.9}{1-.1}$$
$$= \frac{.9}{.9} = 1$$

Therefore, $.99999\ldots = 1$.

27. $\dfrac{1}{3}, \dfrac{1}{9}, \dfrac{1}{27}, \dfrac{1}{81}, \dfrac{1}{243}$

Here, $a_1 = \dfrac{1}{3}$, $n = 5$, and

$$r = \frac{\dfrac{1}{9}}{\dfrac{1}{3}} = \frac{1}{9} \cdot 3 = \frac{1}{3}.$$

$$S_n = \frac{a_1(r^n - 1)}{r - 1}$$

$$S_5 = \frac{\dfrac{1}{3}\left[\left(\dfrac{1}{3}\right)^5 - 1\right]}{\dfrac{1}{3} - 1}$$

$$= \frac{\dfrac{1}{3}\left(\dfrac{1}{243} - 1\right)}{-\dfrac{2}{3}}$$

$$= \frac{\dfrac{1}{3}\left(-\dfrac{242}{243}\right)}{-\dfrac{2}{3}} = \frac{121}{243}$$

29. $-\dfrac{4}{3}, -\dfrac{4}{9}, -\dfrac{4}{27}, -\dfrac{4}{81}, -\dfrac{4}{243}, -\dfrac{4}{729}$

Here, $a_1 = -\dfrac{4}{3}$, $n = 6$, and

$$r = \frac{-\dfrac{4}{9}}{-\dfrac{4}{3}} = -\frac{4}{9} \cdot \left(-\frac{3}{4}\right) = \frac{1}{3}.$$

$$S_n = \frac{a_1(r^n - 1)}{r - 1}$$

$$S_6 = \frac{-\dfrac{4}{3}\left[\left(\dfrac{1}{3}\right)^6 - 1\right]}{\dfrac{1}{3} - 1}$$

$$= \frac{-\dfrac{4}{3}\left(\dfrac{1}{729} - 1\right)}{-\dfrac{2}{3}}$$

$$= \frac{-\dfrac{4}{3}\left(-\dfrac{728}{729}\right)}{-\dfrac{2}{3}} = -\frac{1456}{729} \approx -1.997$$

31. $\displaystyle\sum_{i=1}^{7} 4\left(\frac{2}{5}\right)^{i}$

Use $a_1 = 4\left(\dfrac{2}{5}\right) = \dfrac{8}{5}$, $n = 7$, and $r = \dfrac{2}{5}$.

$$S_n = \frac{a_1(r^n - 1)}{r - 1}$$

$$S_7 = \frac{\dfrac{8}{5}\left[\left(\dfrac{2}{5}\right)^{7} - 1\right]}{\dfrac{2}{5} - 1}$$

$$= \frac{\dfrac{8}{5}\left[\left(\dfrac{2}{5}\right)^{7} - 1\right]}{-\dfrac{3}{5}}$$

$$= -\frac{8}{3}\left[\left(\frac{2}{5}\right)^{7} - 1\right] \approx 2.662$$

33. $\displaystyle\sum_{i=1}^{10} (-2)\left(\frac{3}{5}\right)^{i}$

Use $a_1 = (-2)\left(\dfrac{3}{5}\right) = -\dfrac{6}{5}$, $n = 10$, and $r = \dfrac{3}{5}$.

$$S_n = \frac{a_1(r^n - 1)}{r - 1}$$

$$S_{10} = \frac{-\dfrac{6}{5}\left[\left(\dfrac{3}{5}\right)^{10} - 1\right]}{\dfrac{3}{5} - 1}$$

$$= \frac{-\dfrac{6}{5}\left[\left(\dfrac{3}{5}\right)^{10} - 1\right]}{-\dfrac{2}{5}}$$

$$= 3\left[\left(\frac{3}{5}\right)^{10} - 1\right] \approx -2.982$$

35. There are 22 deposits, so $n = 22$.

$$S = R\left[\frac{(1 + i)^n - 1}{i}\right]$$

$$= 1000\left[\frac{(1 + .095)^{22} - 1}{.095}\right]$$

$$= 66{,}988.91$$

There will be $66,988.91 in the account.

37. Quarterly deposits for 10 years give us $n = 4 \cdot 10 = 40$. The interest rate per period is $i = \dfrac{.07}{4} = .0175$.

$$S = R\left[\frac{(1 + i)^n - 1}{i}\right]$$

$$= 1200\left[\frac{(1 + .0175)^{40} - 1}{.0175}\right]$$

$$= 68{,}680.96$$

We now use the compound interest formula to determine the value of this money after 5 more years.

$$A = P\left(1 + \frac{r}{n}\right)^{nt}$$

$$= 68{,}680.96\left(1 + \frac{.09}{12}\right)^{12(5)}$$

$$= 107{,}532.48$$

The woman is also saving $300 per month, so we use the annuity formula to determine that value.

$$S = R\left[\frac{(1 + i)^n - 1}{i}\right]$$

$$= 300\left[\frac{\left(1 + \dfrac{.09}{12}\right)^{12(5)} - 1}{\dfrac{.09}{12}}\right]$$

$$= 22{,}627.24$$

Adding $22,627.24 to $107,532.48 gives a total of $130,159.72 in the account.

39. Find the sum if $a_1 = 6$ and $r = \dfrac{1}{3}$. Since $|r| < 1$, the sum exists.

$$S = \frac{a_1}{1 - r} = \frac{6}{1 - \dfrac{1}{3}} = \frac{6}{\dfrac{2}{3}} = 6 \cdot \frac{3}{2} = 9$$

41. Find the sum if $a_1 = 1000$ and $r = -\dfrac{1}{10}$. Since $|r| < 1$, the sum exists.

$$S = \frac{a_1}{1 - r} = \frac{1000}{1 - \left(-\dfrac{1}{10}\right)} = \frac{1000}{\dfrac{11}{10}}$$

$$= 1000 \cdot \frac{10}{11} = \frac{10{,}000}{11}$$

43. $\displaystyle\sum_{i=1}^{\infty} \frac{9}{8}\left(-\frac{2}{3}\right)^{i}$

$a_1 = \dfrac{9}{8}\left(-\dfrac{2}{3}\right)^{1} = -\dfrac{3}{4}$ and $r = -\dfrac{2}{3}$.

Since $|r| < 1$, the sum exists.

$$S = \frac{a_1}{1-r} = \frac{-\frac{3}{4}}{1 - \left(-\frac{2}{3}\right)} = \frac{-\frac{3}{4}}{\frac{5}{3}}$$

$$= -\frac{3}{4} \cdot \frac{3}{5} = -\frac{9}{20}$$

45. $\sum_{i=1}^{\infty} \frac{12}{5}\left(\frac{5}{4}\right)^i$

Since $|r| = \frac{5}{4} > 1$, the sum *does not exist*.

47. The ball is dropped from a height of 10 feet and will rebound $\frac{3}{5}$ of its original height.

Let $a_n =$ the ball's height on the nth rebound.

$a_1 = 10$ and $r = \frac{3}{5}$.

Since we must find the height after the fourth bounce, $n = 5$ (since a_1 is the starting point).

Use $a_n = a_1 r^{n-1}$.

$$a_5 = 10\left(\frac{3}{5}\right)^{5-1} = 10\left(\frac{3}{5}\right)^4 \approx 1.3$$

The ball will rebound 1.3 feet after the fourth bounce.

49. This exercise can be modeled by a geometric sequence with $a_1 = 256$ and $r = \frac{1}{2}$. First we need to find n so that $a_n = 32$.

$$32 = a_1 r^{n-1}$$
$$32 = 256\left(\frac{1}{2}\right)^{n-1}$$
$$\frac{1}{8} = \left(\frac{1}{2}\right)^{n-1}$$
$$\left(\frac{1}{2}\right)^3 = \left(\frac{1}{2}\right)^{n-1}$$
$$3 = n - 1$$
$$4 = n$$

Since n is 4, this means that 32 grams will be present on the day which corresponds to the 4th term of the sequence. That would be on day 3. To find what is left after the tenth day, we need to find a_{11} since we started with a_1.

$$a_{11} = a_1 r^{11-1} = 256\left(\frac{1}{2}\right)^{10} = \frac{256}{1024} = \frac{1}{4}$$

There will be $\frac{1}{4}$ gram of the substance after 10 days.

51. **(a)** Here, $a_1 = 1.1$ billion and $r = 106\% = 1.06$. Since we must find the consumption after 5 years, $n = 6$ (since a_1 is the starting point).

$$a_6 = a_1 r^{n-1}$$
$$a_6 = 1.1(1.06)^{6-1}$$
$$= 1.1(1.06)^5 \approx 1.5$$

The community will use about 1.5 billion units 5 years from now.

(b) If consumption doubles, then the consumption would be $2a_1$.

$$2a_1 = a_1(1.06)^{n-1}$$
$$2 = (1.06)^{n-1}$$
$$\ln 2 = \ln(1.06)^{n-1}$$
$$\ln 2 = (n-1)\ln(1.06)$$
$$n - 1 = \frac{\ln 2}{\ln 1.06}$$
$$n - 1 \approx 12$$
$$n \approx 13$$

Since n is about 13, that would represent the 13th term of the sequence, which represents about 12 years after the start.

53. Since the machine depreciates by $\frac{1}{4}$ of its value, it retains $1 - \frac{1}{4} = \frac{3}{4}$ of its value. Since the cost of the machine new is $50,000$, $a_1 = 50,000$. We want the value after 8 years so since the original cost is a_1, we need to find a_9.

$$a_9 = a_1 r^{9-1}$$
$$= 50,000\left(\frac{3}{4}\right)^8 \approx 5006$$

The machine's value after 8 years is about $5000.

55. $g(x) = ab^x$
$(1) = ab^1 = ab$, $g(2) = ab^2$, $g(3) = ab^3$

56. The sequence $g(1), g(2), g(3)$ is a geometric sequence because each term after the first is a constant multiple of the preceding term.

57. The common ratio is $r = \frac{ab^2}{ab} = b$.

58. From Exercise 55, $a_1 = ab$. From Exercise 57, $r = b$. Therefore,

$$a_n = a_1 r^{n-1}$$
$$= ab(b)^{n-1}$$
$$= ab^{1+n-1}$$
$$= ab^n.$$

Section 11.4

1. $2! = 2 \cdot 1 = 2$

3. $\frac{6!}{4!\,2!} = \frac{6 \cdot 5 \cdot 4 \cdot 3 \cdot 2 \cdot 1}{(4 \cdot 3 \cdot 2 \cdot 1)(2 \cdot 1)} = \frac{6 \cdot 5}{2 \cdot 1} = 15$

5. $_6C_2 = \dfrac{6!}{2!\,(6-2)!} = \dfrac{6!}{2!\,4!} = 15,$

by Exercise 3.

7. $\dfrac{4!}{0!\,4!} = \dfrac{4!}{(1)(4!)} = \dfrac{1}{1} = 1$

9. $5! + 2! = 5\cdot4\cdot3\cdot2\cdot1 + 2\cdot1$

$\qquad = 120 + 2 = 122$

11. $10 \text{ nCr } 3 = \dfrac{10!}{3!\,(10-3)!}$

$\qquad = \dfrac{10!}{3!\,7!}$

$\qquad = \dfrac{10\cdot9\cdot8\cdot7\cdot6\cdot5\cdot4\cdot3\cdot2\cdot1}{(3\cdot2\cdot1)(7\cdot6\cdot5\cdot4\cdot3\cdot2\cdot1)}$

$\qquad = \dfrac{10\cdot9\cdot8}{3\cdot2\cdot1} = 120$

13. $(m+n)^4$

$\qquad = m^4 + \dfrac{4!}{3!\,1!}m^3n^1 + \dfrac{4!}{2!\,2!}m^2n^2$

$\qquad\quad + \dfrac{4!}{1!\,3!}m^1n^3 + n^4$

$\qquad = m^4 + 4m^3n + 6m^2n^2 + 4mn^3 + n^4$

15. $(a-b)^5$

$\qquad = \left[a + (-b)\right]^5$

$\qquad = a^5 + \dfrac{5!}{4!\,1!}a^4(-b)^1 + \dfrac{5!}{3!\,2!}a^3(-b)^2$

$\qquad\quad + \dfrac{5!}{2!\,3!}a^2(-b)^3 + \dfrac{5!}{1!\,4!}a^1(-b)^4 + (-b)^5$

$\qquad = a^5 - 5a^4b + 10a^3b^2 - 10a^2b^3 + 5ab^4 - b^5$

17. $(2x+3)^3$

$\qquad = (2x)^3 + \dfrac{3!}{2!\,1!}(2x)^2(3)^1 + \dfrac{3!}{1!\,2!}(2x)(3)^2$

$\qquad\quad + (3)^3$

$\qquad = 8x^3 + 36x^2 + 54x + 27$

19. $\left(\dfrac{x}{3} + 2y\right)^5$

$\qquad = \left(\dfrac{x}{3}\right)^5 + \dfrac{5!}{4!\,1!}\left(\dfrac{x}{3}\right)^4(2y)^1$

$\qquad\quad + \dfrac{5!}{3!\,2!}\left(\dfrac{x}{3}\right)^3(2y)^2 + \dfrac{5!}{2!\,3!}\left(\dfrac{x}{3}\right)^2(2y)^3$

$\qquad\quad + \dfrac{5!}{1!\,4!}\left(\dfrac{x}{3}\right)^1(2y)^4 + (2y)^5$

$\qquad = \dfrac{x^5}{243} + \dfrac{10x^4y}{81} + \dfrac{40x^3y^2}{27} + \dfrac{80x^2y^3}{9}$

$\qquad\quad + \dfrac{80xy^4}{3} + 32y^5$

21. $(mx - n^2)^3$

$\qquad = \left[mx + (-n^2)\right]^3$

$\qquad = (mx)^3 + \dfrac{3!}{2!\,1!}(mx)^2(-n^2)^1$

$\qquad\quad + \dfrac{3!}{1!\,2!}(mx)^1(-n^2)^2 + (-n^2)^3$

$\qquad = m^3x^3 - 3m^2n^2x^2 + 3mn^4x - n^6$

23. $(r + 2s)^{12}$

$\qquad = r^{12} + \dfrac{12!}{11!\,1!}r^{11}(2s)^1 + \dfrac{12!}{10!\,2!}r^{10}(2s)^2$

$\qquad\quad + \dfrac{12!}{9!\,3!}r^9(2s)^3 + \cdots$

The first four terms are

$r^{12} + 24r^{11}s + 264r^{10}s^2 + 1760r^9s^3.$

25. $(3x - y)^{14}$

$\qquad = \left[3x + (-y)\right]^{14}$

$\qquad = (3x)^{14} + \dfrac{14!}{13!\,1!}(3x)^{13}(-y)^1$

$\qquad\quad + \dfrac{14!}{12!\,2!}(3x)^{12}(-y)^2$

$\qquad\quad + \dfrac{14!}{11!\,3!}(3x)^{11}(-y)^3 + \cdots$

The first four terms are

$3^{14}x^{14} - 14(3^{13})x^{13}y + 91(3^{12})x^{12}y^2$

$\quad - 364(3^{11})x^{11}y^3.$

27. $(t^2 + u^2)^{10}$

$\qquad = (t^2)^{10} + \dfrac{10!}{9!\,1!}(t^2)^9(u^2)^1$

$\qquad\quad + \dfrac{10!}{8!\,2!}(t^2)^8(u^2)^2$

$\qquad\quad + \dfrac{10!}{7!\,3!}(t^2)^7(u^2)^3 + \cdots$

The first four terms are

$t^{20} + 10t^{18}u^2 + 45t^{16}u^4 + 120t^{14}u^6.$

29. The rth term of the expansion of $(x+y)^n$ is

$$\dfrac{n!}{[n - (r-1)]!\,(r-1)!}(x)^{n-(r-1)}(y)^{r-1}.$$

Start with the exponent on y, which is 1 less than the term number r. In this case, we are looking for the fourth term, so $r = 4$ and $r - 1 = 3$. Thus, the fourth term of $(2m + n)^{10}$ is

$$\dfrac{10!}{(10-3)!\,3!}(2m)^{10-3}(n)^3$$

$$= \dfrac{10!}{7!\,3!}2^7m^7n^3$$

$$= 120(2^7)m^7n^3.$$

31. The seventh term of $\left(x + \dfrac{y}{2}\right)^8$ is

$$\frac{8!}{(8-6)!\,6!}(x)^{8-6}\left(\frac{y}{2}\right)^6$$

$$= \frac{8!}{6!\,2!}x^2\frac{y^6}{2^6}$$

$$= \frac{7x^2y^6}{16}.$$

33. The third term of $(k-1)^9$ is

$$\frac{9!}{(9-2)!\,2!}k^{9-2}(-1)^2 = \frac{9!}{7!\,2!}k^7 = 36k^7.$$

35. The expansion of $\left(x^2 - 2y\right)^6$ has seven terms, so the middle term is the fourth. The fourth term of $\left(x^2 - 2y\right)^6$ is

$$\frac{6!}{(6-3)!\,3!}(x^2)^{6-3}(-2y)^3$$

$$= \frac{6!}{3!\,3!}(x^2)^3(-8y^3)$$

$$= 20x^6(-8y^3) = -160x^6y^3.$$

37. The term of the expansion of $\left(3x^3 - 4y^2\right)^5$ with x^9y^4 in it is the term with $\left(3x^3\right)^3\left(-4y^2\right)^2$, since $\left(x^3\right)^3\left(y^2\right)^2 = x^9y^4$. The term is

$$\frac{5!}{3!\,2!}(3x^3)^3(-4y^2)^2$$

$$= 10(27x^9)(16y^4) = 4320x^9y^4.$$

Chapter 11 Review Exercises

1. $a_n = 2n - 3$

$a_1 = 2(1) - 3 = -1$

$a_2 = 2(2) - 3 = 1$

$a_3 = 2(3) - 3 = 3$

$a_4 = 2(4) - 3 = 5$

Answer: $-1, 1, 3, 5$

2. $a_n = \dfrac{n-1}{n}$

$a_1 = \dfrac{1-1}{1} = 0$

$a_2 = \dfrac{2-1}{2} = \dfrac{1}{2}$

$a_3 = \dfrac{3-1}{3} = \dfrac{2}{3}$

$a_4 = \dfrac{4-1}{4} = \dfrac{3}{4}$

Answer: $0, \dfrac{1}{2}, \dfrac{2}{3}, \dfrac{3}{4}$

3. $a_n = n^2$

$a_1 = (1)^2 = 1$

$a_2 = (2)^2 = 4$

$a_3 = (3)^2 = 9$

$a_4 = (4)^2 = 16$

Answer: $1, 4, 9, 16$

4. $a_n = \left(\dfrac{1}{2}\right)^n$

$a_1 = \left(\dfrac{1}{2}\right)^1 = \dfrac{1}{2}$

$a_2 = \left(\dfrac{1}{2}\right)^2 = \dfrac{1}{4}$

$a_3 = \left(\dfrac{1}{2}\right)^3 = \dfrac{1}{8}$

$a_4 = \left(\dfrac{1}{2}\right)^4 = \dfrac{1}{16}$

Answer: $\dfrac{1}{2}, \dfrac{1}{4}, \dfrac{1}{8}, \dfrac{1}{16}$

5. $a_n = (n+1)(n-1)$

$a_1 = (1+1)(1-1) = 2(0) = 0$

$a_2 = (2+1)(2-1) = 3(1) = 3$

$a_3 = (3+1)(3-1) = 4(2) = 8$

$a_4 = (4+1)(4-1) = 5(3) = 15$

Answer: $0, 3, 8, 15$

6. $\displaystyle\sum_{i=1}^{5} i^2 x$

$= 1^2x + 2^2x + 3^2x + 4^2x + 5^2x$

$= x + 4x + 9x + 16x + 25x$

7. $\displaystyle\sum_{i=1}^{6} (i+1)x^i$

$= (1+1)x^1 + (2+1)x^2 + (3+1)x^3$

$\quad + (4+1)x^4 + (5+1)x^5 + (6+1)x^6$

$= 2x + 3x^2 + 4x^3 + 5x^4 + 6x^5 + 7x^6$

8. $\displaystyle\sum_{i=1}^{4} (i+2)$

$= (1+2) + (2+2) + (3+2) + (4+2)$

$= 3 + 4 + 5 + 6 = 18$

9. $\displaystyle\sum_{i=1}^{6} 2^i$

$= 2^1 + 2^2 + 2^3 + 2^4 + 2^5 + 2^6$

$= 2 + 4 + 8 + 16 + 32 + 64 = 126$

10. $\displaystyle\sum_{i=4}^{7} \frac{i}{i+1}$

$$= \frac{4}{4+1} + \frac{5}{5+1} + \frac{6}{6+1} + \frac{7}{7+1}$$

$$= \frac{4}{5} + \frac{5}{6} + \frac{6}{7} + \frac{7}{8} \qquad LCD = 2^3 \cdot 3 \cdot 5 \cdot 7 = 840$$

$$= \frac{672}{840} + \frac{700}{840} + \frac{720}{840} + \frac{735}{840} = \frac{2827}{840}$$

11. $\overline{x} = \dfrac{\text{total}}{5}$, where total $= 684,588 + 680,913$
$+\, 654,110 + 652,945 + 652,848.$

$$\overline{x} = \frac{3,325,404}{5} = 665,080.8$$

The average share volume is about $665,081$ (thousand) for the five given days.

12. $2, 5, 8, 11, \ldots$ is an *arithmetic* sequence with $d = 5 - 2 = 3$.

13. $-6, -2, 2, 6, 10, \ldots$ is an *arithmetic* sequence with $d = -2 - (-6) = 4$.

14. $\dfrac{2}{3}, -\dfrac{1}{3}, \dfrac{1}{6}, -\dfrac{1}{12}, \ldots$ is a *geometric* sequence with

$$r = \frac{-\dfrac{1}{3}}{\dfrac{2}{3}} = -\frac{1}{3} \cdot \frac{3}{2} = -\frac{1}{2}.$$

15. $-1, 1, -1, 1, -1, \ldots$ is a *geometric* sequence
with $r = \dfrac{1}{-1} = -1$.

16. $64, 32, 8, \dfrac{1}{2}, \ldots$

$32 - 64 = -32$ and $8 - 32 = -24$, so the sequence is not arithmetic.

$\dfrac{32}{64} \neq \dfrac{8}{32}$, so the sequence is not geometric.

Therefore, the sequence is *neither*.

17. $64, 32, 16, 8, \ldots$ is a *geometric* sequence with

$$r = \frac{32}{64} = \frac{1}{2}.$$

18. $10, 8, 6, 4, \ldots$ is an *arithmetic* sequence with $d = 8 - 10 = -2$.

19. Given $a_1 = -2$, $d = 5$; find a_{16}.
$$a_n = a_1 + (n-1)d$$
$$a_{16} = -2 + (16-1)5$$
$$= -2 + 15(5)$$
$$= -2 + 75 = 73$$

20. Given $a_6 = 12$, $a_8 = 18$; find a_{25}.
$$a_n = a_1 + (n-1)d$$
$a_6 = a_1 + 5d$, so $\quad 12 = a_1 + 5d \quad (1)$
$a_8 = a_1 + 7d$, so $\quad 18 = a_1 + 7d \quad (2)$

Multiply equation (1) by -1 and add the result to equation (2).

$$\begin{array}{ll} -12 = -a_1 - 5d & -1 \times (1) \\ \underline{18 = a_1 + 7d} & (2) \\ 6 = 2d \end{array}$$

$$3 = d$$

To find a_1, substitute $d = 3$ in equation (1).
$$12 = a_1 + 5d \quad (1)$$
$$12 = a_1 + 5(3)$$
$$12 = a_1 + 15$$
$$-3 = a_1$$

Use $a_1 = -3$ and $d = 3$ to find a_{25}.
$$a_n = a_1 + (n-1)d$$
$$a_{25} = -3 + (25-1)3$$
$$= -3 + 24(3)$$
$$= -3 + 72 = 69$$

21. $a_1 = -4$, $d = -5$
$$a_n = a_1 + (n-1)d$$
$$a_n = -4 + (n-1)(-5)$$
$$= -4 - 5n + 5$$
$$= -5n + 1$$

22. $6, 3, 0, -3, \ldots$
To get the general term, a_n, find d.
$$d = 3 - 6 = -3$$
$$a_n = a_1 + (n-1)d$$
$$a_n = 6 + (n-1)(-3)$$
$$a_n = 6 - 3n + 3$$
$$a_n = -3n + 9$$

23. $7, 0, 13, \ldots, 49$
$a_1 = 7$, $d = 10 - 7 = 3$; find n, the number of terms.
$$a_n = a_1 + (n-1)d$$
$$49 = 7 + (n-1)(3)$$
$$42 = 3(n-1)$$
Divide by 3.
$$14 = n - 1$$
$$15 = n$$
There are 15 terms in this sequence.

24. $5, 1, -3, \ldots, -79$
$a_1 = 5$, $d = 1 - 5 = -4$; find n, the number of terms.
$$a_n = a_1 + (n-1)d$$
$$-79 = 5 + (n-1)(-4)$$
$$-79 = 9 - 4n$$
$$-88 = -4n$$
$$n = 22$$
There are 22 terms in this sequence.

25. Find S_8 if $a_1 = -2$ and $d = 6$.

$S_n = \dfrac{n}{2}(a_1 + a_n)$; find a_8 first.

$$a_8 = a_1 + (8-1)d$$
$$= -2 + 7(6)$$
$$= -2 + 42 = 40$$
$$S_8 = \dfrac{8}{2}(-2 + 40)$$
$$= 4(38) = 152$$

26. Find S_8 if $a_n = -2 + 5n$.

$$a_1 = -2 + 5(1) = 3$$
$$a_8 = -2 + 5(8) = 38$$
$$S_n = \dfrac{n}{2}(a_1 + a_n)$$
$$S_8 = \dfrac{8}{2}(3 + 38) = 4(41) = 164$$

27. Find the general term for the geometric sequence $-1, -4, \dots$.

$a_1 = -1$ and $r = \dfrac{-4}{-1} = 4$.

$$a_n = a_1 r^{n-1}$$
$$a_n = -1(4)^{n-1}$$

28. $\dfrac{2}{3}, \dfrac{2}{15}, \dots$

$a_1 = \dfrac{2}{3}$ and $r = \dfrac{\frac{2}{15}}{\frac{2}{3}} = \dfrac{2}{15} \cdot \dfrac{3}{2} = \dfrac{1}{5}$.

$$a_n = a_1 r^{n-1}$$
$$a_n = \dfrac{2}{3}\left(\dfrac{1}{5}\right)^{n-1}$$

29. Find a_{11} for $2, -6, 18, \dots$.

$a_1 = 2$ and $r = \dfrac{-6}{2} = -3$

$$a_n = a_1 r^{n-1}$$
$$a_{11} = 2(-3)^{11-1}$$
$$= 2(-3)^{10} = 118,098$$

30. Given $a_3 = 20$, $a_5 = 80$; find a_{10}.

$a_n = a_1 r^{n-1}$

For a_3, $\quad a_3 = a_1 r^{3-1}$
$$20 = a_1 r^2.$$

For a_5, $\quad a_5 = a_1 r^{5-1}$
$$80 = a_1 r^4.$$

The ratio of a_5 to a_3 is

$$\dfrac{80}{20} = \dfrac{a_1 r^4}{a_1 r^2}$$
$$4 = r^2$$
$$r = \pm 2.$$

Since $20 = a_1 r^2$, and $r^2 = 4$,
$$20 = a_1(4)$$
$$5 = a_1.$$

Now find a_{10}.

$$a_n = a_1 r^{n-1}$$
$$a_{10} = 5(\pm 2)^{10-1}$$
$$= 5(\pm 2)^9$$

Two answers are possible for a_{10}:

$5(2)^9 = 2560$ or $5(-2)^9 = -2560$.

31. $\displaystyle\sum_{i=1}^{5}\left(\dfrac{1}{4}\right)^i$

$a_1 = \dfrac{1}{4}, r = \dfrac{1}{4}, n = 5$.

$$S_n = \dfrac{a_1(1 - r^n)}{1 - r}$$

$$S_5 = \dfrac{\dfrac{1}{4}\left[1 - \left(\dfrac{1}{4}\right)^5\right]}{1 - \dfrac{1}{4}}$$

$$= \dfrac{\dfrac{1}{4}\left(1 - \dfrac{1}{1024}\right)}{\dfrac{3}{4}}$$

$$= \dfrac{1}{3}\left(\dfrac{1023}{1024}\right) = \dfrac{341}{1024}$$

32. $\displaystyle\sum_{i=1}^{8}\dfrac{3}{4}(-1)^i$

$a_1 = -\dfrac{3}{4}$ and $r = -1$, and $n = 8$.

$$S_n = \dfrac{a_1(1 - r^n)}{1 - r}$$

$$S_8 = \dfrac{-\dfrac{3}{4}\left[1 - (-1)^8\right]}{1 - (-1)}$$

$$= \dfrac{-\dfrac{3}{4}(1 - 1)}{2}$$

$$= \dfrac{-\dfrac{3}{4}(0)}{2} = 0$$

33. $\displaystyle\sum_{i=1}^{\infty}4\left(\dfrac{1}{5}\right)^i$

The terms are the infinite geometric sequence with $a_1 = \dfrac{4}{5}$ and $r = \dfrac{1}{5}$.

$$S = \dfrac{a_1}{1 - r} = \dfrac{\dfrac{4}{5}}{1 - \dfrac{1}{5}} = \dfrac{\dfrac{4}{5}}{\dfrac{4}{5}} = 1$$

34. $\displaystyle\sum_{i=1}^{\infty} 2(3)^i$

The terms are the infinite geometric sequence with $a_1 = 6$ and $r = 3$.

$S = \dfrac{a_1}{1-r}$ if $|r| < 1$,

but $r = 3$ so S does not exist, and, thus, the sum does not exist.

35. $(2p - q)^5$

$= \left[2p + (-q)\right]^5$

$= (2p)^5 + \dfrac{5!}{4!\,1!}(2p)^4(-q)^1$

$\quad + \dfrac{5!}{3!\,2!}(2p)^3(-q)^2 + \dfrac{5!}{2!\,3!}(2p)^2(-q)^3$

$\quad + \dfrac{5!}{1!\,4!}(2p)^1(-q)^4 + (-q)^5$

$= 32p^5 + 5(16p^4)(-q) + 10(8p^3)q^2$

$\quad + 10(4p^2)(-q^3) + 5(2p)q^4 - q^5$

$= 32p^5 - 80p^4q + 80p^3q^2 - 40p^2q^3$

$\quad + 10pq^4 - q^5$

36. $(x^2 + 3y)^4$

$= (x^2)^4 + \dfrac{4!}{3!\,1!}(x^2)^3(3y)^1$

$\quad + \dfrac{4!}{2!\,2!}(x^2)^2(3y)^2$

$\quad + \dfrac{4!}{1!\,3!}(x^2)^1(3y)^3 + (3y)^4$

$= x^8 + 4(x^6)(3y) + 6(x^4)(9y^2)$

$\quad + 4(x^2)(27y^3) + 81y^4$

$= x^8 + 12x^6y + 54x^4y^2 + 108x^2y^3$

$\quad + 81y^4$

37. $\left(\sqrt{m} + \sqrt{n}\right)^4$

$= \left(\sqrt{m}\right)^4 + \dfrac{4!}{3!\,1!}\left(\sqrt{m}\right)^3\left(\sqrt{n}\right)^1$

$\quad + \dfrac{4!}{2!\,2!}\left(\sqrt{m}\right)^2\left(\sqrt{n}\right)^2$

$\quad + \dfrac{4!}{1!\,3!}\left(\sqrt{m}\right)^1\left(\sqrt{n}\right)^3 + \left(\sqrt{n}\right)^4$

$= m^2 + 4(m\sqrt{m})(\sqrt{n}) + 6(m)(n)$

$\quad + 4(\sqrt{m})(n\sqrt{n}) + n^2$

$= m^2 + 4m\sqrt{mn} + 6mn + 4n\sqrt{mn} + n^2$

38. The fourth term ($r = 4$, so $r - 1 = 3$) of $(3a + 2b)^{19}$ is

$\dfrac{19!}{16!\,3!}(3a)^{16}(2b)^3$

$= 969(3)^{16}(a^{16})(8)b^3$

$= 7752(3)^{16}a^{16}b^3.$

39. The twenty-third term ($r = 23$, so $r - 1 = 22$) of $(-2k + 3)^{25}$ is

$= \dfrac{25!}{3!\,22!}(-2k)^3(3)^{22}$

$= -18,400(3)^{22}k^3.$

40. The arithmetic sequence $1, 7, 13, \ldots$ has $a_1 = 1$ and $d = 7 - 1 = 6$.

$a_n = a_1 + (n - 1)d$

$a_{40} = 1 + (40 - 1)6$

$\quad = 1 + (39)6$

$\quad = 1 + 234 = 235$

$a_{10} = 1 + (10 - 1)6$

$\quad = 1 + (9)6 = 55$

$S_n = \dfrac{n}{2}(a_1 + a_n)$

$S_{10} = \dfrac{10}{2}(1 + 55)$

$\quad = 5(56) = 280$

41. The geometric sequence $-3, 6, -12, \ldots$ has

$a_1 = -3$ and $r = \dfrac{6}{-3} = -2$.

$a_n = a_1 r^{n-1}$

$a_{10} = -3(-2)^9$

$\quad = -3(-512) = 1536$

$S_n = \dfrac{a_1(1 - r^n)}{1 - r}$

$S_{10} = \dfrac{-3\left[1 - (-2)^{10}\right]}{1 - (-2)}$

$\quad = \dfrac{-3(1 - 1024)}{3}$

$\quad = -(-1023) = 1023$

42. $a_1 = 1, r = -3$

$a_n = a_1 r^{n-1}$

$a_9 = 1(-3)^{9-1}$

$\quad = (-3)^8 = 6561$

$S_n = \dfrac{a_1(r^n - 1)}{r - 1}$

$S_{10} = \dfrac{1\left[(-3)^{10} - 1\right]}{-3 - 1}$

$\quad = -\dfrac{1}{4}(3^{10} - 1)$

$\quad = -\dfrac{1}{4}(59,049 - 1)$

$\quad = -14,762$

43. The arithmetic sequence with $a_1 = -4$ and $d = 3$ is $-4, -1, 2, 5, \ldots$.

$a_n = a_1 + (n - 1)d$

$a_{15} = -4 + (15 - 1)3$

$\quad = -4 + 42 = 38$

$$S_n = \frac{n}{2}\big[2a_1 + (n-1)d\big]$$

$$S_{10} = \frac{10}{2}\big[2(-4) + (10-1)3\big]$$

$$= 5(-8+27)$$

$$= 5(19) = 95$$

44. $2, 7, 12, \ldots$

This is an arithmetic sequence with $a_1 = 2$ and $d = 7 - 2 = 5$.

$$a_n = a_1 + (n-1)d$$
$$a_n = 2 + (n-1)5$$
$$= 2 + 5n - 5$$
$$= 5n - 3$$

45. $2, 8, 32, \ldots$

This is a geometric sequence with $a_1 = 2$ and

$$r = \frac{8}{2} = 4.$$

$$a_n = a_1 r^{n-1}$$
$$a_n = 2(4)^{n-1}$$

46. $27, 9, 3, \ldots$

This is a geometric sequence with $a_1 = 27$ and

$$r = \frac{9}{27} = \frac{1}{3}.$$

$$a_n = a_1 r^{n-1}$$
$$a_n = 27\left(\frac{1}{3}\right)^{n-1}$$

47. $12, 9, 6, \ldots$

This is an arithmetic sequence with $a_1 = 12$ and $d = -3$.

$$a_n = a_1 + (n-1)d$$
$$a_n = 12 + (n-1)(-3)$$
$$= 12 - 3n + 3$$
$$= -3n + 15$$

48. The distances traveled in successive seconds are $3, 7, 11, 15, 19, \ldots$.

This is an arithmetic sequence.

$$S_n = \frac{n}{2}\big[2a_1 + (n-1)d\big]$$

$$S_n = 210, \quad a_1 = 3, \quad d = 4$$

$$210 = \frac{n}{2}\big[2(3) + (n-1)4\big]$$

$$420 = n(6 + 4n - 4)$$

$$420 = 6n + 4n^2 - 4n$$

$$0 = 4n^2 + 2n - 420$$

$$0 = 2n^2 + n - 210$$

$$0 = (2n + 21)(n - 10)$$

$$2n + 21 = 0 \quad \text{or} \quad n - 10 = 0$$

$$n = -\frac{21}{2} \quad \text{or} \quad n = 10$$

Discard $-\frac{21}{2}$ since time cannot be negative. It takes her 10 seconds.

49. $S = R\left[\dfrac{(1+i)^n - 1}{i}\right]$

$$S = 672\left[\dfrac{\left(1 + \dfrac{.08}{4}\right)^{4(7)} - 1}{\dfrac{.08}{4}}\right]$$

$$= 672\left[\dfrac{(1.02)^{28} - 1}{.02}\right]$$

$$= 24,898.41$$

The future value of the annuity is \$24,898.41.

50. Since $100\% - 3\% = 97\% = .97$, the population after 1 year is $.97(50,000)$, after 2 years is $.97\big[.97(50,000)\big]$ or $(.97)^2(50,000)$, and after n years is $(.97)^n(50,000)$. After 6 years, the population is

$$(.97)^6(50,000) \approx 41,649 \approx 42,000.$$

51. $\left(\dfrac{1}{2}\right)^n$ is left after n strokes. So

$$\left(\frac{1}{2}\right)^7 = \frac{1}{128} = .0078125 \text{ is left after 7 strokes.}$$

52. (a) We can write the repeating decimal number .55555... as an infinite geometric sequence as follows:

$$\frac{5}{10} + \frac{5}{10}\left(\frac{1}{10}\right) + \frac{5}{10}\left(\frac{1}{10}\right)^2 + \frac{5}{10}\left(\frac{1}{10}\right)^3 + \cdots$$

(b) The common ratio r is

$$\frac{\dfrac{5}{10}\left(\dfrac{1}{10}\right)}{\dfrac{5}{10}} = \frac{1}{10}.$$

(c) Since $|r| < 1$, the sum exists.

$$S = \frac{a_1}{1 - r} = \frac{\dfrac{5}{10}}{1 - \dfrac{1}{10}} = \frac{\dfrac{5}{10}}{\dfrac{9}{10}} = \frac{5}{9}$$

53. No, the sum cannot be found, because $r = 2$, and this value of r does not satisfy $|r| < 1$.

54. No, the terms must be successive, such as the first and second, or the second and the third.

Chapter 11 Test

1. $a_n = (-1)^n + 1$

$$a_1 = (-1)^1 + 1 = 0$$
$$a_2 = (-1)^2 + 1 = 1 + 1 = 2$$
$$a_3 = (-1)^3 + 1 = -1 + 1 = 0$$
$$a_4 = (-1)^4 + 1 = 1 + 1 = 2$$
$$a_5 = (-1)^5 + 1 = -1 + 1 = 0$$

Answer: 0, 2, 0, 2, 0

2. $a_1 = 4$, $d = 2$

$a_1 = 4$

$a_2 = a_1 + d = 4 + 2 = 6$

$a_3 = 6 + 2 = 8$

$a_4 = 8 + 2 = 10$

$a_5 = 10 + 2 = 12$

Answer: 4, 6, 8, 10, 12

3. $a_4 = 6$, $r = \dfrac{1}{2}$

$a_n = a_1 r^{n-1}$

$a_4 = a_1 \left(\dfrac{1}{2}\right)^{4-1}$

$6 = a_1 \left(\dfrac{1}{8}\right)$

$a_1 = 48$

$a_2 = \dfrac{1}{2}(48) = 24$

$a_3 = \dfrac{1}{2}(24) = 12$

$a_4 = \dfrac{1}{2}(12) = 6$

$a_5 = \dfrac{1}{2}(6) = 3$

Answer: 48, 24, 12, 6, 3

4. Given $a_1 = 6$ and $d = -2$; find a_4.

$a_n = a_1 + (n-1)d$

$a_4 = a_1 + (4-1)d$

$= 6 + (3)(-2)$

$= 6 - 6 = 0$

5. Given $a_5 = 16$ and $a_7 = 9$; find a_4.

This is a geometric sequence, so $a_6 = a_5 r$ and

$a_7 = a_6 r$. Thus, $a_6 = \dfrac{a_7}{r}$ and

$a_5 r = \dfrac{a_7}{r}$

$r^2 = \dfrac{a_7}{a_5}$

$r^2 = \dfrac{9}{16}$

$r = \dfrac{3}{4}$ or $r = -\dfrac{3}{4}$.

$a_5 = a_4 r$ so $a_4 = \dfrac{a_5}{r}$.

$a_4 = \dfrac{16}{\dfrac{3}{4}}$ or $\dfrac{16}{-\dfrac{3}{4}}$

$a_4 = \dfrac{64}{3}$ or $a_4 = -\dfrac{64}{3}$

6. Given the arithmetic sequence with $a_2 = 12$ and $a_3 = 15$; find S_5.

$d = a_3 - a_2 = 15 - 12 = 3$

$a_1 = a_2 - 3 = 12 - 3 = 9$

$a_5 = a_4 + 3 = a_3 + 6 = 15 + 6 = 21$

$S_n = \dfrac{n}{2}(a_1 + a_n)$

$S_5 = \dfrac{5}{2}(9 + 21)$

$= \dfrac{5}{2}(30) = 75$

7. Given the geometric sequence with $a_5 = 4$ and $a_7 = 1$; find S_5.

$r^2 = \dfrac{a_7}{a_5} = \dfrac{1}{4}$, so $r = \dfrac{1}{2}$ or $r = -\dfrac{1}{2}$.

Use $a_7 = a_1 r^6$ to get $1 = a_1 \left(\dfrac{1}{2}\right)^6$, and so

$a_1 = 64$.

$S_n = \dfrac{a_1(r^n - 1)}{r - 1}$

$S_5 = \dfrac{64\left[1 - \left(\frac{1}{2}\right)^5\right]}{1 - \dfrac{1}{2}}$ or $S_5 = \dfrac{64\left[1 - \left(-\frac{1}{2}\right)^5\right]}{1 - \left(-\dfrac{1}{2}\right)}$

$= 128\left(1 - \dfrac{1}{32}\right)$ $= \dfrac{128}{3}\left(1 + \dfrac{1}{32}\right)$

$= 128\left(\dfrac{31}{32}\right)$ $= \dfrac{128}{3}\left(\dfrac{33}{32}\right)$

$= 124$ $= 44$

8. $\bar{x} = \dfrac{271.9 + 199.8 + 198.5 + 183.5 + 179.1}{5}$

$= \dfrac{1032.8}{5} = 206.56$

The average share volume for the five given stocks was about 206.6 million.

9. $S = R\left[\dfrac{(1+i)^n - 1}{i}\right]$

$= 4000\left[\dfrac{\left(1 + \dfrac{.06}{4}\right)^{4(7)} - 1}{\dfrac{.06}{4}}\right]$

$= 4000\left[\dfrac{(1.015)^{28} - 1}{.015}\right]$

$\approx 137{,}925.91$

The account will have $137,925.91 at the end of this term.

10. An infinite geometric series has a sum if $|r| < 1$, where r is the common ratio.

11. $\displaystyle\sum_{i=1}^{5}(2i+8)$

$\quad = [2(1)+8] + [2(2)+8] + [2(3)+8]$
$\quad\quad + [2(4)+8] + [2(5)+8]$
$\quad = 10 + 12 + 14 + 16 + 18$
$\quad = 70$

12. $\displaystyle\sum_{i=1}^{6}(3i-5)$

Use the formula $S_6 = \dfrac{n}{2}(a_1 + a_6)$.

$a_1 = 3(1) - 5 = -2$
$a_6 = 3(6) - 5 = 13$
$S_6 = \dfrac{6}{2}(-2 + 13) = 3(11) = 33$

13. $\displaystyle\sum_{i=1}^{500} i$

Use the formula $S_{500} = \dfrac{n}{2}(a_1 + a_{500})$ with $a_1 = 1$
and $a_{500} = 500$.

$S_{500} = \dfrac{500}{2}(1 + 500)$
$\quad = 250(501) = 125,250$

14. $\displaystyle\sum_{i=1}^{3}\frac{1}{2}\left(4^i\right) = \frac{1}{2}\sum_{i=1}^{3}\left(4^i\right)$

$\quad = \dfrac{1}{2}\Big[4 + 16 + 64\Big] = 42$

15. $\displaystyle\sum_{i=1}^{\infty}\left(\frac{1}{4}\right)^i$

This is an infinite geometric series with

$a_1 = \dfrac{1}{4}$ and $r = \dfrac{1}{4}$.

Since $|r| = \dfrac{1}{4} < 1$, the sum is

$S = \dfrac{a_1}{1-r} = \dfrac{\frac{1}{4}}{1 - \frac{1}{4}} = \dfrac{\frac{1}{4}}{\frac{3}{4}}$

$\quad = \dfrac{1}{4}\cdot\dfrac{4}{3} = \dfrac{1}{3}.$

16. $\displaystyle\sum_{i=1}^{\infty}6\left(\frac{3}{2}\right)^i$

This is an infinite geometric series with

$a_1 = 6\left(\dfrac{3}{2}\right)^1 = 9$ and $r = \dfrac{3}{2}$.

Since $|r| = \dfrac{3}{2} > 1$, the sum does not exist.

17. The amounts of unpaid balance during 15 months
form an arithmetic sequence
$300, 280, 260, \ldots, 40, 20$
which is the sequence with $n = 15$, $a_1 = 300$, and
$a_{15} = 20$. Find the sum of these balances.

$S_n = \dfrac{n}{2}(a_1 + a_n)$

$S_{15} = \dfrac{15}{2}(300 + 20)$

$\quad = \dfrac{15}{2}(320) = 2400$

Since 1% interest is paid on this total, the interest
paid is 1% of $2400 or $24.

The sewing machine cost $300 (paid monthly at
$20), so the total cost is $300 + $24 = $324.

18. The weekly populations form a geometric
sequence with $a_1 = 20$ and $r = 3$ since the colony
begins with 20 insects and triples each week. Find
the general term of this geometric sequence.

$a_n = a_1 r^{n-1}$
$a_n = 20(3)^{n-1}$

We're assuming that from the beginning of July to
the end of September is 12 weeks, so find a_{12}.

$a_n = 20(3)^{n-1}$
$a_{12} = 20(3)^{11}$

At the end of September, $20(3)^{11} = 3,542,940$
insects will be present in the colony.

19. The fifth term ($r = 5$, so $r - 1 = 4$) of

$\left(2x - \dfrac{y}{3}\right)^{12}$ is

$\dfrac{12!}{(12-4)!\,4!}(2x)^{12-4}\left(-\dfrac{y}{3}\right)^4$

$\quad = \dfrac{12!}{8!\,4!}(2x)^8\left(-\dfrac{y}{3}\right)^4$

$\quad = \dfrac{14,080x^8y^4}{9}.$

20. $(3k-5)^4$

$\quad = (3k)^4 + \dfrac{4!}{3!\,1!}(3k)^3(-5)^1$

$\quad\quad + \dfrac{4!}{2!\,2!}(3k)^2(-5)^2 + \dfrac{4!}{1!\,3!}(3k)^1(-5)^3$

$\quad\quad + (-5)^4$

$\quad = 81k^4 - 4\left(27k^3\right)(5) + 6\left(9k^2\right)(25)$

$\quad\quad - 4(3k)(125) + 625$

$\quad = 81k^4 - 540k^3 + 1350k^2 - 1500k + 625$

Cumulative Review Exercises Chapters 1–11

In Exercises 1–4, let

$P = \left\{-\dfrac{8}{3}, 10, 0, \sqrt{13}, -\sqrt{3}, \dfrac{45}{15}, \sqrt{-7}, .82, -3\right\}$

1. The integers are $10, 0, \dfrac{45}{15}$(or 3), and -3.

2. The rational numbers are

$\quad -\dfrac{8}{3}, 10, 0, \dfrac{45}{15}, .82,$ and -3.

3. The irrational numbers are $\sqrt{13}$ and $-\sqrt{3}$.

4. All are real numbers except $\sqrt{-7}$.

5. $|-7| + 6 - |-10| - (-8 + 3)$
$= 7 + 6 - 10 - (-5)$
$= 13 - 10 + 5 = 8$

6. $-15 - |-4| - 10 - |-6|$
$= -15 - 4 - 10 - 6 = -35$

7. $4(-6) + (-8)(5) - (-9)$
$= -24 - 40 + 9 = -55$

8. $9 - (5 + 3a) + 5a = -4(a - 3) - 7$
$9 - 5 - 3a + 5a = -4a + 12 - 7$
$4 + 2a = -4a + 5$
$6a = 1$
$a = \dfrac{1}{6}$

Solution set: $\left\{\dfrac{1}{6}\right\}$

9. $7m + 18 \le 9m - 2$
$-2m \le -20$
Divide by -2; reverse the direction of the inequality symbol.
$m \ge 10$
Solution set: $[10, \infty)$

10. $|4x - 3| = 21$
$4x - 3 = 21$ or $4x - 3 = -21$
$4x = 24$ $4x = -18$
$x = 6$ or $x = -\dfrac{18}{4} = -\dfrac{9}{2}$
Solution set: $\left\{-\dfrac{9}{2}, 6\right\}$

11. $\dfrac{x + 3}{12} - \dfrac{x - 3}{6} = 0$
Multiply by the LCD, 12.
$12\left(\dfrac{x + 3}{12} - \dfrac{x - 3}{6}\right) = 12(0)$
$x + 3 - 2(x - 3) = 0$
$x + 3 - 2x + 6 = 0$
$9 - x = 0$
$9 = x$
Check $x = 9$: $1 - 1 \overset{?}{=} 0$ *True*
Solution set: $\left\{9\right\}$

12. $2x > 8$ or $-3x > 9$
$x > 4$ or $x < -3$
Solution set: $(-\infty, -3) \cup (4, \infty)$

13. $|2m - 5| \ge 11$
$2m - 5 \ge 11$ or $2m - 5 \le -11$
$2m \ge 16$ $2m \le -6$
$m \ge 8$ or $m \le -3$
Solution set: $(-\infty, -3] \cup [8, \infty)$

14. $f(x) = .07x + 135$
$f(2000) = .07(2000) + 135$
$= 140 + 135 = 275$
The weekly fee is $275.

15. Let $(x_1, y_1) = (4, -5)$ and $(x_2, y_2) = (-12, -17)$. Then
$m = \dfrac{y_2 - y_1}{x_2 - x_1} = \dfrac{-17 - (-5)}{-12 - 4} = \dfrac{-12}{-16} = \dfrac{3}{4}.$
The slope is $\dfrac{3}{4}$.

16. To find the equation of the line through $(-2, 10)$ and parallel to $3x + y = 7$, find the slope of
$3x + y = 7$
$y = -3x + 7.$
The slope is -3, so a line parallel to it also has slope -3. Use $m = -3$ and $(x_1, y_1) = (-2, 10)$ in the point-slope form.
$y - y_1 = m(x - x_1)$
$y - 10 = -3[x - (-2)]$
$y - 10 = -3(x + 2)$
Write in standard form.
$y - 10 = -3x - 6$
$3x + y = 4$
Alternative solution: The line must be of the form $3x + y = k$ since it is parallel to $3x + y = 7$. Substitute -2 for x and 10 for y to find k.
$3(-2) + 10 = k$
$4 = k$
The equation is $3x + y = 4$.

17. $x - 3y = 6$
Find the x- and y-intercepts. To find the x-intercept, let $y = 0$.
$x - 3(0) = 6$
$x = 6$
The x-intercept is $(6, 0)$.
To find the y-intercept, let $x = 0$.
$0 - 3y = 6$
$y = -2$
The y-intercept is $(0, -2)$.
Plot the intercepts and draw the line through them.

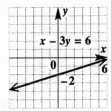

18. $4x - y < 4$

Graph the line $4x - y = 4$, which has intercepts $(0, -4)$ and $(1, 0)$, as a dashed line because the inequality involves $<$. Test $(0, 0)$, which yields $0 < 4$, a true statement. Shade the region on the side of the line that includes $(0, 0)$.

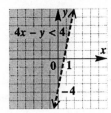

19.
$$2x + 5y = -19 \quad (1)$$
$$-3x + 2y = -19 \quad (2)$$

To eliminate x, multiply equation (1) by 3 and equation (2) by 2. Then add the results.

$$\begin{array}{rl} 6x + 15y = -57 & 3 \times (1) \\ \underline{-6x + 4y = -38} & 2 \times (2) \\ 19y = -95 & \\ y = -5 & \end{array}$$

Substitute -5 for y in equation (1) to find x.
$$2x + 5y = -19 \quad (1)$$
$$2x + 5(-5) = -19$$
$$2x - 25 = -19$$
$$2x = 6$$
$$x = 3$$
Solution set : $\left\{(3, -5)\right\}$

20.
$$\begin{array}{rl} x + 2y + z = 8 & (1) \\ 2x - y + 3z = 15 & (2) \\ -x + 3y - 3z = -11 & (3) \end{array}$$

To eliminate x, add equations (1) and (3).

$$\begin{array}{rl} x + 2y + z = 8 & (1) \\ \underline{-x + 3y - 3z = -11} & (3) \\ 5y - 2z = -3 & (4) \end{array}$$

To eliminate x again, multiply equation (3) by 2 and add the result to equation (2).

$$\begin{array}{rl} 2x - y + 3z = 15 & (2) \\ \underline{-2x + 6y - 6z = -22} & 2 \times (3) \\ 5y - 3z = -7 & (5) \end{array}$$

Multiply equation (4) by -1 and add the result to equation (5).

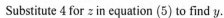

$$\begin{array}{rl} -5y + 2z = 3 & -1 \times (4) \\ \underline{5y - 3z = -7} & (5) \\ -z = -4 & \\ z = 4 & \end{array}$$

Substitute 4 for z in equation (5) to find y.

$$5y - 3z = -7 \quad (5)$$
$$5y - 3(4) = -7$$
$$5y - 12 = -7$$
$$5y = 5$$
$$y = 1$$

Substitute 1 for y and 4 for z in equation (1) to find x.

$$\begin{array}{rl} x + 2y + z = 8 & (1) \\ x + 2(1) + 4 = 8 & \\ x + 6 = 8 & \\ x = 2 & \end{array}$$
Solution set : $\left\{(2, 1, 4)\right\}$

21. $\begin{vmatrix} -3 & -2 \\ 6 & 9 \end{vmatrix} = -3(9) - (-2)(6)$
$$= -27 + 12 = -15$$

22. $\begin{vmatrix} 2 & 4 & 1 \\ 1 & 3 & 6 \\ 2 & 3 & -1 \end{vmatrix}$ Expand about row 1.

$$= 2\begin{vmatrix} 3 & 6 \\ 3 & -1 \end{vmatrix} - 4\begin{vmatrix} 1 & 6 \\ 2 & -1 \end{vmatrix} + 1\begin{vmatrix} 1 & 3 \\ 2 & 3 \end{vmatrix}$$
$$= 2(-3 - 18) - 4(-1 - 12) + 1(3 - 6)$$
$$= 2(-21) - 4(-13) + 1(-3)$$
$$= -42 + 52 - 3 = 7$$

23.   Let $x =$ the number of pounds of $3 per pound nuts.

	Number of Pounds	Price per Pound	Value
$3 nuts	x	3	$3x$
$4.25 nuts	8	4.25	4.25(8)
Mixture	$x + 8$	4	$4(x + 8)$

The last column gives the equation.
$$3x + 4.25(8) = 4(x + 8)$$
$$3x + 34 = 4x + 32$$
$$2 = x$$
Use 2 pounds of the $3 nuts.

24. $(4p + 2)(5p - 3)$
$$= 20p^2 - 12p + 10p - 6$$
$$= 20p^2 - 2p - 6$$

25. $(3k - 7)^2 = (3k)^2 - 2(3k)(7) + 7^2$
$$= 9k^2 - 42k + 49$$

26. $(2m^3 - 3m^2 + 8m) - (7m^3 + 5m - 8)$

$= 2m^3 - 7m^3 - 3m^2 + 8m - 5m + 8$

$= -5m^3 - 3m^2 + 3m + 8$

27.

$$
\begin{array}{r}
2t^3 + 3t^2 - 4t + 2 \\
3t - 2 \overline{\smash{\big)}\, 6t^4 + 5t^3 - 18t^2 + 14t - 1} \\
\underline{6t^4 - 4t^3} \\
9t^3 - 18t^2 \\
\underline{9t^3 - 6t^2} \\
-12t^2 + 14t \\
\underline{-12t^2 + 8t} \\
6t - 1 \\
\underline{6t - 4} \\
3 \qquad \text{Remainder}
\end{array}
$$

Answer:

$$2t^3 + 3t^2 - 4t + 2 + \frac{3}{3t - 2}$$

28. $7x + x^3 = x(7 + x^2)$

29. $14y^2 + 13y - 12$

Look for two integers whose product is $(14)(-12) = -168$ and whose sum is 13. The required numbers are 21 and -8.

$14y^2 + 13y - 12$

$= 14y^2 + 21y - 8y - 12$

$= 7y(2y + 3) - 4(2y + 3)$

$= (2y + 3)(7y - 4)$

30. $6z^3 + 5z^2 - 4z = z(6z^2 + 5z - 4)$

$\qquad\qquad\qquad = z(3z + 4)(2z - 1)$

31. $49a^4 - 9b^2 = (7a^2)^2 - (3b)^2$

$\qquad\qquad = (7a^2 + 3b)(7a^2 - 3b)$

32. $c^3 + 27d^3 = c^3 + (3d)^3$

$\qquad\qquad = (c + 3d)(c^2 - 3cd + 9d^2)$

33. $64r^2 + 48rq + 9q^2$

$= (8r)^2 + 2(8r)(3q) + (3q)^2$

$= (8r + 3q)^2$

34. $\qquad\quad 2x^2 + x = 10$

$\qquad 2x^2 + x - 10 = 0$

$\qquad (2x + 5)(x - 2) = 0$

$\quad 2x + 5 = 0 \quad$ or $\quad x - 2 = 0$

$\qquad x = -\dfrac{5}{2} \quad$ or $\qquad x = 2$

Solution set: $\left\{-\dfrac{5}{2}, 2\right\}$

35. $\qquad k^2 - k - 6 \le 0$

Solve the equation

$\qquad k^2 - k - 6 = 0.$

$\qquad (k - 3)(k + 2) = 0$

$k - 3 = 0 \quad$ or $\quad k + 2 = 0$

$\quad k = 3 \quad$ or $\qquad k = -2$

The numbers -2 and 3 divide the number line into three regions.

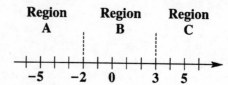

Region A	Region B	Region C

Test a number from each region in the original inequality.

$$k^2 - k - 6 \le 0$$

Region A: Let $k = -3$.

$(-3)^2 - (-3) - 6 \le 0 \quad?$

$\qquad\qquad\qquad 6 \le 0 \qquad$ *False*

Region B: Let $k = 0$.

$0^2 - 0 - 6 \le 0 \quad?$

$\qquad\qquad -6 \le 0 \qquad$ *True*

Region C: Let $k = 4$.

$4^2 - 4 - 6 \le 0 \quad?$

$\qquad\qquad\quad 6 \le 0 \qquad$ *False*

The numbers in Region B, including the endpoints -2 and 3, are solutions.

Solution set: $\left[-2, 3\right]$

36. $\left(\dfrac{2}{3}\right)^{-2} = \left(\dfrac{3}{2}\right)^{2} = \dfrac{3}{2} \cdot \dfrac{3}{2} = \dfrac{9}{4}$

37. $\dfrac{(3p^2)^3(-2p^6)}{4p^3(5p^7)} = \dfrac{3^3p^6(-2)p^6}{20p^{10}}$

$\qquad\qquad\qquad = \dfrac{-54p^{12}}{20p^{10}}$

$\qquad\qquad\qquad = -\dfrac{27}{10}p^{12-10}$

$\qquad\qquad\qquad = -\dfrac{27p^2}{10}$

38. $f(x) = \dfrac{2}{x^2 - 81} = \dfrac{2}{(x + 9)(x - 9)}$

The domain of f is the set of all real numbers excluding ± 9 since division by 0 is not defined. In interval notation, the domain is

$\qquad (-\infty, -9) \cup (-9, 9) \cup (9, \infty).$

39. $\dfrac{x^2 - 16}{x^2 + 2x - 8} \div \dfrac{x - 4}{x + 7}$

$= \dfrac{x^2 - 16}{x^2 + 2x - 8} \cdot \dfrac{x + 7}{x - 4}$

$= \dfrac{(x + 4)(x - 4)(x + 7)}{(x + 4)(x - 2)(x - 4)}$

$= \dfrac{x + 7}{x - 2}$

40.

$$\frac{5}{p^2 + 3p} - \frac{2}{p^2 - 4p}$$

$$= \frac{5}{p(p+3)} - \frac{2}{p(p-4)}$$

The LCD is $p(p+3)(p-4)$.

$$= \frac{5(p-4)}{p(p+3)(p-4)} - \frac{2(p+3)}{p(p-4)(p+3)}$$

$$= \frac{5p - 20 - 2p - 6}{p(p+3)(p-4)}$$

$$= \frac{3p - 26}{p(p+3)(p-4)}$$

41.

$$\frac{4}{x-3} - \frac{6}{x+3} = \frac{24}{x^2 - 9}$$

$$\frac{4}{x-3} - \frac{6}{x+3} = \frac{24}{(x+3)(x-3)}$$

Multiply by the LCD, $(x+3)(x-3)$. $(x \neq \pm 3)$

$$4(x+3) - 6(x-3) = 24$$
$$4x + 12 - 6x + 18 = 24$$
$$-2x + 30 = 24$$
$$-2x = -6$$
$$x = 3$$

But $x \neq 3$.
Solution set: $\emptyset$

42.

$$6x^2 + 5x = 8$$
$$6x^2 + 5x - 8 = 0$$

Use the quadratic formula with $a = 6$, $b = 5$, and $c = -8$.

$$x = \frac{-b \pm \sqrt{b^2 - 4ac}}{2a}$$

$$x = \frac{-5 \pm \sqrt{5^2 - 4(6)(-8)}}{2(6)}$$

$$= \frac{-5 \pm \sqrt{25 + 192}}{12}$$

$$= \frac{-5 \pm \sqrt{217}}{12}$$

Solution set: $\left\{ \dfrac{-5 + \sqrt{217}}{12}, \dfrac{-5 - \sqrt{217}}{12} \right\}$

43.

$$\sqrt{3x - 2} = x$$
$$3x - 2 = x^2 \qquad \textit{Square}$$
$$0 = x^2 - 3x + 2$$
$$0 = (x-1)(x-2)$$
$$x - 1 = 0 \quad \text{or} \quad x - 2 = 0$$
$$x = 1 \quad \text{or} \quad x = 2$$

Check $x = 1$: $\sqrt{1} \overset{?}{=} 1$ *True*

Check $x = 2$: $\sqrt{4} \overset{?}{=} 2$ *True*

Solution set: $\left\{ 1, 2 \right\}$

44.

$$5\sqrt{72} - 4\sqrt{50} = 5\sqrt{36 \cdot 2} - 4\sqrt{25 \cdot 2}$$
$$= 5 \cdot 6\sqrt{2} - 4 \cdot 5\sqrt{2}$$
$$= 30\sqrt{2} - 20\sqrt{2}$$
$$= 10\sqrt{2}$$

45.

$$(8 + 3i)(8 - 3i)$$
$$= 8^2 - (3i)^2 = 64 - 9i^2$$
$$= 64 - 9(-1) = 64 + 9 = 73$$

46. The graph of $f(x) = 9x + 5$ is a line. To find the inverse, replace $f(x)$ with y.

$$y = 9x + 5$$

Interchange x and y.

$$x = 9y + 5$$

Solve for y.

$$x - 5 = 9y$$

$$\frac{x - 5}{9} = y$$

Replace y with $f^{-1}(x)$.

$$f^{-1}(x) = \frac{x - 5}{9}$$

47. Graph $g(x) = \left(\dfrac{1}{3} \right)^x$.

Make a table of values.

x	-2	-1	0	1	2
$g(x)$	9	3	1	$\dfrac{1}{3}$	$\dfrac{1}{9}$

Plot these points, and draw a smooth decreasing exponential curve through them.

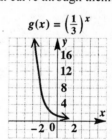

48.

$$3^{2x-1} = 81$$
$$3^{2x-1} = 3^4$$
$$2x - 1 = 4 \qquad \textit{Equate exponents}$$
$$2x = 5$$
$$x = \frac{5}{2}$$

Solution set: $\left\{ \dfrac{5}{2} \right\}$

49. Graph $y = \log_{1/3} x$.
Change to exponential form.
$$\left(\frac{1}{3}\right)^y = x$$
This is the inverse of the graph of
$$g(x) = y = \left(\frac{1}{3}\right)^x$$ in Exercise 47. To find points
on the graph, interchange the x- and y-values in the table.

x	9	3	1	$\frac{1}{3}$	$\frac{1}{9}$
y	-2	-1	0	1	2

Plot these points, and draw a smooth decreasing logarithmic curve through them.

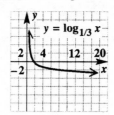

50. $\log_8 x + \log_8 (x + 2) = 1$
Use the product rule for logarithms.
$$\log_8 x(x + 2) = 1$$
Change to exponential form.
$$x(x + 2) = 8^1$$
$$x^2 + 2x - 8 = 0$$
$$(x + 4)(x - 2) = 0$$
$$x + 4 = 0 \quad \text{or} \quad x - 2 = 0$$
$$x = -4 \quad \text{or} \quad x = 2$$
$x \neq -4$ because $\log_8(-4)$ does not exist, so the only answer is $x = 2$.

Solution set: $\left\{2\right\}$

51. $f(x) = 2(x - 2)^2 - 3$ is in
$f(x) = a(x - h)^2 + k$ form.

The graph is a vertical parabola with vertex (h, k) at $(2, -3)$. Since $a = 2 > 0$, the graph opens upward. Also, $|a| = |2| = 2 > 1$, so the graph is narrower than the graph of $f(x) = x^2$. The points $(0, 5)$ and $(4, 5)$ are on the graph.

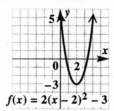

52. $\frac{x^2}{9} + \frac{y^2}{25} = 1$ is in $\frac{x^2}{a^2} + \frac{y^2}{b^2} = 1$ form with $a = 3$ and $b = 5$. The graph is an ellipse centered at $(0, 0)$ with x-intercepts $(3, 0)$ and $(-3, 0)$ and y-intercepts $(0, 5)$ and $(0, -5)$. Plot the intercepts and draw the ellipse through them.

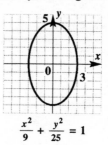

$$\frac{x^2}{9} + \frac{y^2}{25} = 1$$

53. $x^2 - y^2 = 9$
$$\frac{x^2}{9} - \frac{y^2}{9} = 1$$

The graph is a hyperbola centered at $(0, 0)$ with x-intercepts $(3, 0)$ and $(-3, 0)$. The asymptotes are $y = \pm x$. Draw the right and left branches through the intercepts and approaching the asymptotes.

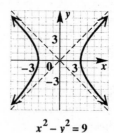

$$x^2 - y^2 = 9$$

54. Center at $(-5, 12)$; radius 9
Use the equation of a circle with $h = -5$, $k = 12$, and $r = 9$.
$$(x - h)^2 + (y - k)^2 = r^2$$
$$\left[x - (-5)\right]^2 + (y - 12)^2 = 9^2$$
$$(x + 5)^2 + (y - 12)^2 = 81$$

55. $a_n = 5n - 12$
$$a_1 = 5(1) - 12 = 5 - 12 = -7$$
$$a_2 = 5(2) - 12 = 10 - 12 = -2$$
$$a_3 = 5(3) - 12 = 15 - 12 = 3$$
$$a_4 = 5(4) - 12 = 20 - 12 = 8$$
$$a_5 = 5(5) - 12 = 25 - 12 = 13$$

Answer: $-7, -2, 3, 8, 13$

56. $a_1 = 8, d = 2$
$$S_n = \frac{n}{2}\left[2a_1 + (n - 1)d\right]$$
$$S_6 = \frac{6}{2}\left[2(8) + (6 - 1)2\right]$$
$$= 3(16 + 10) = 3(26) = 78$$

57. $15 - 6 + \dfrac{12}{5} - \dfrac{24}{25} + \ldots$

This is an infinite geometric series with

$a_1 = 15$ and $r = \dfrac{-6}{15} = -\dfrac{2}{5}$. The sum is

$$S = \frac{a_1}{1-r} = \frac{15}{1 - \left(-\dfrac{2}{5}\right)}$$

$$= \frac{15}{\dfrac{7}{5}} = 15 \cdot \frac{5}{7} = \frac{75}{7}.$$

58. $\displaystyle\sum_{i=1}^{4} 3i = 3\sum_{i=1}^{4} i = 3\left[\frac{4(4+1)}{2}\right]$

$$= 3(10) = 30$$

We used the fact that $\displaystyle\sum_{i=1}^{n} i = \frac{n(n+1)}{2}$.

59. $(2a - 1)^5$

$$= (2a)^5 + \frac{5!}{4!\,1!}(2a)^4(-1)^1$$

$$+ \frac{5!}{3!\,2!}(2a)^3(-1)^2 + \frac{5!}{2!\,3!}(2a)^2(-1)^3$$

$$+ \frac{5!}{1!\,4!}(2a)^1(-1)^4 + (-1)^5$$

$$= 32a^5 + 5(16a^4)(-1) + 10(8a^3)(1)$$

$$+ 10(4a^2)(-1) + 5(2a)(1) + (-1)$$

$$= 32a^5 - 80a^4 + 80a^3 - 40a^2 + 10a - 1$$

60. The fourth term $(r = 4$, so $r - 1 = 3)$ of

$\left(3x^4 - \dfrac{1}{2}y^2\right)^5$ is

$$\frac{5!}{(5-3)!\,3!}(3x^4)^{5-3}\left(-\frac{1}{2}y^2\right)^3$$

$$= \frac{5!}{2!\,3!}(3x^4)^2\left(-\frac{1}{2}\right)^3(y^2)^3$$

$$= 10(9x^8)\left(-\frac{1}{8}\right)y^6$$

$$= -\frac{45x^8y^6}{4}$$